Encyclopaedia of
FOOD CHEMISTRY

Encyclopaedia of
FOOD CHEMISTRY

Volume 1

Manpal Singh

ANMOL PUBLICATIONS PVT. LTD.
NEW DELHI - 110 002 (INDIA)

ANMOL PUBLICATIONS PVT. LTD.
H.O.: 4374/4B, Ansari Road, Darya Ganj,
New Delhi-110 002 (India)
Ph.: 23278000, 23261597
B.O.: No. 1015, Ist Main Road, BSK IIIrd Stage
IIIrd Phase, IIIrd Block,
Bangalore - 560 085 (India)
Visit us at: www.anmolpublications.com

Encyclopaedia of Food Chemistry

First Published, 2007

ISBN 978-81-261-3399-4 (Set)

PRINTED IN INDIA

Printed at Amit Enterprises Delhi.

Contents

Preface

Food Chemistry is a newer sub-discipline, under Food Science. This branch of knowledge enriches us about the composition and quality of food stuffs, we intend to consume. In a scholar's view, we are what we think, while we think what we eat and drink. This is an undeniable fact that our thoughts are determined by the kind of food we take. Food not only serves as a source of energy, it also plays a significant role in the making of one's personality.

Another scholar opines that for a healthy life, a healthy diet is a must. Our body mechanism is controlled by a number of elements, called vitamins, proteins, enzymes, carbohydrates and other simple and complex chemical elements.

All the abovesaid elements find their origin in various food articles, such as, milk, meat, egg, vegetables, cereals etc. Each has its own characteristics. For a healthy diet, different nutrients have to be taken in a proper proportion, because excess of any of these may result in disturbing the system of our body and affecting the health.

A proportionate or a balanced diet is one, which consists of all the necessary nutrients in a proper manner, required for the development and maintenance of the body. It depends upon age, health and profession. For instance, a diet, rich in

carbohydrates, vitamins, mineral salts, fats, and proteins should provide about three hundred calories per day for a normal adult. It comes under Food Chemistry, which has been dealt with, at length in this multi-volume.

Encyclopaedia of Food Chemistry, is an admirable step in this direction. These books throw abundant light on all aspects of food chemistry. It must prove to be a proud possession of students, scholars, academics, intellectuals and of course the general readers, who may like it, equally.

Wise comments and positive remarks for the betterment of these books in future, are solicited.

— Editor

1

Various Consuming Items

Hot Drinks

The word 'Beverage' has been derived from the Latin word 'bever' meaning rest from work. After work one tends to feel thirsty due to fluid loss through perspiration and one is inclined to drink water or other potable beverages.

Beverages are potable drinks which have thirst-quenching, refreshing, stimulating and nourishing qualities. By refreshing one means the replenishment of fluid loss from the body due to perspiration. Stimulation results in increase of the heart beat and blood pressure. This is due to the intake of spirits or tea and coffee. Nourishment is provided by the nutrients in the beverages.

Most of the beverages supply energy in the form of sugar or alcohol. They also provide other nutrients like mineral salts and vitamins. For example, milk gives calcium and citrus fruits give vitamin C. Certain beverages like tea, coffee and alcohol stimulate the heart and kidneys, because tea and coffee contain caffeine and alcohol provides as much as 7 calories per 1 ml.

Alcoholic beverages affect the central nervous system and tranquilise it, thus reducing tension and worries. Initially alcohol acts as a stimulant, being a concentrated source of energy. Later on, however, it acts as a depressant. Alcohol being a concentrated source of energy, it is absorbed directy by the bloodstream from the stomach and therefore provides instant energy to the body and increases the blood pressure for a short while. As mentioned above, alcohol is a depressant, because it relieves the blood vessels of the body from tension. Effects of alcohol remain in the body for about 24 hours.

Major Definitions

1. *Alcoholic Beverage:* Any beverage containing 1/2% of alcohol by volume to 76% alcohol by volume is considered to be an alcoholic beverage. Those with higher than 76% alcohol are known as medicines, such as anaesthetics.
2. *Alcohol:* These are hydroxides of organic radicals. Ethyl alcohol is the only alcohol which is consumable. All other alcohols are poison.
3. *Wine:* According to the 'Wines and Spirits Association of Great Britain'— "Wine is an alcoholic beverage obtained by the fermentation of freshly gathered grape juice, the fermentation of which has been carried out in the district of its origin, according to the local traditions and practices. Nothing artificial is added to it."
4. *Brandy:* Brandy is obtained by the distillation of fermented grape juice and is suitably matured in oakwood casks for a minimum of 3 years. Brandies are never coloured because they obtain a beautiful amber colour from the black oakwood casks. Brandies can be matured upto a period of 70 years (usually used for blending purposes).
5. *Distillation;* Distillation is a process of simultaneous evaporation and condensation of a liquid or a mixture of liquids to purify them.

6. *Liquor:* Liquor is used for all alcoholic beverages in general, but it has a special meaning in the brewery, namely, it is water used for brewing after the removal of unwanted minerals and salts (like calcium carbonate, which forms a precipitate on boiling and spoils the beverage) and addition of required minerals and salts in the manufacture of beer.
7. *Liqueurs:* Liqueurs are sweetened, coloured and flavoured spirits with a base of Cognac, Brandy, Scotch Whisky or sometimes Rum. The colour is obtained from herbs, spices, vegetables or fruit-coloured pigments, either by distillation or infusion. Sweetening is added as and when required.
8. *Whisky:* Whisky is obtained by the distillation of grains, particularly barley malt, and is aged in oakwood casks for a minimum of 3 years for Scotch Whisky, a minimum of 7 years for Irish whisky. Regular Scotch Whiskies are matured for 3-12 years, Premium Scotch or Deluxe Premium Scotch are matured for 12-20 years and the Finest Old 'Blue label' for a minimum of 20 years. Whiskies are coloured by the addition of desired colour caramel. It is distilled either, by the Pot Still method (which gives body) or the Patent Still method (which gives volume).
9. *Rum:* Rum is obtained by the distillation of fermented molasses or sugarcane juice and is matured in oakwood casks for a minimum of 3 years. Rum is coloured by the addition of dark caramel.
10. *Gin:* Gin is a rectified spirit obtained by the distillation of the fermented mash of cereals and grains in a definite ratio. Gin has a typical flavour of juniper berries, coriander seeds and orange peel. It has in all 108 botanicals. Gin is a colourless drink matured in oakwood casks for a minimum of 3 years.
11. *Vodka:* Vodka is the national drink of Russia and goes well with caviar. It is a strong alcoholic beverage

obtained by rectifying spirit made from potato and sweet potato starch and is passed through layers of sand, vegetable charcoal, coal and quarts to make it colourless, tasteless and odourless. Vodka is not matured.

Classification and Division

Beverage

Alcoholic

Fermented
GRAPES: WINES
(a) Table
(b) Fortified
(c) Soarklinn
corn)-Bourbon
(d) Aromatised

MOLASSES: RUM

Food Grains: Beers
(a) BARLEY-BEER
(a) Diet beer
(1-1.5% alcohol)
(b) Light beer
(3-5% alcohol)
(c) Strong beer
(8-10% alcohol)
(d) Draught Beer
(i) Light 3-5%
(ii) Strong 8-10%
(b) MAIZE-STOUT

Distilled
GAINS
(a) Barley malt-whisky
(b) Rye-Rye whisky
(c) Any grain (wheat,
(d) Potato and Sweet
potato starch-Vodka
SUGARCANE and
Scotch Whisky
(a) Light bodied
(Cuba, Mexico
Puerto Rico, Phillipines, Brazil
Venezuela, Argentina,
Peru, Paraguay)
(b) Full bodied (Barbados
Hawaii Jamaica, (Guiana)
(c) Aromatised (Java, Bulgaria
(d) Imitation rum de
FRUITS: BRANDY (from grapes)
-Apple brandy
—Cherry brandy
-Plum brandy etc.
the rocks or with soda or chilled orange juice)
(c) WHEAT-ALE
(d) CORN-PORTER
(e) RICE-SAKEY (Japan)
CHUNG (Tibet)

Compound
(a) GINS
Indian
Imported (Dutch, English,
American)
(b) LIQUEURS:—
Sweetened, coloured and
flavoured spirits
They have a base of either
or Cognac Brandy.
(i) Plant liqueurs
(ii) Nuts Iiqueurs
(iii) Fruit liqueurs
For example: Creme
menthe, Benidictine
Curacao, Cointreau , and
(c) BITTERS
-Fruit
—Aromatic
Angoustra, Grenadine (From Pomegranate)
Campari (Campari is a bitter as well as aromatised
wine. If served as an apertif, if is either served on

Non-alcoholic

Hot
Tea, Coffee, Hot
Chocolate, etc.

Cold

Aerated
Colas, Ginger ale, etc.
LIMCA, SODA

Non-aerated
Cold coffee, nectors, squashes,
juices, milk-shakes
Ice tea.

FRUITS:
(a) PEARS-PERRY (1.5-2% alcohol)
(b) APPLES-CIDER (1.5-2% alcohol)
MISCELLANEOUS:
(a) HONEY-MEAD
(b) CACTUS or PINEAPPLE-PULQUE

- Imitation rums are made in countries where no sugarcane is available. They are made from spirits flavoured with ethyl biterate. 1 ml. alcohol = 7 calories.

Non-alcoholic Beverages (Cold)

Aerated Water: These beverages are charged with Carbon-di-oxide gases.

E.g.(a) Soda Water: Colourless and tasteless.

(b) Tonic Water: Colourless and tasteless.

(c) Dry ginger: Golden straw—coloured with ginger flavour.

(d) Other flavoured waters are:

(i) Fizzy Lemonade

(ii) Orange

(iii) Ginger Beer

(iv) Coco Cola and others.

Non-aerated

Natural Mineral Water/Spring Water: Different varieties of mineral water are:

(i) Appolianaris: It is natural sparkling—Germany

(ii) Contrex: It is Still—France

(iii) Perrier: It is Still Sparkling or in fruit flavours—France

(iv) Vichy Celestine: It is Natural Sparkling—France

(v) Vittel: It is Still—France

Different Varieties of Spring Water are:

(i) Ashboure: It is Still or Sparkling—England

(ii) Badoit: It is Slightly Sparkling—France

(iii) Evian: It is Still—France

(iv) Malvern: It is Still or Carbonated—England

Mineral water has mineral content which is strictly controlled while spring water has fewer regulations apart from those concerning hygiene.

Squashes: The various squashes are Lemon, Mango, Pineapple, Lichi and others.

Juices: Juices can be of two types:

(i) Fresh fruit juices

(ii) Canned fruit juices

Various types of juices available are: Tomato juice, Grapefruit juice, Orange juice, Pineapple juice and others.

Syrups

(i) 'Grenadine (Pomegranate)

(ii) 'Casis (Black Currant)

(iii) 'Citronelle (Lemon)

(iv) 'Gommes (White Sugar Syrup)

(v) 'Framboise (Strawberry)

(vi) 'Cerise (Cherry)

(vii) 'Orgeat (Almond)

Non-alcoholic Beverages (HOT)

Tea

Cold Tea/Russian Tea: Chilled tea liqueur or extract of tea leaves. Served with lemon wedges, garnished with mint, sugar syrup.

Tea: Tea liquor served with sugar and milk.

Lemon Tea: Lemon-flavoured tea.

Coffee

Different types of coffee are:

Cafe Brule: A flambe presentation in which a blend of brandy, sugar and crushed coffee beans are ignited in a larger spoon and stirred into the hot coffee.

Cafe Capuccino: Blending chocolate into coffee and topping with Chantilly cream.

Cafe Viennois: Coffee with hot milk topped with whipped cream.

Irish Coffee: Black coffee, sweetened to the guest's taste,

is reinforced with Irish Whisky and has double cream floated on top.

Cafe Cream: Hot coffee with cream.

Cafe Au Lait: Coffee with milk.

Cafe Glace: Iced coffee or cold coffee. Coffee served with a scoop of ice-cream.

Tea: Tea is the most popular beverage in India. The best quality of tea is produced by India but the maximum quantity is produced by Sri Lanka.

The leaves come from the plant Camellia Sinesis which is a native of China. It is 2-3 ft. high and looks like a shrub. To prevent high growth, pruning is done after 1 year and the height is never allowed to exceed more than 3 ft. This short growth helps in the growth of more shoots (*i.e.* leaves) and less of wood. It is best grown in cool, humid climates where rainfall is plenty (30" of rainfall). But stagnation of water around the plant should be avoided as it may affect the growth. For this reason, the plant is grown on hill slopes.

Manufacturing of Tea: After maturation of the plant, the plucking is done. The pluckers pluck the top leaves and the flower is called Pekoe.

Approximately 40 kilos of tea leaves are plucked in the first shift (of 8 hrs.). After plucking, all the tea leaves are gathered at one place and sent for the withering process. Here, tea leaves are spread on a perforated rack under the sun for 24 hours or till the moisture content comes down to 50%. During the rainy season, leaves are sent to withering rooms and dried by blowing dry warm air. The next step is rolling. Here, tea leaves are passed through a crusher. There they get ruptured and the juice starts coming out and makes the leaves moist again. Then the fermentation is done. The juice has a substance called Tanin. When it comes in contact with air, it gets oxidised and becomes deep cooperish red in colour. It takes the leaves 3 hours to get the deep cooperish red colour. Finally, firing is done where leaves are exposed to dry warm

air till they have 3% moisture content. The leaves acquire black colour. Leaves are then packed and sent for sale. This whole process of making black tea is called the orthodox method. South-Asian countries are famous for producing good grade black tea.

Another type of tea is the Green Tea. It is manufactured in Laos, Indonesia, Burma, Malaysia, China and Japan. It is unfermented tea. After plucking, tea leaves are withered, steamed and then fired and packed. No fermentation is done during its manufacture.

Oolong Tea is another type of tea made in Taiwan. It is semi-fermented tea. During the plucking process, tea leaves are snapped into two, and because of this, a little bit of tanin starts coming out and hence oxidation occurs. Then the leaves are withered and fired.

Tanin contributes to the thickness and flavours of tea. Black tea has a thick consistency but is not very well flavoured. Green Tea has a medium consistency but has the best flavour. Oolong Tea has a watery consistency and is medium-flavoured. The concoction also depends on rolling. The bigger the size of tea leaves, the weaker will be the crushing and hence extraction.

There are 2 methods of rolling:

C.T.C.—Cut-Tear-Curl: Here, the two cylinders have grooves on them. When the leaves are passed through these cylinders, the leaves not only get cut and torn but they also get curled. CTC reduces the fermentation time to 1/3. It is the best method but it can not be applied to Green tea and Oolong tea.

Leg Cut: Here, the leaves are shredded into strips. Therefore maximum tanin comes out and fermentation is very quick.

Coffee: Coffee is a natural product grown in many countries of the tropical and subtropical belt in South Central America, Africa and Asia. It is grown at different altitudes in different basic climates and in different soils and is looked upon as an

international drink consumed throughout the world. Brazil is the world's largest grower of coffee, Columbia is the second, and the Ivory Coast and Indonesia are third and fourth respectively.

The rising popularity of coffee over the past few years may be attributed to the fact that nowadays more and more people spend their holidays in the Continent. Coffee is the Continent's most popular drink and therefore is one of the first habits that tourists pick up abroad.

The trees which produce coffee are the Genus coffee which belongs to the Rubiaceae family. Out of the 50 different species, only two of these are commercially significant. These are known as Coffee Arabica and Coffee Camephora which is usually referred to as Robusta. Arabica accounts for some 75% of the world production.

The coffee tree is an evergreen shrub which reaches a height of 2-3 m when cultivated. The fruit of the coffee tree is known as the 'cherry' and they are about 1.5 cm in length and have an oblong shape. The cherry usually contains two coffee seeds. The coffee tree will not begin to produce fruit until it is 3-5 yrs old and it will then usually yield good crop for about 16 yrs.

Coffee plants require the following conditions for growth: (i) High humidity (ii) Plenty of rainfall (iii) Cooler climate (iv) Care is taken so that water does not stagnate around the plant.

The Blend: Companies who sell coffee have their own blending experts whose task it is to ensure that the quality and taste of their particular coffee brand is consistent, despite the fact that the quality of imported beans does vary from shipment to shipment. Most brands of coffee sold in shops are in fact a blend of two or more batches of beans. Because they have no smell or taste, green beans have to be roasted in order to release the coffee aroma and flavour. The correct roasting gives a uniform colour. The output of different roastings are used to form different blends.

The Common Degrees of Roasting are:

Light or Pale Roasting: Suitable for mild beans to preserve their delicate aroma.

Medium Roasting: Gives a stronger flavour and are often favoured for coffees with well-defined character.

Roll Roasting: Popular in many Latin countries and have a bitterish flavour.

High Roasted Coffee: A strong bitter coffee, although the original flavour is lost.

Commercial coffee roasters can either convert the beans into instant (soluble) coffee or prepare them for sale as roasted or ground beans. When roasted the degree of acidity is less and the bitterness is more in the coffee.

Roasted coffee must be ground before it can be used to make the brew. Coffee is ground to different grades of fineness which suit the many different methods of brewing. The most suitable grinds for some common methods of brewing coffee are:

Method	*Grinding Grade*
Filter/Dip	Fine to medium
Jug	Coarse
Turkish	Medium
Cafetiere	Medium
Vacuum infusion	Medium fine to fine
Espresso	Very fine
Percolator	Medium

Processing of Coffee: The fruit of the coffee plant is called cherry. It has a pulp and two flat seeds inside. Two following methods are used to separate the seeds from pulp:

Fermentation Method: The fruit is dried under sun so that the pulp dries up and the seeds can be removed.

Frying Method: The fruit is allowed to ferment till the pulp falls off and seeds are separated.

In both the cases, one gets coffee seeds. These are freshly roasted and then grinded immediately. The coffee seeds are then roasted at 150°C by hot dry air to give the required colour to the seed. The colour of the roasted seeds are important factors. When the desired colour is achieved, a calculated amount of cold water (to arrest further roasting) is added. This process is called Quenching. Then the coffee is grinded and the degree ranges from fine powder to medium to coarse powder. This is called processing of coffee.

Types of Coffee: There are 3 types of coffee available in market:

Arabica: It is the best quality available. It is available in the finest quality. It is grown in Brazil, Paraguay, Coloumbia, Mexico and India. The crop is very fragile hence the yield is very less. So it is very expensive.

Robusta: The quality is medium, taste and flavour-wise. The yield is far better. It is grown in the West African coast specially in Ghana.

Liberica: It is the cheapest brand of coffee available. It is grown in Malaysia.

Other Type of Coffee Made in the World

French Coffee: In this type of coffee, the coffee seeds are mixed with chicory. The commonly used one is Chicory (Endive plant). It is the root of Endive Plant. It is roasted along with coffee seeds and then powdered. In India, it is available as the 'BRU' Coffee.

Henna Coffee: It is an Austrian coffee. Figs are roasted along with coffee seeds and then powdered. Then they are mixed with boiled water and served with or without milk.

Coffee Pona: It is filtered very finely.

Coffee Espresso: It is steamed coffee with a thick layer of

froth which is sprinkled with cocco powder. When cinnamon powder is spinkled then it is called Capuchino coffee. *E.g.* Cappucino, Coffee Italiano.

Hawaiian Coffee: Black cold coffee with pineapple juice and cream.

Turkish Coffee: Mostly used in the Middle East. Coffee is brewed with very little water with few grains of potassium permanganate in a copper pan. Very strong coffee and copperish red in colour.

Irish Coffee: It is coffee liquor mixed with Irish whisky and whipped cream is poured on top to form a layer.

Instant Coffee: Coffee powder mixed with boiled water and served with or without milk.

Cona Coffee: Coffee made from freshly crushed roasted coffee beans and is prepared in a Cona Coffee apparatus. Very strong coffee served with milk and cream.

Cocco and Chocolate: The best cocco is produced in Assra (in Ghana). 70% of the world's production is done in Ghana, about 24% production is from Brazil and 63% from South West Asia.

Cocco Pod is a big fruit of size 12" x 4" diameter. It has 20-40 seeds. It is used in the manufacture of cocco and chocolate.

Processing starts with fermentation of the fruit. Fruit is fermented so that the pulp leaves the seeds. This produces acetic acid which gives a bitter taste and a brown colour to the cocco products. The seeds are sent for winnowing so that the outer seed is removed and the nib is obtained. It is then treated with alkalines. In presence of alkalis, the seeds develop a typical brownish colour and strong flavour. This is called the Dutch Process. Then the, nibs are crushed (Rolling Process). The rolling or crushing of nibs produces cocco which has 50% cocco and 50% cocco butter. This cocco can be used in two ways.

For preparation of chocolate, cocco mass is mixed with sugar and with butter, then heated to 60-70°C and churned at

a high speed to get the chocolate which is used for finishing of cakes, candies and other commercial food or products. Milk chocolate is manufactured by adding milk powder and cocco butter to the chocolate. The cocco mass is compresed in hydrolic machines to separate into cocco powder and cocco butter. By hydraulic pressure, the fat is separated. This cocco butter can further be used for making imitation chocolates.

Cocoo butter is mainly used in cosmetics.

Breakfast (Petit Dejeuner): The word Breakfast means to break the fast after a long time. It originally started in England where they used to have a lot of courses. The different types of breakfasts are:

1. Continental
2. American
3. English

Continental Breakfast or Cafe Complet or the Complet: This breakfast is an European Breakfast. It comprises of:

I. *Choice of Juice:* Mango juice, Pineapple juice, Tomato juice, Orange juice or Grapefruit juice etc.

II. *Choice of Breads:* Toast (white bread / Brown bread), Rolls, Croissant, Brioche, Muffins, Doughnuts, Danish Pastry served with preserves, Jam, Honey, Marmalade and butter.

III. *Tea/Coffee:* If tea is served with this breakfast, it is known as The Complet. If coffee is served along with this breakfast, it is known as Cafe Complet.

Cafe Simple or The Simple: If the guest orders only coffee then it is called Cafe Simple. If the guest orders only tea it is called The Simple.

Cutlery Required for Continental Breakfast: A side plate or quarter plate with side knife, butter dish with underliner and butter spreader, toast rack, tea /coffee pot, sugar bowl, milk pot/Creamer.

American Breakfast

It Starts with a Glass of Cold Water: Consists of the Table d'Hote breakfast Menu which one would generally find in the menu card of any 5 star hotel.

Choice of Juice: Mango, Pineapple, Tomato, Orange, Grapefruit.

Cereals: Choice of cornflakes, oatmeal, porridge, rice crispies, wheat flakes served with cold or hot milk.

Eggs to Order: Scrambled, poached, boiled, Omelette, served with bacon, ham or sausages.

Choice of Breads: B/F Toast, rolls, Brioche, croissant with preserves like batter, jam, marmalade and honey.

Or

Choice of Indian Breakfast: Puri Bhaji or stuffed Paranthas served with curd, idli, wada/uttapam/dosa/upma served with sambar.

Tea or Coffee

Beverages: Tea, coffee or hot Bournvita, Horlicks, Ovaltine.

English Breakfast: An English breakfast is a very elaborate breakfast. It comprises of ten courses.

Choice of Juice: Pineapple, Tomato, Grapefruit, Orange, Mango etc.

Stewed Fruits: Apples, Prunes, Figs, Pears etc. are cut into small pieces and cooked in sugar syrup flavoured with clove and cinnamon. It is served in a cocktail cup with a quarter plate as underliner and the cutlery provided is a teaspoon.

Cereals: Oatmeal (meal), cornflakes, wheat flakes, rice crispies, porridge are served with cold or hot milk in a soup bowl with a quarter plate as underliner and a dessert spoon is provided as cutlery.

Fish: Herring, Haddock, Kedgres, Sardines, are served.

Eggs to Order: Omelette, boiled eggs, scrambled egg, poached or fried eggs.

Meat: Grilled Bacon, Sausages, Ham, Salami, Kidney or Liver, are served.

Choice of Rolls or Toast: Toast white or Brown or Rolls—like Croissant, Muffins, Brioche, Doughnuts, Danish Pastry are served.

Butter and Preserves: Butter, jam, honey, marmalade, maple syrup etc. provided.

Fruits: Fresh fruits like melon, papaya, mango, orange, grapefruit, pears are served in this course. (When in season)

Beverages: Tea/Coffee or hot beverages like Bournvita, Milo, Horlick, Ovaltine, Cocco etc. can be served.

French ***Petit Dejeuner***	***English*** ***English Breakfast***
Jus de Tomate	Tomato juice
Jus de Pamplamousse	Grapefruit juice
Jus d'orange	Orange juice
Jus d'ananas	Pineapple juice
Compote des Fruits	Fruit Compote
Compote des Figues	Figs Compote
Compote des Pommes	Apple Compote
Flacon de Mais	Cornflakes
Flacon de Ble	Wheat flakes
Souillie d'Avoine	Oatmeal Porridge
Bouillie de Ble	Wheat Porridge

Oeufs Aux Choix (CHOICE OF EGGS)

Mollet	Soft			
Poche	Poached	Jambon	Ham	
Dur	Hard-boiled	Lard	Bacon	(Avec-with)
Brouille	Scrambled	Jaucisse	Sausage	

Frit	Fried	
Hareng Grilles	Grilled Herrings	
Rognon Grille	Grilled kidneys	
Foie d'Agneau Au Lard		Lamb's Liver with bacon
Ris d'Agneau		Lamb's sweet bread
Pain Grilled		Toast
Petit Pain		Bread Rolls
Confiture	—	Jam
Miel		Honey
The		Tea
Café		Coffee
Chocolate		Chocolate
Cacao		Cocco
Pamplamousse		Grapefruit
Melon Frappe		Chilled melon
Demi Pamplamousse		Half grapefruit
Corbeille de Fruits		Fruit basket
Fruit Frais		Fresh fruit

Use of Tobacco

Generally tobacco means a leaf product containing 1-3% of the alkaloid nicotine which produces a narcotic effect when smoked, chewed or snuffed. The plant N Rustica provides tobacco in parts of Europe, but the tobacco of world commerce is N. tabacum. During his voyage to America Columbus discovered the West Indians smoking tobacco in a hollow forked stick. Historians do not know who first brought tobacco to Europe, but most of them give credit to Jean Nicot in 1561. Nicotina and Nicotine incorporate his name.

Main Features: Solaneceous Annual, Alternate, Sessile, Oval or Lanceolate leaves. Rowers produced in a panicle.

Cultivation, Harvesting and Curing: The local soil type and climate within the country, or state determine the type of tobacco grown there. Generally tobacco grows best in warm, even climate on light, well-drained, carefully fertilized soils, that receive weekly moisture from rain or irrigation. Special seedbeds, either open or covered with cloth, glass or plastic provide transplants for the fields. Plants are usually topped (blossom removed) to allow the upper leaves expand. Harvest proceeds by cutting the whole stock or picking leaves successively as they ripen (primed). Primed leaves are supported on strings, wires or sticks. Curing (drying) is done in ventilated barns with natural or artificial heat. In some areas, machine harvested leaves are packed in special frames for curing in heat regulated, forced-air chambers.

Drying time (2-6 weeks) and temperature (70-170°F or 21-77°C) influence the amount of changes that occur in proteins, carbohydrates, organic acids, alkaloids and enzymes in the leaf. Before inserting the leaves in cigars, cigarettes, pipes, chewing tobacco or sauf, lined leaves are fermented by storing them for 6 weeks to 2 years at about 15% moisture and 80-110°F (27-43°C). The methods used for harvesting, curing and fermenting depend on the type of tobacco, intended use and local custom. A flat product made from ground up tobacco plus binding agents is often blended with tobacco leaves.

Economic Importance: Tobacco is economically important in 66 countries and is grown to some extent in all but a few countries.

Diseases Caused: Diseases cause production problems in all countries where tobacco is grown. Annual losses of tobacco products worldwide average 10% or more; in addition, production costs invariably increase. About 25 principal pathogens attack the crop right from the seed sowing stage to marketing.

Root knot, caused by the Nematode Meloidogyne sp. is the most important disease. It occurs in all the warmer countries, particularly in sandy soils. Diseased plants are stunted and

have wilted leaves that yellow prematurely. Knits or galls form on the roots. Infected plants become more susceptible to other diseases such as black shank and bacterial wilt.

Cigars: Cigars are formed by using three types of tobacco leaves:

Fillers: These are broken/crushed leaves which forms the centre of the cigar.

Binders: These are the leaves which are small in size and are bound around the filters.

Wrappers: These are the big leaves which are quite soft and elastic in nature so they can be wrapped to give the final smooth surface. The best wrapper leaves are found in flavours and these are grown in shade.

In order ro bind all these leaves together, a special type of glue is used in manufacturing cigars and cigarettes. The leaves of tobacco are taken and put in bundles and then heated and cured by subjecting them to fire which helps in making the leaves sort of brittle in nature and increase the preserving quality. The bundles are then opened up and graded and used for the particular cigars.

Cigars are usually made by hand but nowadays rolling is also done with the help of machines. The cigar should not be dry, some moisture should be there in order to have full satisfaction and should be stored in cedar wood boxes as they allow that much quantity of air and moisture which can maintain the quality of cigars.

Types of Tobacco

Type	*Major Production Area*	*Use*	*Curing Method*
Flue-cured (bright)	United States (Virginia, North Carolina, South Carolina, Georgia Florida border); Canada (Ontario); Europe; China; Japan; Australia; India	Cigarettes	Primed leaves heat-cured to a bright yellow colour
Light air-	U.S. (Kentucky, Mary	Blending in	Whole plant air

contd......

Type	*Major Production Area*	*Use*	*Curing Method*
cured (Burley,	land, Tennessee); Canada, Europe Maryland)	cigarettes	dried in ventilated sheds
Oriental (Turkish)	Greece (Macedonia), Turkey (Samson)	Blending for aroma in cigarettes	Primed leaves cured in sun on strings.
Cigar filler	Cuba; Phillippines; Puerto Rico; US(Pennsylvania, Ohio); Dominican Republic, Central and South America	Central bulk of cigars	Whole plant air dried in ventilated sheds
Cigar binder	US (Connecticut River Valley, Wisconsin, Pennsylovania)	Binding the filler into shape.	Same as above
Cigar Wrapper	Under Clotheshade; US (Connecticut River Valley, Georgia Florida border); Cuba; Italy; new in Canada & South America	Outer leaf or wrapper of cigar	Primed leaves heated or air dried to a uniform golden brown or a bright green.
Dark air cured and fire cured	US (Kentucky, Tennessee, Virginia)	Pipe, chewing snuff	Whole plant air dried in ventilated sheds.

Strength

1. C Means Colorado—a dark and strong cigar.
2. C.C. Means Colorado Claro—medium cigar.
3. C.C.C. Means Claro—a light cigar.
4. C.M. Colorado Maduro—a very strong cigar.
5. M. Maduro—an exceptionally strong cigar and darker in colour.

Common Cigars

— Esquitos (in tubes)

— Hebeaneoros (in tubes)

— Club No. 4

— Cafe Creme Corona

Casings (sweetening agents) are added to give flavour to the cigarettes. Snuffs are made by fermenting the fire cured leaf and grinding stems and adding salt and flavourings. Cigars are made by wrapping a blinder leaf round a bunch of cut filler leaves and overwrapping with a fine wrapper. Stems and scraps are grounded down to make reconstituted sheets something like the paper-making process which can be used as substitute cigar binders or wrappers. These are also used to supplement natural tobacco in cigarettes.

Names and Sizes of Cigars

1. Corona—A 9 inch long straight cigar with a round end. It gives a slant cut or cut with a cigar cutter in V-shape.
2. Petit Corona—Same shape, 5 inches long.
3. Tres Petit Corona—4½″ long/ same shape.
4. Half Corona—3 ¾″ long.
5. Lansdale—The strongest cigar, 6 ½″ long.
6. Ideals—Torpedo shape cigar 6 ½″ long.
7. Bouquet—Small torpedo shape, 4" long cigar.
8. Panatella—It is opened from both the ends. 3" long with a golden foil to show from which side to smoke.
9. Churoot—Smaller cigar but bigger than bidis and darker, they are tapered at the stroking end and are 6 ½″ long.
10. Whiffs—English cigar open at both ends, 3½″.

Cigarettes: These have long thread-like, shredded tobacco prepared by using machines which are capable of manufacturing 10,000 cigarettes in a minute.

These are never individually made but there is a long cigarette paper in which tobacco is placed and the paper is glued and then cut accordingly and the filter is inserted at the end which has a diameter longer than the cigarette but has been compressed and fitted on the end with a golden brown paper covering.

Filter: This helps in reducing the amount of nicotine that goes along with the smoke to the person's body and is made up of cellulose fibres which are very fine.

There are certain cigarettes like Phillip Moris which have a double filter namely, regular filler and triangles of charcoal.

Synthetic filters are also being used in order to fight the competition in the market like 'Charms'.

Virginia Tobacco: This is the best and the different cuts available are Navy, Silk, Flake and Fine Cut.

How to Judge a Good Cigar: Basically, before they come into the market they are judged by an expert team of connoisseurs. But if a person wants to judge then the foremost point is the reputation of the Company and goodwill, and not an attractive box or packing.

It's odour, flavour, aroma, taste should be checked which can be done by smelling and smoking.

Colour of the Ash: It should be dark greyish, firm and should not fall on its own. The longer it stays the better it's quality is.

Cigars, if taken near the ear and rolled with the hands, produce cracking sounds when dry and means that the cigar are of poor quality.

Older Cigars: You can see spots on the cigar.

Cigarettes: These can be stored for a period of 6 months after which yellow spots appear on the paper and they have a sweet taste when puffed.

Hukka: Desi tobacco is used.

Pipes: Flecked tobacco is used.

Brands

American: Camel, Pallmall, Cheterfiled, Kent.

English: Thon Player's Special, 555, Dunhill, Benson and Hedges.

Mentholated: Cool, Salam

Double Filtered: Phillip Moris (Charcoal and cotton/silk filter)

Internationally Famous Cigars

American	—	Robert Burns
Cuban	—	Romeo Juliet
Danish	—	Ducados
Holland	—	Ambapsado, Rothman's—120 m.m.
Dutch	—	Corona delux, Diplomat
Jamaican	—	Corona delux, Diplomat (delux)
Indian	—	Bengal Tiger

Usually all cigars are derived from N. tobacom except for air-cured Khorka (USSR) which is derived from N. rustica. Its leaf has a coarse texture, it has Nicotine and has citric acid of a higher level than lemon.

Culture

Seeds: The choice of seed is contingent upon the type of leaf desired. Most strained varieties of N. tobaccom are local in their adoptions and are not suited for the production of widely divergent leaf.

Orinoco Group: In production of Flue-curing tobacco.

Pryor Groups: In production of dark, air-cured and fire-cured tobacco.

Havana Seed, Sumatra, Luban represents outstanding varieties used in the production of cigar leaf.

Specification of Seeds: Seedlings are small and are produced in cold farms or outdoor beds covered with thin cloth or something like a straw bed.

Specification of Soil: Soil which is selected should be:

1. Fertile
2. Have good tilth

3. Drainage
4. Be fully exposed to sun

Sterilisation of soil iodine by burning, steaming, using chemicals, such as methyl bromide to destroy used seeds, nematode insects etc. is done 8 to 10 weeks after transplanting the field. Shortly after the flowering, it is topped (disbudded) that is, the terminal growth is removed. Leaves are large and vary from 10-20 in number. Burley, dark air-cured, fire-cured, cigar requires silt loam, clay loam with clay sub soils.

Oriental: Unfertile sandy hill slopes.

Flue-cured: Heavy black clay-like soil.

Use of Fertilisers: NPK Fertiliser are used one ton per acre and can vary in small quantities.

Disease of the Plant

1. Pusarium Whilt
2. Black Root Rot
3. Tobacco Masaic Virus
4. Bacterial Leave Spot
5. Downy Mildew

Insect Pests

1. Green June Beetle Larvae
2. Cutworms
3. Flea Beetles
4. Horn Worms
5. Grass Hoppers
6. Nematodes

These are controlled by various methods including fumigation. The harvesting time is 70-140 days.

Effects of Tobacco: Tobacco can affect the health. Passive smoking places unborn children at serious risk.

Smoking and chewing of tobacco can also lead to cancer.

It is also bad for lungs and heart. Research has found that smoking mothers account for 50,000 miscarriages in the US each years, the culprit being Nicotine. Premature deliveries are also associated with this.

Role of Outlets

One method of classifying food and beverage outlets is outlined above. The basis of this classification is the distinction between commercial catering on the one hand and subsidised or welfare catering on the other. In the case of the former a further distinction is made between those outlets which serve a restricted market and those which serve the general public at large. In the case of the latter a distinction is made between institutional catering and employee catering.

There are two reasons for using this classification here. First, it provides a very broad coverage of food and beverage outlets-broader, for example, than many of the official definitions and classifications of the hotel and catering industry. The 1968 Standard Industrial Classification (SIC), for instance, excludes from the hotel and catering industry, all subsidised catering unless operated by a catering contractor as well as take-aways. The exclusions from the 1980 SIC are much less but it still excludes much of the employee and institutional catering. The second reason for using this classification is that it is based on distinctions which have a significant bearing upon most aspects of the operation of the catering activity. For example, the difference between subsidised catering and commercial catering not only embraces differences of objectives but also covers differences in markets served, differences in organisations involved and differences in marketing and business strategy.

Based on this classification, the aim of this chapter is to list out the different types of food and beverage outlets and to identify their main characteristics. This then serves as a basis for a consideration of issues of relevance to food and beverage service in general.

Major Commercial Sectors: Commercial food and beverage outlets may be defined as those operations in which profitability is the primary concern. Such outlets exist not only in private ownership but also in the publicly-owned sector of the economy where, for example, the British Railway seek to operate their catering outlets on commercial lines. Also, it is worth noting that in the commercial sector, catering may be the main activity of the organisation or company or it may simply be a secondary or additional service to customers, for example, catering in departmental stores or theatres.

A broad distinction can be made between catering for a restricted market and catering for a general market. In the case of the former the market can be restricted in a number of ways: by way of membership criteria as in the case of catering in clubs or by the fact that the catering is only available to those engaged in a specific activity as in the case of transport catering or employee catering. Clearly, there will be some overlap in a few cases. For example, catering at a railway station is normally open to the traveller as well as to the general public.

Commercial Catering for General Market: The provision of food and beverage facilities in hotels ranges from the self-services style often adopted for specific meal periods by transit or motel-type hotels, to waiter silver service used by luxury and resort hotels. In the latter establishments, the food and beverage facilities are usually presented as one of the many features of the hotel because the guests may be staying at the hotel for some time rather than just a night or so as may be the case in transit hotels.

The types of food and beverage found in hotels include silver service restaurants, licensed bars, coffee shops and snack bars, and buffet restaurants, room service and banquet facilities. Some of these facilities are only available to hotel residents, or 'In-House trade', for example room service; while the others, for example banquet, coffee shops and other restaurants, are advertised outside the hotel in order to attract outside customers.

Commercial Catering for a Restricted Market

Transport Catering: Transport catering (*i.e.* road, rail, air and sea) has a number of characteristics not commonly associated with other food and beverage outlets. It usually involves the feeding of a large number of customers arriving together at a catering facility, and who need to be catered for in a specific time period, *e.g.* on board a plane. Sufficient food and beverage supplies are usually carried for a specific number of meal periods, as, if for any reasons this food cannot be served to customers, alternative supplies may not be readily available. The service of food and beverage may be particularly difficult due to the physical conditions within the service area *e.g.* turbulence on board a plane. Finally, there are the problems of staffing these food and beverage facilities; the extra costs involved in the transportation and service of the food and beverages; space restrictions and the problem of security whilst the operation is the transit.

Four main types of Transport Catering may be Identified:

Road: Road catering has progressed from the inns and taverns of earlier days used by those travelling on foot and horseback to the present-day motorway service areas situated along the motorways which traverse the country. These service areas are open twenty-four hours a day and have a particular problem of staffing as some employees have to be brought to and from work over a distance of twenty to thirty miles. Also because of their isolated locations, the hours they are open, and the sheer volume of numbers involved at peak periods, these service areas are also particularly prone to vandalism and littering. They do, however, provide a valuable catering service to the travelling public and the food and beverage facilities usually include self service and waiter service restaurants, vending machines and take-away food and beverages.

Rail: Rail catering may be conveniently divided into two major areas namely, terminal catering and in-transit catering.

Catering at the railway terminals usually comprises licensed bars, self-service and waiter service restaurants, fast food and take-away units, supplemented by vending machines dispensing hot and cold food and beverage. In-transit, catering basically consists of two main types of service. The first is the Restaurant Car service where breakfast, lunch and dinner are organised in sittings and passengers go to the Restaurant car for service where appropriate seating accommodation is provided. The other type of service is the Buffet Car which is a self-service operation in which passengers can go to the car and buy light refreshment over the counter.

Such traditionally organised catering facilities are, however, no longer suitable for the latest development in rail travel which include the Advanced Passenger Train (APT) and the High Speed Train (HST). These trains demand a very streamlined and compact catering service which can provide the customer with food and beverage very quickly. The adoption of the airline service style has been suggested for these new fast trains, and it may be that some thing along these lines will be adopted in preference to the more traditional forms of catering, which are not able to cope as efficiently with the catering demands of these new fast speed trains.

Air: Airline catering has increased and developed considerably over the past twenty-five years. It originally consisted of sandwiches and flasks of tea, coffee and alcoholic beverages, but the progression to today's full and varied service is on par with that of aircraft development itself.

Like the Railways, Airline catering falls into two main areas: Terminal catering and 'In-transit' or 'In-flight' catering. Food and beverage outlets at air terminals usually consist of self-service and waiter-service restaurants, supplemented by vending machines and licensed bars. The in-flight catering service varies considerably with the class of travel, type and duration of flight, etc. At one extreme, the food and beverage portions are highly standardised with the meals portioned into plastic trays which are presented to the passengers and from

which they eat their meals. Disposable cutlery, napkins, etc. are used which reduces the facilities necessary for washing-up and also cuts down on breakages and wastages. At the other extreme there is virtually no portion control; service is from a gueridon trolley, food is portioned in front of the customers and any garnishes, sauces, etc. added according to their requirements. The crockery used may be bone china and this combined with fine glassware and cutlery creates an atmosphere of high-class dining.

A characteristic of airline catering is that this service is often contracted out to a specialist catering firm, which may be supplying a similar service to many airlines.

The provision of a catering service in air travel is normally inclusive in the price of the fare and a particular feature now is the cable service facilities by the different airlines. The increasing use of airlines as a common mode of transport has made competition fierce, and the area of food service is now a particularly competitive aspect of the total service offered by an airline.

Sea: Sea or marine catering varies from the provision of food and beverages on the short sea-route ferries where the catering service does not often feature prominently, to the large cruise or passenger liners where the catering facilities are an important part of the service offered by the shipping line, and are usually included in the price of the fare. On the cruise liners the standard of catering facilities is high because they are an important sales feature in a competitive activity. On the short sea routes, however, price is usually the more important factor and because of the necessity to feed large number of people the catering service provided is usually of the popular and fast-food type.

Clubs: Clubs, as a sector of the hotel and catering industry, are establishments offering food and drink, with at times accommodation, to members and bonafide guests. The types of clubs varying from working men's clubs, to political party clubs, social clubs, sporting clubs, restaurant clubs, to the private

exclusive clubs. Basically, in England and Wales, clubs are of two main types:

Proprietary Clubs: These are licensed clubs, owned by individuals or company and operated by themselves for self-profit, and as such require a Justice's Licence to operate. It is usual for a high proportion of proprietary clubs to resemble licensed restaurants with a substantial part of their turnover obtained from the sales of food.

Registered Clubs: These are registered clubs in which the management is responsible to an elected committee; the members own all the property including the food and drink, and pay their subscriptions to a common fund. As a non-profit making club which belongs to all the members and provides a service to the members, it is a genuine members' club and does not require a Justice's Licence to operate. It requires to be registered. The turnover of the members' clubs is mainly obtained from the sale of drinks which are normally sold at a competitive price as the profit element in clubs is lower than say in public houses.

Function Catering: Function catering may be described as the service of food and beverage at a specific time and place, for a given number of people, at a known price. Examples of function catering include social functions, such as wedding, dinner dances, business functions such as conferences, meetings and work lunches; and those functions that are organised for both social and business reasons such as outdoor catering at a show or exhibition.

Function catering is found in both the commercial and non-commercial sector of the catering industry. In the commercial sector, function catering is a specialised organisation operating in its own function facilities or as an outdoor-catering specialist operating in the clients' facilities or rented facilities or as a separate department within an hotel.

In the non-commercial sector, function catering is rarely the primary reasons for providing the establishment with catering facilities. Such establishments include hospitals,

schools, industrial cafeterias, etc. where the functions are not usually organised on a purely profit basis as they are in the commercial sector, but rather to serve a specific need of the organisation. Typical examples would be Diwali functions, retirement parties, fund-raising events, etc.

Restaurants and Snack Bars: Unlike the hotel facilities commercial restaurants do not offer accommodation and therefore their primary function is the provision of food and beverage. Because these restaurants do not have any in-house trade they are very reliant on passing trade and the reputation they develop from word-of-mouth advertising.

The various types of restaurants include snack bars, cafes, coffee shops, take-aways, steak bars, speciality restaurants, hot cuisine restaurants, etc. These diverse types of restaurants have service-styles, ranging from the self-service catering to the more elaborate methods of table service (e.g. French, Russian, English) found in luxury restaurants, and those particular service techniques specific to speciality restaurants such as Chinese, Polynesian and Indian.

In many restaurants today separate bar areas are provided for pre and after-meal drinks. These have the double advantage of offering the customer a place to sit and relax away from the dining area and they allow a faster seat turnover in the restaurant.

Fast Food and Take-away: This sector of the industry is concerned with the preparation and service of food and beverages quickly for immediate sale to the customer for consumption either on or off the premises. Examples of fast food and take-away operations include the many themed hamburger units found in the high streets of most of the cities today, as well as operations such as fish and chip shops.

There are several identifiable characteristics common to this sector of the industry. Like,

(a) The units are usually themed around a product *(e.g.,* Hamburgers), a range of products (*e.g.* fish), or products of a country (*e.g.* Chinese).

(b) The units are often owned by large chains or are franchised.

(c) The 'Product' is very well-marketed, through the themed product, the decor and atmosphere, the high and consistent standard of the product, advertising on television, local radio and newspapers, and the container boxes for takeaway items.

(d) The pricing of the items is within a fairly distinctive known price range.

(e) The commodities used are often of the convenience type (*e.g.* frozen chips, concentrated beverage syrups).

(f) The method of food production is often partially or fully automated thereby de-skilling the job.

(g) The method of food service is simplified and basic.

Public Houses: Public Houses consist of a varied group of establishments, which mainly offer the general public alcoholic liquor for sale for consumption on and off the premises. The supply of food, at one time ancillary to liquor, has been seen more in recent years as a very important element in the 'product' for the consumer.

The characteristics of public houses are, firstly, that they all require a licence to operate and that the licence is only granted to suitable persons, and secondly, that most public houses are owned by a brewery company, providing an integration of their production with the retail distribution of alcoholic beverages.

In order to become more competitive, to meet customers demands, the catering premises in most pubs have improved considerably in recent years. Some brewer companies have classified their public houses by the level of catering offered. This ranges from those offering just sandwiches to hot and cold snacks, to a cold buffet counter, to a bistro-type operation, to a griddle or steak bar to a full a la carte menu. The range of food items offered mainly being of the convenience food type but at times extending to fresh food items. Several brewery

companies market specific pubs, by the type of catering offered, with a brand image, *e.g.* Beefeater Steak houses.

Major Welfare Sectors: Subsidised or welfare food and beverage establishments may be defined as those operations in which the profitability of the catering facility is the outlet's primary concern. Since the operations are either completely or partially subsidised by a parent body, such establishment's primary obligation is the well-being and care or their customers or patients. Unlike customers frequenting commercial sector operations, these customers often do not have a choice of catering facilities, for *e.g.* in hospitals and schools. Non-commercial operations are usually subsidised by government bodies which dictate an allowance per head, or by parent companies who may have a similar arrangement.

A distinction can be made between institutional catering and employee catering facilities, *e.g.* in hospitals and schools. Non-commercial operations embraces catering in institutions such as prisons, schools, hospitals. An important characteristic of this type of catering is that the market is not only restricted to the inmates of the institutions but in most cases it is also captive. In addition, in most cases, institutional catering is completely subsidised. Employee catering may be in the public or private ownership and covers the provision of food and beverage services to employees. The degree of subsidy in this type of operation varies considerably and also in many cases the market is not entirely captive. In other words, the catering outlet may be competing with the catering facilities provided at nearby restaurants, pubs and take-aways or with food brought in by the workers from their homes.

Institutional Catering: Institutional catering establishments include schools, universities, colleges hospitals, the Services, and prisons. In some of these establishments certain groups of customers do not have to pay for the provision of the food and beverage services as they are completely or partially subsidised by various government funds. This part of the catering industry is commonly referred to as the welfare sector. Very few catering

contractors are found in this sector so the majority are self-operated.

Schools: The school meals catering service was until recently structured on a dietary basis with a daily or weekly per capital allowance to ensure that the children obtained adequate nutritional levels from their meals. Most of the schools used to operate their dining rooms on a family-type-serve or a self-service basis with the traditional 'Meat and two veg' lunch being very much the norm.

There is today, however, a trend of provision of a snack-type lunch as an alternative to a main meal. Some schools now provide sandwiches, rolls, pies, soups, yoghurts, etc., and the children may choose from this selection. This trend has gone one step further in some areas which have drastically cut their school meal service and are providing more of the dining room space for the children to bring in their own lunches from home. Whether this trend will continue and spread in the future, or whether it is merely a current 'fad' is a debatable issue. It does seem likely, however, that now introduced, the snack-type meal will at least remain as an alternative to the traditional school meal.

A further trend since 1983 has been for some local education authorities hire specialist contract caterers.

Universities and Colleges: The public sector universities and local education authority institutions such as polytechnics, colleges of education and colleges of further education provide catering facilities for the academic, administrative, technical and secretarial staff as well as for full-and part-time students and visitors. The catering service in this sector of the industry suffers from an under-utilisation of it's facilities during the three vacation periods and also in many instances at the weekends.

Universities are autonomous bodies and are responsible for their own catering services. They are, however, publicly accountable for the expenditure to the University Grants

Committee (UGC) which allocates the funds on behalf of the Exchequer. The UGC's policy on catering allows for a subsidy on capital costs, *i.e.* "Buildings and equipment, Landlord expenses and rent and rates where applicable". Apart from a few special exemptions to certain universities, they are expected to break even. University catering units have traditionally been of two basic kinds:

(a) Residential facilities attached to halls which serve breakfast and evening meals.

(b) The central facilities which are open to all students and staff and usually serve lunches and snacks throughout the day with beverage. These catering facilities have to openly compete with the students's union services and independently-staffed senior common rooms.

The residential students pay in advance for part or in total for their board and lodgings. This method has been abandoned by many universities in recent years who have provided limited kitchen facilities in the residences to enable students to prepare and cook their own meals if they wished to and also introduced a pay-as you-eat system for residential students. Unfortunately, this introduction has reduced the catering revenue from students who prepare and cook their own meals.

To offset the losses incurred and to achieve a position of breakeven in catering, universities often utilise their residential and catering facilities by making them available at commercial rates to outside bodies for meetings, conferences and for holidays during the vacation periods.

Polytechnics, colleges of education and colleges of further education are financed by the local education authorities. There is no overall objective catering policy and the practices vary from one institution to another depending very much on the local authority and the local political party in power. In general most polytechnics and colleges catering is subsidised with policies usually requiring that food and labour costs be covered for the year from the catering revenue, by accepting losses in

the summer term because of circumstances requiring lower student attendance and by making a gross profit during the rest of the year.

Hospitals: Hospital catering facilities have improved considerably over the past twenty years with the result that new hospitals in particular are benefiting from the well-planned and managed catering services. Hospitals require a specialised form of catering as the customer is normally unable to move elsewhere and choose alternative facilities and therefore special attention must be given to the food and beverage. As with the school meals service, the hospital catering service is structured on a per capital allowance for patients but with staff paying for all of their meals..

Traditionally, a decentralised approach was used in the hospital in which the patient's food and beverage were portioned away from the main production area, at the hospital's kitchens and wards. This, of course resulted however, in the patients receiving cold, unappetising meals because of the length of time between the food being prepared and the patients actually receiving it. Today this method of food service is being replaced by the centralised approach which involves the preparation of the patients' trays in or close to the main production area. From here they are transported by trucks or mechanical conveyors to various floors, and from there directly to the patients so that there is very little delay between the food being plated and served to the patient.

A recent trend, though, has been for hospital catering to be open for tender by contract caterers where in many instances a general production system for several nearby hospitals would have to be operated to be viable.

The Services: The Services include the Armed Forces—the Navy, Army and Air Force; the Police and Fire service; and some government departments. In the Armed Forces they often have their own specialist catering brand for example the Army has the Army Catering Corps, and civil organisation such as the Metropolitan Police force and the Post Office service

also their own catering departments. The levels of food and beverage facilities for the Services vary from the large self-service cafeterias for the majority of personnel, to the high-class traditional restaurants for the more senior members of staff. A considerable number of functions are also held by the Services giving rise to both small and large scale banquet arrangements.

Prisons: The diet for the inmates is based upon fixed weekly quantities of specific named food commodities with a small weekly cash allowance per head for fresh meat and a further separate weekly cash allowance per head for the local purchase of dietary extras of which a proportion is spent on fresh fruit.

The catering within the prisons is the responsibility of the prisons governor with delegated responsibility being given to a catering officer.

Employee Catering: This is the provision of catering services to employees. The activity may be performed either directly by the employer or subcontracted out to contract caterers. A direct catering service which is running smoothly and being well-managed is unlikely to change to using a contractor. Employing the services of a catering contractor involves defining exactly what is required of the contractor in terms of level of service, costs per employee head, revenue, etc. At present contract caterers operate between 25 and 30 per cent of employee catering units and this figure has changed little since contract caterers first began to appear.

In providing a catering service for the employees, the parent company may decide at one extreme to subsidise the facility or at the other to pass all the costs on to the customer. There are various formulas for subsidising prices, but a general one is for the revenue from the catering facility to cover food and labour costs with the remaining costs, such as premises and equipment, fuel costs and management fees to be met by the employer.

In some sectors of the industry the catering service may

be provided virtually free, such as in oil companies, insurance and banking, with the employees making a small token payment per meal.

A variety of catering styles and level of services are found in industrial catering situations. The majority of the market is catered for by popular and fast-food facilities incorporating different methods of service, such as self-service cafeterias, buffet restaurants and vending operations. Management in large companies may also have the additional choice of waiter service facilities. At the top end of the industrial catering market, that is, those facilities catering for directors and executives, the standard of food and service can equal those found in commercial high class restaurants, with silver service being the norm.

The number of food and beverage outlets reviewed here illustrate the diversity of the hotel and catering industry. It is diverse because it caters for a diverse and growing eating-out market. As with all marketing situations, it is prone to change, but although there may be shifts from one sector to another in terms of volume, within the general structure of the industry the future points the way towards growth and expansion.

The Meal/Drink Experience: The 'meal/drink experience' may be defined as a series of events, both tangible and intangible that a customer experiences when eating out. It is difficult to define exactly where a meal/drink experience actually starts, and indeed ends, although it is usually assumed that the main part of the experience begins when a customer enters a restaurant and ends when he leaves the restaurant.

The series of events and experiences the customer undergoes when eating out may be divided into those tangible aspects of the product, that is the food and drink, and those intangible aspects such as service, atmosphere, mood, etc. In order for the product offered by the restaurant to be acceptable to the customer, it is necessary for the tangible and intangible components of the restaurant to combine together and to be seen by the customer as one total product, and not a number

of disjoined parts. It is therefore the duty of the restaurant owners or managers to construct this complete product and to present it to the customer.

The different components of the meal/drink experience that may be identified and which are considered by a catering facility include the following:

Food and Drink: The kind of food and drink that people choose to consume away from home depends mainly on the amount of money that they are prepared to pay for within a given set of circumstances at a specific point in time. Apart from the cost aspect, the factors which are of particular concern to the customer include:

(a) The type of choice of food and drink available.

(b) The quality of the product offered (*e.g.* fresh or convenience food; Chateau bottled or a blended non-Chateau bottled wine)

(c) The quantity of product offered *i.e.* the portion sizes.

(d) The consistent standard of the product. A customer returning to the restaurant and repeating his order of an item would expect the product to be very similar to what he had experienced before.

(e) The range of textures, flavours, aromas and colours offered by a food dish or the taste, aroma and colour offered by a drink.

(f) The appropriate temperature at which the food and drink is served.

(g) The presentation of the food and drink and how it enhances the product offered.

(h) The perceived value for money.

Variety in Menu Choice: The type of menu offered by an establishment and the variety of menu choice should also conform to the requirements of the total meal experience. At the lower level of the market the choice of many items in a restaurant is usually fairly limited for a number of reasons.

Firstly, price. If a customer is paying for a three-course meal the range of menu items that can be made available within the cost limits of such an operation is obviously more limited than in operations where the customer's average spending power is higher. Secondly, the amount of time taker to consume meals at this level of the market varies between less than half an hour and one hour, but rarely exceeds this. Of this amount of time the customer in these establishments takes proportionately little. In studying the menu choice. Thirdly, it may be suggested that customers frequenting this lower level of the market may feel uncomfortable if presented with a very large menu selection and prefer a more limited, but still varied menu choice.

In high-class restaurants where the average spending power is higher, the menu selection is normally much greater. In these establishments which encourage a luxury meal experience, the minimum amount of time customers usually spend on a meal is one and a half hours, and may often be three to four hours depending on the size of the group and the occasion. The proportion of the time that may be devoted to reading the menu and selecting from the menu items is correspondingly greater. The customers frequenting these types of establishments would also expect to be offered a large menu selection, including a number of chef's specialities and other dishes.

The menu choice offered by a restaurant is therefore dependent on a number of interrelated factors—the price the customer is willing to pay, the amount of time available for the meal experience; the level of the market to which the restaurant is catering and consequently the types of customers likely to frequent restaurant. Further considerations affecting the choice of menu from the caterer's point of view are the production and service facilities available, the skills of the staff, the availability of commodities and the potential profitability of the menu.

Level of Service: Broadly speaking, the higher the cost of

the meal the more service the customer expects to receive. In a self-service cafeteria the degree of service he receives is comparatively little. Customer collects his own food from the service counters, carries it to a table, and generally clears his dishes from the table at the end of his meal. As the cost of the meal to the customer increases so will the amount of service he receives. At the top end of the catering market, where the customer pays over Rs. 300 per head for a meal, full waiter silver service is provided.

The actual service of the food and beverage to the customer is described as the 'direct' service; part of restaurant's total service, however, is also comprised of 'indirect' services. These include the provision of cloakroom facilities—somewhere for the customer to leave coats and bags safely; and the availability of a telephone for customer use and this is particularly important in restaurants with a large business lunch trade. It is necessary for a restaurant to identify the level of service it is going to offer in its catering establishment and to extend this standard of service throughout all aspects of the operation.

Thus if a restaurant has a very formal type of food and beverage service, usually associated with high class operations, the other aspects of the restaurant service are equally formal-the speed, efficiency and dress of staff, the degree of personalisation and courtesy the customer receives, and so on. It is important therefore for a restaurant operation to consider not only the service of the food and beverages for which the staff are usually adequately trained, but also to take into account the indirect service aspects of the operation, which the customer particularly notices.

Value for Money: The concept of value for money varies from one sector of the market to another, and indeed from one customer to another. In majority of cases however, a customer will frequent a restaurant not only because of its food and service, but also because he feels it gives him value for money. At the lower end of the market inclusive price meals are often offered for *e.g.* by many of the fast food operations, so that a

prospective client is aware in advance what the total cost of the meal will be, and this will help alleviate any concern the customer may have about being able to pay.

At the other end of the market, menu items are often charged for separately, because at this level of the market the total cost of the meal is not such an important factor to the customer as perhaps are the other aspects of the operation, such as the standard of food or the level of service offered. However, there has been more emphasis over recent years on the set price menu in quality and luxury restaurants, particularly for lunch to attract business account meals where an idea of the cost can be ascertained beforehand.

Today, some establishments include a service charge in the price of their meals, others show it separately, while some operations do not include it in the individual price of the dishes, but add it to the end of the bill. Some schools of thought are of the opinion that by not showing these added extras in the price of the individual menu items, the customer may be encouraged to spend more because the prices will appear very reasonable; while others are of the opinion that the customer prefers to know exactly what he is paying and does not like to see these 'extras' added at the end of the bill. At present no one level of the market adopts one particular approach, although broadly speaking it is more usual for these additional costs to be separated out at the higher average spending level of the market, than at the lower.

Interior Design: The interior design of a restaurant is one of the first physical aspects of a catering operation that a customer comes into contact with. The first impression of the restaurant is very important. A potential customer passing by may like the look of the establishment and decide to come and eat there while the customer who has actually planned to eat in the restaurant and likes what he sees when he enters, feels pleased with his choice of restaurant.

The interior design of a restaurant is composed of many different aspects—the size and shape of the room, the furniture

and fittings, the colour scheme, lighting, etc. As with the previously described aspects of a restaurant, there is a need for a sense of totality in a restaurant's interior design. The colour scheme of the restaurant should blend, balance and be enhanced by lighting arrangements; tables and chairs should be ergonomically designed and allow for manoeuverability for both customers and staff.

The interior design of a restaurant contributes greatly to the creation of its image. A self-service cafeteria in a factory for example usually consists of a very large open area, tables and chairs of a standard design and shape, the colour scheme of the restaurant has few variations and lighting arrangements are purely functional. For this type of catering operation a consistently steady seat turnover is required, and this is achieved by designing the interior of the restaurant, such as pictures or murals to hold the customer's interest.

In luxury restaurants, however, the seat turnover is not so critical and infact customers may be encouraged to stay in the restaurant to spend more. In these types of establishments the interior design of the restaurant is very comfortable—the lighting in the restaurant is quite subdued; the colour scheme has warmth and depth, there may be several particular points of interest in the restaurant, such as pictures or murals to hold the customers interest; tables are farther apart, and may be separated from one another in booths or by partitions; and the chairs are so designed that the customer may sit in them for several hours without feeling uncomfortable.

The importance of an interior design to a catering facility and its success cannot be emphasised enough and it is essential that professional advice is obtained from experienced designers.

Atmosphere and Mood: The atmosphere or mood of a restaurant is a difficult aspect of an operation to define, but is often described as an intangible 'feel' inside a restaurant. Not all restaurants have an obvious type of atmosphere, others try to deliberately create one. For example, luxury high-class restaurants often have a very formal atmosphere which is

created by the dress and attitude of the staff, the decor of the restaurant, the service accompaniments, the type of clientele that frequent these restaurants, etc. Other restaurants such as the currently popular bars, try to create a relaxed informal atmosphere, and one that is very sociable and 'trendy' to be a part of and seen at.

The atmosphere of a restaurant is therefore affected by many different aspects of the operation—the decor and interior design of the restaurant, the table seating arrangements, the service accompaniments, the dress and attitude of the staff, the tempo of service, the age, dress and sex of the other customers, etc. All these various aspects combine together to produce the restaurant's atmosphere and, it is important therefore that all these elements should combine and compliment one another to present a picture of totality to the customer; should one of these aspects be out of keeping with the others, disharmony may result in the customer's image of the restaurant, and the customer will invariably leave remembering that one small aspect.

In a high-class restaurant for example it would be expected that linen napkins would be provided; should paper ones be provided the customer may feel cheated when he considers the price he is paying for the meal.

Expectation and Identification: A single customer or a group of customers arriving at a restaurant for a meal bring with them a series of expectations regarding that restaurant—the type of service they will receive, the price they will pay, the expected atmosphere and mood of the restaurant, etc. The customer's expectation may therefore be varied and numerous, ranging from the restaurant which the customer frequents because he wants to be seen there and participate in its social atmosphere to the small quiet restaurant where the customer may go because of its intimate and personal nature.

Upon arrival at the restaurant, if the product presented to the customer is in harmony with his choice he will have a relaxed and enjoyable meal. Should the customer sense

disharmony, however, between his expectations of the restaurant and the actual product he finds, like for example it is too intimate for the occasion, he may not enter the restaurant but choose another. If the disharmony is not realised until the customer is eating at the table, it is unlikely that he will leave but will have a hurried and uncomfortable meal.

There is therefore a need for the customer to identify and associate himself with a particular restaurant for a particular meal occasion. He may not always identify with the same restaurant, as his needs and expectations may vary from one meal experience to the next.

For example, at a business lunch a customer may require an expensive hot cuisine restaurant with an atmosphere condusive to discussing business. Such a restaurant, however, may not be suitable for the same customer to take his wife and family to for a special occasion. A customer therefore, has different needs and expectations on different meal occasions, and similarly at different times of the day, for example lunch and dinner.

These alternating needs of customers should be identified by a restaurant and catered for, so that the restaurant offering formal business lunches may offer special function catering in the evening, when the demand for business meals is minimal. There is a danger in these situations, however that restaurants may be led into catering for mixed markets, and it is important therefore for a restaurant offering different levels of service at different times of the day, to keep them completely separate, and not attempt to be all things to all people. Those operations that have taken this approach invariably adjust downwards to a lower socio-economic market segments than the one to which they were originally catering.

Location/Accessibility: The location of the restaurant is one of the few, if not the only unchangeable aspect of the operation, and is therefore one which requires considerable thought and attention at the initial planning stage. The

restaurant's location in relation to its present markets should not only be considered, but also its location to possible future markets. For example a restaurant may rely heavily on several large firms in the surrounding area for the majority of its lunch-time trade and should several of these firms leave the area this would seriously affect the restaurant trade.

Accessibility to the restaurant is another important factor, particularly if it is situated out of a town and not within easy reach by public transport. In these cases the majority of the restaurant's clientele would travel by private transport and thought must be given to car parking spaces for the customers.

Staff: Staff employed by a restaurant operation should complement the meal experience of the customers, and they are able to do this in a variety of ways their attitude to customers, their age and sex," their uniform, the tempo of their service and so on.

The number of staff servicing a restaurant is closely related to the prices charged by the establishment and the level of service that it offers. In self-service operations very few service staff are required; in some establishments the ratio may be as high as one member of staff to eight customers. These latter types of operations are, however, charging the customer for this extra attention service and must therefore be seen to have an adequate number of staff.

Not only does the number of staff in a restaurant contribute to the meal experience, but also their attitude to customers and the tempo of their service. In a large industrial cafeteria where the ratio of service staff to customers is low, and speed of staff is important, the staff are required to work at a fast and efficient speed, and where possible leave the customers to serve themselves.

In a luxury restaurant the tempo of the staff is considerably lower and more relaxed because of the high ratio of service to customers. It should be noted that the attitude of the staff

is almost totally influenced by the management attitude and the environment in which the staff are working.

Additionally, such things as the level of skills of the staff, their visual cleanliness, as well as their sex, age and nationality are of importance.

Finally, the uniforms of the service staff should be appropriate for the level of the catering operation, and again this physical aspect of the restaurant must be seen to be a part of the establishment's totality. In some lower market level operations the staff may only be provided with overalls; this is in complete contrast to high class restaurants, particularly in hotels, where there is a very strict demarcation of uniform styles according to the status of the service personnel.

2

Wine and Liquers

Wines and their Application

All wine bottle openers have a corkscrew with a spiral edge and a sharp tip, used to pierce the foil around the cork. The corkscrew should be inserted into the cork absolutely straight and then driven in a clockwise direction. The tip should never pierce the cork as it could dislodge particles of the cork into the wine. The corkscrew is moved clockwise till 3/4 of it is above the mouth of the bottle, and then it is moved anticlockwise to dislodge the cork from the opener. Sparkling wine bottles are always opened with bare hands. The different types of wine bottle openers are:

T-spiral: The oldest type of opener, it is best used for opening bottles of white wines. It consists of a simple T-shaped implement, the spiral comprising the vertical limb of the T, while the handle forms the bar. After removing the tin foil from the bottle, the cork is wiped with a waiter's cloth. The opener is held horizontally, then turned over into the vertical position and inserted into the centre of the cork.

T-spiral with Cap: This opener resembles the T-spiral, but

has a metallic cap around the screw. The corkscrew is driven into the cork till the cap grips the mouth of the bottle. The screw is then moved clockwise till the cork is 3/4 above the mouth; finally the screw is moved anticlockwise to dislodge the cork.

Wooden Double Barrel: This opener consists of a corkscrew encased in two wooden barrels or cases, one above the other. One barrel, when moved clockwise, makes the screw descend, and the other, when turned anticlockwise, makes it rise.

Angel Wings Wine Bottle Opener: The most popular multipurpose wine bottle opener, the angel wings opener has two projecting metallic wings on either side of the screw, with a handle on top. When the handle is moved' clockwise, the wings rise till they are horizontal (which lifts the cork upwards); after this, the wings are pulled down to raise the cork and then the handle is moved anticlockwise to dislodge the cork. The screw has a plastic cap at the bottom to protect the cork.

Waiter's Friend: A multipurpose wine bottle opener, the waiter's friend is a can opener, knife, bottle opener and wine bottle opener rolled into one. After pushing the screw into the cork, the other end of the opener is pressed down over the edge of the mouth of the bottle. It is then pulled up and moved in an anticlockwise direction.

Red Wines are of Three Types: Young (which have been matured for a short period and have no sediment or lees), moderately old (which have a small amount of sediment) and very old red wines (which have heavy sediment)

The equipment required for the service of red wine includes:

(i) The ordered wine bottle
(ii) A wine bottle opener
(iii) Red wine glasses
(iv) A quarter plate
(v) A waiter's cloth
(vi) An ashtray
(vii) A wine salver (14' diameter)

The procedure followed for the service of young red wines is as follows:

(i) The wine list is presented.

(ii) The guest's order is taken and the BOT (Beverage Order Ticket) cut.

(iii) The wine bottle is taken from the main bar or the cellar and the BOT deposited.

(iv) The rest of the equipment is checked , laid on the wine salver, and placed on top of the sideboard.

(v) The wine bottle is presented. A folded waiter's cloth is spread on the left forearm, and the bottle supported on it with the label on top, & the base of the bottle gripped with the fingers. On the right of the host, the neck of the bottle is raised so that the label is clearly visible, and simultaneously the name (and vintage year, if any) of the wine announced.

(vi) A wine glass is placed on the right of the host, just below the water goblet. Glasses for all the guests are placed and the ashtray is placed in the centre of the table. The wine bottle and the rest of the equipment is placed on the right of the host.

(vii) The bottle is opened and the cork smelled. The cork should smell of wine and not cork; a corky wine is spoilt. The cork is placed on the quarter plate and presented to the host for his approval. If the wine is corky, the waiter must inform his superior-a new bottle may be opened.

(viii) A small pad of the waiter's cloth is made and held it in the left hand. The wine bottle is held in the right hand, with the index finger on the shoulder, three fingers on one side and the thumb on the other side. 1/8th of the wine is poured into the host's glass, for approval of temperature. The mouth is wiped with waiter's cloth. The wine may be 'over' (warmer than the desired temperature) or 'under' (cooler than

desired). In either case, the host is asked if the entire wine should be left for 'rooming' or 'chamber' (to attain room temperature), or whether less than half the wine should be served right then. In the latter case, the bottles are placed on the sideboard.

If the wine is 'over' it may be cooled by placing in the refrigerator for a while. To warm the cool wine, an uncorked bottle is placed in the room.

(ix) The guests are served, in the correct sequence, serving the host in the end. If any wine is left over at the end, the bottle is placed on the host's right, with the label facing the host. The quarter plate is removed and the opener is placed on the sideboard.

(x) If the host wants another bottle of wine opened, the entire procedure must be repeated, and the glasses should be changed. The ashtray is changed when needed.

(xi) For moderately old or very old wine, decanting is recommended, *i.e.*, transferring wine from the bottle to a carafe or decanter, for separating wine from the sediment. Decantation must always be done in the cellar, so that the lees are not disturbed while moving the wine bottle.

For decanting, a funnel is placed in the decanter, with a tall burning candle, behind it, for light. Holding the wine bottle carefully, the wine is poured into the decanter until the wine is just above the lees in the bottle. The rounded shoulders of the bottle hold ,the sediment and prevent it from escaping. The original bottle and cork is taken to the host along with the decanter, for his approval.

If the host wants the wine to be opened at the table, the bottle should be brought to the table in a wine cradle, in order to disturb it as little as possible. The bottle is opened in a lying position, and is then presented in the cradle itself. It is taken to the cellar, decanted and served.

All white and rose wines are served cold, at a temperature

of 48-50°F while sparkling wines are served chilled. To maintain the coolness of the wines, wine buckets or wine coolers are used. These are of three types:

1. A small wine cooler, used for white and rose wines, with a quarter plate as an underliner.
2. A large Champagne cooler, used for sparkling wines, with a large plate as an underliner.
3. A wine cooler with a stand (in the absence of which a half plate may be used as an underliner). The stand has a flat disc with a hole in the centre, to allow moisture to drip through it. While the exterior of the wine cooler is highly polished, the interior is made of an insulating material and has a matt finish. This wine cooler may be used for white, rose or sparkling wines.

While serving white wine, the bottle is first placed in an angular position in the empty wine cooler, which is then filled with ice cubes and some water. The water, besides having a better cooking effect, also makes the replacement of the bottle in the wine cooler, much easier. The wine cooler is covered with a waiter's cloth, folded in half, which is used for wiping the bottle when removing it from the cooler. The equipments required for the service of white wine are as follows:

(i) The ordered wine bottle
(ii) Wine cooler with underliner .
(iii) Required number of white wine glasses
(iv) Wine bottle opener.
(v) Two waiter's cloths, one to be used to wipe the bottle and cover the bottle in the ice-filled wine cooler, and the other, placed on the waiter's forewarm, to be used to wrap the bottle (in order to prevent to hands from warming the wine. This is not needed if gloves are used).
(vi) Ashtray and quarter plate.
(vii) Wine salver.

The procedure for serving white wines is as follows:

(i) The wine list, is presented, the order taken, and a BOT (Beverage Order Ticket) cut. All the equipments are collected and brought to the sideboard.

(ii) Glasses and the ashtray are placed on the table. One quarter plate is placed on the side, to be used for presenting the cork. The wine cooler and the wine bottle opener are left on the table. If a wine cooler with a stand is being used, it is placed on the right of the host. The wine salver is removed to the sideboard.

(iii) Holding the mouth of the bottle with the waiter's cloth, it is removed from the wine cooler, simultaneously wiping the bottle with the waiter's cloth covering the cooler.

(iv) The bottle is presented, with the label facing the host, and the name of the wine is announced, along with the vintage year, if any.

(v) The bottle is placed on the right of the host and opened. The cork is presented for the host's approval.

(vi) The waiter's cloth, is folded into four on the neck of the bottle and folded at the two ends diagonally across each other. The doth is pulled downwards, holding the two ends, thereby tightening it around the bottle. The bottle is held with the index finger at the point where the two ends of the waiter's cloth cross.

(vii) A little wine is poured for tasting. To stop the flow, pouring is stopped and the bottle is twisted slightly upwards (this can only be done in the case of white wine. If this was done with a bottle of red wine, the wave thus created would disturb the sediments at the bottom of the bottle. White wines however, do not have sediments).

(viii) If the host is not satisfied with the wine, and if it is over chilled or under chilled, the bottle would be either kept for rooming or replaced in the wine chiller. The host

may also like a part of the wine to be served right then wine. It should again be offered for approval, before being served to the guests. If any wine is left over, the host is asked whether the bottle should be left on the table, or replaced in the wine chiller. If the bottle is to be placed in the chiller, it's neck should be away from the host.

If the host would like another bottle of wine, the entire process must be repeated all over again, changing the glasses as well. White wines are best with white meat, namely chicken or fish. Dry Chablis is specifically suited to oysters.

Champagnes are mostly dry, if not specifically mentioned. They include the following brand names: Dom Perignon, Charles Hiedsieck, Moet et Chandon, Veuve Clicquot Ponsardin (white and rose), Laurent Perrier (white and rose), Alfred Rothschild, Perrier Jouet, Mercier, Bollenger, Lanson, Lanson Black Label, Blanc de Blanc. All sparkling wines are served in the same manner, at 45°F, covered with a napkin, called a 'baby wrap'. Whereas Champagne is served in Champagne flutes, saucers or coupes, other sparkling wines are served in an all-purpose tulip. The bottle is always opened with bare hands; the cork is never given for smelling and the wine is offered for tasting. The host is always served first, followed by the guests.

Champagne may be served in three ways:

1. At any time, of the day with any food, in a champagne coupe.
2. At a celebration—in a Champagne saucer. For this the bottle is agitated, to build up the pressure. The cork flies off and the champagne is poured intermittently into the ready glasses, giving time for the froth to settle before pouring further.
3. With food—in a Champagne tulip/flute. Less surface area is exposed, so the gas escapes slowly—also the flute occupies less space on the table.

If champagne is ordered, it is served from the bottle; but if it is served complimentary, the glass full of champagne is brought from the bar. The bottle is kept closed by using a special stopper.

The procedure for serving Champagne (and other sparkling wines) is as follows:

(a) The wine list is presented and the order taken.

(b) A BOT is made out and all the equipments are bought.

(c) The bottle is placed in the Champagne cooler with ice and water. It is wrapped with a napkin and covered with another napkin when brought to the table.

(d) Ashtray and glasses are placed on the table. The cooler is placed on the right of the host.

(e) The bottle is removed from the cooler, simultaneously wiping it. Holding the neck of the bottle with the other napkin the bottle is presented, announcing the wine. The bottle is placed on the table and wrapped.

(f) The bottle is opened. First, the tourniquet and the foil are removed. The wire loop is twisted and the wire removed. Holding the bottle (wrapped in the 'baby wrap') at 45° angle from the glass, in the right hand with the thumb on the cork, the cork is grabbed with the left hand, twisted off and the wine poured into the glass.

(g) Then the bottle is rewrapped and replaced in the chiller.

Brand Names of Champagne

1. Dom Perignon
2. Charles Hiedsieck
3. Moet et Chandon
4. Laurent perrier
5. Perrier jouet
6. Mercier
7. G.H. Mumm's-Cardon Rouge

8. Pol Roger
9. Veuve Cliequot ponsardin
10. Lanson Black Lable
11. Alfred Rothschild
12. Bollenger.

Various Combinations: The following alcoholic beverages are best served with the combinations mentioned below:

Whisky

(a) Whisky may be consumed straight (neat) in an old-fashioned whisky glass.

(b) On the Rocks-Three or four cubes of ice are put into the glass, and then the spirit is poured over. Putting the ice in first helps cool the whisky at a faster pace and also prevents splashing.

(c) Whisky may be served with water and ice.

(d) It can be served with soda and ice.

(e) It may be served with a mixture of half a soda and half water.

(f) Whisky may be consumed with mineral water.

(g) It may be mixed with tonic water.

Avoid mixing whisky with juices as it spoils the aroma of the whisky.

Gin: May be consumed

(a) Straight

(b) On the rocks

(c) With tonic water, well-chilled

(d) With Lime cordial and soda

(Without soda, it is a 'gimlet'. A sweet gimlet has equal quantities of cordial and gin, with cherry and a dry gimlet has 15 ml cordial, 45 ml. gin and is garnished with a olive or lemon slice. It is stirred with ice in a mixing glass, strained and served).

(e) With citrus drinks like Limca or Citra.

(f) With Lemon juice and soda.

(g) Gin can replace vodka in many cocktails such as Bloody Mary and Screwdriver.

Vodka

(a) Vodka is best consumed neat and well-chilled, drunk in one gulp.

(b) It may be served with soda or water.

(c) Vodka may be served with various juices as a cocktail.

(d) It can be mixed with citrus drinks like Limca.

Rum

May be consumed.

(a) With cola, well-stirred. (The best mix)

(b) With water and ice.

(c) With soda and ice.

(d) As a soothener in winter, with lukewarm water.

Brandy

May be consumed:

(a) Straight (Cognac is served in a brandy inhaler; other brandies are served in an old fashioned whisky glass).

(b) With lukewarm water.

(c) With Ginger Ale at room temperature.

(d) With 1 teaspoon honey, 2-3 drops lemon juice, 1/2 teaspoon ginger juice, mixed with 30 ml. Brandy and water as desired, to obtain a good soothener.

Beer: Beer would always be served chilled but ice should never be added as it absorbs carbon dioxide and makes the beer flat, which can cause dysentry. Beer should be poured into a tilted glass from the side beer of the glass, so that a good head of foam is formed. Beer may be served in tankards, mugs, goblets and various other types of beer glasses. It is most

appropriately served in a glass, so that 2½ glasses can be obtained from 650 ml beer bottle.

Fortified Wines: 60 ml of a fortified wine are served per portion, and it must be served in the appropriate glass at a temperature of about 48-50°F (port at 54°F). If a bottle is opened, it must be opened with bare hands, as most fortified wine bottles (especially port and sherry) have a mushroom cork. Otherwise the service is the same as that of red wines.

There are a rich variety of theories to account for the origin of word 'Cocktail'. All that is know for certain is that Cocktails as they were called in the American Bars came into popularity after the 1914-18 war and reached their zenith in the late twenties and early thirties. Cocktail is taken before any meal to stimulate the appetite and aid in digestion. "It is a mixed drink made of spirits, bitter, flavouring and may be sugar syrups. This definition however is not strictly accurate as all Cocktails may not contain all four of these ingredients.

Qualities of a Good Cocktail

1. It must whet the appetite and not dull it. This first basic requirement of a good cocktail automatically eliminates a host of over-sweetened over-fruit juiced, over-egged, over-creamed cocktails.
2. It should stimulate the mind as well as the appetite. A well-made cocktail is one of the most delicious of drinks. It pleases the senses. The sheer delight of those who usually partake this refreshing drink breaks the ice of formal reserve.
3. It must be pleasing to the palate. In order that a Cocktail may satisfy the palate, it must be dry, yet smooth, indeed in compounding a Cocktail, the first thought should be the production of a drink sufficiently dry to wake up the taste buds yet not so sour or so bitter or so aromatic as to be impalatable.
4. It must be pleasing to the eye.

5. It must have sufficient alcoholic flavour to be readily distinguishable.
6. It must be well-iced.

Cocktail Ingredients: Every Cocktail properly so called, must contain two distinct types of ingredients.

It also may but need not necessarily contain a third type. They are:

1. A base.
2. A modifying agent, smoothing agent.
3. Additional flavouring and colouring ingredients.

A Base: This is the fundamental and distinguishing ingredient of the Cocktail and usually comprises more than 50% of the entire volume of the Cocktail. Strictly speaking the base must always consist of spirits like—Whisky, Gin, Brandy. By common acceptance however combination of various liquors or aromatic wines as a base have also come to be called as Cocktail. For example-A mixture of Italian and French Vermouth with bitter. Normally the Cocktail base will consist of a single liquor and this one liquor being the distinguishing and predominant ingredient determines the type of Cocktail. Thus we have gin-based Cocktail, whisky-based Cocktails, rum-based Cocktail and within certain limits it is possible to combine two liquors as a base.

A Modifying Agent: It is an ingredient in combination with the base which characterises the Cocktail and without this ingredient the base no matter how vigorously shaken, how thoroughly chilled would still not be a cocktail & would remain merely a chilled liquor. Its functions are to smooth down the biting sharpness of the liquor and at the same time to add character to its flavours. The flavour of the modifyer in itself should never predominate but should always remain submerged.

The modifying agents can be further divided into 3 groups.

Aromates: Include aromatic wines such as French and

Italian Vermouths, bitters of various types. For example e.g. Angostura bitter.

Bitter: The generic name of highly flavoured pungent and bitter liquor compounds which are prepared from the fruits, bark, roots, or leaves of various trees and plants, mostly from sub-tropical or tropical counties. They bear the name either of whatever flavour they may be or the registered names of the people who make them i.e. Angostura bitter.

Fruit Juices: All fruit juices which are used such as orange, lemon, pineapple, tomato etc. with or without sugar.

Miscellaneous Smoothing Agent: These include all types of sugar syrups, cream, eggs, milk etc.

All these modifying agents, particularly the aromatic wines and above all the bitter must be used with great discretion. Just how far one should go with each agent one has to learn with experience, relying both on his palate and on the comments of his friends. With bitters the safe rule is that never more than one or two dashes with a special sprinkler top to each drink. It is somewhat more difficult to lay down any general rule as to the use of citrus juices.

In using cream or eggs one must remember that one is preparing a drink to stimulate the appetite and not dull it. The object of using cream or eggs is merely to produce a drink pleasing to the eyes because of its creamy and foamy texture and at the same time make it smooth tasting. A reasonably safe rule for these miscellaneous smoothing agents is a maximum of 1/2 white of an egg, one yolk of egg, 1 teaspoon full of cream, one teaspoon of sugar syrup to each drink.

Additional Flavouring Agents: These include all the various cordials and liqueurs, sometimes the ingredients which are used as a modifying agent in one cocktail may be used for flavouring and colouring in another.

Classification of Cocktails: It is estimated that there are over 10,000 mixed drinks and cocktails in use. It is quite obvious that it is impossible for any one man to know them all. The

greatest problem that faced the bartenders when cocktail became a rage was to ensure uniformity in their composition. The United Kingdom Bartenders Guild (U.K.B.G.) made this one of their main objectives over a period of years. They have compiled a Cocktail recipe book which is universally accepted as an authentic method of dispensing the drinks.

Iced mixed drinks can be roughly divided into four groups. Following is a short selection of the few standard drinks in each group.

Aperitif: These must stimulate the appetite. They must, therefore, be appealing to the eye and have an immediate effect. Broadly speaking, these drinks are usually made from wines mixed with the spirit, e.g. Dry Martini, Manhattan; Rob-Roy etc.

Cocktails: This is the largest group of all, and consists of all short, shaken mixed drinks. They should contain a spirit base, colouring and flavouring ingredient and may be a fruit juice or other modifying agent. Most common ones are Sidecat, White Lady, Bronx etc.

Long Drinks: These are served in tall tumblers such as High Balls, or Collin's glass.

Stimulants: Instead of going to the doctors, if one has to put one self into the hands of the bartender when one has a hangover, and he will prescribe cocktails for one's needs. These are also described as "Pick-me ups." Champagne Cocktail etc.

Aperitifs

(a) *Martini:* When taking orders for Martinidry or sweet, one must always make sure that the customer is asking for a Martini Cocktail and not simply a Martini Vermouth, sweet or dry. The latter is simply a Martini Vermouth with ice, a slice of orange and may be a splash of soda if required and served in goblet type glass.

(b) *Dry Martini:* 45 ml. dry gin, 15 ml dry vermouth (French)

with cubes ice in the mixing glass, strain into cocktail glass, and serve with an olive, or squeeze a twist of lemon peel on top.

(c) *Sweet Martini:* 30 ml. dry gin, 30 ml. sweet vermouth (Italian) stir and strain as before, add a cherry.

(d) *Sweet Manhattan:* 30 ml. Rye whisky, 30 ml. sweet Vermouth, stir and strain into a cocktail glass, add a cherry.

(e) *Dry Manhattan:* 45 mls. Rye whisky, 15 ml. Dry vermouth, a dash of Angoustra bitter. Stir and strain into a cocktail glass.

(f) *Bamboo:* 30 ml. Dry sherry, 1/4 French vermouth, 1/4 Italian vermouth, stir and strain as before. Add a twist of lemon peel.

(g) *Dubonnet:* 30 ml. Dry gin, 30 ml. dubonnet. Stir and strain into a cocktail glass, add a twist of lemon peel.

(h) *Rob-Roy:* 45 ml. Scoth whisky, 15 ml. sweet Vermouth, a dash of Angoustra bitter, stir and strain into a cocktail glass, add a cherry.

Cocktails

(a) *Bronx:* 30 ml. Dry gin, 15 ml. fresh orange juice, 15 ml. French Vermouth. Use a shaker and strain into a cocktail glass.

(b) *White Lady:* 30 ml. Dry gin, 15 ml. Cointreau, 15 ml. fresh lemon juice. Use a shaker and strain into cocktail glass.

(c) *Side Car:* 30 ml. Cognac brandy, 15 ml. Cointreau, 15 ml. fresh lemon juice; Use a shaker and strain as before.

(d) *Pink Lady:* 30 ml. White of egg, 30 ml. dry Gin, dash of Grenadine. Shake well and strain into a frosted glass.

(e) *Bacardi:* 30 ml. Bacardi rum, 15 ml. lemon juice, 15 ml. grenadine.

(f) *Alexandera:* 30 ml. Cognac Brandy, 15 ml. Creme de

Cacao white, 15 ml. French cream. Shake well and strain into cocktail glass.

Long Drinks

(a) *Horse Neck:* Peel the lemon into a continuous spiral strain, hang it from the edge of a high ball glass. Place few cubes of ice in the tumbler and a dash of Angostura bitter, 60 ml of brandy, fill with Ginger Ale.

(b) *High Balls:* A cube or two or ice in the high ball glass with the desired spirit and mineral water or plain water.

(c) *John Collins:* Tall glass or a Collins glass, gin, 15 ml. lemon juice, cubes of ice, 1/2 teaspoon of powdered sugar. Stir with a long spoon and add soda.

Note: John collins is usually prepared with Hollands Dry Gin whereas Tom Colins is prepared with London Dry Gin.

(d) *Gin Fizz:* Pour 60 ml. gin, 15 ml. lemon juice, a little sugar syrup or powdered sugar. Shake well pour into a sour glass (Delmonico) and add soda. A little white of egg can be added to the shaker to retain the froth for a longer time.

(e) *Pimms' Cup:* There are a number of pimms cups:

Pimms Cup No. 1	With Gin
Pimms Cup No. 2	With Whisky
Pimms Cup No. 3	With Brandy
Pimms Cup No. 4	With Rum
Pimms Cup No. 5	With Vodka

A tankard is used, put a little cucumber peel, a little fresh fruit (season wise, grapes, oranges, apple) add some cubes of ice and 60 ml of the appropriate spirit. Fill the tankard with lemonade.

Stimulants (Pick-Me Ups)

Flips: All flips are usually prepared as follows:

Use a shaker, a yolk of egg, a teaspoon of powder sugar,

60 ml. of sherry port or 60 ml. of spirit depending upon the type of flip required. Shake well and strain into a cocktail glass, grate a little nutmeg on top.

Noggs: Into the shaker, one whole egg, a little powder sugar, 60 ml. of sherry, port or 60 ml. of desired spirit depending upon the type of Nogg required. 1/4 pint of milk. Shake well and hard, strain into a tall glass or tumbler, grate a little nutmeg on top.

Champagne Cocktail: Into a Champagne coupe glass, place one cube of sugar saturated with Angostura bitter, one slice of orange and fill the glass with Champagne. A little brandy can be added where desired.

Pousse Cafe: This is an exotic drink consisting of liqueurs of different colours floating one on top of another according to their densities in a straight-sided tall glass.

Examples:

1. *Pousse Cafe Americano:* Equal portions of Marashino liqueur, Curacao (Red), Chartruse (green) cognacy brandy.
2. *Pousse Cafe Finnalandia:* Equal portions of Grenadine, Creme de Methe (green), Marashinao and Curacao.

Cocktails: Recipes

Between the Sheet	1 dash of lemon juice
	15 ml. Brandy
	15 ml. Bacardi Rum
	15 ml. Cointreau
	Shake well and strain into cocktail glass. (6 portion)
Cherry Blossom Cocktail	30 ml. Curacao
	30 ml. Lemon Juice
	30 ml. Grenadine
	150 ml. Cherry Brandy
	120 ml. Brandy

	Shake well thoroughly and serve very cold.
Daiquiri Cocktail	1 teaspoon powdered sugar 15 ml. Fresh Lime Juice 45 ml. Bacardi Rum Shake well and strain into cocktail glass.
Gibson	10 ml. Dry Vermouth 50 ml. Gin Stir well and strain into cocktail glass. Squeeze lemon peel on top and add an onion.
Grasshopper	14 ml. Creme de Menthe (green) 30 ml. Creme de Cacao (white) 45 ml. Rum Shake and strain into a large size cocktail glass.
Paradise Cocktail	1 dash Lemon juice 15 ml. Orange juice 30 ml. Gin 15 ml. Apricot Brandy Shake well and strain into cocktail glass.
Rolls Royce	1 Dash Benedictine 15 ml. Dry Vermouth 15 ml. Sweet Vermouth 30 ml. Dry Gin Shake well and strain into cocktail glass.
Americano	15 ml. Campari 45 ml. Sweet Vermouth Lump of ice Use a wine glass or a goblet.

Fill with soda, stir and add a twist of lemon peel.

Bloody Mary
45 ml. Vodka
90 ml. Tomato Juice
Tomato juice, 2 dashes of Worcestershire sauce,
Tobasco Sauce (optional), salt and pepper.

Cuba Libre Cooler
60 ml. Bacardi Rum
Juice of 1/2 a lemon
Use long tumbler. Add ice and lime wine and fill up with coca cola.

Gin Sling
Dissolve 1 teaspoonful of sugar in water.
60 ml. dry gin.
1 lump of ice.
Serve in a long tumbler which is filled with water or soda.

Negroni Cocktail
15 ml. Gin
15 ml. Sweet Vermouth
15 ml. Campari
Stir well and strain into cocktail glass.
Serve with a twist or orange rind.

Planters Pinch Cooler
30 ml. Lemon or Lime Juice
15 ml. Gommes syrup
1 dash Angoustra Bitters
60 ml. Jamaica rum.
Shake, pour unstrained into tall tumbler, add a slice of pineapple and slice of orange and serve with a straw.

Rickey	Ice cubes Juice of 1/2 a lime or 1/4 lemon 60 ml. of any spirit (Whisky, Gin, Bourbon, Apple, Brandy) Add the spirit into medium sized glass, lime juice, ice and fill the glass with soda and leave rind of lime or lemon glass.
Screw Driver	15 ml. Vodka 2/4 Orange juice Stir well and strain into hi-ball glass Add a slice of orange.
Singapore Sling	Juice of 1/4 lemon 16 ml. Dry Gin 45 ml. Cherry Brandy Shake well, strain into medium-size glass and fill with soda, add cubes.
Brandy Fix	Place in a small tumbler. 1 teaspoonful of sugar. 1 teaspoonful of water to dissolve the sugar juice of 1/2 a lemon, 30 ml. of Cherry Brandy and 45 ml. of Brandy. Fill the glass with fine ice and stir slowly a slice of lemon and serve with a strain.
Smash	Dissolve 1 lump of sugar in a medium-size glass and add 4 leaves of mint, crushing the sugar and mint lightly together. Put a lump of ice in the glass and then add 60 ml. of Bacardi Rum, Brandy, Gin or Whisky. Decorate with a slice of orange and squeeze lemon peel on top.

Sours: (Gin Sour, Whisky Sour)	Juice of 1/2 a lemon 1 teaspoonful sugar 60 ml. of Gin Whisky Shake well and strain into medium-size glass and add a slice of orange and a cherry.
Fizzes: (Gin Fizz)	Juice of 1/2 a lemon 1/2 a tablespoonful powdered sugar 60 ml. Gin Shake well, strain into medium-size glass and fill with syphon soda water.
Brandy Fizz	Juice of 1/2 a lemon 30 ml. tablespoon powdered sugar 60 ml. Brandy Shake well, strain into medium-size glass and fill with syphon soda water.
Sherry Sungaree	2 ml. sherry. 1 teaspoon fine sugar. Till the tumbler with ice and pour sherry and sugar. Grate nutmeg on top.
Port Sungaree	60 ml. Port Wine 1 teaspoon sugar. Till the glass with ice and pour port and sugar grate nutmeg in top.
Flips (Rum Flip)	1 egg yolk 1 teaspoon powdered sugar 60 ml. Rum, Brandy and port wine or Sherry wine.

	Shake well and strain into medium-size glass.
	Grate nutmeg on top. In cold weather a dash of ginger can be added.
Cobblers	Half tilt the glass with pushed ice, add one teaspoon powder sugar, 60 ml. Gim/ Whisky/ Brandy/Port or shemy. Stir well and pour into the glass and decorate with a slice of orange.
Egg Noggs	1 whole egg
	1 teaspoonful powdered sugar
	60 ml. of any desired spirit or 60 ml. of fortified wine.
	Milk 150 ml.
	Shake well and strain into long tumbler. Grate a little nutmeg on top.
Baltimore Egg	1 fresh egg
	30 ml. tablespoon sugar
	30 ml. Brandy
	30 ml. Rum
	30 ml. Maderia
	1/2 a pint fresh milk.
	Shake well and strain into a long tumbler.
	Grate a nutmeg on top.
Rusty Nail	30 ml. Scotch whisky
	30 ml. Drambuie
	Into the mixing glass add the ingredients, put some ice cubes. Stir and strain into a

	Roily-Polly glass, add orange peel as garnish.
Pina Colada	60 ml. gin 60 ml. Pineapple juice 60 ml. Coconut milk Take a cocktail shaker, add the ingredients, some ice cubes and shake it well. Strain into a Collin's glass and add pineapple slice as garnish.
B and B	30 ml. Cognac Brandy 30 ml. Benedictine Mix them well with some ice cubes in a mixing glass and strain into a Brandy Baloon Glass.
Black Russian	30 ml. Vodka 30 ml. Kahlua Stir them in a mixing glass with ice cubes and strain into a parfait glass. Top it up with soda.
Margarita	45 ml. Tequila, 15 ml. Cointreau 10 ml. Lemon Juice. Stir them well with ice cubes in a mixing glass and strain it into a champagne strainer.
Gimlet	60 ml. Gin 15 ml. Lime Juice cordial Stir them well with some ice cubes in a mixing glass and strain it into a cocktail glass. Garnish with lemon slice.
Harvey's Wall Banger	Harvey' Wall Banger 30 ml. Vodka

	30 ml. Galliano 150 ml. Orange juice Stir well and pour into the glass Garnish with orange slice and cherry.
Margarita	45 ml. Tequila 15 ml. Triple Sec or Cointreau 30 ml. Fresh lemon juice Pour all the ingredients into the cocktail shaker & shake well. Pour into the glass which has a salted rim. Put a slice of lemon around the rim of the glass.
Rusty Nail	45 ml. Scotch whisky 45 ml. Drambiue Lemon peel Take a old fashioned whisky glass put some ice cubes and pour scotch and Drambiue in the glass. Serve with a twist of lemon.
Bombay Cocktail	60 ml. Brandy 15 ml. Sweet Vermouth 15 ml. Dry Vermouth 3 dashes of Curacao Stir well and pour into a cocktail glass.
Tequila Sunrise	60 ml. Tequila and orange juice poured into float grenadine syrup to give an effect of sunnise.

The Components: Nowadays the cocktail habit is no longer a craze. People tend to prefer bars which dispense all types of drinks from a good glass of beer to more intricate concoctions. However, whether it is a cocktail bar or any other type of bar, certain essential equipments and ingredients are required to

cater for all types of customers. Some of the essential equipment required are as follows:

One large and one medium-sized mixing glass or Bar Glasses with a lip for pouring. These glasses are used for preparing mixed drinks such as Martini, Manhattan etc.

Bar Spoons	For use with mixing glasses. They are long-handled silver spoons like Parfait spoons.
Hawthorn Strainers	Their coiled spring edge ensures a snug fit while straining from the mixing glasses.
Cocktail Shaker	Of different capacities for preparing all shaken Cocktails. They are silver containers with a lid and in-built strainer. An electrically operated blending machine, for preparing all types of shaken cocktails. Ideally suited for a very busy bar and for preparing well-blended smooth cocktails in very short time.
Boston Shaker	Consisting of two approximately 1 pint silver or stainless steel glasses (cups) which fit into each other. These are used for shaking flips, noggs etc. for large amount of ingredients for large shaken mixed drinks.
"Muddler" Spoons	These muddlers have a round flattened end for crushing mint leaves, fruit or sugar cubes as part of a drink preparation
Bitter Bottles	These bottles are fitted with conical sprinkler tops with a very small top opening to provide droplets or "dashes" as required in some cocktail receipes.

Other equipments include, a sharp knife, cutting board, ice buckets, ice tongs, ice scoop, ice shaver, the chopper, fruit squeezer for a busy bar, an electrically operated fruit squeezer, nutmeg grater, drinking straws, cocktail sticks, tin opener; bottle openers, juice strainer for fruit juices, wooden spoon for use while straining fresh fruit juices, small funnels for transferring liquids from one bottle to the other, salvers, cocktail napkins, coasters, ashtrays, candle stands, candles, match boxes; flower vases, toothpicks, cigar cutter, etc. Other items depend on customer's whims and the location of the bar in a hotel or a bar by itself on a busy street. These might include city maps, city entertainment diaries; time-tables for airways and railways, first-aid supplies music system, telephone, peg measures, optic measures, inverted fixed optic measures etc. All types of glassware including a good supply of cocktail glasses.

Ingredients: These might include, oranges, lemons, cucumber peels, fresh eggs, cream, milk, sugar, coffee, tea, olives, cherries, cocktail onions, cloves, salt, pepper, cinnamon, ginger, mint, nutmeg, sauces, tomato juice, good supply of clean ice etc.

Gin: London Dry, Hollands, Plymouth, Beefeater, Gordons, Old Tom, Indian Gin.

Whisky: Scotch, Rye, Irish, Bourbon, Canadian, Indian Whisky.

Rum: Jamaican Ćuban Demerara, Bacardi, Indian Rum.

Brandy: Cognac, Armagnac Marc, Indian brandy.

Vodka: Imported as well as Indian.

Aromatic: Italian Vermouth (Sweet), French Vermouth (dry) Bianco, Wines, Dubonnet, Lillet, Campari, Cinzano, Nouilly Vermouths: Prat, St. Raphael.

Bitters: Angostura bitter, Amer Picon, Underberg; Fernet Brance; Peach bitter, orange bitter Alsinthe etc.

Syrups: Gommes (ordinary sugar syrup without any flavour); Rashberry; Strawberry, Gooseberry, Lime cordial; Grenadine etc.

Liqueurs: Cointreau; Creme de Menthe, Advocat, Chartreuse, Benadictine Curacao, etc.

Fortified Wines: Port; Sherry, Marsala.

Accompaniments with Cocktails: Cheese straws, crisps, olives, gherkin, (capers), for special functions etc.

Other Spirits

1. *Slivovitz:* This is a spirit which is distilled from plums, mainly from Hungary, Yugoslavia and Romania. It is very similar to the Alsatian Plum Brandy.
2. *Calvods:* A potent spirit distilled from cider. In U.K. the same product is known as Apple Jack.
3. *Aquavit:* A highly rectified colourless spirit, distilled mostly from grains and potatoes and flavoured with carraway seeds.
4. *Grappa:* A crude Italian spirit distilled from the husks, skins, pips and stacks of the grapes after they have been pressed and the wine made.
5. *Zubrowka:* A Russian spirit which is just as raw as Vodka but not so clear. It is in fact Vodka in which some Zubrowka grass has been steeped, so that it has a little colour, aromatic bouquet and a little bitter taste—whereas Vodka is colourless and free from all bouquet and taste.

Bar-equipment: The Bar should be simple, have a non-fussy shape and every drink and piece of equipment should have it's regular place, to which it is returned immediately after use.

It may be tedious to be so methodical, but it saves undignified scrambles in the night, as the professional bartender will testify. Some of the common equipments used in the bar are as follows:

A Lemon Squeezer: No particular design can be selected as the best, since it is a matter of personal taste as to which

can be handled most easily and which extracts the juice most efficiently. If there are to be many guests, plenty of juice is squeezed beforehand for mixed drinks, but not so far in advance that it loses its freshness.

Ice Container: It should have the capacity and efficient insulation. If white wine or champagne are to be served, ice bucket containers in which they can be cooled are also required.

Tongs: These are more efficient than a spoon when ice is to be put into a glass and they do not carry any unwanted water.

Ice Crushers: Comes in various designs for crushed ice (sometime it is described as being shared) required for drinks, daiquiris and a variety of other treats. Crushers can usually be adjusted to produce cracked ice for drinks like the Old Fashion. Otherwise, cubes are taken in a tea towel and cracked with a kitchen mallet. If drinks are mixed in an electric blender, whole ice cubes will damage the blade, but this problem does not arise with mixers made especially for bars.

Jugs: Required for iced water for fruit juice and have an involuted power to hold back the cubes. These types of jugs are also used as mixing glasses and do not require strainers.

A Scoop: It is needed to load crushed or cracked ice into a drink, mixing glass or shakes.

The Bartender's Friend: It is an all-in-one device that opens corked or crown- topped bottles, is preferred and a bottle opener is permanently fixed to the bar.

Measure: It is known as peg measure to differentiate between a large peg and a small peg. The large peg measures 60 ml whereas the small peg measures 30 ml.

Bitter Bottles: These are fitted with caps like those used to pour vinegar over salad oil. This means that a dash of Angostura or orange bitters can be shaken without the need for measurement.

Mixing Glasses: These come in various shapes and sizes

but are always big enough for the preparation of several drinks at once. The mixing glass, sometimes, known as a bar glass, is for drinks that are to be stirred not shaken. These are drinks that are intended to be clear, not cloudy, and which would be particularly harmed by the greater dilution caused in shaking. After being stirred, the drink is strained into a cocktail glass straight up, or on to fresh ice in a large glass.

Muddlers: Comes in various shapes and sizes, large ones for use with a mixing glass and small ones for the drinker's own amusement. They have a bulbous end and are intended for crushing sugar and pounding mint in a drink. A similar device with a paddle-like end is called a swizzle stick. A swizzle stick adds decorative value in a drink.

Strainers: The classic strainer type that is popular, clips on to the mixing glass.

Bar Spoons: These are used to stir drinks in a mixing glass or after they have been served, the wrong end can be used as a muddler.

Shakers: These are used in drinks that contain fruit juice, syrups, very thick liqueurs, or any ingredients that demand a thorough mix. Because the ice gets knocked about in the mixer, there may be considerable dilution, and a clear drink cannot easily be produced.

Knife and Board: Easy to forget, yet obviously necessary. How else could one produce with easy efficiency skivers and twists of lemon or cucumber peels, and slices of orange, which are crisp clean and handsome ?

Useful Glasses

Cocktails: Essential, elegant and neatly proportioned. A stem just long enough to project the small conical bowl from the warmth of the hand. Opening sufficiently wide to display a garnish. Capacity 4 oz.

Old Fashioned: It is used for any cocktail, served on the rocks that also doubles for whisky. The typically old fashioned

glass is less attractive for that purpose than the cut-crystal and faintly tampered tumbler, traditionally used for Scotch. Has a capacity of 6 oz.

Hi-ball: An intermediate size which can serve several purposes. Capacity 10 oz.

Collins: For long drinks, the taller the better. Always narrow often with perfectly straight sides. Has a capacity of 10 to 12 oz.

Sour: A stemmed glass for a drink which is not strictly a cocktail but is not long. A similar glass is sometimes used for a fizz. Has a capacity of 5-6 oz.

Sherry: Those tall schooners are stilted and lumpy and do not hold the bouquet. A dry sherry in a large glass warms before it is drunk. It is better to have a small glass which can be refilled from a cooled bottle. Has a capacity 2 oz.

Pousse-Cafe: A tall narrow liquor glass.

Liqueur: Rich sweet liquors are served in small quantities. Has a capacity 1-2 oz.

Snifter: The traditional brandy glass well-rounded to be warmed in the palm of the hand with a short stem. The rim turns in to hold the powerful bouquet.

Champagne: It can accommodate the garnish on a champagne cocktail better than the narrow tulip glass, which is preferable because it retains the sparkle better.

White Wine: A long stem so that the hand does not warm the drink. A tall narrow bowl to retain the cold. A fine line for a delicate wine.

Red Wine: Solid-looking to match the fatness of burgundy or Claret with a short stem and rounded bowl to be embraced warmly. The rim wide enough to let the wine breathe.

Toddy: Fire-proof glasses with handles are useful for hot drinks.

Bar-trade Tips

1. A sharp knife is used to cut fruits for garnish.

2. The fruit garnish is covered with a damp cloth to maintain their freshness and crispness.
3. The bar equipments are always clean and dried after every use.
4. As soon as the tin juices are opened they are poured are out in clean glass jars to avoid tanning.
5. The ice once used is never reused. The second drink might get spoiled.
6. Some lukewarm water is sprinkled on the ice cubes. It removes the glossy appearance and makes them shiny.
7. Cocktails made with juices, egg-white and cream are always shaken.
8. Cocktails made with Vermouths and Bitters are always stirred.
9. Rubber mats are used on the floor to avoid slipping.
10. The drink is always in the peg measure holding the peg measure on top of the glazes. Any spillage goes in the glass but not out of the glass.
11. Always replace the lid of the bottle after each use no matter how busy one is.
12. The bottle is placed back in its position after each use and helps in remembering as to where the particular brand of bottle is lying.
13. Ice scoop is used to put ice cubes. One should never try to use a glass for this purpose because if the glass breaks one would have to throw away the entire ice.
14. Juices and fresh cream are tasted before using them for cocktail. It helps in minimising wastage of liquor.
15. The cocktail shaker and strainer are cleaned after each use so that the next drink is not spoiled.
16. After polishing glasses they are stocked the upside down so that dust does not collect in them.
17. A small diary should be maintained in which one notes down all the recipes of cocktails alphabetically.

18. Sodas and beers are chilled before service, if they are flat the drink will be spoilt.
19. Ice cubes are always put in the glass first, then the drink poured over ice cubes.
20. Seeds are removed from the fruit garnishes before using them.
21. Remember one short talk with the guest can help one sell one extra drink.
22. The bottles of lime juice, squashes and sauces should be wiped before using.
23. Behind the counter it is ensured that the empty bottles and juice tins are thrown in the dustbin and not be placed on the floor as they cause in convenience.
24. Beer bottle needs an a angle while opening. This helps avoid the beer from fizzing out.
25. One should never open a new bottle unless the first one is finished.

Liqueurs and their Application

Spirits: All top quality spirits are made from barley, but whiskies may be manufactured from any grain-rye, barley, wheat etc.

The grain is first screened, then soaked in treated water for about 2 days, spread on malting floor, and then roasted (the roasting being done preferably on an open fire pit or kiln with charcoal or vegetable charcoal). For Scotch whisky, the roasting is done only with vegetable charcoal, which gives more intense heat and also smoulders, causing a lot of smoke which gets incorporated in the grain and imparts a characteristic smoky flavour. The grain is then crushed into grist, which is boiled with treated water for 36 hours. The wort thus obtained is mixed with sugar (a larger amount than for beer, so that the alcohol content of the spirit is higher), and boiled. It is cooled to 16°C or 60°F, and cultured yeast is added to it for fermentation. After fermentation, it is strained to give crude

beer, which may be distilled by either of the two methods: the Pot Still or the Coffey (patent) Still.

The basic principle of distillation is as follows: Alcohol vapourises at 76°C and water at 100°C.

Pot Still: The Pot Still method is a very slow process and may allow some leakage of alcohol; however, most top quality spirits are distilled by this method, especially malted whisky and Cognacs. In this method, the entire apparatus is made up of copper (which is a good conductor of heat, and also does not react with alcohol) pot or alembic, with a capacity of 1,000-1,200 litres, on brick kilns. Each kiln has four pots. The distillation is done twice. The first few drops of distillation are called the heads and the last few drops are called the tails.

The first distillation yields a distillate of low alcohol content (25-30% of volume) and is called low wine. Once the residues in the alembic have been cleaned, the distillate is again charged, and the distillate thus obtained has an alcohol percentage of 65-70%. However, the alcohol content must not exceed 43% by volume; so the second distillate is analysed for alcoholic content and distilled water is added to reduce the alcohol content to 43%. After distillation, the spirit is matured in oakwood casks for a minimum of 3 years (Regular Scotch is matured for 3-12 years and Premium Scotch for 12-20 years).

Patent Still Method: Also called the Coffey still or Continuous still.

The Patent still consists of two stainless steel, heavily insulated columns, each eighty feet high. Column A is the analyser and Column B is the rectifier. In Column A the spiral copper pipe, is supported by plates at each turn. The column (as well as column B) is heated by hot steam, and the bottom of the column is the hottest. Warm wort travels down through the pipe, and starts boiling by the time it reaches the bottom. The hot wort then travels (pumped up) into the storage tank of the rectifier. (Rectification is the removal of other vapours from alcohol). The wort trickles from the tank as very tiny

droplets, hits the hot copper plates underneath and vaporizes. The vapours travel through a long tube past the condenser, where they appear as the distillate. To get rid of the water in the distillate, it is rectified by passing the distillater through the pipe. This process is used for regular brands from grain spirits, but not brandy; the Still can be used continuously and is much faster. No vapours are lost in the process.

American Proof: According to the American Proof system, if AP (American Proof) is mentioned on the label of a bottle of spirit, the alcohol content will be exactly half of the number mentioned. For instance, if the label reads 90 AP, then its alcohol content is 45%.

Gay-Lussac: According to the Gay-Lussac system, 100° proof has 57.1% alcohol by volume.

Proof: An arbitary system of measuring the alcoholic strength of a spirit. A spirit of 100° proof is one which contains 50% alcohol by volume at 60°F. This is according to the American proof system. According to Gay Lussac a spirit of 100° proof is one which contains 57.1% alcohol by volume at 60°F. There are two types of proof spirits—

(i) Under proof spirit—U.P.

(ii) Over proof spirit—O.P.

In case of Under Proof Spirit it is ahiays subtracted from 100 namely 25° U.P. will be 75° proof spirit. In case of O.P. 100° is added to it i.e. 5° O.P. will be 105° proof spirit.

Calculation:

Suppose it is a 25° U.P. spirit i.e. *75°* proof

According to Gay Lussac:

100° proof spirit is— 57.1%

therefore, 1 proof spirit is— 57.1%

100

and 75° proof spirit is— $\frac{57.1 \times 75}{100} = 42.8 \text{ v / v}$

Brands Names of Whisky

Scotch Whiskies

(a) Super Deluxe /Premium Brands

1. Johnny Walker Blue Label
2. Johnny Walker Yellow Label
3. Johnny Walker Gold Label (Matured for 12-20 years).

(b) Delux/ Premium Brands (Matured for a minimum period of 12 years)

1. Royal Salute
2. Chivas Regal
3. Johnny Walker Black Label
4. Dimple Scotch
5. Glenfiddich
6. King of Kings
7. Black Dog
8. Queen Anne
9. John Haig (12 years)
10. Ballantine (12 years)
11. Grant's Premium
12. Old Smuggler
13. Dancing Doll Scotch
14. Old Parr
15. 100 pipers (12 years old)

Regular Brands

1. Johnny Walker Red label
2. Vat—69
3. J & B Scotch
4. Cutty Sark
5. Ballantine

6. Grant's
7. Teacher's
8. Passport
9. Mckinly's
10. Dewar's
11. John Haig
12. Black & White
13. White label
14. White Horse
15. Bell's

Canadian Whiskies

1. Canadian Club
2. Seagram's Crown Royal
3. Windros
4. Wiser's Delux
5. G & W Special

Irish Whiskies

1. John Jameson
2. Old Bushmill's
3. Tullamore Dew
4. Paddy Irish
5. Dunphy's Original Irish

American Whiskies (Bourbon)

1. Southern Comfort
2. Old Grand Dad
3. I.W. Harper
4. Bourbon Supreme
5. Beam's Choice
6. Old Crow
7. Old Turkey

8. Virginia Gentleman
9. Old Hickory
10. Early Times
11. Jack Daniels
12. Old Forester
13. Jim Beam
14. Old Taylor
15. G.W. Bourbon

Indian Whiskies

DELUXE BRANDS

1. Peter Scot
2. Single Malt
3. Royal Challenge
4. Vintage
5. McDowell's Signature
6. McDowell's Premium
7. Aristocrat Premium
8. Bagpiper Gold
9. Colonel's Special
10. Brahin's Premium

Regular Brands

1. Director's Special
2. Aristocrat
3. Royal Reserve
4. Bagpiper
5. Diplomat
6. McDowell's No. 1
7. White House
8. Hayward's
9. Officer's choice
10. Solan No. 1

Manufacture of Brandy: 'Brandy' is derived from the phrase 'Burnt Wine' and is also called 'Delightful soul of wine' and 'Eau de Vie' (water of life).

The story goes that a Dutch wine shipper, who shipped wines from France to Holland, hit upon the idea of 'concentrating' the wine (so that he could transport more of it) by distilling it, and then diluting the 'concentrated' wine at the destination to obtain wine. However, when he tried this, he was told to stop 'diluting' the burnt wine, as it had a unique bouquet and flavour. Thus, 'brandy' was created. The main brandy producing district in France is Cognac, which is the district above Bordeaux. The brandies made from grapes outside the territorial limits of the Charante Valley are called Armagnacs. Also, Cognacs are made from using the grape pulp the first time; Armagnacs are made from the second extraction. Any spirit produced within the following legal territories is called a Cognac:

Grande Champagne — Petit Champagne

Les Borderies — Fine Bois

Bons Bois — Bois Ordinaries

Bois Commune

Brandy is colourless, but obtains a beautiful garnet red colour from the black oakwood casks in which it is matured. Brandies must be matured for at least 3 years; Cognacs are matured for 7 to 12 years, and may be matured upto 70 years. The brandy keeps on maturing till it is 70 years old, after which it starts evaporating at the rate of 1½ - 2% per annum. Therefore, a stored 70 year old brandy is kept in glass or ceramic jars, and drunk only on very special occasions or used exclusively for blending.

The process of brandy manufacture is:

Ripened Grapes (variety: Pomac) → Sorted and Screened → Destalked → Winepress → Must → Fermentation → Wine → Distillation → Bottling.

Wine is distilled twice by the Pot Still method to yield brandy. The first distillation is known as Premier Chauffe. It yields a distillate with an alcohol content of 25- 30% by volume. This distillate is subjected to a second distillation, called Bonn Chauffe, to raise the alcohol content to 43%. The brandy is subjected to maturation in oakwood casks, during which the colour changes from golden to amber. The brandy may be blended (by addition of another, older brandy) to improve its colour, bouquet, aroma, and also the alcohol percentage.

Brandy is always corked capped, as a loose cap may allow the volatile liquid to evaporate. Instead, if a cork is put, the vapour rising from the brandy keeps the cork moist; thus the cork remains swollen and fits tightly in the mouth of the bottle. The brandy must not touch the cap, otherwise it will eat away the cork.

Brand Names of Cognacs: Remy Martin VSOP, Napoleon (VSOP and XO) Martell, Camus XO, Bisquit VSOP, Hennessey (VSOP and XO), Courvoisier VSOP("Le Co-gnac de Napoleon"). Indian Brandies: Doctor's brandy, Teacher's brandy, Honey Bee, Beehive, Aristocrat, Golconda, Hayward's.

Manufacture of Gin: Gin is a compounded drink with a typical flavour of juniper berries. It was introduced to the world by the Dutch in the 17th century. Dutch and English gins are the most popular.

Queen Anne (1702-14) of England increased the excise duty on beers and spirits in order to popularise gin. The most famous are Dutch/Holland Gins and English Dry Gins (the word 'dry' in this context is a generic term which can be used by any country which produces gin). The grain formulae for gin are:

(a) *Dutch:* Equal quantities of barley, corn and rye, cooked to a mash and fermented.

(b) *English:* 75% corn, 15% barley, 10% any grain (depending upon availability). It is lighter as compared to Dutch Gin.

Gin is a rectified spirit produced in a column still; Dutch gins are made in a Pot Still. The process of manufacture is:

Grains → Screened → Soaked → Green malt → Toasting → Malt → Crushing → Grist → Boiling in water → Straining → Wort → Boiling for 12 hours with sugar → Cooling to 60°F → Fermentation in vats → Subjecting to Pot/Patent Still → Crude spirit with low alcohol content.

At this point, the spirit is flavoured (by the addition of 108 botanicals: the predominant flavours are of juniper berries, coriander seeds and orange peel) in either of the two ways :

(a) A part of the distilled is mixed with the botanicals, boiled and then subjected to the Still, to give gin or gin flavour. This is returned to the rest of the distillate and again rectified.

(b) The entire distillate is flavoured and then rectified.

Gin is matured for at least 3 years in oakwood (white) casks.

Brand Names of Gins: Old Town, John Booth's, Gordon's, Ballantine London Dry Gin, Lamplighter, Tanqueray, Beefeater, Booth's High and Dry, House of Lord's, Seagram's (a very famous port and spirit shipper), Coates Plymouth, Calvert London Dry Boodle's, Hiram Walker, White Satin, Gilbey's, Holland Dry Gin.

Indian brands: Blue Riband, Forbes, High Society, John Booth's, Dunhill, Aristocrat, Blue Angle, Silver Anchor, Mikado, Whitefield.

Manufacture of Rum: Rum is a pure product of sugar, which is made from molasses (a black and viscous liquid derived from sugar) or from pure sugarcane juice, which is used to make top quality rum. This juice is diluted to reduce its sweetness and yeast is added in order to start the fermentation. If rum is made from molasses, the molasses are diluted to reduce its viscosity and then it is fermented. After fermentation, the wort is subjected to a Pot Still (for top quality rum) or a Patent Still (for regular brands). The distillate obtained

is mainly colourless and hash. It is matured for at least three years, and distilled water is added to it to lower the alcohol content to 43%. For uniformity of colour, dark caramel is added. No caramel is added to white rum. The best rums are :

1. *Puerto Rican and Cuban:* Light-bodied and dry.
2. *Demerarian:* Distilled in British Guiana and South America.
3. *Jamaican:* full-bodied and rich.

Brand Names of Rums are as follows:

1. *Puerto Rico:* Don Q, Ron Bacardi, Ron Rico, Ron Merito.
2. *Demerare:* Lemon Hart, Bacardi (white).
3. *Jamaica:* Appleton, Lemon Hart, Myers, Captain Morgan, Red Hart, Western Pearl, Colonel Club.
4. *Cuba:* White Rum.
5. *Others:* Demerara, Old navy, Kelly's White Lable, Sam Weller, Black Diamond, Dregger, Golden Pilsner, London dock, Old Charlie.
6. *Indian:* Old Monk, Sea Pirates, Old Monk Gold, Black bull, Hercules, Old Adventure, Captain Henry, Sikkim XXX, Tropicana, Contessa, Dunhill, Royal Treasure, Old Vat, Whitefield.

Manufacture of Vodka: The national drink of Russia, Vodka (which is always best drunk, chilled and neat) used to be produced from potato and sweet potato starch. Nowadays, it is manufactured from grains and specifically, not potatoes. It is never matured, and is a rectified spirit. To make vodka, a fermented mash of grains is subjected to the Patent Still. The distillate thus obtained is passed through layers of vegetable charcoal, sand and quartz to make it tasteless and colourless. For every 1 gallon of vodka to be thus filtered, 1½ lb of vegetable charcoal is needed for at least 48 hours. After 48 hours, 20% of the charcoal is replaced by fresh charcoal. Russian vodka can be sold at an alcohol strength of upto 69% as it is not governed by British laws.

Brand names of vodka are as follows:

Gordon's vodka, Nikolai vodka, Cossack vodka, Stroganoff, Pushkin, Crown Russian vodka, Imperial vodka, Golubok.

Zubrowka (Zubrowka is an East Polish grass, a blade of which is put in each bottle to impart a bitterish flavour and yellowish tint). It is 80 proof.

Starka (Nearest in character to whisky, it is made of rye and is matured). It has 45% alcohol by volume.

Jarzebiac (Flavoured with Rowantree berries, 80 Proof). Luksuowa (Diluted thrice and served chilled, 70 Proof).

Wisniowka (Sweet vodka, flavoured with cherries, drier than cherry liqueur 80 Proof).

Smirnoff (British vodka made from grains, filtered twice through wood charcoal which removes undesirable flavour. Of exceptional purity, it is sold under two strengths : 80 Proof and 90 Proof).

The following pages shows a photographs of different varieties of Wine Glasses from left to right their names are:

1. Beer glass 2. Champagne tulip/flute 3. Hack & Mosele glass 4. Claret/Red Wine glass 5. Port wine glass 6. Brandy glass Baloon/Inhaler/Snifter 7. Cocktail glass 8. White wine/ A.P. glass 9. Collins/Zumbie glass.

The 2nd photograph is that of different wine bottles. From left to right their names are:

1. Bordeaux wine bottle 2. Champagne wine bottle 3. Burgundy wine bottle 4. Sherry wine bottle 5. Port wine bottle 6. Mosele wine bottle 7. Hack wine bottle 8. Rose d'Anjou wine bottle.

Liqueurs are sweetened, coloured and flavoured spirits, generally having a base of Scotch whisky or Cognac brandy. They are sweetened with sugar and coloured with vegetable pigments. Liqueurs are served, 1 oz, chilled, in liqueur glasses. They are basically digestives, drunk after a meal and generally come in very fancy bottles.

Two methods are used to extract maximum colour and flavour:

Steeping: Used to extract flavour mainly from soft fruits. Fresh or dried fruits, but not nuts, are cut into small pieces and seeds are removed as they often contain bitter oils. The fruit is placed in large vessels, into which Scotch whisky or brandy is pumped from below. The spirit remains in contact with the fruit for 4-6 months, during which, due to the action of the alcohol, the fruit gets mashed up. The solids are separated from the liquid by filteration, or distillation if necessary. Required sweetening and colouring are added to the distillate.

Percolation: Used to derive flavours from plant (Such as stems, leaves, bark and roots) which are hard and from which it is very difficult to extract flavour. The desired herbs and spices are put, along with the base and boiled (percolated) in large vessels. This is distilled along with the herbs spices. The distillate thus obtained is highly concentrated and defined colour and sweetening is added according to the type of liqueur. Honey, especially heather honey, is added to some plant liqueurs. Some liqueurs, like tequila are unsweetened and some have a base of rum.

Some famous liqueurs are:

Advocat: It is Dutch. A sweet yellow liqueur with a brandy-base and egg yolk flavour, served at room temperature.

Amar Picon: A brandy-based, dark and bitter French liqueur from Bordeaux. It is rarely consumed neat, and is served as an aperitif by mixing 2 oz of Amar Picon with 4 oz of cold water, sweetened with grenadine syrup.

Anisette: A brandy-based liqueur flavoured with aniseeds.

Apricot liqueur/Apricot Brandy: A golden-hued French liqueur, flavoured with apricots and served cold. It is often shipped under trade names such as Apry or Abricotine. Abricotine is a brown, sweet and fruity, brandy-based liqueur, served at room temperature.

Absinthe: A green, French liqueur with a flavour of wormwood and ani seeds, with a brandy-base. It is not sweet and is always served with a lump of sugar.

Benedictine: A golden, brandy-based French liqueur which is spicy and sweet and has a flavour of herbs, plants and peels. It was originally made by monks and is served cold.

Chartreuse: A sweet and spicy, brandy-based liqueur made by monks. It is French or Spanish, and yellow (Alcohol strength 75-80 AP) or green (90-110 AP). It has a flavour of anjelica.

Cherry Brandy: Made from 40% Scotch whisky, 40% Cognac brandy, 20% cherry juice and 10% mash of cherries. It is a red liqueur with a flavour of cherries. It is sold under two trade names: Queen's Cherry Brandy and Sportman.

Cherry Liqueur: A blackish liqueur with a brandy-base and has the flavour of black cherries.

Cointreau: A colourless French liqueur brandy-based and flavoured with orange. It is served chilled, but in a dry glass, as even a drop of water can turn Cointreau milky.

Creme de Cassis: A blackish, brandy-based French liqueur-flavoured with black currants.

Creme d'Ananas: A yellow, brandy-based, pineapple-flavoured French liqueur.

Creme de Banana: A French liqueur, golden in colour and flavoured with bananas. It is based on brandy.

Creme de Cacao: A French liqueur, chocolate brown in colour, brandy-based. It has a flavour of vanilla beans and cacao.

Creme de Framboise: A brandy-based liqueur, made in France. It has the colour and flavour of raspberries.

Creme de Fraise: A brandy-based liqueur with the colour and flavour of strawberries.

Creme de Noyaux: A light yellow, bitterish sweet liqueur with a base of brandy. It has a flavour of fruit stones.

Creme de Menthe: A green or white liqueur, flavoured with mint and men thol, and based brandy-based.

Creme de Vanille: A dark Mexican liqueur, brandy-based and having a flavour of the finest Mexican vanilla beans.

Creme de The: A brandy-based French liqueur with the colour and flavour of tea.

Creme de Chocolate: A brandy-based liqueur, with the colour and flavour of chocolate.

Curacao: A garnet red, liqueur-base with a Scotch whisky base, and flavoured with dry orange peels.

Double Orange: A brandy-based French liqueur with the colour and flavour of oranges.

Drambuie: A golden liqueur, flavoured with Heather honey and with a fine Scotch whisky base.

Danziger Goldwasser: A golden, brandy-based liqueur flavoured with orange peel and various other spicy herbs and plants.

Forbidden Apple: A golden, apple-flavoured brandy-based Italian liqueur.

Forbidden Fruit: A Peninsulean liqueur brandy-based, coloured and flavoured with oranges.

Grand Marnier: A cognac based, golden liqueur flavoured with oranges. It is also called 'Golden Curacao'.

Glen Mist: A golden Irish liqueur with a Irish whisky base and with a flavour of honey.

Goldwasser: A golden, brandy-based liqueur with a flavour of aniseed and oranges. It also has gold flecks.

Half om Half: A French liqueur brandy-based and flavoured with oranges. Made by mixing 50% Curacao and 50% Brandy.

Irish Mist: An Irish liqueur, light golden in colour and with a Irish whisky. It is flavoured with Heather honey.

Kiimmel: A brandy-based liqueur flavoured with caraway

seeds and cumin seeds. Occasionally sugar crystals may be added as well. It is medium dark in colour and is also called 'Kiimmel Crystalline'.

Kirsch: An unsweetened Alsation liqueur, brandy-based and with a flavour of bitter almonds.

Kahlua: A dark and very strong Mexican liqueur brandy-based and with a flavour of coffee.

Liqueur d'or: A dark coloured French liqueur brandy-based. It is flavoured with oranges, herbs and plants and has 18 carat gold leaf flecks.

Maraschino: An Italian liqueur, originally from Zara. It is colourless and has a flavour of green Maraschino cherries.

Ocha: A brandy-based Japanese liqueur with a flavour of fresh-green tea. It has a mild green colour.

Silvovitz: A brandy-based liqueur with the colour and flavour of plums.

Tia Maria: A rum-based, dark brown Jamaican liqueur flavoured with coffee.

Triple Sec: White Curacao, made by almost every liqueur producers of the world. It was originally called Cointreau.

B and B: A golden, French liqueur made by mixing half Cognac and half Benedictine.

Peach Liqueur: A peach-coloured, brandy-based liqueur with a flavour of fresh and dried peaches.

Van der Hum: A South African liqueur, dark in colour and based on Scotch whisky. It is a spicy aromatic liqueur with a flavour of fruits (especially tangerines), plants, seeds and bark.

Beer: The basic raw material required for beer are barley malt, sugar, yeast, hops and liquor (treated water, used often with permission from the local corporation).

Contents of Beer

1. Water: 89-91% by weight.

2. Alcohol: 3-5% by weight (in exceptional cases, upto 10-12%).
3. Carbohydrates, sugar or dextrin: 3-4% by weight.
4. Protein: 0.4-0.5% by weight.
5. Carbon dioxide gas: 0.4-0.5% by weight.
6. Minerals, salts: 0.2% by weight.

Types of Indian Beers: Indian beer bottle capacity is 650 ml.

Kingfisher, UB (United Breweries), Golden Eagle, Birdie, Rosy Pelican, Black Partridge, Gymkhana, Guru, Black Label, Thunderbolt and others.

Draught Beer Container Sizes

Pint = 20.45 litres (4.5 gallons)

Firkin = 40.90 litres (9 gallons)

Keg = 45.50 litres (10 gallons)

Kinderkin = 81.80 litres (18 gallons)

Barrel = 163.65 litres (36 gallons)

Hogshead = 245.50 litres (54 gallons)

Draught Beer: Freshly gassed beer.

Lager Beer: Bright, clear, light-bodied beer which is effervescent. It is lagered, *i.e.*, matured for 3-6 months in a cask which is lined with a material called 'Pitch' and then it is carbonated. All Indian and American beers are lagered beers.

Bock Beer: A heavy beer, dark in colour and sweet in nature. It is made during winter, to be used in spring. A Bock Beer Day is celebrated particularly in Germany which lasts for six days.

Ale: It is made from aromatic malt or starchy cereals and brewed. It is usually full bodied and is more bitter than beer. It is fermented at a higher temperature than beer and the yeast comes on the top. It is called the top fermentation beer.

Stout: A very dark type of Ale with a strong malt flavour. It is a little sweet and also has a strong flavour of hops.

Porter: A type of rich beer froth in it. It is made from dark malt and is sweeter than regular stout but is less happy than regular ale and is stronger than stout.

Pilsner: A beer named after Pilsner, who introduced beer to the world, it is bright lagered beer.

Lager Beer: A bright, clear and bight-bodied beer which is sparkling and effervescent. It is a brew of malt barley, hops, sugar, yeast and brew. It is matured in a dark wood cask which is lined with a material pitch for about 3 months so that it mellows. It is then carbonated. All Indian and American beers are lager beers.

Beer	Manufacture Chart
↓	
Barley ↓	(Generally, the entire crop of the village is bought. It is cut, dried naturally and stored in sacks, fumigated, away from the floor
Ripened ↓	and walls, for better ventilation. The barley is screened for any for eign material like straw and pebbles.
Stored ↓	This is done on conveyor belts or floors, mechanically or manually).
Screened ↓	
Soaked ↓	(The barley is soaked in water for 3-4 days, so that the barley can absorb water to its fullest strength. This softens the gluten).
Green malt ↓	(The barley is spread on a 'Malting Floor', with a very small amount of moisture. The barley germinates to form 'Green Malt').
Malting ↓	(The barley is roasted to change its colour from pale to dark brown, in kilns or on conveyor belts. It is then dried till the moisture con tent is 0.5%.

Screening ↓	(Due to malting, the starch becomes soluble, changing to maltose; this is called malted barley, which is then screened to remove unroasted or overroasted grains).
Grist making ↓	(Malted barley is coarsely ground to a grist).
Wort Making ↓	(The grist is mixed with hot water and boiled for about 12 hours to allow infusion to take place, for maximum extraction).
Straining ↓	(The strained liquid is the wort and the solids are the spent barley, which is used for cattle feed).
Copper/Brewing Kettle ↓	(The wort is run into a kettle and a weighed quantity of hops and sugar are added. It is then boiled for about 36 hours to ensure that the liquid gets maximum extraction from hops and the sugar is equally distributed. It is then strained to remove the spent hops, cooled at 16°C or 60°F and run into a fermenting vessel called a vat).
Vat ↓	(A weighed quantity of cultured yeast mixed with tepid water is added, and fermentation is started. Initially, the fermentation is slow, but as it becomes vigorous, heat is evolved. The temperature must be controlled and surface scum must be skimmed off so that the liquid remains in contact with air; some scum may be left as a dust cover. When the yeast stops working on the sugar, the yeast rises to the top and is removed—In India this is called top fermentation).
Filtering ↓	(After filtering, the beer is checked for clarity and consistency. If it is not clear enough, filtering may have to be done).
Cooling ↓	(After cooling, in case of lager beers, the beer is filled in casks and matured. After

maturing, the beer is siphoned off, leaving sediments in the cask).

Pasteurisation (The beer is filled in bottles, subjected to a machine for gassing and cork-crowning and then pasteurised at 140°-135°F for about 20 minutes. The beer is immediately cooled, labelled, packed and sold).

Famous Beers of the World

1. Argentina: Schneider, Rio Sagundo.
2. Australia: Foster's Lager Beer, Abbott's Lager Beer, Crown Lager Beer, Carlton's Draught Beer.
3. Austria: Puntigan
4. Brazil: Brahma.
5. China: Yuchuan beer, Tsingtao beer.
6. Canada: Dow's ale, Labott's ale.
7. Czechoslovakia: Pilsner Urquell (Urquell being the name of the man who first developed beer).
8. Denmark: Carlsberg, Toburg, Special dark lager beer.
9. Germany: Dortmunder Union, Dortmunder Hansa, Lowenbrau, Bicks, Heineken's, Hofbrau, Kronenbourg (French).
10. Holland: Heineken's, Amstel, Special dark beer.
11. Ireland: Harp Lager Beer, Guiness Stout, Guiness Draught Beer.
12. Italy: Peroni Lager Beer.
13. Japan: Asaki Lager Beer, Kirin Lager/Stout/Black beer.
14. Mexico: Carta Blanco.
15. New Zealand: Zealandia half half.
16. Poland: Krakus.
17. Sweden: Three towns.
18. Singapore: Tiger, Anchor Beer.
19. Switzerland: Zurich Export Lager, Cardinal.

20. Thailand: Amrit lager, Singha lager.
21. USA: Budweiser, Falls City Beer, Anchorsteam, Utica Club Pilsner, Utica club Cream Ale, Point Special beer, Uncle Ben's lager beer.
22. UK: Bass ale, White bread ale, Mackerson's stout, Double Diamond beer, Skol Beer, Bass Red Triangle, Bass Blue Triangle Beer.
23. Uruguay: Salus.

India and the USA produce light lager beers. 'Lagering' refers to the maturing (for 3-6 months) of beer in pitch-lined oakwood casks; the pitch retains the bitterness of the beer and also protects it from oxidation.

Storage and Upkeep: Alcoholic beverages are very expensive, and are a good source of revenue; therefore it is very essential to take great care while purchasing and storing wines and spirits.

It is advisable to buy large consignments at one time, to save on freight charges. Indian importers of foreign-made liquors must get permission from the RBI and excise department to import wines and other foreign-made liquor. The following licenses must be obtained in order to sell alcoholic beverages.

L_1: For service of liquor in a bar or a restaurant. Timings must be specified, along with details of whether Indian or imported, regular or delux brands are to be served. A separate license must be obtained if beer is to be served.

L_2: Mainly for room service. If liquor is to be served round the clock, then a separate fee is charged.

L_3: Is for storage of large stocks of liquor. In airport duty-free shops, an excise officer is often deputed to check on the storage and sale of alcoholic beverages.

L_{20}: For sale of liquor at banquets. This license is issued on the name of a particular person, for a particular day and time. The licence becomes null and void as soon

as the specified time is over; the venue, number of persons, type of function, brands to be served must also be specified. Corkage is charged if the guest brings his/her own liquor.

The following points must be considered while storing alcoholic beverages:

Location: The cellar should be situated away from vibrations, such as busy main doors, plant and machinery. Ideally, the cellar should be underground and should be made of limestone or unplastered bricks which absorb moisture well, and keep the cellar moist and cool. Alternatively, the cellar may be located on the ground floor.

Temperature: Ideally the temperature in the cellar should be 54°F. The temperature may be controlled thermostatically. As the colder, heavier air settles at the bottom, the lower part of the cellar is cooler than the upper part. For purposes of aeration and cleanliness, wines should be stored 3" off the wall on floor, on their sides. Beer, requiring low temperatures, are stored in their cases, on the ground, on a plank. Sparkling wine in the lower shelves of the cellar, then, in ascending order, comes white and rose wines; young red wines; medium old red wines and very old red wines at the very top.

Moisture (Humidity): The atmosphere of the cellar must be fairly humid, in order to maintain the coolness of the cellar and prevent the corks from shrinking (dryness may also attract pests and bacteria). Moisture, however, can cause black fungus—this is harmless and is infact, called a 'friend of wine'. It can be easily wiped off.

Lighting: Wines are very sensitive to light; therefore, there should be no direct penetration of sunlight in the cellar. If sunlight enters the cellar, it should be indirect or diffused. There should be diffused lighting or shaded lighting.

Hygiene: Cellars should be swept and mopped daily. Occasionally, spring cleaning should be done whenever needed. The walls should be whitewashed when necessary. The door

should be such that it prevents the entry of rodents. Access to the cellar must be restricted, by setting specific timings. Nowadays, time-lock systems are used on cellar. Self-locking doors must have an opening device from inside aswell.

Some special directions:

1. Shelves should be 6" off the walls and 3" of the ground.
2. All wine bottles should be kept horizontally.
3. Beers and spirits should be kept erect so that the cap is not eaten away by the alcohol.
4. It is preferable to have a separate cellar for wines and beers, and a separate one for spirits.

Different Ways of Stacking of Wines

1. *Wooden Blocks:* The wine bottles are stored on their sides and each variety has got a separate block. If fresh consignment is received, it should be stacked in a new block and previous block bottles should be consumed first. Then the block should be rinsed.
2. *Wire-Blocks:* They can be raised according to requirements and can be spread on floor accordingly.
3. *Wooden Platform:* In this case a wooden platform is made 6" above the ground and at specific points. Some nails or wooden pieces are fixed in order to support the bottles.

In each case a card is attached to the bins stating the

1. Behaviour
2. History
3. Date of Purchase
4. Approximate Life

The current stock is also maintained by making the entries while taking out the bottles or addition of new stock.

Length of Storage of Beverages

Beer: A simple rule 'To keep it away' is followed. This

means that beers are not placed in cellar. Foreign beer can be stocked for about 6 months. They are stored in their boxes.

Sparkling Wines: The imported ones can be stored but in less numbers as they lose effervescence although a class of clientelle values the old champagne. (For this reason, some wine bottles can be stored and sold at high profits)

White Wines: They mature in bottles to a small extent. Due to this reason, they can be stored for years. A moderate stock for 1-3 years should be maintained.

Red Wines: These are the ideal wines to store as and when there is supply in the market. They are stored (young) for a number of years. They mature and the quality improves (if stored) and can be sold at high profit.

Care should be taken to increase the length of storage. One should see that corks do not shrink, bottles should not leak and sediments do not get disturbed and light does not affect its colour.

They should be sold at the prime of their youth, if not they are useless.

Fine and Dry Varieties: All dry varieties start deteriorating after 6 months. Sweeter and darker varieties improve, when aged particularly vintage and crusted port. They don't go bad easily depending upon situation.

Spirits: They hardly go bad or improve when stored in bottles except that they evaporate and one may lose alcohol if stored for a long time.

Liqueurs: They are stored as they improve on aging (In large quantities, they can be stored).

The rest are stored like other spirits.

Bottle and Cask Sizes: The following sizes of bottles and casks refer to wines, in particular Champagne, but the first five can also be referred to for spirits also:

1. Baby/Split/Nip: 6.5 oz-¼ bottle

2. Pint: 13 oz-½ bottle
3. Bottle/Quart: 26²/₃ oz (equivalent to a regular bottle of 750 ml).
4. Magnum: 2 quarts-52 oz (2 bottles)
5. Jeroboam: 2 magnums-104 oz (3 litres)
6. Rehoboam: 6 quarts-156 oz (4½ litres)

   The following casks are very rarely shipped. They are shipped only when demanded for:

7. Methuselah: 8 quarts-208 oz-1.65 gallons (American gallons)
8. Salamanazar: 12 quarts-312 oz-2.44 gallons
9. Balthazar: 16 quarts-416 oz-3.30 gallons
10. Nebechandenzzar: 20 quarts-520oz-4.07 gallons.

The following terms are applicable to Champagne wine only:

Name of term	*Type of Champagne*	*% of sweetening*
Brut	Very, very dry	0.5-1.5%
Extra sec	Extra dry	1.5-3%
Sec	Dry	3-5%
Demi sec	Medium dry	5-7%
Doux	Very sweet	7 and above

Cognac Brandy Abbreviations and their Age

1. VO: Very Old; 10-15 years
2. VOP: Very Old Pale; 15-20 years
3. VSO: Very Superior/Special old; 20-25 years
4. VSOP: Very Superior/Special Old Pale; 25-40 years.
5. XO: Extra Old; 50-70 years (such as Camus and Napoleon)

Age of Cognac, according to stars:

*: 3-4 years

**: 5-6 years

***: 7-10 years

****: 10-15 years

*****: 15-20 years

Wines

Varieties of Grapes Used for the Manufacture of Wines

Vitis Vinifera: Both red and white varieties of grape, used for producing both red and white wines. It is one of the most superior varieties and is best suited for Champagne.

Boal: A red grape, used in the manufacture of sweet Madeira.

Cabernet: Black grapes , best suited for red Bordeaux wines.

Cabernet Sauvignon: Red as well as white varieties of grape, suited for producing Bordeaux wines, and especially wines from the Loire valley.

Gamay: A red grape, used in the manufacture of red Burgundy wines, and especially Beaiyolaise and Cote Maconnaise.

Merlot: A grape which is good for red Bordeaux wines and gives a medium to full-bodied wine, like St. Emillion and Pomeroe. (Red)

Semillion: A grape suited for Sauternes and Barsac. (White)

Resting: Grapes from France or Germany, best suited for white wines.

Furmint: A white grape, used in making a sweet Hungarian wine called Tokay.

Muscat: A white grape, used in French, Italian and Spanish sweet white wines.

Grenache: A black grape, used in French and Spanish red wines.

Pinot Noir: A black grape, used in the manufacture of red Burgundy wines and Pink Champagne.

Pinot Blanc: A white grape best suited for white Burgundy as well as Champagne.

Chardonnay: A white grape, best suited for white Burgundy wines like Chablis, Meursault, Montrachet, Mercury and Ruilly.

Hligothe: A grape used generally for any red or white wine.

Malvasia: A grape used generally for any red or white wine.

Diseases which Attack Grapes

1. *Odium:* A powdery fungus which attacks the leaves and produces brown spots; it also destroys the skin and root of plants. The best remedy is a Sulphur di-oxide (SO_2) gas spray, which is usually done over a vast area by aircraft.
2. *Pylloxera:* Oval-shaped lice which attack the leaves, fruit, stem and root of the plant. This too is prevented by an SO_2 spray.

Wine Manufacture Calendar

January: Pruning starts.

February: Pruning finishes.

March: Ploughing starts, so as to turn the soil upside down for airing of the roots.

April: Ploughing finishes.

May: Sulphur di-oxide liquid spray is done in the vineyard so as to protect the plant and the leaves from frost and diseases like Sodium (a lice which attacks the leaves).

June: Vine flowers appear and the temperature should be about 19-20°C. A spray of powdered sulphur is done to avoid any attack of disease on the flowers.

July: Regular spray of Bordeaux Mixture (Copper Sulphate, lime and water in a definite ratio) is done in the vineyards.

August: Grapes change colour from green to pale or black.

Oakwood casks are inspected and cleaned. Sulphur di-oxide gas spray is done inside and outside the casks to prevent any bacterial growth. The casks are kept in the sun for about 15-20 days.

September: The crop ripens to its full strength and the plucking of the grapes is carried out.

October: Fermentation starts.

November: The long shoots of the plant are cut, collected and used as fuel.

December: The soil of the vineyard is brough back if washed away by the rains, owing to the vineyards being on a sloping surface.

Wine Manufacture Chart

Viticulture
↓
Vintage (Plucking of the grapes)
↓
Wine Press (1. Troddening, 2. Balloon press, 3. Manual press, 4. Electrically operated press yields the most.)
↓
Fermentation
↓
Racking ($C_6H_{12}O_6 \xrightarrow[\text{at 60–80°F}]{\text{By the Action of Yeast}} 2C_2H_5OH + 2CO_2$)
↓
Finning
↓
Filteration
↓
Refrigeration
↓
Sulphuring
↓
Maturing
↓

Wine tasting by connossieurs

↓

Blending

↓

Bottling

↓

(Not in the case of Vintage wines and wines meant for the domestic market)

Process of Manufacture of Wines

Viticulture: Viticulture refers to the growing of grapes. The grape plant is a creeper which has to be grown on sloping, rocky and chalky soil. Rocky soil is essential to prevent the retention of water, which could rise upwards into the vines, diluting the sugar in the grapes or even causing the grapes to burst.

After one crop, soil which has been washed away by the rain must be brought back and prepared for cultivation, by ploughing. The vines too are prepared for the next crop by pruning them and removing the long shoots. The vineyard is sprayed at this stage (either with sulphur di-oxide gassing, or sulphur di-oxide and water mixture spray) to protect the plants from diseases. The sulphur di-oxide spray is repeated when grape flowers appear, and again when the grapes are about to ripen.

Bamboos are placed at regular intervals, with wires stretched between them, to support the vines. These wires should not be stretched too high, as the warmth rising from the soil would not be able to ripen the grapes; they should not be too low, as the blossoms would then touch the soil.

Vintage: Vintage refers to the plucking of grapes. This is done manually, often using special scissors. The grapes are collected in conical cane baskets and taken to the wine press. The grapes are then sorted and graded manually, all rotten, damaged and raw fruit are discarded. The fruit is then shifted to the de-escalating machine to remove the stalk from the grapes (as stalks contain bitter oils).

Wine Press: Grapes are crushed to yield Must (fresh grape juice), by four methods :

Troddening: Troddening refers to the crushing of grapes by treading on them. For this, the grapes are placed in a large, shallow tank and the people treading the grapes, wear boots with spikes, which prevent crushing of pips (as pips contain bitter oils).

Balloon Press: The balloon press consists of an expandable balloon which is fitted in a tank filed with grapes and equipped with an outlet for the Must. On being filled with air, the balloon expands, crushing the grapes against the sides of the tank and squeezing out the juice.

Aid

Manual Press: The manual press works on the principle of hydraulic pressure. As the central disc descends, it crushes the grapes.

Electrically-operated Press: This press consists of a roller with a plate fitted all around it. The grapes are crushed between the roller and plate.

Fermentation

$$C_6H_{12}O_6 \xrightarrow[\text{Between } 60-80^\circ F]{\text{By the Action of Yeast}} 2C_2H_5OH + 2CO_2$$

The process of fermentation takes about 3-4 days, and is the result of the action of natural yeast (saccharomyces ellipsoides) on grape sugar. From white (pale) grapes—white wine is obtained, and from red grapes, white, red and rose wines are obtained. The skins of the grapes contain most of the pigments; therefore if the skins are removed from the Must before fermentation, a white wine is obtained. If allowed to remain in the Must for a short time (usually a few hours), a rose wine is obtained.

If the skins are removed only after fermentation is over, a red wine is obtained.

Nowadays, fermentation takes place in thermostatically

controlled fermentation tanks, keeping the temperature within the range of 60 to 80°F. With the onset of vigorous fermentation, heat is produced. To prevent excessive heat (which would inactivate the yeast and thus halt fermentation), cold water is run through pipes embedded in the tanks; this ha. a cooling effect. If the temperature is too low, hot water is run through the pipes.

Fermentation takes place on its own. However, to produce a sweet fortified wine (a fortified wine has an alcohol percentage of 18-25%), a definite amount of Brandy is added during fermentation. Due to this, the alcohol content of the wine increases and yeast is inactivated.

For a dry fortified wine, Brandy is added after fermentation (when all the sugar will have been used up by the yeast), which increases the alcohol content of the wine.

Grape sugar may be added to wine in certain circumstances. Grapes which have a lot of sugar but less yeast is used for many sweet wines. Whole grapes that have more yeast but are less sugary are used in dry wines.

Racking: The process of removing pulp and yeast particles from wine is called racking. The wines are kept in oakwood casks (with no airspace on top, as air oxidises wine into vinegar) for about a month, to allow the solids or lees to settle at the bottom of the cask, after which the clear wine is siphoned off into another cask. The process is repeated 3-4 times, and so racking can take upto 3 or 4 months.

Finning: Finning is done in order to remove all suspended particles in wines. It may be done using either of the following three finning materials.

(a) Egg albumen (using stiffly beaten egg white).

(b) Icing glass derived from the bladder of a sturgeon.

(c) Blue Icing glass (for the use of which special permission has to be obtained from the Appellation d origin controlee. It is used to remove traces of toxic copper oxide from grapes).

These materials have clinging power; they attract the particles in the wine which then settle at the bottom, after which the clear wine is siphoned off.

Refrigeration: After finning, the wines are refrigerated in order to avoid bacterial attack.

Filteration: Filteration is done if the wine is not clear enough. It is a very slow process and is usually done only in the case of wines meant for the domestic market.

Maturing: Once the wine is clear and free of impurities, it is filled in black or white oakwood casks and kept for maturing, the maturation period varying greatly for different types of wines. Champagne, for instance, is never matured while other wines may be matured for 3, 6 or 9 months or even for a number of years. Champagne and other sparkling wines are bottled, gassed and ready to consume after finning or filteration— the second fermentation takes place in the bottle itself, using the natural sugars and yeast present in the wine, in the presence of the air in the bottle. Once a bottle has been filled, *i.e.*, the wine has been glassed, the wine will not mature further. The wine-filled casks are kept on rocks in the cellar to age, the period depending upon the type of wine. White wines are usually matured for 3 to 6 months and are best consumed when young.

Maturing makes the wine mellow and smooth and removes harshness. Oakwood is porous and allows enough air into the wine to make it mellow, but not enough air to oxidise the wine into vinegar. The wine also acquires colour from the oakwood casks. Oakwood casks, on being repeatedly used, become seasoned, and after a certain period of time, also stop imparting colour to the wine.

Prior to maturing, the casks are subjected to the process of 'sulphuring'. This consists of spraying or gassing the empty casks with Sulphur di-oxide, 15 to 20 days before filling them, and then drying them in the sun. Similarly, bottles and corks are treated with sulphur di-oxide and placed upside down to prevent microbial growth.

In the case of sherry, the Solera system of blending and maturing is used which involves blending of sherry from two or more vineyards.

Solera System of Blending and Maturing of Sherry: Once the top cask gets filled, the Sherry starts trickling downwards into the lower casks. The top cask is continuously filled. The entire process takes from six months to a year, for the wine to reach in the last cask from where the blended and matured Sherry is withdrawn.

COUNTRIES AND BRANDED WINES

Italy

Chianti: A red wine, dry as well as sweet. Asti Spumante: A dry white Sparkling wine.

Spain: There are basically two types of sherries: Fino's and Olorosos.

Fino's: There are three varieties of Fino's—Manzanilla, Fino's and Amantillado.

Manzanilla: A very dry wine, pale in colour, with a light-body; it is matured for approximately six months.

Fino's: A very dry wine, pale in colour and with a medium-body; it is matured for about a year.

Amantillado: A full-body dry wine with a nutty flavour. It is either pale or light brown in colour, and is matured for about 2 years.

Olorosos: Olorosos is of two types: Amorosos and Oloroso.

Amorosos: A medium dry wine, sweet in taste, full-bodied, with a nutty flavour.

Oloroso: A sweet wine, deep golden in colour, full-bodied with a nutty flavour.

Cream Sherry: A sweet sherry, golden in colour. It is very smooth in taste and is full-bodied.

Brown Sherry: A very sweet sherry, which goes well with

the sweet course. It is dark brown in colour, is full-bodied and has a nutty flavour.

Malaga: A sweet wine which goes very well with the dessert course and is white in colour. It has a light-body.

Rioja: It's red in colour and is full-bodied table wine which generally accompanies a meal.

Portugal

(i) Port: There are many varieties of port, which include:

(a) Dow's Vintage Port ....

(b) Croft's Vintage Port ....

(c) Tawny Port ....

Crusted Port: This is a blend of two vintage ports. Generally vintage wines are never blended together, but Crusted Port is an exception. This wine is obtained by blending a wine of one year with a wine of the subsequent years, and is matured by keeping in oakwood casks for a long time. The wine is then bottled and kept in the cellar for a long time, so that it develops a crust on top. This crust is always served with the wine itself.

Ruby Port: It is a blend of two young Ports and is matured for a very short time. It is best consumed when young.

Madeira: A sweet wine, usually served with dessert. At times it is also served as an apertif wine.

Germany

Hock: Amber red in colour, Hock is a table wine and generally accompanies a meal. It is a natural Still table wine, but nowadays one gets sparkling Hock too.

Moselle: A white, sparkling dry wine, it is also known as 'Sekt' (which means 'sparkling' in German). Germans also call it German Champagne.

USA: California produces the second biggest quantity of wines in the world. They imitate all the wines from other countries and have thus gained popularity.

Barbara: Basically this is the name of a type of grape. This wine is dark red in colour and is like Italian Chianti.

Cabernet: Also a type of grape, Cabernet is one of the best Californian wines. It is dry in nature, has a fruity flavour and is matured for about four years. It is red wine.

Gamay Beaujolaise: A red wine with a light body; it is fruity in flavour.

Pinot Blanc: A dry white wine with a distinct grape aroma.

Tinta Madeira: A sweet, red fortified wine, resembling Port wine.

White Reisling: It is a dry white wine, and is consumed when young. It is matured for 3-4 months only.

Australia

Dalwood: A red wine with a light-body.

Lovedale: A red wine with a light-body.

Hunter River: A dry white wine.

Rhine Reisling: A dry white wine, consumed when young.

South Africa

Neederberg: A dry white wine.

Zonne Blcom's Cabernet: A sweet red wine.

Pearl's Rodeberg: A dry fortified wine, similar to Port.

Mymering: A sweet wine, pale in colour, also called the South African Sherry.

Wine Producing Districts of France

From the point of view of wine-making, France is divided into districts which are further subdivided into subdistricts, villages, chateaux and vineyards. Each of these divisions has its own quality control system—'Appelation d'origin controlee Francaise', which ensures that the wine produced is of the very best quality.

The three main wine-producing districts of France are Bordeaux, Burgundy and Champagne.

Bordeaux: Situated on the left bank of the river Rhine, below the district of Cognac, Bordeaux is divided into 5 major wine-producing districts: Medoc, Graves, Sauternes, St. Emillion and Pomerol. Only wines produced within the legal boundaries of the Gironde valley can be classified as Bordeaux wines. Bordeaux is best known for its red wines, called Claret in England.

District: Bordeaux

Name of Subdistrict	*Trade Name of Wine*	*Nature of Wine*	*Colour*	*%of Alcohol*
Medoc	St. Juliene		All are	
	St. Estephe	All are natural	Red	11%
	Chateau Lafite	Still wines, semi-		
	Chateau Latour	sweet in		
	Chateau Margaux	nature		
	Pouillac		All are	11%
Graves	Chateau Haute Brion		Red	
	Chateau Haute Billy	All are natural	Except	
	Chateau Olivier	Still wines,	White	11%
	Chateau Carboniaux		Red	
Sauternes	Chateau Bonns	All are natural	White	
	Sauternes	Still wines,	White	11%
	Preignac	Very sweet in	White	
	Barsac	nature	White	
	Chateau La Tour		White	
	Blanc		Red wine	
	Chateau d'Yquem		Red wine	
St. Emilion	St. Emillion	Natural Still	Red	
	Chateau Cheval Blanc	wines, dry	White	11%
Pomerol	Pomeral	Natural Still	Red	All Have
	Chateau Petrus	wines, sweet in	White	
		Nature		11%

— Area under vine: 105,000 hectares (260,000 acres)

— Yearly production: Nearly 600 million bottles

— Colours: Red, dry white, sweet white, rose and Cremant (sparkling) wines

— About 8,000 Chateaux: From the famous "Great Growths' to 'Petits Chateaux", including branded wines

— A very wide range offering consumers fine value for money.

Red Wines: These are made from a blend of Merlot, Cabernet Sauvignon and Cabernet Franc grapes. Some can be enjoyed quite young, while the Great Growths can age for many years. Account for 3/4 of total production.

White Wines: These are made from Semillion, Sauvignon Blanc and Muscadelle grapes. The dry white wines are fruity and aromatic and are generally put into dark green bottles. They are best enjoyed within 2-3 years after the vintage. Great Growth wines can age longer. They account for 1/9 of total production.

Sweet White Wines: Made from the same grape varieties, are usually bottled in clear glass.

They are made from overripe grapes with concentrated sugar and aromas and can age for several years.

Storage: After purchasing and if the wine is not meant to be consumed immediately, it is best stored at a cool temperature (11-18°C/52-14°F) lying down on its side.

Serving Temperatures: Red wines are generally best at 16-20°C (61-68°F) and White, Rose and Cremant wines between 6 and 10°C (43 and 50°F).

Burgundy: Known as the 'Golden Slope', Burgundy is best known for its white wines. The wines of Burgundy include:

District: Burgundy

Name of Subdistrict	*Trade Name of Wine*	*Nature of Wine*	*Colour*	*% of Alcohol*
(i) Cote d'or	Cote de nuits Nuit St. George Romanee Conti Chambertin Cote de beaune Meursault	All natural Still, sweet as well as dry.	Red White	If dry 12-14% If sweet 11%
(ii) Chablis	Chablis Balmur La Clos Vaudeser	All natural still dry wines	All White	11%

contd.....

Name of Subdistrict	*Trade Name of Wine*	*Nature of Wine*	*Colour*	*% of Alcohol*
(iii) Cote Macounaise	Beaujolaise Givery Mercury Pouilly Fuisse Pouilly Loche	All natural Still, dry wines	Red White	11% 11%
(iv) Cote Chalonnaise	Cote Chalonnaise Rully	Natural still dry Sparkling	Red White	10% 12%
(v) Cote du Rhone (Cote du Rhone)	Cote du Rhon Chateau neuf	sweet in nature du pape'	Red Red	9-10%

Concept and Terminology

1. Age: Maturing of wines and spirits in oakwood casks (oakwood casks have a holding capacity of 1361). White oakwood casks have a capacity of 32 gal lons and are mainly used by the Americans only to mature Bourbon whisky.
2. Aperitif: Alcoholic beverage consumed before a meal as an appetizer.
3. Aroma: Taste of wine in the mouth.
4. Bouquet: Smell of wine in the glass.[1]
5. Baby/Nip/Split: Quarter size bottle (6½ oz.)
6. Bodega: Spanish for ground—level cellar.
7. Body: Refers to the consistency of an alcoholic beverage; *i.e.*, thin wine has less body, and is matured for a shorter time.
8. Blending: Mixing of two or more similar wines, may be of different ages, to obtain a better quality and taste. Vintage wines are always blended with vintage wines only. Blending improves the standard of wine, *i.e.*, blending older wines with younger wines.
9. Brut: Refers to the percentage of sweetening in Champagne—5 to 15%.
10. Cask: A barrel-shaped container made out of oakwood,

used for maturing and storing wines and spirits; its capacity is 36 gallons.

11. Butt: Standard shipping cask for sherry only. Its capacity is 32 gallons.
12. Case: A container of 12 bottles.
13. Chaise: French for ground level cellar.
14. Chambre: Bringing red wine to room temperature *i.e.* 65°F.
15. Crust: Some wines, when poured in glass, throw up sediments in the form of crusts on the top layer.
16. Corky: When the wine attains the flavour of cork due to faulty corking or a faulty cork.
17. Corkage: When the customer brings his own liquor in a licensed place, some amount is charged from the guest for rendering services to the guest.
18. Cocktail: A mixture of two or more alcoholic or non-alcoholic beverages in a definite ratio.
19. Curve: French for blending.
20. Demi-sec: Medium dry; percentage of sweetening is 5-7%.
21. Depot: Sediments which all wines throw, but this is more prominent in red wines.
22. Doux: Very sweet; the percentage of sweetening is 7% and above.
23. D.O.M.: Deo Optimo Maximo. The best wine often dedicated to God. This term is found on a bottle of Benedictine liquer.
24. Eau de vie: Water of life, also known as the 'Delightful soul of wine'.
25. Feints: The heads and tails, *i.e.*, the first and last parts of distillation.
26. Foreshots: End product or residue left in the still after distillation is over.

27. Fine Champagne: A type of Cognac Brandy, classified as VSOP. Brandy is the only alcoholic beverage which is not coloured. It attains its golden straw colour from the cask itself. Fine champagne brandy is made by mixing 60% grapes from Grande Champagne area and 40% grapes from Petit Champagne area.
28. Frappe: Iced or chilled. The term applies to the service of liqueurs.
29. Fliers: Floating crystallised particles seen in overchilled white wine bottles.
30. Full-bodied: Very well-matured wine or spirits.
31. Fino: A term applied to a dry type of sherry.
32. Fine maison: Best brandy of the house.
33. Lees: Sediments which settle down at the bottom of the cask prior to racking (separating the wine from sediments).
34. Optic: An instrument attached to the mouth of spirit bottles only when the bottle is inverted, it dispenses an exact amount (30 ml) of the spirit at a time.
35. Rocking: Removing/siphoning the clear wine from the cask and leaving behind the sediments.
36. Pomac (or cider): Fermented apple juice.
37. Punt: An inverted depression at the base of a bottle which increases the surface area and also helps to withstand the internal pressure of the wine.
38. Sec: French for dry; Percentage of sweetening is 1.5-3%.
39. Sekt: German for sparkling wine.
40. Trocken: German for dry.
41. Ullage: A term used to describe the loss of wine or spirits due to evaporation or leakage.
42. Vin Moussena: French for sparkling wine.
43. Weepes: Leakage of wine through cork due to faulty or crumbled cork.

44. Degorgement: A process used in the production of champagne, namely, removalof sediments from the champagne bottle after secondary fermentation.

Food and Wines: General rules for serving wine with food:

1. Young wines are always served first, followed by mature wines.
2. Dry wines are generally served first, followed by sweet wines.
3. Red wines are, followed by white wines.
4. Light wines first, followed by heavy wines.
5. Whenever food contains anything acidic in nature, no wine is served.
6. With Indian and Chinese food, generally no wine is served, instead, beer and cider are served. Anyhow, with Chinese food, Chinese wines can be served.

Menu and Wines

1. Generally, with hors d'oeuvres, dry Sherry goes the best. Some alcoholic beverages are specifically suited to certain types of hors d'oeuvres like: Vodka with caviar, dry Chablis (or dry Graves) with oysters, and dry Sherry with smoked salmon.
2. With soups, fortified wines like dry Sherry, Port, Maderia or Marsala are served.
3. With the fish course, young white wines or sparkling wines, like white Bur gundy, dry Chablis, dry Sauternes and Moselle may be served.
4. Entree: With Italian pasta products, Chianti or Asti Spumante may be served. Egg preparations are served with white Cote du Rhone and organ meats, such as liver and kidneys go best with Moselle. Red meat served as entree are best with full-bodied, heavy red wines (Bordeaux, Burgundy and Chianti) and white meat with white wines like Burgundy.
5. In the main course, red meat (such as lamb pork and

beef) are served with heavy, full-bodied red wines, like Bordeaux and Burgundy. White wines are served with chicken, fish and other white meat; this includes wines like white Burgundy.

6. The sweet course is generally served with sweet wines like Sauternes, Sweet Graves, Tokay, Barsac, sweet Sherry, sweet Port, sweet Madeira, etc.
7. As far as the cheese course is concerned, soft cheeses are served with white wines and hard, ripened cheeses are served with red wine.
8. Cafe (coffee) is served with chilled liqueurs, Cognac, Sherry or Rum.

Major Categories: Wines may be broadly categorised as 1. Table/natural still wines, 2. Fortified wines, 3. Sparkling wines and 4. Aromatised wines.

Table Wines: Table wines are natural still wines, served along with meals. Their alcohol content is generally low (about 10-12%) but some German Table wines may have 8-10% alcohol. Generally, red meat (like beef, pork and lamb) are served with full-bodied red wines, and white meat (such as chicken or fish) with white wines. Sherry, though not a Table Wine, is served with soup. Light, medium and heavy wines are served with different foods.

Examples of table wines include red and white Bordeaux wines, red and white Burgundy wines, Tokay, Barsac, Hock and Moselle (served with white meat).

Fortified Wines: Fortified wines, with an alcohol content of 18-22% are those wines to which brandy has been added during or after fermentation. The addition of brandy during fermentation halts the fermentation and results in a sweet wine, whereas fortification after fermentation results in dry wine. By the addition of this extra alcohol, the keeping quality of the wine increases (which is why fortified wine can be sold by the glass) and it does not oxidise easily. Examples of fortified wines include Port (served at 54°F), Sherry (served at 55-56°F), Madeira and Marsala.

Sparkling Wines: Sparkling wines are those wines in which carbon dioxides are incorporated. They are effervescent.

Champagne (attributed to the blind monk Dom Peignon, who introduced gas into wine) owes its effervescence to naturally produced carbon di-oxide. After the primary fermentation, the skin, pips and other solids are removed and the wine is bottled. Extra grape sugar is added to the wine, and it is corked, allowing a little space above the wine, so that the air can allow fermentation to occur.

The champagne bottles are placed on a cone-shaped rack on which they are rotated (so that the yeast mixes well with the wine) till they are upside down, and all the sediments collect in the bottle's neck. The neck is than frozen by dipping it in a cold brine solution, and the frozen lump of sediments is removed.

The bottle is subjected to a machine which replenishes any wine loss from degorgement, and checks if the pressure of gas is 70 lb PSI. The bottle is then corked with a mushroom-shaped cork, secured with a wire clip and tin foil.

Other sparkling wines are made by incorporating carbon-di-oxide in wines. They are ready to drink after gassing and are never matured. Bottles of sparkling wine, are opened with bare heads.

Examples of Sparkling wine include Champagne, Moselle, Sparkling Burgundy-Rully, Asti Supamte, Sparkling Hock and Sparkling Chianti.

Aromatised Wines: Aromatised wines are those wines in which aromatic agents (such as herbs) have been added to the base of a fortified wine, which has not reached the specified standards. It is allowed by law to add certain aromatic ingredients. These aromatised wines are generally served as aperitifs and have a high alcohol content (between 10 and 19%). These wines are not served below the 60 ml measure. White aromatised wines dry are served chilled, but sweet red ones are served at room temperature.

Examples of aromatised wines are:

(a) *Campari:* Red and very bitter, either served on the rocks or with soda or orange juice.

(b) *Cinzano:* A vermouth which may be dry and pale (French) or sweet and red (Italian-the Italians discovered Cinzano which they make from white grapes and mature for at least a year) As a drink, 60 ml is served but as a bitter only a dash of it is used.

(c) *Martini:* The term martini may apply to a vermouth (sweet or dry) or a martini cocktail (with gin).

(d) *Dubonnet:* An aromatised wine, red in colour.

(e) *Lillet:* A white aromatised wine made from cactus.

3

Utility of Vitamins

Vitamins are the discovery of 20th century scientists. Carbohydrates, proteins and fats were considered as the nutrients essential for health. In the history of nutrition, the findings of a Polish chemist, Casimir Funk; in 1911 was a turning point. He found out for the first time that a disease syndrome could be produced in an experimental animal if inadequate supply of certain factors in food other than carbohydrates, proteins and fats takes place. He gave the name for the water soluble substance as "vitalamine" which was later known as vitamin. Scientists from various parts of the world were engaged simultaneously in finding out these specific factors which promote growth and maintenance of various organs in our body. Takaki, a physician in the Japanese Navy, first, demonstrated that fatal diseases like beriberi could be treated with rice bran, vegetables, fish and meat. Thus the study of vitamins brought to light about 17 different vitamins. Each one of them has its own history, chemistry, structure, function, sources, requirements and disorder symptoms.

Vitamins are defined as organic compounds which are necessary for good health and vitality. Vitamins are required

in minute quantities and their deficiency results in structural and functional disorders of various organs in the body. Thus vitamins are essential for life.

Classification of Vitamins: Vitamins are classified broadly into two groups: (a) fat-soluble vitamins, and (b) water-soluble vitamins.

Vitamins (Fat-soluble): Fat-soluble vitamins are soluble in fats and fat solvents. They are insoluble in water. So these vitamins are utilised only if there is enough fat in the body.

Vitamin A and carotene, vitamin D, vitamin D_2, vitamin D_3, vitamin E and vitamin K are fat-soluble vitamins. Water-soluble vitamins are soluble in water and so they cannot be stored in the body. Therefore, a day-to-day supply of these vitamins is essential.

Vitamins (Water-soluble)

B-complex Vitamins: 1. Vitamin B (Thiamine), 2. Riboflavin, 3. Nicotinic acid and Nicotinamide (Niacin), 4. Pyridoxin (B_6), 5. Pantothenic acid, 6. Folic acid, 7. Biotin, 8. Choline, 9. P-amino benzoic acid, 10. Inositol, and 11. Vitamin B_{12}.

II. Vitamin C or Ascorbic acid, and

III. Vitamin P. (Bioflavonoid)

Vitamin A: Vitamin A was the first fat soluble vitamin to be discovered. In 1913 Osborne and Mendel found that certain natural fats and oils stimulated growth in rats. Again McCollum and Simmonds (1917) demonstrated that Xerophthalmia was due to the lack of a fat-soluble vitamin. Steenbock (1919) discovered the vitamin A activity of carotenoids. In 1920 Roseheim and Drummond demonstrated that the carotene of plants had similar biological action to that of vitamin A.

In 1930 Moore discovered that when carotene was fed to vitamin A-deficient experimental animals vitamin A was found in their liver. In 1931 Karser obtained the structure of vitamin A. In 1937, Kuhn and Morris synthesised vitamin A. In the same year Holmer and Corbet obtained vitamin A in a crystalline form.

Chemistry: Vitamin A in its pure form is a pale yellow substance soluble in fat. It is unsaturated alcohol which is stored in the body as esters. Vitamin A as such occurs only in animal foods. Retinol is the generally accepted chemical name for vitamin A. It is stable to heat Vitamin A is present in the form of retinol (vitamin A), Retinal (retinene) or vitamin A aldehyde, retonic acid or vitamin A acid, dehydro retinol (vitamin A_2) and carotene (provitamin A). Carotene has alpha, beta and gamma isomers. Vitamin A, is poor in its biological activity. Upon hydrolysis each molecule of beta-carotene yields 2 molecules of vitamin A. But the biological value of beta carotene is only half of vitamin A.

Functions: Vitamin A is essential for the building and growth of all cells, especially skeletal cells. Vitamin A is also needed for proper tooth structure. It also helps in the integrity of the epithelium, especially the mucous membranes, which line the eyes, mouth, alimentary canal, salivary glands, respiratory and genito-urinary tracts. In the deficiency of vitamin A the epithelial tissues of these parts are keratonised.

Vitamin A maintains normal reproductive function in males. Vitamin A is again essential for the maintenance of normal vision in dim light. This vitamin combines with protein to form rhodopsin or visual purple which is present in the retina. Visual purple absorbs light and is decomposed to retinene and protein. The protein and retinene are partially recombined to form rhodopsin in the dark. Vitamin A is essential for this reaction and so continuous supply of this vitamin is required.

Rhodopsin Cycle

(Rods of the retina)

Reteinal dehyde + opsin (Dark) ————————↓

Rhodopsin (Visual purple)

↓

Retinene + protein←————Lights←————

Vitamin A plays an important role in maintaining the

structure of myelin sheath. Vitamin A also helps the synthesis of mucopoly-saccharides.

High serum retinol levels reduce the risk of cancer. This is a new finding in vitamin A studies and further research is going on.

Vitamin A is also known as anti-infective vitamin as it provides resistance power to the body by maintaining normal structure of the organs.

Deficiency: Vitamin A deficiency in man may be due to low intake of vitamin A sources or due to interference with absorption or storage of vitamin A in the body. In certain diseases proper assimilation of vitamin A is affected. Loss of vitamin A occurs in the blood during some infectious diseases.

Clinical manifestations of vitamin A deficiency are retardation of growth and lowered resistance to infection. Deficiency of vitamin A shows in delayed adaptation to darkness on coming from the lighted area because it produces changes in the eyes. In a healthy person the eyes adapt quickly to dim light and bright light. In vitamin A deficiency state rhodopsin is not formed effectively, therefore vision in dim light is not easy. This symptom is known as night blindness. As the deficiency advances a condition known as xerophthalmia develops. The conjunctiva becomes dry and lustreless. The transparent appearance of the eye and its elasticity is lost. The eye becomes grey and opaque. If this condition persists the eye becomes infected and ulcerated. This is a serious condition of the eye which results in blindness. Softening of the cornea brings blindness. Advanced neglected xerophthalmia leads to degeneration of cornea and blindness. This condition is known as keratomalacia.

Bitot's Spots: Grey coloured triangular spots are located in the conjunctiva in severe cases.

Phrynoderma: In vitamin A deficiency, the skin becomes rough because the epithelial tissues are affected. This condition is known as follicular keratosis or toad skin or phrynoderma.

Mucous membraneous structures of respiratory tract, alimentary tract and genito-urinary tract are affected and infectious diseases of these areas result due to a deficiency of vitamin A. Thickening and dryness of the alimentary canal reduce the secretion of digestive juices which leads to diarrhoea.

Experimental animals showed that vitamin A exerts a protective effect against artificially induced cancer and urinary calculus.

Hypervitaminosis: If excess amount of vitamin A is consumed it shows ill-effect in the body. The symptoms are drowsiness, headache, vomiting, itching, skin lesions and loss of appetite.

Sources of Vitamin A: Vitamin A activity is in the form of retinol and carotenoids. The best sources of retinol are colshark and halibut liver oils, and liver of animals like sheep, goat and cow. Butter, ghee, egg, especially egg yolk, and all milk powders are good sources of vitamin A.

Carotenoids are present in green leafy vegetables, yellow vegetables like carrot, pumpkin and yellow fruits like papaya and mangoes. Vegetable oils are available in processed forms (vanaspati) and they must contain specific amounts of retinol as per government rules.

Loss of Vitamin A during Cooking and Storage: Under Indian conditions SO per cent of available vitamin A in food is lost during cooking. However, cooking loss of carotene is negligible. Since most of our dietary vitamin A is from the conversion of carotene, cooking loss is not considerable.

Requirements of Vitamin A: In our diet vitamin A is present as preformed vitamin A or retinol or as carotene which is converted into vitamin A. Retinol or preformed vitamin A is present only in animal foods. In an average Indian diet vitamin A is present in the form of beta carotene. It is converted into retinol. Due to various physiological reasons only one-third of beta carotene is absorbed. But from leafy vegetables and from carrots and papaya the conversion rate is about 50 to 99 per cent.

Considering these factors the ICMR Expert Group recommended a daily intake of 750 mg of retinol/day for an adult. If beta carotene is consumed the recommendation is 3000 mg of beta carotene for an adult/day.

Prophylatic Treatment: In India a majority of the people consume vitamin 'A' deficient diet. As per 1978 studies half of the world's 15 million blind persons are in developing countries. Out of this two out of every three cases are preventable nutritional or infective conditions. The age group most affected in our country is 1-5 years and males are affected more than females.

Low availability of vitamin A in the diet and its poor utilisation due to protein deficiency are the main causes for deficiency. Now the Government of India has introduced prophylatic measures to combat vitamin A deficiency. Intake of one teaspoonful of codliver oil daily provide 3000 I.U. Four ml of red palm oil, supplying oral dose of 1,00,000 microgrammes of vitamin A palmitate through 5 ml of arachis oil are some of the measures.

In severe cases of deficiency large doses of 50,000 mg are administered for a few weeks.

Vitamin D: Vitamin D is otherwise known as 'sunshine vitamin' as it can be synthesised from sunlight by our body. Palm (1890) first demonstrated that a bone deformity, ricket, can be prevented by exposure to sunlight. Mellanby (1919) discovered that codliver oil can cure rickets produced in experimental animals. McCollum (1922) showed that codliver oil contained two vitamins, vitamin A and vitamin D. Recrink and Van Wijk (1931) isolated crystalline vitamin D from activated ergosterol. Schenck (1937) obtained a crystalline vitamin D from activated 7 dehydrocholesterol.

Chemistry: Vitamin D is fat soluble. It is a group of different sterol compounds with antiarchitic properties. They are vitamin D, vitamin D_2 and vitamin D_3. Among these vitamin D_2 or calciferol and vitamin D_3 or cholecalciferol are

biologically important. Ergosterol and 7 dehydrocholesterol act as provitamin of vitamin D. When ergosterol of vegetable origin is exposed to ultraviolet light calciferol is formed. Vitamin D_3 is produced by the action of ultraviolet rays of sunlight on 7 dehydrocholesterol which is present in our skin. If the body is kept in 'purdah' cholecalciferol is not produced by the body.

Functions: Vitamin D itself is in an inactive form. Vitamin D is concentrated in the liver and in minute quantities in skin, lungs, brain, spleen and kidneys. In the liver vitamin D is hydroxylated to 25-hydroxy vitamin D, (25-OH-Dj). It is also known as 25-hydroxyl-cholecalciferol. From the circulating blood vitamin D, reaches the kidney and here it is further hydroxylated to form 1,25-Dihydroxy cholecalciferol (1,25, (OH)D,). This is the active form of vitamin D.

Vitamin D promotes the absorption of calcium and phosphorus from the small intestine. In the DNA vitamin D helps to form active calcium binding protein. It also maintains the concentration of calcium and phosphorus in the blood. Vitamin D helps the deposition of calcium Phosphate in the bone. In rachitic children there may be delay in dentition and malformation of the teeth occurs.

Sources: Fish liver oils, egg yolk, milk, butter and ghee are the best sources of vitamin D. Some brands of butter are fortified with vitamin D. Sunlight is the best natural source of vitamin D.

Vitamin D is absorbed from the intestine with dietary fat from where it is taken to the liver.

Deficiency of Vitamin D: Deficiency of vitamin D results in poor absorption of calcium and phosphorous from the intestine and excessive loss of them in the urine and faeces. Concentration of calcium in the plasma is diminished and it produces tetany which is charaterised by hyperexcitability of the nervous system. This stimulates the parathyroid glands to decalcify bone calcium to raise the plasma calcium level.

Deficiency of vitamin D leads to rickets in children and osteomalacia in adults.

Rickets: The name ricket is given to vitamin D deficiency after a famous English bone-setter by Ell. In mild deficiency of vitamin D, the symptoms shown are restlessness and irritability. Ricket is a disease of infancy and childhood. Calcium and phosphorus are not deposited enough to form strong bones and as a result the leg bones become bent and deformed. Small round unossified areas are developed near skull bones and projections are formed on the osteochondal junction of the ribs. It resembles a rosary and it is known as 'rickety rosary'. The abdomen is distended and chest deformities like depression of ribs attached to diaphragm take place. This is known as Harrison's sukus. This gives an appearance of 'pigeon breast'. In a later stage pelvic deformities, bow legs and knock knees develop. Nutritional rickets occur in infants and children of poor families. Premature infants are more susceptible to rickets.

Osteomalacia occurs among adults. It is often known as 'adult rickets'. Osteomalacia is most common among women who live on a poor diet lacking in calcium and vitamin D. Pregnant women are more prone to this. It is mainly because of the increase in weight during pregnancy accompanied by the deficiency of vitamin D. It is again common among women who observe 'purdah'. Repeated pregnancies and prolonged lactation further deplete calcium stores which results in osteomalacia. Pain, softening and tenderness in the bones and fractures are the common complaints of the patients of osteomalacia.

Prophylatic Treatment: For the treatment of rickets and osteomalacia 1,000 to 5,0001.U. of vitamin D should be administered orally for about a month. If there is marked improvement the dose can be reduced to 8001.U./day for at least six months. A massive dosage of 1,000,0001.U. can also be administered but small dosage is safer.

Storage: Cholecalciferol is converted to 25-hydroxy cholecalciferol in the liver. This is the active form of vitamin

D. Vitamin D is stored in body tissues in the inactive form. Adipose tissues and skeletal muscles are the major storage sites. Brain, lungs, skin and spleen are other places where it is stored.

Vitamin E: Evans and Bishop (1923) discovered that for normal reproduction of experimental animals a fat soluble vitamin is needed. It was later named vitamin E or tocopherol. Female rats with vitamin E deficiency are unable to carry the young ones throughout the gestation period. The word tocopherol is derived from the Greek word tocos (childbirth or offspring).

Chemistry: There are eight naturally occurring tocopherols with vitamin E activity. Among these Alpha tocopherol is the most active. Tocopherols inhibit oxidation; therefore it is known as antioxidant vitamin. Vitamin E is soluble in fats and stable to acids and heat.

Sources: Cereal gem oils are the richest source. Vegetable oils except coconut oil and nuts are fair sources.

Functions: Vitamin E has a prominent role in maintaining the stability and integrity of cell membranes. One latest discovery about its function is its role in erythrocytes survival among premature infants. Experimental animals showed that deficiency of vitamin E results in anaemia in monkeys. Vitamin E deficiency produced muscular dystrophy in rabbits and reproductive failures in rats. One important function of vitamin E is its antioxidant properties. Through this activity it reduces the oxidation of vitamins A and C and sulphur containing amino acids.

Through this activity it stabilises the lipid parts of cell membrane. The presence of vitamin E as a natural component of vegetable fats protects against rancidity. Recent studies on rats have shown that vitamin E may protect lung tissues from damage by air pollution through nitrogen dioxide and osazone. In experimental animals vitamin E has been shown to promote the effectiveness of linoleic acid in preventing essential fatty

acid deficiency. Studies have suggested the possibility of polyunsaturated fatty acids having a role in brain development during infancy. The antioxidant property of vitamin E helps fatty acids from oxidation and thus it indirectly helps the brain development.

Vitamin E Deficiency: Vitamin E deficiency is not common among human beings as it is widely distributed in foods. Experimental deficiency in female rats results in foetal death and in males testicular changes in sterility. Muscular dystrophy and paralysis are the other symptoms of vitamin E deficiency. Premature babies with vitamin E deficiency showed impaired fat absorption. If vitamin E deficiency is present in the pregnant modier transfer of placental blood to foetus is poor and so haemolytic anaemia may occur in the baby after delivery. Vitamin E deficiency in adults results in poor absorption of fat and increased haemolysis and increased excretion of urinary creatinine. Brown muscle pigment deposition catted ceroid occurs with tocopherol deficiency. Liver necrosis, erythrocyte haemolysis and anaemia are the other symptoms.

Requirements: For males 30 I.U. and for females 25 I.U. are recommended. Human milk supplies enough tocopherol for the infants.

Vitamin K: In the 1930's a Dutch scientist Dam and his associates found that "Coagulation vitamin" is necessary for clotting or coagulation of blood. It was named vitamin K, Karrer and coworkers (1939) isolated pure vitamin.

Chemistry: There are two naturally occurring forms of vitamin K, vitamin K_1 and vitamin K_2. Vitamin K_1 is vegetable origin and vitamin K_2 is produced by bacterial sythesis.

Sources: Plant oils are the best sources. Rice bran oil and wheat germ oil, soyabean and cottonseed oils are the best sources. Wholegrain cereals, fleshy foods and organ meats are fair sources. Vegetables and fruits are poor sources. It is a fat soluble vitamin and is easily oxidised by acids, alkalies and light.

Functions of Vitamin K: Vitamin K is essential for the formation of prothrombin which helps normal clotting of blood. Vitamin K is not directly involved in the coagulation but it is essential for the coagulation of blood. It takes part in the carboxylation of glutamic residue.

Requirements: The exact requirement is not known. If lowered prothrombin level is there 1-2 mg of vitamin K/day is recommended.

Deficiency: Dietary deficiency is not common. Deficiency of vitamin K is indicated by a tendency to bleed from skin and mucous membranes. Deficiency occurs due to deficient production of it by the gut, with prolonged antibiotic therapy, due to poor absorption of vitamin K, in malabsorption of fats and due to low prothrombin synthesis in liver disease. Pyloric obstruction and diarrhoea or dysentery can result in poor absorption.

Deficiency of vitamin K among newboms, especially among premature babies, delays clotting of blood. If the mother's intake during pregnancy is poor it can lead to this condition.

Excess of vitamin K due to excessive administration produces hypoprothrombinemia and disorders in liver.

Vitamin K is not stored in large quantities in the body. Only in the liver vitamin K is stored.

Water-soluble Vitamins: There are 13 vitamins in this group. Vitamins C and B complex family, comprising twelve vitamins, are soluble in water.

In 1882 Takaki, the Director General of the Naval Medical Service in Japan, demonstrated that 'beriberi', a fatal disease in the navy, could be cured by including a mixed diet of whole cereals, vegetables and milk. In 1897 Eijkman found that fowls on a poor diet manifested symptoms like beriberi. He suggested that instead of rice, rice Polishings must be included in the dietary pattern to avoid beriberi. In 1911 Casimer Funk described a group of water-soluble substances present in food which can act as anti-scorbutic, anti-beriberi and anti-pellagra

factor. In 1926 McCollum and Kennedy suggested different substances in food with different properties which were later identified as anti-beriberi and anti-pellagra factors.

Thiamine (B_1): The occurrence of beriberi led to the discovery of thiamine. Followed by the findings of Takaki, Eijkman established that rice polishings prevented a disease similar to beriberi in fowls. Dr. Grijins, an associate of Eijkman, suggested beriberi was a deficiency disease and rice polishings contain the factor. Funk, Suzuki and Edic attempted to isolate the 'anti-beriberi' factor. In 1926 Jausen and Donal's isolated crystals of thiamine. In 1936 Dr. R.R. William synthesised thiamine.

Chemistry: It is a white powder which is readily soluble in water. Alkali destroys thiamine. Thiamine is commercially produced as thiamine hydrochloride. It has a yeast-like odour and it is stable in its dry form.

Physiological Functions: Thiamine combines with pyrophosphate and forms thiamine pyrophosphate. (TPP) Thiamine pyrophosphate participates in intermediary metabolism of carbohydrates. Transketolase and enzyme has thiamine pyrophosphate as a co-factor. It acts as a co-enzyme in the metabolic pathway of carbohydrates at different places. Animal experiments showed that thiamine pyrophosphate is essential for protein metabolism. Thiamine is also required for the synthesis of glycine in the body.

Thiamine pyrophosphate is present in the peripheral nerve cells. It has a specific effect on nerve tissues. In thiamine deficiency the peripheral nerves degenerate. Thiamine is also required for the maintenance of normal gastro-intestinal tone and motility. Thiamine seems to help the heart in its noitnal action. Thiamine is otherwise known as "morale vitamin" because it helps the transmission of nervous impulses and maintains the stability of nerves.

Deficiency of Thiamine: It is difficult to identify the deficiency earlier. In mild deficiency loss of appetite, fatigue,

irritability, depression and constipation occur. Gastrointestinal and cardiovascular disorders with peripheral nervous problems initiate the deficiency state of thiamine.

In mild deficiency loss of appetite, loss of weight and strength fatigue, emotional upsets, irritability, depression, anger and fear are manifested. Constipation is common among such people. Headaches, insomnia, and techycarida with slight exertion occurs during thiamine deficiency. Severe deficiency of thiamine causes beriberi. Two forms of beriberi occur in human beings. They are dry beriberi and wet beriberi.

Dry Beriberi: In dry beriberi, the peripheral nerves of the legs and arms are affected first. Degeneration of the myelin sheath of the nerves is followed by the degeneration of the axis cylinder. Calf muscles become lender. Numbness in the toes and ankles takes place. The knee and ankle jerk and later on the sensation is lost. Tingling and numbness of the legs and hands are followed by wasting of muscles and difficulty in walking.

Wet Beriberi: All the symptoms of dry beriberi are seen in wet beriberi also. In addition to these oedema in legs and in between cardiac muscle fibres is observed. These changes in the heart muscles lead to enlargement of the heart. Later palpitation, oliguria, and dyspnoea may develop. Oedema on the legs will be prominent. The patient may be bed-ridden and cardiac failure may take place. If untreated it leads to death within hours.

Infantile Beriberi: It affects infants within six months. The early symptoms are restlessness, sleeplessness and loss of appetite. The affected child is pale, oedematous and ill-tempered. Vomiting and green-coloured diarrhoea are common. If treatment is delayed it leads to death.

Requirements: Thiamine requirement is related with energy intake. The Nutrition Expert Group of ICMR and FAO/WHO recommended 0.4-0.5 mgs of thiamine per 1000 kcals. For infants 5.9 mgs/kg and from 1 year onwards proportionate kcals requirements are recommended.

During deficiency state 5 mgs of thiamine orally three times a day is recommended. It is readily absorbed from the small and large intestines. It is not stored in the body.

Sources: Rice polishings, dried yeast and wheat germ are rich sources of B_1. Whole cereals like wheat, oats, legumes, oil seeds and nuts are good sources. Milled cereals, vegetables, fruits, milk, meat and fish are fair sources. On milling, thiamine is lost from cereals. Commercial preparation of B_1 is extracted from natural sources or it is synthesised. Nowadays cereals are enriched with thiamine. Flours are enriched and 'abelled to show the amount of enrichment.

Riboflavin: Scientists from various parts of the world reported that a growth promoting factor substance remained even after the beriberi preventive factor is destroyed from yeast A yellow green fluorescent compound was known to be in the whey from milk. In 1932 the German scientists Warburg and Christian discovered the first flavoproteins from yeast. It was first known as 'Yellow enzyme'. In 1933 Kuhn and Gyorgy isolated riboflavin and Booker recognised the growth promoting property of it. It was synthesised by Kuhn and Karrer in 1935.

Chemistry: Riboflavin is dimethyl-iso-alloxazine 'attached to ribose.' It is slightly soluble in water and stable in acid medium. In alkaline medium it is easily destroyed. Bright light also destroys riboflavin.

Functions: Riboflavin is part of food co-enzymes as riboflavin monophosphate or flavin mononucleotide (FMN) and flavin adenine dinucleotide. Riboflavin is closely related to biological oxidations in cells. Flavoprotein enzyme system is present in different parts of our body. It takes part in cell respiration. Cellular growth cannot evolve in the absence of riboflavin. Riboflavin through the enzyme system possesses the ability of transferring hydrogen in chemical reactions.

Thus, it takes part in carbohydrate and protein metabolism. Riboflavin is essential for the health of skin, mucous membraneous structures of alimentary canal and eyes. For

normal vision riboflavin is essential. The retina has free riboflavin which is converted by light to a compound that stimulates optic nerves.

Deficiency: In mild deficiency of riboflavin weakness, anorexia, apathy, burning at the angles of mouth, in the eyes and over the skin occur. Normal metabolism of carbohydrate and protein is affected. Riboflavin deficiency symptoms are generally known as ariboflavinosis. Ariboflavinosis consists of angular stomatitis, with cheilosis, glossitis, skin changes, seborrheic dermatitis and vascularisation of the cornea.

The angles of the mouth become pale and crust is formed in the corners. Fissures occur at the corners of the mouth and it is known as cheilosis. Crust is formed on the surface of lips and removal of it makes it red in colour. Papilae of the tongue are peeled off and the tongue shows a 'magenta' or red tint on the surface. This is referred to as glossitis. On the nasolabial folds, ears, vulva in the female reproductive organs and on the scrotum in the male, scaly desquamation of the skin takes place. These are known as nasolabial seborrhoea and scrotal lesions. The lesions are cured by the administration of riboflavin.

Even in a mild deficiency state burning and watering of the eyes, sensitivity to light and dimness of vision take place. Vascularisation of the cornea and abnormal pigmentation of the iris are the other symptoms manifested. Vision is blurred and conjunctivitis and itching may occur in riboflavin deficiency.

Requirements: Various studies on the requirement of riboflavin showed that 1.4 mg per day for a sedentary man, 1.7 mg for a moderate worker and 2.3 mg for a heavy worker are essential. For females the requirement is 1 mg, 1.1 mg and 1.5 mg, respectively. During pregnancy and lactation additional 0.2 mg and 0.3 mg riboflavin are required. During infancy (0-6 months) 0.71 mg/kg and 6-12 months, 0.65 mg/kg. For 1-3 years 0.7 mg, 4-6 years 0.9 mg, and 7-9 years, 100 mg. of riboflavin are required. During adolescence 15 mg in the earlier stages, that is, 10-12 years and 1.6 mg during later adolescence

for boys had 1.4 mg in early adolescence and 1.4 during later adolescence for girls are recommended.

Sources: Fleshy foods like liver, egg, skimmed milk powder and other milk products are rich sources of riboflavin. Milk, meat, fish, whole cereals, legumes, dais, oil seeds, nuts and leafy vegetables are good sources of riboflavin. Milled cereal and flours, roots and tubers and other vegetables are fair sources of riboflavin.

Riboflavin is readily absorbed from the intestine and the excess is excreted through urine.

Niacin (Nicotine Acid): Niacin is also known as nicotinic acid or nicotinamide. In 1725 a Spanish physician Casal described the occurrence of red sickness. In the eighteenth century, the Italian physician Frapolt named the disease as pellagra, meaning rough skin. In 1911 Funk isolated niacin from rice polishings. In 1937 Elvegem and co-workers showed that the deficiency disease (black tongue in dogs) was cured by an isolated substance from liver. Goldberger cured and prevented pellagra by adding meat, vegetables, fruits and eggs in the diet. Thus, pellagra preventing factor (P-P-factor) was identified from foods. Further investigations showed that nicotinamide was equally effective as nicotinic acid in curing pellagra.

Chemistry: Niacin is a white crystalline compound soluble in water, stable to heat, light, acids and alkalies. In the body niacin is converted into niacinamide.

Physiological Functions: Like thiamine and riboflavin niacin takes part in the metabolism of carbohydrates, proteins, and fats through the enzymatic action. It is part of co-enzyme I and co-enzyme II which are essential for the conversion of lactic acid to pynivic acid. They are otherwise known as NAD and NADH. More than 40 biochemical reactions depend on these enzymes. Dehydrogenation reaction in Kreb's cycle, oxidation of alcohol to aldehyde, of glucose to gluconic acid, of malic acid to oxaloacetic acid, of lactic acid to pyruvic acid,

glucose -1-6-phosphate to phosphohexouric acid and reactions in the synthesis of fatty acids, cholesterol and for the conversion of phenylalanine to tyrosine are some of them. It takes part in tissue oxidations. Nicotinic acid is essential for the normal functioning of the skin, intestinal and nervous systems.

Effects of Deficiency: In a mild deficiency state weakness, loss of weight, loss of appetite, lassitude, irritability, burning tongue, and constipation occur. Pellagra is the characteristic disease resulting from niacin deficiency. It is otherwise known as three 'D's disease—dermatitis, diarrhoea and dementia. The dermatitis of pellagra is different from other skin diseases. It occurs mainly on the exposed portions of the body. Lesions are seen on the face, neck, surfaces of the hands, elbows, feet and other parts of the body which are exposed to sunlight.

Another peculiarity of dermatitis is that the skin lesions are symmetrical in shape, and they always occur on both sides of the body simultaneously. The skin is cracked and scaly. Burning sensation and itching are common. The lesions leave a reddish, dark brown colour. The mucous membrane of the alimentary canal is affected and glossitis, abdominal pain, and loss of appetite are some of the symptoms. Nausea, vomiting and diarrhoea are the other symptoms of niacin deficiency. Delirium is the common mental disturbance in pellagra. Earlier symptoms of nervous disturbances are depression, dizziness and insomnia. It leads to death.

Pellagra is common among jowar eaters. Jowar has a very high leucine content and this excessive leucine interferes with the utilisation of nicotinic acid. Leucine precipitates mental changes but isoleucine promptly reverses them to normal.

Dr. Gopalan and his co-workers in India have done extensive studies for a decade (1960-69) in this connection. Among maize eaters pellagra *is* common because of the high leucine content in them. Leucine interferes in the conversion of tryptophan to niacin and also the conversion of niacin to NAD and NADP in the tissues.

Sources: Cereals, especially whole cereals, dried yeast, liver,

groundnut, legumes, fleshy foods and fish are good sources. Milk, eggs, and vegetables are fair sources.

Tryptophan-niacin Relation: Tryptophan is an essential amino acid and it is a precursor of niacin. Vitamin B1 or thiamine, B2 or riboflavin and pyridoxine are essential for this conversion of tryptophan to niacin. Another essential amino acid leucine interferes with the conversion of tryptophan to niacin and niacin to co-enzymes. Milk is a rich source of tryptophan. Animal foods rich in protein are good sources of niacin.

In a healthy person's body 60 mg of tryptophan is equivalent to 1 mg of niacin. Thus the dietary allowances are usually made in terms of niacin equivalents.

Niacin equivalents (mg) = Niacin content (mg.) + tryptophan content (mg)/60

Requirements: The FAO/WHO expert group has recommended an allowance of 6.6 mg niacin equivalents per 1,000 kcals. For all age groups this requirement pattern is suggested.

Pyridoxin (B_6): Three naturally occurring compounds with similar activities are together known as Vitamin B_6. They are pyridoxine, pyridoxal and pyridoxamine. It is also known as adermin. In 1934 Gyorgy found a factor leading to a type of dermatitis in rats. In 1938 it was isolated and named pyridoxine.

Chemistry: It is a white crystalline substance soluble in water, quite stable to heat and acid. In an alkaline medium it is easily destroyable.

Physiological Function Vitamin B_6 acts as a co-factor for several enzymatic reactions in all the three metabolisms. In protein metabolism, transamination takes place with the help of this enzyme system. In carbohydrate and fat metabolisms, there are intermediary reactions where vitamin B_6 acts as a co-factor.

It is part of the transaminase system which has a number of enzymes, e.g.,

1. Glutamic acid + oxalo acelate - alphakete-glutarate + aspartate. Alanine + alpha ketoglutarate = glutamic acid + pyruvic acid.

2. Amino acid decarboxylase- Amino acids to amines

   e.g., Histidine = Histamine + CO_2

   Tyrosine - Tyramine + CO_2

Tryptophan conversion to niacin has a series of biochemical reactions. In pyridoxine deficiency this conversion step will not materialise and niacin is not formed.

Removal and transfer of sulphur groups from sulphur-containing amino acids like cysteine requires pyridoxine.

In carbohydrate metabolism pyridoxal phosphate is essential for the action of phosphorylase enzyme which converts glycogen to glucose.

Vitamin B_6 helps in energy transformation in brain and in nervous tissues. In its deficiency, convulsive seizures occur in animals and in human beings. Especially infants fed on evaporated milk develop nervous irritability and convulsive seizures.

Deficiency: Separate deficiency of B_6 is not common. But as a part of general deficiency of B complex, vitamin B_6 deficiency also occurs. In experimental cases anorexia, nausea, restlessness, lethargy, seborrhoea, hyperpigmentation, pellagra like dermatitis, cheilosis, glossitis and angular stomatitis were observed. Changes in the periphery of nerves with sensory impairment and motor abnormalities were also observed. In infants convulsions are seen. Immune response is impaired in B_6 deficiency.

Sources: Dried yeast, rice polishing and liver are rich sources. Milieu like cholam, cambu, pulses, beetroot, cabbage and fleshy foods are rich to fair sources, respectively.

Pantotnenic Acid: In 1933, R.J. Williams reported that a factor present in food is essential for growth in yeast. It is

named pantothenic which is derived from a Greek word 'Panthos' meaning everywhere. In 1939, it was isolated and synthesised in 1940.

Chemistry: It is soluble in water and it contains the amino acid alanine. It is stable to heat but prolonged exposure to dry heat destroys it. It is also destroyed by acid and alkali.

Functions: Pantonenic acid forms a complex compound co-enzyme A (CoA) and acyl carrier protein and thus it takes part in the metabolism of carbohydrates and fats. It is also necessary for the formation of acetyle choline, which is a precursor of heme which in turn is essential for haemoglobin synthesis. CoA is also essential for the synthesis of cholesterol and sterob.

In our body pantotnenic acid is present in the liver, adrenal gland, kidneys, brain and heart.

Deficiency: Deficiency of pantothenic acid is rare. Experimental production of its deficiency showed loss of appetite, fatigue, weakness, headset, insomnia, nausea, emotional instability, dizziness, imparted motor coordination, muscle cramps, burning sensations in the feet and changes in the heart beat rate and in blood pressure.

Burning feet syndrome is associated with neurological and mental disturbances and it is cured only by pantothehic acid. Gopalan (1946) suggested the treatment for burning feet as administration of capantothenate. He also found that other B-complex vitamins like thiamine, riboflavin and niacin were not effective in its cure.

Sources: Dried yeast, liver, rice polishings and wheat germs are the rich sources. Whole cereals, legumss, nuts, fleshy foods, eggs and fish are fair sources.

Requirements: For adults 10 mg/day, for adolescents 8-10 mg/day, children 5-8 mg/day and for infants 1.5-2.5 mg/day was recommended by the FAO/WHO expert committee. During pregnancy and lactation 10-15 mg/day was recommended.

Vitamin B_{12}: Minot and Murphy in 1926 reported that pernicious anaemia could be controlled by feeding liver in large quantities to patients. Castle found that beef muscle is also effective in controlling pernicious anaemia. He established that two factors were involved in curing pernicious anaemia, the intrinsic factor of normal gastric secretion and the extrinsic factor vitamin B_{12}. In 1948, two British scientists, Smith and Parker, and Dickes from the USA isolated B_{12}. Its structure was determined in 1955.

Chemistry: Vitamin B_{12} contains the mineral cobalt which gives its characteristic dark red colour. Cyanocobalamin and Hydroxy cobal are the two forms of B_{12}.

Functions: B_{12} is essential for the maturation of red blood corpuscles and it combines with globulin. B_{12} takes part in the enzymatic reaction. Especially co-enzyme B_{12} is very prominent in the body. Niacin, riboflavin and magnesium are essential for this conversion of B_{12} to co-enzyme B_{12}. B_{12} co-enzyme is required for the action of dehydrase reaction.

Vitamin B_{12} acts as a necessary substance for the lobo protein action in the nervous tissues. It is also necessary for the synthesis of myelin. Within the bone marrow B_{12} cc-enzyme participates in the synthesis of DNA.

Co-enzyme B_{12} is essential for carbohydrate, protein and fat metabolism. In experimental animals vitamin B_{12} has been found to stimulate growth. B_{12} co-enzyme is also essential for the synthesis and transfer of single carbon units such as the methyl group, eg., methionine and choline synthesis.

Deficiency: Due to the deficiency of vitamin B_{12}, pernicious anaemia occurs. To prevent anaemia two factors are required: an intrinsic factor produced by the gastric parietal cells and the extrinsic factor of vitamin B_{12}. If due to any cause enough intrinsic factor is not secreted, B_{12} is not absorbed and utilised properly. B_{12} is otherwise known as extrinsic factor. Congenital intrinsic factor deficiency and low absorption of B_{12} or its deficiency leads to pernicious anaemia. In pernicious anaemia

the life span of red blood cells reduced to 60 days instead of the normal 120 days and their size and shape also is changed. Skin changes show a yellowish pallor.

Soreness and shine on the tongue are other manifestations of skin lesions. Gastric secretion is reduced. Stomach mucosa shows atrophic changes. Nervous tissues are degenerated, especially the posterior column of the spinal cord is affected. Peripheral neuropathy and later on mental changes are common. Tingling, numbness, loss of sensation in the limbs, exaggerated knee jerks and ankle jerks are the other symptoms. Depression and other psychosis symptoms are common in later stages.

Sources: Organ meats like kidney, liver, brain, meat, poultry, egg, fish and milk are good sources of B_{12}. Vegetable foods lack B_{12}.

Requirements: B_{12} requirements for a normal person is 1 mgm (microgram) per day. During pregnancy and lactation it is increased to 1.5 mgm (microgram). Infants and children up to 10 years require 0.2 microgram and adolescents need 1 microgram.

Choline: In, 1934, Best and Huntsman discovered fatty liver in choline deficiency among rats. Choline is a component of phospholipid lecithin arid it forms acetyl choline with alcohol and takes part in biochemical reactions. It is involved in the transmission of parasympathen'c and other nerve impulses to effective organs. It promotes growth in animals.

In its absence neutral fats and cholesterol esters accumulate in the liver which produce fatty liver. Choline is otherwise known as a lipotropic factor as it prevents fatty liver. Choline enhances the oxidation of fatty acids and promotes the formation of lipo proteins in the liver. Phospholipids take part in the transport of fatty acids and cholesterol in the blood and from the liver and their deposition and removal in adipose tissues. It is essential for the transfer of nerve impulses. It takes part in the formation of methionine from homocystine. In its deficiency production is reduced.

Sources: Egg yolk, wheat germ, organ meats like kidney, brain, heart, meat, yeast, soyabean and peanut and skimmed milk are good sources of choline. The amino acid methionine acts as a donor of methyl group to convert cltanolamine to choline.

Requirements: Choline requirements of human beings is not known and its deficiency is also uncommon.

Biotin: The importance of biotin was first observed by Boas in 1927. He found mat raw egg white produces dermatitis, loss of hair, and muscular incoordination in experimental animals. Therefore, it was first called anti-egg white injury factor. In 1931, Gyorgy gave die name vitamin H or Haunt, which means skin. It is sparingly soluble in cold water and alcohol Biotin is present in human tissues and is necessary for bacterial growth. It is also necessary for normal gestation and lactation in sals. Biotin plays an important role in maintaining the skin soft and for the health of nervous system. Biotin also acts as a co-enzyme for carboxylatioa of acetyl-CoA to form malonyl-CoA in fatty acid metabolism.

Egg White Injury Factor (Avidin): Rats kept on a diet of uncooked egg white developed dermatitis, muscular incoordination and spastivity of legs and loss of hair. This is produced by a protein present in egg white called avidin. Avidin combines with biotin which is unavailable to the body. This avidinbiotin compound is excreted in the stcol. Since avidin is present in egg white it is also known as egg white injury factor. Since heat destroys avidin, cooking of egg destroys it.

Deficiency of biotin among human beings is not common. Sydenstricker and co-workers in 1942 produced biotin deficiency in man by giving 30 per cent of the total calorie intake from fresh egg white. Dermatitis of varying types on the necks, hands, arms and legs were seen. It was followed by lassitude, muscle pain, anorexia, nausea and anaemia. Injection of biotin brought rapid cure.

Sources: Dried yeast, rice polishings, liver, groundnut,

soyabean, whole cereals, legumes and fleshy foods are good sources of biotin.

Requirements: A normal balanced diet supplies the requirement of biotin. Its deficiency is not common. The exact requirement of biotin is not known.

Inositol: Wolley and co-workers in 1940 reported that deficiency of inositol produced alopaecia in the mouse. Inositol is a component of phospholipids.

Inositol is essential for the growth of fibroblasts. It is present in phospholipids and it exerts a lipotropic effect. It is present in different tissues, especially in heart and skeletal muscles. It also enables peristalisis of the intestine.

Effects of Deficiency: In experimental animals deficiency of inositol produced retarded growth, loss of hair or alopaecia. It also produces spectacled eye or loss of hair around the eye.

Sources: It is widely distributed in nature. Fruits, milk, vegetables, nuts, meat, wholegrains, yeast and bacteria are good sources of inositol.

Para-amino-benzoic Acid: In 1941, Ansbacher found out that para-amino-benzoic (PABA) is essential for normal growth of rats. For normal lactation and for the prevention of greying of hair para-amino-benzoic acid is necessary. It is present in folic acid.

Para-amino-benzoic acid inhibits the bacteriostatic effect of sulphonamides. It also inhibits the production of thyroid hormones.

Requirement of PABA for human beings is not known. Average diets provide adequate amounts of PABA.

Vitamin C: In the eighteenth century James Lind, a British naval surgeon, demonstrated that citrus juices cured scurvy among sailors. Hoist and Frolich produced scurvy in guinea-pigs. The study of ascorbic acid was thus actually initiated in 1907 by these scientists. Later in 1928, Szent-Gyorgy isolated a factor from orange and cabbage juice. It was hexuronic acid

which prevented scurvy. Wangh and King in 1932 showed that hexuronic acid was identical with vitamin C and they isolated it from lemon juice.

Chemistry: Ascorbic acid is a white crystalline substance readily soluble in water. It is present in two forms: L-ascorbic acid and D-ascorbic acid. L-ascorbic acid has anti-ascorbulic property while D-ascorbic acid has no such property. It is easily destroyed by oxidation, heat and alkalies. Even when it is exposed to air it is oxidized to dehydroascorbic acid. During cooking much of it is destroyed. Iron and copper act as catalysts and cooking in these vessels increases the loss of vitamin C.

Ascorbic acid is also lost by the action of an enzyme, ascorbic oxidase, present in vegetable oils. When vegetables are cut into fine pieces more enzyme is released and it causes more loss. This can be minimised by immersing vegetables in boiling water as heat destroys the enzyme. Adding sodium bicarbonate to preserve the colour of vegetables destroys vitamin C. As it is readily soluble in water washing vegetables after cutting, cooking in too much water and cooking in cold water enhances the loss of vitamin C.

Ascorbic acid is readily absorbed from the intestine.

Physiological Functions: Due to its great affinity for oxidation it takes part in oxidation and reduction reactions in the tissues. Ascorbic acid is essential for the formation of collagen present between cells. It is otherwise known as the cementing material or intercellular cement of the body. If collagen is defective tissues in the capillaries, teeth and bone matrix are net formed properly. Ascorbic acid is also necessary for the formation of osteoblasts. Vitamin C is essential for cholesterol metabolism.

It is also necessary for the conversion of phenylalanine to tyrosine and for the oxidation of tyrosine. It takes part in the conversion of tryptophan to 5-hydroxy-tryptophan. Ascorbic acid reduces the ferric iron to ferrous iron and iron is absorbed

only in this form. It is also essential for rapid healing of wounds as it helps in the formation of connective tissues. Vitamin C is also essential for the utilisation of folacin.

Deficiency of Vitamin C: Prolonged deficiency of Vitamin C produces scurvy. In mild deficiency fatigue, weakness, irritability and frequent infection occur. Pain in bones is common.

Scurvy among infants is known as infantile scurvy. The infant cries on moving its legs and arms. Early symptoms include listlessness, fatigue, fleeting pains in the joints and muscles, bleeding gums, internal haemorrhages and shortness of breath. In severe deficiency of vitamin C scurvy occurs. Swollen gums, loose teeth, soft or malformed weak bones, anaemia, degeneration of muscle fibres including heart muscles are some of the symptoms of vitamin C deficiency. Separation of ribs through beading is also observed. This can be distinguished from rickets from its shape. The skin becomes rough and dry. Pyrexia, rapid pulse and susceptibility to infection are common in scurvy.

In adults, weakness, spongy bleeding gums, loose teeth, swollen joints and haemorrhages in various tissues are the symptoms.

Propylaxis for infantile scurvy include 50 mgs of vitamin C with orange juice three or four times a day. In the case of adults 100 mgs of ascorbic acid twice daily is recommended. Citrus fruits, sprouted pulses and soaked cereals must be liberally used for vitamin C.

Dietary Sources: Citrus fruits like lime, orange, pineapple, ripe mango, papaya, cashew fruit and tomato are good sources. Amla or Indian gooseberry is the richest source of vitamin C. One hundred grammes of amla contain 700 mg of vitamin C. Guava is also rich in this vitamin C. Leafy vegetables are good sources of vitamin C. But this vitamin is readily destroyed by oxidation during cooking and storing. About half of the vitamin C in raw food is lost during cooking.

Requirements: The ICMR (1984) recommended 40 mgs of vitamin C as sufficient for an adult considering its cooking loss. During pregnancy and lactation the requirement was reviewed and the committee recommended that no additional allowances were needed during pregnancy but 80 mgs of vitamin C is required during lactation. For infants 20 mg per day and for children 40 mg per day is enough.

Folic Acid: Folic acid has been called folacin, pteroylglulamic or vitamin Bc. In the early 1930s a search for anaemia-preventing factor was done by many scientists. Dr. Lucy Wills (1934) showed that there was a factor in yeast and crude liver extracts which cured tropical macrocytic anaemia. This factor was known as "Wills factor". In 1946 Keresztesy and Silverman isolated folic acid from liver. The word folic acid is derived from the Latin word folium (leaf) as it is found in green vegetables. Many bacteria synthesise folic acid but lactobacillas grows well only if folic acid is supplied. So folic acid is otherwise known as Lactobacillas.

Chemistry: Folic acid is a yellow crystalline compound widely distributed in nature. Both folic acid and folinic acids occur in foods as conjugates of glutamic acid. It is slightly soluble in water, relatively unstable to heat and easily destroyed by exposure to light. Folacin occurs in different forms. Most of it occurs as pterolypolyglutamates. It is sensitive to light and cooking loss is reduced by anti-oxidant agents like vitamin C.

Physiological Functions: Folic acid is converted to active folinic acid. Folic acid acts as a co-enzyme. Co-enzyme takes part in metabolic process when it transfers single carbon units from one compound to another. Folacin co-enzymes are involved in many methylation reaction. Folic acid is essential for nucleoprotein synthesis and it is required for cells division and also for maturation of erythrocytes. The role of folacin in protein synthesis is made in cancer therapy. Folacin inhibitor like methortrexale are used in chemotherapy to inhibit tumor growth. It is essential for reproduction in animals. Folic acid also helps the hair growth and health of skin.

Deficiency: Lack of folacin results in megaloblastic anaemia. It is otherwise known as macrocytic anaemia. This mainly affects pregnant women. A study in Madras showed that 20 per cent of maternal deaths occurred due to anaemia. Poor vegetarian diet supplies very little folacin. This type of anaemia is due to inadequate formation of nucleoproteins.

This prevents the megaloblasts in the bone marrow from maturing into erythrocytes. In megaloblastic anaemia the number of red blood cells produced in the bone marrow is reduced. The haemoglobin level is also reduced to 6-9 per cent and R.B.C. count to two to three million per mm. Along with inadequate diet it also results from malabsorption syndromes like tropical sprue. Often B12 deficiency after a surgical removal of some part of the stomach also produces megaloblastic anaemia.

Requirements: 100 mgm for adults and 300 mg during pregnancy are recommended by the expert committee. Breast-fed infants will receive about 25 to 30 mg of folate daily. Requirement for children is not known.

Sources: Organ meats like kidney, liver and dark green leafy vegetables, soyabean and groundnuts are rich sources of folic acid. Other vegetables, legumes, eggs, wholegrain cereals and fruits are good sources.

Different Sources of Vitamins

Vegetables and Fruits are the classes of foods reflect the origins and methods of storage of the principal vitamins in these natural sources.

The water soluble factors are synthesised in the first place by plants and the quantities produced in excess of a plant's requirements are laid up in its own normal storage organs, such as underground roots and tubers, fruits and seeds. These are subsequently the foods to which we look for our supplies of the B vitamins, vitamin C and carotene. The oily fractions of seeds also contain vitamin E.

Vitamins A and D are found in foods of animal origin, having reached their final form in the animal body.

We would like to study first vegetables as sources of supply, then fruits and cereals, and finally dairy products, meat and fish.

Green Leafy Vegetables: Vegetables refer to edible parts of plants other than fruits and seeds. There are certain anomalies in any classification and this is no exception. Tomatoes, for example, cucumbers and marrows are structurally fruits, but are often eaten with meats and savouries, and are frequently included in the term 'vegetable.' By contrast rhubarb, which is eaten in the same way as fruits, as a desert, is the coloured stalk of the rhubarb leaf and is no more a fruit than is celery.

Vitamins synthesised by plants remain stable as long as they are in living undamaged cells, but injury to the cell walls exposes the contents to air and permits oxidation and vitamin destruction to occur. During harvesting, transport and storage, vegetables are subjected to conditions that may considerably lower their vitamin content. Gathered freshly from the garden they not only look and taste more attractive but are of greater nutritive value.

In present day civilizations vegetables have to be transported sometimes for long distances by rail, road or water and measures have to be taken to preserve as much as possible of their nutritive value.

Careful packing is essential to avoid bruising and mechanical injury and this is even more important for delicate leafy vegetables such as spinach than for firmer green vegetables such as cabbage or root vegetables such as parsnips and potatoes. We generally see green vegetables packed fairly tightly in open work sacks. This avoids overheating in the interior of the pack, and also protects them from being jostled and bruised.

Exposure to the sun causes not only oxidation but also loss of water and consequent wilting, which is another factor introducing loss of vitamins.

Finally, high temperatures reduce the content of certain vitamins to a considerable degree. Vegetable produce carried by sea is generally packed at refrigeration temperatures.

After travelling for hours or even days, vegetables are often exposed in markets or shops for another period of varying duration before they are taken home to be eaten or cooked.

Storage of Fresh, Unprocessed Vegetables: Even green vegetables have to be kept for short or long periods and most seasonal crops are stored in order to be available after the season has ended.

Root crops can be stored in bulk, as for example, in potato caves (or clamps) for several months. Home storage takes place in cellars, lofts, cupboards, larders and refrigerators. Large-scale storage is also essential in catering establishments and ships.

The two vitamins which are found in highest concentration in vegetables are A (as carotene) and C. Of these carotene withstands storage better than vitamin C.

Carotene in Raw Vegetable Foods: We cannot assume that a yellow colour always indicates the presence of carotene because there are other yellow pigments, e.g. xanthophyll in some vegetables. Carrots contain a high proportion of carotene; swedes do not. Carotene is present in its highest concentration in green leaves, but is masked by the more dominant green of the chlorophyll. It is generally true that the greener the leaf, the higher the concentration of carotene. So good sources are broccoli leaves, watercress and spinach, and among animal feeding stuffs — grass. The outer deep green leaves are better sources than the inner pale leaves.

Vegetables fail to yield all their carotene during digestion and this fact must be taken into consideration while estimating the vitamin A values of diets where the vitamin A is provided from vegetable sources. From 40-50 per cent of the carotene of green leafy vegetables is available, but only about 25 per cent of that in carrots. Utilisation of the carotene is best when it is

present, as in lettuce, along with vitamin E — another example of vitamin interactions.

The following table shows the carotene content of some vegetables which are generally eaten raw.

Carotene in Raw Vegetables (i.u. per ounce)

	Up to 200		*200-1,000*		*Over 1,000*	
Name	*i.u./oz.*		*i.u./oz.*		*i.u./oz.*	
Celery	Peas	60	Tomato	720	Watercress	1,200
Cucumber	Onion	160	Lettuce	900	Carrut	3,000-4,000
Radish	Cabbage	180				

These days in agriculture it is a common practice to cut young grass instead of making hay. The grass, that is very green and contains a good supply of carotene, is finely divided and dried rapidly by a special process. It is used as a source of carotene in animal foods. It is bright green when fresh but may lose as much as 40-50 per cent of its potency during storage, or more if permitted to become damp.

Vitamin C in Raw Vegetable Foods

Please note that these are all values in *one ounce* of each vegetable. While it is easy to eat several ounces of raw tomatoes or celery, shredded raw cabbage cannot be eaten in *quantities* similiar to those of cooked cabbage and both watercress and mustard and cress are very light in weight. Thus it generally happens that a liberal helping of a vegetable with a low vitamin C potency supplies more than a small helping of one with a high potency.

Effects of Short Periods of Home Storage: Cabbage, left whole and stored in a cool, dry place, loses only little of its vitamin C in a week and only about one-third of its content in a month, where as minced or finely cut cabbage may lose much of its vitamin C in a matter of minutes.

This is important and means that raw cabbage for use in salads should be shredded as near as possible to the time of

serving, spinach and asparagus, both of which wilt quickly may lose 40-80 per cent of their content if kept for several days before being used.

The following table shows recorded vitamin C values in stored vegetables. From these, losses in 4 days and in 1 week can be read.

Vitamin C (in mg./oz)

Vegetables	*Fresh*	*Stored 4 days*	*Stored 1 week*	
Brocolli tops	49	37	30	room temperature
Brussels sprouts	38	34	—	" "
Cauliflower	24	—	17	" "
Cabbage	14	—	13	0-3°C.
			11	room temperature (20°C.)
Spinach	14	—	13	0-3°C.
			2	room temperature (20°C.)
Asparagus	3	0.5	—	" "
Peas	3	—	1.5	" "

The green vegetables are better sources of vitamin C than root vegetables. Potatoes/parsnips, onions and carrots contain relatively little, but as the quantities eaten are greater, they contribute what may be quite a valuable quota to the daily intake. In many diets, potatoes are the chief source of vitamin C.

These vegetables of hard texture are generally better stored at low temperatures. Potatoes may lose up to 50 or 60 per cent of their vitamin C in six months. Onions may lose as much in three months.

Methods of Preservation of Vegetables: From olden times the housewife has practiced methods of preservation to have vegetables available when out of season. Her methods: pickling, salting and drying have been followed by modern procedures such as canning, freezing and dehydration. All these, both old and new, are still used.

Salting: This is, the oldest method of food preservation

that is still in use. The domestic preservation of vegetables like beans, by packing them in jars with salt is very destructive to vitamins. Little or no vitamin C remains after three months' storage and 50-80 per cent of the original carotene is lost.

Pickling: Methods of pickling differ much and figures available indicate that the effect on vitamin C is also variable. Homemade chutney is considered to retain in some instances as much as 94 per cent of the vitamin C in the fresh ingredients—other samples contain only 40 per cent. Sauerkraut has a good vitamin C value when freshly made but loses it rapidly on storage.

Canning: Carotene is relatively stable to canning but vitamin C losses vary. It may be destroyed by the preliminary blanching of vegetables and is not well retained by non-acid vegetables like beans, peas and spinach, which lose about half their original content. Canned foods are best kept in a cool dry place. Storage of canned vegetables at 50°F. Causes losses of 5-15 per cent of the vitamin C content, while at 80°F. This loss is increased to 20-35 per cent.

Freezing: In 1929 a quick method of freezing foods was developed and it has proved effective for the preservation of vitamins.

Although preliminary blanching has to be used to destroy oxidative enzymes in vegetables, ultimate losses in vitamin C are only from 20-24 per cent. The best temperature for storage seems to be about 0° F.

These 'quick frozen' foods do not readily lose vitamins on thawing but do on being allowed to stand at room temperatures. They ought therefore to be used as soon as they have been thawed.

Cooking: Whatever the method of harvesting, storage or preservation, most vegetables are eaten cooked, and the vitamin value of the food as it arrives at the table is the figure that is of the greatest interest to those engaged in the planning of diets and the preparation of meals.

Carotene is stable to cooking and there is very little difference between estimates made on raw and cooked vegetables, but vitamin C, one of the least stable of all the vitamins, is readily lost and every possible measure should be taken to preserve its activity.

The following points should be kept in mind:

1. Vitamin C is readily soluble in water. It tends, to pass into solution in the water in which the vegetable is being cooked. This happens more readily if the vegetable is already cut into small pieces or shredded. If the cooking water is to be used, the leakage of vitamin C into it does not mean loss, but in those cases where the cooking water is not to be used, it should be kept to as small a quantity as possible.

2. Vitamin C is also lost by oxidation. Vegetables should therefore to be plunged into boiling water which has been actively boiling for some little time so that air has already been boiled off. Rapid boiling does less damage than slow simmering.

3. Vitamin C is less stable in alkaline than in neutral or acid solutions. The practice of adding soda to leafy vegetable so as to retain the green colour is therefore not good for the preservation of vitamin C.

Pressure Cooking: This is the method where by the largest quantities can be vitamin C retained, the losses being only about 10-40 per cent. This is, of course, not the only consideration when a method of cooking is being chosen, but is important in view of the increasing use of pressure cookers.

The following diagram shows a comparison between the percentages of vitamin C and carotene retention in three different methods of cooking vegetables.

The following table gives what information is available on the average values of cooked vegetables.

Carotene in Cooked Vegetables (i.u. per ounce)

None or Very Little		
Beetroot	boiled	
Cauliflower	boiled	
Celery	boiled	
Marrow	boiled	
Mushrooms	boiled	
Parsnips	boiled	
Potatoes	boiled	
Swedes	boiled	
Turnips	boiled	

Contd...

None or Very Little		
Up to 200 i.u./oz.		
Artichokes	boiled	28 (estimated)
Asparagus	boiled	37
	press cooked	40
Beans	baked	47
French beans	boiled	162 (estimated)
Runner beans	boiled	197
Brussels sprouts	boiled	85 (estimated)
Cabbage.	boiled	161
	press, cooked	173
Leeks	boiled	90 (estimated)
Lentils	boiled	57 (estimated)
Onions	boiled	150 (estimated)
Peas	boiled	51 (estimated)
	processed	48
Over 200 i.u./oz.		
Carrots	boiled	3,000
(approx.) of which	750 available for conversion	

	press cooked 3,200 800 available for conversion
Kale.	boiled 3,300 1,320 available for conversion
Spinach	boiled 2,300 1,020 available for conversion
Turnip tops	boiled 5,700 2,280 available for conversion

Vitamin C in Cooked Vegetables (mg./oz.)

None		
Beans	baked	
Lentils	boiled	
Marrow	boiled	
Peas	dried	
	processed	
	split, boiled	
1-4 mg./oz.		
Asparagus	boiled	2
Broad beans	boiled	2.5 (estimated)
French or Runner beans	boiled	1.6 (estimated)
Beetroot	boiled	1.7
Carrots	boiled	up to 1
	press cooked	2
Celery	boiled	1
Leeks	boiled	3 (estimated)
Onions	boiled	1
Parsnips	boiled	1.5

Contd...

None		
Peas	boiled	2.0
Potatoes, old	boiled	1.4-3

	press, cooked	4.5
	roast	2-3
	in jackets	3.3
	fried	2
Pumpkin	boiled	0.6
Tomato	fried	1.5-3.0
	canned	4
Turnips	boiled	2.8
5-9 mg./oz.		
Artichokes, globe	boiled	5.6 (estimated)
Broccoli, tops	boiled	6.3
Brussels sprouts	boiled	8-14
Cabbage	boiled	3-10
Potatoes, new	boiled	6.4
	press, cooked	6.9
Kale	boiled	5
	press, cooked	
Spinach	boiled	7.2
	press, cooked	8.6
Swedes .	boiled	6.5
20-20 *mg./oz.*		
Cabbage	press, cooked	13
Cauliflower	boiled	11
	press, cooked	15
Turnip tops	boiled	11.6
Over 20 mg./oz.		
None		

The usual practice of warming up vegetables or cooking them a second time is thus destructive of vitamin C and is justifiable only when there are abundant alternative sources of the vitamins in the diet.

In large scale cookery, vegetables are frequently prepared some hours or even a day before they are cooked and are left standing in water. This is not recommended for green vegetables but potatoes suffer merely slight losses.

It is clear from the above discussion that vegetables can be good sources of both carotene and vitamin C. The following measures help in keeping the values as high as possible.

1. Using as soon after gathering as possible.
2. Storing in a cool dry place, away from direct sunlight.
3. Rapid cookery with the use of as little water as possible.
4. Avoiding cooking too much and having to re-cookleft-overs.

Vitamin Rich Fruits: Various considerations about vegetables also apply to fruits. These, too, are sources chiefly of carotene and vitamin C. Although small quantities of the B vitamins may be present, these reach a higher concentration in seeds.

Hard fruits, such as hard vegetables, withstand transport better than the softer varieties. Many fruits that are imported from abroad are packed while still unripe to prevent damage from bruising and crushing during transport.

Carotene in Raw Fruits: Apricots, peaches and tomatoes are the only fruits having a carotene content comparable with that of vegetables. Raw apricots contain approximately 200 i.u. per ounce, peaches slightly less and tomatoes 280-300 i.u. per ounce.

The following values may be useful for reference. Carotene in Raw fruits (i.u. per ounce)

50-200 i.u./oz-Apricots	
Blackberries	196
Gooseberries	85
Peaches	57

Plums	185
Tangerines	100

Other fruits contain less than 50 i.u. per ounce and are not in general worth considering as sources of carotene. They are important, however, for their vitamin C content.

Among such, apples, pears and bananas, that bear transport well and can be stored without mechanical damage, are not good sources of vitamin C. Those a high value, which can be stored are the citrus fruits.

These may lose up to 20 per cent during one month's storage. Black currants, which exceed all the others, have to be preserved so as to be available when not in season. Rose hips that are also very high in their vitamin C content are not eaten raw.

Methods of Preservation of Fruits: The different methods used for preservation of fruits are drying, bottling, canning and jam making.

Dried fruits contain no vitamin C. Currants, sultanas, raisins, prunes, dried bananas may be valuable for their content of mineral salts and sugars but no reliance can be placed on them as sources of vitamins.

Jam: This, is a useful domestic method of preserving fruit with its vitamin C. The rapid boiling required does not destroy the vitamin C and, although there is progressive loss on storage, jams kept for reasonable periods still contain measurable quantities, particularly those made from fruits like black currants, strawberries and gooseberries that have high initial values.

Orange peel also contains vitamin C, so that marmalade, if made by modern methods (e.g. pressure cooking), is a fairly good source of vitamin C. The old-fashioned custom of leaving the cut-up orange to stand for twenty-four hours before being cooked exposes the fruit to air and to consequent loss of vitamins.

Vitamin C in Jam (mg. per ounce)

Black currant	1
Strawberry	6
Gooseberry	3
Raspberry	2.5
Marmalade	2
Plum	1

Canning and Bottling: Exclusion of air is more efficient in canning than in bottling and this helps in the retention of vitamin C. However, home-bottling is a useful method and, if the bottled fruit is kept away from light, the vitamin value is retained moderately well.

Losses in canning or bottling fruit juices and sliced fruits are about 25 per cent.The addition of ascorbic acid in weights not more than 0.2 per cent of the weight of the fruit is said to stabilise the colour of canned fruit without affecting its flavour and the vitamin is retained for long periods in 30-60 per cent of sugar as syrup.

Little vitamin C is retained in canned fruit. Black currants may contain as much as 26 mg./oz. and strawberries, grapefruit and gooseberries from 5-7 mg./oz. Other fruits contain less than 4 mg./oz.

Cooking: Fruits cooked in pies, stewed or baked contain very little vitamin C- in most instances less than 1 mg. per ounce. Exceptions are those fruits which have very high initial values, e.g. black currants, strawberries, raspberries and gooseberries. These, even with a loss of about 50 per cent, still contain appreciable amounts.

On the whole, vegetables are better sources of both carotene and vitamin C than fruits. The latter, with one or two notable exceptions, need to be eaten raw and as fresh as possible if they are to contribute an appreciable amount of vitamin C to the daily food intake.

Cereals and Nuts: Though the term cereal should be applied only to grain products but in practice other foods of a starchy nature like sago (from a stem) and tapioca (from a root) are included because of their similar uses in cookery.

Nuts, too, may be either seeds, whole fruits or parts of fruit, but for the purpose of considering them as sources of supply of vitamins they will be grouped together.

All these are storage organs of plants. They are not structures where in we find vitamin C. Carotene likewise is found in only one or two exceptional instances like yellow maize and some samples of rye and millet. Cereals and nuts are, however, the storehouses for vitamins of the B complex and therefore they form a group of very interesting products — interesting because of their geographical occurrence, regional use and widespread consumption.

They are eaten as whole grains, as flour or meal from whole grains or as products prepared from parts of grains or other plant structures.

To understand the composition of grain products it is essential to be quite familiar with the structure of a single grain of any cereal, e.g. wheat, oats or rice. All these are members of the grass family and every grain is alike in being a single-seeded fruit where in the wall of the seed is fused with that of the fruit forming a single outer layer or integument. This is generally very hard, consisting of cells with a thick tough wall. Inside is a compact white mass, the 'endosperm.'

The cells of this part of the seed have thin walls but are packed so tightly with starch grains that the cell walls can be ignored. At one end of the grain, generally the more pointed, is the embryo or germ, the part that would grow into a new plant if the seed were sown. This is very small and can be made out as a slight indentation on the otherwise smooth surface of the grain. It occupies only about one-fiftieth part by weight of the whole grain but a section examined by microscope shows it to be made up beginnings of a shoot with rudiments of

leaves and the growing point of a root. This embryo is separated from the endosperm by a layer of cells, the scutellum through which nutrients from the endosperm pass to the embryo when growth begins.

The starchy endosperm is totally devoid of vitamins but members of the B group are present in the outer cells close to the integument, in the scutellum and in the embryo itself. Whole grains therefore contain fair quantities; products made from the germ contain more still, but foods prepared solely from endosperm are poor sources.

Whole wheat was previously ground between stones to produce flour for bread making and a few stone-grinding mills still exist and work in Great Britain. The meal produced by the first grinding is very coarse and is sifted to produce fine wholemeal flour. About the year 1870 steel roller mills were introduced. These can be worked so as to extract from flour whatever proportion of the grain is desired. If only about 70 per cent (chiefly endosperm) is used, the flour, which is white, is said to be of 70 per cent extraction. The highest extraction rate used is about 90-92 per cent and includes most of the grain with the germ but without some of the hard integument or husk that goes to form bran. Various grades are also milled between 70 and 90 per cent.

Other grains used whole are oats, barley and maize, but these are coarsely ground to produce meal rather than flour. Finely ground flour from maize endosperm (with a low extraction rate) is cornflour and this is also used for custard powders.

Pearl barley and rice are freed from their husks before being offered for sale as food, and this process removes not only husk but embryo with the result that the final product contains very little vitamin B.

In primitive regions like parts of the Sudan, sorghum (Kaffir corn) is prepared by village women. They soak it until the grains are soft enough and then pound these to a coarse pulp

that can be made into flat cakes and baked. The resulting cakes or biscuits are called Kisra and contains most of the vitamin B of the fresh grain.

All the B vitamins are not distributed in the same way in whole grains; for example, oats, which has a high content of B_1 is low in nicotinic acid and the converse is true of barley.

The values of grains which are used whole are listed in the following table.

Micrograms per oz. *Whole grains (raw)*	B_1	*Riboflavin*	*Niacin*	B_6
Barely	30	10	700	130
Maize	94	37	430	200
Millet	130	10	460	90
Oats	180	31	340	–
Rice, polished	18	20	300	92
Rye80	30	340	100	
Sorghum	130	23	740	—
Wheat	135	60	1,070	110

The Preservation: Grains being hard and dry and not easily damaged, grains withstand storage well. Loss of vitamins occurs if they are allowed to become damp or overheated. Moreover they are liable to be attacked by parasites — larvae of moths, weevils and moulds which may cause deterioration. But grain kept dry and clean retains its B vitamins for long periods of storage.

Most cereals are used in food as meal or flour, and in this form they provide the staple foods to people in many countries, the choice depending on climate, soil, custom and economics. The various rye breads of the European continent, the polenta and spaghetti of Italy, the oatcakes and porridge of Scotland, and the wheaten bread of Australia, the North American continent and Great Britain are examples. In India and China

whole rice is still eaten to greater extent than bread made from any kind of flour.

Storage and Upkeep: Ground products of whole grains keep only for limited periods because they contain in addition to starch, small quantities of fat (in the germ) and moisture. In conditions of temperature conducive to enzyme action, the fats are split producing fatty acids and the presence of these makes the entire meal rancid.

It was this tendency of whole meal flour to become rancid that made the roller mill such a welcome innovation to millers, because the white flours thereby produced were for the first time free from the oil of the germ and kept for long periods without rancidity. They were also whiter and this pleased public tastes, satisfying a desire that can be traced back to the time of the Roman occupation of Britain. At the time when modern white flour first became readily available, i.e. in the latter half of the 19th century nothing was known of the effects that might be produced by a resulting shortage of B vitamins. Thus white flour became popular because it could be stored without deterioration and made a loaf light in colour and in texture.

Vitamin-B Values of Meals, Flours and Pellets

Micrograms per oz.	B_1	*Riboflavin*	*Niacin*	*Pyridoxin*
Arrowroot	—	—	—	—
Barley meal	—	—	—	—
Cornflour \	—	—	30	—
Custard powder	—	—	30	—
Wheat flour				
70 per cent	24	10	300	72
81 per cent	68	23	450	—
92 per cent	130	60	800	104
Lentil flour	—	—	321	—

Contd...

Micrograms per oz.	B_1	*Riboflavin*	*Niacin*	*Pyridoxin*
Maize meal				
60 per cent	14	9	140	—
96 per cent	90	370	430	16
Millet meal	130	10	460	90
Oatmeal	128	40	300	70
Rye flour	77	30	340	98
Sago	—	—	60	—
Tapioca	—	200	330	—
Taro (Hawaiian)				
Corms	60	8	170	—
Leaves	66	132	460	—

These are not all prepared from cereal grains.

Arrowroot is obtained from the underground stems of a species of Maranta. The starch is extracted and grated to a very fine powder. With boiling water or milk it readily forms a smooth jelly-like paste and was previously used in invalid cookery, but now it is clear that it has no vitamin content and any value it possesses is by virtue of its calories and digestibility only.

Custard powders consist chiefly of cornflour, coloured and flavoured.

Lentil flour, from a leguminous plant, contains some protein but little vitamin value except from the presence of some nicotinic acid.

Maize meal is unique among ground cereals in that it contains carotene. Whole yellow maize contains 28 i.u. per oz. Italian polenta is made by boiling yellow maize meal with a small amount of water to form a thick paste. It is eaten either alone or with cheese.

Sago and *tapioca* resemble each other in being composed chiefly of starch, with no vitamin B_1 Tapioca contains fair quantities of riboflavin and nicotinic acid but sago very little. Tapioca is prepared from starch found in the tuberous roots of the cassava, and sago is obtained from the starchy pith in the stems of several species of palm.

Taro is a plant of some interest. Both its corms and leaves are used in the preparation of food in Hawaii and both contain good amounts of the B complex.

These are other cereals are mostly eaten cooked or at any rate processed on a large scale.

Processed Cereals: A great many cereal foods are cooked during large scale manufacture but not further cooked in the home. These include all those generally grouped as 'breakfast cereals.' They vary widely in their vitamin B content according to the grain from which they are formed, the part of the grain used and the temperatures to which they are subjected during preparation.

The values of some of these are listed below.

Cereal Products, Processed but not Further Cooked

Vitamin	B_1	*Riboflavin*	*Niacin*	*Pyridoxin*
Bemax(wheat germ)	450	200	1,700	450
Bran, processed	180	140	5,000	?
Cornflakes	trace	_	—	—
Puffed wheat	—	30	1,300	?
Shredded wheat	73	40	1,400	?
Wheat flakest	38	50	1,800	?

Manufacturers' data. Lampitt and others, 1945.

Bicknell and Prescott, 1947.

Bread: The domestic cooking of bread has mostly given way to large-scale preparation. Breads are thus intermediate products between processed cereals and cooked cereals. They cannot be stored as long as the cereals which are specialty

packed but can be kept for several days and in this respect are unlike cereal dishes that are eaten immediately after being cooked. The various crisp-breads properly packed can be kept in the same way as breakfast cereals and perhaps starch-free breads like Energen rolls should be included here also.

Wheaten bread has been eaten in Great Britain since the latter part of the 18th century. Before that, most of the bread was made from barley or rye and rye bread is still popular on the continent of Europe.

By the middle of the 19th century the consumption of bread was very high and has been estimated as about 21 ounces per head daily in 1840. This would all have been from stone-ground flour. By 1934, when most of the bread was white, the daily consumption had fallen to 11.6 oz. per head and during the period when bread was rationed the level permitted was 9 oz. Since rationing ceased, the level has again risen and in 1951 was 12.3 oz. per head. The effect of this difference in quantity consumed and type of flour is seen in the variation caused in vitamin B_1 intake from bread.

Year	*Estimated mean consumption of bread per day oz.*	*Type of bread*	*Vit. B_1 mg./oz.*	*Vit. B_1 eaten as bread mg.*
1840	21.3	Stoneground wholemeal	0:13	2.8
1934	11.6	White, 70 per cent extraction	0.024	0.28
1945	9	90 per cent extraction (ration)	0.13	1.17
1951	12.3	81 per cent extraction	0.048	0.59
1954	?	white, restored	0.048	
	?	or 81 per cent extraction (National flour)	0.048	?
		or wholemeal, 90-92 per cent extraction	0.09	?

The bread cannot be exactly equivalent in nutrients to bread of a high extraction rate. There are many known and no doubt still some unknown factors in wheat germ, all of which are not added. Nothing has so far been said about the vitamin E of grains. It is all situated in the oil of the germ and is, removed with the other germ factors in the preparation of white flour.

We can argue that there are no signs of vitamin E deficiency in man and therefore removal of vitamin E is of no importance, but this is by no means unquestioned. All these large-scale changes in foods consumed by any population are in the nature of nutritional experiments and time alone shows their effects.

Among breads should perhaps be considered the crispbreads and various forms of biscuit. Most of these are subjected to temperatures high enough to make the final products brown and crisp throughout. This is destructive of vitamin B_1 but not of the heat-stable fractions. Bread contains more B_1 in the central uncharred portion than in the crust and slices of bread, toasted, lose some B_1 in the charring that occur on the outside.

Vitamins of Breads and Biscuits

		Micrograms	*per oz.*
	B_1	*Riboflavin*	*Niacin acid*
Biscuits, water type	37	30	300
Biscuits sweet	18	10	200
Bread, wheaten 70 per cent	24	10	200
Bread, wheaten 81 per cent	48	17	330
Bread, wheaten 92 per cent	90	50	600
Bread rye crisp	70	—	—
Sorghum, Kisra	—	12	800

Cooked Cereals: Cereals, cooked before purchase or else in the home form part of almost every meal and contribute the bulk of the calories in any daily diet. It is therefore easy to see that the intake of 3 vitamins is largely controlled by the type and quantity of the cereals eaten.

The cooked polished rice of the Far East is still responsible for the prevalence of beriberi in India and China, and white bread, unfortified as it was in Great Britain before 1940, doubtless caused minor degrees of deficiency, for example, digestive and cardiac disturbances.

A great many cereals in various forms are now available for domestic cookery—flours for making into cakes and pastries; meals such as oatmeal, maize meal ground rice, semolina and ground almonds for the making of porridge and puddings — grains, whole and split such as barley, rice, lentils and peas

Raw Cereal			***Cooked***		
	Moisture per cent	***B_1 mg./oz.***		***Moisture per cent***	***B_1 mg./oz.***
			Bread from		
81 per cent extraction flour	12.3	.068	81 per cent extraction flour	36.0	.048
Oatmeal	8.9	.128	Porridge . .	89.1	.015

Other Vitamins: The newly discovered

for puddings, soups and savouries, pellets such as tapioca and sago for puddings and other dishes, and finally, made-up cereal preparations such as macaroni and spaghetti.

These can be classified into two main classes, those mixed with different amounts of fat and only small amounts of water and usually baked into pastry and cakes, and those cooked with larger amounts of water which is afterwards drained off, sometimes to be used, sometimes not.

In all those processes the B_2 vitamins withstand heat well, but being water-soluble may be dissolved out into the cooking water and lost if this is discarded.

Vitamin B_1, being less stable, is retained in the inner portions which are not oxidized but is gradually lost in prolonged or excessive heat or by surface charring.

It is important, too, to remember that cereals cooked in water take up large quantities and that while this increases the bulk, the weight of the cooked food is to a large extent that of its moisture content.

Thus it requires only 2 oz. of rice to make 24 oz. of rice pudding and only 1 oz. of oatmeal of make 6 oz. of porridge.

These points are illustrated in the following comparisons between the vitamin B_1 content of equal weights of raw cereal and pudding or other dish prepared from it.

members of the B-complex are present in most cereals but values are not yet available.

Vitamin E is obtained mainly from cereals, the oils of which are rich sources, although some is present in leafy vegetables and in animal products. Values are known for a limited number of foods and some are listed below:

Vitamin E in Some Cereal Foods

		mg./oz.
Wheat and wheat products	whole wheat	0.3-1.0
	wholemeal flour	0.6
	white flour	0.3
	macaroni	0.3
	wheat germ (Bemax)	8.0
Oats	whole oats	0.3-0.6
	oatmeal	0.6
Other grains	barely	0.9-1.4
	maize	2.8
	rice	0.1-0.2
	rye	2.9

Vitamins in Nuts: The only vitamins in nuts are those of the B-complex and vitamin E, but some nuts are relatively high in these.

The following table provides the information at present available.

	1*Microgrammes per oz.*			
	B_1	*Riboflavin*	*Niacin*	*mg./oz.*
Almonds, whole	24	90-152	334	—
Brazil	—	—	—	—
Chestnuts, raw	80	—	—	—
roasted	—	—	60	—

contd.....

	B_1	1*Microgrammes per oz.* *Riboflavin*	*Niacin*	*mg./oz.*
Hazel	168	—	60	—
Coconut, fresh	10	30	100	0.1-0.3
desiccated	trace	—	—	—
Peanut	179	80	2,600	3.0
Walnut	51	37	320	—

Green Walnuts also Contain Vitamin C: Since the B vitamins and particularly vitamin B_1 are obtained mainly from cereals, there is a real danger of deficient intakes when large quantities of refined cereals are eaten because no foods of any other kind will make up for what these lack.

When the population can choose freely it is essential that everyone should have some knowledge of these facts so that the choice of food should be governed not only by preference but also by knowledge of nutritional value.

Dairy Produce: Modern methods have resulted in the production of farm animals very different in type from their wild ancestors: cows which yield milk when it is not needed by their calves, and hens which lay more eggs and at most seasons of the year. Man no longer has to hunt wild animals for his food. He can to a great extent control the production of both vegetable and animal food for his own use.

The cow eats carotene in grass, changes it into vitamin A and passes this into its milk from which it is available not only in the form of liquid milk but as butter and cheese. Hens eat grain and poultry mashes and pass different vitamins, e.g. A, D and riboflavin, into the yolks of their eggs.

Dairy produce is thus a source of vitamins very valuable in human nutrition and particularly so in relation to vitamin. A because it becomes available as the performed vitamin rather than as the less readily assimilated carotene of vegetables.

Vitamin D is also present in eggs and butter though in much smaller quantities than in fat fish.

These dairy foods are available in the fresh state/when they can be eaten raw or cooked, They are also treated to make them suitable for storage, after which again some may be eaten without further cooking but some, e.g. dried eggs, are suitable only for use in cooked dishes.

Dairy Foods: Truly speaking, only fresh milk, whole or skimmed, raw cream, and fresh eggs are the dairy foods which are freshly used.

Milk

Fresh milk may be:

(a) T-T, or *Tuberculin tested*, when it is from cattle regularly tested and proved to be free from tuberculosis and kept from contact with other herds,

or

(b) Accredited, when it is from cows subject to veterinary inspection every three months. This milk must pass a certain standard indicating freedom from bacterial contamination.

(c) *Cream:* Raw cream is made up of the milk fats that rise to the surface when milk is left to stand. It contains less water than whole milk and a higher concentration of vitamins.

The vitamin content of milk and cream depends partly on the food of the cow and partly on its breed. Animals fed on young spring grass yield milk with the highest concentration of vitamin A in the fat.

Eggs: The concentration of vitamins in eggs depends upon the hen's food, the vitamin — A content being at its highest when the hens are fed on grass or green leaves and that of vitamin D when they are given cod liver oil or some other concentrated source of vitamin D in their food or when they live in sunshine. Laying hens kept intensively are sometimes exposed to ultra-violet light so as to increase the quantity of vitamin D in the eggs.

The colour of the yolk is no guide to its vitamin content. A pale yolk may be relatively high in performed vitamin A.

The colour can be made deeper by the feeding of foods rich in carotene or even by feeding various pigments. Deeply coloured yolks are preferred in some areas and pale yolks in others. The vitamins of eggs, except riboflavin and vitamin B_6, are present only in the yolk.

Vitamins in Fresh Milk and Eggs 1 oz. of each

	A	*D*	*E*	B_1	*Ribo-flavin*	*Niacin*	B_6	*C*
	i.u.	*i.u*	*mg.*	μg.	μg.	μg.	μg.	mg
Milk, whole	20-40	3	0.03	13	40	—	20	0.4
skimmed	—	—	0.01	13	40	—	—	0.4
Cream, raw.	450	14	?	13	40	—	?	0.3
Egg whole	250	17	0.6	37	170	—	56	—
yolk	200-800 upto 50	1.5	120	100	—		—	
white	—	—	—	—	100	—	30	—

Milk can be subjected to certain processes while still fresh, the most important of these being pasteurisation, to that destroy bacteria that are specially injurious to health, e.g. those responsible for diphtheria, typhoid and tuberculosis.

The milk is subjected to a temperature of 145°-150°F. (62.8-65.6°C.) for 30 minutes or else to 162° F. (72.2°C.) for 15 seconds and cooled

Pasteurisation does not affect the content of vitamins A, D or riboflavin. *Cream* may be clotted by being heated to A better comparison a half pint of milk (10 oz.) is perhaps between and one egg (2 oz.):

> The vitamins of scalding' temperature and then being cooled in shallow pans. This causes the fat to be caught in the meshes of clotted protein and the whole mass thickens to a stiff paste.

	A	D	E	B_1	Ribo-flavin	Niacin	B_6	C
	i.u.	i.u	mg.	µg.	µg.	µg.	µg.	mg
½ pint milk	300	30	0.3	130	400	—	200	4
1 egg	500	34	1.2	74	340	—	112	—

The vitamin content remains unaffected.

Yoghourt is milk soured by the culture in it of a specific lactic-acid-producing bacillus. The temperature of the culture is not high and there is nothing in the process to destroy the natural vitamins of milk. Vitamin C may be slightly reduce but is low in milk even when this is fresh.

Preservation of Dairy Foods: Milk can be preserved by concentration or by being converted into cheese; cream can be canned and the fats of milk can be made into butter that keeps much longer than liquid milk.

Evaporated milk is prepared by subjecting whole milk to heat to destroy any bacteria that are present and then concentrating it to one-quarter or one-fifth of its original bulk. It is treated to reduce the size of its fat globules, canned, and sterilised.

Condensed milk is concentrated rather less, down to one-third of its original bulk and may be sweetened by the addition of cane sugar. This helps to keep it sterile.

Dried milk is a fine powder, produced either by spraying or by passing liquid milk over heated rollers, thus reducing its moisture content to under 3 per cent. Dried milk may be full-cream, half cream, or skimmed. Dried skim-milk does not contain the fat-soluble vitamins.

Butter: Butter is produced by separating the milk fats from the rest of the milk. The separation is more complete than in the preparation of cream, butter containing about 85 per cent of the milk fat and only traces of moisture. The natural colour of butter depends on its carotene content, this being derived from green foods, but some pale butter may have a high concentration of performed vitamin A.

Butter thus contains the fat-soluble vitamins of milk in higher concentration but not the water-soluble vitamins. Its keeping properties depend on the amount of moisture present. Water in butter causes the development of rancidity.

Cheese: Cheese contains most of the solids of milk, a gallon of milk being required to produce about a pound of cheese of the Cheddar type. The milk is soured either by the action of its natural enzymes or by rennet coagulation after which the whey is extruded and the cheese is permitted to ripen. Different moulds and yeasts are used to produce cheese of characteristic type and flavour. Most of the common cheeses are made from full-cream or partly skimmed cow's milk but sheep's and goat's milk are also used and in some countries, the milk of the reindeer, buffalo and llama.

Cheeses contain only small quantities of vitamin D, but are rich sources of vitamin A and contain also some riboflavin. They form the best means of storing the vitamin A of milk. The riboflavin may even increase in the surface layers during ripening, probably owing to synthesis by micro-organisms.

Eggs: Eggs can be preserved whole in water-glass (Sodium silicate) or by being dipped in a cellulose varnish. Dipping them in oil also preserves them for short periods. The riboflavin quite stable during storage but 50 per cent of the pyridoxine may be lost in 12 months. Eggs are preserved commercially by drying, the whites and yolks being beaten together first and then sprayed through warm air forming a powder.

They are reconstituted by the addition of water. Some vitamin A may be lost by prolonged storage.

Other Vitominous Foods: Other foods that are similar to dairy foods must be included here, i.e. margarine, lard, cooking fat and dripping. Margarine was first produced during the Franco-Prussian war from emulsified purified beef suet. Modern margarines can be obtained from animal fats, hardened vegetable oils or mixtures of both, and in Great. Britain margarine is now fortified by the addition of concentrates of vitamins A and D.

Neither lard nor cooking fat contains any vitamins but dripping may contain small amounts.

Vitamins of Prepared Dairy Produce per ounce

	A	*D*	*E*	*B_1*	*Ribo-flavin*	*Niacin*	*B_6*	*C*
	i.u.	*i.u*	*mg.*	*μg.*	*μg.*	*μg.*	*μg.*	*mg*
Milk, evaporated condensed	105	1	.09	18	100	100	?	0.6
Sweetened	105	1	?	30	100	100	15-30	0.9
Dried	300	3	.014	85	330	200	90	2.0
Butter Cheese	800	17	0.5	—		—	—	—
Cheddar	370	4	0.3	9	140	100	18	—
Roquefort	700	?	?	8	100	350	?	—
Processed	310	?	?	3	110	40	?	—
Eggs, dried	850 or less	?	?	10	—	—	—	—
Margarine	560	56	15	—	—	—	—	—
Lard	—	—	—	0.7	—	—	—	—
Cooking fat	—	—	—	—	—	—	—	—
Dripping	28	?	9-12	—	—	—	—	—

The Effect of Cooking on the Vitamins in Daily Foods: Milk is extensively drunk raw but is also added to hot drinks such as tea and is itself heated in the preparation of beverages like coffee and cocoa. It is also an ingredient in many cooked dishes like custards, milk puddings, soups and sauces.

The vitamins that would be destroyed by these cooking processes are fortunately present only in small amounts in milk, e.g. vitamins B_1 and C.

Vitamin A and riboflavin are both stable to heat and are therefore reduced only slightly if at all in cooking.

This stability of vitamin A and riboflavin is important and means that cooked dairy produce can be relied upon as sources of these vitamins. Individuals who are advised to take one or two pints of milk daily sometimes have difficulty in consuming this quantity and are apt to assume that it must all be taken

as liquid milk. Provided that vitamin B_1 is adequately supplied by other foods there is no reason why milk should not be taken cooked in a variety of ways, e.g. half a pint in coffee for breakfast or at a mid-morning break, another half-pint at lunch or dinner partly in soup, partly in pudding, and some in an evening drink.

Eggs also retain their stable vitamins when used in the making of cakes and puddings.

Meat and Fish: Animals, whether mammals or fish, store fat-soluble vitamins in their body fat and also in glandular organs, e.g. the liver. Small amounts are also found in muscle. Thus when these parts of the animal body become human food, they provide not only protein or fat but also vitamins to the diet.

Vitamin A is stored in the liver and the human liver may contain more than a year's supply that is one reason why gross vitamin-A deficiency is not common in Great Britain (Tomaszeski, 1947). It occurs only when individuals have been in short supply for fairly long periods.

Vitamin D is found not only in the livers of fish but also in the body oils so that fat fish can be a source of supply of this vitamin for those who do not take fish liver oils.

Vitamin E is stored partly in muscle but more in the liver. Its concentration in animal foods is, however, much lower than in cereals.

Of the various water-soluble vitamins, vitamin C is found in the animal body in only very small quantities, the only organ containing appreciable amounts being the adrenal gland. This is very small and is not generally eaten. Vitamin B_1 is found in the liver and smaller quantities in muscle, but except for pork and bacon, muscle meat contains very little. The pig is exceptional in its ability to store vitamin B_1 in its muscles.

Some of the B_2 vitamins are, by contrast, found in large quantities — riboflavin in liver and kidney and nicotinic acid in these glandular organs and also in muscle.

Most of the vitamins obtained from meat and fish are heat-stable and are thus available when the food is cooked.

Preservation: Curing of meat is an old process. Both curing and corning are effective for the preservation of the B vitamins other than B_1. Corned beef retains only 20-50 per cent of its vitamin B_1 but as much as 90 per cent of its riboflavin.

Hams prepared commercially by being placed in pickling solution for 12 days and then being smoked for 18 hours were found on test to contain 73 per cent of their original vitamin B_1 92 per cent of their riboflavin and 84 per cent of their nicotinic acid.

Generally nitrites and nitrates are used in the curing of meat and, if the same meat is used with vegetables in stews, these salts will rapidly destroy the vitamin C of the vegetables.

Canning is a satisfactory method of preservation particularly for fish containing vitamins A and D. Unless subsequent storage is unduly prolonged, there is very little loss. Canned fish such as herrings and sardines are thus valuable sources of vitamin D.

Vitamin A in Meat and Fish (i.u. per oz. approx.)

Less than 20 i.u. loz.		*20-100 i.u. loz.*		*100-500 i.u. loz.*		*Over 1,000 i.u./oz*
Veal	11	Bloater	32	Oyster	120	Calves' liver
Beef		Herring		Sheep's kidney		1,100-1,200
Mutton		Mackerel	42	250-300		
Cod	about 14	Kipper	50			
Haddock		Sheep's heart	60			
Hake		Canned salmon	70-80			
		sardine				

Vitamin D in-fat Fish (i.u. per oz. approx.)

Cod roe	fried	20
Herring	fried	250
Kipper	grilled	250

Mackerel	fried	100
Salmon	canned	170
Sardines	canned	280

Vitamin E in M 250 Meat and Fish (mg. per oz. approx.) (only a few values available)

Bacon	15
Beef lean	15
Chicken	.06
Calf liver	.04
Mutton, lean	.22
Pork, lean	.20
Cod roe	1.5-2.2
Haddock, fresh	11

Although there the B vitamins, foods is great variation in the distribution of do fall roughly into groups.

Foods poor in B vitamins	*(Micrograms per oz. approx.)* Vitamin B_1	*Riboflavin*	*Niacin*
Beef, corned roast or stewed	10-15	40 60-70	50 1,000
Sweetbreads Tongue }	12-20	30-60	up to 800
Chicken, roast	30	20-70	1,000-2,000
Mutton, roast boiled }	20-25	40	1,600
Better sources of the B vitamins			
Bacon, fried	120-200	100-150	1,000
Ham, boiled	110	30-100	500
Pork, roast	300-400	35-50	1,200.
Cod roe	200	200	400
Salmon, canned	300	60	1,800
Liver	110	890	3,600

Cooking: During cooking, vitamin is B_1 may be lost. In stewing as much as 50 per cent may be lost and in roasting

40-60 per cent. But by the ordinary methods of cooking meats: roasting, grilling, frying, braising or stewing, the losses of riboflavin and nicotinic acid amount to less than 15 per cent.

Comparing these as sources of the B vitamins with foods in the cereal class, we find that one slice of bread made from 81 per cent extraction flour provides 3 times as much vitamin B_1 as one ounce of roast beef but only one-third as much nicotinic acid.

Miscellaneous Foods

Sweets and Preservers: The only sweet that may contain appreciable quantities of performed vitamin A is *ice-cream,* good samples of it may contain as much as 400 i.u. per ounce. Small quantities of carotene are present in chocolate (20 i.u./oz.), lemon curd (13 i.u./oz.) and jams (6 i.u./oz.).

Chocolate contains no vitamin C, but enough of the B complex (B_1 30 ug., riboflavin 70 ug., niacin 300 ug. per oz.), carotene 20 i.u. per oz. and traces of vitamin D.

Molasses, that is obtained from the crude liquor from which sugar is crystallised, contains the B vitamins in addition to cane sugar, glucose, fructose and mineral matter. Cane molasses contains about:

14 micrograms of vitamin B_1
55 micrograms of riboflavin per oz.
and rather over 400 micrograms of nicotinic acid

During further purification these are destroyed and neither sugar nor golden syrup contain any vitamins.

Honey contains small quantities of:

vitamin B_1	5 micrograms
riboflavin	20 micrograms
nicotinic acid	100 micrograms per oz.
vitamin B_6 up to	80 micrograms
and vitamin C	1-6 mg.

Jellies contain no vitamins but are convenient for 'setting' fruit. The flesh of a large orange or grape-fruit can be removed with very little damage from half a fruit and can be set in jelly in a small container. This is an easy way of carrying the fruit for picnic lunches and obviates the sticky fingers caused by peeling an orange. Although geltine contains no vitamins, it is not without value. Eaten with bread, the amino acids of gelatine together with those of wheat contribute the equivalent of good protein to the diet.

Puddings and Deserts: These depend for their value on the ingredients used and the method of cookery.

The value can be estimated easily by reference to the values given for cooked cereals, fruit, eggs and other ingredients.

Spices, Condiments and Flavouring Agents: These are all eaten in small amounts and, with few exceptions contribute little of nutritive value to the diet.

Vitamins in 1 Teaspoonful (½ oz.)

	A *i.u.*	B_1 µg.	*Riboflavin* µg.	*Niacin* µg.	C *mg.*
Paprika (green)	—	—	—	—	4
Paprika pods.	—	—	—	—	2-100
Marmite	—	112t	225	2,150	—
Bovril	—	9	60	2,000	—
Ham & beef paste	—	4	5	17	—
Salmoon & Shrimp paste	—	3.5	3	35	—
Dried thyme	—	40	—	—	—
Dried mint	—	—	—	—	1-14
Ginger	—	—	—	25-30	—
Sage	trace (carotene)	—	—	—	1

Manufacturer's data.

Freshly prepared. [There is progressive loss on keeping.]

Booth and others, 1944.

Fish Liver Oils

	per oz. Vit. A i.u.	Vit. D i.u.
Halibut liver oil	600,000	60,000
Cod liver oil	30,000	3,000

Beverages: Drinks prepared with milk contain the vitamins of milk according to the amounts used. This may be so little as to be negligible in the case of tea and may be very little in cocoa.

White coffee contains more. Coffee contains no vitamins and tea only a trace (6-11 ug./g. of riboflavin) but cocoa contains traces of vitamin A and of the B complex.

Fruit drinks and squashes vary in their content of actual fruit juice and neither flavour nor colour are reliable indicators. Where, however, the makers state the proportion of genuine fruit juices present, rough estimates of vitamin value can be made.

Special fruit juices and purees are now available where in the content of vitamin C is stated on the label. These are generally preparations of black currants, rose hips and orange.

Black currant puree	18 mg. vitamin C/oz.
Black syrup (Ribena)	20 mg. vitamin C/oz.
Rose-hip syrup	43 mg. vitamin C/oz.
Concentrated orange juicet	45 mg. vitamin C/oz.
Orange and rose-hip syrup combined (Rosena)	56 mg. vitamin C/oz.

Manufacturers' data. Min. of Food, 1945.

These can all be diluted.

Alcoholic beverages contain no vitamin C but some contain members of the B complex.

Micrograms per oz	B_1	*Riboflavin*	*Niacin*
Ale, strong	3	10 or more	400-700
Beer, mild	1	10	40
Beer bitter	1	8	100-200
Stout	1	10	150-300
Wine	less than	1 2-8	?

Min. of Food, 1945.

Randoin, 1949.

Cailleau and others, 1949.

The vitamin B_1, even in a pint (20 fl. oz.) of beer is thus negligible but the riboflavin content may be as high as 150-200 ug. and nicotinic acid 2,000 or more ug.

The absence of vitamin B_1 in alcohol coupled with the fact that alcohol contributes a considerable quota of calories to a diet is one reason why individuals who consume large quantities of alcohol, and often consequently fail to take enough of the protective foods may suffer a shortage of vitamin B_1 This deficiency is in some cases severe enough to cause symptoms.

Vitamin Scale and Vitamin Quantities

The isolation of pure vitamins led to further work on their chemistry and the development of chemical methods of analysis which had previously been impossible. With the evolution of these techniques we have become less dependent upon the use of experimental animals, but a difficulty exists in that many chemical methods of estimation give accurate results only when no interfering substances are present and are not applicable to food analyses, since many other factors are always present in foods together with vitamins.

A brief description of some of the commoner chemical methods is given, followed by outlines of the biological and microbiological techniques which are still used.

Chemical and Physico-Chemical Methods of Estimation

Vitamin A: Vitamin A activity is measured either chemically or spectroscopically.

Chemical Method: This depends on the fact that a solution of vitamin A in chloroform when mixed with a 25-per-cent solution of antimony trichloride in chloroform produces a blue colour, whose intensity depends on the quantity of vitamin A present. The colour develops gradually, reaching a maximum after ten seconds and then gradually fades. Its intensity is measured by a colorimeter.

Spectroscopic Method: This depends on the production of the same blue colour, but it is measured by a spectrophotometer instead of visually.

Vitamin B_1

Chemical Estimation: The method is based on the fact that vitamin B_1 can be converted by alkaline oxidation into another compound, thiochrome, that gives a blue fluorescence in ultra-violet light. The food or other material containing the vitamin is first extracted with water and then heated to get an adsorbate on Kaolin. This is oxidized to thiochrome and the intensity of the fluorescence produced, that is measured by special apparatus, is proportional to the amount of vitamin B_1 present.

The method used for testing pharmaceutical preparations is suitable only for high concentrations. It involves precipitating the vitamin by an alkaloid reagent, silicotungstic acid and determining the weight of vitamin precipitated.

Vitamin C: This is the vitamin that is most easily measured chemically. The method is based on the fact that vitamin C is a reducing agent, so it readily takes oxygen from other compounds and thereby reduces their oxygen content (or its equivalent) and becomes itself oxidized. A dye, methylene blue and another compound derived from indo-phenol are blue in the oxidized state but colourless when reduced. By means of a chemical titration it can be found out how much

vitamin C (ascorbic acid) is required to reduce a given quantity of one of these dyes. This is compared with a solution of the substances under test and the strength of the latter can then be calculated. The practical details are a bit difficult. Various modifications of the test solution have to be made before titration, so as to remove any other reducing compounds.

Vitamin D: A chemical procedure has been published for the estimation of vitamin D, but unfortunately this can be used only for highly concentrated solutions. It is therefore still necessary to use biological methods in spite of the fact that these are less accurate than chemical analyses.

Biological and Microbiological Methods of Estimation: The methods used now, have been developed from the earlier procedures used but, the careful design of each test and the selection of animals has led to greater accuracy, and a further advance has been the discovery that micro-organisms can be used in some tests instead of rats, mice or guinea-pigs.

Yeasts, moulds, and bacteria need vitamins for their multiplication and fail to develop if even one vitamin essential for their multiplication is absent. These organisms cannot, of course, be observed singly but a colony can be used to replace a single animal of higher type.

Growth of a colony occurs either in a liquid containing the necessary nutrients in solution or on the surface of a nutrient jelly. Generally test-tubes or flasks are used for liquid media and shallow dishes for jelly cultures. The obvious saving of space by the use of such test organisms is a great advantage. Compare, for example, the space required for two or three dozen tubes or flasks with that required for cages of rats —> and the time involved in giving the necessary attention to each.

Whatever the organism used, whether rat, chicken, mould or bacterial colony, the principle underlying the assay is the same and the methods are comparable. The commonest feature used for measurement is growth-rate, but there are others such as, for example, the heart-rate of rats, and the production of carbon dioxide by yeasts.

In a test, groups of animals or cultures of micro-organisms are given graded doses of the pure vitamin and their responses give a standard with which test substances can be compared.

Simultaneously, other groups of cultures similar to the ones on standard are given graded doses of the test substances and their responses are measured in the same way.

Since no two animals and no two cultures of bacteria are ever exactly alike, there is always some variation in response, even to the same dose. This is why *groups* of animals and two or more tubes or flasks of bacteria are used at each dose level.

There is a definite mathematical relationship between the response and the dose, but this is not a simple ratio and has to be worked out by statistical methods that are beyond the scope of this book.

However if care is taken to use experimental material which is as uniform as possible and if conditions are the same for all groups of material then the results generally show the graded responses to be expected from graded doses. Having measured the responses obtained by given doses of standard and the responses from given weights of test substances, it is not difficult to calculate the amount of vitamin in the dose of test substance used and hence the weight in say 100 g. or 1 oz.

Vitamins which are estimated by biological or microbiological methods are those of the B group, vitamin D and vitamin E.

Vitamin B_1

Biological Assay: The first methods to be used depended on growth rate or on the cure of head retraction in pigeons, but a more accurate method now in use makes use of the fact that rats deprived of vitamin B_1 develop in contrast to man a slower and slower pulse rate (a slow pulse rate = bradycardia: a rapid pulse rat = tachycardia).

Small animals such as rats, mice and birds have, when

healthy, very rapid heart-beats, the normal speed for a rat's heart being 500 or more contractions, or beats, per minute. Rats are used for test when they have been on a vitamin B_1-deficient diet long enough for their pulse rate to be reduced to 350-400 beats per minute. Doses of either pure vitamin B_1 or foods containing the vitamin cure" this bradycardia, but the cure is only temporary and its duration depends on the amount of the B_1 given. When this feature of vitamin B_1 deficiency is used in the assay of foods, the calculation is based on the fact that the duration of cure bears a mathematical relationship to the weight of vitamin given.

The pulse rate is too fast to be measured by hand and is accurately recorded by an electrocardiograph.

Microbiological Assay: The microbiological method for vitamin B_1 is not as satisfactory as that used for some of the other vitamins of the B group, but it is being developed and gradually perfected. One procedure depends on the fact that the organism, lactobacillus fermenti, depends for its growth on the presence of vitamin B_1.

Other Vitamins of the B Complex Microbiological Methods: Most of the other vitamins of the B group are measured satisfactorily by microbiological methods commonly. Two types of organisms are most used: bacteria that during multiplication produce lactic acid and moulds that produce an extensive mycelium. The lactic acid bacteria are permitted to develop in tubes of liquid media to which graded doses of, for example, riboflavin are added. When the cultures have been incubated for a long enough period for acid to be produced, definite volumes of the liquid cultures are titrated against alkali of known strength to determine how much acid has been formed. From this it is possible to calculate the weight of vitamin present.

When moulds are used, the mycelium develops on the surface of the medium, where it can clearly be seen showing the gradation between none or very little at one end of the series to a dense growth at the other end. These mycelial

growths can be lifted off the surface of the medium, dried and weighed and again we get a graded series of weights according to the amount of vitamin present in the culture.

The following is a list of vitamins which can be estimated by microbiological methods and some of the many organisms which may be used.

It may be noticed that four of these use mutants of neurospora. These are varieties produced by exposing the fungus (mould) to X-rays, a procedure that makes the organisms sensitive to lack of these particular vitamins.

Vitamin D

Biological Estimations: Methods for the measurement of vitamin D are less accurate than those for most of the other vitamins. Test animals still have to be used, pending the perfecting of chemical procedures and, in spite of great improvement, the error is still considerable.

Rats and chicks are both used but the method involving the least error is that wherein bone calcification in chicks is considered the criterion.

Vitamin E and fertility in the rat

Chicks grow quickly and X-ray negatives of their bones show the effects of lack of vitamin D before any gross signs of rickets can be detected. The method involves feeding to groups of chicks a basal diet containing no vitamin D, to which has been added graded quantities of standard oil or test oil and taking X-ray photographs of the legs at the end of a six-week period. The joint chosen for examination is the hock joint where the tarsal bones form a junction with the metatarsus. At this point there is a very easily seen centre of ossification (bone formation) separated from a very clearly defined metatarus. The distance between these two is very definite and can be measured to a fraction of a millimeter by means of a micrometer guage and magnifying eye piece. This is called the tarsometatarsal distance (or T.M.T.) and is found to differ with

the vitamin D supplies given to the chicks, being narrow in normal chicks but wider in chicks with poor calcification. In this case there is an *inverse* ratio between the distance and the amount of vitamin D in the ration.

Because of the variation between chicks even on the same ration, large numbers are required for the test but where this is possible results obtained are better than by any other method. This is the procedure used to measure the vitamin D potencies of cod liver and halibut liver oils used in infant feeding.

Vitamin E

Biological Method: This depends on the fact that the female rat, when fed a diet lacking in vitamin E fails to carry her young to term. They are formed, but part way through pregnancy the maternal weight begins to drop due to the resorption (re-absorption) of the embryos. No abortion and no birth of young occur. Vitamin E is measured by finding the least dose of any test food which will prevent resorption and allow the birth of living young. There are also chemical and physical methods but the details of these are rather beyond the scope of this book.

Chemical Combination

Vitamins are chemically unrelated compounds that are grouped together because each is essential in the diet in minute amounts and is required for specific metabolic reaction within the cell. Several of the vitamins though conveniently considered *as* single substance actually are a group of structurally related compounds which tends to behave alike physiologically. In general however physiological function tend to be limited to one active form to which the related forms are converted.

The discovery of the vitamins unhered in a new era in nutrition. It was one in which *small* things in the diet assumed *great* importance. The percentage of body weight attributable to vitamins is minute. Nevertheless, the amount, even though small, is indispensable for normal functioning.

Early evidence of the existence of vitamins came from the search for treatments of certain human diseases that persistently plauged people. One of these diseases was scurvy. In was one of the hazards explorers and sea voyagers encountered. In the year 1747 relief from scurvy was enjoyed by a few of the British navy. Freedom from the illness came about in this way. Limited amounts of lemons and oranges were aboard ship and were given to some of the sailors; the recovery was remarkable. In another part of the world, another disease was making the sailors of the Japanese navy ill. This disease was beriberi. The diet of the men consisted mainly of white rice. With the addition of vegetables, meat, and fish, a change instituted in 1884, the incidence of beriberi decreased.

Despite the improvement in health brought about by changes in the diet, the realisation that a lack in the diet was responsible for scurvy and beriberi was slow. In fact it was almost 150 years after the experiment at sea with lemons and oranges before there was acceptance of scurvy as being caused by a nutritional deficiency; and it was not until the early 1900's that scientists and physicians accepted beriberi as the result of a deficiency in the diet. That deficiencies in the diet could be responsible for disease was a new concept, which was adopted with reluctance.

Another avenue of approach to the study of the dietary deficiency diseases was the use of animals in the experimental laboratory. In the final analysis, it was the well-controlled laboratory experiments that eventually gave conclusive evidence that certain diseases are caused by the lack of dietary nutrients. In the experimental studies a diet of pure chemical compounds was fed, pure protein (such as casein or albumen), pure fat (such as lard), pure carbohydrate (such as dextrin), and minerals. The purified diet was really crux of the experiment, for the natural food more than likely would have provided some of the unknown factors. As it was young animals on the purified diet failed to grow; mature animals failed to maintain body weight.

Mc Collum in his study of the history of nutrition found Jean Dumas (1800-1884) to be the earliest man of science to question the adequacy of a diet composed only of protein, carbohydrate, fat and salts. The experience that led Dumas to such a prophetic belief was not with experimental animals, however, but with citizens of France. At the time of the siege of France by the Germans (1870-1871) food was scarce; there was no milk for children. Scientists attempted to prepare artificial milk using protein, fat, carbohydrates, and salts. The combination was disastrous to the young. It led Dumas to conclude that there was something lacking in a combination of the then known nutrients essential for life.

Animal experiments that led to the discovery of vitamins included those of Pekelharing in 1905 at the University of Utrecht, who made the important observation that a diet of caseim, egg albumen, rice flour lard, and salt kept his rice alive only for four weeks. When the added milk to the mixture the rice assumed healthy.

Sir Fredick Hopkins (England) is considered as one of the most advanced thinkers of his time as animal nutrition in 1960 he made an observation,"... no animal can live on mixture of carbohydrates, proteins and fats, and even when the necessary inorganic material is carefully supplied, the animal still can not flourish."

McCollum in 1907, believed with many others of his time that the failure to survive on purified relation of protein, fat, carbohydrate and minerals, was due to unpalatability of the mixture and the consequent refusal to eat a sufficient amount of the ration. He tried to improve the flavour by adding sugars and other variables in rations, but the animal still failed to survive. Even though his experiments appeared to be a failure, it was successful in that it led to his discovery of vitamin A, a few years later.

McCollum and Davis discovered that there was a substance present in butterfat or egg-yolk fat that made difference between

moderate success in the nutrition of young rats on certain diets and prompt failure in its absence. Osborne and Mendel at the Connecticut Experiment Station found cod liver also to contain this essential substance. Thus the first vitamin was discovered and called fat-soluble vitamin A.

Historical Background: The discovery of vitamins is a long story that started with observations made centuries ago in China where the disease now recognised as beriberi was certainly known. It is probable also that liver was used in very early times for the treatment of diseases which may have been related to a vitamin deficiency. Experience in the Middle Ages showed that certain foods were definitely connected with the prevention and cure of scurvy. All these and other observations led ultimately to the scientific investigations of the present century.

Many a names are associated with this history of exploration into the field of nutrition, but in general our knowledge today is the result of the careful observations, experiments and experiences of many individuals in different countries.

Vitamins have existed in foods as long as the plants that produce them and the animals that eat them, but our knowledge of them as such dates only from the beginning of the present century. It is now recognised that without vitamins in their daily food both man and animal become unhealthily and at length succumb to various diseases. These have come to be called the 'Vitamin Deficiency Diseases' and several are familiar by name, for example *rickets* in which the bones of a child or young animal become soft and bend under the weight of the body, and *scurvy* in which the gums bleed and the joints become swollen and painful on account of internal bleeding into them. These and other deficiency diseases have been known for many years but it is only within living memory that we have learnt of their cause and of how to prevent theme.

Descriptions two to three thousand years old have been found in China of the disease beriberi, now known to be due to the lack of an essential vitamin, and in the Middle Ages

many sailors died of scurvy because the food they were able to carry on long voyages was short of vitamins. Vasco da Gama who sailed from Portugal and found a route to India and the East via the Cape of Good Hope in 1497, lost 100 men out of his crew of 160 from scurvy.

One hundred years later, when Drake and other seafaring men from Great Britain were exploring the seas and finding new sea-routes, Sir Richard Hawkins was able to write in his *Observations on his Voyage to the South Sea* published in 1593, this description of the disease and its remedy:

> "The signes to know this disease in the beginning are divers, by the swelling of the gummes, by denting of the flesh of the legs with a man's finger, the pit remaining without filling up in a good space; others show it with their lasinefse, others complain of the cricke of the backe ... all of which are for the most part, certain tokens of infection. That which I have seen most fruitful for this sickness is sower oranges and lemmons."

This is a very good description of scurvy. Hawkins knew that it was caused by lack of fruit and vegetables and he knew mat he could prevent it by giving his crew oranges and lemons.

In the 18th century Captain Cook was also careful to take a supply of fruit and vegetables with him to protect his crew from the same serious disease, and in 1804 regulations were introduced into the British Navy providing all ratings with a daily ration of lemon juice. These measures were all taken as a result of experiences and observations. Nothing was known about vitamins and we were as ignorant in 1900 as Hawkins was three hundred years before, about *why* oranges and lemons were effective remedies for scurvy.

Scientific discoveries are sometimes made by a deliberate and planned search, by experiments designed to give solutions to definite problems, but sometimes the 'discovery' follows years of careful observation and recording; one could cite

many instances, of which the use of fruits and vegetables for scurvy is one, in which the practical application came first and the underlying reason was discovered later.

Towards the end of the 19th century, Dr. Eijkman, a military medical officer stationed in the Dutch East Indies, noticed that poultry, when fed on rough unpolished rice remained healthy but that when fed on the polished rice which was usually reserved for the men in the camp, they developed a paralytic condition in which they were unable to stand and could not hold up their heads (head retraction).

Soldiers at the station were also suffering at the time from beriberi, a disease in which great weakness was accompanied by a condition resembling paralysis of the limbs and frequently causing death. Dr. Eijkman found that he could treat his men in the same way as he treated the hens by giving them either rough unpolished rice or rice polishings and that this was effective (Eijkman, 1897). He did not, at the time, recognise that beriberi was a deficiency disease.

He thought there was some toxic substance in the starch of the polished rice and that there was an antidote in the polishings and it was not until several years later that another Dutch scientist, Grijns, put forward the suggestion that beriberi was caused by a lack of some vital ingredient which was present in the polishings of rice but not in the starchy part of the grain.

The existence of such 'accessory food factors' was finally established by Frederick Gowland Hopkins at Cambridge. Between 1906 and 1912 he was carrying out of series of experiments on proteins and other known nutritional factors in foods. He fed groups of rats on whole milk, and others on all the known constituents of milk and found that while the first group grew normally and were quite healthy the second group soon began to lose weight and became ill. When, however, he gave milk to these also, they recovered and began to grow normally.

Hopkins concluded that there must be, in whole milk, unknown substances necessary for life and growth but in such small quantities that they had so far defied chemical analysis. He quoted the well known diseases, scurvy and rickets, as examples of such deficiency diseases and foresaw that there were 'many more nutritive errors' affecting health, but at that time quite unknown.

There was a need for a name for the new factors in nutrition. The name which became *vitamin* originated in this way. Funk, a Polish biochemist, coined the term "vitamine" in 1911 to designate the anti-beriberi factor that he was investigating at the time. He combined *vita,* to designate "essential for life," and *amine* to indicate the chemical structure. The term was accepted and used for all of the unknown factors. In 1919 the final "e" was dropped, for by then it was abundantly clear that there were several unknown dietary nutrients, and that the known ones did not all have the amine structure. The change in spelling avoided any implication of chemical for the as yet unidentified dietary factors.

There seemed to be only one factor needed to be supplement the purified diet at the time of discovery of vitamin A. In a couple of years, however, it was recognised, that not one, but two factors were lacking in the purified diet for normal growth, one soluble in fat, the other soluble in water. Later of course, a number of new fat soluble and some more water soluble factors were identified.

Here we give a summary of the isolation and synthesis of various vitamins:

The scientists who made these first discoveries about vitamins were wise in not attempting to give each one a name as nothing was known of their chemical nature. All that was certain was that there were several different factors in a number of different foods and that each played its own part in the maintenance of health and the protection of the body from particular diseases.

Year	Isolation or Identification	Synthesis or Structure
1925	vitamin D by Irradition	
1926	Vitamin B1	
1925	Iusitol	
1931	Vitamin A	
1932	Vitamin C	
1933	[Riboflavin 1	Vitamin C
	Pantothenate	
1935	Biotin	Riboflavin
1936	Anti-egg-white factor	Vitamin D
	Vitamin E	Thiamine
1938	Nicotinic acid Pyridoxine	
1939	Vitamin K	Vitamin K
1940	P- Amino benzoic acid	Pentothenatic
1942		Biotin structure
1945	Folk acid	Folic acid
1948	Vitamin B_{12}	Folioric acid
1955-56	Vitamin B_{12}	structure

McCollum and Davis, two other Americans, adopted the custom of referring to the vitamins by the letters of the alphabet (McCollum and Davis, 1915), and this proves so useful and easy that we still use the letters today.

A = the growth factor in butter fat

B = the anti-beriberi factor in rice polishings and wheat germ

C = the anti-scorbutic factor in fresh fruit and vegetables.

It was not until after the World War that vitamins D and E were discovered.

Mellanby (1918-19), showed by feeding tests on puppies that rickets was a deficiency disease and, moreover, that it could be cured or prevented by the same foods, i.e. butter and cod liver oil, which contained the growth-promoting vitamin A. It was thought at first that growth and the prevention of rickets were both due to the same factor, However, later work showed that the growth-promoting factor could be destroyed by heat and oxidation while the anti-rachitic property remained, and also that although butter was weaker than cod liver oil in both properties it had *relatively* higher proportion of the growth factor A. It was therefore concluded that there must be two factors: vitamin A controlling growth and vitamin D protecting against rickets.

By this time it was known also that vitamin A prevented a disease characterised by drying of the eyes leading to blindness and known as xerophthalmia.

It would be a mistake to pass on without mentioning the notable work of Herriett Chick and her colleagues, who went to Vienna in the years 1918-21 and did much to improve the health of children who at the time were dying of malnutrition. Supplies of cod liver oil were used for the infants who had both xerophthalmia and rickets, and the rather surprising discovery was made that rickets could be cured by exposure to sunlight without any cod liver oil or butter but that one or other of these was needed to cure the condition of the eyes.

This observation led to recognition of the fact that in sunlight the body can make its own vitamin D, and this is one reason why exposure to sunlight is beneficial. Vitamin D is sometimes called the 'sunshine vitamin.'

It will be noticed that both A and D occur in fats and oils of animal origin/and B and C in plant products. This led to an early distinction between fat-soluble and water soluble vitamins, the latter being found in the cell-sap of plants or in the non-fatty storage organs.

In 1922 came the discovery of vitamin E. Evans and Scolt

(1922), found that rats, even when given a good diet containing protein, salts and all the other known essential factors, failed to breed unless certain natural foodstuffs were added. The foods they found effective were fresh lettuce leaves, wheat germ and dried alfalfa. These, therefore, were sources of the fertility or anti-sterility factor which was called vitamin E.

The best source of vitamin E was found to be the oil extracted from wheat germ. Vitamin E was therefore the first example to be found of a fat-soluble vitamin occurring in a vegetable oil.

Thus by the early 1920s five definite vitamins had been recognised and something was known of their occurrence in food and of the functions each.

This knowledge is summarised in the following Table:

Vitamin	*Solubility*	*Function*	*Best Sources*
A	Fat-soluble	Promotes growth Prevents or cures xerophthalmia	Cod liver oil, butter.
B	Water-soluble	Prevents or cures beriberi in man, polyneuritis in birds	Wheat germ, rice germ, yeast.
C	Water-soluble	Prevents and cures scurvy	Citrus fruits. fresh vegetables
D	Fat-soluble	Prevents and cures rickets	Cod liver oil, butter.
E	Fat-soluble	Prevents and cures sterility in female rats (not in male	Fresh lettuce leaves, wheat germ oil, dried alfalfa. rats)

During the second quarter of the present century there have been discoveries not only of other vitamins but of the chemical nature and distribution of all those known and the physiological actions of most. Many have been isolated from natural sources and methods have been evolved by which many can now be synthesised. The following is a brief outline of the various discoveries which have been made since 1925.

The first step of the importance was the recognition (Goldberger and others, 1926), that pellagra was associated

with lack of a factor closely related to, but not the same as, the anti-beriberi factor. This he called P-P or pellagra-preventing. The following year a distinction was drawn betweer them on account of their stability to heat and they were renamed vitamins B_1 and B_2, the former or anti-beriberi factor being less stable to heat than the anti-pellagra factor B_2.

Shortly afterwards B_1 was isolated in crystalline form (Jansen and Donath, 1926), and it was shown that B_2 was not a single substance but was in itself a 'complex.'

The next step was the recognition, between 1920 and 1929, that carotene, the yellow plant pigment, was a precursor of fat-soluble A (Moore, 1929).

This discovery was followed in 1930 by the first preparation of vitamin D, calciferol (Askew and others, 1930).

In 1933 vitamin C was synthesised (Ault and others, 1933), and riboflavin was isolated from the rest of the vitamin B complex (Booker, 1933).

In 1935, a vitamin concerned in the coagulation of blood was discovered by a Danish scientist (Dam, 1935) who named it vitamin K (Koagulations vitamin) and the next year riboflavin was synthesised. In 1935, too, another factor was found in lemon rind and juice which appeared to be different from ascorbic acid (Szent-Gyorgyi, 1936). This was called vitamin P and its status as a vitamin is still uncertain. More will be said about this later.

In 1936, vitamin E (tocopherol) was isolated from wheat germ oil (Evans and others, 1936), and a series of researches began on its chemistry and very complicated behaviour.

In 1937, ten years after its isolation, vitamin B_1 was synthesised and given the name aneurin and about the same time (Williams and Clive, 1936), it was recognised that the so-called P-P or pellagra-preventing factor was nicotinic acid, another factor of the vitamin B_2 complex (Harris. 1937).

A second form of vitamin D was also distinguished: D_3 or

irradiated 7-dehydrocholesterol, the form in which vitamin D is present in fish liver oils.

In 1938 came the isolation of another vitamin B factor — B_6 (Harris and Folkers, 1939), needed to protect rats from dermatitis. The terms vitamins B_3, B_4 and B_5 had been tentatively given to factors thought to exist, but these had not proved to be genuine new vitamins and the terms did not persist, B_6, however, was established as a real and necessary vitamin and it was synthesised in 1939. Vitamin B_6 is pyridoxin.

In the same year two other substances with vitamin activity were isolated: vitamin K or menaphthone and pantothenic acid. Choline and para-aminobenzoic acid were not newly isolated or synthesised but were for the first time found to be necessary for certain vital physiological processes and were therefore included among the vitamins.

In 1941, a factor, known since 1932 as H was identified as a newly isolated chemical compound biotin (Du Vigneaud and others, 1941). In the same year folic acid was isolated from the green grass and inositol, a known constituent of wheat germ, was recognised as a vitamin.

It took chemists until 1945 to discover the nature of biotin and to synthesise it and it remains one of the rare Vitamins.

The most recent vitamin to *he* added to the list is vitamin B_{12} (Rickes and others, 1948; Smith, 1948), isolated from liver after six years of intensive research and since proved to be the active factor which is used in the treatment of pernicious anaemia.

There are various factor whose nature and functions are still being studied, i.e. B_{12}, B_{14}, B_{15} and U, but in the summary which follows are included all those vitamins which are the present identified and about whose functions we have some definite knowledge.

A Few Dates in Vitamin History

B.C. Beriberi known in China.

Middle Scurvy common among sailors. Discovery made ages that oranges and lemons
18th were protective. Capton Cook took vegetables with him to protect century
century his crew from scurvy.

1804 Regulations introduced into the British Navy providing for a daily ration of lemon juice for all ranks.

1897 Eijkman, in the Dutch East Indies, discovered that beriberi in man and polyneuritis in birds were both caused by eating polished rice and cured with rice polishings.

1906-10 Gowland Hopkins demonstrated the existence of Accessory Food Factors.

1912 The word 'Vitamin' first used by Funk.

1915 McCollum and Davis demonstrated two growth factors: fat-soluble A and water-soluble B. The anti-scurvy factor became water soluble C.

1918 Mellanby showed that the anti-rachitic factor was different from vitamin A and called it D.

1920 J.C. Drummond suggested the present spelling 'vitamin' for all these factors.

1922 Evans and Scott described vitamin E, the anti-sterility vitamin.

1926 Recognition of the P-P (pellagra-preventing) factor. Distinction between B_1 and B_2. Isolation of crystals of vitamin B_1.

1928 Carotene, the plant pigment, recognised as a precursor of vitamin A.

1930 Vitamin D, calciferol, prepared by the irradiation of ergosterol.

1933 Synthesis of vitamin C. Recognition of riboflavin as part of vitamin B_2.

1936 Isolation of vitamin E- tocopherol.

1937 Discovery of vitamin K. Synthesis of riboflavin.

Synthesis of vitamin B_1- aneurin. Identification of nicotinic acid as the P-P factor. Recognition of vitamin D_2 as the anti-rachitic factor on fish liver oils.

1938 Isolation of vitamin B_6.

1939 Synthesis of vitamin B_6- pyridoxin. Preparation of menaphthone, a form of vitamin K. Pantothenic acid isolated.

1940 Choline suggested as a member of the B_2 complex.

1941 Isolation of biotin and folic acid. Recognition of inositol as a vitamin.

1945 Synthesis of biotin.

1948 Isolation of vitamin B_{12} from liver.

From the historical development of vitamins presented in the preceding pages it is obvious that vitamins were discovered not by their presence in the diet but because of their absence. A lack of most, but not all of vitamins known to be required by results in typical clinical syndromes or symptoms that can be cured by adding necessary vitamin to the diet. This negative approach to the functioning of vitamins is a useful tool in research. The investigator obtains clues to the metabolic rate of the nutrient when he observes the functional failure that results nutrient is absent from diet.

As a preface to the discussion of individual vitamins an explanation of the general meaning of the term should be made. Vitamins are organic compounds, necessary for growth and maintenance of life, which must be provided in the diet. The body is not able to synthesise them, at least in amounts sufficient to meet its needs. An exception is vitamin D; exposure to ultraviolet rays develops the vitamin from its precursor present in the skin. Vitamins are regulatory substances, each performing a specific function. They are carried in the blood stream to all parts of the body.

4

Significance of Spices

Down through the ages India has been known as the land of spices. Sweet or pungent, aromatic and seductive spices have played a special part in the life, legend and ritual of the country. With the march of time, although there have been several changes in the vise of spices, their qualities for enhancing food still reign supreme.

The primary function of spices in Indian food is to improve the flavour of the dish. Many spices such as cloves, coriander, cumin seeds, cinnamon, etc., because of their volatile oil contents, impart various flavours to the food. These appeal to the sense of smell and render the food more palatable. The primary quality the common man looks for in a food product is its organoleptic quality (or the sense of taste), rather than its nutritive value. As a result even highly nutritious food is not accepted unless it is adequately spiced. It is only where pleasure to the eye and palate meet that food becomes fully acceptable. The best chefs of the world consider flavourings and seasonings absolute necessities for achieving this objective.

The success of cooking depends largely on their aid. However, spices must be used with skill and, above all, sparingly. All palates do not crave highly spiced foods, yet most people demand that food should be adequately spiced, for the flavour of insipid food can be improved very much by the use of some suitable spice. Even the people of the West, particularly the English and Americans who have a tradition of bland food, are now introducing a variety of spices into their dishes and appreciating the added piquancy of the food. Western palates, however, are sensitive, and therefore can appreciate the subtle flavours more readily than we do, as unfortunately, owing to the continuous consumption of over spiced foods, we have lost this power. Whenever possible, only such spices should be used as bring out the natural flavours of the main ingredient, rather than imparting a new one. Intelligence, carefulness, thorough, sound judgement, a steady hand and a keenly perceptive palate are qualifications every cook must possess in order to prepare food that is appetizing and pleasant to the taste.

Importance in Food

Besides enhancing the flavour and aroma of food, spices have a physiological action beneficial to our system. They act as a stimulus to the digestive system and help digestion in many ways.

Every spice used in the making of a curry is a preservative. All have some antiseptic value and many are carminatives, i.e. they tend to reduce flatulence as in the case of omum (ajwain) water given to babies. The traditional combination of spices used for certain dishes must have been made with this factor in mind, e.g. the use of ginger or mint with peas. Peas are flatulence-forming and ginger or mint counteracts this effect. Asafoetida, commonly used with pulses (dais) of various types functions in a similar manner.

Some spices are used to give colour to food, to improve eye appeal, e.g. turmeric in yellow rice, red chillies in vindaloo,

rattanjog in roganjosh, etc. Spices also serve as a thickening agent in the preparation of curries, e.g. onion and poppy seeds.

Different Uses

The therapeutic value of spices is well known. Spices and herbs play a very important role in some of the simple but effective home remedies. Almost every spice has medicinal properties. Investigations are being conducted scientifically to elucidate the therapeutic properties of the various spices. Many have already been accepted and are being used today, e.g. aniseed (saunf) which has been used from prehistoric times as a flavouring for cough mixtures and as a herb tea to soothe the nervous system and induce sleep. Scientists have recently probed the effect of cloves on the digestion. Studies of clove oil have shown that it stimulates the flow of gastric juices and is non- injurious to the lining of the stomach. Garlic has been accepted in both India and China as one of the treasured spices and medicinal agents. It has been indicated in the treatment of numerous diseases such as haemorrhoids, rheumatism, dermatitis, abdominal pains, coughs, loss of appetite, etc. Although garlic is claimed to be useful in the treatment of such a variety of diseases, it is not popular because of its strong smell and pungency. Research has now made it possible to administer the active principle of garlic in a fairly concentrated form. Ginger tea is used commonly to ease an ailing stomach. Nutmeg is used medicinally to cure biliousness, diarrhoea and headaches. The ancients prized the seeds of fenugreek because of its stimulating effect on the digestive system.

To enhance the flavour and aroma of foods, spices may be used whole or in pieces, as in pulaos and biryanis, or in powdered or ground form. Indian cooks are artists in the use of spices. Money may limit the variety of spices used by some, but others are limited only by their own creative imagination. While freshly ground masalas (mixed curry spices) are preferred to powdered masalas, owing to the shortage of domestic labour there is a definite trend today towards using spices mainly in

powdered form. Many combinations of spices to suit different tastes are marketed today as various forms of curry powders. When using spices in powdered form it is essential that the powders be extremely fine. Coarse powders do not blend well with the rest of the ingredients; this causes a separation thus allowing the spices to float. The fine powders should be mixed into a paste with vinegar or water and then fried well before meat or vegetables are added. Besides curry powders, practically all the dry spices such as coriander, turmeric, cumin, fenugreek, cloves, cinnamon, mustard, etc., can be cleaned, dried and powdered.

Kinds of Spices

As a result of modem techniques, moist spices such as onion, garlic and ginger are now obtainable in powdered form. This process, while retaining the original flavours, improves the keeping quality of spices, and makes transportation more economical and easy. Onion powder is used commercially in the manufacture of processed meats and sausages, and for flavouring vegetables and meat. Garlic powder has many commercial uses and is also now being used by the housewife who wants a clean and convenient method of using this popular flavour. All powdered spices, however, must be kept in airtight containers to prevent spoilage and deterioration in aroma and flavour, caused by the evaporation of volatile oils. Ground masalas are also available in the market today. They are packed in plastic bags and sold, but the keeping quality of these is poor as they are susceptible to spoilage.

The liquid spices now available in the market as straight liquid spices, such as liquid red chillies, green chillies, ginger, coriander, cumin, garlic, nutmeg, cinnamon, cloves, cardamom, asafoetida, saffron, curry leaves, etc., or as ready-to-use compounds such as liquid masala for curries, pulao, mutton, fish, etc., have revolutionised the use of spices. These liquid spices are extracts of indigenous natural spices and are the result of years of research. By the separation of the active

principles from the residual matter, concentrations are prepared. These are easy to use and are being used widely in the commercial manufacture of sauces, fruit juices, ketchups, salad dressings, tinned pulaos, curries, etc. They have a better keeping quality than curry powders, and mix readily with the rest of the ingredients, thus bringing a uniformity of colour and flavour to the food.

They are clean, compact, suitable for transportation and storage and are available in glass bottles with droppers, ranging from 15 ml (oz) to 500 ml (1 lb). They impart original flavours and colours. Thickening agents such as ground onion, poppy seeds, coconut and coconut milk should still be added to maintain the desired texture of the food. For those spices where flavour is the primary function as in cardamom, cloves, cinnamon, nutmeg, etc., the liquid form can be used most effectively. It also makes it possible to be able to use out-of-season flavours. For example, when ginger or green chillies are not available, as in the summer months, the same flavours can be readily obtained by using liquid spices. Provided they are used carefully, liquid spices do not in any way increase costs, while at the same time they save labovir and time.

Allspice: This is the berry of 'Eugenia Pimenta', a small tree grown in the West Indies. The berries are gathered when green and unripe and dried in the sun. Then they turn black. Large quantities of these are used for the manufacture of sauces and pickles sold commercially. The berries combine the flavour of cloves, cinnamon and nutmeg. This spice is also known as 'Jamaica Pepper'.

Cloves: A very old and important spice, the clove was known before the time of the early Egyptians. The word is taken from the Latin 'Clavus' and the French 'Clou' both meaning nail, which describes its shape. The clove tree (of the order of the myrtle) grows to a height of 9 metres. The unopened flower buds are carefully harvested. Upon browning, they lose half their weight. Cloves are used in both Western and Indian dishes, both savoury and sweet, e.g. pulaos, cooked ham, sauces

such as bechamel sauce and bread sauce, meat dishes, fruit salads, baked apple, etc. The oil of cloves is used in dentistry to soothe toothaches, and its antiseptic properties are well known. It is grown on the islands of Zanzibar, Pemba and the East Indies, West India and other tropical regions.

Nutmeg and Mace: These are the only known cases of two different spices from the same fruit. Nutmeg is grown in the Dutch East Indies and Grenade in the British West Indies. The tree begins bearing fruit at the age of 8 years and continues to yield for half a century or more. In its wild state the tree grows to a height of about 22 m. or 70 ft. The nutmeg fruit resembles an apricot in size and shape. What we call the nutmeg is the seed. It is protected by a thin shell; the shell has a coat of orangy flesh which dries into mace. Although the aroma of mace is similar to that of nutmeg, separate uses exist for each. Nutmeg is used in Western and Indian dishes in puddings, sweetmeats, for flavouring egg and milk and in some curries. Mace is also used to flavour sauces, stocks, sweet dishes such as halwas as well as in betel nut mixtures. Nutmeg is used grated fine and mace coarsely crushed. When buying nutmegs choose those which are round, compact, of oily appearance and heavy for their size.

Mustard: A product of great antiquity and grown in most parts of the world, mustard is used in both Western and Indian preparations. It is used in the West in powdered form, as a table condiment and for flavouring sauces, etc. The pungency is more fully brought out when mustard is moistened with water (prepared mustard). In India it is used whole for tempering dishes, and in pickles. It is also an important ingredient in certain masala powders, e.g. vindaloo.

Cinnamon: This is taken from the bark of an evergreen tree belonging to the 'laurel' family and is chiefly cultivated in Ceylon and the East Indies. First the outer bark is stripped of the three-year-old branches, and then the inner bark is loosened and dried. The best cinnamon should not be too dark in colour and should be as thin as paper. It has a fragrant odour and

its taste is pleasant and aromatic. It is used in both Western and Indian cookery, in cakes, buns, ketchup, pickles, in pulaos with cardamom and cloves, etc. It is used in the preparation of garam masala for curries.

Cardamom: This is the fruit of a reed-like plant, native of the mountains of the Malabar Coast, but cultivated in Jamaica as well. The fruit is a small pod and the seeds within the pods have a strong, sweetish flavour. The pods vary from 0.5-2 cm. 1/4-1") and more in length. The small-sized ones are considered the best. Rich curries like khorma include cardamom. Powdered cardamom is a big favourite in cakes, puddings, halwas, etc. Pulaos and biryanis also depend on this spice for their popular flavour. Tincture of cardamom is used in stomach medicines.

Pepper: This is obtained from the seeds or berries of the plant 'piper nigrum' which grows in Malabar and other parts of India. Black pepper is made from dried whole berries, and white pepper from the same berries with the husk dried and removed. Pepper is one of the most universally popular spices, and is used in every type of savoury dish to improve flavour and to add to its piquancy. It is also one of the table condiments.

Mignonette Pepper: This is ordinary white pepper coarsely crushed.

Long Pepper: It is a spice similar in taste to ordinary pepper but not so pungent.

Cayenne Pepper: Dried fruits of the capsicum plant grown in the Cayenne Islands. This is red in colour, resembling the red chillies of India without being as pungent. This is used in Western cookery where pungency is required, e.g. for devilled prawns.

Krona Pepper: Bright red pepper from the Hungarian paprika. It is much milder than cayenne pepper and forms a palatable seasoning for Western savoury dishes.

Red Chillies: Chillies are grown in large quantities in India and are used extensively in Indian savoury preparations. They also form an important ingredient of curry powders. Apart

from being used as a condiment, they are also used in several medicinal preparations for the treatment of lumbago, neuralgia, rheumatism, etc. Although there are several varieties they fall under two main groups: Capsicum Anum and Capsicum Frutescens. The variety acuminatum (Capsicum Anum) is the principle source of commercially used red chillies.

Coriander: The fruit of a plant called 'Corianderum Sativum' of the parsley family. It is extensively used in the preparation of Indian savoury dishes and forms an essential ingredient of curry powders. It is used in the preparation of frankfurters (sausages) and by confectioners and distillers. The leaves are used for flavouring curries, in salads and chutney and as a garnish.

Turmeric: This is the aromatic root or rhizome of a plant grown in India and the West Indies. It is sold in root form and as powder. The hard resinous flesh of the dried root varies from a dark orange to deep reddish brown, but in the powdered form it exhibits a characteristic yellow colour. Turmeric is a member of the ginger family but has a different flavour. It is used in curry powder not only for imparting colour and flavour to curries but also because of its preservative qualities. Game and meats that are dried are soaked in liberal quantities of turmeric and salt. Turmeric is used freely in Indian medicines too. Plasters of it are applied in cases of bruises; as a dry dressing on open wounds; paste of turmeric and neem leaves is used to heal smallpox marks, and so on. Turmeric has sometimes been used to replace the more expensive saffron as a colouring agent.

Ginger: This is the tuberous root of a plant 61 cm. (2ft.) or more in height and is one of the few spices that grow underground. Ginger is mainly produced in Jamaica and to a less extent in Sierra Leone, China, Japan and India. Because of its stimulating and digestive properties it is employed medicinally for dyspepsia and colic. It is also frequently used to disguise nauseous tastes in medicines. The most common use of ginger is in making masalas, soft drinks, pickles and

preserves, and in confectionery such as ginger biscuits, gingerbread, snaps, etc. Crystallised ginger is made from the young roots and is a delicacy. The best grades selected are called 'stem ginger'.

You can make instant pickle by chopping fresh ginger and green chillies and adding lemon juice together with salt. It is delicious specially one day after preparation.

Cumin seed: This is a herb of the caraway type, producing seeds of a pleasant smell and flavour which are very popular. It is most widely grown in several European countries, India and Mexico. The cumin plant is less than 30 cm high. The seeds come from the fruit and their slightly bitter taste is said to stimulate the appetite and digestive system. Cumin is added to curry powders and used whole in savouries, biscuits and fried rice. It is also used in the manufacture of liquors, pickles and sausages.

Fenugreek: These are the dried ripe fruits of an annual plant grown in many European countries and India. The fresh leaves and tender stalks are used widely as a vegetable. Because of their stimulating effect on the digestion they are administered medicinally even to horses and cattle. The seeds are used in making curry powders and fried whole for seasoning pickles and vegetables. When ground with water, fenugreek becomes a slimy paste and so is used to give batters smoothness.

Aniseed: The seed is small, oval in shape, somewhat similar to caraway and it grows in China, Mexicc, Peru, Argentina, Spain, Malta, Syria, Lebanon, India, Pakistan, Hongkong and Egypt.

The seed is small, oval in shape, somewhat similar to caraway and is usually a greenish brown in colour. It is used in Western preparations and in Indian dishes. It is identified with the licorice flavour. It is used in making licorice confections, in baking, in the cordial anisette and other liquors. Its digestive properties are well known and it can be eaten plain after a meal.

Fennel Seed: The dried fruit of a perennial herb of the parsley family, fennel is grown -in Europe, India, Lebanon and Argentina. The seed has a licorice-like flavour resembling anise. Fennel is used both whole and ground. Bakers use it whole in Italy to stud breads and rolls. It is also used in sweet pickles, in the seasoning of soups, fish dishes, sauces, and in South Indian meat curries. It is also attributed with medicinal properties.

Caraway Seed: Fruit of a perennial plant of the parsley family. It has a pleasant, slightly sharp flavour with a sweet undertone. It is a native of Europe. Caraway seed is mostly used whole and is the principal flavouring in rye bread, and is widely used in rolls, biscuit and cakes. It is also used in cheeses, sauerkraut, pickles, soups, meats and stews. It is the chief ingredient of 'Kummel Cordial'.

Aromatic Herbs and Plants: The ones commonly used are parsley, bay leaves, thyme, marjoram, sage, tarragon, chervil, chives, onions, shallots, garlic etc. The bouquet garni which is the mainstay of French cookery consists of sprigs of chervil, chives, thyme, bay leaves, tarragon and parsley.

Parsley: It possesses a wonderful quality of masking the taste of a stronger flavour; so do not use too much of it when flavouring a dish. It plays an important part in cookery. It not only gives a finishing touch to stews and soups and sauces, but it is also used a great deal for garnishing purposes. Powdered parsley is an excellent herb which imparts a delicate flavour. To make powdered parsley, infuse some parsley leaves in boiling water for a few seconds. Drain and put into a hot oven to dry. Put through a sieve and use as required.

Tarragon and Chervil: The leaves of chervil possess a flavour which is appreciated by many people. Tarragon belongs to the same family as wormwood and the botanical name is 'Artemisia Dracunculus'. Of all the pot-herbs, tarragon and chervil give the strongest smell. They are used a great deal in French cookery for entrees, sauces, soups, salads, chad-froids, etc. Tarragon leaves are also used for flavouring vinegar.

Thyme: It belongs to the same family as mint. The leaves are used fresh or dry for stuffings, soups, etc.

It possesses a highly aromatic flavour and should, therefore, be used sparingly. Lemon thyme is a smaller variety and has a strong perfume like the rind of lemon. Leaves of ajwain (omum) can be used as a substitute for thyme.

Burnet: This is not used much in cookery now except for salads. When bruised it smells like cucumber.

Capsicum: There are several kinds of capsicum cultivated in the East and West Indies and in America. The capsicum fruit is both pungent and stimulating. In Mexico the pods are called chili and they are used to make hot-pickle and chili-vinegar. It is the powder of the dried seeds and pods that goes to make cayenne pepper (as a substitute use Kashmiri chilli powder). Capsicums are considered to be very wholesome.

Savory: There are two varieties of this herb; one is called the "summer variety". It is taken from a flowery herb. Both varieties are used extensively in Western cookery.

Marjoram: There are four kinds of marjoram. The sweet or knotted marjoram, originally a native of Portugal, is the one which is commonly used. It is used for soups, sauces, stews, etc., and used either fresh or dry.

Mint: This belongs to a family of plants called Labiatae. The spearmint or the 'Mentha Viridis' commonly cultivated in gardens has a better flavour and is the one used popularly in cooking. It has the property of removing flatulence.

Bay Leaves: The leaves of the common laurel are used in cookery to give a kernel-like flavour to stocks, mirepoix sauces, custards, puddings, blancmanges, and to the milk which is used for mixing cakes. They are generally dried and used.

Basil: A favourite herb in French cookery. It has a flavour similar to that of cloves. It is used for making such things as mock turtle soup and clear soups made out of shellfish. It is also used for flavouring vinegar.

Onions: The name onion is given to all plants of the onion tribe, e.g., leeks, garlic and shallots. The onion is, next to salt, the most valuable of all flavouring substances used in cookery. When onions, shallots or garlic are used in cookery they should be well blended with other flavours. Whenever onion is used as a condiment or seasoning and the article properly treated, the strong smell is removed.

Garlic: This consists of groups of several bulbs called cloves, all enclosed in one membranous skin. When used carefully and sparingly, garlic is an excellent condiment. It is considered to be very wholesome and is a stimulant.

Shallots: This bulbous root resembles garlic. It is a native of Palestine and was introduced to England by the Crusaders.

Carrots and turnips: Next to onions, these are considered to be the most important flavouring vegetables for soups and sauces. Besides being used for flavouring, carrots and turnips are largely used for garnishing certain dishes such, as ragouts, boiled meat, etc. They are also used as a vegetable and as a puree for soup.

It plays an important part in sauces. The rind, juice and the essential oils contain valuable properties. The rind or peel is used for flavouring a variety of dishes. Generally the rind is grated, but the best way is to peel the rind with a very sharp knife as thinly as possible without touching the white part. Lemon rind is preserved in sugar.

When sour limes are in season buy them in large quantities. Extract the juice and mix one cup of salt to every four cups of juice. Shake well and keep in an airtight bottle—under refrigeration it keeps for months. Use whenever fresh lime is required.

Vanilla: This was first discovered by the Spaniards. It is the fruit of an orchid. The best variety is grown in Mexico. It has a delicious flavour and is largely used for flavouring puddings, custards, liquor chocolate, etc. For flavouring purposes it is better to use the vanilla pod or vanilla sugar

rather than the essence of vanilla, since the odour of the essence escapes rapidly.

Curry Powders: For some basic curry powders, see the recipes at the beginning of the Indian Cookery Section of this book.

Beverages : Particular Role

To Prevent Water Depletion: Water being one of the essential constituents of beverages it is very useful in making good the loss of water content in the blood stream, specially after heavy physical exertion. Adequate supply of water is essential for the healthy functioning of all the tissues in the body. For an adult doing no manual work and when the weather is pleasant enough not to cause any excessive perspiration, three litres of water daily are necessary for good health. Two litres are obtained from food so one litre must be taken in the form of water and other beverages.

Refreshing and Restorative: Cold drinks in summer and hot drinks in winter have a greatly refreshing and restorative effect on the body particularly when fatigued.

Supplies vitamins: Certain beverages like lime juice, Ovaltine, etc., are a source of vitamin supply.

Stimulating: Alcoholic beverages are stimulating and have a definite value in case of illness, but if taken regularly even in moderate quantities, lessen the power of resistance to disease, and if taken in excess are very harmful.

Beverages such as tea, coffee and cocoa besides being refreshing are also stimulating but they do not have any aftereffects.

5

Management of Food Items

Kitchen equipment is expensive and to justify the expense it is essential that maximum use is made of it.

This can be done only if the equipment works efficiently and this depends upon care and maintenance. The type of fuel to be used is an important factor. Though solid fuel and oil both have their place under certain circumstances, for professional cookery the choice lies between electricity and gas. In India, salt coke is used in small establishments sometimes. Firewood is used for tandoor.

The routine use, care and cleaning of all items of equipment are important and this should be appreciated and understood.

When selecting equipment, capacity, trade name, good quality, simple design and ease to clean should be the criteria. The purpose and price is also considered.

Kitchen equipment may be divided into three categories:

1. Large equipment:

   Ranges, steamers, boiling pans, fish fryers, sinks and tables.

2. Mechanical equipment:

   Peelers, mincers, mixers, refrigerators, dishwashers.

3. Utensils and small equipment:

Pots, pans, whisks, bowls, spoons, palate knives, graters.

Basic Tools

Stoves: A large variety of stoves is available fired by gas, electricity, solid fuel or oil.

Solid tops should be washed clean, or wiped clean with a cleaning pad. When cool, the stove can be more thoroughly cleaned by washing and using an abrasive such as emery-paper. After any kind of cleaning, a slid top should always be lightly greased.

On the open type of stove, all the bars and racks should be removed, immersed in hot water with a detergent, scrubbed clean, dried and put back in place on the stove.

All gas jets should then be lit to check that none are blocked.

All enamel parts of the stove should be cleaned while warm with hot detergent water, rinsed and dried.

The insides of ovens and oven racks should be cleaned while slightly warm, using detergent water and a mild abrasive. In cases of extreme dirt or grease being stuck on to the stove or oven, a caustic jelly may be used, but thorough rinsing must take place afterwards.

Oven doors should not be slammed as this is liable to cause damage.

The unnecessary lighting or the lighting of ovens too early can cause wastage of fuel, which is a waste of money.

Steamers: Steaming ovens that work from a main steam supply need little maintenance. The door controls should be greased occasionally. The steamer trays and runners should be washed in hot detergent water.

This type of equipment is usually fitted with a gauge

which registers steam pressure (½1b. per square inch), also an overflow valve which gives a warning whistle, if the pressure reaches danger point. These should be periodically checked by a qualified engineer to ensure that they are working correctly. A constant supply of water should be maintained in the generating tank.

Steamer trays and the inside of the steamer should be cleaned with detergent water and rinsed.

Boiling Pans / Steam-jacket Kettle: Many types are available in different metals and various sizes (10, 15, 20 and 40 litres or more capacity).

They may be heated by gas, electricity or steam from the main supply. As they are used for cooking large quantities of food, it is important that they do not allow the food to burn.

It is for this reason that the steam-jacket type (double walled) boiler is most suitable.

Many of these boilers are fitted with a tilting device to facilitate the emptying of the contents.

After use, the pan and lid should be thoroughly washed with a mild detergent solution and then rinsed well. Any moving parts should be greased occasionally and checked to see that they are in good working order.

If gas fired, the gas jets and pilot should be inspected to ensure correct working. If a pressure gauge and safety valve are fitted, these should be checked to see that they are working correctly.

Deep Fat Fryers/ Fritures: These are among the items of equipment that are commonly used in catering establishments.

An unskilled or careless worker can cause money to be lost by food or fat being spoilt through misuse of the friture.

Modern fryers are heated by gas or electricity. Most incorporate a thermostatic control in order to save fuel and prevent overheating. A fairly recent development is the "cool zone".

This is where the heating elements are at the sides of the fire but the lower part is kept at a reduced temperature. This stops particles that may fall off the food being fried from dropping to the bottom of the friture, burning and so spoiling other foods cooked. This form of heating also saves fat. Frying baskets can also be used.

Deep fat fryers should be cleaned daily after use by:

1. Turning off the heat and allowing the fat to cool.
2. Draining off and straining the fat.
3. Closing the stopcock, filling the fryer with hot water containing detergent and boiling for 10-15 minutes.
4. Draining off the detergent water, refilling with clean water plus vinegar and water, and reboiling for 10-15 minutes.
5. Draining off the water, drying the fryer, closing the stopcock and refill with clean fat.

Hot Cupboards and Bain-marie: Hot cupboards (commonly referred to in the trade as the hot plates) are used for heating plates and serving dishes and for keeping food hot. Care should be taken to see that the amount of heat fed into the hot cupboards and a thermostat is necessary in maintaining this.

Hot cupboards may be heated by gas, electricity or steam. The doors should slide easily and essential greasing may be necessary. The tops of most cupboards are used as serving counters and should be heated to a higher temperature than the inside.

These tops are usually made of stainless steel and should be cleaned thoroughly after each service. (Method as for cleaning stainless steel.)

Bain-marie in this case are open walls of water used for keeping foods hot and are available in many designs, some of which are incorporated into the hot cupboards, some in serving counters, and there is a type which is fitted at the end of the cooking range.

They may be heated by steam, gas or electricity, and sufficient heat to boil the water in the bain-marie should be available.

Care should be taken to see that a bain-marie is never allowed to burn dry when the heat is turned on.

After use, the heat must be turned off, the water drained away and the bain-marie cleaned thoroughly inside and outside with hot detergent water, rinsed and dried. And drain off tap should then be checked for any blockage and then closed.

Grills and Salamanders: The salamander, heated from above, probably causes more wastage of fuel than any other item of kitchen equipment, through being allowed to burn unnecessarily for long unused periods. Most salamanders have more than one set of heating elements or jets, and it is not always necessary to have them all turned on.

Salamanders are heated by gas or electricity or a combination of the two.

Bars and draining trays should be cleaned regularly with hot water containing grease solvent such as soda. After rinsing thoroughly, they should be replaced and the salamander lit for a few minutes to dry the bars.

For an UNDER FRIED GRILL to work effectively, it must be capable of cooking food quickly and it should reach a high temperature (within 15-20 minutes) after lighting and the heat should be turned off immediately after use. When the bars are cool, they should be removed and washed in hot water containing a grease solvent, thoroughly rinsed and dried, and replaced upon the grill.

Care should be taken with the fire-bricks if they are used for lining the grill as they are easily broken.

Contact Grills: These are sometimes referred to as a double-sided or infra-grills, having two heated surfaces facing each other.

These grills are electrically heated and are capable of

cooking certain foods very quickly. The electricity should be turned off after use. When the grill is cool, the cooking surface should be cleaned with a stiff wire-brush. The surfaces are then wiped clean with a damp cloth and lightly oiled to prevent rusting.

Fry Plates, Griddle Plates: These are solid metal plates heated from below. They are used for cooking individual portions of meat, eggs, bacons, etc. They can be heated to a high temperature and are suitable for rapid and continuous cooking. When cooking is commenced, a light film of oil should be applied to the food and the griddle plate. After use, the heat should be turned off and the plate scraped clean, allowed to cool and left lightly oiled.

Sinks: Different materials are used for sinks according to the purpose for which they are intended:

1. Heavy galvanised iron for heavy pot wash.
2. Teak or other hard wood for glass or china wash. (If hard wood sinks are left unused for long periods of time, they should be filled with cold water to prevent shrinkage of the timber.)
3. Stainless steel for general purposes.
4. Glazed earthenware for general light purposes.

Sinks, drainers, waste and overflow outlets should be cleaned with a suitable abrasive power cleaner, thoroughly rinsed with plenty of clean water and left to dry.

Tables: WOODEN tables should be scrubbed clean with hot soda water, rinsed and wiped dry as soon as possible to prevent warping.

FORMICA or STAINLESS STEEL topped tables should be washed with hot detergent water, rinsed with hot water and dried.

MARBLE SLABS should be scrubbed with hot water and rinsed. All excess moisture should be removed with a dry cloth.

No cutting or chopping should be allowed on table-tops; chopping boards must be used.

Hot pans must not be placed upon tables, triangles must be used to protect the table-tops. The legs and racks or shelves of tables are cleaned with hot detergent water and then dried. Wooden table legs require scrubbing.

Butcher's Block: A scraper should be used to keep the block clean. After scraping, the block should be sprinkled with a few handfuls of common salt in order to absorb any moisture which may have penetrated during the day.

Do not use water or liquids for cleaning unless absolutely necessary as water will be absorbed into the wood and may cause swelling.

Storage Racks: All types of racks should be emptied and scrubbed or washed periodically.

Handling of Equipments: When a new item of equipment is installed, it should be tested by a qualified fitter before being used by the catering staff.

The manufacturer's instructions must be displayed in a prominent place near the machine. The manufacturer's advice regarding servicing should be followed and a record book kept showing when and what maintenance the machine is receiving.

Before cleaning, the machine should be switched off and the plug removed from the socket.

Potato Peelers

1. Potatoes should be free of earth and stones before loading into the machine, otherwise damage to the machine will result.
2. Before any potatoes are loaded, the water spray should be turned on and the abrasive plates set in motion.
3. The interior should be cleaned daily and the abrasive plate removed to ensure that small particles are not lodged below.

4. The peel trap should be emptied as frequently as required.
5. The waste outlet should be kept free from obstructions.

Refrigerators: In order to maintain a refrigerator at peak efficiency, the following points should be observed:

1. Defrost weekly. The control should be turned to defrost, the racks should be emptied and racks and interior surfaces washed, rinsed and dried. If the refrigerator is not defrosted regularly, excess frost accumulates on the cooling system, acts as an insulator and causes the refrigerator motor to work longer than is necessary, thus shortening the life of the components.
2. The door or doors should be kept closed as much as possible. If too much warm air is allowed to enter the refrigerator plant, it overworks and excess frost can accumulate on the cooling system.
3. Food should be stored sensibly and in such a way that the cold air can circulate all around. Excessive packing of food into a refrigerator should be avoided.
4. A qualified engineer should be called in at the first sign of any defect in the machinery operating a refrigerator.

Food Mixer: This is an important labour-saving, electrically operated piece of equipment used for many purposes, for example, mixing pastry, cakes, mashing potatoes, beating egg whites, mayonnaise, cream, mincing or chopping meat and vegetables.

1. It should be lubricated frequently in accordance with manufacturer's instructions.
2. The motor should not be overloaded. Overloading can be caused by an obstruction to the rotary components. For example, if dried bread is being passed through the mincer attachment without sufficient care, the rotary cog can become so clogged with bread that it is unable to move. If the motor is allowed to run, damage can be caused to the machine.

3. All components as well as the main machine should be thoroughly washed and dried. Care should be taken to see that no rust occurs on any part. The mincer attachment knife and plates will rust if not given sufficient care.

Mechanical equipments are slicer, grinder, bowl cutter, food processor, etc.

Food Slicers and Choppers: Food slicers available are both manually and electrically operated. They are labour-saving devices which can be dangerous if not operated with care. Because of this, the working instructions should be placed in a prominent position near the machine.

1. Care should be taken that no material likely to damage the blades is included in the food to be sliced and chopped. If a piece of bone comes into contact with the cutting blade, severe damage will result.
2. Each section that comes into contact with food, should be cleaned carefully and dried after use.
3. The blade or blades should be sharpened regularly.
4. Moving parts should be lubricated, but oil must not be allowed to come into contact with any food.
5. Extra care must be taken when blades are exposed.

Chopper (Hand or Electric): The manual type should be washed and dried after- use. Care should be taken with the interior blades; they should be cleaned with a folded cloth.

The electric chopper should be thoroughly cleaned and dried after use, particular attention being paid to those parts which come into contact with food. Care should be taken that no obstruction prevents the motor from operating at its normal speed. Moving parts should be lubricated regularly in accordance with the manufacturer's instructions.

Masher (Hand or Electric): The hand type should be washed immediately after use, then rinsed and dried.

The electric masher should have the removable sections

and the main machine washed and dried after use, extra care being taken over those parts which come into contact with food.

Lubrication as for all electrical appliances.

Tea Making Appliances: There are two main types of boilers. Bulk boilers from which boiling water can only be drawn when all the contents are boiling and automatic boilers which provide a continuous flow of boiling water.

Bulk Boilers: Used mainly where large quantities of boiling water is required for a given time. They should be kept scrupulously clean, covered with the correct lid to prevent anything from falling in and when not in use they should be left filled with clean water.

Automatic Boilers: These boilers have automatic water supplies and can give freshly boiled water at intervals. It is important to see that the water supply is efficiently maintained, otherwise there is a danger of the boiler burning dry and being damaged.

Pressure Boilers: This is the type that operates many still sets, consisting of steam-heating milk boilers and pressure boilers providing boiling water.

Care should be taken with the pilot light to see that it is working efficiently. As with all gas-fired equipment, it is essential that regular inspection and maintenance is carried out by gas company fitters.

Power Economy

Energy can be saved in the kitchen by

1. Turning gas flames once the food is cooked.
2. Using the lowest possible gas flame at all times, to correctly achieve the task.
3. Placing lids on pots when boiling water or cooking.
4. Turning off gas flames, ovens, and hot plates when work is finished.

5. Cooking food no longer than is absolutely necessary.
6. Preheating ovens only when needed, don't let them be on standby all day.
7. Checking the effectiveness of oven's doors and the seals to prevent loss of oven heat.
8. Keeping oven door closed to prevent loss of temperature.
9. Keeping refrigerators and deep freeze compartment doors closed.

Prerequisites for Safety

Safety Factors:

(a) Excessive haste should be prevented.

(b) Distraction should be avoided.

(c) Failure to apply safety rules.

Excessive haste may cause people to take chances which inevitably lead to mishaps. Disconnect the power.

Accidents may be caused by not concentrating on the job in hand.

It is the responsibility of everyone to observe the safety rules.

Ensure operational procedure for all machines and equipments which should be in good working order.

Prevention of Cuts and Scratches:

Knives

1. Use the right knife for right job.
2. Use a sharp knife.
3. Handles should be free from grease.
4. When carrying these the points must be downwards.
5. Keep these flat on the board or table.
6. When cleaning, keep the sharp edge away from the hand.

Cutting Blades on Machines: Guard should always be in place when the machine is in use.

Never place hand under the blade. Before the guards are removed for cleaning, stop the revolving blades.

Cut from Meat or Fish Bones: Much care should be taken and cuts from bone should not be neglected. Always seek medical help.

Prevention of Burns and Scalds

1. Roll down the sleeves of jackets and overall. Use aprons of a suitable length so as to give adequate protection.
2. Trays containing hot liquid should be handled carefully. (Put one hand on the side and other on the end of the tray.)
3. A good thick dry cloth is very useful.
4. Put some white thing, e.g. flour on handle and lid to indicate that it is hot.
5. Handles of pans, etc. should not protrude over the edge of the stove.
6. Carry the large full pan correctly.
7. Take extra care with certain foods, e.g. when boiling sugar.
8. Frying, specially deep frying, needs special attention.

   The pan should not be more than 2/3 full. Drain the wet food first before putting in the fat. If the fat in pan bubbles over on to a gas stove, then turn off the gas taps.

   Fire blankets and fire extinguishers should be provided in the kitchen.
9. Keep the steamer in proper working condition. Clean the drain hole. Never open the door immediately the steam is turned off. It is better to wait for about half a minute before doing so.
10. When pouring boiling liquid, keep your face away as far as possible.

U

V

W

Y

❑❑❑

Index

Machinery

1. See that the machine is in correct running order before use.
2. The machine should be operated only by the person who is engaged for it.
3. Use a particular machine for a particular type of job.
4. When using mixing machine, never put your hand in the bowl until the balance hook or whisk have stopped revolving.
5. Remove the plug first before cleaning.
6. There should be proper earth connection of the electrical machines.
7. In case of any doubt, always inform the maintenance department.

Handling of Gas Cylinders: The risk of explosion from gas is considerable. To avoid this, it is necessary to ensure that gas is properly lit. Never allow the gas to collect in the oven, or on range. See that there is no leakage on the gas pipe or any connection with gas supply.

Floors: Accidents are caused by grease and water being spilled on floors and not being cleaned up. See that the floor is always clean. Never keep anything on the kitchen floor.

Never keep liquid items in containers on the shelves or racks beyond the eye-level specially when hot.

Keep the kitchen well lit, well ventilated and dry.

Safety in Restaurant: It is essential that people working in the restaurant are capable of using the tools and equipments in a manner which will neither harm themselves nor those with whom they are in contact. They should be aware of the causes of accidents and be able to deal with any which occurs.

Accident may come in various ways.

1. Excessive haste.
2. Distraction/carelessness.
3. Failure to apply safety rules or lack of knowledge.

Everybody must take greatest care to observe the rules of safety to avoid great pain and loss of time.

The waiter when engaged with carving, filleting, must use the correct sharp knife and must see that the handle is free from grease. When carrying a knife, he must keep the point downward and try to avoid movement with the knife. When not in use, keep the knife on its side on the table or sideboard preferably in the drawer. After every use, wipe the knife with the edge away from the handle.

During carving and filleting meat, fish, take care of bones which may be dangerous. Try to use a glove when dealing with such items.

Use a good dry thick folded cloth for carrying. Soup, roast gravy, etc. should be handled carefully.

When preparing flambe dishes, do not bring your face near to the pan when flaming the dish.

When dealing with gas stove, always check the gas first. Also check if there is any leakage in the pipe. Never allow the gas to collect on the stove by switching on the stove in advance. This will cause explosion.

Keep the floor as clean as possible. Never keep anything on the floor which may cause accident.

Fire Safety

Fire Precautions: Fires in hotels and catering establishments are common and all too often can result in injury to the employee and in serious cases either injury or loss of life to employees and customers.

Fire Prevention: A basic knowledge regarding fire should assist in preventing fires and handling them if they do occur. Three components are necessary for a fire to start, if one of the three is not present, or is removed, then the fire does not happen or it is extinguished. The three parts are:

1. Fuel—something to burn.

2. Air—oxygen to sustain combustion (to keep the fire going).
3. Heat—gas, electricity, etc.

Methods of Extinguishing a Fire

To extinguish a fire, the three principal methods are:

1. Starving—removing the fuel.
2. Smothering—removing the air (oxygen).
3. Cooling—removing the heat.

Therefore one of the sides of the triangle is removed.

The Fire Triangle: The fuel is that which burns, heat is that which sets the fuel alight and oxygen is needed for fire to burn. Eliminate one of these and the fire is put out. Oxygen is present in the air, so if air is excluded from the fuel and the heat then the fire goes out. For example, should the clothes of someone working in the kitchen catch fire then the action to be taken is to quickly wrap a fire blanket round the person and roll them on the floor. In doing so, the flames have been cut off from the source of air. (The oxygen has been taken away from the triangle.) In the event of fire, windows and doors are to be closed so as to restrict the amount of air getting to the fire. Foam extinguishers work on the principle that the foam forms a 'blanket' thus excluding air from coming into contact with the fuel.

Should fat or oil in a pan ignite, then the pan should be quickly covered with a lid or other item or fire blanket so as to exclude air. It is also essential to turn off the source of heat, gas or electricity, etc. so that the heat is taken away from the triangle.

Water extinguishes fire by dousing the flames thus taking the heat out of the triangle provided the fuel is material such as wood, paper, etc. If fat or oil is alight, water must not be used as it causes the ignited fat to spread thus increasing the heat. Water extinguishers must not be used on live electrical equipment because water is a conductor of heat.

In the event of a small fire in a store, it may be possible to remove items in the store to prevent fire from spreading.

Fire doors are installed for the purpose of restricting an area so that in the event of a fire, the fuel is limited.

Procedure in the Event of a Fire:

1. Do not panic.
2. Warn other people in the vicinity.
3. Do not jeopardise your own safety or that of others.
4. Follow the fire instructions of the establishment.
5. If it is a small fire, use appropriate fire extinguisher.
6. Do close doors and windows, turn off gas, electricity and fans.
7. Do not wait for the fire to get out of control before calling the fire brigade.

It is important that in all catering establishments, passageways are kept cléar and that doors open outwards. Fire-escape doors and windows should be clearly marked and fire-fighting equipment must be readily available and in working order. Periodic fire-drills should occur and be taken seriously since lives may be endangered if there is a fire. Fire-alarm bells must be tested at least four times a year and the staff should be instructed in the use of fire-fighting equipment. All extinguishers should be refilled immediately after use.

All fire extinguishers should have been manufactured in accordance with ISI specifications; they should be red with an additional colour code to indicate the type and with operating instructions on them.

Red	—	water
Cream	—	foam
Black	—	carbon dioxide
Blue	—	dry powder
Green	—	halons (vapourising liquid).

Various Fire Extinguishers

Foam (Cream): Foam puts out fires by forming a blanket of foam over the top of the fire. It is particularly good for putting out fat fires because the foam stays in position and so stops the fire-reigniting. Foam can also be used on fires of natural materials.

Carbon Dioxide (CO_2) (Black): Carbon dioxide gas is used on fires on inflammable liquids and has the advantage that it does not conduct electricity.

Disadvantage: CO2 gas has limited cooling properties and therefore is not the most efficient way of putting out a fat lire.

Dry Powder (Blue): Dry powder is commonly used for fat fires. It does not conduct electricity, and some all-purpose powders can be used on fires in natural materials. Powders based on bicarbonate of soda are used in most extinguishers.

Disadvantage: Dry powders usually have limited cooling properties.

Fire-hoses: Fire-hoses are used for similar fires to those classified under water fire extinguishers. It is necessary to be familiar with the instructions displayed by the fire-hose before using it.

Water Sprinkler Systems: These consist of sprinklers from the main water supply fitted in the ceiling. The system is designed to automatically spray water over the whole area when the temperature rises above a pre-set level, e.g. 75° (167°F).

Each extinguisher should be fixed on a suitable bracket, be properly maintained and should be always available for use. It should be immediately refilled after use. It is important that the staff must learn how to use them.

Research and development by the manufacturers of fire-fighting equipment inevitably leads to changes and increased efficiency in the various appliances as it is important that the best fire extinguishers are always available. Always consult the Fire Prevention Branch of the Fire Brigade.

6

Background of Catering

The Indian sub-continent is vast and has a heterogeneous population of almost 800 million. Apart from Hindi there are several other major languages as well as hundreds of different dialects that are spoken and the people of India belong to a variety of religions.

The country is divided into several states and each state has its own culture with traditions that go back thousands of years. Tradition in food too differs from state to state, just as the European food of Italy is totally different from that of neighbouring France or Germany.

India has a richly varied cuisine. In the cold climate of Kashmir, the basic food stuffs are often warming and filling. In the hot South, foods tends to be cool and much of it is steamed, a method of cooking seldom used in the north. The Pulao's are by origin a Muslim dish from the Middle East and most of the states have adopted them in their own traditional ways of cooking. In the South, rice is eaten with almost every meal in one form or another. In the North, bread such as Chapatis or parathas or nans, are popular.

Indian cooking has more combinations of flavours and seasoning than any other nation in the world ranging from hot and sour, hot and nutty, sweet and salty to the freshness and sweetness of highly aromatic curry leaves and the pungency of asafoetida. Spice shelves often contain more than 30 seasonings and it is the combination of these seasonings that differentiates eating habits from state to state. The mixture of these numerous seasonings makes Indian cuisine unique in the world of cooking.

The introduction of western cooking by the British during the days of the Raj did no favours to the country. European food doesn't adapt extremely well to the Indian taste and this is reflected in the large number of Indian restaurants that have flourished in the Western World during the past 20 years.

Indians have much to be praised for their style of cooking and the fantastic range of fabulous food stuffs from around the country. The heritage of regional cooking dates back to thousand of years and can be equalled in authenticity only by the Chinese cuisine which has a long and varied history.

The recent history of the growth of catering and the hotel industry in the country which has roots in ancient times, has been enumerated below.

Food Sciences in India

The hotel and the catering industry started late in the 19th century with the development of the major cities, improved transportation and the coming of the railways. In ancient times, travellers were varied but could always rely on a meal whilst passing through. As the centuries progressed, travellers, mostly pilgrims, would be cared for in the temples or monastries. During the latter part of the 18th century and the formation by the British of the East Indian Company, it was apparent that catering expertise on a more formal basis were required and with the advent of the Railways in the mid 19th century, small hotels and clubs became part of everyday life for those who could afford to eat out. For the ordinary man, street-side catering

was a way of life and for many hundred of years, the science of cooking delicacies like pakoras and samosas and their spicy titbits was a way of ensuring that anyone with a few paise to spare for food would never go hungry.

Prior to the formation of hotel companies the "Club" were restricted unfortunately to European or upper class Indians. However, it did give many Indians the opportunity to work in restaurants. These skills were often passed down from father to son. The vast Indian armies too, required catering on a different scale and the Officers Mess then, resembled high class restaurants with usually a very good kitchen attached.

The princely palace of many royal Indian Maharaja's and families were also run on hotel lines of the present day at least as far as food and drinks were concerned. Kitchen with Indian and European chefs were considered normal and the banquets of yesteryears far outshone anything that is provided today even in the best of five star hotels.

India has a long culture heritage of catering and hospitality. The diversity of the regional dishes and talent to produce them has long been inherited and it is because of this history that India with its numerous hotels and restaurants ranks among the world's leading hospitality and tourist venues.

Indian Regional Dishes: The cuisine of India differs from State to State as the foods of Europe differ from one country to another. However, the features that all the states have in common are the vast variety of spices and seasonings.

The taste of India's real food is in the millions of private homes scattered across the country. From the temperate northern state of Kashmir, where food is eaten to warm the body, to the tropical southern states of Tamil Nadu and Kerala where the foods are of necessity cool, the variety is incredible. It is unfortunate that most hotels and restaurants do not experiment more with Indian cookery and keep to the old tried and tested recipes, which although good, does tend to become rather boring. All Indian food is served with either rice or

bread or both. In the North, it is whole wheat breads, such as chapatis and parathas, that are commonly eaten, and in the South it is rice. The traditional Indian bread is flat, baked on cast-iron griddles. The Muslims introduced ovens where sour doughs and plain bread, such as nan and roti could be baked.

According to the ancient Indian system of ayurvedic medicine, all spice and herbs have medicinal properties. Turmeric for example is an antiseptic (often put on fish before frying). Asafetida is a digestive (hence it is often put in pulses). Garlic is good for circulatory ailments; coriander and tamarind for constipation, cloves for toning up the heart and black pepper for giving energy.

Delhi-Punjab and Haryana: Delhi, India's capital since the early part of this country, has been invaded, pillaged and conquered by Parthians, Turks, Afghans, Mughals and Britons. As the years passed much of their cuisine was absorbed into the cities life-style—others changed. Pulaos and samosas are both derivations of middle-eastern cuisine, split-peas cooked with ginger and cummin (Dal); meat cooked with fried onions, (Do-piaz); chicken and almonds braised in youghurt (Dumpukht)—all these and more have been added over the years to the Northern Indian and particularly Delhi cuisine.

These mostly middle-eastern dishes were just added to the Hindu dishes already in existence thus making the repository more extensive. Pakoras of every type, mint and coriander chutney's, aloo-ki-tikiyas, (Potato cakes) eaten with sour tamarind chutney, chapatis eaten with potato, peas and cauliflower combined with turmeric, ginger and chillies, Rogan Josh, cubes of goat meat cooked with spices-further enriched Indian cusine.

Tandoori food was brought to Delhi after the Partition in 1947 and this style of cooking not only 'caught on' in Delhi but most of the world. Indian restaurants specialise in Tandoori (especially, chicken). Many Punjabis are farmers and consequently some of Punjab's best food is country-food, based on butter milk, ghee, and freshly harvested produce. Like,

layered parathas (griddle bread) stuffed with grated radish and pomegranate seeds; buttery mustard greens, (sarson da sag) to be eaten with fresh corn bread, spicy chana bhatura, or aloo subzi with puries.

Thus these states have a wide range of both indigenous and international cuisine well-blended into indigenous the life-styles of the people in these states.

Uttar Pradesh and Rajasthan: One of the larger communities are Marwaris-originally from Rajasthan who have settled all over the North, but particularly in Benaras. Marwari kitchens may be entered only by a professional Brahmin cook called a 'Maharaj'. For religious reasons the food is served directly from cooking pots into thalis. A thali might contain for a meal-a salad of shredded, fresh ginger with radish and chillies and aubergine, kalonji, potatoes and cauliflower; a dal, youghurt karhi, resullas, kachoris, rotis and papad.

Lucknow has a strong Muslim influence. Kakori Kababs are a Lucknow speciality. This is meat pounded with poppy seeds and cloves. Breakfast in Lucknow could consist of kulchas-flat, sour-dough breads eaten with nahari-stew or Rogani Roti (rich whole wheat bread) eaten with spicy fried liver. More formal meals have a taste of the middle-east, whole marinated legs of lamb, cooked with almonds and poppy seeds or whole chickens stuffed with keema.

To the west of UP stretches the desert of Rajasthan. The ancient Rajputs hunted and ate wild boar. The fat and skin of the animal was and still is cooked with fried onions, coriander, cumin, ginger and garlic. It is also used to flavour moong dal.

In Rajasthan, meat, including poultry and fish, are mostly marinated, skewered and grilled over fire to make Soola Kababs.

To the South of UP lies the central Indian state of Madhya Pradesh, with its capital at Bhopal—where the food is traditionally Muslim in character, spicy and aromatic. Korma is served with an equally aromatic Baghare Chaval (rice) flavoured with ginger and garlic, and Baigan ki Boorani, an

Afghan dish (fried slices of aubergine topped with garlic flavoured youghurt.)

Fish from the lakes Bhopal are also very popular. White fleshed fish such as Sanwal are cut, dipped in seasoned youghurt, and then in a coriander, chilli and garlic paste and fried.

Youghurt, coriander, chillies, and ginger form a large part of the cuisine in Bhopal.

Gujarat: The modern state of Gujarat, was formed only in 1960. Gujarati food belongs to Gujarati-speaking people, whether they happen to be in Ahmedabad, the state capital, in Saurashtra, the western portion of the state, in cosmopolitan Mumbai or in one of the thousand of villages throughout the area.

For many Gujarati vegetarians, much of the food is based on grains, beetroot and vegetables grown on family farms. Millet is staple here, a grain that has more protein, vitamin E, and iron than wheat. It is ground, mixed with spinach and green chilli to make a dough, and then formed into breads—Dhebras—on a cast-iron griddle. This bread is often eaten with Chhundo, a sweet chutney made with green mangoes, cardamon, cloves and youghurt. Batata Shaak also popular are diced potatoes sauted with mustard and cumin seeds; and of course the dhoklas—savoury steamed cakes made from split-pea butter.

Much of the state's sugarcane production is used to make jaggery, which is raw, unprocessed lump sugar. Most of the jaggery is made into loaf-like lumps. Gujaratis often like to sweeten their spilt-peas and vegetables with jaggery.

One of the special breads is Batloo, a flat griddle bread of millet flour with tiny depressions which looks like the cracked earth of the surrounding country.

Ahmedabad has one distinction that sets it apart from other provincial capitals. It is the only city where the best of regional fare may be sampled in a fine restaurant. This

vegetarian restaurant, Vishala is set up like a village with banquet style dining. Hadvas are served here, savoury spilt pea cakes, encrusted with seasame seeds; rice steamed with tiny stuffed aubergines; corn cooked with drumsticks; and Khandvi— tiny rolled crepes.

One of the oldest Muslim communities in Gujarat is that of the Bohris disciples of the 11th century missionary, Abdullah, and converts from Hinduism. They are among the very few communities in the country with a passion for serving soups-hot soup, cold soup, breakfast soups. A favourite is Sarki, a cold summer soup that uses to our dal for the basic/Stock and then adds diced cucumbers, tomatoes and spring onions, before garnishing with youghurt.

Parsis are another Gujarati-speaking minority. They are Zoroastrians who fled Iran in the 8th century and landed just north of Bombay. Parsi food is a delicious blend of western influences, a Gujarati love of sweet and sour mixes, and the Persian genius for combining meat with dried fruits. One of their delicious chutneys is "Meva Nu Achar" made with carrots and dried fruit. Sali Jardaloc Margi is a sweet and sour dish of chicken and dried apricots. One of their more well-known meat dishes is Dhansak, which is basically mutton cooked with dal.

Maharashtra: Starting around Maharashtra, one can almost begin to draw a line that divides the basically wheat-eating north from the rice-eating south. Maharashtrian eat both wheat and rice, but the latter is given a place of prominence. It is used, in every form imaginable from steamed rice flour dumplings (Modaks), to delicate rice flour biscuits called Anarsa. Anarsa typifies the best of the Maharashtrian food. This cuisine from India's heartland aims to satisfy and delight. It depends, for its final taste, not so much on rare ingredients as on the expert blending of sweet, salty, and sour flavours.

While the food of certain Brahmin communities around Poona maybe restrained and vegetarian, one has only to look at the extravagance and richness of the diet of the Maharashtrian head quartered in Kolhapur or the spicy exhuberance of that

of the coastal fisherfolk to realise that even within the state of Maharashtra, there are several different types of cuisines.

The Maharashtrian, belonging to a caste of working-class agriculturists, have traditionally eaten and enjoyed whatever came their way. On hunts, they marinate quail with youghurt, cloves and black pepper, wrap them in wet clay and bake them. Maharashtrian eat their food hot and unsweetened.

Ratnagiri on the Konkan coast has the distinction of producing one of the most sought after mangos. It is called Alphonso. It is plump with a clean yellow skin touched with red. It is juicy, and sweet. One of the most delicious products of this mango is Amba Barfi.

Mumbai, the state capital and western India's principal seaport is a vibrant blend of Gujarati Hindu merchants, Parsi industrialists, Maharashtrian intellectuals— Baghdadi Jews, movie stars, Goan lawyers and the original Koli fisherfolk. But the one equaliser in Mumbai to which everyone succumbs is Bhel-Poori. The place to eat bhel-poori is the Chowpatty beach. Some puffed rice is put in a bowl. Wheat flour crisps are crumbled over, generous sprinklings of chopped onion and potato added, two chutneys are spooned in a sharp, hot green one made from chillies and coriander, and a thick sweet and sour one made from tamarind and dates. Some sev which is vermicilli made from chickpea flour finishes off the snack.

Kashmir: The beauty of Kashmir is awesome. One of the most beautiful sights is acre upon acre of mauve crocuses— flowers that provide Kashmir and all of India with an ancient crop, Saffron. Local people use the saffron, sparingly of course as it is very expensive, to flavour rice and other dishes.

For their cuisine, the Hindus and later on the Muslim inhabitants of the land drew on what they could find, raise or grow, acknowledging the culinary whims of their changing overlords—Afghans, Mughals, Sikhs and Britishers. The upland forests of Kashmir are carpeted with juicy morels, mushrooms which grow everywhere and can be harvested at no cost. There

is plenty of wild asparagus. There are enough pastures to raise sheep and goats for milk. Growing food is more of a problem. Terraces are used for the staple-rice. Wheat, too, is grown to make superb breads such as the Kuecha, and the sesame encrusted Tsacharu.

With the abundance of mountain lakes, Kashmiris harvest specially the lotus roots which taste like artichoke hearts, called Nedr. They are cooked with fish, Wastahaak (tender spring greens) and with lamb. Best of all they are fried in mustard oil to make crunchy chips. The lake waters also produce edible lotus seeds, water chestnuts, and fish.

The day begins in Kashmir with tea called Kahva. It is China-like tea mixed with cardomoms and crushed almonds. The bread of Kashmir are unique. Related more to the bread of Afghanistan, Central Asia and the Middle-east rather than the traditional Hindu bread. They are generally baked in clay or brick ovens and bought every day. They range from shortbread like buns, to discs of flaky pastry, to spongy, chewy creations something like a doughnut. They can be sweet or salty-some are encrusted with sesame seeds, others with poppy seeds.

Vegetables, mostly season greens, are staple and invariably cooked in mustard oil and water with the addition of chillies. Hindus use asfoetida also as flavouring while the Muslims use garlic. Lamb is, the main meat, cooked with youghurt and the flavouring depends on the religion. Hindus use asfoetida and ginger or fennel; Muslims use garlic and onions.

The Kashmiri onion is different than other Indian onions. It is a cross between a shallot and a spring onion (called Pasan). While summer and autumn in Kashmir produce an abundance of vegetables there is a constant awareness of the dearth during winter. All vegetables that can be cut and dried, like turnips, green leaves, tomatoes, chillies are preserved. The fishermen do likewise. Their dried fish, Hogaad, has to last until spring.

Kashmiris make exquisite chutneys using walnuts, or sour cherries, or yellow pumpkin, or white radishes. Other specialities include Jeddar (mushrooms cooked with tomatoes, ginger and fennel), Shikar (lightly vinegared duck cooked with garlic and chillies) and Gurd Muj (flesh cooked with white radish). One of Kashmir's most delicious meat dish is rib chops braised in broth and then fried till crisp.

Bengal: No meal in Bengal is considered complete without fish-however, out of all the varieties available the freshwater variety is much prized. Bengalis not only love fish, they are exceedingly particular about it when they shop. They examine the eyes of the fish for clearness and the inside of gills for redness; they know that Lata fish is good for invalids, and must be bought live. Hilsa is the supreme Bengali fish and is bought whenever the availability of fish coincides with the availability of extra cash. Hilsa may be cooked in many ways. It can be boned (not an easy task as Hilsa is full of tiny bones) and smoked, and it can be prepared as elish bhapa. Cut pieces of hilsa are mixed with a paste of grounded mustard seeds, mustard oil, red and green chillies turmeric and salt. This is then cooked either wrapped in banana leaves or steamed.

Mustard oil and mustard seeds are very much part of Bengali cooking. Many Bengal dishes required mustard seeds oil, for tempering.

Calcutta, Bengal's capital city of about 13 million people lived the longest under the continuous and very direct foreign rule, and has therefore a mixed culinary history of Portugese, Mughal and British influences. But the Bengali however westernised never gave up his language neither did he give up his passion for fish, rice and sweets.

Date jaggery is one of the sweet delicacies made by tapping the date palm tree. However, India's premier sweet is Rasgulas, which is curdled milk, moisture squeezed out, rolled into balls and dropped into boiling syrup where they puff up. Sondesh is made with the same dough mixed with sugar and cooked over a low flame until all moisture evaporates. In

spring, jaggery is used instead of the traditional sugar giving a Nutan Gurer sondesh which has a caramel flavour and colour.

A lunchtime favourite is Sukto. It is served at the meal and is a mixture of diced and fried vegetables like bitter gourd, white radish, potato, seem (a hard-skinned flat bean) plus leaves and stems popular with Bengalis. Split-peas are added and then the whole mustard seeds and cumin. Panchphoran is a spice mixture used only in Bengal, and consists of whole cumin seeds, whole fennel seeds, whole fenugreek seeds and whole radhuni (like parsley seeds and mustard seeds). This Panchphoran is added to the sukto.

Another famous Bengali sweet dish is Mishti doi, a thick, sweetened, youghurt set, in earthen cups.

Hyderabad: The cuisine of Hyderabad, which was once a home to a wealthy Muslim court in the heart of Hindu South India, combined the very best of Muslim foods—Kababs, pulaos, kormas, and youghurt dishes with the aromatic, pungent and creamy flavourings of the south. Mustard seeds, cinnamon, curry leaves, red chillies, peanut, tamarind and coconut milk are used widely. A sauce of roasted and ground sesame seeds is mixed with green chillies and tamarind paste to make Til ki Chutney. Dried beans and lamb, which are often stewed together in Persian cookery, are perked up here with tamarind, cumin, red chillies and curry leaves to become the delicious Dalcha of Hyderabad.

Some of Hyderabad's best dishes include; Lukmi which is ravioli-like squares of pastry filled in the centre with spiced meat and deep fried, Chippa ka Gosht, which is chunks of meat marinated in a paste of onion, green chillies, coconut, garlic, garam masala and youghurt and then cooked very slowly in a clay pot. There is Ambaday ki Bhaji-a dish of sorrel-like greens, Baghare Baigan-small whole aubergines, slit, browned and cooked in a nutty sauce of sesame and peanuts, and aromatic tomato puree with tamarind, curry leaves and garlic; Murgh Methi which is chicken cooked with freshly

sprouted fenugreek greens; Kacchi Biryani, a rice and meat mixture and of course Kacche Dahi ke Koftay, in which minced meat spiced with coriander, mint, garlic and garam masala is ground to a fine paste, formed into balls and fried, put into youghurt beaten with garlic, mustard seeds, curry leaves and chillies.

Tamil Nadu and Karnataka: These two states, even though separated by different languages, share a common heritage of food, especially food that are basically vegetarian. There are of course many areas specially in Karnataka which are renowned for its Bisi Bela Huli Anna— a hot, tart porridge of rice, split-peas and vegetables; garlic flavoured idlis as well as Ahay Pad Wada, which is large wadas, Soppu Palaya, which consists of a dish of steamed spinach enriched with milk and flavoured with mustard and cumin.

Rice is the king of grains in both states. No meal is possible without rice making an appearance in one form or another. Most people are vegetarians therefore grains and split peas are the core of the diet. These basic grains and split peas are transformed into Sevel (Fresh noodles) and sweet jellies such as the Hulbai of Mysore.

South Indians are coffee drinks. In Madras they make strong filtered coffee, but mix it with lots of hot milk. Morning coffee is served with dosas, which are thin round, golden crisp and smooth on one side and made out of the same batter. Both dosas and idlis are the breads of the south, nourishing and digestible. They may be eaten with butter and honey, or with chutneys, or stuffed with a spicy blend of potatoes and onions.

Sambar is to our dal boiled with turmeric, vegetables, tamarind, curry leaves and sambar powder and tempered with mustard seeds and whole red chillies. Rasam is made of to our dal boiled with turmeric. When soft, the thin broth on top is ladled off and is tempered with mustard seeds, curry leaves and lot of crushed peppercorns and seasoning. Sambar powder, a ground spice mixture having chillies, channa dal, coriander seeds and fenugreek are mixed in along with curry

leaves. For the soupy rasam, the dal is mixed with tamarind water, ground black pepper, red chillies and cumin with coriander leaves sprinkled over and tempered in hot oil.

Kerala: Kerala is the spice area of India. There is peppercorns, nutmeg and with it mace, cinnamon, cloves and tamarind, cardamom, ginger and turmeric, along with the special curry leaves and the thing that makes most Kerala food unique, the coconut.

Stews made of meat or chicken simmered in coconut water are eaten with uppamas. Uppamas or Appams are rice flour pancakes designed to have thick soft, spongy centres and thin, golden crisp edges. Toddy, from the coconut plant is used as a fermenting agent for the batter and kept overnight. Uppamas, ferment and thin-down in the morning.

Fish is often prepared with chilli pastes and left for several days to mature in terracotta pots.

Vegetables are cooked simply. Green papaya and jackfruit are prepared with a variety of seasonings and coconut milk.

A speciality desert is Prathaman; moong dal boiled with coconut and flavoured with raw palm sugar, cardamom and ginger.

Hotels in India

The diverse cuisines of Indian states finds expression in the hotel industry across the country. India has several renowned chain of hotels and their history represents an important milestone in the growth and development of the hotel and catering industry.

Taj Group of Hotels: The Taj Group of Hotels is one of the largest chain in India with several hotels abroad also. The parent hotel, the Taj Mahal Hotel, Mumbai is rated among the 10 best in the world.

The founder of the House of Tatas, Mr. Jamshedji Ratanji Tata in 1897, formed the Indian Hotel Company, and built the

exquisitely beautiful Taj Mahal Hotel in Bombay. The door opened in 1903 and has been an important landmark opposite the Gateway of India ever since.

In 1971, the 220-room Taj Mahal hotel in Bombay was converted into a 325 room hotel, and a multi-storey structure was built adjoining the original property.

In 1972 the Lake Palace at Udaipur and Rambagh Palace in Jaipur were linked to the Taj and a chain was born.

In 1974, a new company was floated, which created the Taj Coramandel in Madras. In the same year the chain broadened with acquisition of the Fort Aguade Beach Resort in Goa.

In 1976, the Fisherman's Cave was built. It is a 30 minutes drive from Madras on the Bay of Bengal with a private beach.

In the same year the Taj Group opened the Taj Flight Kitchen in Bombay catering to both domestic and international flights.

In 1978 the glittering Taj Mahal Hotel was opened in Delhi and this was followed two years later in Varanasi by the Taj Ganges.

The 1980's saw the rapid expansion of the group—the gigantic Taj Palace in New Delhi; the Shiv Nivas Palace at Udaipur; another Flight kitchen in New Delhi airport; the Taj Residency in Bangalore, at Ooty; the Raj Mahal Palace at Jaipur and the Taj View Hotel in Agra, Sri Lanka. And the chain continues to expand.

Oberoi Hotels: Rai Bahadur Mohan Singh Oberoi, Chairman and founder of Oberoi Hotel was born in August 1900 in Bhaun, a small village in what is now Pakistan.

Soon after his marriage in 1922 he arrived penniless in Simla, and found a job as a Front Office Clerk in the Cecil Hotel at a salary of Rs. 40 per month. Some years later he moved to Clerker hotel (also in Simla) and gained experience in all aspects of the hotel operations. In 1934 he bought Clerker hotel by mortgaging all his assets and selling his wife's jewellery.

In the early 1930's a cholera epidemic struck Calcutta and the Grand Hotel, the largest in the city was forced to close. In 1938 Mr. Oberoi acquired the Grand and converted it into a profitable business venture. The Oberoi Grand remains to this day Calcutta's leading hotel.

In 1943 Mr. Oberoi took over the Associate Hotel of India (AHI) with eight hotels in Northern India including the Cecil in Simla, Maidens and Imperial in Delhi, and 4 hotels now in Pakistan. The Oberoi chain grew rapidly with hotels in Gopalpur, Darjeeling, Chandigarh and Srinagar.

The Oberoi International in New Delhi, which opened in 1965 was the first modern luxury hotel in the capital. The 35-storey Oberoi Towers opened in Bombay in 1978, the tallest building in India.

Mr. Oberoi's dedication to the industry is evident from the Oberoi Centre of Learning and Development based in Delhi.

Oberoi is now part of the International Chain of Hotels with has properties not only in India but also in Australia, Indonesia, Egypt, Iraq, Sri Lanka, Singapore, Zanzibar and Nepal.

Mr. Oberoi was named as 'Man of the World' for 1983 at the annual convention of the International Hotel Association in New York and selected as the outstanding hotelier of the year by the Hotels and Restaurants International.

India Tourism Development Corporation: ITDC (India Tourism Development Corporation) that runs the largest hotel chain of India has largely contributed in the development of tourism industry in India. It was founded in 1965.

ITDC after its formation put in exceptional effort to promote the hotel industry as well as tourism by developing new hotels and tourism spots.

ITDC launched it's own project for a five star hotel in Bangalore that is Bangalore Ashok and soon launched another hotel in Delhi in the name of Akbar hotel and thus started the

unending story of grouping upcoming new hotels all over India and abroad which has now made ITDC one of the largest international hotel chains.

The Ashok group of hotels is but one of the services provided by the state—owned by the Indian Tourism Development Corporation. It has the largest network of hotels in the country. ITDC has promoted the concept of joint venture hotels, both with the state government and private entrepreneurs for setting up, running and marketing hotels under the Ashok brand name.

ITDC runs duty-free shops at all international airports, both in the arrival and departure halls. Its range of products on sale and the competitive prices make the ITDC duty-free shops a real bargain in Asia.

The corporation also runs a few restaurants, with specialised cuisine. It has restaurants, in Agra, en route Agra, in distant Moscow, the Soviet capital city and in Wellington, New Zealand.

ITDC has also entered into agreements for running hotels overseas, and the first of its hotels is under way in Cuba. Other hotel and restaurants project have also been taken over at other places all over the world.

Ashok Group of Hotels: The Ashok Group of Hotels has been India's gracious and regal host to leading national and international visitors. The group is known for its unique mix of traditional Indian facilities that are beyond compare. The group has 39 hotels in 32 major destinations. The range of hotels varies from five star deluxe to first class to budget, depending on tourist needs.

The hotels have now been classified into three categories—Elite, Classic and Comfort. The six Elite hotels are Ashok group's five star deluxe properties that cater to the upmarket clientele. The Classic group comprises of sixteen units that meet the expectations of the business travellers and mid-market guests. The Comfort category provides home comforts at three locations across the country.

The hub of the Ashok group's activities is in New Delhi, India's ancient capital. It has been apparently called the 'Conventional Capital of Asia' largely due to the presence of the Ashok group, which offers over two thousand rooms in its various hotels strategically located across the Capital.

The Ashok Hotel, the group's flagship, offers the largest conventional hall in India with a capacity of seating 2500 people and is named after the legendary philanthropist emperor Ashok. The other hotels of the group are the Samrat, Qutab, Kanishka, Janpath, Lodhi, Ranjeet and Ashok Yatri Niwas their names reflecting India's rich heritage.

Six Ashok Group Hotels are situated in west India-Jaipur, Udaipur, Aurangabad, Kolhapur and Pune. Udaipur studded with palaces, lakes and temples has the Laxmivilas Palace Hotel. Jodhpur, another beautiful city of Rajasthan, has Ratanada Ashok. Aurangabad, situated in the heart of Maharashtra, is the base for visiting Ajanta and Ellora. Shaline Palace Ashok in the princely town of Kolhapur and Executive Ashok in Pune complement the Ashok group in the west.

In central India, the Varanasi Ashok situated in the holy Hindu city, provides a base for tourists to explore the city. Located atop Shamlan Hills in Bhopal is Hotel Khajuraho Ashok built in the shadows of the famous, thousand year old Khajuraho temples. Hotel Jammu Ashok, in the north of the country, provides a convenient base for excursions into the beautiful valley of Kashmir. Manali, in Himachal hills has Manali Ashok. The Bharatpur Forest Lodge is strategically positioned in the haven of indigenous and migratory binds. Agra, the city of Taj Mahal, has Hotel Agra Ashok.

Even within a nation as varied as India, the diversity of culture and landscape of eastern India encapsulates the mystery and beauty of the Orient. There are Ashok Group Hotels at Ranchi, Calcutta, Puri and Bhubaneswar. The latter is an excursion base for the Sun Temple at Konark and Chilka bird sanctuary. Puri combines the pleasure of a seaside resort with the privilege of being one of the most sacred pilgrim centers of India.

The soundproof Ashok at Dum Dum airport in Calcutta is ideal for transit passengers. From Brahmaputra Ashok at Guwahati one can explore Kaziranga the home the of one-horned rhinoceros. In Shillong, the capital city of Meghalaya, the Ashok group is represented by Pine Wood Ashok. Japfu Ashok named after one of the peaks of Nagaland is positioned at Kohima. Imphal Ashok nestles in Manipur's wooded hills.

The Ashok group offers a complete Indian experience with an ever-increasing network, taking care of its guests with a combination of dependable comfort and traditional Indian hospitality:

Welcomgroup: Welcomgroup is the Hotel Division of ITC Ltd. ITC entered the field of hoteliering in 1975 with the opening of Chola in Madras. Two more hotels, Mughal in Agra in 1976 and Maurya in New Delhi in 1977 followed in rapid succession, firmly establishing Welcomgroup as one of the fastest growing groups in India. After 1977 Welcomgroup went into the management operation and marketing of hotels owned wholly by other entrepreneurs, or jointly with ITC.

Welcomgroup is a completely new concept in hoteliering in India. The chain's philosophy of individuality was set up by its first three hotels, each of which took as its themes a famous dynasty from India's past namely, Chola, Mughal and the Mauryan.

ITC Sheraton Management signed an exclusive agreement with Welcomgroup in 1979, in the field of Marketing and Reservations for the first three hotels of Welcomgroup. In 1987 Welcomgroup Windsor Manor and Sea Rock were Sheratonised. Welcomgroup Adyar Park has been included in this agreement renaming the hotel as WG Park Sheraton. Recently opened Welcomgroup Rajputana Palace Sheraton Hotel in Jaipur has also been brought under the Sheraton banner.

All Welcomgroup hotels are equipped to meet the most stringent international standards of efficiency service and state

of art technology. In every hotel, local Indian elements are blended beautifully with the most sophisticated comfort and facilities.

1974	The hotels divisions of ITC Ltd. established as: Welcomhotels and Individuals
1975	Opening of Welcomgroup Chola Sheraton (Madras)
1976	Opening of Welcomgroup Mughal Sheraton (Agra)
1977	Opening of Welcomgroup Maurya Sheraton Hotel and Towers (New Delhi)
1978	Opening of Welcomgroup Rama International (Aurangabed)
	Opening of Welcomgroup Umaid Bhawan Palace (Jodhpur)
	Opening of Welcomgroup Usha Kiran Palace (Gwalior)
1978-79	Rajendra Villas Palace (Mysore)
	Hotel at Balrampur
	Hotel at Gorakhpur
1978-86	WG Banjara (Hyderabad)
1978-87	Welcomgroup Mansingh (Jaipur)
1978-89	WG Druk (Bhutan)
1979	Opening of Welcomgroup Sea Rock Sheraton (Bombay)
	Opening of Welcomgroup Royal Castle (Khimsar)
	Marketing agreement with Sheraton Management Corp. the second largest hotel chain in the world
1979-81	Fernhill Palace (Ooty)
1979-88	Welcomgroup Cida de Goa (Goa)
1980	Welcomhotels and Indovilles become Welcomgroup
1980	Welcomgroup High Land (Kargil)
1980-93	Welcomgroup Maurya Patna (Patna)

1981	Opening of Welcomgroup Bay Island (Port Blair)
1981-82	Palace Hotel at (Bhopal)
1981-82	Palace Hotel at (Bharatpur)
1982	Opening of Welcomgroup Windsor Manor Sheraton (Bangalore)
	Expansion of Welcomgroup Maurya Sheraton Hotel and Towers (New Delhi)
1982-84	Welcomgroup Mumtaz (Agra)
1983	Welcomgroup Gurkha House Boats (Srinagar)
1983-90	Welcomgroup Nilambagh Palace (Bhavnagar)
1985	Opening of Welcomgroup Park Sheraton (Madras)
1986	Opening of Welcomgroup Vadodra (Baroda)
	Opening of Welcomgroup Manjarun (Mangalore)
1988	Welcomgroup Valley View (Manipal)
	Welcomgroup Graduate School of Hotel Administration (Manipal)
1989	Welcomgroup South Park (Trivandrum)
1989	Welcomgroup Nedous (Srinagar)
1992	Welcomgroup Rajputana Palace Sheraton (Jaipur)
1992-93	Span Resorts (Manali)
1993	Welcomgroup Central Park (Bangalore)

No more with Welcomgroup.

The Railway and Airlines have and still do play a major role in the growth of the hotel and catering industry.

Railway Catering: In the mid nineteenth century the railway network began in India with an operation that was to extend the length and breadth of the vast subcontinent. With travel made easier, people travelling from one part of the country to the other, required food and drink on route. At most of the larger stations catering Refreshment Rooms were opened. The trains would halt for an appropriate length of time so that the passenger could get a simple meal.

Third class passengers, unable to afford the luxury of eating in the Refreshment Room could get snacks from the numerous vendors on station platforms.

Railway Companies (as most part of each region had its own) even went to the extent of setting up hotels attached to the station so that passengers who were travelling from one region to another could spend the night before or after in relative comfort. The luxury of Sleeping Cars and Restaurants Cars were a much later development.

At turn of the century, the railway ministry decided to contract out the catering requirements to attract companies with a catering background so that the traveller could be more professionally served during long journeys.

Role of Spencers: The association of the Spencers, with Railways began as early as 1910. Spencer was a company of repute with a large network all over India. Catering was done on a contractual basis. The catering and hotels division of Spencers was responsible for these operations.

The Spencers were catering under their own name to the M and SR (Madras and Southern Railways) and under the name of "Brodons" they were catering to the G.I.P.R. (Great Indian Peninsular Railways) whose headquarters were in Bombay, Gujarat etc. Under the name of "Kellners" they created the BNR (Bengal Nagpur Railways) and also the N.W.F.R (North West Frontier Railway) and between Delhi, Amritsar and Lahore. Even at that time Spencers had as many as 180 refreshment rooms all over India. The size of the Restaurants, the staff and the menu were structured to meet most of the demands the people passing through that particular area.

Even the smallest Restaurant had:

A Manager

A Cook-cum-Waiter (known as Butler in those day)

A Cleaner

The biggest Refreshment room had a bar attached since

prohibition was non-existent then. Still larger facilities even dispensed such items as simple medicines, aerated drinks and tobacco.

In minor trains, the guard on checking tickets would take orders from the passengers of first, second and even third class passengers and telegraph the same to the station ahead, depending on the time of the train. Food, crockery, cutlery and glassware were brought in baskets which would hold food for 2-3 persons. Food for one person was usually prepared while for 2 or more the food was served in ceramic dishes. Obviously, the size of the basket was quite large, measuring 2 feet by 1½ feet. The most interesting part of the operation was that of the halting of trains was adjusted to suit the meal timings. In faster trains, Express and Mail they had pantry cars and restaurants on wheels, fitted with a regular kitchen, a pantry and an eating area. The cooking fuel was coal. Menus were quite elaborate and reflected British rather than Indian taste. Breakfast would consist of porridge, butter, toast with preserves and tea or coffee. Lunch would include soup, chicken or mutton curry before concluding with dessert.

Catering was considered as an amenity promised to traveller, rather than as a business. The food served was cheap and wholesome. This was possible because Spencer were given free supplies of coal and free freight for all provisions and perishables. Aerated water were supplied to refreshment room by the representative zonal factories.

After independence Spencer lost the contract, being a foreign company.

Present Setup: The present Railway Catering service is managed both Departmentally (The Indian Railways) and through licensed contractors.

Catering facilities are available at 2995 stations and in 88 pairs of trains, the majority (2877 station) being catered by contractors.

With hundred and thousands of people travelling by train

everyday throughout the country the turnover is enormous for the department. Recently the Indian Tourism Development Corporation (ITDC) has been appointed as consultant for the improvement of meals. Disposable aluminium foil casseroles has been introduced on some of the major routes. The Railways also owns and operates Railway Hotels at number of railway stations.

To give some idea of the enormous quantity of food stuffs required annually by railways catering services the following list has been included.

Black Gram Dhuli	—	1,25,000 kg
Bengal Gram Dhuli	—	50,000 kg
Toour Dhall	—	1,00,000 kg
Green Gram Dhuli	—	7,500 kg
Raw Rice	—	5,00,000 kg
Boiled Rice	—	85,000 kg
Sugar	—	1,50,000 kg
Nescafe	—	25,000 kg

Airline Catering: Unlike the other transport systems like rail or ocean liners, air transport has to carry pre-prepared food on board, as there are no facilities available on any airline for cooking or presentation. Most airlines design and plan flight catering establishment/kitchen to meet their own requirements.

Civil Aviation progressed rapidly after the Second World War when a large number of surplus DC-3 and C-47 aircraft were available for disposal. From 1946 private commercial airlines began with subsequent food requirement. Initially, food was provided in boxes, prepared, as the majority of aircraft were without facilities of in-flight catering services. There were no flight kitchen, nor were there any public/ private flight kitchens in existence to meet their requirements.

This system was to continue until the industry was

nationalised and was divided into two separate corporations one for domestic routes and the other for international.

Indian Airlines and Air India: Modernisation and expansion of the aircrafts increased the carrying capacity of the passengers in both airlines. It was necessary to increase and expand the catering services to passengers with a service compatible with cost of the airline ticket.

Several flight kitchens were established in Delhi, Bombay, Madras, Calcutta to cope with the need of international carriers. Air India floated a subsidiary company known as HCI (Hotel Corporation of India) with a view to operate flight kitchens known as the Chefair and also to accommodate travellers in five star category hotels at major airports (The Centaur group of Hotels). Indian airlines also launched a subsidiary company known as Allied Service in order to establish flight kitchens at airports where there are no HCI flight catering establishments.

In Madras, plans are currently foot to introduce a flight kitchen and hotel in the new domestic airport complex. This will be part of the Allied Services chain. At present, meals are taken from the Taj and Welcomgroup, so meals are of a consistently high quality.

Most airline meals are deep-frozen, often cooked in a blast freezer with units which freeze hot within seconds to prevent loss of flavour as well as bacteria from breeding. The food items, in individual casseroles are stored in freezer units until required. They are transported in portable freezes or in dry ice to the aircraft. They are then heated as required with the aid of microwave ovens and served to the passengers.

Catering in Operation

Classification of Catering Operations

Luxury Hot Cuisine (French Service)	Speciality Restaurants (For eg. Steak House)

Haute Cuisine
(Russian Services)

↓

Counter Or Bar Service

↓

Fast Food Joints

↓

Take Away, Drive In
(*e.g.* → KFC, &
Hamburger/Hot Dog stand)

Definitions

1. Hospitality. *means providing facilities and making the customer feel at home. Ambience, the quality of staff service, warmth all contribute to comfort factor.*
2. Catering: *Meeting the basic needs of food, drink and accommodation.*
3. Commercial Sector: *This includes those establishments which provide food, drink, accommodation and refreshment in exchange of money. They are run as businesses, operating to provide these products, e.g., hotel, restaurant, take away, pubs etc.*
4. Catering Service Sector: *This sector exists to provide hospitality and catering services in a manner similar to the commercial sector, the difference being that these services are secondary or indirect.*
5. Secondary or Indirect Services: *are those which are not the main priorities of organisation but are provided to support main activities e.g. Hospitals, Prisons, Public Institutions Employees.*
6. Subsidised: *Part of the operational costs are met from another service.*
7. Breakeven Analysis. *Enough sales must be generated to meet all the costs associated with sales.*
8. Organisation: *A company which owns various outlets.*
9. Outlet: *The unit of management or establishment with each sector usually on one side e.g. Hotel.*

10. Facilities: *are operations provided by the hotels which earn revenue for organisation.*
11. Operations: *Activities or group of activities which may earn revenue for the organisation or which may provide support to services which do not directly earn revenue.*
12. Turnover: *The income earned from the sale of products and services, sometimes also called revenue.*
13. Tender: *An offer to complete certain tasks or provide goods and services within a certain time and for a fixed amount of money.*
14. Captive Customers: *Those customers that have no choice of outlet.*

Types of Catering Establishments

1. Hotel is a place where the customer is provided boarding and lodging.
2. Motel is a hotel situated on a highway along with services of garage and station.
3. Boatel: A houseboat hotel.
4. Floatel: A luxury hotel on a luxury liner.
5. Lotel: Hotel with helipad facilities.
6. Rotel: Hotel on wheels.
7. Metel: Totally mechanised hotel.
8. Kiosk: Stall on roadside for snacks, open on all sides.
9. Soda Fountains: Outlets selling aerated drinks.
10. Milk Bar: A restaurant dispensing milk and milk products.
11. Restaurants: A place where one can eat, pay for the same and go.

Types of Food Services Facilities

1. Fast Food Services
2. Mid-priced Restaurants
3. Upscale Restaurants

4. Take-out food Service facilities
5. Hotel/Motel Food Services
6. Food Services at Leisure facilities
7. Food Services in Health care facilities
8. Food Services in Business and Industries
9. Food Services in Educational Institutions
10. Community Food Services

Fast Food Services: The term fast refers to the services rather than to the food. Fast food services does not mean that quality of food is inferior or loaded with calories of food but is for faster services. These food services rely enormously on automation. On the other hand since there is the need for faster service, menu choices are limited. Hamburger chains may also offer—Chicken, ham, fish salads. In order to eliminate misconceptions about food served by the fast food chains regarding the calorie or nutritional contents the food services provide nutritional information or label the foods they offer.

1. Fast food services are not suitable for special occasions.
2. Fast food services are planned to facilitate a fast turnover rate.
3. Seating ranges from less comfortable to moderately comfortable.
4. Profit is divided from high turnover rather than high profit margin.
5. Cleanliness is more important than decor.
6. Fast food services have a crowded atmosphere.
7. Fast food services control a specific segment of market.
8. The fast food services provide quality food for limited money.
9. Method of sales promotion activity are adopted by issuing discount coupons, games, prizes etc.
10. In order to attract kids as captive customers, electronic animated cartoons and special shows are shown.

11. In the fast food services, drive-in facilities are considered an added convenience.
12. There is a specific time standard set for service.

Mid Priced Restaurants

1. These food services offer mid-priced items.
2. Services is not as fast as fast food outlets.
3. All age groups are target consumers.
4. Generally the consumers demands low calorie food.
5. It generally has a varied menu, modest decor and pleasing atmosphere in order to generate repeat business.
6. Special entertainment is sometimes offered.
7. Generally this group covers Restaurants, Coffee Shops, Diner, Pancake Houses and Theme restaurants.
8. The menu has food items which is suitable for all age groups.

Upscale Restaurants

1. Consumers look for a dining experience that is unique and memorable.
2. Consumer's expectation are high quality food with high prices.
3. There should be elegant atmosphere and decor.
4. Dishes on the menu reflect class and style of the restaurant.
5. Gueridon service may be provided.
6. There is a need for a sophisticated blend of all desirable attributes.

Take-Out Food Services Facilities

1. The menu generally includes all food items that are feasible for take-away.
2. Consumes prefer drive—in facilities.

3. Free home delivery is generally provided.
4. Special packing material require an add-on expense.

Hotel/Motel Food Services

1. Consumers are mainly houseguests and outsiders.
2. Courteous services required.
3. It is a matter of convenience required by consumers who prefer services at the same location.

Food Services at Leisure Facilities

1. The group of restaurants which fall in this category are located at amusement parks, zoo, sports arenas, resorts areas, vacation land and other tourist attractions.
2. Consumers are captured patrons and guests who do not like to waste time hunting for eating places.
3. These food facilities don't allow food to be brought in from outside.
4. Consumes are of all age and sex.
5. Cleanliness and quick service are important for repeat business.
6. Variety and calorie are not important for consumers.
7. Open air Restaurants are preferred.
8. Consumer's have limited expectations except in resort areas.
9. Aerated drinks are preferred.

Food Services in Health Care Facilities:

1. Health-care facilities include hospitals, nursing homes and convalescent centres.
2. Menus are planned extremely carefully.
3. There is an emphasis on specialised diets.
4. Goals and objectives are different.
5. Dieticians are specially hired.

Food Services in Business and Industry

In this category fall cafeterias and restaurants in plants and industries.

1. Worker's motivation is the prime reason for providing catering.
2. There are nominal charges at these places.
3. There is a demand for high quality and large quantity of food.
4. Since there is, limited time a quick meal service is desired.
5. Variety in menus is offered.
6. These facilities are not open to the general public.

Food Services at Educational Institutions

1. These services include schools, colleges, universities.
2. Dietary goals are given high importance.
3. Variety in menu is preferred.
4. Teenagers prefer fast food items.
5. Resident Hall facilities are a must.
6. Generally there is a nominal charge for these food services among students.

Community Food Services

1. Several government-sponsored programmes that provide food or food services for special segments of the population *e.g.* WIC (Women, Infant, Children).
2. It is a meal on wheels program.
3. Consumers belong to a different socio-economic group.
4. Convenience food, snacks and readily prepared food are served by food services.

Some specialised services that are sometimes offered in various food facilities are as follows:

Carrousel: It was introduced for the first time in Britain

in 1982. It comprises of large rotating arrangements of shell, approximately 2 meters in diameter with food and drinks arranged. The customer remains stationary as the carrousel revolves once every minute to enable them to select items. Only ½ of the Carrousel are in service so that items are removed from other half. The shell can be replenished in the running area behind the unit. The cutlery, napkin and beverages are usually separately available.

Advantages

1. It is a method of self service. It has customers through out between 8 to 10 persons per minute which is higher than the traditional line counters.
2. It requires less space than many other self service layouts particularly service of operations.

Disadvantages

1. The operator must be well-trained to reach quickly and efficiently to maintain a steady flow rate of customer.
2. Customers may be delayed, waiting for a complete revolution of the unit. In this respect the Carrousel service is mostly suited to operations where there is repetition on a frequent basis.

Salad Bars: It is a self-service set in which each guest is given the opportunity to prepare his or her own salad from an attractive array of fresh vegetables and salad fruits that has been cleaned and sliced or quartered into bowls. Salad plates are available at one end of the salad bar. The guest prepares their own tossed green salad and helps themselves to a variety of prepared salad accompaniments such as dips, bread sticks and salad dressings. More elaborate salad bars provide cheese and different kinds of bread. Expensive salad bars include dishes such as pickled Harrin, Sardins, Chilly sliced varieties ham, Tuna salad. A clean panel through which the guest can see the selection is mounted between guest and salad. This panel is called Snuz Gard and helps keep the food germ-free.

The duties of the waiter is to firstly take the guest's meal and wine order and then to inform them when and how to begin with the salad bar. They also assist the kitchen staff, in upkeep of the salad selection by informing them when the food at salad bar needs replenishing. Servers remove salad dishes from the table as they accumulate and keep beverage fresh for this course.

Oyster's Bar is a buffet featuring oysters on half shell, along with various sea food sauces and mustard that compliment the oyster. Sometimes boiled shrimps and other appetisers like sea food are included selection. The chef may want to display sea food and exotic Ice ships, net and diving relics to add to the appeal.

Dessert Table: A tantalising display of tarts, cakes, creams, eclaires, fresh fruits, soft cheese are displayed in a buffet fashion. Dessert plate and fork are at hand on the dessert table. Usually desserts are presented in position and guests can help them selves. The displayed desserts are brought to the guest table on a tray or a dessert cart at the end of meal and served portions from the selection at hand.

Small Gas Board buffet features large selection of food with many Scandanavian selections such as cheese and Sherry. In many places the set price includes self service buffets of any kind food. Usually guests may also come back to the Small Gas Board table and refill their plates as often as they desire.

Bibliography

Achaya, K.T. : *Indian Dairy Products,* Asia Publishing House, Bangalore, 1974.

Adam, H. : *Food Science,* John Wiley and Sons, New York, 1982.

Austin, A. and Ram, A. : *Studies on Chapati-Making Quality of Wheat,* Indian Council of Agricultural Research, New Delhi, 1971.

Bachman, W. : *Food Chemistry,* Maclaren and Sons Ltd., London, 1955.

Balsdon, J.P. : *Life and Leisure in Ancient Rome,* Bodley Head, London, 1966.

Beersteeker, E. : *The Biochemistry of Vitamins,* Reinhold, New York, 1950.

Bharathi, V.V. : *Anxiety Behaviour in Children,* Discovery Publishing House, New Delhi, 1992.

Board, R.G. : *A Modern Introduction to Food Microbiology,* Blackwell Scientific Publications, London, 1983.

Burton, T.L. : *Recreation Research and Planning,* Allen and Unwin, London, 1970.

Cambell, A.M., Penfield, M.P. and R.M. Griswold : *Experimental Study of Foods,* Houghton, London, 1990.

Cameron, Spencer : *Food Science,* Pargamon Press, New York, 1984.

Challoway, D.H. : *Nutrition and Physical Fitness,* W.B. Saunders Co., Philadelphia, 1979.

Cherry, G.E. : *Town Planning and its Social Context*, MacGraw-Hill, London, 1970.

Cherry, J.P. : *Protein Functionality in Foods*, American Chemical Society, Washington D.C., 1981.

Chichester, C.O. : *The Chemistry of Plant Pigments*, Academic Press, New York, 1972.

Dale, E. : *Management Theory and Practice*, McGraw-Hill, New York, 1973.

Dharamraj, E. : *Food and Hotel Legislations and Policies*, New Age Publishers, New Delhi, 1998.

Dumazedier, J. : *Towards a Society of Leisure*, Free Press, New York, 1967.

Dyke, S .F. : *The Chemistry of Vitamins*, Interscience Publishers, New York, 1965.

Edmunds, F. : *Environmental Administration*, McGraw-Hill, New York, 1973.

Eley, G. : *Wild Fruits and Nuts*, E.P. Publishing Ltd., Yorkshire, 1976.

Feiffer, M. : *Giving Places*, MacMillan, London, 1985.

Furia, T.E. : *Hand Book of Food Additives*, C.R.C. Press, Ohio, 1977.

Geiss, I. : *Proteins : Structure, Function and Evolution*, Menlo Park, California, 1983.

Gell, D. L. : *International Hotel Industry Policy*, Van Nostrand Reinhold, New York, 1990.

Gyorgy, P. and Pearson, W.N. : *The Vitamin*, Academic Press, New York, 1967.

Haschemeyer, A.H. : *Proteins : A Guide to Study of Physical and Chemical Methods*, John Wiley & Sons, New York, 1973.

Hudson, F. : *Development in Food Proteins*, Applied Science Publisher, London, 1982.

Inglett, G. : *The Quality of Foods and Beverages : Chemistry and Technology*, Academic Press, New York, 1981.

Jackson, R. : *The Geography of Recreation and Leisure*, Hutchinson, London, 1972.

Jasscler, D.K. and M.A. Joslyn : *Fruit and Vegetables Juice,* The AVI Publishing Co., Connecticut, 1971.

Johnson, A.H. : *Encyclopedia of Food Science,* The AVI Publishing, Connecticut, 1978.

Kacharoo, P. : *Pulse Crops of India,* ICAR, New Delhi, 1970.

Kadans, J.M. : *Encyclopedia of Fruits, Vegetables, Nuts and Seeds,* Parker Publishing Co., New York, 1973.

Kalyani, T. : *Home and School Environment,* Discovery Publishing House, New Delhi, 2003.

Kinton, R. : *Practical Cookery,* Edward Arnold, London, 1981.

Krause, M.V. and L.K. Mahann : *Food Nutrition and Diet Therapy,* W.B. Saunders Co., Philadelphia, 1979.

Kumari, P. : *Role of Anganwadi Workers in Integrated Child Development Services Programme,* Discovery Publishing House, New Delhi, 1994.

Lal, H. : *Total Quality Management : A Practical Approach,* New Age Publishers, New Delhi, 2000.

Law, C. : *Urban Tourism : Attracting Visitors to Large Cities,* Mansell, London, 1993.

Less, R. and E.B. Jackson : *Sugar Confectionery and Chocolate Manufacture,* Leonard Hill Books, Burks, 1973.

Lickorish, L. J. : *Hotel Industry and International Balance of Payments,* International Institute of Scientific Travel Research, Geneva, 1953.

Lloydryall, A. and W.T. Pendzer : *Handling Transportation and Storage of Fruits and Vegetables : Fruits and Peanuts,* The AVI Publishing Co., Connecticut, 1974.

Mackinney, G. and A.C. Little : *Colour of Foods,* The AVI Publishing Company, Connecticut, 1972.

Malhotra, R. K. : *Food Service and Catering Management,* Anmol Publications, New Delhi, 1998.

Marton, I.D. and A.J. Macbod : *Food Flavours,* Elesvier Scientific Publishing Co., New York, 1982.

Masefield, G.B. : *The Oxford Book of Food Plants,* Oxford University Press, London, 1975.

Mate, S.A. : *Bakery Technology and Engineering*, The AVI Publishing Co., Connecticut, 1972.

———— : *Cereal as Foods and Feeds*, The AVI Publishing Co., Connecticut, 1959.

Mayo, J. : *The Psychology of Leisure Travel*, CBI Publishing Company, Boston, 1981.

Mcwilliams, M. : *Food Fundamentals*, John Wiley & Sons, New York, 1979.

Medlik, S. : *Hotel Management: Past, Present and Future*, Heinemann, London, 1976.

Mellor, J. D. : *Fundamentals of Freete-Drying*, Academic Press, London, 1978.

Meyer, L.H. : *Food Chemistry*, Reinhold Publishing Corporation, New York, 1961.

Michael Eskin, N.A. : *Plant Pigments, Flavours and Textures*, Academic Press, New York, 1979.

Murthy, R.B. : *Breeding Procedures in Pearl Millet*, ICAR, New Delhi, 1977.

Nath, P. : *Vegetables for the Tropical Region*, Indian Council of Agricultural Research, New Delhi, 1976.

Oberleas, D. : *Trace Elements in Health and Disease*, Academic Press, New York, 1976.

Palmer, T. : *Understanding Enzymes*, John Wiley & Sons, New York, 1981.

Pamela, V.P. : *Directory of Wines and Spirits*, North Wood Publication, London, 1974.

Paul, P.C. and Palmer, H.H. : *Food Theory and Application*, John Wiley & Sons, New York, 1972.

Peckham, G.G. : *Foundations of Food Preparation*, MacMillan Publishing Co., New York, 1979.

Pigman, W.W. and Horton, D. : *The Carbohydrates : Chemistry and Biochemistry Plants*, C.S.I.R, New Delhi, 1969.

Pigram, J. : *Outdoor Recreation and Resource Management*, Croom Helm, London, 1993.

Potter, N.M. : *Food Science*, The AVI Publishing Co., Connecticut, 1978.

Pudney, John : *The Thomas Cook Story*, Michael Joseph, London, 1953.

Puruthi, J.S. : *Spices and Condiments*, National Book Trust of · India, New Delhi, 1976.

Ranganna, S. : *Manual of Analysis of Fruit and Vegetable Products*, Tata McGraw-Hill Publishing Co. Ltd., New Delhi, 1984.

Rani, Usha D. : *Nutrition and Religion*, Discovery Publishing House, New Delhi, 2003.

Rao, D. V. : *Community and School Nutrition Education*, Discovery Publishing House, New Delhi, 1998.

Roberts, D.V. : *Enzyme Kinetics*, Cambridge University Press, 1977.

Rogers, C.A. : *Profitable Poultry Keeping in India and the East*, Treasure House of Books, Bombay, 1966.

Ross, G. F. : *The Psychology of Hotel Management*, Hospitality Press, Melbourne, 1994.

Schapira, J. : *The Book of Coffee and Tea*, St. Martin's Press, New York, 1975.

Schultz, G.E. and R.H. Schirmer : *Principles of Protein Structure*, Springer-Verlage, New York, 1979.

Shreedevi, V. : *Nutrition Education*, Discovery Publishing House, New Delhi, 1997.

Smartt, J. : *Tropical Pulses*, Longman Group Ltd., London, 1976.

Stadlmare, W.J. and. Cotterill, O.Y. : *Egg Science and Technology*, The AVI Publishing Co., Connecticut, 1973.

Stumbo, C.R. : *Thermobacteriology in Food Processing*, Academic Press,, New York, 2000.

Swern, D. : *Bailey's Industrial Oil and Fat Products*, John Wiley & Sons, New York, 1964.

Tandon, G.L. : *Preservation of Fruits and Vegetables*, Indian Council of Agricultural Research, New Delhi, 1960.

Tandon, H.P. : *Egg and its Care*, Ministry of Agriculture, Govt. of India, New Delhi, 1973.

Thangam, P.E. : *Modern Cookery for Teaching and Trade*, Orient Longmans Ltd., Bombay, 1983.

Thankamma, J. : *Food Adulteration*, S.G. Wasani, New Delhi, 1976.

Tressler, D.K. : *Food Technology the World Over*, The AVI Publishing Co., Connecticut, 1963.

Uma, K. : *Poverty, House Hold Fold Security and Nutrition in Rural Areas*, Discovery Publishing House, New Delhi, 2000.

Underwood, E.J : *Trace Elements in Human and Animal Nutrition*, Academic Press, New York, 1977.

Veldhnis, M.K. : *Food Science and Technology*, The AVI Publishing Co., Connecticut, 1977.

Vidya, D.B. : *A Textbook of Nutrition*, Discovery Publishing House, New Delhi, 1996.

Weiss, T. J. : *Food Oils and their Uses*, The AVI Publishing, Connecticut, 1970.

Whitaker, J.R. : *Principles of Enzymology for Food Science*, Marcel Dekker, New York, 1977.

Williams, P.N. : *The Chemical Constitution of Natural Fats*, Chapman and Hall, London, 1964.

Winitz, M. : *Chemistry of Amino Acids*, John Wiley & Sons, New York, 1961.

Yasoda, P. : *Determinants of Nutritional Status of Pre-School Children*, Discovery Publishing House, New Delhi, 1998.

Yukic, T.S. : *Fundamentals of Recreation*, Harper Row, London, 1963.

1

Primary Techniques

Application of physical and chemical methods of processing foods from the time of harvest, slaughter or catch, during storage and till the time of consumption. The objectives of such processing are:

1. Improvement in aesthetic quality leading to better acceptability.
2. Enhancement of nutritive value.
3. Prevention of deteriorative changes.

Normal culinary practices are not included except those practised on a large scale or on a commercial basis.

Fundamental Objectives

1. Cooking partly sterilises food. Above 40°C (104°F) the growth of bacteria falls off rapidly and in general it ceases above 45°C (113°F). Non-sporing bacteria are killed at item apertures above 60°C (140°F) for varying periods of time, e.g., to make milk safe, it is pasteurised at 63°C (145°F) for 30 minutes or at 72°C (161°F) for 15 seconds. Boiling kills living cells, with the exception

of spores, in a few seconds. Spore- bearing bacteria take about 4 to 5 hours of boiling to be destroyed. To destroy them in a shorter period of time higher temperatures must be used.

2. Cooking helps to make food more digestible. Complex foods are often split into simpler substances during cooking. This helps the body to absorb and utilise the food more readily than if consumed in its raw form.
3. Cooking increases palatability.
4. Cooking makes food more attractive in appearance and, therefore, more appetizing.
5. Cooking introduces variety. Many different types of dishes can be prepared using the same ingredient.
6. Cooking helps to provide a balanced meal. Different ingredients combined together in one dish make it easier to provide a balanced meal.

Objects of Cooking

Different raw materials are used in cooking to produce a complete dish. Each ingredient in a dish has a special part to play and a knowledge of what each food does is necessary to understand cooking thoroughly. Grouping the foods according to their use, makes it possible to remember the facts better. The materials are thus classified according to the part they play in making up a dish. Given below is a classification of raw materials.

1. Foundation ingredient.
2. Fats and oils.
3. Raising agents.
4. Eggs.
5. Salt.
6. Liquid.
7. Flavourings and seasonings.

8. Sweetening agents.
9. Thickening agents.

Foundation Ingredient: Every dish has a foundation ingredient on which the other ingredients are based. The foundation ingredient can be a liquid or a solid, e.g., flour in bread, meat in roast, milk or stock in soups. When examining a recipe for proportions, all foods which serve the same purpose should be reckoned together. It is not only necessary to know the proportions of various ingredients but also the composition of the different ingredients and the action of heat on these. Heat may not affect the whole material but only certain constituents.

Food is composed of the following five constituents:

1. Carbohydrates.
2. Fats and oils.
3. Proteins.
4. Minerals.
5. Vitamins.

Action of Heat on Carbohydrates: Generally starch swells up and becomes softer. Starch is enclosed within granules. These starch granules cannot be easily digested but heating makes them swell up and then burst and release the starch. Starch itself becomes gelatinised.

Sugar is changed into invert sugar (by a process where sucrose is broken down into glucose and fructose) when heated with fruit juices or other weak acids. In the presence of invert sugar, the solubility of sucrose is increased, and it tends to recrystallise more slowly. Inversion raises the boiling point and lowers the freezing point of a sugar solution. Invert sugar is slightly more easily digested and absorbed.

When sugar is boiled it passes through regular stages till it becomes caramelised at 163°C (325°F). The following are the stages thorough which it passes:

	°C	°F
Small thread	102°C	216°F
Large thread	103°C	218°F
Pearl	106°C	222°F
Soft ball	114°C	237°F
Hard ball	120°C	247°F
Soft crack	143°C	290°F
Hard crack	156°C	312°F
Caramel	163°C to 177°C	325°F to 350°F

Action of Heat on Protein: Protein coagulates or sets when heated. The white of an egg is a good example. The coagulation process helps in keeping the nourishment within the food; it is sealed in by the formation of a coat on the outside. Hence, when boiling protein foods such as meat and fish, put them in hot water. When roasting meat, put it in an oven which is at a fairly high temperature, and then reduce the heat. Another action of heat on meat is the softening of the muscle fibres. Normal cooking methods cause the elastin in the connective tissue to shrink, and the collagen is converted into gelatin by heat in the presence of water, and as a result the muscle fibres separate and meat becomes tender.

This is done more satisfactorily at a medium temperature than at a high temperature. Meat also shrinks during cooking. The higher the temperature the greater the shrinkage. Cooking at medium temperatures renders protein foods more digestible. The myoglobin in the muscle tissues and haemoglobin in the capillaries give meat its red colour. Both these decompose on heating causing the brown colour which is generally found in cooked meat. The change takes place at a temperature of 65°C (149°F). At high temperatures the protein itself gets denatured thus making it less nutritive.

Action of Heat on Fats: Fat melts when it comes in contact with heat. At high temperatures fats decompose into fatty

acids and acrolein. Darkening of fat is caused by cooking at high temperatures and also by the presence of carbonised bread crumbs and small particles of cooked food. The temperature at which decomposition sets in, when an amount of fat is heated, is known as its smoke point. The specific temperature at which this takes place varies with different fats and is lowered by repeated heating.

With repeated use, fat deteriorates, because of excessive temperature, moisture, air and the presence of carbonised crumbs and small pieces of food, to a point where it will just bubble in the pan, too weak to perform the function of frying. Abnormal absorption caused by frying for too long at too low a temperature also contributes to the breakdown. Fat also begins to thicken and become gummy. This condition is known as polymerisation, and fat that has reached this stage is no longer fit for use.

Action of Heat on Minerals: There is no appreciable loss of minerals due to cooking, excepting when cooking liquor is thrown out. Some minerals are made readily available by cooking.

Action of Heat on Vitamins: Vitamin C is the only vitamin which becomes unstable when heated, although careful cooking helps to retain some of it. Vitamin A and D are not destroyed by ordinary methods of cooking. Vitamin B may be destroyed during cooking if cooked at high temperatures, e.g., baking of biscuits and manufacture of breakfast cereals, by the addition of baking soda, or away.

Fats and Oils: Fats are solid at normal temperatures and melt when heated. Those used in cooking are butter, margarine, lard, suet, dripping and hydrogenated fat.

Oils are liquid at normal temperatures but solidify at lower temperatures. The commonly used cooking oils are coconut oil, sesame oil, cotton-seed oil, olive oil, peanut oil, mustard oil, corn oil, and sunflower oil.

Hydrogenation of Oils: Whale oil, cottonseed oil, soya

bean oil, peanut oil, etc., are sold as shortenings of varying consistencies. The conversion of oil into fat is brought about by a process known as hydrogenation. This consists of treating oil under pressure and at a suitable temperature with hydrogen; in the presence of a catalyst, usually nickel. Under these conditions the unsaturated fatty acids present in the oil combine with the hydrogen. This chemical process brings about a physical change, the liquid oil becoming a solid fat. The unsaturated fatty acids are chiefly those of the oleic type but when hydrogenated, are converted into stearic acid which is solid. The varying consistencies are due to the process of hydrogenation being stopped at different stages. Fats are used in cooking either as shortening agents or as frying media.

Fats are used in confectionery to enrich the food and to impart to them shortening qualities. Collectively they are referred to as shortening agents. Their effect is to break down or destroy the toughness of gluten, so that instead of being hard and tough to eat, foods containing fat break off short and readily melt in the mouth. The factors to look for in shortenings are:

(i) *Creaming Value:* This affects the cake volume.

(ii) *Shortening Value:* The shortness or shortening power it gives to the result. Shortness is necessary for the baking of biscuits, cookies, wafers and pastries.

(iii) *Stability:* Its keeping quality. If the stability is not good, the keeping quality of the product will be decreased.

(iv) *Consistency:* Hardness or softness depending on the purpose, e.g. hardness is needed for puff pastry, softness for cake-making.

(v) *Water Absorption Power:* Correlated to emulsification value of shortening.

As shortening agents, fats add to the nutritional and satiety values of flour mixtures and contribute to their characteristic flavour and texture. The type of fat and the way in which fat is incorporated affect the texture, as in short crust pastry, flaky

pastry and puff pastry. As heat melts fat, cooking must be done at the correct temperatures so that the flour can absorb the fat as it melts. If the heat is insufficient, the fat runs out of the mixture and is wasted, besides leaving the food tough and dry. Fat which has been broken up into very fine particles, as in creaming, is more easily absorbed than fat left in large pieces. In general, the richer the pastry the hotter the over, and the richer the cake the cooler the over.

In general, the richer the pastry the hotter the oven, and the richer the cake the cooler the oven.

Fat as a frying medium functions in three ways. It serves to transfer heat to the articles to be fried, it adds to the nutritive value of the food and it contributes to the flavour of the food. Fat used as a frying medium should have a high smoke point, low congealing point, low moisture content, high stability and acceptable flavour.

Rendering of Fat: Rendering of fat is the process of melting to extract fat from fatty tissues. A good supply of dripping can be obtained by rendering down suet (fat surrounding the kidney of a cow or sheep) and pieces of fat from meat. This can be done in two ways:

1. Cut the fat into small pieces and place in a baking tin in a slow oven until the fat has melted and there are only crisp brown pieces of tissue left. Strain through a fine cloth into a clean basin pressing the tissues to squeeze out all fat.
2. Cut the pieces of fat as for above method. Place in a pan without a lid and with very little water. Boil until the water has evaporated and then heat very gently until the fat melts and leaves only crisp brown pieces of tissue. Strain as before.

In both cases the temperature should not be too high or else the fat will decompose.

Clarification: This is a method of cleaning used fat. For

this process, put fat in a pan with enough water to cover it. Bring to boil without covering. Strain and cool. Lift off the cake of fat floating on top, turn it upside down and scrape off any foreign particles sticking to the bottom. The fat is then heated gently till it stops bubbling, to remove all water particles.

Varieties of Fats and Oils

Fat / Oil	**Butter**
Source	Cream
Properties	Salted or unsalted. Difficult to handle when chilled, unique flavour and enriching qualities.
Uses	Not suitable for deep frying. For shallow frying, add 1 table spoon oil to raise smoking point.
Fat/Oil	**Margarine**
Source	Groundnuts; palm, coconut and fish oils, milk whey.
Properties	Not easy to spread when chilled. Distinctive flavour not to everyone's taste.
Uses	As above but not used for its flavour. Economical for baking.
Fat/Oil	**Soft-blend margarine**
Source	As above
Properties	Will spread when chilled.
Uses	Excellent for all-in-one cake and pastry making.
Fat /Oil	**Low-fat spread**
Source	Blended vegetable oils.
Properties	Half the calories of butter, less fattening than margarine.
Uses	Can be used for most baking, but not for pastry.
Fat/Oil	**Cholesterol-free spread**
Source	Vegetable oils.
Properties	Spreads when chilled.
Uses	As above.
Fat / Oil	**Aerated white vegetable fat**
Source	Blended vegetable oils.
Properties	More flavour than lard. Easy-to-mix, creamy texture.

Uses	Can be used for most purposes, but rather expensive for deep frying. Excellent for fork-mix pastry.
Fat / Oil	**Lard**
Source	Pork
Properties	White solid fat, very economical.
Uses	Good for short-crust pastry when combined with margarine. Does not cream with sugar. Good for shallow frying.
Fat / Oil	**Suet**
Source	Lamb or beef.
Properties	Hard white fat. Available shredded or solid.
Uses	Best used for suet pastry, pudding and stuffing.
Fat / Oil	**Spreading blend**
Source	Buttermilk, butter oil, vegetable oil.
Properties	A mixture of butter and margarine. Low fat content. Just over half the calories of butter. Slight butter flavour. Spreads when chilled.
Uses	Suitable for all cooking, except frying. Baking results acceptable. Do not use for pastry.
Fat/ Oil	**Corn oil**
Source	Corn.
Uses	Suitable for all purposes, especially deep frying.
Fat / Oil	**Sunflower oil**
Source	Sunflower seeds.
Uses	As above.
Fat / Oil	**Sesame and almond oils**
Source	Sesame seeds, bitter almonds All purposes.
Properties	Delicate flavour.
Uses	All purposes.
Fat / Oil	**Olive oil**
Source	Olives.
Properties	Distinctive flavour.
Uses	Suitable for most purposes, but not for deep frying.
Fat / Oil	**Soya bean oil**
Source	Soya beans.

Properties	Distinctive flavour.
Uses	Suitable for all purposes, but not ideal for dressings. Low shelf life.
Fat / Oil	**Coconut oil**
Source	Coconuts.
Properties	Strong aroma and flavour.
Uses	Used mainly in Kerala.
Fat/Oil	**Soft-blend margarine oil**
Source	Groundnuts.
Properties	Distinctive flavour.
Uses	Suitable for all purposes.
Fat / Oil	**Blended oil**
Source	Vegetable and animal oils.
Uses	Can be used for the same purposes as corn oil but needs straining after use if it is to be refused for deep frying.
Fat / Oil	**Solid vegetable oil**
Source	Palms.
Properties	Suitable for all purposes, except dressings because it solidifies when cold.

Raising Agents: The function of a raising agent is to puff up the food so that it spreads and rises, thus making it light instead of close and heavy. The tiny air spaces caused by the raising agents are retained during the process of cooking. The leavening of the flour mixture is accomplished by the expansion of incorporated air and by the internal production and expansion of water vapour and carbon dioxide. When the product is heated, the air expands and part of the water vapourises. The formation of carbon dioxide requires the presence of suitable micro-organisms or chemical agents. During the first part of heating, gas production is accelerated and the gas formed expands as the temperature rises.

Gluten, a substance found in flour, develops when moisture is added, and helps to keep the raising agent within until the food sets, thus helping to make the food light. Gluten is a sticky, elastic substance which stretches as the air or gas expands

and prevents these from escaping. The air or gas trapped in the mixture expands further when heated and makes the cooked food light. Too much raising agent raises the food too rapidly, breaks the surface and allows the gas produced to escape, and as a result the product sinks and becomes heavy.

Air as a Raising Agent: Air is incorporated by sifting flour, by creaming shortening, by beating eggs or by beating the mixture itself.

Water Vapour as a Raising Agent: Water vapour is formed in quantities sufficient to raise the mixture when liquid and flour are in equal volumes.

Chemicals as Raising Agents: Chemicals such as sodium bicarbonate, cream of tartar, baking powder, ammonium carbonate, etc., are added to mixtures to make food light. In each case the result is the production of carbon dioxide.

Baking powders are mixtures of sodium bicarbonate with some suitable acid such as cream of tartar, diluted with corn-flour to give a product the desired strength. The corn-flour also serves to separate the acid and the base, thereby increasing the stability of the mixture during shelf life. General proportions used are 1-2 tsp baking powder per 450 gm of foundation. The richer the mixture, the less the baking powder required. When the mixture is of an acidic nature, (as when sour milk, buttermilk, treacle, vinegar or jams made from acid fruits are used) the addition of cream of tartar or tartaric acid is not necessary, since the soda will act upon these other foods and cause the required gas to be given off. Four tsp baking powder is equal to approximately one tsp of soda in a carbon dioxide yield. In certain flour mixtures variations from neutrality are deliberately sought by including a reaction with either soda or cream of tartar, e.g., to get the dark, rich colour in chocolate cakes which many people prefer, sodium bicarbonate is used.

By the Addition of Yeast: Carbon dioxide is produced either from sugar by yeast, or from a carbonate, usually sodium bicarbonate, by action of an acid.

Yeast consists of microscopic, unicellular plants which are capable of rapid multiplication when conditions are favourable and which obtain energy by breaking down sugars to carbon dioxide and alcohol. This process is known as fermentation, and is brought about by the enzyme zymase found in yeast. Yeast also produces enzymes which are able to split disaccharide sugars.

Dried Yeast: This is a mixture of yeast and corn-flour or cornmeal pressed into cakes and dried. The yeast continues to live but in an inactive state. When furnished with warmth and moisture, it begins to develop and multiply, but this process is slow. Dried yeast has to be soaked in lukewarm water and mixed with very soft dough (sponge) for a preliminary period of development before all the other ingredients are added.

Activated Dry Yeast: This develops more rapidly than dried yeast and is the type now available. It can be used in straight dough mixing. It is less perishable than compressed yeast. The shelf life of both dried yeast and activated dried yeast is longer at refrigerated temperatures than at room temperature.

Compressed Yeast: This is a moist mixture of yeast plants and starch. The yeast remains active and will grow and multiply rapidly when added to dough. It has to be kept at refrigerated temperatures and keeps well for only a few days. If held in the freezing compartment, it retains its activity for a longer period.

Eggs: Although hens' eggs are the most popular, eggs of turkeys, guinea fowls, ducks and geese are also used. Eggs are used in various forms in cooking:

1. As a leavening by foam formation to make the dish light, as in souffles or meringues.
2. For binding, as in egg croquettes.
3. To improve flavour and colour, as in cakes.
4. To add to the nutritive value:

(a) The protein of eggs is unexcelled in quality, and is used as a standard against which other food proteins are measured.

(b) Eggs make a substantial contribution of vitamins A and D and also contain a small amount of B complex vitamins and minerals.

(c) The essential fatty acids are more finely emulsified in eggs than in any other food thus assuring easy, complete digestibility and ready utilisation.

5. To decorate and garnish dishes.
6. To enrich a mixture as in the case of rich cakes.
7. As a thickening, as in custards and cooked salad dressings.
8. For emulsification, as in mayonnaise sauce.
9. For coating, as in fried foods.
10. As the first semisolid food taken by infants.
11. As a nourishing and easily digestible food for invalids.
12. As a quick-cooking and nourishing dish for breakfast or main meals.

The action of heat on proteins accounts for the principal changes that occur when eggs are cooked. Both yolks and whites of eggs coagulate when heated. The temperature at which coagulation begins, the rate of coagulation and the firmness of the gel will depend on:

(i) Intensity of heat,

(ii) Length of heating period,

(iii) The presence of added material such as water, milk, sugar and salt.

Acid and salt speed up the process of coagulation, e.g., to prevent disintegration of eggs when poaching, vinegar and salt are added.

Sugar increases the temperature at which coagulation takes

place. When eggs are diluted by adding milk or water, the coagulation temperature is raised. If water is added instead of milk, coagulation results in a flocculent rather than a gelled structure.

An undiluted whole egg begins to coagulate at 57°-66°C (135°-150°F) and to gel at 70°-74°C (158°-165°F). The white of the egg alone starts coagulating at 58°-62°C (136°-144°F) and begins to gel at 62°-70°C (144°-158°F). In all cases, the higher the cooking temperature, the more rapid the coagulation and the firmer the gel. Changes associated with the cooking of eggs can however, be produced at temperatures much below boiling, by a longer heating period. The egg that is cooked for a longer period at a lower temperature is superior in nutritive value, as well as in flavour, to an egg cooked at a higher temperature for a shorter period.

The blue ring round the yolk of an egg, sometimes found in a hard-boiled egg, is due to the formation of ferrous sulphide. This is brought about by the union of iron released from the yolk of the egg and hydrogen sulphide from the white, which in turn is released from sulphur found in the white. The amount of such discolouration depends on time and temperature of cooking, and the age of the egg. Cooking for 30 minutes in water at 85°C (185°F) followed by rapid cooling, gives very little discolouration. Eggs which are six weeks or more old discolour rapidly. In any case, boiled eggs used for stuffing, garnishing etc., should be cooled quickly after boiling, after cracking or without cracking.

Characteristics of Fresh Eggs: The shell of an egg may be either brown or white depending upon the breed of hen producing it. But, irrespective of colour, the shell of a fresh egg always has a delicate velvety appearance, called "bloom", that is due to the protective mucous coating which the shell has when the egg is laid. The inner air cell of a fresh egg is small. The contents when broken out from the shell show an upstanding, well-rounded yolk covered with a clinging layer of thick egg white, which, in turn is surrounded by a relatively

smaller amount of thin egg white. The yolk may be either golden or light yellow, depending upon the amount of carotenoid pigments contained in it, and the white is practically colourless or possibly slightly opalescent. There is no adore other than that characteristic of eggs, and when cooked, the flavour will be excellent.

Deterioration of Eggs: As soon as the egg is laid, changes begin to take place which, if not checked by suitable storage conditions, result in undesirable deterioration. An account of these changes and their effects follows:

As Shown in the Shell: With deterioration, the mucous protective covering of the shell soon disappears, leaving the shell shiny and more porous, so that moisture and carbon dioxide can pass more easily through it from the egg contents. This causes a loss in weight of the egg, a contraction of egg contents, and a resulting enlargement of the air space between the two membranous linings of the shell.

As Shown in the Egg White: Accompanying the loss of carbon dioxide through the shell, the reaction of the egg contents, especially of the egg white, changes from nearly neutral in the fresh white, to alkaline, becoming progressively more alkaline as the egg ages. It should be said, however, that factors other than the loss of carbon dioxide may be responsible for the increase in this alkalinity.

As egg deterioration progresses, the thick white gradually loses its firm, jelly-like consistency and decreases in volume, while the volume of thin egg white increases. These changes continue until the entire white portion becomes watery and drains away from the yolk when the egg is broken out of the shell. The cause of this change in thick egg white is not, at the present time, entirely clear, although various reasons have been advanced to account for it.

As Shown in the Egg Yolk: Upon aging, the yolk of the egg also becomes more liquid and enlarges as water passes into it from the white through the enveloping vitellin membrane.

As a result, the increasing volume of the yolk stretches the surrounding membrane and weakens it so that, in the broken-out egg, the yolk is no longer upstanding and well-rounded, but instead becomes wide and flat. Indeed, with continued aging, the membrane which encloses the yolk may become so weakened that it breaks when the egg is opened, and, as a consequence, the white and yolk cannot be separated from each other.

As Shown by Substances with Off-odours: Finally, in aging eggs certain chemical reactions occur among the various egg components which yield compounds that impart characteristic and familiar off-odours and flavours; and if these reactions continue long enough, such eggs become inedible.

Among the compounds that are made from these reactions is the foul-smelling, gaseous substance, hydrogen sulphide. When eggs are heated for some time, as when hard-boiling them in the shell, this gas, formed chiefly in egg white, passes through the vitellin membrane and reacts with iron compounds in outer yolk areas to form ferrous sulphide which can be seen as a black, film-like deposit enveloping the yellow, hard-boiled yolk. Although this film can be encountered to a certain extent in a fresh egg, its formation can be largely avoided—if the aging of the egg has not been carried too far—by taking care not to prolong the cooking period unnecessarily, and by being careful to use the hard-boiled egg immediately.

Boiling Eggs: Put enough water into a pan to cover the egg and bring it to the boil. Lower the egg into the water with a metal spoon.

Reduce the heat and cook gently for 3-4 minutes for a lightly cooked egg. For a moderately soft egg, allow 4-5 minutes. For a hard-boiled egg, allow 10 minutes.

Duck's eggs must be boiled for not less than 14 minutes.

Put the hard-boiled egg into cold water and leave to cool to prevent the formation of dark rings round the yolk.

Salt: Salt, although used in small quantities, is an essential ingredient for both sweet and savoury dishes. Salt helps to bring out the flavour of other ingredients. If too much is added, food becomes inedible, and too little makes food insipid. Gauging the right amount to be added is learned by experience. Salt adds to the nutritive value of food by providing an essential mineral, sodium chloride.

Salt has physical effects on the gluten of flour. In reasonable quantities it strengthens gluten and increases its resistance to the softening effects of fermentation. Too much salt on the other hand will remove the power of gluten to hold gas. Salt also acts as a preservative; it speeds up coagulation of eggs and lowers the freezing point of food.

Liquid: Liquid is necessary to bind dry ingredients together, to dilute food, to act as a cooking medium and to thin down a gravy or sauce. Milk, water, stock and fruit juices are the most commonly used liquids. Eggs may be used for binding.

When too little liquid is added, the food does not get cooked, or as in the case of cakes, it retards the action of raising agents, resulting in a hard, heavy cake. Gravies or soup containing insufficient liquid are 'stodgy'. Too much liquid results in a watery product, or, as in the case of cakes and puddings, makes them soggy.

Flavourings and Seasonings: The taste and acceptability of food depends to a very great extent on the correct amounts of flavourings and seasonings. These are variable ingredients and the types and amounts necessary for different dishes must be carefully studied. To add just enough and no more should be the aim of every cook.

Those which combine taste and smell such as essences, cardamom, nutmeg, basil, thyme, etc., are called flavourings. Those which only enhance taste are seasonings, e.g., red chillies, mustard, pepper, etc.

Sweetening Agents: These are not as necessary as salt.

Their lack will not affect the texture of the food or the lightness of a dish. A bad cook generally adds too much sweetening to cover other faults.

The type of sweetenings used are:

(i) Sugar,
(ii) Treacle,
(iii) Jaggery or molasses,
(iv) Syrups such as golden syrup, etc.
(v) Jams,
(vi) Honey,
(vii) Fruit juices.

If too much sweetening is used, food becomes cloying and if too little is used, sweet dishes become tasteless.

Theory

Sugar	
Type	Granulated sugar (white)
Description	Refined sugar medium-size.
Uses	All-purpose but for many tasks another type may be better or quicker.
Type	Castor sugar
Description	Refined white sugar with the smallest crystals.
Uses	Because it dissolves quickly it is especially good for cake-making, whisked sponges and sauces. Best for caramels and syrups.
Type	Icing sugar
Description	Made from sugar—relatively large crystals ground to a fine powder. Contains a small percentage of calcium phosphate to prevent clogging.
Uses	Finest and smoothest of all sugars. Best for icing for cake decorations, for meringues and for sprinkling and coating. Sometimes takes on a slight greyish tinge if sieved through a metal sieve or if a metal spoon is used.

Type	Preserving sugar
Description	Refined white sugar—relatively large crystals, makes it ideal for preserving.
Uses	Dissolves slowly and does not clog bottom of pan or form as much scum as granulated sugar when used for preserving.
	Can also be used (coloured) to decorate cakes and breads.

Thickening Agents: Thin foods such as milk, soups, gravies, curries, etc., are often improved by the addition of a thickening agent. The thickening used also increases the nutritive value.

Starchy foods and eggs are commonly used as thickening but sugar will thicken a syrup, and gelatin will set a thin liquid if chilled. Ground onion, coconut and poppy-seeds are used for thickening curries in Indian cookery.

Process of Measurement

This is a very important aspect of cookery, particularly for bulk of food production. To obtain a standard product with a standard yield, as is essential for commercial foods, weighing and measuring must be done accurately.

A set of scales, measuring jugs, and standard measuring cups and spoons must be provided in every kitchen. The scales should denote both grams and ounces and the measuring jug both liters and pints. Weighing is more accurate than measuring but for expediency, measuring can be used. When measuring, it is better to use level measurements, as this is less subject to error. Spoons and cups vary greatly in size. Therefore test them carefully and always use the same utensils for measuring.

Component Making

Solid foods which are to be mixed have to be reduced into sizes which will allow them to combine readily. A certain amount of preparation is thus necessary.

Washing: This is done to remove superficial dirt. Wash vegetables, meat and fish in cold water before any preparation is done. If they are soaked for a long period or washed after cutting there is greater loss of water soluble minerals and vitamins.

Peeling and Scraping: Spoilt, soiled and inedible portions are removed. Skins of vegetables like potatoes, carrots, etc., or of fruits, are removed by either peeling or scraping. When peeling, remove as little of the fleshy part as possible.

Paring: Remove surface layers by cutting as in paring an apple. A circular motion is used.

Cutting: Reducing to small parts by means of a knife or scissors. When the reduction is done by a chopping knife or a food chopper it is known as chopping. Cutting into even-sized cubes is called dicing. Cutting into very fine pieces is mincing. Shredding is cutting into fine long pieces with a knife or shredder. Slicing is also cutting into thin long pieces, but these are not as fine as in shredding.

Grating: Reducing to fine particles by rubbing over a rough, sharp surface.

Grinding: Reducing to small fragments by crushing in a mill, a grinding stone or an osteriser.

Mashing: This is a method of breaking up soft foods such as cooked potatoes or vegetables.

Sieving: Passing through a mesh to remove impurities, to break down to even portions, or to enclose air.

Milling: used for cereals, to remove husks, etc.

Steeping: Extracting colouring and flavouring by allowing ingredients to stand in water generally at a temperature just below boiling point.

Centrifuging: Promotion of separation through the application of a whirling force, e.g., separating cream from milk.

Emulsification: Blending one liquid with another in which it is insoluble, e.g., oil and egg yolk in mayonnaise.

Evaporation: Removal of excess moisture by heating.

Homogenisation: Subdivision of large drops into smaller ones by forcing them through a small opening under great pressure, e.g., fat in cream.

Mixing the Food

Beating: This method can be used with thin mixtures or liquids. This should be done carefully with the aim of enclosing air. This term is used synonymously with whipping, e.g., beating of eggs.

Blending: Mixing two or more ingredients thoroughly.

Cutting: In Usually the incorporation of fat in flour and other sifted ingredients with a knife. This method produces a relatively coarse division of the fat and does not result in blending, e.g., cutting fat into pastry mixture.

Creaming: Softening of fat by friction with a wooden spoon, generally followed by the gradual incorporation of sugar as in cake making.

Folding: Mixing by a careful lifting and dropping motion. The edge of the spoon is used and the mixture is lifted, turned completely and gently replaced. All the movements in this method, though deft, should be so gentle that the different ingredients are almost coaxed together.

Kneading: Manipulating by alternating pressure with folding and stretching. The food is pressed with the knuckles. The dough is brought from the outside of the bowl or basin to the centre and at the same time the bowl is moved so that a different section is kneaded each time. This ensures a thorough distribution of ingredients, e.g., bread dough, chapati dough, etc.

Pressing: This is done to shape foods like cutlets and

sometimes as a method of subdivision to separate liquids from solids by weights or mechanical pressure, as for paneer.

Rubbing: In Rubbing fat into flour using the tips of the fingers and thumb and lifting the hand out of the basin as in the case of short crust pastry. Rub until the mixture looks like bread crumbs.

Rolling: In Rolling butter or fat in a soft dough, e.g., puff pastry.

Stirring: Mixing foods with a suitable tool such as a spoon by a circular motion in contact with the pan. Generally this is a gentle movement but is changed to suit different dishes, as when used to prevent sticking or burning in hallways and toffees. If used too vigorously, it tends to drive out air or other gas previously enclosed as raising agent.

Texture: Texture is the term used to describe the characteristics of a finished food product. The order in which the ingredients are added, the way of mixing and the method of cooking affect the resulting product.

A good cook should not only be able to distinguish between one texture and another but also be able to produce what he or she wants. Only by observation, experience and perseverance will a person be able to know what the correct texture of a particular product should be. A brief description of some commonly found textures and their correct occurrence is given below, but it must also be borne in mind that the difference between one texture and another is very fine.

Firm and Close: The air bubbles made by the raising agents are many but small, and the mixture is not in the least spongy. The fat included prevents the mixture from being too hard, e.g., in biscuits or plain short pastry.

Short and Crumbly: This is similar to firm and close, but more fat is added, e.g., in shortbread or nankhatais.

Spongy: A soft and elastic texture showing inclusion of air, e.g., Swiss rolls, sponge cakes and idlis.

Light and Even: Holes are plentiful and of a fair size. The food is firm but not hard or tough. It is neither as short as pastry nor as spongy as sponge cakes, e.g., Madeira cake, Queen cake.

Flaky: This is caused by the method of adding fat. Thin crisp layers are formed, separated by air pockets. The flakes themselves should not be tough, e.g. flaky and puff pastry, chiroti, etc.

There are some textures which are incorrect in any dish, e.g.

Coarse: Holes are large and uneven, and the food is sunken in the centre. This is brought about by the addition of too much raising agent or too little liquid.

Tough: Coarse mixtures are also tough. Toughness is caused by too much liquid or through incorrect mixing. This will also result if too little fat is added.

Hard: A bad fault brought about by the addition of too much liquid or too much pressure while mixing. Hard mixtures are usually heavy since the air enclosed is removed.

Style of Cooking

While the correct preparation of ingredients and correct mixing are necessary, greater skill is needed in the actual cooking of the food. The different methods of cooking are:

1. Roasting
2. Baking
3. Frying
4. Boiling
5. Poaching
6. Steaming
7. Stewing
8. Braising
9. Broiling
10. Grilling.

Roasting

Spit Roasting: The food to be cooked is brought into direct

contact with the flame of a clear, bright fire. The food is basted with fat and also turned regularly to ensure even cooking and browning. This method, now known as spit roasting, is not often used as only good quality meats are suitable for it. Roast meats, however, have a very good flavour, and are still served in large hotels and in special restaurants and hotels, e.g., as barbecued meat.

Oven Roosting: This has now taken the place of spit roasting, because of its convenience, although only first class meat, poultry and vegetables are thus cooked. It is done in a closed oven with the aid of fat. The joint is raised out of the fat by means of bones or a trivet to prevent the meat from frying and becoming hard. Frequent basting, however, is essential. The food is put into a fairly hot oven for 5 to 10 minutes and the temperature is lowered to allow the joint to be cooked through. Cooking in a moderate oven for a longer time produces a better cooked joint than cooking at high temperature for a shorter period.

There is also less shrinkage. Therefore, more portions can be obtained. Aluminium foil is now used in oven roasting. The joint is larded or browned in fat. A matignon [finely minced 2 medium carrots, 2 onions, 2 sticks of celery (heart), 1 tbsp ham, 1 sprig thyme,½ a bay leaf crushed and stewed in butter] may or may not be used to cover the joint which is then wrapped tightly in aluminium foil and cooked in the oven till done. This method is an improvement on oven roasting as the meat retains its moisture and flavour.

Pot Roasting: This method is used to cook small joints and birds if no oven is available, but a thick heavy pan is essential. Enough fat is melted to cover the bottom of the pan. When the fat is hot the joint is browned. It is then lifted out and 2 or 3 skewers are put into the pan, on which the joint is placed. This is to prevent the joint from sticking to the pan. The joint should just touch the fat. The pan is then covered tightly with a well-fitting lid and cooked over a very slow fire. The joint could

be basted if lean and turned occasionally to ensure even cooking. Prepared root vegetables and potatoes can also be cooked around the meat.

Baking: The food to be cooked is surrounded by hot air in a closed oven. The action of dry heat is modified by the steam which arises from the food whilst cooking. Bread, cakes, pastry, puddings, vegetables and potatoes may be cooked by this method.

Frying: This is a method of cooking whereby the food to be cooked is brought into contact with hot fat. Food cooked in this way is said to be indigestible, but if the method is correctly and carefully carried out, the food is quite suitable for normal people. The advantages of frying are:

(i) Fried food is very appetizing,

(ii) It is a quick method of cooking,

(iii) The keeping quality of fried food is strengthened. It is however, an expensive method of cooking meat as only the best parts are suitable (chops, liver, etc.). For other foods and for reheating of food, frying is a good means of providing variety.

There are two types of frying:

(i) shallow fat frying;

(ii) deep fat frying.

Shallow Fat Frying: Only a little fat is used and the food is turned over in order that both sides may be browned. Generally this method is applied to pre-cooked food unless the food takes very little time to cook (omelette, liver, etc.). Some foods contain sufficient fat in themselves (e.g., bacon, sausages, etc.) and additional fat is not necessary although some cooks prefer to use a little fat. Fat absorption is greater when food is shallow fried then deep fried.

Deep Fat Frying: The food is completely immersed in hot fat and, therefore, a large quantity of fat is required. The quantity of fat requires some time to heat. Special care must

be taken to prevent overheating of fat, as this spoils both the food and the fat. The fat decomposes at high temperatures.

If the fat is not hot enough the food breaks up and absorbs extra fat thus making the product unfit for consumption. Use a frying basket wherever possible to remove fried foods easily.

Almost all foods require coating before frying, since not only are the juices and flavour of the food to be kept in, but the fat must be kept out. Materials used are:

(i) Eggs and bread crumbs;
(ii) Flour and milk (not suitable for deep fat frying);
(iii) Batter (flour, milk and eggs);
(iv) Thinly rolled pastry;
(v) Oatmeal or vermicelli;
(vi) Lamb or pig's caul;
(vii) Began (gram flour or chick pea flour) batter.

Both sweets and savouries may be cooked by this method without the flavour affecting the fat or the food, provided correct principles are applied. Food cooked by deep fat frying has a much better appearance than that cooked by shallow fat frying as it is evenly browned.

General Rules for Frying

(i) Have the fire clear and hot.
(ii) Make the food into suitable sizes and shapes and see that it is free from cracks.
(iii) Apply coating evenly. The bread crumbs used should not be coarse. Remove any excess and firmly press on loose crumbs.
(iv) When frying chicken dry off all moisture before dipping in flour.
(v) Use fat with a high smoke point. The fat must be quite still and at the right temperature required, before the food is put in.

(vi) Do not put in too many pieces of food at the same time as this will lower the temperature.

(vii) See that the temperature is not increased for increased output.

(viii) Follow a time and temperature chart.

(ix) Fry to a golden brown on both sides turning over the food if necessary.

(x) Drain well on absorbent paper and serve attractively,

(xi) Cover fats left in the fryer between frying periods and maintain at a temperature not higher than 200°F.

(xii) Once fat has been used for frying, strain and store in cans in a refrigerator.

(xiii) Replenish the original volume with fresh fat after each frying period.

(xiv) Fat begins to thicken, getting to be what is known as gummy or syrupy, with continuous use. This condition is known as polymerisation and such fats are more viscous than fresh fat. Since there are no known food additives for inhibiting this condition, fat that has reached this stage is no longer fit for use.

(xv) Fats that are used for frying should have high stability. Normally fats themselves do not get rancid during the frying process but rancidity in the finished product affects the keeping quality and acceptability of the product. For commercial purposes fat of over 100 hours stability is desired for continuous trouble-free frying.

(xvi) Darkening of the fat is caused by cooking at too high a temperature which carbonises loose bread crumbs and small particles of fried food. Such fat should be strained and replenished with fresh fat before being used again.

Boiling: Food is cooked by surrounding it with boiling or simmering liquid (stock or water). Only just sufficient liquid should be used to cover the food to be cooked. To retain nourishment and flavour in food, plunge into boiling liquid.

Allow to reboil and then to simmer. The liquid thus obtained is known as Tot Liquor and contains some nourishment and flavour. This should not be wasted. It can be used as a substitute for stock when stock is not available or cooled and used to prepare stock. If a well-flavoured stock is required, vegetables and meat should be put into cold water and allowed to cook gently. Salted or pickled meats should always be started in cold water.

Generally speaking, vegetables grown above the ground are cooked in boiling salted water and vegetables grown below the ground are started in cold salted water with the exception of new potatoes and new carrots. Dry vegetables are started in cold water. Salt is added only after the vegetables are tender.

Fish should be put into hot liquid and allowed to just simmer.

Advantages of Steaming

(i) Food cooked by this method is easily digested.

(ii) All nourishment and flavour are kept in the food.

(iii) Food cannot be easily overcooked.

(iv) Pudding basins and other containers need not be filled to the top, thus allowing room for food to rise and so making food light.

General Rules for Steaming

(i) Prepare the vessel and have the water boiling fast.

(ii) If a basin is to be used as a container inside the steamer, see that it is greased and covered with a well-fitting lid or with greaseproof paper. This is to prevent condensed moisture from falling on the food.

(iii) Keep a kettle of boiling water handy so that the water in the steamer can be replaced as it boils away. Frequent refilling however, should be avoided as this causes a draught on food.

(iv) Never allow the water in the steamer to go off the boil,

except when preparing dishes with eggs as the main ingredients, such as custard. For this, water must be kept at the simmering point otherwise the custard will curdle.

(v) Dish food quickly and serve hot. An appropriate sauce must accompany steamed food to make it appetizing, as steamed food is generally bland.

Stewing: This is a very gentle method of cooking in a closed pan using only a small quantity of liquid. The food should never be more than half covered with the liquid and the food above this level is thus really cooked by steam.

Advantages of Stewing

(i) Cheap cuts of meat, old fowls and tough or under-ripe fruits may be prepared by this method as the slow, moist method of cooking softens fibres, rendering the food tender.

(ii) Meat and vegetables may be cooked and served together, making an appetizing dish, while saving fuel and labour.

(iii) Stewed food may be cooked in the oven after other food is cooked, or it may be cooked on the side of the fire or on a very small gas or oil flame, thus again saving fuel.

(iv) All nourishment and flavour are retained so the food is very appetizing.

General Rules

(1) Have ready a pan with a well fitted lid.

(2) Prepare the food and cut into pieces conveniently for serving.

(3) Use tepid liquid and only sufficient to half cover the food.

Braising: This is a combined method of roasting and stewing in a pan with a tight-fitting lid. To be strictly correct this

method requires a special pan but a casserole dish or stew pan makes a good substitute. The meat should be sealed by browning on all sides and then placed on a lightly fried bed of vegetables (generally root). Stock or gravy is added which should cover two- thirds of the meat. The flavourings and seasonings are then added. The lid is put on and the food is allowed to cook gently on the stove or in the oven. When nearly done the lid is removed and the joint is frequently basted to glaze it. This latter process is always done in the oven.

Broiling: Broiling is cooking by direct heat and is synonymous with grilling. In pan broiling, the food is cooked uncovered on hot metal such as a grill or a frying pan. The pan or grill is oiled slightly to prevent sticking. Excess fat accumulated while cooking should be poured off.

Grilling: Used synonymously with broiling. This is cooking by dry heat. The food is supported on a grid iron over the fire, or on a grid placed in a tin under a gas or electric grill, or between electrically heated grill bars.

Grilling over the Heat: This is cooking on greased grill bars with the help of fat over direct heat. Only first class cuts of meat, poultry and certain fish can be prepared this way. The grill bars are brushed with oil to prevent food from sticking, and can be heated by charcoal, coke, gas or electricity. The bars should char the food on both sides to give the distinctive flavour of grilling. The thickness of the food and the heat of the grill determine the cooking time. Grills are typical a la carte dishes and are ordered by the customer to the degree of cooking required, such as rare, medium rare, medium or well done.

Grilling under the Heat-salamander: Cooking on grill bars or on trays under direct heat. Steaks, chops, etc., are cooked on the bars but fish, tomato, bacon and mushrooms are usually cooked on trays. This method can also be used in the preparation of foods au gratin and when glazing is required.

Grilling between Heat: Food is cooked between electrically heated grill bars.

Infrared grilling This is cooking by inferred radiation. This method reduces cooking time considerably, e.g., a steak can be grilled in one minute.

Soups

Aim of Soup Making: To prepare an appetizing, economical and easily digestible dish by extracting nourishment and flavour from the solid to the liquid.

Stock is the foundation of all good soups. Stock is made by putting solid food into cold water and by a long slow process of cooking extracting the nourishment into the liquid. The best stocks are made from meat and bones together, e.g., white consomme. Since this is expensive for household purposes as well as for institutional feeding, bones and leftovers are used. The following stocks are adequate as well as economical.

Bone Stock: Use any kind of bones, cooked or uncooked. Place in a pan. Cover with cold water, add a little salt and bring to the boil. Simmer gently to allow the scum to form on the surface then gently remove it. A little cold water could be added occasionally to accelerate the process. Skim well and for each 600 ml of liquid add one onion, one carrot and a bouquet garni. Put the lid on and simmer very slowly for at least 2 to 3 hours. Strain. Cool quickly and use.

Meat and bones not only contribute flavour but body. Body is largely made up of gelatine and other soluble products extracted from meat and bones.

Depending on the size of the bones and the way they are cut, stock from meat bones is simmered from 4 to 10 hours and poultry from 2 to 6 hours.

The best bones to use for stock are those from young animals. These contain more gelatine and the marrow contains more flavouring ingredients. Shin and knuckle bones are best with neck bones next. The bones are best cut into 4 lengths and split.

To make 5 litres of good stock use 2 kg. of meat bones, 6 litres water, 12 kg. of flavouring vegetables (mirepoix).

In a brown stock, the meat, bones and vegetables are first browned.

In a white stock, they are not browned. Also avoid using highly coloured vegetables such as carrots, beans etc.

Vegetable Stock: Use any mixture of vegetables, the outside leaves of cabbage, cauliflower stalks, celery stalks and leaves, shells of peas, etc. All these must be washed thoroughly to free them of grit, sand and worms. Shred vegetables and add to cold water. Cover and boil for 20 to 30 minutes. Strain and use.

Fish Stock: Cover the fish bones, skin, head and other trimmings with water or milk and water. Add 1 small onion, 2 cloves, ½ a bay leaf and a sprig of parsley. Bring to boil. Cover and simmer for 20 minutes. Strain and use.

Emergency Stock: Dissolve meat or yeast extract in boiling water or in pot liquor.

Pot liquor is the water in which meat, poultry, fish, vegetables, macaroni, etc., have been boiled for the table where the purpose is to retain as much nourishment and flavour in the solids as possible. This is done by putting the solids into hot water but still some nourishment and flavour escape into the liquid. This liquid (pot liquor) can be used as a substitute for stock.

No stock will keep for very long. Because stock is a good medium for bacterial growth, great care must be taken in its preparation and storage. In hot climates stock should be stored in a refrigerator and boiled to prevent spoilage. Cool quickly by emptying into shallow and clean bowls. Stock which contains starchy foods should not be used after 12 hours.

Consomme: A clear soup prepared from beef, chicken or game stock, garnished with a variety of ingredients. This can be served hot or chilled. Consomme takes its name generally from the garnish, e.g. Consomme Julienne—consomme

garnished with boiled matchstick-like strips of vegetables such as carrots, beans, turnips, etc. Consomme Royale consomme garnished with savoury custard cubes.

Broth: A good stock (beef, mutton or chicken) cooked with diced meat, vegetables and rice or barley and served with the solids. Some of the nourishment is retained in the solids by putting them into tepid liquid. Broths are also different from ordinary soups in having the thickening (rice, barley, macaroni, etc.) put in at the beginning and cooked with the other ingredients. Examples: scotch broth, minestrone, escudella-catalina, etc.

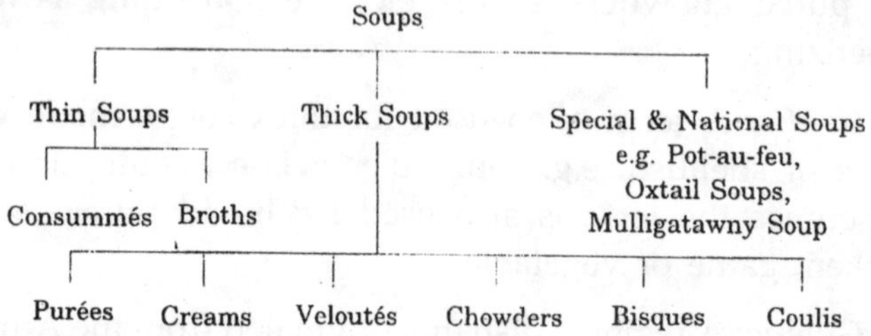

Potage or Soupe: These terms cover a wide variety of soups, e.g., potage bortsch, potage minestrone, potage mulligatawny, soupe au gumbo, soupe aux rognons, etc.

Puree: A soup thickened by its main ingredient and passed through a sieve. Milk or a little flour blended with milk is added to prevent the puree from separating, but not as a thickening. The consistency of the soup should be like thick cream. Generally served with croutons (fried cubes of bread). Examples: puree-de lentils, puree de haricot blanc, puree Parmentier, puree de tomatoes, puree de pois frais.

Cream: A soup of cream consistency which is generally made with vegetable puree mixed with bechamel or white sauce. It can be finished with cream if desired, e.g., creme de celeii, creme de tomato.

Veloute: A thick soup made from white stock and roux, finished with a liaison of yolks and cream.

Chowder: An American soup resembling a stew made of

meat, fish or vegetable with milk, salt pork and various seasonings. Crackers are generally added just before serving. Examples: prawn chowder, oyster chowder, pork chowder, vegetable chowder.

Bisques: These are thickened fish soups generally made from shellfish and fish stock and thickened with cream. Diced fish is served in it. Example: bisque de homard.

The value of any soup will depend on the ingredients used. Clear soups are of value only to stimulate the appetite at the beginning of a meal while thick soups such as cream, veloute and puree, chowders, broths, etc., are nourishing as well as appetizing.

Coulis: A term often used for thick soups made with a puree of shellfish, e.g. coulis d'ecrevisses, coulis de crabes. Sometimes the term is also used for liquid purees such as chicken, game or vegetables.

Gumbo: A regional dish that originated from the American South. Contains okra, rice and a seasoning called fille, which is powdered young sassafras leaves.

2

Different Techniques

Methods of Heat Transfer

Heat can be transferred by three methods—conduction, convection and radiation. Two or three of these methods may be combined. For example, in baking a cake, converted heat cooks the mixture, then the cake comes in contact with the cooking utensil. Heat is also transferred by conduction. The browning of the cake is by radiation.

Conduction: Conduction is the passage of heat through a solid or one solid to another provided they are in contact—pot on a hot plate. Some materials retain heat better than others. Copper, for example, heats up very quickly and evenly but does not retain heat well. Stainless steel, iron, pyrex and ceramic cookware, on the other hand, retain heat well. Heat conduction include shallow frying, sauteing and stir-frying.

Convection: Involves the transfer of heat in liquids and gases. In heating liquids or gases, convection currents are produced which distribute the heat. For example when water is heated in a saucepan, the water at the bottom is heated first.

It rises to the top because it is lighter and the heavier cold water takes its place at the bottom of the pan.

Radiation: It is the passage of heat in straight rays from a hot object. Any object in the part of the rays becomes hot e.g. gas and electric grills.

Basic Methods

Cooking helps to preserve colour, get a variety of texture, enhance flavours and make the nutrients easily available, though a few of the nutrients, which are water soluble are lost or oxidised.

Roasting

The term roasting was originally applied to the cooking of large pieces of meat, on a turning spit over an open fire, a practice which would be reasonably called grilling or perhaps barbecuing today. Traditional methods of roasting, practised at present are (1) pot roasting, (2) oven roasting, (3) spit roasting, (4) tandoori method of roasting.

Pot Roasting: Cooking is done in a covered pot or pan. Only good quality meats, i.e. small joints and birds are used.

The meat, poultry or game bird is tied to retain the shape. It is placed in the thick-bottomed pot and skewers placed at the bottom to prevent the meat from sticking to the pan. Fat should cover the bottom of the pan. The joint should touch the fat. Cover the pot with a well-fitting lid and cook on slow fire. The joint should be put in the pot for flavouring after 20 minutes. When the joint is cooked, remove the joint of excess fat. Add stock, so that the juices mix and get the flavour from the aromatic vegetables, and from the basis of the accompanying sauce or gravy, which is called *Jus Roti.*

Oven Roasting: It is cooking in a roasting tray in an oven with the aid of fat and first-class meat; poultry and certain vegetables are used. The raw meat joint is trimmed and tied

if necessary to retain the shape. It is put in hot fat and all sides are browned to seal the juices. Salt could be sprinkled later on. Baste the meat often, so as to help the meat colour without drying. The average time for cooking is 15 minutes per lb (455 gm). The time could be adjusted to the size of the joints. For small joints the heat must be constant, and for large joints at first it should be strong heat (450°F) and then reduced to 300°F. After 15 minutes aromatic vegetables should be added for flavouring during the process of cooking.

Pork and veal should be well cooked, whereas beef should be just done or underdone. To ascertain the point of cooking, the finger should be pressed against the meat and if it offers no resistance, then it is cooked. Another test is to pierce the meat with a fine skewer and if no pinkish liquid oozes out, it is cooked.

The cooked meat should be removed from the roasting tray and the excess fat poured off. The roasting tray should be placed on the fire and the juices allowed to caramelise and the brown stock added and brought to boil. (The aromates and meat juices give flavour and taste to the stock). Thus the brown gravy called *Jus Roti* is prepared and the actual process is known as deglaze.

Spit Roasting: It is the original from of roasting. The same principle of seasoning, basting and cooking with a fierce heat is done. It is cooking by direct heat, basted with fat. A spit is constantly revolved to ensure even colouring and cooking. First-class quality of small pieces of meat are used and the meat pieces should be succulent (Juicy). In the case of small game, fuel should be wood, and it should produce low flame.

Basic Rules

(a) Items of meat, poultry and game for roasting must be tender and of top quality.

(b) Food items must be raised off the bottom of the roasting tray to prevent the frying of the lower part.

(c) The juices from the process of roasting meat, poultry and game must be used to make the accompanying gravy.

(d) To facilitate the slicing of roasts, and other cuts of cooked meat, piece should be allowed to remain at room temperature, after it has been removed from the heat for 15 to 30 minutes. This sets the juices, allowing them to soak back into the muscle and it is easier to slice because the meat becomes firm.

(e) The thicker the meat, the lower should be the cooking temperature.

(f) Meat is considered rare when It reaches the temperature of 60°C (140°F); this is the internal temperature of the cooked spots. For well done meat, the internal temperature is 82°-85°C (180°-185°F).

Poeling: It is similar to pot roasting; the difference is that it is cooked entirely in butter. Usually poultry and game are peeled. The poultry is wrapped with thin slices of pork fat, covered with or without buttered paper, placed in a pot and basted with melted butter while it is cooking, covered with a tight-fitting lid. Cut aromatic vegetables are added for flavouring, and later on moistened with Madeira wine (if not available, then red wine could be used.)

After the liquor has absorbed the aroma of the vegetables and the cooked food item is removed, it is then strained, grease removed and reduced a little, seasoned and served as the accompanying sauce.

Special poelings are known as "Encasserole" and "Encocotte". The poeling is done is special earthenware utensils and served in the same.

Encasserole: The food item is peeled, only in butter, without the addition of vegetables. When the cooking is done, the food item is withdrawn for a moment and some excellent brown stock is poured. The piece is put into the earthern utensil and kept hot in a bain-marie and served in the same dish.

Encocotte: The procedure is the same as encasserole, except that the food item is garnished with vegetables such as mushrooms, the bottoms of artichokes, small onions, carrots, turnips, etc. which are turned (shaped) and cooked in butter before being used.

Basic Rules

(a) The size of the pot with tight-fitting lid should be in keeping with the item to be peeled.

(b) Sufficient butter should be used to allow for adequate basting and no liquid must be used during the process.

(c) The lid should be removed when the items are two-thirds cooked to colour the food item.

(d) The vegetables, juices and sediments must be used to make the accompanying sauce.

Grilling

Cooking by the grilling method began years ago, when meat was first placed near or over an open fire. It has steadily advanced through the years to the most modern gas, the electric grills and infrared models. The food is placed on grill bars, and subjected to the action of radiation which could be from below or above. The source of heat may be from charcoal, coke, gas or electricity. In an authentic grill the heat is located below the grill bars, and if the source of heat is above the grill bars, the grilling equipment is called salamander.

The grill bars should be cleaned and oiled, or the bars will rust and food will stick. Usually small items are grilled, the meat is cut, trimmed and slightly flattened with a bat. Then the meat should be seasoned with salt and pepper and brushed on both sides with oil. The meat is then grilled on the grill bars, turned when half cooked, brushing oil occasionally and cooked to required degree. The fierce heat seals the cut ends of the meat and browns the surface, so that all the juices are retained in the meat. It could be termed as cooking along with concentration of flavour.

French Grilling Terms

Au bleu — rare, very underdone
Saignant — underdone
A point — just done (medium)
Biencuit — well cooked

Basic Rules

(a) Items for grilling must be small in size, flattened and of top quality.
(b) The items should be oiled and seasoned. Fish must be floured, oiled, and seasoned.
(c) Heat should be intense to enable sealing at once and colouring; a brown crust should form, which seals the juices of the food.
(d) Grill bars should be cleaned and oiled

Broiling

Is cooking by direct heat and is used synonymously with grilling, It could be done by source of heat above or below. It is dry cooking, and can be done on a grid (iron bars) or pan broiled, where the food is cooked uncovered. The pan or grill is oiled slightly to prevent sticking, excess fat removed.

Stirred eggs (oeuf sur le plat) is an example of broiled eggs. The egg is slipped on to an egg dish, cooked on top of the stove, till white coagulates, then it is placed in a broiler, heat from above. Tenderest and most expensive cuts of beef and fish are most often broiled. Grid markings on steaks add attractiveness. Whole spices broiled on a hot tawa bring out the flavours (pan broiled).

Basic Rules

(1) Tawa or pan or grill may be oiled slightly to prevent sticking.
(2) Dry cooking should take place.

Baking

Baking is cooking of food by the action of dry heat in an oven. The degree of dryness of heat may be modified by the amount of steam produced from the items baked. Bread rolls, cakes, pastries, puddings, potatoes, vegetables, farinaceous dishes are baked.

During baking, the mixture or moisture in the product expands by air, steam or carbon dioxide.

Basic Rules

(1) Items should be placed in the right position in the oven, the cool spots should be avoided.

(2) The oven should have the correct temperature according to the baked items, e.g. rich cake should have a lower temperature than pastries, which should have a high temperature. Baked yeast goods should have a hot oven 220°C (425°F).

Frying

Food is fried when it is placed or immersed in oil or fat at a sufficiently high temperature; used to brown the surface atleast and partially or completely cook the interior of the food. Frying adds flavour to otherwise bland and tender foods; it also creates a crisp crunchy texture.

There are two types of frying (1) deep frying, (2) shallow frying. Deep frying is cooking food by immersing it in fat, while shallow frying is cooking it in a little fat in frying or saute pan or on a griddle (hot plate).

Deep Frying: It is a quick method of frying and a popular method for Indian cuisine. The holding temperature range for deep-fryer is 90°C-110°C.

As the food is immersed in hot fat, it is essential that the fat is of a good quality having a high smoking point. The frying pan should have a sufficient depth. The size and shape

of the food to be fried should be uniform. Fried food have a good appearance. Some of the fried food are: rissoles, cromesquies, chicken and mutton cutlets, fish cakes, potatoes, vegetables, fritters, beignets, croquettes, pakodas, macchi pakodas, vadas, moong dal ladoos, samosas, bondas, poories, kachories, etc.

Basic Rules

(a) The fat must be hot enough to seal the outside of the items being fried. This prevents the food from absorbing any fat and to prevent any loss of flavour. The exceptions to this rule are for the frying of Dauphine, Chamonix potatoes for which the temperature is approximately 145°C to allow the slow expansion in size while cooking. Frying temperature varies from 160° to 195°C.

(b) Prevents the escape of juices; all items, except potatoes must be seasoned and coated with either batter, flour, egg or egg and breadcrumbs before being fried.

(c) Too many items should not be put at a time as it will reduce the temperature to the point where the food will absorb fat, lose its juices, become soggy and heavy or break up.

(d) The fried items should have a uniform golden colour.

(e) Drain well on paper and serve at once on doilys.

(f) The fat used for frying should be strained and stored in a cool place.

(g) If fat, that has been used for cooking, is used again, it should be replenished with fresh fat for reuse in the ratio of 1:2.

(h) If possible, a frying basket should be used to give uniform colouring and heating to the items fried.

(i) Wet food items, e.g. potatoes should be well dried before being fried.

Shallow Frying: Tender, good quality items, like fish, meat

and poultry are generally used for frying, as this method completes the cooking of the dish. The presentation side should always be fried first and then turned, both sides are browned evenly. For bacon and sausages as they have sufficient fat, do not add additional fat.

The food items, that are shallow fried are fish cakes, medaillons, eggs, parathas, dosas, cheelas, alloo ki tikkies, etc.

Sauteing comes under shallow frying—the word sauteing comes from the French word *sautoir* which means "to jump". Food items, e.g. thin meat pieces, fish, game and vegetables are cut into very small thin pieces. They are placed in a heavy-bottomed pan with a little hot fat and the pan with the items is shaken and flipped occasionally over the fire.

Basic Rules

(a) Items should be seasoned before being shallow fried.

(b) Food items must be placed to cook in hot fat, so as to seal the outside.

(c) Drain well the food items after cooking, or the food will be greasy, e.g. bacon rashers.

(d) Before sauteing, food having a high water content should be salted and allowed to drain and then dried on paper towels. (Salt draws out the moisture).

(e) The saute pan and frying pan should have a thick bottom and straight sides.

Solar Cooking: It is similar to what we use in the kitchen to cook food. Solar cooker works only on *solar energy,* it emits no smoke, no soot spoils the cooking utensils. It keeps the surroundings clean. Above all, it conserves the precious energy resources of the country and saves us money.

Principles: Solar cookers can be broadly divided into two categories:

(1) Cookers utilising solar heat with little or no concentration of rays. It may be termed as a "hot box".

(2) Cookers which deflect solar energy from a large surface to a smaller area to produce high temperatures.

The first type of solar cooker commonly known as "hot box" consists of a well-insulated box, the inside of which is painted dull black and is covered by one or more transparent covers. The purpose of the transparent covers is to trap heat inside the solar cooker. These cookers allow the radiation from the sun to come inside but do not allow the heat from the black absorbing plate to go out of the box. It is because of this that the temperature of the blackened plate inside the box increases and can heat up the space inside to temperature up to 140°C which is adequate for cooking.

The second type of solar cooker uses a lens or a reflector suitably designed to concentrate the solar radiation over a small area. This cooker is able to provide high temperature on its absorbing surface when suitably designed.

The cooking time is about 2-2½ hours for cooking depending upon the kind of food and the season. The time taken for cooking is less in summer than in winter.

Advantages: The solar cooker has a number of advantages over the traditional cooking devices. These are:

1. It cooks up to four items at a time.
2. It preserves the nutrition value of the food because the cooking is done at low temperature.
3. It does not require constant attention.
4. It is safe and does not burn.
5. Saves fuel.
6. Keeps food hot for a long time.

Disadvantages

1. It needs the sun and will not give results on a cloudy day.
2. Only simmering method is adopted.

Solar Cooker

1. Outer box of a solar cooker may be made of wood, iron sheet or filter fibre reinforced plastic having suitable dimensions.
2. The inner box can be made from galvanised iron, or mild steel or aluminium sheet. All the four sides and the bottom of the inner box which are exposed to the sun are coated with blackboard paint which absorbs solar radiation.
3. Thermal Insulation: The space between the outer box and the inner box is fitted with insulating material such as glass wool, thermocole, etc. This prevents heat losses from the cookers.
4. The double glass cover: Generally a double glass cover is provided for a solar cooker. These covers have length and breadth slightly bigger than the inner box and can be fixed in a wooden frame maintaining a small spacing between the two glasses. This space contains air which acts as an insulator and prevents heat escaping from inside. The frame is attached by means of hinges. A rubber strip affixed all around prevents any heat leakage.

Mirror: It is used in a solar cooker to increase the radiation input on the absorbing surface. Sunlight rays on the mirror gets reflected from it and enter the box after passing through the glass covers. This radiation is in addition to the radiation entering the box directly and helps to quicken the cooking process by raising the inside temperature of the cooker.

Inferred Cooking: Dry heat is excellent for cooking meat, fish and other items which are fairly tender to begin with. The food is placed a few inches above or below the heat source. Some of the heat which is given off by the heat source is carried to the food by means of convection, the rest by means of radiation. The radiation is in the form of inferred rays. Inferred rays are a form of radiation which are just a little longer than

those waves which are seen as red and shorter than radio waves and microwaves.

Infra-Red Waves Penetrate Food Slightly: Inferred wavelengths are small and range in length from about 0.07 to 1000 millionths of a metre. Inferred waves are energy waves just a little longer than light waves.

Wavelengths from 1.4 to 5 millionths of a metre are the most effective for cooking food, because they pass through the vapours and act on the food and the moisture contained in the food. Shorter wavelengths are not well absorbed by the water in the food.

Catering equipment which relies on inferred waves for energy transfer should produce inferred waves from 1.4 to 5 millionths of a metre.

As temperature increases in a heat source, the energy radiated shifts to shorter-wavelengths. As the temperature is doubled, the radiated energy shifts maximum to half the wavelengths.

A temperature of 649°C (1200°F), for example produces a maximum energy wavelength of 3 millionth of a metre effective in inferred cooking.

However, the surface of food cannot stand temperatures much above 260°C (500°F) without burning (unless they are very moist). With high temperatures, food must be placed some distance from the heat source. High temperatures are desirable to produce inferred waves of a maximum cooking effectiveness but are so hot that if food is placed near this heat source, it will burn.

Inferred heating is fast because the energy is transmitted from the heat source to the food almost instantaneously, not having to be carried by means of hot air to the food, as in the case of convection heating. The inferred rays penetrate the blanket of vapour surrounding the food and enter the food a few millimetres which also speeds up the cooking process.

Vapour Blanket Food: Some of the new grills produce a great amount of inferred energy and hence are much faster than the old style grill which produced less inferred energy.

The closer the food is placed to the source of radiation, the faster it would cook. Radiant heat energy is transmitted directly to the food passing through the vapour blanket.

The amount of energy which reaches food from an inferred source varies proportionately to the distance the food is away from the source.

A steak placed 5 cm away from a heat source will cook four times as fast as the same steak 10 cm from the source.

Inferred rays travel in straight lines in all directions from the emission source. Aluminium foil should not be placed on the food as the rays penetration will be less.

Inferred lamps are also used to keep food hot and to give them a rich red colour on buffet tables and cafeteria lines. Inferred lamps transmit heat instantaneously to food exposed to them and make the food appear more red in colour.

The amount of heat absorbed by food from an inferred source depends upon the reflectivity of the surface of the food; the darker the colour of the surface of the food, the greater the rate of absorption.

Inferred lamps and inferred grills are designed so that most of the energy emitted is inferred.

One inferred grill cooks a 340 gms steak in about 6 minutes— less than half the time required for grilling in a conventional grill. Infra-red energy is produced by electric cooking as well as by gas cooking.

The ultra-ray is gas fired and is a recent invention and produces 100% inferred radiation. It consists of three layers of Incaloy metal screens sandwiched together. Air and gas readily pass through the screen, produce a blanket of flame about 1 cm in thickness.

Temperatures range around 621°C (1150°F), reaching maximum capacity within six seconds.

With such a grill, time required for grilling is determined by the distance from the energy source and its temperature.

Haute Cuisine has some special methods of cooking which is done when required.

Sous Cloche (Under bell Cooking): The name is derived from a bell container, made of glass or silver. It has two parts, a base and a lid. The dish is cooked in the cloche and it is served in the same container. The bell-shaped lid is opened in front of the customer.

Champignons Sous Cloche: The ingredients are: Mushrooms, white wine and butter. Heat the butter in the base of the container, add sauted mushrooms, wine, seasoning. Place the lid on top, cook for 2 minutes. The container is placed in front of the customer and the waiter removes the bell-shaped lid (all the essence and goodness is retained).

En Papillote: This is a traditional method of cookery, where the article of food is placed inside oiled grease-proof paper cut according to the shape of the food.

It is generally applied to items that need less cooking time, e.g. tender cuts (prime quality of meat), fish or vegetables that need less cooking time.

Meat when cooked by this process is set rapidly by shallow frying, without cooking fully. This is done to seal natural juices. Only tender meat is used for this method such as minute steak, kidneys; liver, etc. Generally a garnish is placed on the oiled grease-proof paper which is wrapped round the article of food and the edges sealed by folding them over. The bag is then placed on the service dish and cooked in the oven. The steam that rises within the paper container makes the top surface of the paper rise, e.g. Cote de Vean en Papillote. Prepare as follows:

Fold the grease proof paper as illustrated, well covered

with oil. Place on the paper, a thin slice of ham (pear-shaped) the same size as of a cutlet. Spread duxelle (finely chopped mushrooms, onions, carrots) in the ham slice. Then lightly saute the prepared seasoned veal cutlets for 2 minutes on each side. Place the cutlet on the duxelle that has been put on the ham slices, then place a slice of ham on top.

Take grease-proof paper, fold into two, cut heart shape with scissors; it should be larger than the food item. Open the cut paper, on one side and place the food item as explained above. Seal the edges of the paper from point A in the diagram and fold at 20 mm intervals, flattening each fold, before the next fold (see diagram).

Sous La Cendre: Meat slice is larded, seasoned and marinated in white vinegar. Wrap in a paste covering the meat. Place in the oven after making ½ inch hole. (This is done to let off the steam.) When half baked, put demi-glaze in the hole, finish baking. Serve hot with Madeira sauce.

Forced-Air Connection Oven: The forced-air connection oven is best suited for use in a kitchen either as a cooking agent or a reheater prior to presentation of the food on a counter service. The advantage is that an even temperature is maintained throughout the cooking process but a high degree is reached in a relatively short time, i.e. 390°F in 15 minutes. This is achieved by a centrifugal fan drawing air across the heating elements and distributing it evenly. The movement of air strips away the dead air which normally surrounds a product when first put into the oven and promotes faster heating up. At the same time because of the air movement taking place, the temperature throughout the whole oven remains constant, whereas in conventional static atmosphere units there is a tendency for temperature gradients to be developed.

The oven can roast, grill, bake, thaw and cook quickly and efficiently.

The advantages of convection oven is two fold, it is not

only speedier, it reduces the weight loss incurred in the longer cooking cycle with conventional ovens.

Forced air convection ovens are also eminently suitable as end cooking units for prepared frozen meals coming individually or in bulk. The food can be placed in the oven straight from the zero cold room and end-cooked within 20 to 30 minutes.

The smaller type of forced air convection oven is suitable for such installations as snack-bars, bakeries and in public houses where there is a demand for hot dishes.

This system could also be applied in the kitchen having satellite service points, snack-bars, bakeries where there is demand for hot dishes.

Microwave: Although a form of energy, it is not heat. The principle of microwave involves a type of electronic heating and is concerned with the production of high frequency waves from an oscillator tube called a magnetron. The electromagnetic waves have energy and this energy is transported to the desired load for heating. The energy created penetrates articles of food to a depth of 3" and causes the molecules that make up the food to rub together and create friction, supplying the heat which cooks or reheats the food.

Microwaves do not heat the oven cavity. Microwaves are reflected by metal items and therefore are unsuitable for use with this equipment.

The advantage lies in the speed of the operation, although browning is not possible. It is excellent as a reheater. Microwave ovens allow direct customer participation with the use of time tokens.

Electronic Cookers: The hotter a substance becomes, the more rapidly its molecules or atoms begin to vibrate. The common methods of applying heat by conduction, convection or radiation, all involve supplying heat energy to excite.

Electricity can be changed into microwaves using a special

valve called a magnetron. The microwaves pass through air or glass, china, plastics and paper without heating them, but when they strike solid and liquid foods, they are absorbed. Their energy is transferred to the food molecules which vibrate rapidly. The resultant heat cooks the food. Microwaves are absorbed by most types of food in varying degrees dependent on the cellular structure of the food.

Affecting Factors

(a) *Density:* The denser the food, the more difficult it is to cook. A joint of red meat will not be easily and uniformly penetrated by microwaves whereas liquids, fish, white meats, etc. offer far less resistance.

(b) *Quantity:* The heating is proportional to the quantity of food placed in the oven. For instance if one portion takes 30 seconds to heat, two portions will take 45 seconds, whereas three portions will take 60 seconds. For each additional portion, half the time taken by the first portion is added.

(c) *Shape:* The thicker the product, the longer it will take to cook. Shape as far as possible should have uniform thickness, e.g. truss chicken.

(d) *Containers:* They should be of glass, china, plastic or paper to allow transmission of microwaves.

1. Microwave heats the food by agitating the food molecules.
2. It is a rapid method for small quantities and small portions of uncooked food.
3. There are no food odours and less steam.
4. Excellent method for reheating, frozen cooked food, canned food for reheating assembled food on a plate or pre-cooked snack meals.
5. Rapid service and needs less staff.

One disadvantage is that food does not brown; browning elements can however be fitted, if required.

Rechauffe: Rechauffe means to reheat. Frequently many cooks come across a few left-overs in their kitchens. In the interest of economy, a sound knowledge of the left-over foods is necessary. These left-overs should be used in some form or the other. Many tasty and interesting dishes can be prepared, but care should be taken that the food is thoroughly and carefully reheated. It is also important that only sound food is used. "If there is any doubt, throw it out." If meat is over-cooked, it will toughen the fibres and harden the proteins thus making food indigestible and unpalatable. Meat and fish lose some of their nourishing components in cooking and this loss could be supplemented by other food. In addition to these points, the importance of suitable seasonings and flavourings and suitable accompaniments adds and supplements any deficiency in food value, and also determines the success and value of a rechauffe dish. Care must be taken that frequent repetition does not occur.

Always the use of minced meat or mashed potatoes will make it monotonous. In producing a dish with left-overs, it should not be obvious that the main ingredient of such a dish is not fresh. Rechauffe dishes can be made attractive, if prepared and served in small individual dishes or patty cases and fish in small scallops or coquille shells. The dishes must be decorated so as to look attractive and pleasing to the eye. It should be combined again to enhance the flavour with a sharp sauce as the natural flavour is lost to some extent.

Food may be warmed in a moderate hot oven carefully covered. The vessels containing food, if placed in a tray of water in the oven, prevents over-drying. Steaming is one of the best methods of reheating as the heat does not dry or over-cook the food. Condensed steam should be prevented from falling in the food.

General Directions

1. To bring down the food cost, utilise the left-overs, such as vegetables, eggs, rice, paneer, sauce or gravies, etc.

2. The left-over vegetables could be mixed with salad dressings, used as salads and served cold.
3. Sauces could be made sharp by adding pepper, mustard, vinegar or tomato puree, etc.
4. Panada or thick binding sauces will help to bind mashed or chopped ingredients for croquettes. Mashed potatoes could be used without sauce, fish cakes, aloo ki tikkis, aloo ka parathas, etc.
5. Cooked left-over rice could be made into croquettes, lemon or curd rice.
6. Curries could be improved by adding fresh ingredients such as beaten curds or ground cashewnuts or coconut paste. Onions, mushrooms, curds, curry paste improve the flavour and taste when added.
7. It is important that the excess fat, gristle is removed or it will make the rechauffe dish greasy.
8. Trimmings and bones could be used for brown stock.
9. White sauce is appropriate for fish and poultry, and brown sauce for the meat.
10. Marinating of meat or fish improves the flavour.
11. To improve the taste for fish, white meat, lemon juice, parsley, nutmeg could be added.
12. The flavourings and seasonings, when mixed with finely minced meat, blend better and heat penetrates quickly to all parts and this reduces the reheating time. Fish can be flaked and blended with spices.
13. The addition of moisture in the form of gravy or sauce is necessary for the reheated meat dish, whether it is incorporated inside or served separately as it gives moisture to the dish and improves the taste.
14. If raw ingredients are to be added to the rechauffe dishes, they must be cooked first, as the short time necessary for reheating does not allow time for raw food to be thoroughly cooked, e.g. onions in croquette

mixture, potatoes in fish cakes, rice in cutlet mixture.

15. Protection of the food from direct heat of an oven or from hot fat can be done by giving protective coating by enclosing with batter, egg and breadcrumbs or covering with mashed potatoes as in shepherd's pie.
16. Left-over stale bread can be used to prepare breadcrumbs and puddings, cheese toast.
17. Rechauffe dishes should be dished neatly, decorated and garnished.
18. It is important that the dishes should be piping hot.

Reheating Dishes

Meat: Any meat or mixture of meats—white or red—poultry, game or rabbit can be used with suitable flavourings and seasonings for croquettes, cutlets, rissoles, kromesky, salads, cottage pie, mirotan of mutton, shami kababs, keema pulao, etc.

Croquettes: Make the panada sauce, chop meat finely, otherwise the mixture cannot be smooth, add seasonings and flavourings, add to the panada and mix well. Turn the mixture to get cold and firm. Make rolls of even size of the mixture. Coat with egg and breadcrumbs. Press the coating well and coat a second time. This improves their appearance. Fry in hot deep fat, drain well and serve with fried parsley.

For Cutlets: Form minced meat into cutlet shapes, season, egg and breadcrumb twice, put a piece of macaroni in the end of each, and deep fry. Serve with fried parsley.

For Rissoles: Roll out short crust pastry very thin and cut into rounds about 4 to 4½" in diameter. Place a little of the prepared mixture on each, roll and seal edges, coat with egg and crumbs, fry in deep fat.

Kromeskys: These make a dainty small entree dish. Just the scraps and trimmings of white meat, ham or tongue can be used. Prepare coating batter, mince meat and mushrooms finely, add seasoning, grated lemon rind and mix with sufficient

panada. Mix well, form into small rolls and wrap each in a small slice of bacon or caul. Dip the rolls in batter. Fry in moderately hot fat till crisp and golden brown in colour. Garnish with lemon and parsley.

Chicken: Left-over chicken cut into joints can be reheated carefully in sauce and if the skin and bones are removed, the chicken can be used with bechamel sauce in patty cases or croquettes. Cold roast chicken could be used to prepare chicken sandwiches and canapes, chicken chat, etc.

Fish: Cooked fish is more insipid than meat and requires to be seasoned well and flavoured. Lemon juice, parsley, anchovy essence are the flavourings used.

Cold fish may be used in may interesting dishes, e.g. fish mornay, curried fish, fish cakes, fish cutlets or croquettes, fish pie or fish salads.

Fish Cakes: Flake the cooked fish, add boiled potatoes (two-third fish and one-third potatoes), chopped parsley, lemon juice and seasoning. Add a little melted butter, egg to bind and cool the mixture. Make round cakes, egg and breadcrumb twice and deep fry. Garnish with lemon and fried parsley.

Fish Croquettes or Cutlets: Flake fish well, season, mix with panada to bind and then roll, egg and breadcrumbs and deep fry.

Fish Scallops: Butter some china scallop shells or small deep round bowls, place the flaked fish that is seasoned and mixed with lemon juice. Cover with white sauce or sprinkle breadcrumbs and melted butter on top. Pipe the mashed potato all round the fish.

Reheat in a quick oven or gratinate, till the surface is nicely browned.

Fish Pakodas: Dip fish in besan (gram) batter and deep fry.

Vegetables: Cold cooked vegetables such as peas, beans,

cauliflower, potatoes, mixed with vinaigrette or mayonnaise are used as a salad. Cold boiled potatoes can be used for potato salads, tikkies or for saute potatoes. Mashed potatoes are used for fish or potato cakes, or for making borders for salads, or covering meat cutlets. Rice, cooked spaghetti or macaroni can be used for hors d'oeuvres with chopped onions, tomatoes, cooked meat, vegetables mixed with vinaigrette or mayonnaise.

Bread: Trimmings of crust, etc. should be kept until dry, lightly browned in the oven and made into breadcrumbs. Stale bread can be used for bread pudding, shahi tukra, bread halwa, pakodas, etc.

To control the cost of food, the left-overs in the kitchen are—rice, mutton, gravy, vegetables, etc.

The day should start with the chef surveying the walk-in deep freezer. He should prepare dishes using the left-overs. The left-overs should be reheated before use. The left-over items could be mixed with fresh items and prepared.

Boiling

Boiling is cooking by immersing the food in a pan of liquid, which must be kept boiling all the time, i.e. quite a number of bubbles should be seen on the surface. Boiling is restricted for meat and poultry for the first few minutes in order to seal the pores, to retain natural juices and then gentle boiling must take place which is known as simmering at 82°-99°C (180°-210°F). Boiling temperature is 100°C (212°F). Vegetables are boiled at 100°C (212°F). Blanching is used when food needs to be precooked or partly cooked.

Basic Rules

(a) The food items should be completely immersed throughout the process; if excess evaporation takes place the liquid must be replenished.

(b) The flavour of meat and poultry is enhanced by the addition of herbs and vegetables to the cooking liquor.

(c) Scum that arises during the boiling must be removed or it will discolour and spoil the taste.

(d) The liquid in which the food is boiled is known as pot liquor, as it contains some nourishment and flavour. It should not be thrown away but used for gravies or sauces. Use correct ratio of liquid to food.

(e) Fast boiling should be done to green vegetables as it preserves the green colour and prevents excessive loss of vitamins and mineral salts. The pan should be uncovered while boiling.

(f) Root vegetables, with the exception of new potatoes, must be placed in cold water to boil. This helps to improve their flavour, as the acrid taste of root vegetables are leeched out.

(g) Stocks, sauces and soups should be simmered gently. The stock will get cloudy as the scum in fast boiling may get mixed in the pot liquor.

(h) Always poach fish and never boil, or the fish will break up. Fish can be wrapped in a cloth and then poached.

Blanching: Place food item in cold water and bring to boil, remove. It seals the juices and flavours.

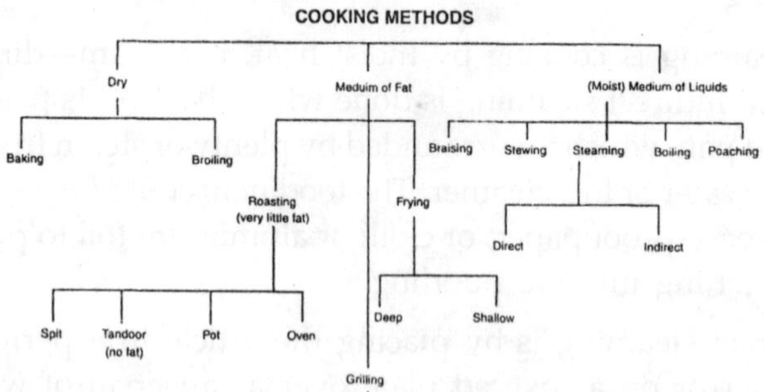

Poaching

Poaching is cooking gently in water which is just below boiling point (93°-95°C). The liquid is half as close to boiling

point as possible without there being any perceptible movement of the liquid.

Basic Rules

(a) Food items should be completely immersed in the liquid and left in throughout the cooling process.

(b) As soon as the liquid, with the food Item begins to boil, lower the temperature to allow poaching.

(c) Whole fish must be placed into cold liquor to poach. Cuts of fish must be placed into boiling liquor, temperature lowered and allowed to poach.

(d) The liquid for poaching eggs at one side of the pan should be simmering and fresh eggs must be added one at a time; this will help to get the rough shape. Placing them into liquid without any movement will result in flat eggs.

(e) Adding vinegar and salt helps in quicker coagulation and prevents disintegration.

(f) It is important that fresh eggs should be used for poaching.

Steaming

Steaming is cooking by moist heat, i.e. steam—direct or indirect. Indirect steaming is done when the food Is placed in a closed pan, which is surrounded by plenty of steam from fast boiling water or in a steamer. The food item could be protected with grease-proof paper, or cloth or aluminium foil to prevent water getting into the pudding.

Direct steaming is by placing the article in a perforated container or on a covered plate over a saucepan of water.

Basic Rules

(a) Only small suitable food items should be steamed, as it is cooked slowly. Green vegetables are not suitable for steaming as they lose the colour.

(b) Puddings in basins must be covered with grease-proof paper to prevent condensed steam from making the pudding wet.

(c) Root vegetables including potatoes must be placed on perforated tray that will allow condensed steam to circulate, and cook them.

(d) It is a good method for invalids, as the steamed foods are easily digested. Nourishment and flavour are retained in the food and usually the food cannot be easily overcooked.

Stewing

It is a time-consuming method of cooking, but the advantage is that the coarser, older and cheaper types of poultry and meat are used as they are unsuitable for grilling and roasting. Cheaper cuts of meat and certain fish dishes (bouillabaisse, matelotte) are prepared by this method, as it renders it tender and palatable.

Stewing is gentle simmering in a small quantity of water, stock or sauce, until the cut food items are tender, and both liquid and food are served together. Stewing is done in covered saucepans or casseroles on top of the cooker, or in the oven—but always at a low temperature.

Basic Rules

(a) While stewing, the liquid should cover the food items.

(b) It is a slow process in the liquid or sauce and the long process of cooking will give a concentrated flavour to both the food and the sauce.

(c) Items suitable for stewing are those which are of a tough nature.

(d) The liquid should not reach too high a temperature, as the slow process of cooking by gentle heat, converts the connective tissues into gelatin so that the meat fibres fall apart easily and become digestible. The

proteins are coagulated without being over hardened and the soluble nutrients and flavour pass into the liquid which is served.

Braising

This is a combined method of roasting and stewing. In a pan with a tight-fitting lid or casserole, braising is done to prevent evaporation, so that the food retains its own juices together with the article added

Blanching is used when food needs to be precooked or partly cooked for flavouring, e.g. bacon, ham, vegetables and herbs. The meat and poultry used for braising is mature, it is a long method of cooking and tenderises the less tender cuts and grades of meat.

The meat is first browned in its own fat or small amount of hot fat. This seals the pores and thus the juices are retained. It is then placed on a bed of root vegetables. Stock or demi-glaze, should come to two-thirds of the meat. The flavourings and seasonings are also added. The lid is put on and it is allowed to cook gently on the stove or in the oven. When nearly done, the lid is removed and the joint is frequently basted to glaze it. This latter process is always done in the oven, to have an even colouring. The above method is the basic method of braising. The advanced method of braising is by first of all marinating the lardons in the wine and aromatic flavours and then larding the joints, and placing it on a bed of prepared aromatic vegetables in a braising pan and then adding the marinade along with the cooking liquor for braising. Brandy Is used for game.

Basic Rules

(a) Braising pan (with handles on either side and tight-fitting lid) should be in keeping with the size of the food item to be braised. This is to ensure even cooking, to avoid hot spots and cold spots.

(b) Fresh vegetables and herbs, with aroma, should be used as a bed for the joint to be braised.

(c) Food items, i.e. meat, poultry and game, should be evenly coloured while frying before being braised.

(d) Braising should be done slowly so as to enable the interchange of flavour between the item and the liquid.

(e) An appropriate liquid should half cover the food item.

(f) Items of meat, poultry and game suitable for braising are those of a tough nature, rather than those of a tender nature.

(g) Braised vegetables are better if served with a good sauce made separately. The braising liquor is strong and similar in taste to the vegetables.

3

Food Preparation

Brief Description

Bake: Baking is a method of cooking by dry heat usually in an oven.

Barbecue: A preparation of meat where meat is basted with highly seasoned sauce.

Batter: A mixture of flour and liquid which can be used for coating or for frying.

Beat: Mixing any substance by vigorous motion wherein air is incorporated.

Blanch: Dipping any food article into boiling water for a few minutes.

Blend: Combining two or more ingredients.

Broil: To cook by exposing food directly to the heat or through a grill.

Browning: A substitute added to darken certain preparations.

Brushing: Thin coating of beaten eggs or milk to pies or any dish to provide a glossy appearance to the product.

Caramel: A brown product of sugar on heating, usually used for coating or decorating or flavouring dishes.

Caramelise: Heat sugar to brown colour.

Cisel: Finely chop vegetables.

Coat: To cover with a thin layer.

Concass: Chop roughly.

Condiments: Spices and seasonings.

Consistency: Texture of the mixture.

Cream: Beat fat with a wooden spoon until it is light and fluffy.

Croutons: Small diced bread pieces fried or toasted for decorating soups.

Custard: Steamed or baked mixture of milk and egg.

Cut and fold: Mix flour very gently into a mixture.

Cutlet: Minced meat shaped into small round or oval shapes which are usually baked, grilled or fried.

Dice: To cut into small even cubes.

Dough: Mixture of flour and liquid in a stiff form.

Dredge: Cover food with a thin sprinkling.

Dust: To sprinkle lightly with powdered sugar, milk or flour.

Escalopes: Fried sliced meat dipped in egg and bread crumbs.

Farce: Stuffing of any kind.

Fillet: Fish with bone removed or slices from the breast of birds.

Flan: A thin pastry case.

Flute: A long crisp roll of bread used for decoration.

Fondant: Sugar boiled to 234°F and smoothened into a sweet.

Fritters: Foods like meat, vegetables, fish or fruits first coated with batter and then fried.

Frosting: Icing or coating sugar for decoration.

Fry: Cooking food in hot fat.

Gelatin: A product made from cow's hooves and generally used for jellies. Now available from vegetable sources also.

Gild: Cover an object with beaten eggs.

Glace: Giving a glossy appearance for certain dishes with icing, frozen items or with jelly.

Hors-d'oeuvre: Small savoury tit bits used as appetizers before a meal.

Junket: Milk coagulated by the addition of rennet.

Kabab: Minced or small meat pieces braised or curried.

Knead: Forcefully mixing the flour using the knuckles of the hand.

Lard: Fatty substances of meat.

Macaroni: Paste of flour forced through a tube. Thus long strands are obtained which are known as macaroni.

Macaroons: Small cakes with coconut.

Marinate: Soak meat or fish in mixture of vinegar and spices.

Mask: Coat a dish with sauce or moulded jelly.

Mayonnaise: By mixing egg yolk, oil, vinegar and mustard a thick sauce is made which is used for dressing salads.

Mince: Finely slice or shred.

Muffin: Moulded baked batter used with butter.

Noodles: Flour paste in different shapes usually used for Chinese cooking.

Parboil: Partially cooked food.

Pare: To peel.

Pluck: Remove feathers from poultry.

Poach: Cooking on low temperature in hot liquid in an open pan.

Puree: Smooth mixture obtained by sieving cooked fruits or vegetables.

Raising agents: Substances which produce gas when acted on by heat. Examples are baking soda, baking powder and yeast.

Saffron: A yellow colouring removed from a flower used for appearance and flavour.

Sauce: A pouring mixture used for enhancing the flavour and taste of another dish.

Saute: Cook in shallow fat to brown.

Scald: Immerse food in boiling liquid.

Shortening: Fat used for baking.

Shred: Cut in fine strips.

Sift: Put dry ingredients through a sieve.

Simmer. Cook food in liquid on low temperature just below the boiling point.

Souffle: Lightly baked or steamed pudding puffed with egg whiles.

Spaghetti: An Italian paste finer than macaroni and coarser than vermicelli.

Steam: Cook in steam.

Steep: Soak in liquid.

Stew: Cook by simmering in little liquid.

Stock: Flavoured liquid from meat, vegetables, fish or poultry used for soups, stews or sauces.

Weight and volume equivalents are essential to work out new recipes. To scientifically impart the knowledge of cooking

weights and volume equivalents are used. Some common weights and volume are given below:

1 teaspoons	...	5 millilitres
3 tablespoons	...	3 teaspoons or 1/2 ounce liquid
1 Cup	...	16 tablespoons or 8 ounces or 235 millilitre
1/2 Pint	...	10 ounces or 295 millilitres (approximate)
1 Pint	...	590 millilitre (approximate)
2 Pint	...	1.18 litre
1 Ounce	...	30 grammes
16 Ounce	...	455 grammes
1 Pound	...	455 grammes
2.5 Pound	...	1 kilogramme
1 Ounce	...	30 millilitre
34 Ounce	...	1 litre

Inducements and Aims

1. Cooking partly sterilises food. Above 40°C the growth of bacteria decreases rapidly and in general it ceases above 45°C. Non-sporing bacteria are killed at temperature above 60°C for varying periods of time. For instance, to make milk safe it is pasteurised at 62.8°C for 30 minutes. Boiling kills living cells, with the exception of spores, in a few seconds. Spore-bearing bacteria take about 4 to 5 hours boiling to be destroyed. To destroy them in a shorter time higher temperature must be used.
2. Cooking helps to make food more digestible. Complex foods are many a time split into simpler substances during cooking. This helps the body to absorb and utilise the food more readily than the raw food.
3. Cooking increases palatability.
4. Cooking makes food more attractive in appearance and therefore more appetising.

5. Cooking introduces variety. Many different types of dishes can be prepared with the same ingredients.
6. Cooking helps to provide a balanced meal. The different ingredients combined together in one dish make it easier to provide a balanced meal.

Over-heat: This is cooking on greased grill bars with the help of fat over direct heat.

Only first class cuts of meat, poultry and certain fish can be prepared this way. The grill bars are brushed with oil to prevent food sticking and can be heated by charcoal, coke, gas or electricity. The thickness of the food and the heat of the grill determine the cooking time.

Under-heat: Cooking is done on grill bars or on trays under direct heat. Steaks, chops and similar foods are cooked on the bars but fish, tomato and mushroom are usually cooked on trays.

Between Heat: Food is cooked between electrically heated grill bars.

Infrared: This is cooking by infrared radiation. This method reduces cooking time considerably, for instance, a steak can be grilled in one minute.

Microwave Cooking: Microwave heating is used in cooking in all advanced countries. Since the heating is very quick it has its own advantages. A simple microwave oven consists of a metal cabinet into which a quick magnetron is inserted. A magnetron is a kind of electron tube within a magnetic field which generates high frequency radiant energy. Alongwith a magnetron a metal fan is also installed and the fan distributes the microwave throughout. Foods placed in the oven is heated by microwaves from all directions and it helps in easy cooking.

Valuable Parts of Food

Tarragon: It is a native of Southern Europe. It is particularly common in France. The leaves of the plant are used.

Hop: Found in the Northern Temperate Region and Southern Europe. The flowers and shoots of the plant are used either fresh or dry.

Basil: It is an annual plant native to India. Its use in the West is relatively recent. Leaves are used fresh or dried. Fresh leaves have a taste similar to that clove. When dried, they taste more like curry leaves. It has mainly used in Italian dishes like Italian Pasta Sauce.

Sage: There are many varieties of this herb which are native to the Northern Mediterranean Coast. Leaves are used fresh or dried, mainly in stuffings.

Bay Leaf: The tree was originally found in Asia and is now grown in major parts of Europe. Leaves are used fresh or dried in stocks as bouquet garni, and as flavouring in many Indian dishes.

Thyme: There are many varieties of this perenninal herb which belong to Southern Europe and the Mediterranean. The leaves have many uses and forms the basic part of bouquet garni. It has a strong, sharp taste, hence, lemon thyme which is less strong with a lemon tange is often used instead.

Chive: Is a native of the coolest parts of Europe. It is now available in many parts of the world including America and Canada. Its stems are generally used. Their bright green colour and onion-like taste makes them an ideal garnish for salads and soups.

Oregano: Refers to wild Majoram. It is native of Asia, Europe and North Africa. It is now available worldwide. The leaves are usually used dried and are the best known for there use in Italian dishes, specially pizzas.

Rosemary: It is an aromatic herb native to the Mediterranean and is unavailable in most parts of the world. The leaves are used fresh or dried, most notably with lamb.

Dill: This plant belongs to the parsley family and is a native of Southern Europe and Western Asia and is now also

grown in Northern Europe. The leaves and seeds are used fresh or dried. The most important use of the leaves is probably in dill pickles, though they can also be used in salads or in cooked vegetables, and are essential to dill vinegar.

Chervil: It is an annual plant from Southern Russia, and the Middle East. It is now grown mainly in France. Fresh or dried leaves are used.

Fennel: A plant native to Southern England, and is now grown in many countries including America. The fresh or dried seeds or the dried roots are traditionally used with fish.

Grated Horse Radish: The grated outer part of the pungent root of the horse radish plant, a native of Eastern Europe, is usually added to cream, vinegar or mayonnaise to make a relish.

Cayenne Pepper: Refers to the dried pods of red chilli pepper. It is a native of Central America. It is ground to produce this hot pungent spice.

Paprika: It is made from a type of pepper native to South America. Paprika varies in taste from very hot to mild and sweet, and in appearance from rosy brown to scarlet.

Caper: Capers are unopened buds of a shrub native of the Mediterranean region. They are used only in their pickled form.

Olive: The fruit of a tree native of the Mediterranean Coast. They are used only in the pickled form. There are a variety of olives like pickled olive, black olive, dried olive, Spanish olive, olives with almonds, olives stuffed with dried pepper.

Gherkin: These are small pickled cucumber-like vegetables and can be eaten with cold meat, sausage and salad.

Application of Menu

A list of items served by a particular a Food and Beverage outlet is called a Menu.

A menu must be planned with utmost care. Menu planning

is a process by which the menu is planned for a F & B outlet keeping in mind the type of clientle the outlet wants to cater for. The main advantage of a well-planned menu is that it leads to consumer satisfaction. It also helps to motivate employees for a responsible and successful service.

Menu is the statement of F & B items provided by food establishments primarily based on consumer demand and designed to achieve organisational objectives. It represents the focal point around which components of food service systems are based. The basic consideration one has to keep in mind (while planning menu) are two different sets of people and how each set has a different view point. One set is the management or people involved in servicing the guest. The other set is the consumer or the guest.

Menu Management

Organisational Goal and Objective: The menu should be such that it doesn't hurt the goal and objectives set by the management.

The Menu Shouldn't Overshoot the Budget: The management should have some idea about how much a guest can spend. In other words one must conduct a survey on how much is the cost spending power. The factors on which spending power depends are:

Type of Clientele: The selling price should be such that one can keep a good profit margin after taking into account food cost, and overhead costs. Thus if dish price is very high it will not sell much. A guest will only spend Available Disposable Income (ADI) or customer spending power and food sale largely depends on it. Hence food budgeting is very important.

Budgets are normally a projection of:

(i) Established costs for serving actual number of covers.

(ii) Per capita allowance for expected number of covers.

(iii) Forecasting the number of consumers during a specific period of time.

Food Marketing Condition: The management must take advantage of the comparatively low rate of seasonal vegetables and fruits. Seasonal menus should be planned in order to avoid higher food cost.

Facilities and Equipments: While planning a menu the availability and capacity of equipment must be kept in mind as far as possible, labour-saving devices must be used to cut down on labour costs and overheads.

Personnel Skills: The manager must bear in mind, that the restaurant brigade might not be skilled in the service of particular type of food or kitchen staff might not be skilled in the preparation of particular foods, hence such items should be excluded from the menu.

Menu Pattern and Different Types of Menu

Menu pattern is the outline of food items and choices to be included in each menu.

The different types of menu are:

(i) Fixed or Set or Non Selective or Table D'Hote

(ii) Selective or A La Carte

(iii) Partially Selective

Table D'Hote: This is based on the pattern followed in the older days of guests taking whatever food the host had to offer. The term Table D'Hote if literally translated to English means table of the host. So whatever the host had for himself. *i.e.* Breakfast, Lunch or Dinner he could offer the same to the guest, therefore the choice was negligeable. The Table D'Hote menu is followed in factories and educational institutions and in commercial hotels. As the menu is pre-determined the covers are set beforehand. The guest has to pay for all dishes whether he has consumed a particular dish or not. There is no choice or limited choice available. The production of food becomes easier and food cost can be forecasted beforehand. There is however very little potential for preferences and also the menu is not flexible.

A La Carte: In an A La Carte menu there is a wide choice of dishes and each item is priced individually. More service skill is required and a lot of equipment is needed. An A La Carte menu is planned according to time, season and climate. The food is kept in a semi-prepared form and cooking is done as and when the order is received. The portion size is large. The choice of dishes are elaborate and the guest pays for whatever he has ordered. The portion is laid according to the order. This type of menu is very popular in restaurants.

Partially Selective Menu: This is a combination of Table D'Hote and A La Carte menu, the choice being limited.

Difference between A La Carte and Table D'hote

A La Carte	*Table D'Hote*
Food is kept in a semi-prepared form and takes time to serve.	Food is kept in fully prepared form and can be served immediately.
Portions served are large.	Portions served are small.
Food items are individually served and guests pay for what the orders.	Menu is collectively priced and the customer has to pay for the full menu whether he consumes a certain dish or not.
There is a vast choice. The menu is elaborate.	There is limited or no choice. The menu is comparatively small.
Silver is laid according to the dishes ordered.	Silver for the whole menu is laid in advance as the menu is known in advance.

Cyclic Menu: In case of a cyclic menu, the menus are repeated on a weekly or monthly basis. While planning it the seasonal availability and price is kept in mind.

Banquet Menu: It is an elaborate Table D'Hote menu offering more than four courses. The dishes are rich in quality and often fairly expensive as banquet is a formal feast. It is generally given to highlight important events like the visit of Head of States, marriage etc.

The banquet menu is prepared by the F & B Manager or the Banquet Manager in consultation with the Chef. A set

menu of various price ranges is made and offered to the client who organising the party after ascertaining prices per head.

Suggestions for extra dish/dishes by the clients are often considered, provided it doesn't imbalance the menu. The following points are kept in mind:

(i) Menu should not be thrown off balanced.

(ii) Costing of menu should not be affected unless the client is prepared to pay additional charges.

(iii) Work capacity and equipment available in the kitchen should be borne in mind.

(iv) Facilities available in the restaurant and capacity of service staff should also be considered.

Mock Menu for Practical

MENU

Salade De Russe

——— x ———

Consomme Xavier

ou

Creme de Tomate

——— x ———

Poulet a la Kiev

ou

Cotelettes des Legumes

Pommes de terre frites

Petit pois ou beurre

——— x ———

Gateaux Moka (Moccha)

——— x ———

Cafe

Description of Menu

1. Salade de Russe (Combination of par-boiled potatoes, green peas and carrots bound with mayonnaise)
2. Consomme Xavier (A thin, amber-coloured, crystal clear soup garnished with threaded egg).

3. Creme de Tomate (Puree of tomatoes garnished with fresh cream and crou tons)
4. Poulet a la kiev (Breast of chicken de-boned and flattened, stuffed with garlic flavoured butter, dipped in egg, rolled in breadcrumbs and deep fried to a golden brown colour)
5. Cotelettes des Legumes (Boiled, mashed potatoes, carrots, beans and green peas made into cutlet-shape, dipped in egg, rolled in breadcrumbs and fried to a golden brown colour)
6. Pommes de terre frites (Finger-size potatoes deep fried to a golden brown colour).
7. Petit pois au beurre (Boiled fresh green peas tossed in butter)
8. Gateaux Moka (Moccha) (Coffee cake)

MENU II

Consomme Juliene

ou

Puree Solferino

——— x ———

Chop de Porc Charcutiere

ou

Oeuf Poche florentine

Pommes de terre Bretonne

Petit pois Bonne Femme

——— x ———

Bavaroise Rubane

——— x ———

Cafe

S. No.	*Name of Course*	*Name of Dish in French*	*Name of Dish in English*	*Table Appointments*
1.	Soup course	(a) Consomme Juliene and small knife or (b) Puree Solferino	(a) Consomme Juliene or (b) Potato and tomato soup	Dessert spoon, quarter plate

contd.....

S. No.	Name of Course	Name of Dish in French	Name of Dish in English	Table Appointments
2.	Main course	(a) Chop de Porc Charcutiere or	(a) Porked chops or	Large knife and a large fork
		(b) Oeuf Poche Florentine	(b) Poached eggs	
		(c) Pommes de terre Bretonne	(c) A Potato dish	
		(d) Petit pois Bonne Femme	(d) A Vegetable dish	
3.	Dessert course	Bavaroise Rubane	Tricoloured Rubane	Dessert spoon and dessert fork.
4.		Café	Coffee	Coffee cup with coffee saucer & coffee spoon, placed on the right hand side. (from the sideboard)

S. No.	Sideboard Appointments	Pantry Appointments	Description of Dish
1.	Service set (Service fork and service spoon) and soup spoons	Soup bowls with under-liners, quarter plate, bread boat and butter strainer.	(a) Thin, amber coloured crystal clear soup garnished with julienes of carrots and turnips. (b) Thick soup made out of puree of potato and tomato, garnished with carrot balls and potato balls.
2.	4 service sets 2 oval flats-1 small, 1 medium, crumbing pan and brush, 1 small knife.	Hot large plates, Entree dishes-double compartment, triple compartment, oval flat.	(a) Shallow fried pork chops, half the cooking finished in the oven, served with Charchutiere sauce. (b) Poached eggs served on a bed of spinach, put in short crust pastry flan. (c) Dices of potatoes cooked in consomme with chopped onions, garlic and dices of tomatoes.

contd.....

S. No.	Sideboard Appointments	Pantry Appointments	Description of Dish
			(d) Boiled peas, button onions, shredded lettuce tossed in butter with seasoning.
3.	1 services set and 4 half plates	Sweet salver	(a) Cold pudding with chocolate bavarois on top, vanilla in the middle and strawberry in the bottom.
4.	Coffee cup, saucer and coffee spoon	Coffee tray with coffee pot milk jug	(a) Instant coffee served with or without sugar and and sugar pot milk.

MENU III

Consomme Andalouse

ou Puree Parmentière

x

Poulet Saute Chasseur

ou Omelette Espagnole

Pommes Chateau Petits pois a la flamande

x

Charlotte Royale

x

Cafe

S. No.	Name of Course	Name of Dish in French	Name of Dish in English	Table Appointments
1.	Soup course	(a) Consomme Andalouse	Consomme Andalouse	Dessert spoon,
		Ou or	side plate	
		(b) Puree Parmentiere	Potato Soup	with side knife
2.	Main course	(a) Poulet Saute Chasseur ou	(a) Sauted chicken with mushrooms tomato and white win	Large knife, large fork side plate

contd.....

S. No.	Name of Course	Name of Dish in French	Name of Dish in English	Table Appointments
		(b) Omelette Espagnole (c) Pomme Chateau (d) Petits pois a' la flamande	Or with side (b) Spanish omlette (c) Potato dish (d) Vegetable dish	knife
3.	Dessert course	Charlotte Royale	(a) Charlotte Royale	Dessert spoon and a Dessert fork

Sideboard Appointments	Pantry Appointments	Description of Dish
Service set, soup spoon plate and bread boat	Soup bowls with underliners, quarter	(a) Thin amber-coloured soup with tomato puree garnished with dices of tomato, boiled rice, julienes of ham, threaded egg. (b) Thick soup made out of puree of potatoes, garnished with croutons.
4 service sets, 2 oval flats-1 small 1 medium, crumbing pan and a brush, and small knife.	Hot large plates, oval flat and entree dishes	(a) Sauted chicken filled with white wine and brandy. Sauce is made out of butter, chopped shallots, demiglaze, tomato concasse and sliced mushroom sprinkled with chopped parsley (b) It is fluffy omellete made out of egg with dices of tomatoes, capsicum, minced onions cooked like a pancake. (c) Blanched olive-shape potatoes. Cooking is finished in the oven, roasted till it achieves a golden brown colour. (d) Boiled green peas and dices of carrots tossed in butter and flambed.
Service set and cold half plates.	Sweet salver	Surrounded with a layer of swiss roll, with a base of straw-berry jelly and topped with lemon jelly.

Menus Presentation: The success of catering operations depends to a greater extent on the way, the menus are planned and designed. Menus are the backbone of any catering operation. It is the menu which determines or sets the pattern for so many other factors which are important to the operation in other words the catering policies are determined by the menu. The menu should reflect the management policy and thinking. A good menu should ensure profits.

Menus have a lot of sale value. To make the best or most effective use of menus as a means of advertising and selling is called Menu Merchandising. It is that piece of advertising which is sure to be read. For this reason, every detail of the menu deserves closest scrutiny. The menu must fit the market, the facility, the ability of employees if the operation is to succeed. It should help the establishment to sell what it wants to sell more and it should also help the guest to choose from the menu what he wants without much loss of time. Due consideration should be given to the paper on which it is printed, the format, and the layout. In order to increase the merchandising value of menus the following points should be considered:

Clean: The presentation of dirty spotted, worn and old menu is the poorest way to start a meal. Cleanliness is a must in the hotel business. A soiled menu has a very poor merchandising value; it may raise doubts regarding whether the food is being prepared hygienically. Menus in continuous long use are likely to get soiled. Soiled menus must be replaced with new menus. To keep the menus in good conditions the first important thing is that the paper selected should be of reasonably good quality. The cover should be attractive and glazed. The front cover may also be protected by plastic folders. The inner contents should be changed as and when required. It is a very unattractive practice to change the prices on the menu with ink.

Legible: This means that menus should be easy to read. The type selected should be attractive and easy to decipher. It should be of good size so that most people, including the elderly or those with glasses, can read it with ease. As far as possible the menus should be printed. Typewritten menus should be avoided. They are invariably carelessly typed with mispelled words and smeared carbon. Menus should be written in double spacing as single spacing causes confusion and overcrowding. Single spacing tends to give a menu the quality of a telephone directory. Clear readability with plenty of wide

space is essential. The managers must supervise the lighting arrangements at all tables. Customers are annoyed when menus are indistinct or light or so poor that they cannot read it.

Format: The format of the menu should suit the contents of menu. As far as possible different formats should be used for different meals. The format of the menu also depends upon the type of establishment, its standards and nature e.g. hotels, cafeteria, clubs. The menu should not run into many pages. A two-paged menu is considered to be suitable. The hotel address is included. The back of the menu can be used as a wine list. But a separate wine list is more advisable.

Organisation: A menu should be well-organised. Similar items should be grouped together and attractive headings may be assigned to the groups. The purpose of a well-organised menu is that the customer should be able to find out what he or she wants quickly without reading the entire menu. Most customers start reading anything at the top left, so placing cocktails and appetizer in this position would be best. All the categories should be grouped, in an easy manner to find like Appetizers, Sandwiches, Salads, Entrees, Desserts, Beverage.

Restricted Menus: The menus should be short as far as possible. The number of dishes on the menu should be limited. Having a long menu is a poor merchandising policy. There are some operators who may believe that more items on the menu means more business. But actually, it usually works in the reverse.

The following are the disadvantages of very long menus:

1. They tend to confuse the customers and stretch the selection time for the guest.
2. They overburden the kitchen and serving staff.
3. It prevents the chef from concentrating on the presentation and making of the dish.
4. It leads to increased wastage as there will be more left over.

5. They slow down the service as a waiter has to wait for a very long time for the customer to make his selection.
6. They create bottlenecks in the preparation and service area.
7. Buying, receiving and storing of the food becomes difficult and more space is required for storage.
8. They hamper the effectiveness of control process.

It may be well realised that if the menu has many items, there would still be some guests who would asks for something else. The basis for limiting the menu is to learn about the items which are most popular. The person responsible for planning the menus should from time to time analyse the sale of different dishes and the items which do not sell reasonably should be removed. Limited menus does not mean no choice or change at all. Establishments that have permanent regular clientele do change the menus and provide for ample choice. Establishments which have more of tourist business, and where the customers change constantly don't need to change the mentis frequently.

Easy to Change: Even the most carefully planned menus must be changed from time to time. The change may be necessitated because of the change in prices, need to add or drop some items for a variety of reasons. The tendency to mark out items with pencils or pen, whiting items, or worst of all scratch out the old prices and write in the new is undervalue. The changed prices on the menu has a poor merchandising effect. The menu format and design should be such that, it can be changed with the needs of time.

Type of Operation: To serve good food well and promptly one must have a menu designed to fit the place. It must match the size and kinds of equipment, their capacity, and also the skill of the personnel. One should never blindly follow competitors nor include any item unless one can effectively prepare and sell it.

Merchandising Effect: The menu is designed to sell the items that are the speciality of the hotel, or an item that can

be served fast and is profitable. Such an item should be tactfully located. Different typefaces can be used to emphasise one item over another, or attractive borders can be used to make something stand out. While listing the items, the item that one wants to sell fast can be placed at or near the top, since many people read only until they find an item that appeals to them, rather than studying the entire list. Accompaniments should be restricted and specialised.

Language: The language on the menu should be easy for the customer and the staff. Many guests are embarrassed to ask what a foreign term means and will pass on to something that they understand. Where a menu is written in French, the English equivalents should be given. It is a good idea to describe the dishes in the language which is understood by the majority.

Effective Descriptions with Descriptive Headings: Descriptive headings are inserted in the menu for various groups of food which attract the attention of the customer and indicate the nature of the dish more clearly. The headings assigned to different groups should be interesting and should capture the imagination of the guests.

Some examples:

Sea Coast Delights
Gourmet Selection
From the Sizzling Platters
Oyster Bar
Dessert Delicacies
Broiled To Your Order
Chef's Specials.

The explanation of dishes in detail should be avoided, but a brief description would help the customer in making a prompt and suitable selection. Dishes should be described in such a way that a common customer should be able to understand how he is going to get his food and should be able to visualise it. The aim should be to inform the guest regarding the form of food served (method of cooking, cut, hot or cold etc.).

A careful thought given by personnel incharge of making the menu regarding the description location and the inclusion and omission of certain words help considerably in making a menu attractive. It is good merchandising policy to add whatever extra are being given to the customer for e.g. instead of writing simply 'Fillets de Sole Bonne Femme' it is better to feature it as Fillets de Sole Bonne Femme served in white sauce with fresh mushrooms and potato border. Such a description adds to the value of dish. Associating the names of places of origin white describing the dishes are informative and has better sales appeal.

There are words that should not be used on the menu as they are uninviting and lack sales appeal. These are plain, boiled, minced, mashed. Instead of writing plain omlette it could be written as golden or natural. For eg. Listing spinach as garden leaf, boiled potatoes as natural instead of boiled.

For merchandising appeal fresh tender, green, young, spring, new, chilled, special, choice can be used. The word fresh is not exclusively used for vegetables, fruits etc. and can be used when the customer might otherwise suppose the food to be canned or stored.

Some examples are:

Fresh Orange Juice
Grilled Spring Chicken
Fresh Mushrooms
Tender Lamb chops
New Spring Broccoli
New potatoes
Fresh Fruit Salad
Young Turkey
Melon Frappe
Chilled Fruit
New (Bermuda potatoes, cabbage vegetables)
Green (Peas, Cabbage, Apple, Olives)

Young: Chicken, vegetables, Turkey, Lobster, Beets, Carrots

Spring: Vegetable, chicken lamb salad, onion

Garden: Vegetable salad

Tender: Young carrots, buttered beets, baby spring, Roast lamb chops, Spring lamb.

Baby: Sole, Lobster, Lima beans, beets, carrots.

Crisp: Salad, Potatoes

Items should be listed in bold type or at the top of the list. It is not sound merchandising to arrange the meals according to price, and to put lowest priced items first, lower priced next, and highest price last. This focuses too much attention on the lowest priced items, which, if they were listed first would get more orders.

If the business is divided between A La Carte and Table D'Hote and one wants to increase the A La Carte menu, then lesser space should be devoted to Table D'Hote meal listing is. This will attract more attention to the A La Carte choices.

Specials: The inclusion of Specials on the daily menu helps considerably in increasing the sale volume of the establishment. Some of the popular type items are put on the menu under the heading of Specials. The popular type of Specials are as follows:

(1) Daily Special, (2) Chef's Special (3) House Special (4) National Specials (5) Local Special. The special must be choosen with care and priced correctly and supported by good salesmanship or enthusiastic waiters. The Special dish should reflect the occasion or should justify their inclusion on the menu. Since Special's should be specially or exceptionally good, recipes for Specials, are checked over thoroughly before they are introduced or started.

The sale of Specials are watched for a period of time and improvements and revisions are carried out immediately. If a particular item does not sell well or the sale decreases one can

check the recipe to see as to whether this is being followed or needs improvement. Sometimes by a slight change, in the recipe or service, the sale can be increased. Many factors affect the inclusion of Special's on the menu like clientele, leftovers of previous day and different days, of the week.

Special's are normally printed in different colours.

Special's are sometimes priced at break even cost to attract the clientele. Gradually the prices are increased and the customers willingly to pay the increased prices.

Blue Plates: It is also known Preplated service/Quick service. It is an American term for the preplated service of a full meal. It is another way of merchandising by including Blue Plates as popular quick price meals. This also helps in quick and speedy service and is generally used by establishments near to the business centres offices.

Special blue plates are particularly effective at dinner or in the business localities where a average class businessman wants the quick lunch or dinner. Blue plates are generally priced proportionately lower than the cost of an entire dinner. These blue plates appeal to customers who seek a lower price and to those who do not wish to have a complete big dinner and who prefer not to order A La Carte' as their prices are generally much higher.

Menu: A well-designed menu in a proper format is a very useful as a souvenir menu. It is good source of advertisement. Many people have the tendency of collecting the menus of the restaurants they visit. When a menu is demanded as a souvenir perhaps it indicates that a favourable impression has been created. Depending on the policy, the menu can be given to the guests as souvenir. Customers should encourage the guests to carry after the party is over with them, the Banquet Menu.

Cyclic Menus: These are a series of fixed or semi-fixed menus which are repeated in a set period. The length of the period is usually related to the length of the menu.

For example:

21 days for a menu of three main courses.

14 days for a menu of five main courses.

Thus, the greater the choice of menu items, the shorter is the length of the cyclic menus. In practice, cyclic menus cover various periods, usually from 10 to 28 days.

Cyclic Menus are often used in institutional catering establishments (canteens, hospitals etc.) and in restaurants offering Table D'Hote menus.

The main advantages of Cyclic menus are:

(a) Through the use Cyclic menus a pattern emerges which shows clear trends in customer's likes and dislikes of menu items, particularly when offered with specific range of alternatives. With an established pattern of customer demand for a menu item, volume forecasting, purchasing and preparation tend to become more accurate.

(b) The staff requirement can be worked out very accurately when the production requirements are known. This assists not only in the preparation of staff rotas of duty, days off, holidays etc. but may result in the total number of staff being reduced, thus saving in the kitchen and restaurant wage bill.

4

Food Cooking

Some foods like fruits, vegetables and nuts are eaten raw. It is good that they are consumed raw as in the uncooked condition they retain most of their nutritive value. However, most foods are cooked before they are accepted. Cooking of food is the use of heat to bring about desirable changes in foods being consumed. The source of heat may be the result of combustion of wood, coal, gas or oil; by electric heating elements contained in hot plates or electric ovens; or by a microwave cooker.

Kitchen Methods

Cooking of food produces improvement in flavour, texture, and appearance, and this makes the food more palatable and easily digestible. Most foods in the raw state contain harmful micro-organisms and they are destroyed during cooking.

Improvement of Food Quality: Cooking improves the natural flavour of food, e.g., the flavour of meat is developed and enhanced by heating; the flavour of bread develops in the crust during baking; coffee flavour is formed during the

roasting of green coffee, and so on. If a blend of flavours is required, several foods could be cooked together to bring about this. If the object is to change the flavour of a food, cooking helps achieve this by addition of required flavouring material. Cooking could also destroy the flavour of a food. In such a case, if the original flavour of a food is to be retained, there should be minimum cooking. Overcooking results in the vitalisation of flavour substances and a less desired product may be obtained. Colour and texture also influence flavour. Cooking helps develop the desired colour and texture in foods.

Destruction of Micro-organisms: Micro-organisms are ubiquitous in their distribution; they are found everywhere in every environment, and adapt themselves to almost every growth condition. They make their presence known in many ways. Their action may be beneficial as in the production of cheese, pickles, leavening of bread, etc., or harmful as in food spoilage.

The latter type of micro-organisms are not normally present as food contaminants. When present they cause infections or produce toxins. Infective types of bacteria, such as *Salmonella* and *Shigella,* multiply in the intestinal tract and cause disease through the infection of the host. In contrast, bacteria like *Staphylococcus aureus* and *Clostridium botulinum* produce the toxins present in the food at the time of consumption and are the direct causes of illness. Food storage under conditions that prevent bacterial growth do not provide safety against infective pathogens but do so against toxinogenic food poisons.

Molds also produce toxins. The mold *Aspergillus flavus* produces a toxin (aflatoxin) which, when present in food, is a health hazard. There are many other molds producing toxins referred to as mycotoxins. One of the most important methods of protection of food against harmful micro-organisms is by the application of heat. Cooking food to the required temperature for a required length of time can destroy all harmful micro-organisms in foods.

Process of Cooking

Heat may be transferred to the food by conduction, convection, radiation or by the energy of microwaves (electronic heat transfer). Conduction is the method of transfer of heat by contact. Convection is the transfer of heat as a result of the flow of a liquid or gas travelling from the hotter to a less hot part of an oven or saucepan. Radiation is the emission of heat in the form of waves from hot objects. Microwaves are a form of electromagnetic radiations similar to radio, TV, radar, light and infrared waves.

Conduction: In conduction, heat flows from the source to the material absorbing it. Certain materials are better conductors of heat than others. Copper utensils are the best conductors; aluminium ones conduct heat more slowly, and steel ones still more slowly. Glass is a very much less efficient conductor of heat than any of the metals. For efficient conduction to take place from a hot surface to another surface, such as the bottom of a saucepan, it is important that there is as large an area of contact as possible. Hence, the bottoms of pans should be flat and thick.

Convection: When a liquid or air is heated, the portions nearest to the heat become warm and less dense. They rise and are replaced by the denser material, i.e., convection currents flow from a more dense to less dense areas. The convection currents usually flow in a vertical direction and they are hindered by solid materials. Roasting is mainly accomplished by convection. The heat source at the bottom of the oven heats the air which rises and is continuously replaced by cold air. These convection currents create uniform temperature in the centre of the oven where the food is kept, resulting in roasting.

Radiation: Radiation waves travel at the same speed as light rays. Like light rays they are absorbed by dull black or rough surfaces and reflected by smooth, white or metallic surfaces. Radiation waves travel through gases, clear liquids or glass without heating them, or through vacuum. In cooking,

when heat radiations reach the food, only the surface is heated by them. They do not penetrate the food. The rest of the food is cooked mostly by conduction and also to some extent by convection. Radiation heat is used in broiling. Toasting of bread is essentially by radiation heat.

Microwave Heating: Microwaves are absorbed and penetrate the food. The energy of these electromagnetic radiations excite the water molecules (in food) which bear a positive electrical charge in one portioa and a negative charge at another position of the molecule (dipole). When the electric field of the microwave interacts with the water dipole, the water molecules begin to vibrate very rapidly in food. This vibration produces friction that creates heat within the food, thereby cooking it. The excitation of water molecules occurs as far within the food as the microwaves are able to penetrate. This is different from heating by convection methods which transmit heat from the edges of the vessel to the centre. Heating of food beyond the depth of penetration of the microwaves occurs as a result of the conduction and/or convection of the heat created by microwaves.

Use of Cooking Apparatus

Cooking can be carried out in various media or no media at all. Air, water, steam and fat, or combinations of these, are used as cooking media. Microwave heating involves generation of heat within the food; it is not a cooking medium.

Cooking in Air Grilling: Roasting and baking take place in air. Grilling consists of placing the food below or above a red-hot surface. When under the heater, the food is heated by radiation only. This results in the browning of many foods. Then the heat is more slowly conducted through the surfaces of the food downwards. As heating is mostly superficial, gritted foods are usually reversed or rotated in the oven. If the food is above the heater, heat is transmitted to the food through convection currents, as well as radiations with consequent increased efficiency.

Roasting and baking are essentially the same. They are carried out in an oven between temperatures of 120° and 260°C. Generally, the term roasting is applied to meat cooking, while baking is used for breads, cakes and biscuits. The food is cooked partially by dry heat and partially moist heat, if the food is high in moisture content. In baking, the oven atmosphere should be moist initially so that the moisture condenses on the cold surface. This helps in heat transfer and plays a part in the formation of crust. Roasting and baking involve heat transfer from the heat source in the oven by radiation, conduction and convection. Heat is transferred directly onto the container of the food through which it is conducted to the food. Convection currents of air help keep the temperature of the oven fairly uniform. Rough and black surfaces absorbing more radiated heat are more efficient for baking than the bright shiny ones.

Broiling is the cooking of food by exposing it to direct heat. In this case, cooking takes place by conduction through direct contact of food with the hot broiler. Some radiation cooking also takes place in broiling.

Cooking in Water: Boiling, simmering or stewing involves cooking in water. In these cases the medium transferring heat is water. Water receives heat by conduction through the sides of the utensil in which the food is cooked and passes on the heat by convection currents which equalize the temperature and become very vigorous when boiling commences. Water is a poor conductor of heat and its heat capacity is high, i.e., it requires more heat than any other liquid of the same weight to raise the temperature. The boiling point of water is 100°C and it is altered at high altitudes and in presence of electrolytes. Simmering and poaching are methods of cooking of food by immersing in a hot liquid maintained at a temperature just below the boiling point.

Cooking in Steam: Steam is the medium of cooking in steaming, "waterless" cooking and pressure cooking. Cooking

by these methods involves moist heat. In steaming, food is cooked by steam from added water, whilst in waterless cooking the steam originates from the food itself. Cooking food wrapped in aluminium foil or in a plastic bag is a form of waterless cooking. In this case, there is the advantage of preventing the transmission of flavour from or to the sealed food. Pressure cooking is a device to reduce the cooking time by increasing the pressure so that the boiling point of water is automatically raised. While water boils at 100°C at normal atmospheric pressure, it boils at 121°C at a pressure of 1.07 kg/cm^2 which is the pressure at which food is cooked in a kitchen pressure cooker. In cooking by steam, the food is heated as a result of steam condensing on the food, and the release of the large quantity of heat (latent heat) contained in the steam. This continues until the heated food reaches the same temperature as steam.

Cooking in Fat: Fat is used as a medium for cooking in sauteing, shallow fat frying (pan frying) and deep fat frying. Sauteing is the cooking of food in a slightly greased pan. Only thin pieces of food (dosas, for example) are cooked this way. This prevents food from sticking to the pan. The heat is transferred to the food mainly by conduction. The food is to be turned from one side to the other to complete the cooking. In shallow fat frying, food is cooked in a larger amount of fat, but not enough to cover it. Heat is transferred to the food partially by conduction by contact with the heated pan and partially by the convection currents of the food. This prevents local burning of food by convectioning away the intense heat of the frying pan. As in sauteing, even in this case, the food must normally be turned over to ensure some degree of uniform cooking. Both sauteing and pan frying are really a type of baking.

Deep fat frying is similar to boiling. Food is cooked by vigorous convection currents and cooking is uniform on all sides of the food. As fat can be heated to a much higher temperature than the boiling point of water, cooking can be

rapidly completed in deep fat frying. In most foods, this high temperature results in rapid drying-out of the surface, and the production of a hard, crisp surface, usually brown, and the absorption of a fair amount of fat, which raises the calorific value of the food substantially. Fats should not be heated to the smoking point as they decompose at that point to fatty acids and glycerol, followed by the decomposition of glycerol to acrolein, which causes irritation to the eyes and nose.

Foods may also be prepared by a combination of media. Preparation of *uppuma,* for example, involves the use of a combination of fat and water media. Some components (onions) of the preparation are first browned in a small quantify of fat followed by the cooking of semolina in water.

Cooking in Microwave: Microwave cooking, which was first recognised as valuable for cooking food in 1947, is becoming more and more popular because it is a convenient and very quick method of cooking food. Microwaves do not require any medium for transfer of heat in cooking. They are generated by an electric instrument called magnetron. Microwaves have a high frequency of 2,450 megacycles (or 2,450 million times) per second. The microwaves can be absorbed, transmitted or reflected when they impinge on substances. They pass through paper, china, glass and some plastics without absorption, and are reflected by metals and absorbed by food. When food is kept in the cavity of a microwave oven for cooking, microwaves generated by the magnetron strike the food and the metal wails of the oven. Microwaves that strike the metal wall are reflected and bounce back so that they disperse through the oven.

This is important for accomplishing a uniform heating of food. Microwaves penetrate food to a depth of 2.5 to 7.5 cm. Up to this limit of penetration of the microwaves, the food gets heated and cooked. Thus, food will heat up inside and outside at the same time. When the thickness of food is large, microwaves will heat it up throughout at the same time to the depth of their penetration and the portion of food beyond it

will be heated more slowly by conduction. Food, for cooking in a microwave oven, should be kept in containers that transmit the microwaves and should not absorb or reflect them. This is achieved by-using paper containers, such as paper plates or cups, or utensils made of plastic, glass or chinaware, which do not contain metallic substances.

Advantages of Microwave Cooking: The most important advantage of microwave cooking is the speed with which cooking may be accomplished resulting in a considerable saving of time in cooking. Microwave cooking is ten times faster than cooking by the conventional methods; a cake cooks in just five minutes. The length of time of cooking will be increased if the quantity of food to be cooked is increased. Microwave heating also saves time in heating refrigerated or frozen foods; the latter can be defrosted in minutes instead of hours. Another advantage of this method of cooking is that only the food is heated during cooking. The oven does not get heated and even the utensils do not become hot, except when the cooking period is relatively long and some heat is conducted from the food to the container. The flavour and texture of prepared food do not change when reheated by microwaves. This is another advantage of this type of cooking. Electrical energy required for microwave cooking is less than that required by the traditional method.

Limitations of Microwave Cooking: There are some limitations to microwave cooking. As the process of cooking does not rely on the conduction or convection of the material being heated, it is normally possible to heat the whole of the food simultaneously. Due to the short period of time of cooking, the food does not become brown on the surface as many foods do when they are roasted or baked in a conventional oven. Baked products without a brown surface do not have appeal. To overcome the problem created by lack of browning a microwave oven is combined with a browning unit. Another disadvantage of microwave cooking is that it cannot be employed for simmering or stewing for tenderising foods, as

also for deep-fat frying. Also, the short cooking time may not give a chance for the blending of various flavours to develop as in the conventional methods of cooking.

Recent Trends in Cooking: When food is subjected to heat many changes occur; there is some destruction of proteins, lipids and vitamins, which is detrimental to the nutritional value of food. However, there are also some beneficial changes. Cooking is required if we have to obtain the maximum nutritive value of some foods and maintain a safe and wholesome food supply. Heat treatment is also one of the most common and effective methods of food preservation and may be used alone or in combination with other preservation techniques. Some of the changes that take place in the constituents of food during cooking are discussed below.

Changes in Proteins: The principal effect of heat on protein is de-naturation. This results in the destruction of micro-organisms and inactivation of microbial and natural enzymes within the food. Cooking also destroys the toxic proteins and peptides, enzyme inhibitors, antivitamins and other natural toxicants in food, which can seriously affect their nutritive value. Legumes contain trypsin inhibitors, haemogglutinins and other toxic substances which affect the digestibility and availability of sulphur-containing amino acids. These are destroyed by heat. Approximately 40 per cent of the growth-depressing effect of uncooked soyabeans is due to the presence of trypsin inhibitors. Of the remaining proteins of soyabeans digestibility increases after cooking. Cereal grains also contain trypsin inhibitors and natural toxicants. Heat destroys these antimetabolites in rice, wheat and oats, but has little effect in other cereals.

Cooking can also result in the interaction of protein with nonprotein components of the food system; there can be interaction of protein with carbohydrate or lipid oxidation products'. There can also be interprotein and intraprotein reactions in the presence or absence of oxygen. These changes in cooking result in nutritional unavailability of proteins.

There is significant loss of lysine and the sulphur-containing amino acid cysteine after heating proteins. On prolonged heating, tryptophan, methionine and the basic amino acids are also lost. Charring and the presence of off-odours during cooking is due to destruction of amino acids and proteins. These changes affect the palatability of the final product.

Interaction between the free amino groups of proteins with reducing sugars or carbonyl groups formed by lipid oxidation results in non-enzymatic browning (Maillard browning). This reaction is of importance because it is responsible for many of the specific tastes, aromas and colour of foods. In this reaction, the amino group of the essential amino acid lysine reacts with carbonyl groups of sugar and fatty-acid oxidation compounds, which results in a decrease in the nutritive value of proteins.

Changes in Carbohydrates: Monosaccharides, oligo and polysaccharides undergo many transformations when cooked in an aqueous medium. The sugars are subjected to degradation and epimerisation and, of over 100 compounds that are formed by such transformations, hydroxymethyl furfural is the most important compound. This compound and furfural are also products of non-enzymic browning reactions. But these compounds have no adverse effects.

Starch molecules which are the main source of calories in many diets, when heated in an aqueous or moist environment, swell and rupture, and this permits greater enzymatic digestion by enzymes like amylases. Cooking thus increases the digestibility of carbohydrates. Starch, when subjected to dry heat at a temperature of 200°C or higher, breaks down resulting in the formation of dextrin and volatile compounds like furfural and hydroxymethyl furfural.

Changes in Lipids: Lipids undergo hydrolytic, oxidative, polymeric or other degradative changes which modify not only the physical properties of the lipid but also their biological properties when heated. The hydrolytic and oxidative changes

result in rancidity. Hydrolytic rancidity is catalysed in food at high temperature and pressure in an aqueous medium in the presence of acids, alkalis and lipolytic enzymes (Upases). Hydrolytic rancidity by itself does not bring about any significant change in the nutritive value of the food. However, the objectionable flavour imparted by free fatty acids lowers the consumption of food.

Oxidative rancidity is responsible for more losses in the quality and nutritive value of lipids than any other change. This rancidity results in the formation of hydroperoxides as primary products owing to the attack of oxygen on the unsaturated centres of lipids. The products of oxidation exhibit strong unpleasant flavours even when present in extremely low amounts. The oxidation of a fat at high temperature, in addition to volatile and non-volatile compounds associated with lipid oxidation at normal temperature, gives isomerisation products (trans- and conjugated double bond products), cyclic compounds, dimers and polymers. While lipids oxidized at normal temperatures exhibit no toxicity, thermally oxidized lipids have shown various detrimental effects. But the lipid or food system would be unpalatable long before the concentration of the toxic compound reached a hazardous level.

In addition to the effect on the biological properties of lipids, thermal effects bring about physical and chemical changes also. In sauteing and shallow fat frying, the quantity of oil used is small, cooking time is short and there is generally no reuse of fat or oil and thus there is little concern over the nutritional effects of lipids absorbed from such cooking. In contrast, there has been a great deal of concern over deep fat fried foods. If the deep fat frying is continuous, oxidative changes are small because the fat absorbed by the food is constantly replaced. In discontinuous deep fat frying, as is common in homes, there is liberation of fatty acids due to the addition of water to the oil from the food, decreased unsaturation and increase in peroxides, conjugated double

bonds and polymers. Such fats absorbed by foods could be toxic when consumed.

Changes in Vitamins and Minerals: These are lost primarily by leaching, oxidation of the water-soluble nutrients and thermal destruction.

The loss of water-soluble vitamins ranges from 0 to 60 per cent as a result of leaching and thermal destruction. In addition to losses by water and heat, ascorbic acid is lost by oxidation due to exposure to air of food during cooking. Vitamin A and carotene are water insoluble and, as such, are not lost as a result of leaching and their destruction due to oxidation is very slight. Frying and roasting cause their loss up to 40-60 per cent. Minerals also are lost on account of leaching and their losses are smaller (0 to 35 per cent).

Changes in Colour: Colour factors in foods, such as anthocyanins, carotenoids, chlorophylls, myoglobin, etc., are affected by heat. In addition to heat, the acidity or alkalinity of the cooking medium, oxygen and presence or absence of metals, also contribute to colour changes when heated. In some cases, the colour changes that take place in foods on cooking are desirable (as in baking) while in some other cases the changes may be undesirable (as in the prolonged cooking of cabbage). The cooking condition should be so organized as to obtain the desired colour qualities in the cooked food.

Art of Cooking

For every human being, the food is essential and nourishes the body. It is defined as anything eaten or drunk that can be absorbed by the body to be used as an energy source, building, regulating or protective material. Food can be defined as anything solid or liquid, which when swallowed, digested and assimilated in the body keeps it well. Food is a prerequisite of nutrition. Nutrition is defined as food at work in the body. In other words, food being processed and used for various functions in the body.

The Functions

1. Food provides materials for tissue building, growth and body repair mainly through proteins and minerals, eg., proteins and minerals.
2. Food provides energy to the body through nutrients like carbohydrates and fats (lipids), eg., carbohydrates, fats.
3. Protective foods are essential for safeguarding against the diseases, eg., Vitamins, minerals.
4. Regulatory foods are needed for the normal working of the body which includes heart beating, body temperature maintenance, muscle construction, blood clotting, waste products removal from the body and control of water balance.
5. Food also serves the social function. Special foods are served in public gatherings, ceremonies and religious functions, eg., water, roughage.

Various Groups: A balanced diet is one that contains different types of foods in such quantities and proportions from which the need of calories, minerals, vitamins and other nutrients is adequately met. These foods are divided into various groups as follows:

Group I: It includes cereals, millets, roots and tubers. All these primarily supply energy or calories, eg. rice, wheat, jowar, maize, bajra, ragi and tomato, potato, sweet-potato, colocasia, yam. These are an important source of thiamine, niacin and iron.

Group II: The food stuffs in this group are primarily sources of protein, eg. Dals, grains, peas, beans, groundnuts, cashewnuts, almonds, coconut, milk, curd, buttermilk, paneer, mava, eggs, fish, mutton, chicken, pork and other flesh foods. Milk and milk products are important sources of calcium and riboflavin. Meat, fish and eggs rank first for their protein, iron and niacin content.

Group III: These food stuffs supply energy or calories, eg., vegetable oils, vanaspati ghee, butter, cream, sugar, jaggery. This group constitutes about 1/6th of the energy value of the diet. Cooking oils include mustard oil, coconut oil, til oil, groundnut oil, palm oil and sunflower oil supply essential fatty acids.

Group IV: These are mainly suppliers of minerals and vitamins. This includes protective vegetables and fruits.

(a) Green leafy vegetables, palak, methi, gogu, drumstick, coriander, fenugreek and arvi leaves, kulfa and similar greens supply vitamin B_2, folic acid, calcium, iron, fibre and carotenoids.

(b) Yellow or orange fruits and vegetables - papaya, mango, carrots, yellow pumpkin provide carotenoids.

(c) Fruits and vegetables rich in vitamin C - amla, guava, lemon, orange, phalsa.

These are the only important sources of ascorbic acid, contribute half of the vitamin A requirement, supply 1/5th of the iron required.

Group V: Other vegetables and fruits are included in this group. These provide variety in taste and texture and furnish roughage in the diet. eg. fruits, stems, leaves and flowers of plants, lady's finger, brinjal, bitter gourds and other gourds, cabbage, cauliflower, drumstick supply carotenoids, folic acid, calcium and fibre in the diet. Various fruits included in this group are bananas, melons, grapes, apples, sapotas, berries, pears, plantain, lichis, pomegranate etc.

Simple Techniques of Cooking

Food preparation is an important step in meeting the nutritional needs of the family. It is not enough that food be nutritious, it has to be pleasing in appearance and taste so that it is eaten.

Principles of Cooking

1. All foods must be cooked in a way that 'keeps the flavour in'.
2. Sometimes the flavour of the food is 'drawn out' into the gravy or broth.
3. The preservation of the maximum nutritive value can be ensured by using correct methods of cooking, suited to the particular foods.

The process of subjecting foods to the action of heat is termed as cooking. Food can be cooked by various methods to make it edible.

Roasting: Food is brought into direct contact with the heat source. The food is periodically coated with fat and turned over the fire from time to time to cook it evenly. It can be done on live coals, under a grill, in the tandoor or any other oven. Temperature to be maintained is 350° C to 375° C. The term roasting is applied to meat cooking. It is of three types.

(a) *Spit Roasting:* The food to be cooked is brought in contact with direct flame in front of a clear bright fire. The food is basted over with fat and is also turned regularly to ensure even cooking and browning, eg. Barbecued meat.

(b) *Oven Roasting:* This is cooked in a closed oven with the aid of fat. The food is put into a very hot oven for 5 to 10 minutes and the temperature is lowered to allow the joint to be cooked. Cooking in a moderate oven for a longer time produces a better cooked joint than cooking at high temperature for a shorter period as the meat retains its moisture and flavour.

Broiling or Grilling: The food is placed on a metal grill directly over or below the source of heat either gas or electricity. Under the heater, the food is heated by radiation only. This results in browning of various foods. Then the heat is more slowly conducted through the surfaces of the food downwards. If the food is above the heater, heat is transmitted to the food

through convection currents, as well as radiations with consequent increased efficiency. This method is usually used for tender cuts of meat, poultry, etc. and when foods need to be browned as made crisp on the surface, eg. pizzas, sausages, bacon, etc. When food is cooked uncovered on heated metal or a frying pan, it is known as pan-broiling.

Baking: It requires an oven or tandoor as any equipment in which hot air circulates around the food placed in the action of dry heat is combined with the steam generated from the food during cooking. Foods baked are generally brown and crisp on the top and soft and porous in the centres, eg. cakes, breads, puddings, vegetables, biscuits, pastries, potatoes, meat dishes in sauce etc. Temperature maintained in this is 120° C to 160° C. Baking involves heat transfer from the heat source in the oven by radiation, conduction and convection.

Frying: In frying the food is cooked in hot fat. Fat has very much higher boiling point than water. Frying can be done by different ways:

(a) *Sauteing:* This method involves cooking food in just enough fat or oil to coat/base of the pan in which the cooking is done eg. Dosa. The food is tossed occasionally or turned over to enable the pieces to come in contact with the oil or fat and evenly cooked and prevents food from sticking to the pan. The heat is transferred to the food mainly by conduction.

(b) *Shallow frying:* Sufficient quantity of fat is used in the pan and food is turned to cook both sides equally as it is done for paratha, omelette, pancake, and tikki. Heat is transferred to the food partially by conduction and partially by convection currents.

(c) *Deep frying:* It is done in a deep saucepan or kadai that contains excess quantity of fat or oil to immerse the food fried. Potato chips, bonda, pakoras and puris are deep fat fried preparations. For satisfactory results, food should be fried in fats and oils heated to 320° C.

Fat can be heated to a much higher temperature than the boiling point of water, so cooking is rapidly completed. In most foods, this high temperature results in a hard, crisp surface, usually brown and the absorption of a fair amount of fat which raises the calorific value of the food substantially.

(d) *Fricasseeing:* It is cooking in a small amount of fat and then serving with sauce.

Moist/Wet Heat Methods

1. *Boiling:* Cooking foods in water that is bubbling constantly over the food in a pan. The temperature of the water is 100° C when it boils, eg. potatoes, eggs, sweet potatoes, rice, etc.
2. *Simmering:* It is done at 85° to 90° C temperature when foods are cooked in water or liquid which is not bubbling vigorously as in boiling, it is simmering. The very tiny bubbles coming to the surface and breaking. This method takes longer to cook food than boiling eg. cuts of meat used for stews or stock preparation.
3. *Blanching:* In this method of cooking, the food is usually dipped in boiling water for 5 seconds to 2 minute, depending on the texture of the food. The purpose of this method is to remove peels of fruits, vegetables, nuts, etc. easily without changing their texture too much. The process of subjecting foods to boiling temperature for short period is known as blanching.
4. *Steaming:* It requires food to be placed in a vessel in which steam can enter and cook the food. The steam is generated by boiling water in a pan and another pan is placed in water. The second pan contains the food to be cooked, e.g. Idlis, khaman, dhokla and other fermented products. This food is easily digestable, nutritious and full of flavour. While in waterless cooking the steam originates from the food itself. Cooking food wrapped in an aluminium foil or in a plastic bag is a form of waterless cooking.

5. *Pressure Cooking:* This is a method of cooking food by steam under pressure. While water boils at 110° C at normal atmospheric pressure, it boils at 121° C at a pressure of 1.07 kg/cm^2, which is the pressure at which food is cooked in a kitchen pressure cooker (food is cooked at 110°-112° C). It is a quicker method than pan steaming generally done in pressure cooker. The storter cooking time enhances nutrient retention and palatability eg. curries, soups, broths, stews, etc.
6. *Poaching:* This method involves cooking in a small amount of liquid at a temperature just below the boiling point (simmering). Foods generally poached are eggs, fish, fruits, etc.
7. *Stewing:* This is a gentle method of cooking in a pan with a tight fitting lid, using small quantities and the other half in the liquid is cooked at simmering temperature (98° C). It is usually used for cooking meat, dais, etc.

Other Methods

Braising: This is a combined method of roasting and stewing. It is cooking over direct heat or in an oven in a small amount of liquid at a low temperature with the pan tightly covered.

Solar Cooking: It is a method of cooking food by converting the solar energy into heat energy. Solar cooker is placed at such an angle that the mirror reflects the sun's rays into the food placed in the container. The containers are blackened to absorb the maximum of the sun's heat.

Infra-red Radiation or Microwave Cookery: The food to be cooked is placed in an electronic oven where it is exposed to the penetration of microwaves produced by a magnetron tube. The microwaves cause agitation of the molecules within the food so that heat is generated. The microwaves can be absorbed, transmitted or reflected and they can pass through paper, china and glass. Cooking time is shortened to ten times less than may be needed by conventional methods.

Preservation of Food: Food spoilage is brought by the action of enzymes present in foods or due to the action of micro-organisms such as mold, yeast and bacteria or due to infestation with insects and worms. The environment unfavourable to the action of enzymes or to the growth of micro-organism is the main objective of food preservation. Preservation of food also helps in increasing the shelf life of foods, making seasonal food available through out the year, adding variety to the diet, and improving the nutrition. Preservation is defined as the state in which any food may be retained over a period of time without being contaminated by pathogenic organisms or chemicals and without losing optimum qualities of colour, texture, flavour and nutritive value.

The different methods of food preservation (household and commercial) are of two types:

(a) *Bactericidal Methods:* Bactericidal method are those which destroy bacteria. These include applying heat by cooking, canning, sterilisation, irradiation and preservation.

(b) *Bacteriostatic Methods:* Bacteriostatic method are those in which the growth of bacteria, yeasts and moulds is retarded by dehydration, freezing, treatment with antibiotics, salting and pickling.

Bactericidal Methods

Heating: Food is commonly preserved through the application of high temperatures. A temperature considerably above that of the body may result in either a pasteurised by being held at a temperature from 60° C to 66° C for thirty to forty minutes, during this time most of the organisms, although not all, are killed. This method is used for temporary preservation of milk and for preservation of fruit juices and other fruit products of delicate flavour.

Canning: It is the preservation of food in sealed containers, after the application of heat, through steam under pressure.

If the effectiveness of pasteurisation and sterilisation is to extend over a period of time of practical significance, the material thus treated must be protected from fresh contamination by micro-organisms. The first foods preserved by canning were sealed in glass.

Drying: Removal of moisture is of benefit in preserving food. Drying alters greatly the character of the food and requires some time for preparation both before and after the process. Meat and fish have been dried for centuries. Other foods that are preserved through dehydration method are figs, apples, prunes/plums and raisins.

Sun drying is a common method of food preservation. Dried foods include papdi, papads, macaroni, bari, potato wafers and so on. Fruits like apricots, bananas, dates, figs, grapes, raw mango, peaches, pears, pomegranate seeds are also preserved by sun drying.

Smoking: Foods can also be dried by exposing them to smoke by burning some special kinds of wood in this method, while the heat from the smoke helps in removal of moisture, exposure to smoke imparts a characteristic flavour to the food. Fish and meat are the foods usually preserved by this method.

Mechanical Drying: Dehydrators, rollers dryers and spray dryers are the mechanical devices used for drying food. Solid foods such as papads, vegetables such as green peas, onions, potatoes are spread in thin layers on metal trays inserted in the dehydrator and then dried by heat. Food in the form of a fine dry powder can be obtained by using either roller drying or spray drying. In roller drying, the finely ground wet suspension of the food is spread as a thin layer on a revolving drum which is heated. The resultant dry powder is then scraped off with a blunt knife and packed. In the spray drying process the food in the form of a liquid or a finely ground suspension is pumped through fine jets and falls as a spray into a chamber through which hot air is circulated. The fine droplets of the liquid quickly lose their moisture by evaporation and fall on the floor of the chamber as a fine powder.

Sterilisation by Cooking: Sterilisation by cooking is a method of preserving foods by preventing the growth of micro-organisms through the application of high temperatures. Boiling temperature (212° F or 100° C) if maintained for a sufficiently long period of time, helps the heat to completely penetrate the foods and kill the bacteria.

Pasteurisation: Pasteurisation may be brought about by the holding process in which milk is heated to at least 62° C (143° F) and kept at that temperature for at least 30 minutes; or by the high temperature short time method in which milk is heated to 71° C (160° F) and kept at that temperature for at least 15 seconds.

Milk may be sterilised either by boiling for a period of time or by the application of heat as in the preparation of evaporated milk. Sterilisation depends on the colour of milk and gives it a slightly caramelised flavour, while pasteurisation does not change the colour or flavour of milk.

Preservation of Foods by Irradiation: In this process, gamma rays or high speed electrons are used to destroy micro-organisms. These radiations are termed as ionising radiations.

Low level irradiation presents the germination potatoes, onions and carrots. Metabolic processes can be controlled by this treatment in foods thus delaying their ripening. Pork and beef can be made free of tapeworms and trichenella by irradiation. Heavy losses in the granary due to insect infestation can be checked by irradiation.

Bacteriostatic Methods

Refrigeration: Micro-organisms, although not readily destroyed by severe cold, and much less active at low temperatures i.e. 0° C to 5° C, refrigeration is widely used both in homes and in commercial plants, as a means of maintaining the low temperature found satisfactory in the storage of perishable foods. Fresh milk, meat and similar foods are kept just above the freezing point. Certain fruits and vegetables also keep better when cold.

Wrapping certain fruits and vegetables in paper or cellophane, or coating them with wax, improves the keeping quality. Pears and apples of high quality to be kept for winter use and oranges, lemons and grapefruit are often wrapped. (The fruit or the paper in which it is wrapped may be treated to retard mould formation.) Green peppers, tomatoes, oranges and other citrus fruits, cucumbers and cantaloupes are among the foods whose keeping quality is improved when they are coated with wax.

Freezing: Freezing, (-18° C to -40° C) like cold storage, does not destroy the micro-organisms and enzymes present in the foods. Foods can be quick frozen in about 90 min or less by placing them in contact with the coil through which the refrigerant flows, by blast freezing in which cold air is blown across the food and by dipping in liquid nitrogen.

Scalding to destroy the enzymes is essential for successful preservation of almost all vegetables by this method. Hence, vegetables to be used fresh and crisp in salads cannot be preserved by freezing. Fish, poultry and game, as well as beef, veal, pork and lamb are preserved by freezing.

Freeze-drying: During this process the frozen food is placed under vacuum to remove the water, and it is then packaged in the presence of an inert gas such as nitrogen. The advantage of this method is that the product retains its original volume and shape and rehydrates easily. Products which have been subjected to freeze drying are coffee, beef, pork, chicken, soups and several food mixtures. Products of freeze-drying have a long shelf life, possess a light weight and can be stored and refrigerated, but they are considered inferior to frozen or canned foods.

Dehydro-freezing: This process is a combination of partial dehydration and freezing. The vegetables and fruits which are subjected to dehydrofreezing are first heated in order to lose about half their water content, and then frozen. Later, when the food is cooked water is added to reconstitute it.

Chemical Preservation: Certain chemicals like benzoic acid or sodium benzoate and sulphur dioxide are helpful in preserving foods either by retarding or preventing the growth of micro-organisms. Sodium benzoate upto a concentration of 0.1 % is used for all coloured fruits and vegetables. Potassium metabisulphite is used as a source of sulphur-dioxide and is used only for colourless fruits like apple, lichi, and raw mango. Meats may be cured with smoke that contains phenols.

Sugar, a common home preservative, is often used in such quantities as to increase the concentration of the food and make it an unfavourable medium for the organisms. Salt and acid, in the form of vinegar or lemon juice, are other substances added for their preservative action. Frequently, the fermentation process is used to produce, within the food itself, acetic or lactic acid that exerts the same preservative action as is obtained by the addition of vinegar. Salt is used as an aid in the preservation of vegetables by fermentation. It aids both in drawing out the juices and in delaying the action of certain spoilage agents. Formerly, spices were often used in such quantities as to have a definite preservative effect.

Oils and spices alongwith salt and sugar provide a medium that resists the activity of the micro-organisms in food. Spices such as chillies, fenugreek, mustard and pepper are used in pickling. Certain preservative such as borax, boric acid, sulfites, and formaldehyde, once sold as canning powders, are now considered injurious to human beings, and their use is usually prohibited.

Preserving and Pickling: Preserving and pickling are methods of preservation frequently employed. Long before canning was known as a means of preserving foods, the use of preservatives was practised. Salting of foods was an early method of preservation. Salt is a valuable preservative, both as an antiseptic and as an agent for removing water. Placing foods in brine of certain concentration promotes fermentation and the foods develop an agreeable flavour. It is a good method of preserving vegetables and fruits like tamarind, raw mango,

amla, fish and meat. Sugar in large amount is likewise a favourite preservative because the high concentration of sugar solution exerts a high osmotic pressure and withdraws water from the micro-organisms, thereby preventing their growth. It is the one used in making preserves, jams, jellies and marmalades.

Dehydration: Dehydration of foods to prevent spoilage does not necessarily mean complete removal of water, but it does mean concentration to such a point that the liquid is denser than the body fluid of the organisms. When the liquid outside the cell wall is more dense than that inside, the liquid within tends to be removed from the cell and the body processes are delayed or prevented. Hence, although the moisture is present, it may not be available to the organism which might cause spoilage. Scalding to destroy the enzymes is highly beneficial in prolonging the storage life of dehydrated vegetables. A variety of dehydrated food products available in the market are dehydrated soups, pre-cooked peas and cereals.

Antibiotics in Food Preservation: Very often antibiotics are used in the feed of animals or in treating them. Sometimes cows and poultry are fed on antibiotics. The Food and Drug Administration, specifies that milk from such cows should not be used for human consumption for at least three days following the treatment of the animal with the antibiotic.

Crops are sometimes sprayed with antibiotics. Since no residues of antibiotics are present in the foods eaten, there is no hazard to the health of the consumer. When fish and poultry are treated with antibiotics, their shelf life increases two to three times, because of the reduced growth of the micro-organisms. Similarly, antibiotics are used in the ice-crush used for packing raw fish and shellfish. The cooking of fish and poultry destroys the antibiotic residues, rendering them harmless for human consumption.

Food Analysis: Food has been a basic part of our existence.

The study of food science involves understanding the nature, composition and behaviour of food materials under varying conditions of storage, processing and use. Many physical and chemical reactions occur due to the interaction with the medium of cooking, and the environmental conditions such as heat, cold, light and air to which they are subjected during cooking.

Cooking: Foods like fruits, vegetables and nuts are eaten raw but most of the foods are cooked to bring about desirable changes. The process of subjecting foods to the action of heat is termed as cooking.

Preliminary Preparations: It means the tasks done before the final preparation of food which includes cleaning, peeling and stringing, cutting and grating, sieving, soaking, processing, coating, blanching, marinating, sprouting, fermenting, grinding, drying and filtering.

Cleaning: Many food products like vegetables, fruits etc. may have portions to be discarded, eg., Withered or discoloured leaves in cabbage and green leafy vegetables.

Other aspect of cleaning is washing that is applicable to fruits, vegetables, cereals, pulses and non-vegetarian foods. Washing fruits render them dirt free, insecticides, sprays, and chemical free. Washing in warm water helps to kill the worms in cauliflower. Washing flesh food products helps to remove blood, dirt and unwanted impurities. Washing cereals or dals help to remove husk, mud and any other unwanted matter but may cause the loss of B complex vitamins especially, thiamine.

Peeling and Stringing: Both these methods involve the removal of non edible or fibrous portion of fruits or vegetables, eg. Peeling of banana, potato, bottle gourd, stringing of beans. Peeling brings about loss of nutrients present under the surface of food, hence the product should be boiled or blanched and then peeled, eg., Vitamin C in potatoes.

Cutting and Grating: This is dividing the food into smaller pieces, thus helping in easy cooking. Various terms used under this are:

- *Cut:* to divide into pieces or to shape with knife
- *Chop:* to cut into no specified shape
- *Mine:* to chop very fine
- *Dice:* to cut into small uniform cubes
- *Slice:* to cut into uniform slices

Cutting and grating is done to make the food easily consumable and spoiled portion of the food can be discarded. But if it cut into small pieces, there is a greater loss of nutrients.

Sieving: It is done to remove coarse fibres, dirt, stones and insects eg. sieving flour with baking powder for cake preparation.

Soaking: It is done in water either plain or salted with sodium chloride or sodium bicarbonate to hasten the process of cooking.

Processing: It includes all the things to get food ready for cooking and serving by enhancing the flavour of the product. The various processes included under this are:

- *Mix:* to combine ingredients in such a way that the parts of each ingredient are evenly dispersed in the total product.
- *Blend:* to mix two or more ingredients so completely that they lose their separate identities. Ingredients in ice creams and milk shake blend so well that individual ingredients cannot be seen.
- *Bind:* to cause a mixture of two or more ingredients to cohere as a homogenous product. Binding agents like starches and eggs are used in making custards. Another example is dough making.
- *Beat:* to move an instrument back and forth to blend ingredients together to achieve a smooth texture.
- *Whip:* to beat with a rapid lifting motion to incorporate air into a food e.g. eggs.
- *Fold:* to mix a whipped ingredient lightly with an other

ingredient by gently turning one over the other with a flat implement eg. egg white in souffle preparation.

- *Mashing:* crushing the food product to a smooth structure eg. mashing vegetables in cutlet preparation.
- *Stuffing:* Introducing or filling a food stuff eg. stufing mashed potatoes in parathas.
- *Coating:* It refers to covering a food with layer of crumbs, flour or other fine substances before cooking it. The different ways of coating are:
- *Dredging:* Passing food through a fine dry or powdery substance in order to coat it Substances often used are flour, powdered almonds and bread crumbs.
- *Breading:* Three steps are involved. The product is first dredged in flour, then dipped in egg mixture liquids then in crumbs.
- *Battering:* This is dipping the food product in batter. Batter is a semi liquid. It usually consists of an egg liquid mixture thickened with flour to achieve a smooth consistency. Bengal gram flour or rice flour can also be used for this purpose.

Blanching: This is plunging food into boiling liquid and immersing in cold water. This destroys enzymes present in food hence used as preparation or preservation. Foods normally blanched are tomatoes, potatoes, almonds, carrots and beans.

Marinating: It is soaking a food in a marinade to add flavour or to tenderise it or both. A marinade is any liquid made up for purpose of marinating. Vegetables marinated are brinjal, onions, radish, bittergourd, potatoes and chillies. Meat marinade made up of oil, flavour builders and acid.

Sprouting or Germination: All kinds of grams like green gram, bengal gram, peas and cereals like ragi and wheat are generally sprouted to increase the digestibility and water soluble vitamins.

Fermentation: It is process of breaking down of complex

matter into simpler ones with the aid of enzymes and bacteria. This can be done under aerobic or anaerobic conditions. Fermented foods are more nutritious than their unfermented counterparts. eg. Idli, bread and dhokla.

Grinding: This includes both wet and dry grinding to enhance the taste and flavour. Wet grinding includes the grinding idli, dosa batter and preparations of chutneys. Dry grinding is grinding spices for masala powders and wheat for wheat flour.

Drying: It is the removal of moisture from food products and prolong the shelf life of the food. Cereals, pulses and spices that are normally used are dried products only. Food stuffs generally dried are mango and gongura for pickles.

Filtering: This process is generally done to remove dirt, unwanted particles or to remove moisture from food stuff. In the preparation of cottage cheese or paneer, whey water is extracted. Food stuffs filtered are coffee, tea, rice, soups, fruit juices and tamarind water.

Roasting: This process should be grouped under actual cooking methods but certain recipes, however, demand roasting as prepreparation. eg. rava and vermicelli.

Cereals and Cereal Products: The word cereal is derived from ceres, the Roman goddess of grain. The principal cereal crops are rice, wheat, maize or corn, jowar, ragi and bajra. 80% of dry matter of cereals is carbohydrate, mainly the starch. Cereals contain 6 to 12% protein and provides more than 50% of the daily protein requirement and lipid content is 1-2%. About 95% of minerals are the phosphates and sulphates of potassium, magnesium and calcium. Whole grain cereals are an important source of B vitamins in our diet. There are various by - products of cereals especially wheat includes whole wheat flour which contains the finely ground bran, germ and endosperm of the whole kernel. In maida the bran and germ are separated. Semolina is coarsely ground endosperm. Macaroni products are also called pasta or alimentary pastes

include macaroni, spaghetti, vermicelli and noodles. Macaroni is a tube form, spagetti may be either tube or rod, vermicelli is a tiny rod and noodles are flat strips. Malted wheat is prepared by steeping grains in cold water for 36 hours, allowed for germination and making flour after drying thoroughly. Amylase Rich Food (ARF) is germinated cereal flours which are extremely rich in the enzyme alpha-amylase. Glutamic acid is derived from wheat which is a salt like product generally used to bring out the flavour of other foods or seasonings. Wheat bran prevents constipation and may lower the risk of colon cancer.

Rice products include milled rice and parboiled rice i.e. soaking paddy in water for a short time followed by heating once or twice in steam and drying before miling. Other products are rice starch granules which are quite small and are embedded in a protein matrix and it is used in puddings, ice creams and custard powder. Rice bran is normally finely granulated light tan in colour and it has bland flavour and can be used in preparation like bread, snacks, cookies and biscuits. Rice bran oil is rich in vitamin E. Parched rice is a crisp product with a greyish to brilliant white colour and is sold either salted or unsalted. Puffed rice, flaked rice and aromatic rice are other rice products.

Millets are hardy plants capable of growing in areas where there is low rain fall and poor irrigation facilities. Maize and sorghum are major millet crops and bajra is a pearl millet while ragi is a finger millet.

Maize products are degerminated flour which consists mostly of the endosperm, corn germ oil which can be refined to produce a high quality vegetable oil for cooking; popcorn is a method of starch cookery; and corn starch which is very inexpensive. Jowar, ragi and bajra are usually milled and made into flour for consumption purpose. Ragi is more suitable for malting which forms a high calorie-density weaning food with excellent nutritional qualities. Bajra is also suitable for malt preparation.

Pulses: Pulses are the edible fruits or seeds of pod bearing plants belonging to the family of leguminous. Major pulses important in our diet are red gram dal, bengal gram dal, black gram dal, green gram dal and masoor dal. Some are used as whole grams.

Cow pea, rajmah and dry peas also come under leguminous family. Pulses give 340 calories per 100 g. and are important sources of protein (17 to 25 gms/100 gms) i.e. chiefly globulins. Pulses contain 55 to 60% starch, 1.5% lipids. They contain calcium, magnesium, zinc, iron, potassium and phosphorus. Legume seeds are excellent sources of B complex vitamins particularly thiamine, folic acid and pantothenic acid.

Legumes are generally milled to remove the outer husk by wet or dry process. Whole grams which have hard outer covering need soaking prior to cooking. Whole legumes can be germinated by soaking overnight, draining water completely and tying in a loosely woven cotton cloth for a day or two and sprinkling water twice or thrice a day. By this method, digestibility increases, cooking time decreases and improves availability of nutrients.

Nuts and Oil Seeds: Nuts are seeds or fruits consisting of an edible fat containing kernel and surrounded by a hard or a brittle shell. Nuts are also rich in protein and contain a high level of fat. Groundnuts are particularly rich in thiamine and nicotinic acid.

The meal obtained after extraction of the oil from the seed is richer in protein than the seed itself as fat is removed. Oil seed cake is mostly used for cattle feed. But now improved method of extraction and careful handling of cake have helped production of edible grade de-oiled meal which is used for the development of various products like multi purpose food that are used in feeding programmes.

Coconut is not a nut but rather a stone fruit. Coconut water is a refreshing drink. The white flesh is rich in calories. From grated wet coconut kernel, good quality oil as well as protein

and carbohydrate can be extracted to prepare coconut honey. The white flesh, when dried called copra has a high content of oil.

Groundnut is also known as peanut and monkey nut. The chief product is the oil, which can be used either as cooking oil or for making margarine and soap. The secondary product is the residue or cake left after the expression of the oil which is purified and used in supplementary mix. Ground nut kernels are ground to a smooth paste and made into milk which is used as supplement to the diets of pre-school and school children. Peanut butter also is prepared.

Soyabean contains about 40% protein and also upto 20% fat. The main problems with soyabean are that it has an undesirable beany odour, has a poor cooking quality and has hemoagglutinins and contains trypsin inhibiting factor. These are rendered inactive by suitable heat processing. Apart from being used as oil, soya bean is used in sausages, biscuits, breakfast foods and other cereal products. Soya protein is an important constituent in some infant foods and milk substitutes. Soyabean milk is also extracted. Tofu is made out of soyabean which is a substitute for cottage cheese in cookery.

Milk and Milk Products: Milk is a complex mixture of lipids, carbohydrates, proteins and many other organic and inorganic compounds. Milk fat or butter fat is of great economical and nutritive value. The flavour of milk is due to milk fat. Milk is a true emulsion of oil-in-water. Milk fat is a mixture of several different glycerides. Casein is a milk protein which constitutes 80% of the total nitrogen in milk. Other than this, lactalbumin, lactoglobulin, lactoferrin, serum albumin and serum trasferrin are present which are together called whey proteins. The chief carbohydrate present in milk is lactose or milk sugar is a disaccharide. Milk has, good quality protein, easily digestible fat and significant amount of calcium.

Milk is not only used as such but many non-fermented and fermented products are used in cookery. Skimmed milk is

made by removing fat through centrifugation method and is fortified with vitamins A and D. If 50-60% of the water is evaporated from milk, it is called vaporated milk. Sweetened condensed milk is made from pasteurised milk that is concentrated and sweetened with sucrose. Dry milk can be made with whole or skimmed milk.

Milk powder can be dehydrated to about 97% by spray drying and vacuum drying. Khoa is a compact mass of very small uniformly sized granules formed due to the coagulation of milk proteins by vigorous boiling.

Basundi is prepared by concentrating milk to half of its original volume by open pan concentration and adding sugar and other condiments. Ultra high temperature processed milk (UHT milk) is prepared by heating milk at temperatures higher than those used for pasteurisation for 2-6 seconds.

Rabri is a prepared, concentrated and sweetened product comprising of several layers of clotted cream. Channa is a major heat and acid coagulated product used in preparation of rasogullas, sandesh and rasmalai. Ice cream is a frozen dairy product consisting of whole milk, skim milk, cream, butter, condensed milk products or dried milk products.

Standardised milk has 4.5% fat and 8.5% SNF. Toned milk is prepared by mixing milk reconstituted from skim milk powder with buffalo milk containing 7 % fat. Double toned milk is prepared by admixture of cow's or buffalo milk with fresh skimmed milk and has 9% SNF and not less than 1.5% fat. The other types of milk are recombined milk, sterilised milk, filled milk and flavoured milk. Fermented milk products are cream, butter, cheese (made of casein) and curd.

Eggs: Eggs are rich source of all nutrients except ascorbic acid. It contains 12-14% proteins which are well balanced with respect to all the essential amino acids. Egg is one of the richest sources of lecithin-a phospholipid which forms a part of the structure of every cell wall in the body. Calcium, zinc and

sulphur are the minerals present in egg. It is a rich source of vitamin A, riboflavin, folic acid and B.

Flesh Foods: The term refers to muscle of warm blooded four legged animals. The chief ones being cattle, sheep and pigs. Organ meats include liver, kidney, heart, thymus and pancreas, brain and sausages which are made of ground or minced meat. Meat contains 15-20% protein of outstanding nutritive value. Fish protein has higher biological value than meat protein. The fat content varies from 5 to 40% with the type, breed and age of the animal. Carbohydrates are found in very small quantities in meat. The mineral elements occur either as separate ions or in a variety of compounds within muscle. Meat is an excellent source of some of the vitamins of the B complex and a good source of iron and phosphorus.

Poultry includes chickens, ducks, geese, turkeys and pigeons. Poultry meat has a high protein content about 25%. Chicken fat is more unsaturated than that of red meat. Poultry flesh is a good source of B vitamins and minerals.

Fish contains around 20% protein which has a high biological value. Fish contains less amount of fat compared to meat and poultry. It is rich in calcium. Fish liver oils are excellent sources of fat soluble vitamins.

Vegetables and Fruits: Vegetables are plants or parts of plants that are used as food. They are called protective foods as they are rich in minerals and vitamins. Green leafy vegetables are good sources of beta-carotene and folic acid. Roots and tubers give more calories compared to green leafy vegetables. They are fairly good sources of vitamin C. Other vegetables contain high amount of moisture, poor in all nutrients and contribute to the fibre content of the diet.

Fruits are produced from flowers and they are the ripened ovary or ovaries of a plant together with adjacent tissues. Fruits are very poor sources of protein and fat; contain high amount of moisture; poor sources of iron but good sources of carotenes and vitamin C.

Sugar and Related Products: Sugar and jaggery increase the palatability of food. Sugar provides only energy to the body, with less protein, minerals and vitamins. Sugar related products are corn syrup-contains 75% carbohydrate and 25% water; molasses - residue that remains after sucrose crystals have been removed from the concentrated juices of sugar cane; maple syrup made by evaporation of the sap of the sugar to a concentration not more than 35% water; honey-contains 17% water and carbohydrates like fructose, glucose, maltose and sucrose with small amounts of minerals, vitamins and enzymes; jaggery- obtained from sugar cane and it is rich in iron.

Fats and Oils: Fats that have a relatively high melting point and are solid at room temperature are called fats, whereas those that have lower melting points and are liquid at room temperature are called oils. These are concentrated sources of energy, excellent sources of fat soluble vitamins A, D, E and K; and provide essential fatty acids.

Spices: Aromatic food substances which enhance flavour are classified into spices, herbs and seasonings. Spices are dried roots, barks or seeds used whole, crushed or powdered. Herbs are usually the fresh leaves, stems or flowers of herbaceous plants. Seasonings are the bulbous group almost invariably used fresh like onion, garlic and shallots. Ani seed or somfu is used to counteract flatulence, asafoetida - as antimicrobial agent; bay leaves -flavouring agents for pulav, soups, meat preparations and fish; cardamom-favouring agent for sweet preparations; chillies-used in making chutneys, pickles and as seasoning agent; cinnamon-used for spicing sauces and pickles; clove-has anti mutagenic effect; coriander seed- used in flatulence, vomiting and intestinal disorders; cumin seeds - used as stimulant and carminative agent; fenugreek seeds - aid in maintaining the blood glucose levels in non-insulin dependent diabetes; garlic-used to treat various digestive disorders, as antifungal agent, reduces fat content in blood and reduce blood pressure; ginger- reduces inflammation and pain in joints; kokam and mango powder - sour agent in cookery;

mint leaves-anomatic culinary herb with pepper mint oil used for flavouring gum, confectionery, tooth paste etc. and when fresh used for salad dressing and garnishing. Mustard seeds are rich in sulphur and used for seasoning of various preparations. Nutmeg has antimicrobial property and nutmeg and mace are used as flavouring agents in meat and fish dishes, pickles and preserves.

Onion is used as a flavouring agent in food preparations and has antibacterial properties, lower blood cholesterol and lipid levels and useful in preventing heart diseases. Pepper is used along with hot milk for throat infections and as garam masala. Poppy seeds are used as thickening agents in preparation of gravies. Saffron is used mainly for its yellow colour and has a pleasant aroma, essential oil crocin and colouring principle crocerin used in various sweet dishes, soups and sauces; used as a sedative and used for eye infections. Tamarind is used as souring agent in chutneys, chat, pickles and as thickening agent in gravies. Turmeric is a colouring agent used in curries which has a colouring substance curcumin. It is used to relieve sore throat, cough, cold and against flatulence. Vanilla imparts flavour and is used in ice creams, custards, puddings and cakes. Monosodium glutamate (MSG) is popularly known as Ajinomoto used as a flavouring agent.

Beverages and Appetisers: A beverage is any material used as drink for the purpose of relieving thirst and introducing fluid to the body, nourishing the body and stimulating or soothing the individual. Refreshing beverages are plain water, carbonated beverages not containing fruit juices, fruit juices, iced tea and butter milk with salt and lime juice. While nourishing beverages are different types of milk and milk products like butter milk, milk shakes, eggs, fruit juice, glucose and lemonade. Stimulating beverage includes egg nogs made with whisky, brandy, coffee; coffee or tea; cocoa or chocolate beverage. Soothing beverages are warm milk and hot tea. Soups, fruit juice and alcoholic drinks in limited quantity are appetizers. Alcoholic beverages include toddy, beer and wine.

Methods of Food Processing

Preservation and processing begin when the soil is prepared for planting. Fruit processing is continuous from the time fruit leaves the growing plant till it is securely protected from microorganisms or macro-organisms. Harvesting, sorting, washing, peeling, preparation, heating, refrigeration, dehydration, concentrating, treating with chemicals and packing are steps in the chain of processing events. Amchur, dried chillies, ginger, kokum; apple juice, canned peaches, pears, canned mushroom, squashes, nectars, jams, jellies, marmalades are some examples for vegetable and fruit processed products. These products have limited demand due to their excessive costs, lack of traditional tastes and lack of their availability in remote areas.

The fundamental principle of preserving foods by heat is known as 'processing'. It consists basically in the application of heat in varying degrees to the food in closed containers, for a sufficiently long time to sterilise the contents before these are hermetically sealed. The method of processing varies from food to food.

The term processing as used in canning industry refers to heating or cooling of canned foods to inactivate bacteria. Processing consists in determining just the temperature as well as the extent of cooking that would suffice to eliminate all possibilities of bacterial growth. Almost all fruits can be processed satisfactorily at a temperature of 100° C. Vegetables require to be processed at higher temperatures of about 115° C to 121° C. The temperature and time of processing vary with the size of the can - the larger the can, the greater the processing time and vice-versa.

Processing Methods: There are generally three types of cookers used for canning of fruits. These are: open cookers, continuous non-agitating cookers, and continuous agitating cookers.

Open Cookers: These are very simple in construction and

consist essentially of wooden tubs or galvanised iron tanks of any desired capacity. The sealed cans are placed in crates of galvanised iron and immersed in the tank containing boiling water, which is kept boiling by letting in jets of steam through perforated pipes placed underneath the false bottom of the tank.

Continuous Non-agitating Cookers: In these cookers, the cans travel in boiling water in crates carried by overhead conveyors in a single file on continuously moving belt.

Continuous Agitating Cookers: These cookers are of various designs and are generally used in big production units. The sealed cans move in the non-agitating cookers, but are at the same time rotated by special mechanical devices to agitate the contents of the cans. By this technique, the processing time is reduced considerably.

Different operations involved in processing of any kind of fruit are washing, grading, peeling and coring, blanching, canning, filling, exhausting, sterilising, and cooling while for vegetables washing, preheating, peeling if needed like for potatoes, inspection and trimming, size grading, filling into the cans, sealing and processing and cooling the heat-processed cans.

Sorting and Grading: After preliminary sorting, the fruits and vegetables are graded. This is necessary to obtain a pack of uniform quality as regards size, colour etc. It is done by hand or with the help of grading machines.

Washing: The graded fruits and vegetables are washed with water in different ways, such as soaking or agitating in water, washing with cold or hot water sprays etc.

Peeling, Coring and Pitting: The washed fruits and vegetables are prepared for canning by peeling, coring, blanching etc. Fruits and vegetables are peeled in a variety of ways: (1) by hand or with knife, (2) by machine, (3) by heat treatment and (4) by lye solution i.e. caustic soda solution.

Blanching: Treatment of fruits and vegetables with boiling water or steam for short periods, followed by cooling prior to canning, is called 'blanching'.

Can Filling: The cans are washed with water or subjected to a steam jet to remove any adhering dust or foreign matter.

Syruping or Brining: The cans are filled with hot sugar syrup for fruits and with hot brine for vegetables. The syrup or brine should be added to the can at a temperature of 79° C to 82° C, leaving suitable head space.

Lidding or Clinching: Lid is partially seamed to the can by a single first roller action of a double seamer. The lid remains sufficiently loose to permit the escape of dissolved as well as free air from the contents and also the vapour formed during the exhaust process.

Exhausting: Before sealing the cans finally, it is necessary to remove practically all air from the contents which is called 'exhausting'. Containers are exhausted either by heat treatment or by mechanical means.

Sealing: After exhausting, the cans are sealed by special closing machines known as double seamers.

The processing time for different kinds of fruits and vegetables has to be relatively increased with further increase in the altitude. As the altitude above sea level increases, the pressure required to maintain the specified processing temperature would also increase. The pH value has a great influence upon the destruction of micro-organism. The lower the pH, the greater is the ease with which a product can be processed or sterilised. Similarly any increase in the temperature, after closing of the can, develops pressure inside the can. After processing, the cans are cooled rapidly to stop the cooking process, especially in cans at the centre of large stacks where they may remain hot for several hours. Cooling is done by (a) immersing or passing the hot cans in tanks containing cold water (b) spraying with jets of cold water

0000(c) turning in cold water into the pressure cooker in the case of canned vegetables or (d) exposing the cans to air in small lots when water supply is scarce. Once cooling is done, batches of finished cans should be finally tested for leak or imperfect seals by tapping the can top with a short steel rod. A clear ringing sound indicates a perfect seal, while a dull and hollow sound implies a leaky or imperfectly sealed can. The other method used for testing the vacuum inside the cans is using a vacuum tester which is portable and handy.

After the cans are tested for defects, they are then labelled by hand or by machine, and packed in strong wooden cases or corrugated cardboard cartons, in standard packing. Till they are dispatched, they should be stored in a cool and dry place.

5

Objects of Cooking

Aims and Objectives

The aim or the intention of cooking is to see that the food cooked undergoes a physical change, sometimes a chemical change and is acceptable.

The object of cooking is to achieve certain results such as:

To facilitate and hasten digestion, so that the cooked food is absorbed by the digestive system and subsequently assimilated by the body. This is largely determined in the manner the food is cooked. During the cooking process, it breaks down the cellulose in plant food, softens some of the connective tissues of meat, breaks down and gets starches present. The alteration is brought about in texture, by physical and chemical changes thus assisting mastication.

A physical change occurs when a substance changes its form, colour or size, but still remains that same substance, like water that changes to ice. A chemical change occurs when a substance changes its form, colour or size, combining so as to

form an entirely new body, e.g. green marrow changes its colour and milk changes to curd.

Cooking Requirement

The effect of cooking upon the three chief constituents of food— proteins, carbohydrates and fats—is visible in their increased digestibility.

Proteins: The protein of meat (myosin), of egg (albumen), of wheat (gluten), of pulse (legumin) is coagulated by heat. Avoid high temperatures as the protein hardens, denatures and shrinks and the food becomes indigestible. The connective tissue is converted into gelatine which is soluble in water and rendered digestible. The proteins' biological value is improved by moderate heating.

Carbohydrates

(1) Starch in food is greatly affected by heat. By moist heat, it is converted first into a soluble form and then by extreme heat into a new substance, sweetish in flavour—dextrin as in the crust of bread. Moist heat causes the starch grains to swell; it gelatinises at a temperature below boiling point of water, the degree of heat varying with the kind of starchy food.

(2) Cellulose is softened by the application of moist heat.

(3) Sugar when heated in water dissolves, then colours, upon further heating, turns brown and becomes a caramel and emits a lovely flavour, but does not crystallise.

Moderate heat does not cause much loss of mineral salts and vitamins, except vitamin C.

Fats

(1) If heated to a very high degree for a long time, fats undergo partial decomposition and fatty acids and glycerol are produced. Glycerol further decomposes

into acerolin which is an irritating compound to the digestive system.

(2) Cooking pleases the eye and is receptive to the palate and helps to stimulate the digestive juices, thereby creating an appetite. On application of heat, the red colouring matter such as haemoglobin in meat decomposes and changes the red colour to brown. Flavours are developed, which simulate the digestive juices. The colours of the vegetables are brightened. Green vegetables fast boiled without a lid, improve the green colour. Cauliflower when boiled, in acidic medium, gets a white refreshed look.

(3) Cooking sterilises the food partially. Cooked food can be stored for a longer time and it prevents food poisoning and diseases when stored properly. Some of the disease producing germs are killed by cooking. They are killed because of high temperature during the cooking process. A temperature of 60° (140°F) applied over 30 or more minutes, kills most of the pathogenic germs. However, there are some spore germs which are not destroyed by the usual cooking temperature: they take about 4 to 5 hours to get destroyed at a high temperature. High temperature penetrates to the centre or interior of food, specially in pork, beef and mutton.

(4) Cooking retains, as far as possible, the nutritive and flavouring ingredients. The flavour depends upon the amount and kind of extractives present, and the acids developed. Nutritive value is enhanced if the fat proportion in the meat is more. While cooking, the nutrition could be preserved by using the cooking liquor.

(5) Cooking gives a variety to the menu, as one food item could be cooked in various ways and given different textures, e.g. mutton in a soup, roast joint, croquettes, stews, steaks, keema, sookha meat, boti kababs, etc. Different methods of cooking when used, make the

menu interesting and enhance variety. It is, therefore, easier to plan balanced diet.

(6) Cooking preserves food for a longer time. The high temperature destroys bacteria and limits spoilage. It is economical as the cooked leftovers could be utilised and interesting new dishes could be prepared.

Raw Material Used: For the preparation of good, tasty, colourful dishes, it is essential to have a basic knowledge of the raw materials, their characteristics and the special part they play. This knowledge helps to substitute materials when necessary, or rectify the texture and taste if something goes wrong. Also it helps to improve the quality and get standard end product.

The raw materials are thus classified according to the part they play in making up a dish:

(1) Salt

(2) Liquids

(3) Sweetenings

(4) Fats and Oils

(5) Raising Agents

(6) Thickenings

(7) Flavourings and Seasonings

(8) Eggs

Salt: It brings out the flavour of other ingredients. The other name for salt is sodium chloride: it is readily available in all parts of the world in a solid (rock salt) or in a solution form (sea salt).

Salt, having a distinctive taste, transforms an insipid dish to a wonderful dish. It should be used skilfully, or too much of it could spoil the dish.

Salt is available in 3 forms: 1. Table salt (fine) containing phosphate: (2) Coarse or Freezing salt for culinary purposes; (3) Celery salt. It is a blend of celery root and ordinary salt

and is purchased ready prepared. It is used for flavouring certain dishes as an alternative to fresh celery or celery seed.

Uses of Salt

(1) Use of the correct amount of salt improves the flavour of the savoury dishes and when a little is added to sweet dishes, it enhances the flavour.

(2) It has a physical effect on the gluten of flour and strengthens gluten and increases its resistance to the softening effects of fermentation.

(3) Cauliflower, when put in salted water, makes the insects come out.

(4) It has a controlling effect on the activity of yeast in bakery products. It controls fermentation and hence it has marked effect on crumb, crust and colour of baked products.

(5) Salt added to water, for cooking green vegetables, helps in colour retention and enhances the taste.

(6) Salting is one of the oldest popular methods of preserving ham, bacon, fish etc.

(7) Salt is essential for good health.

Liquids: Liquids are important as they are used for the purpose of cooking, binding and coating, etc. Milk, water, stock and fruit juices are the most commonly used liquids. They prevent food from burning, bind dry ingredients and help in maintaining the right consistency.

Water and milk are used for preparing poaching liquor *[court bouillon)* soups, sauces, gravies, cakes and pastry mixtures and kneading of doughs, etc.

Buttermilk is used for curries, kadi, etc.

Stock is a liquid containing soluble nutrients and flavours, which are extracted by prolonged and gentle simmering (except fish stock— 20 mins.). They are used as the foundation of soups, sauces, gravies, curries, poaching liquors and many important kitchen preparations.

It is important that the correct amount of liquid should he used, or too much would make the food stodgy, soggy or watery.

Sweetening: When sweetening is used with other foods it enhances the combined sensations of odour and flavour of the dish produced. It also adds its own sweetness, and is a versatile food product. Its uses in the kitchen are varied. Substitution of one sugar (castor, icing) for another in a baking formula, gives allowance for the difference in the sweetening effect.

Sweetening is available in various forms—granulated, fine-grained, powdered and in a solution form.

Sugars vary in their sweetening quality and are available in the following forms—granulated sugar, castor sugar, icing sugar, lactose (milk sugar), cane syrup, maple syrup, treacle, honey and golden syrup.

Functions of Sweetening

(1) Low concentration of sugar speeds the effectiveness of baker's yeast by providing an immediate, fast cooking course of nourishment for its growth—thus hastening the leavening process.

(2) The ability of sugar to crystallise, gives a delightful variety in cookery.

(3) Sugar gives puddings, breads, buns and bread rolls a good flavour (caramel), the characteristic golden brown colour, and a tender and even texture.

(4) Fermentation is aided by sugar in brewing, wine-making and in the distillation of spirits.

(5) Sugar has long been used as a preservative, it is an effective preservative and can dissolve in water at very high concentration.

(6) It is an energy food and can be stored for a long time without getting spoilt.

Role of Bakery

Stock Sugar Syrup

500 gm Cube sugar	120 gm
Glucose	1 litre Water

Boil all the ingredients together in a clean Sugar boiler. Strain through a cloth into a basin and keep covered.

Method and Temperatures of Cooking: Take the required amount from the Stock Sugar and boil to the correct temperature between 116° and 160°F.

Keep the side of the pot clean during the boiling, by washing the sides frequently with water.

Do Not Stir: Add a few drops of lemon juice and shake well.

Pulled Sugar: Pour on to a slightly oiled marble slab, allow to cool slightly, with a knife turn the sugar overfolding together, work the sugar with the hand to a pliable consistency.

Aniline powder is best for this type of work, being very strong in colour and easily dissolved in warm water.

The colour is added to the cooked sugar when poured onto the marble slab and before working with hands.

Degrees and Stages of Cooking the Sugar

116°C	Soft Ball	Suitable for Marzipan
119°C	Ball	Suitable for Fondant
121 °C	Hard Ball	Suitable for Nougat
140°C	Small Crack	Suitable for Italian Meringue
153°C	Crack	Suitable for Dipping Fruit
160°C	Hard Crack	Suitable for Pulled Sugar

Using Fats and Oils: Fats and oils are nutritionally useful and in some form, economical sources of energy and give a satiety value to the dish. They also contribute characteristic

palatability, qualities of flavour and texture. They are popularly used as the medium of cooking.

Fats are solid at ordinary temperature and melt when heated. Oils are liquids at ordinary temperature. Only coconut oil solidifies at low temperature.

Various fats used in cooking are—lard (pig's fat), suet (fat around kidneys), dripping (Tallow Beef fat), butter, margarine, ghee, hydrogenated fat, cocoa butter (for confectionary).

Oils are extracted from coconut, palm, sesame, cotton seed, olive, peanut, mustard, corn and sunflower. Salad oil is a deodorised vegetable oil and is used for salad dressings, etc. as olive oil—the best for the purpose—is very expensive and scarce.

Fats and oils are used for various purposes, the major culinary part played by them are as—(1) spreads, (2) shortenings, (3) salad dressings, (4) frying media and (5) tempering.

Spreads: Butter and margarines are used for spreads, and their function is to add to the flavour, nutritional value and satiety value of breads.

Shortening: These are fats which shorten the gluten strands, surround them and make them more easily broken (short). When added to bread, it gives a bit of tenderness, richness and a sheen to the crumb.

Tempering: Dals, curries, rice dishes, etc. are tempered. The fat or oil is heated to which cumin seeds or mustard, or fenugreek seeds, etc. is added and poured over the dals.

Salad Dressings: Fat is used for the various salad dressings—Hot animal fat dressings, which consist of bacon fat, vinegar and seasonings, served hot, are used on green hot salads. Cooked dressing is a cooked mixture of egg, vinegar, fat, starch and seasonings. French dressings are emulsions of oil, vinegar or lemon juice and seasoning. Mayonnaise is an emulsion of oil, acid, egg yolk and seasoning.

Frying Medium: Fats and oils are used as a medium of cooking, i.e. pan roasting, frying, and sauteing. When fats or oils are heated, a temperature is reached at which visible fumes appear, which is defined as smoke point. Fats with high smoke point are suitable for frying. Different fats do not have the same smoke point. The highest frying needed for any food is about 199°C (390°C). Hydrogenated fats have a high smoke point and are good as a frying medium compared to other fats.

For the selection of a good fat, the following points should be remembered. A frying fat should have a high smoke point, low congealing point, high stability, low moisture content, and should have an acceptable flavour.

Fats and oils have a high percentage of unsaturated fats and have more shortening power, than saturated fats. Animals fats such as butter, lard and suet make the food short, i.e. break off (short) and readily melt in the mouth. For the selection of shortening, the following factors should be observed:

(1) Shortening should cream well or it will affect the cake volume.

(2) It should have no moisture; then it will be short. Shortness is correlated to baking biscuits, cookies, pastries, wafers, mathies, nankhatais, nimkies etc.

(3) The shortening should be stable and should not get rancid so that the cooked product has a good keeping quality.

(4) The right consistency of fat, i.e. solidified for puff and flaky pastry, soft for cakes and biscuits help to get a good texture. To acquire good results in baking, because of fats and other ingredients, certain rules should be followed— The richer the pastry, the hotter the oven, and the richer the cake, the cooler the oven."

Rendering of Fat: Animal fat is heated and melted and this renders fat from fatty tissues. Tallow, suet and lard are usually rendered and used for cooking. The fat is cut into small pieces and placed in a pan and put in the oven or on slow fire, until

the fat melts, and there are crisp brown pieces of tissues left. This should be strained through a fine cloth into a clean bowl.

Clarification of Fat: Used fat should be clarified and then used for better results in cooking. Strain the used fat and then mix double the quantity of water in a pan and bring it to the boil. Strain again, cool and place it in a refrigerator. The fat will solidify and float on top. Lift the cake of fat, turn it upside down and scrape off the foreign particles that have collected. Heat the fat on slow fire, till the water evaporates and then strain and store it in a cool place.

Various Leavening Agents: Leavening is increasing the surface area of a dough or batter by creating within myriads of gas bubbles puffing up, thus increasing the volume and making it light.

The expansion of these gases during baking increases the volume of the product and gives a desirable porous structure. The aeration of flour products is affected by the following:

(1) Biological (yeast)
(2) Chemical (baking powder)
(3) Mechanical (whisking, beating)
(4) Lamination (folding, rolling)
(5) Combination of the above.

Yeast: It is a living micro-organism and is a form of plant life. It requires for its growth, food, moisture, warmth and air. The primary function of yeast is to change sugar into carbon dioxide gas, so that the dough, in which it is generated, is aerated. It also assists in mellowing and ripening the gluten of the dough, and it contains vitamin B complex.

When yeast is dispersed in water at a suitable temperature and mixed to a smooth paste with flour and all the food necessary for fermentation are present, e.g. sugar, skimmed milk, the sugar acts as food and soluble proteins to form the structure and building material of the new cells. Activity starts within the yeast cell, yeast exudes a substance known as an

enzyme which changes a solution of sugar (sucrose) and water into a simple sugar (dextrose) thus absorbed within the yeast cell and changed to carbon dioxide and alcohol, together with other by-products.

Other enzymes in flour and yeast change some of the soluble starch to sugar, which in turn is assimilated by the yeast and changed, so that aeration goes on from this continued production of gas.

When using yeast, these points should be remembered:

(a) It remains dormant in cold.

(b) It rises in warmth.

(c) It is killed by heat (if temperature is over 127° F).

(d) It is fed by sugar, and in the dough it caramelises on the outside, forming a brown crust.

(e) It is fed by flour, the starch gelatinises with the moisture inside the dough and cooks into a light spongy mixture which is digestible.

(f) Strong or hard flour is the best to use in bread-making.

(g) If salt is added in the correct proportions, it gives a good flavour, controls fermentation and improves the colour of the finished goods.

Commercial yeast is of two main types compressed in cake form (also in a liquid form) and dry yeast in granular form.

Compressed yeast is a moist mixture of yeast plants and starch. The yeast remains active and multiplies rapidly if added to the dough. It should be kept in the deep freezer, because it retains its activity for several weeks, or even months. Liquid yeasts and starters are sometimes used in household bread-making. They are mixtures of active yeasts from compressed form with yeast food, such as sugar and potato flour or dough. They have a short storage life.

(i) Dry yeast: It is a mixture of yeast with corn meal or starch pressed into cakes and dried. Such yeasts

continue to live, but are in an inactive state. When furnished with food and moisture they begin to develop and multiply, but do so slowly. They can be stored for a long time and are inexpensive.

(ii) Activated dry yeast: It develops faster than dry yeast and is less perishable than compressed yeast. It is used in straight dough mixing. It should be stored at refrigerator temperature for a longer life.

Chemicals: Chemical aeration is brought about by the production of carbon dioxide from the action, in solution, of an alkali and an acid in the presence of heat. The acid and alkali in correct proportions (which is harmless to human digestion) is baking powder. With this method of aeration, the greater amount of gas should be generated after heat has been applied, i.e. when the goods are in the oven.

The carbon dioxide which is generated passes into the air cells, already created and is held by the network of gluten in the dough. This generation of gas and the consequent expansion causes an increase in the volume, which is held by the coagulation of the gluten and other proteins that are present.

The dough cooked becomes light and digestible. Baking powder consists of two parts acid and one part alkali. This can be bought readymade or it can be made as per the following recipes:

Recipe I:	Cream of tartar	60 gms
	Bicarbonate of soda	32 gms

Corn-flour (optional, increases the stability of the mixture.)

Method: Sieve together several times. It has to be kept in an airtight tin. The addition of one part rice flour or Corn-flour will help to keep dry.

Recipe II:	Tartaric acid	30 gms
	Bicarbonate	30 gms

Method: Same as above.

The latter recipe is suitable, but once mojstened it reacts rapidly, so that much of the gas is lost, if delayed in reaching the oven.

Other acids which may be used, when combined with bicarbonate of soda will produce carbon dioxide such as vinegar and sour milk.

When sodium bicarbonate is heated, it gives off part of its carbon dioxide, but leaves sodium carbonate as a residue. This compound is unpalatable and hence should always be neutralised with an acid which leaves a relatively tasteless and harmless salt.

If no sodium bicarbonate is added, the cookies are pale in colour; when it is added, the intensity of colour increases with the quantity of soda. Cream of tartar used in angel cakes increases their activity and tends to make them white and fine grained.

Mechanical Aeration: It is incorporating air by whisking, beating and sieving. When sugar and eggs, fat and sugar, fat and flour or any combination of these are beaten or whisked together, or Hour is sieved, it works as aeration. Whichever way the air is introduced into the mixing, be it by hand, whisk., spatula or by machine, it is still termed mechanical aeration.

Air is incorporated into the mixing by one of the above means and is held there by the fat, eggs, or both; aeration is brought about by the expansion of the air in the hot oven, together with the water vapour pressure within the air bubbles.

One of the best examples of mechanical aeration is the sponge cake, where a foam is produced from the eggs and sugar, the flour is then folded in the sponge and baked. The egg and flour proteins coagulate and the starch cooks thereby making the sponge set.

Water Vapour: Lamination acts as a raising agent. In pastries— folding and rolling helps to give the lift. In idlis, khaman dhoklas, etc. steam formed by the heat, helps to puff

up, as liquid and flour are present in equal quantities. As the popcorns have moisture inside the grains, when heated, they expand in volume.

The fundamental objective when making puff pastry is to build up a structure of fat and dough consisting of many layers so that when subjected to heat in an oven, it will expand and lift evenly, to produce goods with short eating properties.

In the oven, the pastry, which consists of thin layers of dough separated by films of fat, comes under the influence of heat, and the gluten in the dough layers is caused to expand and blister. The fat melts and the dough layers are insulated and the fat takes on a higher temperature, dough layers are cooked, the gluten coagulates and becomes almost rigid and the pastry does not collapse.

It is the expansion and blistering of the gluten in the dough layers as a result of steam pressure from the water in the dough that is responsible for the lift.

Combinations: Danish pastries are a combination of aeration by yeast and by lamination. For this, a rich yeast dough is made and butter is incorporated as for puff pastry.

In this, the pastry expands in volume, because of the process of aeration, by lamination, and at the same time is aerated by the action of yeast.

Various Thickening Agents: Thickening agents give body, consistency, and palatability, when used. They improve the nutritive value. Flavoured liquids are thickened and converted into soups, sauces, gravies, curries, bavarois, mousses, puddings, etc.

Binding agents are used to form a mixture of ingredients into a cohesive mass.

The thickening agents are starch, agar, eggs, gelatin, coconut, tamarind, curd, poppy seeds, onion paste, coriander power, etc. Starch is the reserve carbohydrate of plants and is abundant in common foods: corn, rice, wheat, potatoes, arrowroot, and

tapioca; rice has about 80% starch, wheat 70% potatoes about 19% starch.

Starch is Classified Into: Cereal starches, root starches and instant starches.

Cereal Starch is Found in: Rice, wheat and maize. When cooked, corn, wheat, or rice, begins to gelatinise at about 71° (160°F) but requires about 85-96°C (185-205°F) for maximum thickening. Cereal starches when cooked and cooled tend to gel, while most root starches do not, as they have more thickening power than root starches. Root starches are tapioca, potato and arrowroot which thicken at a lower temperature of 65-71°C (150-160°F) than cereal starches. They thicken completely below boiling point. Instant starches are pre-cooked starches, also known as pre-gelatinised starches. Instant puddings, etc. in the market contain instant starches.

Agar is a rigid transparent gel, and is used for salads, sweet dishes. It is a seaweed, widely found in China.

Eggs give a firm gel when baked, and thicken, but are soft when stirred.

Gelatin (unflavoured) is used for salads, cold sweets, cold soups. When set, it is transparent, firm and quivery.

Coconut is used for curries, and gives the gravy a whitish colour, thickens and enhances the taste.

Tamarind is used in a pulpy form and it gives a thickening to the consistency of the gravy and gives a brownish appearance and an acidic taste.

Curd is used for thickening curries and it gives a whitish appearance, and a good flavour.

Poppy seeds, onion paste, coriander powder give body to the curries, and slightly thicken the gravy and blend with the flavour of the curry.

Spices and Herbs: Spices and herbs give flavouring and

seasoning to the dishes. To get effective results, not only should the food please the eye, but should also flatter or stimulate the palate. The success of cooking largely depends upon the help we obtain from flavouring and seasoning. The spice we use for this purpose should be used sparingly, as well as with skill. All palates may not crave for highly spiced food, yet majority of people demand that the food be moderately flavoured with the right constituents.

To use flavouring and seasoning rightly is a great accomplishment; the dish could be spoilt by being overseasoned. Seasoning should bring out the natural flavours of the main ingredients and blend with them.

Seasonings as such have little or no nutritive value but are valuable for they give variety to the dishes and have medicinal value.

Importance of Spices and Herbs

Help in Digestion: From prehistoric times, spices have been used. Clove oil stimulates the flow of gastric juices; garlic, aniseed and asafoetida are taken for better digestion, and to reduce the chances of hypertension.

Used for Medicinal Purposes: For toothache, clove oil relieves pain. Turmeric and oil applied on swellings and hurts, cures, as it is believed to have antisceptic qualities, etc. Garlic and saunf help digestion, ginger added to tea helps to cure colds. Turmeric added to milk is given to a person who is in a state of shock.

Enhance Flavour: Dishes would be insipid and bland if the spices were not added, because they give a good flavour and stimulate appetite. The flavourings added should not be overpowering, but there should not be a suspicion that they have been added. This mental search for the flavour makes the dish more interesting.

Monosodium glutamate is a flavour enhancer in meat and fish dishes. Cinnamon, bay leaf, nutmeg, saffron, pepper, cloves, etc. also enhance flavour.

Monosodium glutamate is salt like crystals known as MSG. It is marketed under various brand names and is produced by a special process from wheat, soya and sugar beet. It enhances flavour in the cooking of savoury dishes.

Improve Appearance: Some of the spices give colour to the food and improve the appearance of the dish. Turmeric, saffron, coriander leaves, poppy seeds, ratanjot and colour to the food, which makes the dish interesting.

Improve Palatability: Salt is one of the important seasonings that enhances the taste of the food. It also brings out the flavour. The other seasonings that improve the palatability are pepper, chillies, poppy seeds, coriander seeds and paprika, etc.

Act as Preservatives: Many food are preserved for a longer time with the help of spices. Salt is used extensively for preserving—brine solution. Other spices that have the preservative quality are— turmeric, cloves, mustard, ginger, garlic, asafoetida. Pickling is one of the forms of preserving.

Spices that have flavourings and seasonings are: Garlic, ginger, cloves, cinnamon, cardamom, cumin seeds, mustard seeds, poppy seeds, nutmeg, asafoetida, coriander powder, mace, pepper, fenugreek, chillies, saffron, aniseeds, turmeric, paprika, caraway seeds, allspice, sesame.

Various herbs are parsley, celery, coriander leaves, thyme, tarragon, rosemary, mint, marjoram, sage, bayleaf, basil, chervil, caripatta, etc.

Spices and herbs are available in different forms: fresh, dehydrated and powdered, liquids and some are essences (concentrated) of spices, e.g. clove, cinnamon, etc.

Many combinations of spices in a powdered form are marketed today, e.g. chana masala, rasam powder, sambar, masala for onions, garam masala, jal zeera masala, meat masala, etc. This process retains original flavours and improves keeping quality and makes transportation economical and easy. Spices are also available in liquid form.

Spices and Herbs

Spices and Herbs	Characteristics	Eggs and Cheese	Meat and Poultry	Fish and Shellfish	Vegetables/ Fruits	Desserts/ Baked Goods
1	2	3	4	5	6	7
Allspice	It is a berry, which combines the flavour of cloves, cinnamon and nutmeg. It has a spicy sweet, mild, very pleasant flavour. Native of West Indies.	Cottage cheese, eggs. casserole.	Beef stews. meat loaf. baked ham and lamb. Barbecue sauce.	Boiled fish. oyster, stew.	Mashed sweet potatoes, spinach, turnips. aubergine and carrots.	Cakes, milk pudding, apple pies, stewed fruit.
Aniseed	Fruit of an annual plant of the parsley family. It has a delightful sweet liquorice. flavour, Grown in India, Pakistan, China, Mexico, etc.	Cheese	Curries, pickles	Fish and shellfish cookery. Fish pakodas.	Pickles, curries.	Rolls, cakes. biscuits, pastry for pies and tarts. Baked apples.
Capers	Flower heads of Mediterranean caper. Picked, dried and stored in lime solution.	Canapes	Hors d'oeuvres	Poached fish	—	—

contd....

Spices and Herbs	*Characteristics*	*Eggs and Cheese*	*Meat and Poultry*	*Fish and Shellfish*	*Vegetables/ Fruits*	*Desserts/ Baked Goods*
1	**2**	**3**	**4**	**5**	**6**	**7**
Carraway seeds	Elongated seeds, give a strong flavour. Grown in Holland and Germany	Cheese spreads, cottage cheese, omelettes.	Soups, goulash, pot roasts, stews. liver dishes.	Poached fish. tuna fish, casserole	Apples, beets, cabbage, coleslaw. asparagus.	Dumplings, biscuits, rye bread, cookies, cakes, pastry. confectionery.
Cardamom seeds	Fruit of a reed-like plant, seeds within the pod are used. Has a strong, sweetish flavour. Small ones are better flavoured. Native of Malabar Coast, also grown in Jamaica.	—	Pickling, curries, beef dishes, stews. sausages.	Fish curries	Bhajees, potato dishes, baked apples, iced melon, pulaos. seyvian.	Rice puddings, bread, buns, biscuits. pastry, kheer, rich Mughlai biryanis.
Cinnamon	Thin inner bark of the cinnamon tree. has a fragrant odour and the taste is pleasant and aromatic.	Paneer curries.	Ham glaze. pot roast, and kaliyas. kormas. shorbas.	—	Squashes, sweet potatoes. apples and peaches.	Breads, cakes, rolls. cookies and doughnuts.

contd....

Spices and Herbs	*Characteristics*	*Eggs and Cheese*	*Meat and Poultry*	*Fish and Shellfish*	*Vegetables/ Fruits*	*Desserts/ Baked Goods*
1	2	3	4	5	6	7
	Cultivated in Sri Lanka and East Indies. Obtainable in a powder form.				biryani pulao.	
Garlic	A bulb that has several cloves, all enclosed in one membranous skin. It is stimulating.	Eggs Portugaise, cheese pizza.	Curries, bhunava meat, sauces, soups.	Fish pickles.	Hors d'oeuvres. vegetarian curries, pizzas	—
Chilli powder	Pods of chillies are dried and powdered. pungent, red in colour. Kashmiri chillies give colour and not piquancy. Grown extensively in India.	Cheese spreads. panner pokodas.	Curries, koftas. all Mexican meat dishes. stews. dhansak.	Prawn curries. fish curries.	Potato and corn drips, rice dish.	—
Cloves	Dried flower heads of the clove tree.	Paneer curry.	Ham, pork. beef, stews,	Poached fish. fish curries	Apples, canberries, bhajees,	Bread, cookies, fruitcake and gingerbread.

contd....

Spices and Herbs	*Characteristics*	*Eggs and Cheese*	*Meat and Poultry*	*Fish and Shellfish*	*Vegetables/ Fruits*	*Desserts/ Baked Goods*
1	2	3	4	5	6	7
	Native of Zanzibar and also grown in Pamba, East and West Indies.		Sausages, sauces, soups, curries.		curries.	Gingerbread.
Ginger	Stem and root have a hot flavour. Grown in India and Jamaica.	—	Steak, cheese dishes, seekh kabab tikkies curries, koftas.	Fish, prawn.	Chutneys, pickles, bhajees. rice dishes. apples, sweet potatoes, melon. grapefruit.	Gingerbread, cakes, cookies and pies.
Mustard	It is a pungent spice, used as table condiment.	Cheese spreads and boiled eggs.	Soups, ham. pork, beef, chicken, sauces, curries.	Boiled or poached seafood curries.	Curries, sauces, tempering, pickles.	—
Nutmeg and Mace	It is the seed of the nutmeg tree and has a coat of orangy flesh, which is dried	Eggnogs, custards and cream cheese.	Meat balls, sausages, beef. Mughlai dishes.	Poached and fried eggs.	Potato dishes, rich dishes. halwas.	Bread, cakes, custards and cookies.

contd....

Spices and Herbs	*Characteristics*	*Eggs and Cheese*	*Meat and Poultry*	*Fish and Shellfish*	*Vegetables/ Fruits*	*Desserts/ Baked Goods*
1	**2**	**3**	**4**	**5**	**6**	**7**
	into mace. Grown in Dutch, East Indies and Kashmir.					
Paprika	Dried and ground red pimentos.	Poached eggs.	Veal and chicken, goulashes, stews.	Poached fish.	All vegetables, pizzas. crackers.	Meat filling for patties.
Pepper	Black and white berries of the pepper plant. It improves flavours and adds piquancy— used as a table condiment.	Devilled eggs, cheese spreads. canapes. sauces.	All poultry. beef and meat dishes and stuffings, sauces, soups.	All fish dishes.	All vegetable dishes.	Meat filling for patties.
Poppy seeds	Poppy seeds are of the large poppy plants. Seeds contain very little opium.	Cheese mixture, egg dishes.	Meat and poultry stuffings, creamed dishes, veal and poultry dishes. curries, kormas.	Baked and poached fish.	Gravies, bhajees.	Bread rolls, sweet roll fillings. cakes, chickies (praline).
Fenugreek	Dried ripe fruit	—	Curries.	Fish dishes.	For tempering	—

contd....

Spices and Herbs	*Characteristics*	*Eggs and Cheese*	*Meat and Poultry*	*Fish and Shellfish*	*Vegetables/ Fruits*	*Desserts/ Baked Goods*
1	2	3	4	5	6	7
seeds	of an annual plant. Grown in European countries and India. Fresh leaves are used as a leaf vegetable (Methi Saag).			prawn curries.	pickles.	
Saffron	Dried stigma. from the flower of the crocus plant. Deep orange colour. Very expensive. Grown in Mexico and Kashmir.	Egg halwa. paneer dishes.	Veal and poultry, Mughlai curries.	Creamed fish.	Saffron rice, biryanis. curries.	Srikhand, Indian sweets, shahi tukra, halwa. etc.
Turmeric	Aromatic root or rhizome of a plant grown in India and West Indies. It has a characteristic yellow colour.	Egg curries.	Meat, biryani curries.	Prawn curries, fish masala.	Curries, yellow rice.	
Basil	The herb has a	Sprinkled on	Lamb's liver,	Shredded	Green leaves.	Added to

contd....

Spices and Herbs	*Characteristics*	*Eggs and Cheese*	*Meat and Poultry*	*Fish and Shellfish*	*Vegetables/ Fruits*	*Desserts/ Baked Goods*
1	2	3	4	5	6	7
	mild and sweet flavour.	egg canapes, egg sandwiches, scrambled eggs.	bacon, lamb chops, beef stews.	leaves sprinkled over fried fish, used for making turtle soup.	asparagus. broccoli and all tomato dishes.	dough while mixing.
Bay leaves	Leaves of the common laurel tree. It has a very mild sweet distinctive flavour.	Egg Creole.	Boiled ham. pickled ox tongue, stews. soups, sauces, lamb, boiled chicken, curries, kormas. kaliyas, shorba.	Fish chowders, soused herrings, fish casseroles, shrimps, fish curries.	Tomatoes, rice. beans, carrots, vegetables. hors d'oeuvres, dals, bhajees. pulaos.	—
Dill	Leaf herb resembling fennel, bitter flavoured herb. Helps in the preparation of dill vinegar and manufacturing of gin.	Cottage cheese.	Lamb chops.	Fish dishes	Cabbage, cauliflower dishes.	
Marjoram	A sage like plant	Souffles.	Beef and lamb	Baked or	Carrots,	—

contd....

Spices and Herbs	Characteristics	Eggs and Cheese	Meat and Poultry	Fish and Shellfish	Vegetables/ Fruits	Desserts/ Baked Goods
1	2	3	4	5	6	7
	mostly used in Italy. Leaves and flowering shoots are used.		roasts, meat loaf, stuffings and sausages.	creamed fish.	onion, Brussel sprouts, peas, potatoes and zucchini (like tori).	
Mint	Leaf herb, has a pleasant flavour.	Cream cheese	Lamb, veal.	Poached and boiled fish, patrani machi	Peas, potatoes. carrots, chutney, raitas.	
Parsley	Green leaf herb. Rich in vitamin A & C and iron. Curly and broad parsley leaves are available.	Egg dishes, cheese spreads.	Accompaniments, stews. soups, casseroles, poultry. beef, pork. chicken dishes.	All fish and shellfish dishes.	In or sprinkled on vegetables and rice dishes.	
Sage	—	Cream and cottage cheese.	Pork, lamb, stuffings.	Chowder and fish stuffings.	Egg-plants, onion. peas, and tomatoes.	—

contd....

Spices and Herbs	Characteristics	Eggs and Cheese	Meat and Poultry	Fish and Shellfish	Vegetables/ Fruits	Desserts/ Baked Goods
1	2	3	4	5	6	7
Thyme	Cultivated fragrant leaves. Many uses in the kitchen. Used fresh or dried.	Eggs, cottage and cream. dishes.	Beef, pork and lamb stews, liver paste, chicken and turkey.	Crab, scallops and shrimp. Food items.	Beets, onions, Beans, tomatoes and mushrooms.	
Tarragon	It belongs to the same family as common wood. It gives a strong flavour.	—	Chicken, duck. turkey, goose and liver paste.	Boiled salmon, crab, meat. lobster, shrimps. poached fish.	Asparagus, beans, beets. tomatoes.	—
Fine herbs	A combination of finely chopped tarragon, chervil and parsley.	Cheese, omlettes.	Soups, sauces, roast dishes.	Sauces, boiled fish.	Carrots, potatoes.	—

Healthy Food

Introduction: Today the term "Convenience Food" is widely used in the catering industry. This is probably due to the rapid progress in food technology over the past decade.

Many convenience foods have been in use in the industry for many years, without being given this title. This was probably due to the fact, that they were used in small proportions during the preparation of meals.

Considerable amounts of convenience food are now used in all levels of catering, packaged and graded specially for the industry. This varies from individual types of food to complete meals.

The term "Convenience Food", strictly translated, refers to any type of food, where some stage of preparation has been completed beforehand.

Processes: The following are some of the facts and methods used at present. Undoubtedly further progress in new developments will bring forth many more.

Accelerated Freeze Drying (A.F.D): By using high-vacuum techniques it is possible to produce specific conditions of temperature and pressure which create ideal conditions for the dehydration of food without the collapse of cells. Foodstuffs so treated have greatly improved texture retention and rehydration potentialities. There are only a limited number of food in the market, as production costs are relatively high. (See Sublimation of dehydrated peas, cauliflower, spinach, etc.)

Additives: Non-nutritive substances added to food (generally in small quantities) to improve its appearance, flavour, texture and storage properties. Most additives are legally controlled.

Aseptic Canning: A newer development in canning designed for those products which are adversely affected by heat within sealed containers. Products like fruit juices are

sterilised, then placed into sterile cans under microbe-free conditions.

"Boil-in-the-bag" Ready Food: A convenience entree or vegetable commodity. Portioned items are packaged in special plastic bags, sealed, then frozen. To reconstitute, the unopened bag is dropped into boiling water for a predetermined period.

Dehydration: A method of preserving food by reducing the moisture content to such a level that microbes cannot grow and reproduce. The methods used depend on the food, but popular techniques include tunnel, roller and spray-drying. Nutritionally the products lose little value, but the texture, especially in fruit and vegetable, is suspect. Newer methods (dehydro-freezing, freezing dehydration and quick dehydration) are overcoming this problem. A large variety of food are now available, e.g. pulses and leafy vegetables.

Dehydro-freezing: A method of preservation where the moisture content of certain food is reduced by 70 per cent, then quick frozen. Because the cells do not collapse, the product has good texture retention. Product must be stored in a frozen state.

Freeze Dehydration: (See Accelerated Freeze Drying).

Freezer Burn: In any frozen storage room there is a steady removal of moisture from unprotected food which may result in unpleasant irreversible changes to the colour, texture, flavour and nutritive values.

Freezing: Food freeze over a wide range of temperatures, although their individual freezing points are unidentifiable. Because of the high water content in many food, they freeze at temperatures between 32°F and 25°F. *Quick-freezing* has been defined as the process whereby food passes through this zone of maximum ice crystal formation (25-32°F) in 30 minutes or less. This ensures that minimum damage is done to the cell structure. Those caterers wishing to freeze their own produce should realise that slow freezing in deep-freeze cabinets might

result in the production of large ice crystals, with subsequent breakdown of tissues, especially in fruits and vegetables.

"Free-flowing" Products: Many quick-freezing methods (e.g. fluidised bed and flash freezing) produce commodities that have no surface frost, and therefore do not adhere in the pack. Advantageous where thawed appearance is important, e.g. fruit.

Instant-freezing Preservation: Method of preservation using liquid nitrous oxide at 128°F below zero. By gently flushing food through nitrogen, freezing is almost instantaneous, producing very tiny ice crystals, so important for good texture retention.

Irradiation: A novel method of preservation, as it produces a raw sterile article with only a slight rise in temperature. The reaction depends on the absorption of energy by foodstuffs from a material or machine producing radiation. Process may be used for complete sterilisation, pasteurisation or as a pest and fungi controller, depending on the dose of radiation produced. Nutritionally there is little lost, and the dangers from radiation are negligible. However, this process is still at the experimental stage.

Preservation: Stored food tends to deteriorate under the influence of its own enzymes, chemical action and through the growth and reproduction of micro-organisms. The techniques developed for preservation aim to retard these reactions or destroy the cause of deterioration. Methods include: refrigeration, quick freezing, canning, sterilisation, pasteurisation, irradiation and the addition of various chemicals.

Spray Drying: A method of dehydrating food by spraying a slurry of the commodity into a hot-air chamber, which dries the material immediately forming a powder. Good rehydration properties and nutrient retention in milk, dried soups and instant coffees.

Sublimation: Under certain conditions of temperature and

pressure it is possible for water to change from ice to vapour without becoming a liquid at any time. This allows food to be dehydrated without the collapse of tissues," given products with excellent texture and rehydration characteristics, (see A.F.D.)

Smoked Preservation: This process is applied to a wide variety of raw and prepared foods. When used with fish, usually the fish is salt cured before smoking. Further examples relating to smoked sausages, meats, etc. need to be given.

Different Types

Apart from the different methods already mentioned which produce the greater part convenience food in dry, quick frozen and canned food, there are products which are very suitable for large-scale catering and produce quite an acceptable article. When used with imagination and sometimes methods from conventional cookery, a much wider variety of dishes can be produced. (Lecturer to quote examples and draw attention to display board on "Convenience Food", e.g. instant soups, salami, sausages (tinned), canned rasgullas, condensed milk, haricot beans in tomato sauce, tinned cream, mayonnaise, meat balls in gravy, etc.)

Ingredients Used: Many techniques are used for food preparation before cooking and they are done according to the requirements of the various dishes. This helps to improve appearance, texture, palatability and flavour, and food combine readily. The techniques are divided into two:

(1) Sub-division and fractionalisation.

(2) Combining and mixing in the preparation of food.

Mise-en-Place: "Everything in its place", preparation prior to a task or service.

Sub-division and Fractionausation

Washing: This is necessary to remove superficial dirt. Meat, fish, vegetables and fruits are washed in cold water before any

preparation, i.e. peeling or cutting. If cut and soaked for a long period or washed after cutting, there is a great loss of water soluble vitamins and minerals. The more cut surfaces are exposed, the more nutrition is lost.

Peeling: Spoilt, soiled and inedible portions are removed. Skin of vegetables or fruits is either peeled or scraped. The outer layer can be stripped by steam, or caustics such as lye solution.

Paring: Paring is removing the surface layer in circular motion by pressure of a knife-edge all round the object.

Cutting: Reducing to small pieces with a knife. When a similar result is obtained with a chopping knife, or with a mechanical food chopper, the process is called chopping (small pieces). Dicing—cutting into cubes is known as dicing, as in dicing potatoes, carrots, etc.

Mirepoix: Coarsely cut root vegetables. Desired shape for turning vegetable is barrel.

Mincing: Cutting into very fine pieces, e.g. mutton, onions, etc.

Meringue: Whipped egg white and sugar.

Mandolin: A kitchen utensil to cut vegetables and fruits to various shapes and sizes—Julienne, Batons.

Macedoine: 6 to 8 mm cubes of vegetables or fruits.

Shredding: Cutting into long narrow pieces by means of a shredder or knife, e.g. cabbage.

Slicing: Cutting into thin pieces, but not so fine as shredding.

Slitting: Making a slit in the middle lengthwise, e.g. lady's fingers, green chillies, etc.

Grating: Reducing to small particles by rubbing on a rough surface, as in grating lemon peels, cheese, etc.

Grinding: Reducing to small fragments by crushing, as in grinding spices, or coffee in a mill or on a grinding stone.

Mashing: This is a method of breaking up of soft food with pressure, with a potato masher or with a fork.

Pureeing: Food is strained and a Mouli should be used (veg masher).

Pressing: Separating liquid portions from solids by weights or mechanical pressure, as in making cider from apples, paneer, etc.

Sieving: Passing through a fine wire mesh to remove impurities. It also helps in enclosing air and mixing ingredients evenly, like sieving of flour for cakes.

Refining: Freeing any material from impurities, as in refining cane-sugar.

Skimming: Removing a floating layer by passing a utensil under it (ladle) as in skimming cream from milk.

Rendering: Separating fat from connective tissues by heat as in rendering lard (dripping).

Filtration: Separating solids from liquids, through fine meshed materials, as in filtering fruit juices for jelly through a cloth bag, or fine wire mesh strainer.

Flavouring: A bundle of herbs and vegetables to impart flavour to stock am sauces is Bouquet Garni.

Folding: Method applies combining different ingredients together so that no reduction in lightness occurs.

Flotation: Separating on the basis of difference in specific gravity as in the elimination of the over mature peas in a batch by use of brine of appropriate strength.

Evaporation or Reduction: Removal of water, commonly accelerated by heating without lid.

Homogenisation: Sub-dividing large drops into smaller ones by forcing them through a small aperture under great pressure as in homogenizing the fat in cream.

Emulsification: Dispersing one liquid in another in which it is insoluble. If the dispersion is to be temporary, a stabiliser,

which coats the droplets of the dispersed phase, must be incorporated, e.g. in mayonnaise.

Food by Mixing: Food preparation often involves the combining and mixing of different food or food materials. Important effects of the methods of combining food or ingredients are those related to palatability. Texture and flavour are often controlled to an important degree by the skill and method employed in combining component materials.

Different Styles

Beating: Mixing materials briskly, lifting and dropping them with an appropriate tool. Sometimes used synonymously with whipping as defined below. This is done to thin mixture of liquids. The aim is to mix well and incorporate air.

Blending: Mixing two or more ingredients thoroughly, e.g. blending milk into white roux for bechamel sauce.

Cutting: Usually the incorporation of fat in flour and other sifted dry ingredients with a knife, a method which produces relatively coarse division of the fat and does not result in blending as in cutting the fat into a pastry mixture.

Creaming: Softening fat by friction with a spoon, usually followed by gradual incorporation of sugar as in cake-making.

Folding: Mixing materials with palate knife or wooden spoon, by a careful lifting and dropping motion as in folding whipped egg whites into a cake mixture. Palate knife is to life.

Kneading: Manipulating by alternating pressure with folding and stretching as in kneading bread dough. A method of combination to the extent that it combines water and flour proteins to make gluten.

Marinating: Coating the surface of food materials, a marinade, which is usually a mixture of oil and acid as in marinating the components of a vegetable salad with French dressing.

Sealing: Sauteing or pre-cooking roast, to develop colour and flavour.

Stirring: Mixing materials with an appropriate tool, such as a spoon by a circular motion, as in stirring white sauce, while cooking.

Whipping: Rapid beating with a wire egg beater or mechanical beater usually to incorporate air, as in whipping egg white.

Whisking: Whisking is done when a mixture needs a lot of air, items need to be mixed together so that they do not separate.

The Features

Texture: The texture, as related to food is not an entity in itself, but rather the accumulated effect of several characteristics or qualities to create an individuality such as:

(1) Appearance
(2) Feel to touch
(3) Softness
(4) Mouth feel

Texture is the term used to describe the characteristics of a finished food product. Variety includes some hard and soft food in a meal, so that the amount of chewing required is varied. This is one of the points that is considered while planning a menu. The menu includes dishes that have different textures as soft, crisp, hard, smooth, etc. All the above mentioned factors contribute to the concept of texture.

Appearances: It is the first factor in the appreciation of food. The size, shape and distribution of cells (holes) are of prime importance. Ideally the holes should be comparatively small, slightly oval or elongated and evenly distributed. Large cells (holes) produce coarseness to the eye, as in cakes, bread, etc.

Feel to Touch: The perceptiveness to touch should be exact as it is desired to be—a sponge cake should be light and spongy.

Softness: It is the characteristic of texture and the product should possess the resiliency (springiness) to gradually return to its normal shape—idlis, khaman dhokla, bread, etc.

Mouth Feel: Texture involves feels or bite tenderness—dry, soft, wet, hard, firmness, crumbliness and short.

Observation, experience, preparing foods, by adding ingredients and incorporating and applying the right principles (combining, mixing, fractionation, etc.) has a definite bearing in the result.

Variety of Textures

Firm and Close: Raising agents added raise the food, the volume is increased, the holes are small and many. The products are crisp and not spongy, e.g. biscuits, tartlets, etc.

Short and Crumbly: The product is short and it just melts in the mouth, when eaten. The right quantity of fat added gets this effect. More fat is added, as it prevents the mixture from becoming hard and short. It is very similar to firm and close texture, e.g. shortbread, nankhatai, biscuits, short crust pastry, etc.

Light and Even: There are plenty of holes of more or less the same size. It is less short than pastry and less spongy than a sponge cake. The product is firm to touch, e.g. Madeira, queen cakes, etc.

Spongy: The holes (air cells) are small and evenly distributed, as air has been included. It is soft and elastic to touch as in idlis, khaman dhokla, Swiss rolls, etc.

Flaky: The products have thin crisp flakes and they are formed by air pockets. The crispness is due to the method of rubbing fat with the flour. In order to get a good flaky texture, the right amount of ingredients, proper mixing and correct temperature is essential, e.g. vol-au-vent, patties, bouchees, paratha, tikona mathis, etc.

Smooth: When a dry ingredient is added to a liquid and

the blending results in a smooth texture, e.g. sauces, batters, gravies, phirnees.

Incorrect textures spoil a dish and should be avoided. Some of the incorrect texture are:

Coarse and Open Texture: This may be due to insufficient creaming of the fat and sugar, or careless mixing of flour, when added.

Too much of raising agents also give a coarse and open texture.

Hard Texture: The air enclosed has been driven off, may be due to the addition of more liquid than required, or has not been mixed properly. Low temperature of the oven also spoils the texture.

Soggy Texture: This occurs where there is the presence of too much moisture, as it combines the ingredients and the starch gets overcooked and tends to stick, e.g. while preparing rice.

Lumpy Texture: This happens when the liquid and solid is not mixed properly at the same temperature or even if they are at the same temperature. Once the lumps are cooked, they are difficult to remove. This texture may occur in sauces, gnocchi, suji halwas, etc.

6

Conservation of Food

It is often convenient to preserve foods such as fish, eggs, fruits and vegetables while they are cheap and plentiful, for use when they are not. Careful preservation of food eliminates waste which may happen otherwise during a time of glut. It also helps to prevent undue fluctuation of prices.

Although preservation in one form or another has been practised throughout the world, it developed scientifically into a technology only after the First World War when it was found convenient to use preserved food for the armed forces. The technology has continued to progress, and preserved food is used by rich and poor in everyday life in the West. Easy transportation, variety in food and convenience in use have all added on to the popularity of preserved foods. The development of this technology has now made it possible to have preserved food of practically the same nutritive value and palatability standards as fresh foods.

The object of food preservation is to destroy or inactivate spoilage by micro-organisms, toxins produced by them and the chemicals present which are responsible for the deterioration

of food. This has to be done without affecting the palatability of the food to be preserved, its colour, flavour and nutritive value.

Different Kinds

Preservation can be divided into short or long period preservation.

Short Period Preservation: All moist foods are inclined to decompose rapidly and it is often necessary to use some temporary method of preservation to prevent food from deteriorating before use. In this type of preservation the spoiling agents are not killed but their growth is retarded or inactivated for a short period. This is done in the following ways:

Asepsis: The number of microorganisms is reduced by

(i) *Washing:* Along with the dirt some of the microorganism are removed,

(ii) *Careful Handling:* Crushing and bruising of fruits and vegetables is avoided and damaged ones are discarded.

(iii) *Packing in Sanitary Packages:* This reduces the number of microorganisms to a great extent. Nature preserves nuts, etc., by this method.

Low Temperature (chilling or cold storage): This is keeping food at refrigerated temperature. This is not a drastic method. The food is not altered but the bacteria is held dormant. This method keeps food whole only for short periods.

Some fruits such as bananas however, spoil under refrigeration so they must be kept at 12°C (about 54°F).

Apples and hard fruits will keep for a long time in cold storage as long as they are wrapped in greaseproof paper.

Cold storage has been practised for thousands of years. In old temples and caves (e.g. Kanheri Caves, Bombay) one can see niches surrounded by running water to keep them cool. In Pompeii in Roman times each house had a food storage compartment cut out of rock with running water playing on it.

By Heating: Cooking of food and boiling of milk helps to preserve these longer than if left fresh.

By Mild Antiseptics: Sugar and salt can be sprinkled over fruits and vegetables respectively that are to be used for the next day.

By Removal of Air: If butter is spread over potted meat or egg is dipped in a wax emulsion it can be kept for a longer period than if untreated as the air that is necessary for growth of microorganisms is not available.

Waxing fruits by means of sugarcane wax is another old method which is now being revived in India and is being developed on a commercial scale in Nagpur. The purpose is to exclude air by covering fruits and vegetables with wax emulsion. The results have been very encouraging.

Long Period Preservation: The storage life of foods is increased for a longer period by these methods:

Sun Drying: This is perhaps the oldest method of food preservation and has been used by hunters and travellers for generations. The method of preservation is the removal of moisture to render germs inactive. It has been extensively used in India and fish, meat, vegetables and fruits have been preserved by this method; e.g., dried fish, such as Bombay Duck, dried game meat, lentils, peas, chillies, etc., and dried fruits and nuts such as apricots, dates, raisins, cashewnuts, etc.

Dehydration: Whilst sun-drying is a method of dehydration, by using mechanical devices this process has been speeded up considerably. Dehydration is the removal of moisture to the extent that mould growth does not flourish under carefully controlled conditions of air flow, temperature, and humidity, in a special piece of apparatus known as a dehydrator. By using dehydrators a more standard product is also obtained. Milk and eggs, for example, are roller-dried or spray-dried. The products can be successfully reconstituted. Dehydrated vegetables such as potatoes, peas, carrots, cabbage, etc., are

extensively used. All vegetables must be blanched before being dehydrated (except onion, garlic, etc). Dehydrated food can be easily transported and does not need refrigerated storage. The storage space required is also reduced.

Salting: This very often accompanies dehydration. When foods are just salted, this must be done at as low a temperature as possible. In tropical countries it must be done under refrigeration. Use salt that is as pure as possible, or else the impurities cause a bitterness in the fish or meat being salted.

Two methods are used:

(i) Rubbing the salt on the flesh of fish or meat.

(ii) Soaking in a brine tub. This method is used particularly for tongue, beef, pig's trotters and pig's head. Sometimes the brine solution is injected into the meat. Microbes cannot grow in a strong salt medium.

Smoking: This is also a very old method of preservation. Fish or meat is exposed to the smoke of a slowly burning wood fire thus imparting a smoked flavour and assisting in the preservation of the food stuff.

Bacterial growth is stopped by the drying of the food surface and by the antiseptic action of certain constituents of the smoke, such as creosote, formaldehyde and acetic acid.

The wood must not contain any substance that will impart a foreign flavour to the food, nor must the temperature be too high, or else the food will become too dry, and the smoke will not be able to penetrate the food thoroughly.

Deep Freeze: This does not kill germs, but prevents them from being active. This method is being used more and more in recent years, with the advancement in refrigeration machinery.

Freezing is a relatively simple method of preserving food and is generally accepted as one of the most satisfactory ways of storing perishable foods. Research done in this field has shown that freezing food is the simplest and safest method of preservation provided one follows the five cardinal points:

careful selection, proper packaging, freezing at -18°C (0°F) or lower, storage at -18°C (0°F) or lower with the minimum of fluctuation, and avoiding long storage.

Meat, poultry, products, game, fruits and vegetables and prepared and cooked foods can be preserved by this method.

(i) *Careful Selection:* Careful selection is the first step in successful preservation as freezing does not improve the natural product, except in the case of beef. If quality products are to be taken from the freezer, only quality products should be frozen. Slime or bruises must be trimmed carefully from meat before it is frozen.

(ii) *Proper Packaging.* Experience has demonstrated that special packaging materials are necessary. Inferior or low grade materials should be avoided. A good packaging material should be:

(a) Clean and sanitary.
(b) Odourless, flavourless, and impervious to odours.
(c) Mechanically practical.
(d) Attractive and easily labelled.
(e) Protective against desiccation (freezer burn).
(f) Resistant to oxidation.
(g) Tough and not brittle at low temperature,
(h) Greaseproof and stain-proof.

Numerous papers, foils, plastic films, and waxes meet most of these requirements. Only materials that adequately protect the product should be used. There is quite a range in the prices of these materials. More expensive materials increase packaging costs, but the increased protection usually justifies the extra cost. Even the best available wrapping material must be applied properly. The following suggestions are offered for best results in packaging meat:

(a) Use only approved wrapping paper or containers.
(b) Cut meat to suit needs. Avoid packages weighing more than 4.5 kgs. 2.5 kg. roasts are convenient.

(c) Be sure all sharp edges of bones are removed.

(d) Avoid rolling or folding steaks. Keep all packages as flat and square as possible.

(e) Interleave all pieces of meat with tissue paper.

(f) Maintain as complete contact between the meat and the paper as possible.

(g) Eliminate air from the packages, wrap tightly and seal.

(h) Mark each package clearly with name, locker number, cut of meat, weight, and the date frozen.

(iii) *Freezing at -18°C (about 0°F) or Lower:* Foods should be frozen at -18°C (about 0°F) or lower, and packages should be loosely arranged while freezing, but when completely frozen they may be tightly stacked in the home unit or locker. "Sharp freezing" often refers to freezing at temperatures between —18°C (about 0°F) and -28°C (about 20°F), while "quick freezing" applies to still lower temperatures.

(iv) *Storage at -18°C (about 0°F):* The palatability and nutritive value of food are best preserved at -18°C (about 0°F) or lower. The storage temperature should not fluctuate above this, but experiments indicate that fluctuation below that temperature is not serious. Fluctuation of the storage temperature above —18°C fosters dehydration even with the best of wrapping material. Ice or snow on the inside of the package usually indicates fluctuation above this temperature.

(v) *Avoid Long Storage:* Many people think that because freezing retards bacterial and enzymatic activity a frozen product will keep indefinitely. This is true only to a degree, as there is progressive deterioration in a frozen state. The rate is slow, but nevertheless specific; therefore time in storage should be limited. It is best to plan on a normal seasonal turnover. When meat is carefully handled, properly wrapped, sharp frozen,

and stored at -18°C (about 0°F), it can be stored for a reasonable period of time, but deviation from the ideal in any one of the above factors may cause rapid deterioration.

It is recommended that beef be consumed within twelve months, fresh pork within six months and ground meat, especially pork sausages, within four to six months. Pork and pork products tend to develop rancidity early and cured pork products even earlier; therefore, it is a good practice to stock only a six months supply of pork at one time.

These are the five cardinal points for successful storage. They are not over-exacting and not beyond reach. It is believed that home freezer owners will be far wiser to have all their bulk processing done at the nearest commercial freezer plant and use the home unit only as a home reservoir. This will leave all processing to the professional at the plant, where adequate facilities and equipment are available ad thus relieve the householder. Where small quantities of food a to be handled, as a day's supply of peas beans or strawberries from the garden, the may be handled safely and without materially influencing the storage temperature; but large quantities should be done in units intended for volume. The home unit is essentially a storage unit and not a processing unit.

Hermetically Sealed Containers: The absence of air from any container is a principle of food storage. This method is used for biscuits, cereals and preserved food in bottles.

Canning and Bottling: In this method preservation of food is done in hermetically sealed containers sterilised by heat.

The word 'canning' is of American origin but is now used internationally in connection with preserving foods in tin containers. In its broader sense it also includes preservation by application of heat in glass containers as well. Tins are standardised. Al T A 2½ and A_{10} are the commonly used sizes

in India. A_{10} has a capacity of about 3 litres and is used for the trade as it is an economical size.

The canning of vegetables, fruits, meat, fish, dairy products and most edible products cooked and uncooked, has reached a very high degree of efficiency. Each food item uses a special technique. Colouring agents if added in canned and bottled products are strictly controlled by the Food and Drugs Act. In India colouring agents are specially used for canning cherries and green vegetables.

By sugar: Food products containing 60% sugar do not ordinarily spoil. Jam, jelly, marmalade, preserves, candy fruits, fruit syrups and confectionery are preserved by this method.

Gas Storage: Oxygen (air) is replaced by inert gases e.g. aerated water.

Acid: Acids preserve foods by virtue of their hydrogen ion concentration, which produces a toxic effect on microbial protoplasm, e.g., pickles in vinegar.

By Spices: Essential oils of spices produce a toxic effect on micro organisms, e.g., as in pickles.

By Vacuum Packing: Coffee powder, milk powder, etc.

By fermentation Alcoholic beverages, etc.

By Chemical Preservatives: Sodium benzoate and potassium metabisulphite are the two preservatives permitted by the Food and Drugs Act.

Irradiation: Preservation by irradiation is a new technique which is being explored at present. This is done by radiation of various frequencies ranging from the low frequency electric current to the high frequency gamma rays. In 1957 it was reported that 100 food items have been cleared for human testing. The obvious advantage of food preserved by this method is that food can be stored without refrigeration. Bacon, beans, and peaches have been kept for one year at 22°C (about 72°F) storage temperature without spoilage after irradiation treatment.

Ionized radiation produces some destruction of nutritive value, but not sufficient to be nutritionally significant. Experiments on irradiated powdered milk showed no loss of vitamins. This may be due to a lack of moisture.

Colour changes were noticed in strawberry and milk powder amongst the foods studied. There was some flavour change in the foods, the most noticeable being in ground beef. It is believed, however, that acceptability of the strangeness in odour will be only a matter of time.

Work done hitherto does prove, however, that irradiation as a form of food preservation although still in its infancy, has great possibilities.

Preservation by Antibiotics: This is another method of preservation now under study. Much work has been done on Penicillin, Streptomycin, Aureomycin, Tyrrannising, Chloromycetin, Subtlin, Tetracyclin, Oxytetracyclin, Chlortetracyclin, etc.

Eviscerated chicken dipped in Aureomycin and stored at 4.4°C (about 40°F) was found fresh 7 to 14 days longer than untreated chicken.

Chlortetracyclin (aureomycin) is recognised as one of the most effective of the broad range of antibiotics.

This method is still in the experimental stage. Although food bacteriologists realise the advantage of the preservation of raw foods by means of a non-toxic antibiotic or the use of one in combination with reduced amounts of heat in the processing of canned foods, they are aware of certain problems that could arise with the use of antibiotics as preservatives.

(i) The effect of an antibiotic on a micro-organism is known to vary with the species or even with the strains of the organism. Hence the antibiotic may be effective against some spoilage organisms but not others, or against part of the population in a culture but not all.

(ii) Organisms are known to be able to adapt themselves

to increasing concentrations of antibiotics, so that new resistant strains may develop. There is also the possibility that other organisms not yet significant in food spoilage but resistant to the antibiotic might assume new importance in food spoilage.

(iii) Effects of the antibiotic on the consumer, sensitisation to it, changes in his intestinal flora and the development of strains of pathogen in the body resistant to that antibiotic, are all possibilities. It is, therefore, recommended that the antibiotics used for food preservation be other than those used for therapeutic reasons.

Food Planning

It is the manager's responsibility to have a pattern and a plan for all operations so that it is possible to proceed with the least expenditure of time, effort, material and money, to supervise each area and at the same time direct the overall organisation. One of the most important aspects of organisation and planning in the food industry is 'Menu Planning'. The best results are obtained if the person in charge of the food production plans the menus in consultation with the chef.

Menu is a French word meaning minute details as applied to the kitchen bill, of fare, table card or programme of a meal. Its object is to present a list of dishes and courses, eatables and beverages.

In olden days the menu was not given. The banquet or dinner consisted of two courses. 10 to 40 dishes were set upon the table before the diners entered (hence the word Entree) which when consumed were "Relieved" or 'Removed' by 10 to 40 other dishes; hence the words Relives and Removes.

Backdrop of Menus

Menus were first heard of as being adopted for table use in 1541, when at a banquet given by Duke Henry of Brainsick, His Grace had a sheet of paper by his side to which he

occasionally referred. When questioned by one of his guests about what he was looking at, he answered that it was a programme of what they were eating and that it helped him to reserve his appetite for those dishes he liked. The idea was admired so much that it became generally known and adopted.

Another story is that at a dinner given in the year 1498 Count Hugo de Monfort was noticed to have a piece of written parchment near his plate, which he frequently consulted and when asked what it was explained that he had ordered the head cook to write on it the various dishes that were being sent from the kitchen.

The old fashioned bills of fare were usually written on large cards, gaudily ornamented. Menu planning is the term used to denote the planning in advance of a dietary pattern for a given period of time. By planning a menu, we can come to know what foodstuffs we are going to consume, and in what form, at scheduled meal times such as breakfast, lunch, tea, or dinner.

Menu planning is an art and menus are compiled rather than written and they present not only the picture of a truly balanced appetizing meal but also the complete operation of changing raw materials into cooked and served delicacies.

There are two classes of menus: *Table d'hote* and *a la Carte.*

Table d'hote will be a set menu, often with a choice in each course, at a set price that is charged whether or not the full menu is consumed. The food is usually cooked in advance.

A la Carte menus are those where each dish is priced separately. The customer can choose from the card as many dishes as he wants. The dish will be cooked to order and it is recognised that the customer will wait while the dish is cooked.

Factors that Influence Menu Planning are:

1. Money to be spent
    (a) customer.
    (b) food services.

2. Type of customer.
3. Type of food service.
4. Availability of foods.
5. Seasonability of foods.
6. Balance.
7. Equipment.
8. Labour.
9. Leftovers in hand.
10. Appearance of the final product.
    (a) palatability (shape, flavour.)
    (b) variety (colour, texture.)
11. Menu-planning pattern.
12. Recipes to be used.
13. Food habits of the customer.
14. Policy of the establishment.

Prerequisites for Nutrition

The nutritional needs of the people being catered for are an important consideration especially when planning set meals.

Nutritional needs differ for persons of various ages, sex and activity groups.

It is not necessary that we should have a balanced meal every time we eat. The day's meals together must be considered to see that the right quantity and quality are provided.

The number of meals per day may vary from 3 to 5.

The primary consideration should be quality and not quantity. As a rule our appetites will give us a rough idea regarding quantity but when calculating quantities for institutional or large-scale catering, it is safer to calculate the total quantity, allowing 2400C (calories) per day per adult man, and using this unit to determine the amount required for women and children. The occupation of the people you are

catering for must also be considered. 300 calories more per hour for very hard physical work, 150 to 300 calories for medium work and 75 calories for ordinary physical work should be added to the original amount. Allow about 150 calories per person per day for wastage. No extra food is necessary for people with sedentary occupations.

Mere variety is not enough. A knowledge of the different constituents of food and their functions in the body and the sources from which these foods are obtained should be studied.

Elements of Food

The different constituents of foods are carbohydrates, proteins, fats, minerals and vitamins. These are divided into:

(i) *Body Warmers and Workers:* These elements form the bulk of the food and are obtained from carbohydrates and fats. Although proteins can also be used for this purpose, it is wasteful to use a more expensive and less available food.

(ii) *Body Builders:* These are proteins and some minerals; meat, fish, milk and milk products, eggs and soybean are sources of complete or first-class proteins and therefore, are considered to be superior. Peas, beans, lentils and nuts are sources of second-class or partially complete proteins. A judicious mixture of these provides the best nutrition. Special stress must be laid on ensuring that when planning meals for vegetarians, a third of the proteins is obtained from the first category. When there is a shortage of first-class proteins it is a wise policy to distribute them though the three meals as proteins cannot be stored in the body. Minerals can be obtained from vegetables and fruits.

(iii) *Body Protectors:* These are vitamins, minerals and proteins. It is very easy to overlook the body protectors, particularly vitamins, as they are found in food in small quantities. So special attention must be paid to

them. Cooking and processing destroy some of the vitamins, particularly Vitamin C. This vitamin is highly unstable and will be destroyed by oxidation, leaching, heat or ageing. Vitamin B is lost in the cooking liquor, if the liquor is thrown out, and destroyed when soda is used for cooking. Therefore it is essential that raw fruit, fresh vegetables, salads and some nuts or pulses be included in the daily diet. In addition to these, beverages and roughage must be provided.

Another way of dividing food is as follows:

(i) *For Bones and Teeth:* Calcium is necessary and it cannot be utilised without Vitamin D. Foods rich in calcium are milk and cheese; vegetables contain smaller quantities. Vitamin D is found in animal fat.

(ii) *For Flesh and Muscles:* These need proteins which are obtained from milk and milk products, meat, fish, poultry, eggs, soybean and a second-class variety from peas, beans, lentils, and nuts.

(iii) *For Blood:* This needs iron which besides giving red colour to the blood helps to transport oxygen to the various parts of the body.

(iv) *Food for Protection:* Such as minerals and vitamins. Vitamin A for growth, respiratory organs, skin and eyes. Vitamin B for healthy digestive system, nervous system and good appetite and for metabolism of carbohydrates. Vitamin C for cells of the body, clear skin, bright eyes and general good health. Vitamin D for fixing calcium and phosphorus in bones and teeth.

(v) *Food for Energy and Warmth:* Fat and carbohydrates. It is a good plan when planning meals to put in body-building and protective food first and then add other foods to satisfy hunger.

Child's Food Planning

While planning meals for children, besides a knowledge of their nutritional needs, the fact must be borne in mind that

partaking of food is a significant social and educational experience. Food should be plain, well-cooked, easily chewed, free from strong flavours and odours. There should be interesting variety in colour, flavour and texture. Meals should be served well and meal-time should be pleasant, free from undue stress and resentment.

Old people should have less calories (not more than 1500). Food should be easily chewed. Protective food such as vitamins and mineral should form the chief part. There should be less carbohydrates and fats.

Foods served in commercial restaurants, cafeterias, industrial canteens and hotel dining rooms may constitute only a part of the daily consumption of foods. They should be planned to meet a third of the calorie requirement. Such institutions should so plan their meals that patrons will be able to readily select foods necessary for their bodily needs. This is facilitated by the arrangement of foods for a given meal into menus classed as light, medium and heavy. The inclusion of leafy vegetables and salad plates is a nutritionally sound principle.

Budget The following points must be considered:

(i) Money available for food,

(ii) Cost of raw material,

(iii) Proper ratio between high and low-cost dishes.

Type of Service: Menus should be planned in relation to type of service e.g. cafeteria or seated service, etc., and the distance over which food must be transported for service. When meals are served in decentralised units from a central kitchen, a master menu with modifications should be used.

Availability and Seasonability: Seasonal foods are, besides being economical, also better in quality. These must be studied and included. It is also important to consider availability of food when planning menus.

Balance: The following points must be considered:

(i) Light to heavy and back to light,

(ii) Vary the sequence of preparation, e.g. cold or hot.

(iii) Change seasoning, flavouring and presentation,

(iv) Ensure that garnishes are in harmony with their main dishes.

Equipment and Work Space: Menu should conform to limitation of equipment and physical facilities.

Labour:

(i) Skill and number of employees must be considered,

(ii) Avoid too many last-minute processes,

(iii) Avoid too many long jobs especially for one cook,

(iv) Avoid overly complicated processes or emphasis on wrong kinds of food. Plan meals that the staff can do well.

(v) Have records of time and yield,

(vi) Allow time for portioning and garnishing.

Leftovers in Hand: For wise and economic management, leftovers should be considered and the menu altered or the necessary adjustment made.

Appearance of the Final Product: Food should be palatable to be acceptable and hence recipes should be tried out and standardised before a dish is put on the menu. The tastes of the people catered to should be studied. New foods should be introduced gradually.

To make food attractive and to break monotony, variety in colour, texture and flavour must be introduced, e.g. Cream of tomato soup, Fish colbert with French fried potatoes, Chicken fricassee with fresh peas and fruit salad or Puri and chole, Mughlai biryani, Palaksag, Tomato cucumber, Onion raita and Phren. As eye-appeal stimulates the appetite, relate the colours of the garnish with the basic ingredient. Natural colours should be used for food. Do not over-garnish.

Food Habits of the Customer: This is important when

catering in industrial canteens, schools, institutions, hospitals, etc. For example Jains and most Brahmins are strict vegetarians. Sikhs may eat only 'Jhatka meat'. Orthodox Jews may need Kosher meat. Bengal Brahmins may eat fish but not meat.

Policy of the Establishment: Certain establishments may not serve certain foods because of their policy, e.g. beef, or ham, bacon and pork. In order to have well-planned menus it is necessary to have a well worked-out menu pattern which is suitable to the clientele, and workable as it relates to staff and equipment.

Menu on Plan

(i) Plan regular uninterrupted time for making your menus when you are not too fatigued. Choose a quiet place.

(ii) Have available all records, charts and books which will be helpful.

(iii) Have a well organised, classified and usable menu chart. This will be a guide to menu making and will, if kept up to date, suggest ideas and help to prevent monotony and repetition in menu making. Ideas for menu charts can be collected from books, magazines, visits to restaurants and many other sources.

(iv) Formulate a workable menu-pattern. This means that the menu-pattern should be planned to meet the needs of the various groups of individuals served, and should be based on definite information.

(v) Make menus one week in advance. Use menu charts and other sources to get interest, variety and nutritional adequacy, and to avoid repetition. Recheck menus daily to use up leftovers and to meet other problems.

The arrangement of an elaborate dinner is an art in itself and consists of selecting the various courses so that the dishes harmonise with each other. The following points are helpful:

(i) Each dish should be different in composition and mode of cooking,

(ii) The choice of individual dishes must be excellent, each dish should be well cooked, tastefully dressed and yet distinct in character,

(iii) The harmonising effect is obtained by so arranging the dishes that each one is distinct from the other, bearing no relation in appearance to the preceding or following dish,

(iv) A judicious selection of the raw materials with special regard to the season of the year must be made in all cases,

(v) The various meats and materials forming one course must not be repeated in the same menu from one course to another,

(vi) The various kinds of sauces used in the preparation and as the accompaniment must each be distinctly different in colour, taste and flavour,

(vii) If there are two soups, one ought to be clear and the other cream. If the cream soup is made of vegetables the clear soup must not contain any vegetables,

(viii) When two or more types of fish are to be served the first should be boiled—generally a large fish—and the second fried, grilled or broiled,

(ix) The entree should always be classified so that light dishes such as rissoles, bouchees, croquettes, quenelles, chromesthesia, etc., are served first,

(x) If there are two releves, poultry is served before butcher's meat. The most important point is to have dressed releves before plain roasts. This is followed by the sweet (entremet), savoury, dessert and cheese. Sometimes the savoury is served before the sweet.

Having compiled the menu, it must be written down. Use correct words and phrases. Choose the language easily understood by the type of client being served.

First-class establishments use the French language. Ordinary establishments use English with a few common and easily understood foreign terms. Canteens use descriptive, easily understood English.

Do not mix languages. Always call an Irish stew by that name, or by the name used in the country of origin. Well known terms like hors d'oeuvre, etc., may be used when writing menus in English.

Some cooks give new names to old dishes, or give dishes names which have no bearing on the preparation. This causes confusion and disappointment so this practice must be avoided.

When writing the menu it is very important to see that words are properly spelt. Call dishes by their proper names. Do not mask or disguise. Use the following formula:

(i) Name the article—pomfret, lamb, duckling,
(ii) Add the place of origin—Bombay Pomfret, Punjab Lamb, Kerala Duckling,
(iii) Name the part used—Fillet of Bombay Pomfret, Saddle of Punjab Lamb, Breast of Kerala Duck,
(iv) Name the method of cooking—poached, roasted, braised.
(v) Add the garnish—in white wine, boulangre, and green peas,
(vi) Then having finished up with rather a long list of words, streamline the name leaving out the least important,
(vii) Finally add the sauce where necessary, Apple Sauce for the duck, Redcurrant Jelly for the saddle, etc.

Precautions

(i) Never write the word "gravy" on the menu,
(ii) Add after the main name any sauce which is served separately,

(iii) Add after the main name any accompaniment which is served separately viz: Yorkshire pudding, Cheese straws, etc.

(iv) Parsley is not put on the menu except where it is part of a dish, e.g. Parsley Potatoes.

(v) Watercress may be added either after the main word or by adding it with the word and.

(vi) Sauces and garnishes go directly after the dish to which they belong. Add the vegetable after the main dish and its garnish or sauce; and add the potatoes after the vegetable; add the salad after the roast or the cold dish.

(vii) Do not translate literally, e.g. Chasseur means hunter, Bonne Femme means good woman—in these cases either keep the original word or describe the dish as "with mushroom sauce" etc.

(viii) When something is grilled it cannot also be "a la meuniere". "A la" is only used to denote "in the fashion of, "after or in the style or. This means that it may be used when in the style of a certain town, country, place, etc., such as "a l'anglaise", "a la francaise", "a la portugaise", etc. It might also mean "in the" as in "a la creme". But where a dish is named after somebody it will not have the words "a la" but go straight on to the name, e.g. Peach Melba. In the case of Vichy carrots the word "vichy" belongs to the water there and the correct term is, therefore, "Carrottes Vichy",

(ix) Do not translate words which are very well known like Mayonnaise, Hollandaise, Consomme, Yorkshire pudding, etc.

(x) Do not add fancy names not properly understood, like Grilled Plaice a la meuniere.

(xi) Do not write 'soup and a roll': this looks clumsy.

(xii) Do not write Smoked Salmon Hollandaise. Salmon Hollandaise is hollandaise sauce mixed with poached fresh salmon.

(xiii) Do not repeat the garnish as a vegetable. The word 'Bruxelloise' means with brussels sprouts. Vichy means with carrots, Clamart means with peas, Boulangre means with-baked potatoes, and so on.

(xiv) Avoid over-descriptive words.

Correct Sequence of Courses: Make sure that the sequence of courses is correct. Take as a sample those given later under respective headings.

A few alterations are allowed sometimes; a savoury may be served before the sweet. The vegetable course is sometimes put before the roast. Cheese is sometimes served before the sweet.

Make sure that alternative dishes follow the same order of appearance as in a full menu, e.g. if it is fish or meat, then write the fish course first.

Correct Sequence of Garnishes and Accompaniments: First comes the main name of the dish, followed by the garnish which is to be served separately or the sauce which is served separately. Then the vegetables and then the potatoes.

Card of Menu

(i) Use good quality paper or paper mache or parchment or silk,

(ii) The print should be good and clear, in indelible ink, clean and elegant looking. Words should be correctly spelt. The prices should be up-to-date. A card with old prices crossed out looks untidy,

(iii) A medium-sized menu card is preferable to a too large or small one.

(iv) A menu card should have an attractive layout; one of topical interest is a good choice,

(v) There should be variety and contrast without confusing the customer.

Suggested Food

Breakfast Menu			Notes
1st	Course: Fruit	a. Fresh grapefruit, melon, juices. b. Stewed compote of figs, prunes, apples, mixed fruit. Baked apple.	
2nd	Course: Cereals	Cornflakes, grapenuts, shredded wheat, etc., and always—porridge. Sometimes add the words—Hot or cold milk or cream.	
3rd	Course: Eggs	Scrambled, fried, boiled, poached, Scrambled, fried, boiled, poached, or back), tomatoes, mushrooms.	Where this is the main course, it will follow the fish.
4th	Course: Fish	Grilled mackerel, kippers, herrings, bloaters, fish cakes, haddock, kedgeree. Fried plaice sole, whiting, meuniere. Usually plain, sometimes with tartare sauce.	There are usually just single names.
5th	Course: Meats, Hot	Grilled bacon (back and streaky), ham, gammon, sausage, lamb's kidney, calves' liver, with sauteed or fried potatoes or potato fritters. Bubble and squeak, mushrooms, tomatoes.	
6th	Course: Meats, cold	Ham, tongue, pressed beef, rarely, poultry or butcher's meat; with plain or sauteed potatoes where not included above.	Salad may be served but not put on the menu.
7th	Course: Preserves	Marmalade, jam, honey.	
8th	Course: Fresh Fruit	Apples, pears, peaches, grapes, bananas, oranges, etc.	
9th	Course: Beverages	Coffee, tea, drinking chocolate.	
10th	Course: Breads	Rolls, crescents, toast, brioche, etc.	

Generally speaking, a breakfast will consist of three courses:
Fruit or cereal.

Eggs, fish or meat with or without potatoes.
Preserves and toast, rolls, coffee.

Luncheon Menus: These do not follow as strict a pattern as those for dinners—they are generally smaller, with less

courses, but more choice within each course. Whereas in dinners there are many dishes considered beneath a dinner standard, most dinner dishes can be served at lunch. The emphasis at luncheon is placed on stews, grilled meats, cold buffet and steamed puddings, all of which are seldom or never used at a dinner.

1st Course:

(a) *Cocktails*: Fruit or shellfish.

(b) *Fruit:* Grapefruit, melon, fresh figs, and fruit juices.

(c) *Shellfish*, etc.: Caviar, oysters, seagulls' eggs, snails,potted shrimps.

(d) *Smoked Foods:* Salmon, trout, ham, salami and sau sages.

(e) *Hors d'oeuure.*

2nd Course: *Soup*: clear, petite marmite, consomme base, veloute, creams, purees, bisques.

3rd Course: *Farinaceous*: macaroni, spaghetti, ravioli, gnocchis, etc.

4th Course: *Eggs:* All kinds excluding boiled eggs, but including omelettes.

5th Course: *Fish*: Steamed, poached deep or shallow but without complicated garnishes, grilled, meuniere, fried, prime herrings, skate, etc., shellfish, hot mussels and scallops, cold lobster.

6th Course: *Entrees*: Stews (brown), banquette, navarm fricassee, hot pot, pies, puddings, oxtail, goulash, boiled meats, braised meats, salt meats, veal scallops,

veal cutlets, lamb cutlets and noisettes, pork cutlets, liver, calves' head, pig's feet, tripe, Venae steak, Hamburg steak, sausages (braised), minces, rechauffe dishes, braised game, hare, salamis of game, pilaws. All served with vegetables and potatoes.

7th Course: *Roast*: Mostly butcher's meat, rarely poultry or game. With vegetables and potatoes.

OR

8th Course: *Grills:* All the grilled meats including chicken with garnishes. With a vegetable or grilled vegetables and fried potatoes.

OR

9th Course: *Cold buffet:* All the cold joints, poultry, game, pies, terranes, cold fish—salmon, lobster, trout, crab, etc., in aspic or with a salad and sometimes a boiled or mashed potato.

AND/OR

10th Course: *A vegetable dish:* hot or cold asparagus, globe artichokes, seakale.

OR

11th Course: Souffle: Cheese, spinach, mushroom, etc.

12th Course: Sweet: Steamed pudding, pancakes (not as often as at dinner), milk puddings, fruit—stewed, baked, salad, flans, apple dumplings, pies, tart, fritters, trifle, jelly, egg custards, bavaroise,

	gateaux savarin, babes,, charlottes, profiteroles. Various ices and coupes (not featured as much as at dinner), French pastries.
13th	Course: *Cheeses:* All varieties with celery, radishes, biscuits and butter.
14th	Course: *Dessert:* All the fresh fruits
15th	Course: *Coffees.*

Tea menus In hotels, cooked meals are seldom provided at teatime, other than boiled or poached eggs, but in commercial (public) restaurants the emphasis is on "high teas" and like conteens, they serve lighter snacks more on the lines of a snack bar.

Sandwiches: Very small, dainty, white and brown bread, well buttered. Eggs and cress; tomato, lettuce, cucumber; chutney, savoury paste filling, sometimes chicken; foie gras (not often), smoked salmon (not often).

Hotplate Foods and Buns: Buttered buns, scones, teacakes, sally lunns, scotch pancakes.

Doughnuts, waffles, samosas, sausage rolls, pakoras, mutton patties.

Bread and Butter: White, wholemeal, currant, fancy and proprietary brands.

Preserves: Jams, lemon curd, honey, etc. (not marmalade).

Toasted Items: Toast, teacakes, scones, crumpets, buns, etc. Cheese toast, farinaceous and savoury toasted items as in a snack bar.

Eggs: Boiled, fried and omelette.

Pastries: All varieties, afternoon tea pastries (small), gateaux, etc.

Ices: Plain ices: all varieties of sundaes.

Tea: Indian, Ceylon, China, Russian and iced.

7

Art of Preservation

Foods gradually undergo deterioration or spoilage from the time they are harvested, slaughtered or manufactured. Some foods spoil rapidly; others keep for longer but limited periods. The useful storage life of some fruits, leafy vegetables and animal foods is less than 1-2 days at 21°C. The deteriorative changes in some foods is so rapid as to render them virtually useless in a matter of hours. These result in physiological, chemical and biological changes in food rendering them unfit for human consumption.

Preservation of Food

A number of causes are responsible for food deterioration. These include micro-organisms (bacteria, yeasts and molds); activities of enzymes present in food; insects, parasites and rodents; temperature (heat and cold); moisture; oxygen, light and time. These factors are not isolated in nature. At any one time, many forms of deterioration may take place depending upon the food and environmental conditions. The deterioration owing to individual causes is discussed below.

Microbial Spoilage Bacteria, yeasts and molds spoil food after harvesting, during handling, processing and storage. But not all micro-organisms cause food spoilage; e.g., the use of lactic acid producing bacteria in the making of cheese and fermented dairy products, and yeasts as leavening agents and for the production of wine and beer has already been discussed. However, except the micro-organisms that are specially cultivated under controlled conditions for their beneficial effects, others that multiply on or in foods are frequently the major causes of food deterioration. They are found everywhere; they, however, are not generally found within healthy living tissues, but are always present to invade the flesh of animals and plants when there is a break in their skin or if the skin is weakened by disease or death.

Bacteria are minute unicellular plant-like micro-organisms. The length of a bacterial cell is about 1 μm and somewhat smaller than this in diameter. Bacteria are classified according to the shape of their cells. Cocci are spherical in shape, bacilli are elongated cylindrical forms and, spirilla and vibrios are spiral. They can penetrate the smallest opening; many can pass through the natural pores of an eggshell. Bacterial spores are seed-like and they are more resistant to most processing conditions than yeast or mold spores. Bacteria, with few exceptions, cannot grow in media as acid as those in which yeasts and molds thrive. They multiply by cell division. Under favourable conditions bacteria can double their number every 30 min. Some bacteria cannot tolerate oxygen (anaerobes) and some require oxygen for growth (aerobes). Some can grow in an atmosphere devoid of oxygen but manage also in air (facultative anaerobes).

Yeasts are unicellular plants (fungi) widely distributed in nature and they grow well in a slightly acid medium in the presence of sugar and water. They are found in fruits, cereals and other foods containing sugar. They are also found in soil, air, on the skins and in the intestines of animals. They are somewhat larger than bacteria. The individual cell length is

of about 10 μm and the diameter is about a third of this size. Most yeasts are spherical or ellipsoidal. They have been used for centuries for leavening of bread and to bring about the fermentation of fruit juices. They can be harmful to foods if they bring about undesired fermentations.

Molds are multicellular, filamentous fungi having a fuzzy or cottony appearance when they grow in foods. They are larger than yeasts which they resemble in their nutritional requirements. They are strictly aerobes and require oxygen for growth and multiplication. They tend to grow more slowly than bacteria. Molds frequently thrive under conditions of acidity or of osmotic pressures that are inhibitory to most bacteria. This explains why they are frequently found growing on the surface of jams and jellies. Molds require less free moisture for growth than yeasts and bacteria. The absence of bright light and presence of stagnant air favour their rapid development.

Bacteria, yeasts and molds multiply best between 16° and 38°C. Some grow even at 0°C and others at a temperature as high as 100°C. They alter food constituents. Some can hydrolyse starch and other complex organic matter like cellulose, lignin and pectin; others can hydrolyse Jipids and produce rancidity; still others digest proteins and produce putrid and ammonia-like odours. They can also bring about decolonisation of foods, produce acids, gases or toxins. Thus, bacteria, yeasts and molds are the most important agents in causing the deterioration of foods.

Bacterial spoilage of foods could be of two types: due to food infection or intoxication. The former type of spoilage is on account of organisms present in the food at the time of consumption which then grow in the host and cause disease, while the latter type is due to toxins produced by bacteria prior to consumption which cause disease upon ingestion. The toxin of *C. botulinum* is produced only in an aerobic condition, such as found in canned products and meats in airtight packages. In a frozen stage, *C. botulinum* survives but

does not grow. Hence, frozen foods do not present the hazard of botulism.

Food Enzymes: Enzymes present in plant and animal foods continue to be present and are even intensified after harvest and slaughter. They are responsible for facilitating many changes during storage, such as the changes in colour, texture and flavour noted in fresh produce after it has been harvested. Some of these changes, such as the continued ripening of tomatoes after they are picked and natural tenderising of meat on ageing, are desirable. But these changes can proceed too far resulting in food deterioration if not halted at the appropriate time. The enzymes responsible for deteriorative changes are to be inactivated by a suitable method at the appropriate time to prevent food deterioration.

Insects, Parasites and Rodents: Insects are particularly destructive so far as cereal grains, fruits and vegetables are concerned. The loss of food due to insect destruction varies from 5 to 50 per cent depending upon the care taken in the field and storage. Insect eggs may persist in the foods even after processing as, for example, in flour. It is virtually impossible to produce and transport grains and other food commodities completely devoid of insects and insect pests. Therefore, a certain level of insect contamination in such foods is inevitable but it should be kept to a minimum by controlling the level of insect infestation. Insects in grains, dried fruits and spices are generally controlled by fumigation with fumigants like methyl bromide, ethylene oxide and propylene oxide. Apart from the loss due to the food eaten, insects cause greater damage due to the bruises and cuts they make in foods exposing them to microbial attack resulting in total decay.

Parasitic food spoilage occurs in some foods. Hogs eat uncooked food wastes. The parasitic nematode penetrates the hog's intestine and finds its way into pork. The live worms can infect man if the pork is not thoroughly cooked. A parasite worm belonging to the genus *Anisakis* is found in fish. If such fish is eaten raw it can infect man. Another common parasitic

contamination of foods is *Entamoeba histolytica* responsible for amoebic dysentery. This organism contaminates foods when raw human excrement is used as fertilizer for crops. Infected water and poor hygiene also spread the parasite. Cooking kills most of these parasites.

Rodents contribute substantially to food shortages in countries where they are not controlled. Rats live up to 3 years and may have 3-8 litters. Apart from the fact that they consume large quantities of food, they contaminate food. Rodent's urine and droppings harbour several kinds of disease producing bacteria, and rats spread such human diseases as typhus fever, plague, typhoid fever, etc.

Temperature: Heat and cold, apart from their role in food preservation, contribute to food deterioration if not controlled. The rate of chemical reaction doubles itself for every 10°C rise in temperature. Excessive heat brings about protein denaturation, destroys vitamins, breaks emulsions and dries out food by removing moisture. Uncontrolled cold also will deteriorate food. The freezing and thawing of fruits and vegetables destroy their structure. Skin rupture leaves the food susceptible to microbial infection. If milk is frozen, fat will separate and milk protein will be denatured causing it to curdle. Several fruits and vegetables are damaged even at the temperature of refrigeration (4°C). The deterioration includes off-colour development, surface biting and various forms of decay. Bananas, tomatoes, lemons and squash should be maintained above 10°C for maximum quality retention.

Moisture: Presence or absence of excess moisture in food leads to food spoilage. Moisture is required for chemical reactions and microbial growth. Foods with a high percentage of water deteriorate fast. Perishable foods have a high water content. Leafy vegetables, young shoots, juicy fruits, meats and milk deteriorate rapidly. Even if the moisture content is not uniform throughout the food, it exerts its influence. Surface moisture changes due to changes in relative humidity can cause lumping and caking, surface defects, crystallisation and

stickiness in foods. Condensation of even small amounts of moisture can result in the multiplication of bacteria and molds. This condensation need not come from outside. Fruits and vege tables can give off moisture from respiration and transpiration even when packed in a moisture-free package. This moisture trapped within the package is enough for the micro-organisms to grow. Even non-living foods in moisture-proof packages can give up moisture and change relative humidity in the enclosed space, which can result in microbial growth. Control of moisture in foods is thus very important from the point of view of their preservation.

Oxygen, Light and Time: Air and oxygen bring about a number of destructive changes in food components such as destruction of food colour, flavour, vitamins A and C and other food constituents. Oxygen is necessary for the growth of molds. Oxygen is therefore to be excluded from food in the course of processing, by deaeration, vacuum packing, or flushing containers with nitrogen or carbon dioxide and, in some case by the use of oxygen-absorbing chemicals.

Light destroys vitamins B-2, A and C. It also deteriorates many food colours. Not all wavelengths of natural or artificial light are absorbed by food constituents or are equally destructive. Surface discolourations of meat pigments are different under natural and fluorescent lights. Foods may be protected from light by impervious packing or keeping them in containers that screen out specific wavelengths.

All the food deteriorative factors considered so far are time-dependent. The larger the time, the greater the destructive influences. All foods, fresh or processed, have their qualities at a peak at some time after they are obtained. For best quality, they should be consumed before deterioration sets in. Deterioration with time takes place with most foods except certain foods like cheese, wine and other fermented foods which improve with ageing up to a point.

Food Safety in the Home: Despite the fact that food is free

from harmful levels of micro-organisms when it is brought home, there is still a danger of food spoilage due to the causes discussed above. In drder to avoid it in the home, standards of hygiene should be maintained in all aspects of food preservation. Personal hygiene and kitchen sanitation practice should be maintained. Foods should be appropriately stored. The temperature range favourable for growth of micro-organisms should be avoided. Foods should be kept at a temperature colder than 4°C. Hot foods should be cooled as quickly as possible to the point where they can be placed in the refrigerator.

Processing of Food

The holding of food supplies from periods of plenty for supply during seasons of little or no production is one of the most fundamental problems tackled by man and also animals. For thousands of years, man has extensively practised food preservation. It was necessary to ensure a year-round food supply. Storing of food has become increasingly important today in the progress of economic life. Without the hoarding of food, life as we know today would be impossible.

The earliest methods used for food preservation were smoking, drying and salting. The use of ice and snow to preserve perishable foods was known to early man. Fermentation of agricultural produce also was known from early times. Barley, rice and fruit juices were used for alcoholic beverages. Fermented soyabean products have been used in the East from 2000 BC. Modern methods of food preservation have developed through the centuries on the basis of practices employed in the past. Industrial processing of food started on a small scale in the 19th century, and during this century there have been tremendous developments.

The underlying principle of all preservation techniques, whether carried out in the home or commercially, is to restrict food spoilage so that food can be used safely in a palatable form at a later time. Preservation methods are designed to

inhibit the growth of organisms, the removal of organisms and killing the organisms. Food preservation should also arrest the biochemical breakdown of tissues and the transformation of its cell contents. This is achieved by heat, cold, drying, fermentation, radiation and chemicals. Food deterioration also is brought about by any one of these techniques. The technique employed should be such that micro-organisms are killed or inhibited and biochemical attacks lowered and the food is still left unaltered.

Preservation and Processing by Heat: Of the various methods of food preservation, heat finds very wide applications. The application of heat to food is so universal that "processing" and "heating" are considered as synonymous. This underlines the importance of heat in bringing about physical, chemical and biological modifications in foods. The purpose of heat treatment in food preservation is to kill the micro-organisms and inactivate the enzymes that become active in the subsequent storage of packed food. Various methods have been developed to achieve this purpose without bringing about other undesirable effects in foods.

Effect of Heat on Micro-organisms: Most bacteria, yeasts and molds show a growth optimum between 16°-38°C (mesophiles). Such organisms, in general, do not grow below 5°C and above 45°C. Some organisms grow in the range of 66°-82°C (thermophiles). Most bacteria are killed in the range of 82°-93°C. However, bacterial spores and certain other heat-resistant forms can withstand prolonged exposure to 100°C. Some organisms endure this temperature for 5½ hours. However, they last only minutes at 120°C and are destroyed in moist heat at 100°C in about 15 min. The thermal death of micro-organisms in dry air is due to oxidation. Therefore, a high temperature is required for their destruction. In wet heat, cell death is due to coagulation consisting of reactions between proteins and water; this is also accelerated by raised temperatures.

The complete destruction of microogranisms is known as

sterilisation. This is achieved when organisms are exposed at 121°C to wet heat for 15 min. This can be accomplished in a pressure cooker or autoclave on a small scale or, on a commercial scale, in pressure retorts. It should be remembered, however, that for storage a food need not be completely sterile. In many cases it would be enough if it is "commercially sterile" (appertized), i.e., foods may contain a very small number of resistant bacterial spores, but these will not normally multiply in the food during storage. For most foods, a theoretical reduction to $1/10^{12}$ of the original population is considered satisfactory. Canned foods which are commercially sterile have a shelf-life of two years or more.

Since the heat required for complete sterilisation of food also adversely affects the properties of foods, the heat treatment given should be just enough to destroy the pathogenic organisms and toxins, and ensure desired storage life. Thus, to preserve foods by heat with safety, a knowledge of the time-temperature combination required to inactivate the most heat resistant pathogen in a particular food, and the heat penetration characteristics of the particular food including the can or container if the the food is packed, is required.

Thermal Death Time (TDT) Curve: An interesting aspect of the thermal death of micro-organisms (this applies to bacterial spores also) is that, at a given lethal temperature, the death rate is proportional to the number of micro-organisms still living. This is referred to as the logarithmic order of death, which means that under constant thermal conditions the same percentage of microbial population, will be destroyed in a given time interval, regardless of the size of the surviving population. Thus, if 90 per cent of an original microbial population is killed in the first minute, 90 per cent of the remaining will be killed in the second minute and so on. The logarithmic death rate curve for an organism provides the death rate of the organism in a specific medium or food at a specific temperature. From the death-rate curves at different temperatures, a death-time Curve is obtained. A thermal death-

time curve for a specific organism in a spfecific medium or food provides data on the destruction time for a defined population of that organism at different temperatures.

A thermal death-time curve indicates that greater the initial microbial population in a food, the greater is the heat treatment required for its destruction. This strongly emphasises the importance of good sanitation in food industries, even though heat is a destructive force for micro-organisms. In food preservation by heat, in addition to heat destruction of micro-organisms as indicated by the thermal death-time curve, a number of other factors influence microbial population. The acid pH of the food, for example, will increase the killing of organisms by heat. Foods high in acid (having a pH of 4.6 or less), such as tomatoes or orange juice, need not be heated severely because the acid increases the killing power of heat. With acid foods, in many cases, temperature at or below 100°C for a few minutes constitutes adequate heat treatment. Certain spices and food chemicals act synergistically with heat in killing micro-organisms and so reduce the heat treatment that must be used. As will be discussed subsequently, many other food constituents have an opposite effect on heat sensitivity of micro-organisms and protect them against heat. Thus, thermal death-time curves, to be valid, should be established for the specific food for which a heat process is being applied. To provide a substantial margin of safety it is better to work out the heat treatment by assuming that the food contains a heat-resistant spore and its population is large.

Environmental Factors: The data obtained from thermal death-time curves indicate the time and temperature required to destroy the micro-organisms. For proper preservation of food, every pdrticle of food (within the container if the food is canned) has to reach the lethal temperature for the required time. That is, heat should penetrate into the mass of food or throughout the can. Thus, the extent of heating depends upon the size and shape of the food or can andlalso on the nature and consistency of food. This involves the heat transfer into

the food by conduction or convection. Convection heating is far more rapid than conduction heating and so a liquid food reaches the required temperature more rapidly than a food containing free liquid and solid, which in turn is heated more rapidly than a solid food.

When heat is applied from outside the food nearest to the container (can) surface will reach the sterilisation temperature sooner than the food in the centre of the can. The point in the can or mass of food which is last to reach the final heating temperature is known as "cold point". To ensure necessary sterilisation of food, heating must be continued till the time the cold point attains the lethal temperature and remains at that temperature for the required interval. Since the time required for penetration of heat throughout the food varies depending upon the type of sterilizing unit used, size and shape of food container and the composition of food, it is obvious that the required heat treatment will be different for each specific case. This has to be determined to complete the sterilisation requirement of food.

The required lethal effect at the cold point of the food can be achieved by various time-temperature exposures. For example, destruction of *Clostridium botulinum* in a low-acid medium is complete in 2.78 min. at 121°C or 10 min. at 116°C. Also, the temperature rise during heat penetration accomplishes some degree of microbial destruction. Cooling of sterile food is not instantaneous and some additional microbial destruction takes place during the cooling period. These factors are to be taken into consideration in evaluating the temperature and time required for food preservation by heat.

Several constituents of food protect micro-organisms to various degrees against heat. Sugar in high concentration decreases their susceptibility to heat. Thus, canned fruit in a sugar syrup requires a higher temperature or longer time for sterilisation than the same fruit without sugar. Fats and oils decrease the penetration of wet heat into the cells of micro-organisms and heating becomes more like dry heat. Food

constituents, in addition to the above types of direct protection of micro-organisms, also contribute indirectly to the penetration of heat. Fats are poorer conductors of heat than water. Food consistency determines whether conduction or convection heating will take place. If starch is added as a thickener to a food, it may result in the conversion of the convection heat system to a conduction system thus slowing down heat penetration within the container or mass of food and this too will protect micro-organisms.

Canning: The basic process of canning consists of placing food in a sealable container, closing, heating and cooling. This method of food preservation has been in use since early 1800s though the understanding of the process came much later. Large quantities of foods are canned for preservation. In advanced countries canned foods form a significant part of the diet of the people. Most fruits and vegetables, a wide variety of meats and meat products, fish products, soups and many other items are all canned. A typical canning process includes the following steps:

1. Receiving, cleaning, grading and inspecting of raw commodity.
2. Blanching to inactivate enzymes.
3. Placing in the container with added brine or syrup and deaeration of the product.
4. Heating in a retort, under 1.05 kg/cm^2 pressure using steam for metal cans or pressurised water for glass containers.
5. Partial cooling under pressure in the retort.
6. Additional cooling by water sprays or in a cooling tank.
7. Labelling, racking and distributing.

First, the raw commodity is inspected for any treatment required before canning. Many of them require special treatment, such as washing, trimming, shelling, size grading and so on. In commercial canning, most of these operations

are carried out mechanically. The food may then be blanched to reduce surface contamination and inactivate enzymes. The blanched food is placed in the containers. Brine (1.1 to 1.6 per cent salt concentration) is added in the case of vegetables, meat and fish, and syrup in the case of fruits. In some cases, only water is added. Oils are commonly added to fish products.

The containers generally preferred for most heat-processed foods are tin cans. Glass jars are easy to clean, corrosion free and transparent, but they have the disadvantage of requiring more processing time than cans of the same size. There is also the problem of breakage. As heating time is roughly proportional to volume of the container, larger containers must be heated for a longer time and food canned in such containers will be of lower quality than those placed in smaller containers.

Heating time can be reduced by placing the food in a flexible, relatively thin pouch instead of a can or jar. It has been shown that foods processed in this manner are of comparable quality to frozen foods. The pouches can be heated for serving by simply placing them in hot water, and damage in distribution would be significantly less than with can or glass jar. Pouch canning is gradually gaining importance.

The container with the food is ready for deaeration. This can be done by heating the filled cans in steam or hot water or the use of a vacuum closing machine when the air is sucked out by a pump and the lid is sealed on while vacuum is maintained. Air and gas removal is necessary to prevent the bulging of the can due to internal pressure, oxidation of the contents and inside corrosion of the tin plate.

The sealed containers are subjected to heating in a retort, a chamber in which canned foods are processed. Several types of retorts are used. Still retorts are the simplest type in which the cans remain still while they are being heated. In this case, the heating time to bring the cold point to the sterilising temperature is relatively long. The time taken for sterilisation

can be markedly reduced by using agitating retorts. The cans are agitated during the heating process. The mixing inside the container helps the conduction and convection processes so that heating is achieved in significantly less time and food quality is improved. Whether still or agitating retorts are used, the high temperature required for sterilisation is commonly obtained from steam under pressure. The heating may be batchwise or continuous. The heated containers are partially cooled under pressure in the retorts to decrease the internal pressure of the container. Next, the containers are cooled in water or under a spray. Finally, they are labelled and packed for distribution.

Aseptic Canning: In normal canning, heat transfer from the outside of the container to the inside will require many minutes or even hours depending upon the container size to reach the sterilisation temperature. The time of sterilisation can be shortened to seconds or even fraction of a second in aseptic canning. The basic principle of this method is that food is pumped continuously through a plate-type or tubular heat exchanger which heats it very quickly to a high temperature, holding it at that temperature for the time required and then cooling. Food temperature employed may be as high as 150°C and sterilisation takes place in 1 or 2 s. Such a rapid sterilisation at high temperature is referred to as ultra-high-temperature sterilisation. The sterile food is quickly cooled, placed in aseptic containers and the lids are sealed while in a sterile environment. The food canned aseptically retains the nutrients and the sensory attributes will be good.

With acid foods, foods sterilised as in aseptic canning, while still hot, can be filled into clean but not necessarily aseptic containers. The heat of the food and some holding time before cooling the closed container renders the containers commercially sterile. This type of canning is known as "hot pack" or "hot fill".

With low-acid food (above pH 4.6) the conventional hot pack processing is not possible. In such cases, *"flash 18"* process

(pressure canning) is employed. Low-acid foods are heated above 100°C under pressure for sterilty. If food at that temperature is poured into containers for sealing at atmospheric pressure, there will be violent boiling. This is eliminated by carrying out the canning process in a chamber under a pressure of 1.05 to 1.40 kg/cm^2. Thus, low acid foods can be presterilised by high temperature for a short time. They are then transferred to cans in a pressurised room and heated for an appropriate number of minutes so that commercial sterility of non-sterile cans are obtained and finally the cans are sealed and cooled.

Pasteurisation: When foods are heated in containers or by other methods to a temperature below the boiling point of water for a definite period, the process is known as pasteurisation. This process serves two objectives: it destroys pathogenic organisms associated with the food and extends the product's shelf-life by decreasing the microbial population and inactivating some enzymes. The choice of temperature and time of pasteurisation will be influenced by the consideration of the purpose of the process and the chemical and physical composition of food. Thus, milk is pasteurised at 62.8°C for 30 min. to inactivate pathogens and whole egg (mixture of yolk and white) at 64.4°C for 2-5 min. to control the dissemination of *Salmonella* spp. With sweetened condensed milk, a mild heat treatment is acceptable because of the moderate heat resistance of yeasts, the only organisms that could grow under conditions imposed by sugar concentration. Pasteurisation is also employed in the case of beer, wines, fruit juices and certain other foods mostly to increase shelf-life.

Pasteurised products are not sterile. They contain vegetative organisms and spores still capable of growth. Many pasteurised foods must be stored under refrigeration. Pasteurised milk under refrigeration can be stored for a week or more in good condition. However, at room temperature pasteurised milk may be spoilt in a day. High-temperature short-time treatment

(HTST) using a temperature of 130°C and above for short times, i.e., a few seconds to 6 min. are used to obtain a germ-free stage.

Blanching: Blanching is a heat treatment like pasteurisation. The term is usually used in conjunction with vegetable processing, where the goals are to inactivate degradative enzymes and deaerate the product before further processing rather than kill micro-organisms. If omitted, off-flavour, vitamin losses and colour changes occur in the frozen storage. Blanching is usually performed by dipping the products in boiling water for two to three minutes.

Preservation and Processing by Cold: As stated earlier, preservation of food by freezing and cold storage was known to ancient people. Tribes and nations in the temperate and cooler climates froze their harvest, thawed it when necessary and consumed it. Frozen foods were even transported to short distances to consuming centres. With the development of mechanical refrigeration systems, cold preservation of food, food processing, storage and distribution have become widespread. Refrigeration has also influenced agricultural practices. It has become possible to transport perishable foods for long distances from production to consumption centres and make available seasonal foods at all times of the year.

Effect of CoW on Micro-organisms: While most bacteria, yeasts and molds grow best at temperatures of 16° to 38°C, some micro-organisms continue to grow even at low temperatures. These are known as psychrophilic (cryophilic) organisms. These grow even at 0°C, the freezing point of water. However, lowering the temperature of foods and food products de creases the growth rate of micro-organisms and the growth and multiplicity completely stop when water freezes. In some foods water does not freeze even at -10°C or lower temperatures because of the lower freezing point due to dissolved salts, sugars and other substances. The retardation of microbial and biochemical activities at low temperatures is the basis of preservation of food by cold. Cold temperature

treatment, including severe freezing, only reduces the microbial activities and population but does not kill all bacteria. Foods frozen, even for two years or more, when removed from cold storage have shown bacterial activity when thawed.

Types of Cold Preservation: Cold preservation of food may be categorised into two groups: refrigeration and freezing. Household and commercial refrigerators usually run at 4.4°-7.2°C. Commercial refrigeration (cold or chill storage) may use a slightly lower temperature, depending upon the nature of food refrigerated. Freezing refers to foods maintained in a frozen condition. For good freezing a temperature of -18°C or below is required. Chill storage will preserve perishable foods for days or weeks depending upon the food. Frozen storage (deep freezing) will preserve foods for months or even years.

Refrigeration has certain advantages over freezing. It takes less energy to cool a food to just above its freezing point than to freeze it. Refrigeration storage requires less insulation and refrigeration capacity, and refrigerated products do not suffer from texture and flavour losses caused by freezing. Finally, refrigerated products do not have to be thawed before use—a process that may take as much as 48 hours with large frozen foods.

Further distinction between refrigeration and freezing temperatures relates to microbial activity. Most spoilage organisms grow above 10°C. Some food poisoning organisms grow slowly up to 3.3°C. Cryophilic organisms grow slowly from -4.4°C to -9.4°C provided the food is not frozen. These organisms, however, do not produce toxins or disease but cause food deterioration. There is no growth of organisms below -9.4°C though there may not be a total destruction at that temperature.

Chill Storage: Chill storage is useful as the principal method of preservation and as an adjunct to other methods of preservation. The shelf-life of food is directly linked with the microbial rate of growth as influenced by temperature. Storage

at - 1° and - 4°C can provide stability, particularly in the presence of food preservative or modified atmosphere.

Production of Low Temperature: Chill storage requires controlled low temperature. This is achieved by taking heat away from the storage area. Heat is transferred from the storage area by the principle of latent heat of vaporisation. If the state is changed from liquid to gas or from gas to liquid without affecting its temperature an amount of heat must be added or removed. Liquids like ammonia and freon (Cl_2F_2) are used as refrigerants. Liquid refrigerant is circulated through an expansion valve or capillary tube and then through a heat exchanger, called an evaporator, within the storage area. In passing through the capillary tube, the refrigerant goes from a relatively high pressure to a relatively low one, causing the liquid refrigerant to evaporate and thus reducing the temperature of the liquid-gas phase of the refrigerant. In the evaporator, the remaining liquid evaporates absorbing heat from the storage area. The temperature of the vaporisation of the refrigerant is some 5°-6°C below the storage temperature. A compressor then pumps the refrigerant from the low-pressure cold condition to a high-pressure hot one. The hot gas Clows through a condenser where heat is extracted from the refrigerant and transforms it back to a room temperature, high-pressure liquid ready to flow again to the expansion or capillary tube. Air, water or a combination of air and evaporating water is used to cool the condenser.

For the construction of commercial chill storage rooms, the "refrigeration load" is to be taken into consideration. This is the quantity of heat that must be removed from the product and the storage area in order to go from an initial temperature to the selected final temperature and then maintain this temperature for a specific time. This depends on the storage area and other factors that may generate heat within the storage area or influence the removal of heat from the area. These include the light and electrical installations, number of people working, how often the doors which permit entrance

of warm air are opened and the amount and nature of the food product stored in the refrigerated area. Also, the specific heat of food and the rate of respiration of such foods as fruits and vegetables have a bearing on the extent of refrigeration required. The heats produced from the respiration of fruits and vegetables vary considerably. Products like green beans, peas, spinach, sweet corn and strawberries have a high respiration rate even at 0°C.

Air Circulation and Humidity: The heat produced due to respiration in the vicinity of the food surface in cold storage is to be transferred towards the refrigerator. This is achieved by air circulation. The humidity of the circulated air is to be controlled. If it is too moist, moisture will condense on the cold food and molds will grow on food surfaces at the common refrigeration temperature. On the other hand, if the air is too dry there will be desiccation of food. The optimum relative humidity of the circulated air depends upon the moisture content of the foods and the ease with which they dry out. Most foods store best at a refrigeration temperature when the relative humidity of air is between 80 and 95 per cent. Dry and granular foods store well at about 50 per cent relative humidity.

When foods are to be chill stored for a long time they are protected by packing to prevent loss of moisture. Plastic sacks or moisture-resistant coating are used for this purpose. Cheese ripened for a long time in cold storage are wax coated. Loss of moisture from stored eggs is prevented by dipping the eggs in some oil, to seal the minute pores of the eggshell. If packing is not possible as in the case of tenderising beef by ageing in cool rooms, ultraviolet light is employed to retard mold growth.

Modified Atmosphere: Continued chill storage, and biochemical changes of fruits and vegetables even under chill storage result in overripening, softening and general loss in quality. These changes take place at the normal chemical composition of air (78 per cent nitrogen, 21 per cent oxygen and 0.03 per cent carbon dioxide) and they can be arrested by

using refrigeration in conjunction with "modified atmosphere" or "controlled atmosphere". By raising the carbon dioxide and lowering the oxygen levels, foods can be stored for an extended period of time. In a modified atmosphere of 3 per cent oxygen and 3 per cent carbon dioxide, apples can be preserved in good condition for about a year at 2.8°C and 87 per cent relative humidity. For each food there is an optimal storage temperature, relative humidity, oxygen level and carbon dioxide level which delay ripening process. Modified atmosphere cold storage in gas-tight containers has enabled shipment of even highly perishable foods like lettuce over long distances. Ramification of controlled atmosphere can have other desirable effects on cold-stored food. For example, the presence of diphenyl vapours can inhibit mold growth in citrus foods, and use of ethylenegas can speed up the ripening and colour development of citrus fruits and bananas.

Deep Freesing: Tremendous advances have been made in freezing food for preservation, storage and distribution. Many foods can be frozen for twelve months or more without major changes in size, shape, texture, colour and flavour. Frozen foods have thus attained wide acceptance by the public. At present, no form of food preservation is as well suited to provide maximum convenience as freezing. Complete meals on individual plates can be frozen and are ready for use with a single thawing-heated operation. An unlimited number of items may now be frozen in their final serving forms.

Changes During Freezing and Thawing: Uncontrolled freezing of food can result in the disruption of texture, breaking up of emulsion, denaturation of protein and many other physical and chemical changes. Substances like sugar, salt and proteins dissolved in water lower the freezing point of water, Thus, when food is kept at a freezing temperature, where the temperature is considerably below the freezing point of water, the water component freezes first and leaves the dissolved solids in a more concentrated solution which requires a still lower temperature to freeze it. Thus, different foods with

varying levels of water as well as the amount and nature of dissolved substances will have different freezing points and under a given freezing condition will require different times to reach a solidly frozen state.

When food containing water is placed in a freezer it does not freeze uniformly. The portion of food nearest to the container freezes first and the first ice crystals are pure water. As water Continues to be frozen, the concentration of dissolved solids increases and finally a central core of highly concentrated unfrozen liquid remains. If the temperature is sufficiently low this central core also freezes solid ultimately. In the case of solid foods like a piece of meat at about -4°C, they appear to be solidly frozen but they still contain about 3 per cent water in an unfrozen condition. Even at -18°C not all of the water is completely frozen.

Small quantities of unfrozen water in frozen foods result in the deterioration of food with respect to texture, colour, flavour and other properties. The high concentration of dissolved substances in the remaining water can cause effects of various kinds. The dissolved substances may precipitate or crystallise imparting a gritty, sandy texture to the food; if they remain in solution, the high salt concentration or drop in pH (if the solutes are acidic) might result in protein denaturation. The disturbances in anionic and cationic concentration can disturb colloidal substances, and finally the concentration effects can cause the dehydration of adjacent tissues resulting in loss of tissue turgor. Other effects of unfrozen water are the possibilities of the growth of psychrophilic micro-organisms and the greater action of enzymes.

Formation of ice crystals during freezing can affect the texture of many frozen foods. There is water within and between the cells of food. When water freezes rapidly it forms minute ice crystals and clusters of crystals. In slow freezing, water within and between cells freezes causing physical rupture and separation of cells. If freezing is rapid, the minute crystals formed are only within the cells but not between the cells and

physical damage of the cell is less severe. Rapid cooling also minimises concentration effects by decreasing the time of contact of solutes with food tissues during the transition from the unfrozen to the fully frozen state. For these reasons, rapid freezing is desirable for better product quality.

From the point of view of quality and economic aspects, it is best to freeze foods to -18°C or lower and maintain them at that temperature during storage and transport. In order to achieve the best advantage of rapid freezing, many foods are frozen to temperature somewhat below -26°C but this increases the cost. A freezing temperature of -18°C is very safe from attack by micro-organisms as no pathogenic organisms grow below about 3.3°C and spoilage organisms below about -9.4°C. At -9.4°C, most foods retain considerable frozen water and long storage at this temperature can result in the enzymatic deterioration of food. Even at -18°C some enzymes retain activity. However the rates of reactions they catalyse will be extremely slow. In the case of fruits and vegetables, the enzyme activity even at -18°C is sufficient to bring about spoilage. In these cases, the enzymes are inactivated by blanching before freezing. Properly packed foods frozen and stored at -18°C have high-quality storage life of the order of 12 months and longer.

The kind of damages that occur to food during slow freezing also occur during thawing. Repeated freezing and thawing is therefore very detrimental to the quality of food. Even a fluctuation of 3°C in freezing temperature above and below -18°C at which the food is frozen and stored can damage many foods. Upon refreezing, water melted from small ice crystals tends to bathe unmelted crystals causing them to grow in size. Also, if thawing of frozen foods is slow there is loss of quality due to concentration effects. In slow thawing, there is more time for food constituents to be in contact with concentrated solution thus intensifying their damaging effect. Quick thawing is also desirable from the point of keeping the microbial population in check. Large volumes of frozen food

can take from 20-60 hours for thawing. Since bacteria survive the thawing process, long periods and rise in tempera-. ture of products will be opportunate for bacterial multiplication. Use of microwave heating reduces the thawing time.

Methods of Food Freezing: There are three basic methods of freezing in commercial use. These are freezing in air, freezing by indirect contact with a refrigerant and freezing by direct contact with the freezing medium. Air freezing may be by the use of still air ("sharp freezing") or an air blast.

Indirect contact freezing consists of keeping food or food packages on a surface cooled by refrigerant. In direct-contact freezing, the food or package is submerged in a cold liquid or the cold liquid is sprayed on the food or package.

Air freezing: Still-air freezing is the oldest and least expensive method of freezing. This is the type of freezing carried out in home freezers. In this method, food packs are placed in the freezer in such a way that air can circulate between the packages. The freezer is maintained in the range of -23° to -29°C. There is some air movement due to convection and in some cases gentle air movement may also be promoted.

In air-blast freezing, food packages are carried on an open mesh belt through a tunnel through which air at temperatures of -29°C to -46°C is forced at velocities of 10 to 15 m/s over, under or through the product. Under these conditions, a food packet that takes 72 hours to freeze in still-air can be frozen in about 12—18 hours. A drawback of air-blast freezing is the dehydration of food in an unwrapped condition, which is known as freezer burn. This occurs due to ice changing directly to water vapour molecules without going through the liquid state. Freezer burn results in discolouration, changes in texture and off-flavour. Freezer burn can be minimised by prechilling food with air at high humidity at about -4°C and the prechil-led food is then moved into the colder zone where it is quickly frozen.

Indirect Contact Freezing: Solid foods in consumer size

flat packets are placed on hallow metal plates chilled by circulating refrigerant, so that the food is in direct contact with the cold metal wall but in indirect contact with the refrigerant. The efficiency of freezing depends upon the extent of contact between the plates and the food. For this reason, the packages should be well filled or slightly overfilled to make good pressure contact with the plates. The freezing of liquids and purees by indirect freezing is carried out by pumping them through tubes on the outside of which the refrigeration flows. Through appropriate mechanical devices the frozen food is scrapped to keep the mass in motion thus enabling the continuous bringing of new portions of food into contact with the cold wall. In this method freezing occurs in a matter of seconds.

Immersion Freezing: Immersion freezing has a number of advantages. There is intimate contact between the food or package and the refrigerant. This is particularly important in freezing irregularly shaped food pieces and immersion freezing minimises the contact of food with air during freezing, which is desirable for foods sensitive to oxidation.

The refrigerants used for immersion freezing should be non-toxic, pure, clear, free from taste, odour, colour or bleaching action Low-freezing-point liquids'cooled by indirect contact with another refrigerant or cryogenic liquids like liquid nitrogen, carbon dioxide or freon, are used as refrigerants.

The low-freezing-point liquids used are solutions of sugar, salt or glycerol. A temperature as low as -21°C can be reached with a 23 per cent sodium chloride or 62 per cent sucrose solution. With a 67 per cent glycerol-water solution, one could go down to -47°C.

Cryogenic liquids are liquefied gases of extremely low boiling point, such as liquid nitrogen (BP, -196°C), liquid carbon dioxide (BP, -79°C) and freon 12 (BP, -30°C). The most commonly used cryogenic liquid is liquid nitrogen. Its low boiling point helps very quick freezing of even large food

pieces and it can give a quality unattainable by other non-cryogenic freezing methods. Further, since the cold temperature results from evaporation of liquid nitrogen, freezing by liquid nitrogen does not require a primary refrigerant to cool this medium, and liquid nitrogen is non-toxic and inert to food constituents. The operating cost of liquid-nitrogen freezing, however, is high. With liquid nitrogen it is possible to freeze food down to -196°C, but it is seldom frozen to a temperature below -46°C. Generally, liquid-nitrogen freezing produces less dehydration loss during freezing and less drip loss during thawing than other freezing methods.

Freezing with liquid carbon dioxide is carried out in a manner similar to freezing with liquid nitrogen. Some foods frozen with this refrigerant are equal in quality to those by liquid-nitrogen freezing. In such cases, use of liquid carbon dioxide for freezing is very economical. Of late, freon freezing is coming into use. The installation and operational costs of freon freezing are high but this can be overcome by the efficient recovery of the freon which becomes vapour on contact with food, for reuse.

Packaging for Freezing: Choice of packaging material is of special concern in the manufacture of high-quality frozen foods. The material should be moisture-proof and impermeable to oxygen and flavour compounds both at freezing temperature and after thawing. It should be resistant to chemical attack from the constituents of foods. Most foods expand on freezing. Therefore, the packaging material in which the food is frozen should have good mechanical strength, have a degree of flexibility and should not be completely filled. Many packaging materials such as cans, metal foils, waxed papers, plastic-coated cardboards and plastic foils, cellophane, parchment paper, etc., are all satisfactory for frozen foods. Glass is not generally satisfactory for frozen foods due to the possibility of breakage from expansion and thermal shocks.

Drying: Food preservation by drying is one of the methods practised from ancient times. Drying is the method nature

resorts to preserve foods. Grains in the field dry sufficiently on the stalk by exposure to the sun, which requires no further drying for preservation. This is also true of legumes, innumerable seeds and some spices. The observation of natural drying was adopted by early man to dry fruits, fish and meat by exposing them to the sun. Sun drying is still in use in many parts of the world for preserving certain foods, such as fruits and nuts. However, it is limited by the fact that it is feasible only under climatic conditions of high heat and low humidity.

Drying of food involves complete removal of water under controlled conditions in such a way that the food is not altered and results in minimum changes by the drying process. Dried foods contain moisture to the extent of 1-5 per cent, and they have storage stability at room temperature of a year or longer. On reconstitution with water, dried foods are very close to and virtually indistinguishable from the original foods used in their preparation. Removal of moisture |gpm a solid with minimum change in food material is not an easy problem. Removal of only a part of water of foods, perhaps 1/3 to 2/3 of the water as in the preparation of syrups, evaporated milk, tomato paste, condensed soups, etc., is not considered as drying. Partial removal of water is known as concentration.

Advantages of Drying: Micro-organisms require water for growth and when they are growing on food they get water from food. If water is removed from food, the multiplication of micro-organisms will stop. The highest water contents at which microbial spoilage does not occur in the case of traditional dried foods are: dehydrated whole egg, 10-11 per cent; wheat flour, 13-15 per cent; dehydrated fat-free meat, 15 per cent; dehydrated vegetables, 14-20 per cent; and dehydrated fruits, 18-25 per cent. Drying of foods, thus, is primarily carried out to preserve food by controlling, micro-organisms. Partial drying as in concentration will be less effective than total drying in food preservation. However, concentration is quite sufficient to arrest the growth and multiplication of micro-organisms in some foods.

Drying, in addition to preservation, helps decrease the weight and bulk of food, e.g., 237 ml of orange juice on dehydration yields just 28 g of solids. In some cases the drying process may be chosen to retain the original shape and size. However, in such a case the volume may not be affected but there is reduction in weight. Drying thus results in great economy in storage, packaging and transport of food. Drying also results in the production of convenience foods, such as instant coffee, instant rice, etc. In these cases, cooking steps are completed before the products are dried.

Drying Rate: The amount of water removed from a given weight of material during a given time interval is called the drying rate. The most impcrtant stages in dehydration are penetration of heat into the product and removal of moisture out of the product. Various factors contribute to these two stages. A large surface area of the food to be dehydrated provides more surface area for contact witn the heating medium and for moisture to escape. Therefore, generally foods are divided into small pieces of thin layers for dehydration. This also helps reduce the distances heat must travel to reach the centre of the food and the moisture should travel to reach the sjurface to escape. Also, the greater the temperature differences between the heating medium and the food to be dried, greater is the drying rate. For any given temperature, the rate of water removal from the food is greater in vacuum than under atmospheric pressure. When the heating medium is air, it gets saturated with the moisture from the water driven from the food and this slows down subsequent moisture removal. Air in motion sweeps away moist air from the drying food surface. Thus, air velocity is also a factor that determines the drying rate.

When food is dried in air, the drier the air, the more rapid is the rate of drying. Moist air absorbs and holds less moisture than if it were dry. Dryness also determines the extent to which a food can be dried. Dried foods are hygroscopic. Each food has its own equilibrium relative humidity; i.e., the

humidity at a given temperature where food neither loses moisture to the atmosphere nor picks up moisture from the atmosphere. Below this atmosphere-humidity level food cannot be further dried. For example, potato at 100°C and 46 per cent relative humidity level can be dried to 4 per cent moisture. If we wish to dry it down to 2 per cent moisture at the same temperature in air, then the relative humidity of air should be 15 per cent. Thus, the equilibrium relative humidity data of a food are important both from the point of view of its drying and storage.

There is a relationship between evaporation and temperature in drying. As water evaporates from the food it results in the cooling of the food surface. This is due to the latent heat of phase change from liquid to gas. There are several important consequences of this phenomenon in food drying. Regardless of the drying temperature or heating surface, the temperature of the food will not be higher than a particular temperature as long as water is evaporating rapidly.

Thus, in a spray drier the incoming air may be at 204°C and the exit air at 121°C, but the food particle while drying may be at no more than about 71°C. As the moisture content of food decreases and evaporation slows down the particles rise in temperature. When there is no water, the incoming air and the exit air will have the same temperature (204°C). As foods are heat-sensitive, they are to be removed from the drier before this high temperature is reached.

There is also a time and temperature relationship in food preservation by drying. As food constituents are heat-sensitive, a high temperature for a short time does less damage to food than a low temperature for a long time. Also, when foods are dried they do not lose water at a constant rate all the way down to dryness. At the beginning of drying, and for some time thereafter, water generally evaporates from the food at a rather constant rate and then leaves with a falling-rate period of dryness. For example, 90 per cent of water from carrot dice

can be removed in four hours and the removal of the remaining 10 per cent requires another four hours.

Thus, varying conditions, such as temperature, humidity, air velocity, direction of the air, thickness of the food and others affect the drying of food. Removal of water below 2 per cent without damage to the product is exceedingly difficult.

More than the physical factors, some of the properties of food components and the changes they undergo during drying determine the drying rate. Most foods are not homogeneous at the molecular level. For example, a piece of meat will have lean and fat interlaced or marbled together.

Therefore when a piece of meat is dried it will give up water at different rates in the regions of fat and lean. The drying rate in such a case can be increased by orienting the meat portion relative to the source of heat such that moisture escapes in a line parallel to the layers of fat rather than having to pass through them.

When sugar and low-molecular-weight substances are present in foods, they elevate the boiling point of water and thus affect the drying rate. Also, during drying, concentration of solutes become greater which slows down drying. The escape of water from food depends upon the condition in which it is present. Free water easily evaporates. Water loosely held by force of adsorption to food solids evaporates slowly. Removal of water which enters into colloidal gels, such as when starch, pectin or other gums are present, is more difficult. It is more difficult to remove chemically bound water as in hydrates of salts. The cells of blanched or cooked foods are more permeable to moisture and dry more easily than their fresh counterparts, provided cooking does not cause excessive toughening or shrinking.

In many cases, foods are pre-treated before drying with a view to make the structure more porous so as to facilitate transfer of moisture and thereby speed the drying rate. Porous sponge-like structures are excellent insulating bodies, and

generally slow down the rate of heat transfer into the food. The net effect of food porosity on drying rate thus depends on whether the effect of porosity is greater on speeding the escape of moisture from food or slowing of heating transfer rate. Apart from its effects on the drying rate, food porosity has the advantages of quick solubility on reconstitution, but suffers from the disadvantages of increased bulk and shorter storage stability.

Changes during Drying: Drying brings about changes in the final product quality. One of the most obvious changes during drying of cellular as well as noncellular foods is shrinkage. The type and extent of shrinkage depend upon the rate of drying. For example, on the slow drying of a vegetable dice, first there is surface shrinkage. With continued drying, water is removed from deeper layers and finally from the centre causing continuous shrinkage towards the centre with a concave appearance and the product is dense. With quick high-temperature drying of food the surface becomes dry and rigid long before the centre dries out. Thus, when the centre dries and shrinks it pulls away from the rigid surface layer causing internal splits, voids and honeycomb effects and the product is less dense. Such a product absorbs water, reconstitutes quicker, more closely resembles the original material and is thus favoured by consumers.

With foods that contain dissolved sugar and other solutes in high concentration, drying can result in the shrinkage and sealing off of the surface pores and cracks. This results in trapping much of the remaining water with the food and drying rate dropping off severely. Dried food pieces may also contain voids, cracks and pores of various diameters. The shrinking and pore clogging by the solutes is known as core hardening. Gradual drying with low surface temperature can minimise core hardening.

Some foods which lack structure and are rich in sugar and other materials soften when heated at the elevated temperature of drying. Even after all the water has been removed the foods

will be in a thermoplastic tacky condition, giving an impression that they still contain moisture; they stick to the drier. However, on cooling, the thermoplastic solids harden into crystalline or amorphous glass form. In this more brittle condition they can be easily removed from the drier.

In addition to physical changes described above, drying can bring about a range of chemical changes. These contribute to the quality of both the final product and the reconstituted food in terms of colour, flavour, texture, viscosity, nutritional value and storage stability. Complete prevention of these changes is virtually impossible. They can be minimised by appropriate technology.

Methods of Drying: Methods used for food preservation by drying should produce maximum drying rate with minimum product damage and economic drying costs. A number of drying methods are available; some are particularly suited for liquids, others for solid foods or mixtures containing food pieces. The common drier types used for liquid and solid foods may be categorised as the air-convection drier, drum or roller drier and vacuum drier. In the air-convection method, hot air supplies the heat for evaporation. If liquid, the food may be sprayed or poured into pans or on belts. Drum or roller driers are limited to be used with liquid foods, puree and mashes that can be applied as thin films. Vacuum driers are employed to lower the temperature of drying. Freeze driers are vacuum driers where, at an extremely low temperature, water vapour directly forms ice without going through the liquid state. The above division of driers is not rigid since many driers are combinations, e.g., a drum drier can be operated in vacuum or by blowing high-velocity heated air.

Air-convection Driers: There are some common aspects of the different types of air-convection driers. They all have an insulated enclosure, a means of circulating air through the enclosure and a means of heating this air. They also have means of supporting foods to be dried and devices to collect dried food.

Air may be heated directly or indirectly. In the former method, air is in direct contact with combustion gases. In indirect heating, the air is heated in contact with a hot surface, which is heated by a convenient method. A direct heating method may contaminate the food product which is not the case with indirect heating.

The simplest type of air-convection drier is the kiln drier. In this case there is direct heating of air. This kind of drier does not generally reduce moisture to below 10 per cent. A step more advanced drier is the cabinet drier. Food loaded in trays or pans in thin layers are placed in the cabinet and air heated by an indirect method is blown across food trays. This method is used for small-scale operations. Foods commonly dried by this method are fruits and vegetables. The drying time is of the order of 10-20 hours.

For larger operations a tunnel drier is used. Wet food carts are moved through the tunnel on a continuous belt. Drying air is blown in a direction opposite to the movement of food carts (counter-current principle). The conveyer belt itself in some cases may form a trough (belt trough drier). The belt is usually of metal mesh and heated air is blown through the mesh. Another type of air-convection drier is the fluidised bed drier. In this case, heated air is blown up through food particles with just enough force to suspend the particle in a gentle boiling motion.

The most important kind of air-convection drier is the spray drier. More food is dehydrated by using this drier than all other kinds of driers put together. Spray drying is limited to liquids, low-viscosity pastes or purees. Food in the form of a fine spray or mist is introduced into a tower or chamber along with heated air. The small droplets make intimate contact with hot air, blast off their moisture, become small particles and drop to the bottom of the tower from where they are removed. Drying takes place in a matter of seconds. The air temperature will be 204°C but the food particles never reach a temperature above 82°C as evaporation cools the food. This

method of dehydration can produce a high-quality product even with heat-sensitive materials like milk, eggs and coffee.

Drum or Roller Drier: Liquid foods, purees and mashes are dried by this method. The food to be dried is applied, as a continuous thin layer, on to the surface of a revolving drum or between a pair of drums moving in opposite directions generally heated by steam. The dried thin layer of food on the drum is scraped by a scraper blade positioned at a point on the drum. Some foods when dry are sticky and cannot be scraped off from the drum when hot.

Such a sticky food becomes brittle when cold, which facilitates scraping. Providing a cold zone on a drum thus helps the drying of such foods. The drum temperature may be kept above 100°C by using steam under pressure. Generally, the drum is kept at 150°C when a film of food 1.6 mm in thickness gets dried in one minute or less. For heat resistant food products, drum drying is one of the least expensive dehydration methods. Drum-dried foods generally will have a somewhat more "cooked" character than the same material spray-dried.

Vacuum Driers: This method of drying of foods is expensive but gives high-quality foods. The drying system consists of a vacuum chamber that can withstand external air pressure and contains shelves or other supports to hold food. This kind of vacuum drier is for batch-type operation. For continuous operation a belt drier is used. The shelves are heated electrically or by circulating a heated fluid. The food gets heated by conduction and also by radiated heat from shelves above and below. The drying chamber is evacuated by a suitable device maintained outside the vacuum chamber.

The water that evaporates from the food is suitably condensed. The temperature of the food and the rate of water removal are controlled by controlling the degree of vacuum and the intensity of heat output. Liquid foods dehydrated by vacuum drying have a puffed structure and are easily dissolved

in water. Because of the low temperature used, there is minimun flavour change and other kinds of heat damage in this method of drying.

Freeze-drying is used to dehydrate sensitive high-quality liquid foods such as coffee and juices, and also solids of high value like strawberries, shrimps, etc., having delicate flavour, colour and taxtural attributes which cannot be preserved by any other drying method. In freeze-drying, water evaporates from ice without passing through the liquid stage, i.e., water from ice directly becomes water vapour. This happens at a temperature of 0°C or below, and at a low pressure of 4.7 mm of mercury or less.

Frozen food is dried by keeping in a vacuum chamber maintained at a pressure of 0.1 to 2 mm of mercury and maintaining the temperature of the chamber just below the melting point of ice. Under such conditions water from the frozen food evaporates at the maximum rate. Since the frozen food remains rigid the evaporation of water leaves voids behind it, resulting in a porous sponge-like dried material. After the food is dried the vacuum is broken with inert nitrogen gas and packed under nitrogen. If the vacuum is broken by admitting air the dried product absorbs the air into its pores resulting in impaired storage stability.

Foam: Drying (puff dry) at atmospheric pressure can be employed to dehydrate foods and obtain products with qualities approaching vacuum-dried ones. This method of drying is very economical compared to vacuum-drying method and can be applied to liquid foods and purees which can be pre-foamed before drying.

As foaming exposes an enormous surface area for quick moisture escape, drying of foams can be rapid at atmospheric pressure at a reduced temperature. With foods that do not whip readily, some whipping agents (vegetable proteins, gums, monoglycerides) are added prior to being whipped. Stable foams are cast in thin layers into trays or belts and dried by various heating schemes.

Concentration of Food

Foods are concentrated for the same reasons that they are dried—preservation and reduction in weight and bulk. Nearly all liquid foods are first concentrated before drying because of the technical and economical advantages of such processing. Many foods are better recognised as concentrates than dried ones; examples are fruit-juice concentrates, canned soups, condensed milk, etc.

Concentration may not be effective in all cases as a means of food preservation. The levels of water in most concentrated foods are sufficient for microbial growth. Concentrated non-acid fruit juice and vegetable purees undergo spoilage unless they are further processed. On the other hand, concentrated sugar syrups, sauces and jellies are relatively free from spoilage. The solutions of sugar and salt dissolved in the remaining water give concentrates which exert high osmotic pressure and draw water from the microbial cells or prevent normal diffusion of water into the cells and thus prevent microbial contamination. Solutions containing 70 per cent of sugar or 18-25 per cent salt prevent the growth of all micro-organisms.

Methods of Concentration: As in food drying, one of the simplest methods of concentration is by utilising solar energy. Salt has been manufactured by the concentration of sea water by this method from the earliest times.

Kettle Evaporators: A simple method of concentration is by the use of open kettles and pans, which may be heated by direct flame or steam. High temperature and long concentration times damage most foods. However, kettle or pan concentration is used in the manufacture of syrups as high heat is desirable to produce the colour from caramalised sugar and develop typical flavour.

Flash Evaporation: Purees are concentrated by flash evaporation. Superheated steam (150°C) is injected into the food pumped into a vertical tubular steam-jacketed evaporator where boiling occurs. The boiling mix then enters a separator from which the concentrated food is drained off.

Film Evaporators: Foods are also concentrated using thin film evaporators. In this case, food is pumped into a vertical cylinder which has a rotating element that spreads the food into a thin layer on the cylindrical wall. The cylinder wall is usually heated by steam. Water quickly flashes from the thin food layer and the concentrated food is simultaneously wiped from the cylinder wall. The product temperature may reach 83°C but since the concentrated food will be in contact with the heated cylinder for less than a minute, the damage is minimal.

Vacuum Evaporators: Heat-sensitive liquid foods are commonly concentrated at low temperatures in vacuum evaporators. Evaporation systems can be made that operate many times more effectively than single-effect evaporation. Several vacuumised vessels in series constitute multiple effect evaporators. In this process, the first evaporator under vacuum is heated with steam. The water vapour boiled off from the food is sent to the next evaporator kept at a higher vacuum than the first. The water vapour acts as a heat source to evaporate food in the second evaporator. The water vapour from the second can next be passed to a third evaporator having a higher vacuum than the second and so on. Grape juice and tomato juice are concentrated in this way. This method results in energy conservation.

Freeze Concentration: Concentration by freezing has been used to concentrate fruit juice. When liquid food is frozen, not all the water is converted into ice at the same time. Before the food freezes, the initially formed ice crystals are separated. By repeating this process several times on the concentrated unfrozen food a high percentage of water can be removed. The disadvantage of this method is the unavoidable entrainment loss.

Ultrafiltration and Reverse Osmosis: Another method of concentration of liquid food is by ultrafiltration and reverse osmosis. The process consists of pumping liquid foods at a high pressure against selective membranes that allow water

molecules to pass through them while retaining macromblecules, salts, sugar and other organic molecules. Membranes used in ultrafiltration are "less light" and may allow small molecules to pass through them under moderate pressure while in reverse osmosis the membranes are "tighter" and permeability does not allow even these molecules to pass through. In osmosis, there is the movement of water through membranes from a region of lower concentration to a region of higher concentration. In reverse osmosis this is reversed; under pressure there is flow of water through the membranes from a region of higher concentration to that of a lower.

The concentration of foods results in cooked flavour and darkening of colour. Concentration also causes changes in organoleptic and nutritional properties of foods. Some foods cannot be concentrated beyond a certain point. When foods containing sugar are concentrated sugar crystallises out as concentration increases. Concentration also causes an increase ir. the level of salts and minerals which result in the precipitation of protein. This is the cause of gel formation of evaporated milk after a few weeks' or months' storage.

Food Fermentation: The methods of food preservation so far studied, such as application of heat, cold, removal of water and other methods used, have the common objective of decreasing the number of micro-organisms in food or holding them in check against further multiplication. In food preservation by fermentation, in contrast, multiplicatioin of micro-organisms and their metabolic activities are encouraged. However, the organisms that are encouraged to grow and multiply are the select groups whose metabolic products help food preservation.

Natural fermentations have played a vital role in the preservation of foods from early times. Fruit juices exposed to air underwent natural fermentation and acquired an alcoholic flavour. Milk on standing became slightly acidic. These changes were therefore used as a means of preservation of otherwise perishable foods. At the same time the changes

in the texture and taste of fermented foods were found desirable. These are the principal reasons for the continued use of fermentation in food processing and preservation. The list of foods produced by fermentation is extremely long and includes the following: cheese, curd, butter, all alcoholic beverages, pickles, sauerkraut, vinegar, bread, idli, soya sauce, coffee, tea and cocoa.

The term fermentation means the breakdown of carbohydrate material by micro-organisms (or enzymes) under anaerobic conditions. In common usage, the term fermentation refers to both the anaerobic and aerobic breakdowns of carbohydrates and carbohydrate-like materials. Thus the conversion of lactose to lactic acid by *Streptococcus lactis* which takes place under anaerobic conditions and conversion of ethyl alcohol to acetic acid by the bacteria *Acetobacter aceti* under aerobic conditions are both referred to as fermentation. The term is used in a still broader and less precise manner to describe fermented foods.

In the fermentation of foods, a complex mixture of carbohydrates, proteins, fats, etc., undergo modifications simultaneously under the action of a variety of micro-organisms and enzymes present. Thus, the carbohydrates and carbohydrate like materials undergo fermentation, proteinaceous material undergo proteolysis or "putrefactive" breakdown and lipids "lipolytic" breakdown. The nature and extent of these changes depend upon the food, types of micro-organisms present and conditions affecting their growth and metabolic pattern.

The preservative effect of fermentation is caused by the chemicals excreted by the micro-organisms. The principal chemicals involved are acids (especially lactic acid) and alcohol. These inhibit the growth of common pathogenic organisms in foods. The toxin producing *Clostridium botulinum*, for example, cannot grow at pH values below 4.6.

Benefits of Fermentation: In addition to contributing to

preservation, fermentation of foods has additional benefits. Fermentation produces flavour and textural changes and fermented foods are actually more nutritious than their unfermented counterparts. Microorganisms, in addition to breaking down complex compounds, synthesise several vitamins and other growth factors. In fact, the industrial production of some vitamins is largely by fermentation processes. The digestible carbohydrates and proteins of grains and seeds are enclosed in cellulosic and hemicellulosic structures which are indigestible materials. Fermentation, especially by certain molds, breaks down the indigestible protective coating as they are rich in cellulose-splitting enzymes. The altered structure of the cells makes permeability of digestive juices easy. An additional advantage of such splitting of cellulose, hemicellulose and related materials is that such materials are converted into simpler sugars and sugar derivatives which increase the nutritional value of foods.

Microbial: Activities in Foods Microbial flora associated with foods have a broad complement of enzymes and, as already stated, bring about fermentative, proteolytic and lipolytic changes among others. Proteolytic organisms break down proteins and other nitrogenous compounds resulting in putrid and rotten odour and flavour. Lipolytic organisms hydrolyse lipids giving rise to free fatty acids resulting in rancidity. These changes are un desirable in foods. On the other hand, fermentative organisms converting carbohydrates and related compounds to alcohols, acids and carbon dioxide are not offensive to food tastes. When the concentration of their fermentative products increases, they control the growth of organisms of proteolytic and lipolytic types. However, fermentation of food is complex and only one type of change to the total exclusion of other changes is impossible. Thus, to obtain a product of desired quality, controlled fermentation becomes necessary.

Control of Fermentation in Foods: The factors that influence the growth and metabolic activities of micro-

organisms in food fermentation are the acidity, levels of alcohol, use of starters and levels of oxygen and salts.

Acid: Acids when present in food either as natural constituents, added or formed by fermentation, exert an inhibitory effect on harmful micro-organisms. However, the preservative effect of acid is lost if oxygen is available and surface molds grow which result in the fermentation of the acids. This happens during the ripening of cheddar cheese and constitutes a defect. Some yeasts tolerate moderately high acidity and produce ammonia from the breakdown of protein. This results in the neutralisation of the acid present and permits the subsequent growth of proteolytic and lipolytic bacteria as happens in the ripening of some types of cheese.

These types of changes also occur when raw milk is allowed to ferment. Raw milk contains many types of organisms. First, *Streptococcus lactis* grows producing lactic acid and its growth is inhibited when the acid concentration reaches a particular level. *Lactobacillus* bacteria which are still more acid resistant then grow and produce more acid until the new level becomes inhibitory to their growth. This results in milk clotting and curd formation. In the high-acid environment, acid resistant yeasts and molds become active. The molds oxidize acid and the yeast produces alkaline end products. These result in a decrease of the acid level flavouring the growth of proteolytic and lipolytic spoilage bacteria resulting in an off-odour and gassy condition in the curd.

Alcohol: Alcohol is also a preservative depending upon its concentration. Yeast converts sugar to alcohol but it cannot tolerate alcohol concentration of over 12 per cent. When the alcohol content is below this level, as in beer, pasteurisation of the food becomes necessary to prevent spoilage.

Use of Starters: When a particular type of micro-organisms is present in large numbers and is multiplying at a good rate, it keeps down the growth of other micro-organisms. This is the reason why in curd making or wine making a part from

the previous batch of curd or wine is added to the fresh milk or grape juice. However, in commercial food fermentation, pure cultures are used as starters to obtain products of desired quality.

Temperature: Temperature has an effect on all microbial reactions and hence has to be controlled. In mixed fermentations, as in foods, control of temperature is all the more necessary as different organisms have different temperature sensitivity. For example, in the production of sauerkraut three major types of organisms which grow well at different temperatures, are involved in fermentation. In sauerkraut fermentation, therefore, an initial low temperature (about 21°C), which then is increased somewhat in the later stages of fermentation, is employed.

Oxygen: The level of oxygen determines the extent of fermentation of some foods. Thus providing or removal of air or oxygen is used to encourage or inhibit a particular micro-organism. Also, an organism may have different oxygen requirements for growth than it has for fermentation activity. For example, bakers'yeast (S. *cerevisiae)* and wine yeast (S. *ellipsoideus)* both grow well and produce greater cell masses under aerobic conditions, but they ferment sugar more rapidly under anaerobic condition. Thus, for production of bakers' yeast, the yeast is grown under the aerobic condition by bubbling air through the fermenting medium. In vinegar production there are two steps; the first step involves conversion of sugary substances to alcohol and the second the conversion of alcohol to acetic acid. Alcohol production is an anaerobic process and its conversion to acetic acid is an oxidative fermentation. Thus, in vinegar production, initially fermentation should be anaerobic and this should be followed by generous supply of air.

Salt: Salt can also be used for controlling food fermentation. Lactic acid producing micro-organisms used in the fermentation of pickles, sauerkraut, etc., are tolerant to salt solutions of the order of 10-18 per cent. Many proteolytic and

other spoilage organisms that can infest lactic bacteria are not tolerant to salt above 2.5 per cent. Similarly, salt added to cheese curd controls proteolytic organisms during the long ripening periods.

Radiation: Preservation of food by ionizing radiations is a recently developed method but has not yet gained general acceptance. As is well known, the electromagnetic radiations suppress the growth of most micro-organisms. The possibilities of employing nuclear radiation to sterilise food have been extensively studied since World War II. The harmful effects on the human body from radiations from nuclear explosions have given rise to suspicion in the minds of many people about the safety of the use of irradiated foods. Much work has been done on the safety and wholesomeness of irradiated products but more careful investigations are required before allowing the irradiated food to be used on a large scale.

Kinds of Ionizing Radiations: There are several kinds of radiant energy emitted from different sources which differ in wavelength, frequency, penetrating power and in the various effects they have upon biological systems. Irradiation of foods is carried out by exposing them to high-energy radiations which penetrate them and bring about changes within them. The principal radiations used in food irradiation are gamma rays (wavelengths $<2Å$) emitted from excited radioactive elements ^{60}CO) and accelerated negatively charged particles, electrons (β-particles), which are emitted from a hot cathode. Gamma rays are very similar to x-rays, except that they have much greater penetrating power. An electron beam can be accelerated to very high speeds and increased energies so that it can penetrate foods, by passing it through electronic devices.

Ultraviolet radiations are also employed in preservation; but they have a very low degree of penetration and are employed to inactivate micro-organisms on the surface of food and to treat air, water and the surface of food processing equipment to reduce the number of micro-organisms.

Measurement of Radiation: Ionizing radiations are

measured in terms of rads (and kilorad or megarad). A rad is 10 μJ of energy absorbed by 1 g of material. The radiation effects are related to dose by the relation:

$$n = n_0 \exp(-D/D_0)$$

where n = the number of live organisms following irradiation; n_0 = initial number of organisms; D=dose radiation received (rads), and D_o = a constant depending on type of organism and environmental factors.

Different organisms are sensitive to irradiation to different extent as indicated below:

$10^{3\text{-}4}$—10^7 rads —micro-organisms are killed

10^3 —10^5 rads —insects are killed

10^3 —10^4 rads —sprouting of potatoes, onions, carrots, etc., are inhibited

10^2 —10^3 rads —dose is lethal for humans.

In the case of micro-organisms, the approximate sterilising dose is: bacterial endospores, 3.0×10^6 rads; yeasts and fungi, -5.0×10^4 rads. There are, however, some micro-organisms with exceptional resistance to irradiations; for example, *Micrococcus radiodurans* is resistant up to 55 times greater dosage than gram-negative organisms like *Escherichia coli.*

Mode of Action of Radiations: When gamma rays and electron beam pass through foods, there are collisions between ionizing radiations and food particles at the atomic and molecular levels, resulting in the production of ion pairs and free radicals. The reactions of these products among themselves and with other molecules result in physical and chemical phenomena inactivating micro-organisms in foods. Thus, radiation of foods qan be considered to be a means of achieving "cold sterilisation" of food—a food is freed of micro-organisms without the need for high temperatures.

Water is a ubiquitous ingredient in foods. Of all the reactions brought about by ionizing radiations, the one with

water is the most important in producing foods free of spoilage micro-organisms or pathogens or containing a greatly diminished content of spoilage organisms. Ionizing radiations split water to produce hydrated electrons (e^{-}_{aq}), highly reactive hydrogen (H) and hydroxyl (.OH) radicals, excited water (H_2O) and ionized water molecules (H_2O). Interactions between these products, or these products and other components of food containing oxygen result in the formation of highly reactive species such as hydrogen peroxide.

$$OH + OH \rightarrow H_2O_2$$

$$H + O_2 \rightarrow HO_2$$

$$HO_2 + HO_2 \rightarrow H_2O_2 + O_2$$

Hydrogen peroxide is a strong oxidizing agent and a toxicant to micro-organisms.

Uses of Radiations Ionizing radiations may be used for sterilisation of foods in hermetically sealed packs, reduction in the size of the spoilage flora on perishable foods, elimination of pathogens in foods, control of infestation in stored cereals, prevention of sprouting of potatoes, onions, etc. and retardation of the development of pickled mushrooms. The destruction of the growing point causing tissue darkening of the sprouting areas of vegetables such as potatoes by radiations, eliminates the risk of sprouting during storage.

Irradiation does not very much affect the nutritional properties of foods. The destruction of various amounts of nutrients are of the same degree as in heat processing. Ionizing radiations can also hydrolyse and modify proteins, starch and cellulose. Thus, irradiation can help to improve nutritive value of certain plant materials. Significant levels of toxic or carcinogenic substances are not produced in foods irradiated with sufficient dose of radiations. At appropriate dose of radiations, sterilised or pasteurised foods, are safe from microbial standpoint.

Irradiation of foods can also result in certain undesirable

effects. Ionizing radiation, in excessive doses, can alter the structure of organic and biochemical compounds in food resulting in food damage. Degradation of carbohydrates may result in loss of texture and colour. Protein degradation results in undesirable changes in colour and odour and brings about liquefaction. Egg-white proteins, on irradiation, become thin and watery. Irradiation of fats results in the development of off-flavour and off-odour and the loss of the natural antioxidants. Foods irradiated with large doses of radiation might themselves become radioactive and prove harmful when consumed.

In the irradiation of foods for preservation, the control of the radiation process is very necessary. The radiation dose must be carefully controlled to destroy micro-organisms and inactivate many food enzymes. It should be sufficient to destroy the pathogenic and spoilage causing organisms. Apart from the intensity of radiation, the amount of radiation absorbed and the length of time of radiation are to be controlled. The longer the food is in the radiation path, the more it will absorb. Radiation energy must be provided in such a manner that it reaches every particle of food to ensure adequate killing of all micro-organisms.

As already stated, the use of ionizing radiations for food preservation has been adopted very slowly. This is largely due to the unacceptable flavour of some foods irradiated to sterilisation stage. Also, the fear that radioactivity might be induced in a food has come in the way of its extended use. However, recent reappraisal by the WHO and the Internatinal Atomic Energy Authority has resulted in the recommendation that irradiation dose up to 1M rad is not hazardous.

The disinfection of grains, the preservation, the prevention of sprouting of potatoes and other vegetables are of great economic value. In less developed countries lacking refrigeration facilities, losses from insects and spoilage of food account for more than half the total food production. In such cases radiation of foods will be of great significance. This

requires radiation facilities at a cost that developing countries can afford.

Deterioration and Safeguards

The great bulk of food is comprised of chemicals such as carbohydrates, fats, proteins, vitamins, minerals, water, etc. Besides the natural components of food stuffs, additional chemicals may be incorporated, either directly or indirectly during the growing, storage or processing of food.

Additives of Food: A food additive is defined as a substance or mixture of substances, other than a base foodstuff, which is present in a food as a result of any aspect of production, processing, storage or packing. This definition includes both intentional and unintentional additives. The unintentional additives, which are not added to achieve an effect in the food but which may accidentally enter into foods as a result of their use in agricultural production, raising animals, food processing or packing, are not additives in the technical sense of the term, but they are food contaminants.

An expert committee on Food Additives made up of representatives of FAO and WHO has defined food additives as non-nutritive substances added intentionally to food, generally in small quantities, to improve its appearance, flavour, texture or storage properties. This definition excludes substances added primarily for their nutritive value, such as vitamins and minerals.

A broad definition of food additive is any substance the intended use of which results, directly or indirectly, in its becoming a component of or otherwise affecting the characteristics of any food and which is safe under the condition of its use. Any substance present in food, either by intentional addition or an unintentional contaminant, has its effects on the safety of food.

The Need and Upkeep:For centuries, man has recognised the effects of food additives and has used whatever was

available—marigold for colour, wood ashes for leavening, the lining of calves stomach for cheese making, etc. They were used for effects without knowing the changes they brought about. As long as the ingredient did not make one sick immediately it was alright to use it. Today, thousands of compounds are used as food additives, whose chemical identity and structure are known. They can be obtained in a very high state of purity and, when used, bring about the desired effect in foods.

The use of food additives is imperative in the complex and integrated society in which we live. The areas of food production are separated from the areas of consumption. Additives have provided protection against food spoilage during storage, transportation, distribution or processing. Also, with the present degree of urbanisation, it would be impossible to maintain food distribution without the processing and packing with which many additives are involved.

A number of factors have led to the demand for foods with built-in preparation—"convenience" foods. A can of vegetable soup is a convenient food, since various vegetables have been cut, blended, supplemented with spices and so on. It just requires warming up before use.

There is a great demand nowadays for "instant", heat and serve, and "ready-to-cook" convenient foods. These foods make up about 60 per cent of the food that Americans buy. Such foods result in saving considerable amounts of time and effort in food preparation at home, restaurants, etc. The convenience food revolution would not have been possible without food additives.

Additives permit the variety of foods that we deem desirable and which certainly are objectively important in maintaining good nutrition. Vitamins and minerals are important, among other things, for good health. Many of these chemical additives can be manufactured so that foods can be "fortified" or "enriched". Potassium iodide, for instance,

added to common salt can eliminate goitre, enriched rice or bread with B-complex vitamins can eliminate pellagra and adding vitamin D to cow milk prevents rickets.

Many foods, particularly those with high moisture contents, do not keep well. All foods are subject to microbial attack. Fats or oily foods become rancid, particularly when exposed to humid air. The conservation of the quality of foods against agents causing such deterioration of foods requires the addition of preservatives. Additives are also used to colour foods, add flavour, impart firmness and retard or hasten chemical reaction in food.

There is then the need for the use of food additives to maintain the nutritional quality of food, to enhance stability with resulting reduction in waste, to make food more attractive and to provide efficient aids in processing, packing and transport. The amount of additives used should be kept to a minimum; it should conform to a standard of purity and be safe. On the other hand, food additives must not be used to disguise faulty processing and handling techniques, to deceive the customers, or if it reduces the food's nutritive value, or when the desired effect can be achieved by good manufacturing practices that are economically feasible.

Over 3000 different chemical compounds are used as food additives. They are categorised into different groups. A few types of additives are indicated below.

Antioxidants: An antioxidant is a substance added to fats and fat-containing substances to retard oxidation and thereby prolong their wholesomeness, palatability and, sometimes, keeping time. An antioxidant should not contribute an objectionable odour, flavour or colour to the fat or to the food in which it is present. It should be effective in low concentrations and be fat soluble. Also, it should not have harmful physiological effect.

Some antioxidants used in foods are butylated hydroxyanisole (BHA), butylated hydroxytoluene (BHT),

propyl gallate (PG) and tertiarybutyl hydroquinone (TBHQ), which are all phenolic substances. Thiodipropionic acid and dilauryl thiodipropionate are also used as food antioxidants. Naturally occurring substances that act as antioxidants are the tocopherols. The tocopherols act as biological antioxidants in plant and animal tissues, but they are rarely used as additives because they are more expensive than synthetic antioxidants.

Antioxidants function by interrupting the free radical chain mechanism involved in lipid oxidation. They are effective in small concentrations (0.01-0.02 per cent). Mixed antioxidants sometimes act synergistically. The presence of metallic ions, particularly copper and iron, promotes lipid oxidation through a catalytic action. Acidic compounds, like citric acid, complex with iron and thus, when present along with phenolic antioxidants, enhance the latter's activity.

Browning of cut fruits and vegetables is due to enzymic oxidation of phenolic substances. Antioxidants prevent this discolouration. Ascorbic acid is used as an antioxidant in this case. Acids, such as citric and phosphoric, increase the effectiveness of ascorbic acid in preventing browning.

Chelating Agents: Chelating agents or sequestrants are compounds that form complexes with metal ions. Many metals exist in food in a naturally chelated form, such as, magnesium in chlorophylls, iron in ferritin and haemoglobin, and copper, zinc and magnesium in enzymes. When metallic ions are released due to hydrolytic or other degradative reactions, they are free to participate in reactions that lead to discolouration, oxidative rancidity, turbidity, and flavour changes in foods. Addition of chelating agents results in the complexing of these metal ions and thereby the stabilisation of foods.

Compounds containing two or more functional groups, such as hydroxyl, sulphydryl, carboxyl, phosphate, etc., chelate with metals under favourable conditions. Citric acid and its derivatives, phosphates and salts of ethylene-diamine tetraacetic acid (EDTA) are the most popular chelating agents used in foods.

Chelating agents are not antioxidants; they serve as scavengers of metals which catalyse oxidation. They, however, are antioxidant synergists. Citric acid and its esters in propylene glycol solution are effective synergists in lipid systems.

Polyphosphates and EDTA are used as chelating agents in canned seafoods. Seafoods contain substantial amounts of magnesium ions which sometimes react with ammonium phosphate with the formation of glossy crystals (struvite). Iron, copper and zinc containing seafoods react with sulphides that lead to product discolouration. These reactions are prevented by the addition of chelating agents.

Citric and phosphoric acids are used as acidulants in soft drink beverages. These chelate with metals which otherwise promote the oxidation of flavour compounds and catalyse discolouration reactions. Chelating agents also stabilise fermented malt beverages by complexing with copper, which otherwise catalyses the oxidation of phenolic substances which subsequently interact with proteins to form haziness or turbidity.

Colouring Agents: These include colour stabilisers, colour fixatives, colour retention agents, etc. They consist of synthetic colours, synthesised colours that also occur naturally and other colours from natural sources. Even though colours add nothing to the nutritive value of foods, without certain colours most consumers will not buy or eat some foods. Thus, colours are frequently added to restore the natural ones lost in food processing or to give the preparations the natural colour we expect.

Originally, many colour additives were natural pigments or dyes. For example, spinach juice or grass, marigold flower, and cochineal were used to obtain green, yellow and red colour respectively. This gave place to synthetic dyes obtained from coal tar. Synthetic colours generally excel in colouring power, colour uniformity, colour stability and cost. Further, in many cases, natural colouring materials do not exist for a

desired hue. Carbonated beverages, gelatin dessert, candies and bakery goods are some foods that are coloured with coal tar dyes. As a number of coal tar compounds have been shown to be potent carcinogens, the use of coaltar dyes as food additives is restricted. Many countries have severely restricted the number of coal tar dyes for use in foods, while some other countries have completely banned their use. Food colours used also include some inorganic materials such as iron oxide to give redness and titanium dioxide to intensify whiteness.

A number of natural food colours extracted from seeds, flowers, insects and foods are also used as food additives. One of the best known and most widespread red pigment is bixin, derived from the seed coat of *Bixa orel-lana,* the lipstick pod plant of South American origin. Bixin is not considered to be carcinogenic. The major use of this plant on a worldwide basis, however, is for the annatto dye, a yellow to red colouring material extracted from the orange-red pulp of the seeds. Annatto has been used as colouring matter in butter, cheese, margarine and other foods. Another yellow colour, a carotene derived from carrot, is used in margarine. Saffron has both flavouring and colouring properties and has been used for colouring foods. Turmeric is a spice that gives the characteristic colour of curries and some meat products and salad dressings. A natural red colour, cochineal (or carnum) obtained by extraction from the female insect *(Coccus cacti),* grape skin extract and caramel, and the brown colour obtained from burnt sugar, are some natural colours that are used as food additives.

Curing Agents: These are additives to preserve (cure) meats, give them desirable colour and flavour, discourage growth of micro-organisms and prevent toxin formation.

Sodium nitrite has been used for centuries as a preservative and colour stabiliser in meat and fish products. The nitrite, when added to meat, gets converted to nitric oxide which combines with myoglobin to form nitric oxide myoglobin (nitrosylmyoglobin) which is a heat-stable pigment. The curing

also contributes flavour to the meat. In addition, nitrite curing inhibits the growth of *Clostridium* and *Streptococcus*, and also lowers the temperature required to kill C. *botulinum*.

It has been discovered that cooking nitrite cured meat products results in the formation of small amounts of N-nitrosamines, which are potent carcinogens. The nitrosamines are formed by the reaction of secondary and tertiary amines, through the following type of reaction.

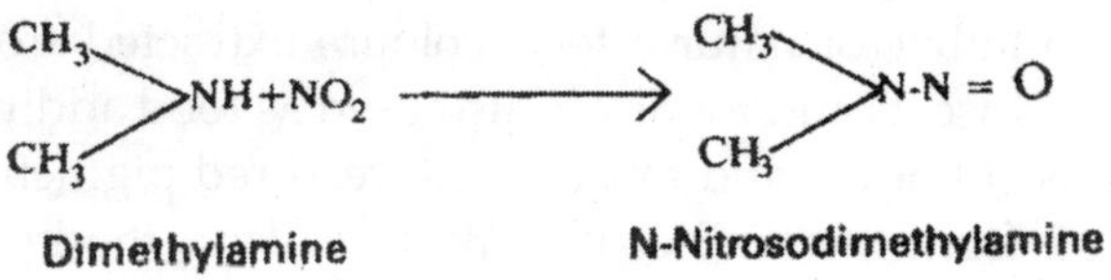

Nitrosation may also take place in foods during storage or processing, and nitrosamines may be ingested as such. There is some evidence that nitrosamines are formed under the strong acid conditions in the human stomach, when nitrite cured meats are ingested.

In recent years, use of nitrite for curing has become controversial. Although the levels are low, the production of carcinogens during food manufacture or preparation cannot be ignored. As the derivation of nitrosamines is due to NO^{-}_{2} in meat, a reduction in the concentration of the curing agent used helps alleviate the problem. However, the reduction in the concentration of NO^{-}_{2} enhances the risk of food poisoning due to *Clostridium botulinum.* Recent results have shown diminished concentrations of nitrite can be used in the presence of another preservative that acts synergistically. In the presence of isoascorbate even very dilute concentrations of nitrite significantly decrease spoilage and toxin production.

Other agents used for curing meat are ascorbates and several phosphates. Ascorbate and isoascorbate react with nitrite to give nitric oxide and thereby accelerate the rate of formation of nitrosylmyoglobin. They also stabilise colour

and flavour. They inhibit the formation of nitrosamines. Through chelating iron, they may contribute also to the antimicrobial stability of cured meats.

Polyphosphates, such as sodium tripolyphosphate ($Na_5P_3O_{10}$) and sodium hexametaphosphate $(NaPO_3)_n$ where n = 10 to 15, are used in meat curing. These compounds, through enhancing water retention, aid tenderness, juiciness and flavour of cured meats. They also influence texture and, by chelating metal ions, act as antioxidants. The latter property also contributes to the antimicrobial properties of cured meats.

Emulsions: Emulsifiers are a group of substances used to obtain a stable mixture of liquids that otherwise would not mix or would separate quickly. They also stabilise gas-in-liquid and gas-in-solid mixtures. They are widely used in dairy and confectionery products to disperse tiny globules of an oil or fatty liquid in water. Emulsifying agents are also added to margarine, salad dressings and shortenings. Peanut butter contains up to 10 per cent emulsifiers.

One of the most widely used emulsifiers is lecithin which is found in milk, egg and soyabean. Lecithin keeps, in milk, the butterfat and water phases more or less uniform. Commercial vegetable lecithin is obtained principally from soyabean. Lecithin is employed in the preparation of cocoa butter and chocolate candy. The texture and keeping qualities of bread and other fermented baking products are improved by the use of lecithin. Lecithin is a more effective emulsifying agent in combination with monoglyceryl stearate and ascorbic acid. A number of mono- and diglycerides and their derivatives are good emulsifying agents. In these cases, the ester groups make the molecule fat-soluble, while the alcohol group lends water solubility to another portion of the molecule. As a result, the molecule can serve as a bridge to keep fat molecules suspended in water. In addition to these natural emulsifiers, there are a number of synthetic ones. These include propylene glycol monostearate, sorbitan monostearate and polysorbates.

Flavours and Flavour Enhancers: Flavouring additives are the ingredients, both naturally occurring and added, which give the characteristic flavour to almost all the foods in our diet. Flavour enhancers are not flavours themselves but they amplify the flavours of other substances through a synergistic effect. Flavour and flavour enhancers constitute the largest class of food additives. These are about 2,100 approved natural and synthetic flavours of which more than 1,600 are synthetic ones.

Natural flavour substances, such as spices, herbs, roots, essences and essential oils have been used in the past as flavour additives. The flavours of such materials are not uniform. They vary with the season and area of production. In addition, the natural flavours are in short supply and the amount of flavour substances in them is very tiny. It would take about a tonne of many spices to produce 1 g of the flavour substances and in some cases only 0.1 g could be extracted. Natural food flavours are thus being replaced by synthetic flavour materials.

The agents responsible for flavour are esters, aldehydes, ketones, alcohols and ethers. These substances are easily synthesised and can be easily substituted for natural ones. Typical of the synthetic flavour additives are amyl acetate for banana, methyl anthranilate for grapes, ethyl butyrate for pineapple, etc. Generally, most synthetic flavours are mixtures of a number of different substances. For example, one imitation cherry flavour contains fifteen different esters, alcohols and aldehydes.

One of the best known, most widely used and somewhat controversial flavour enhancers is monosodium grutamate (MSG), the sodium salt of the naturally occurring amino acid glutamic acid. This is added to over 10,000 different processed foods. This has been in use in Chinese and Japanese cooking for centuries, and was extracted from seaweeds and soyabean. About 65 years ago, a Japanese named Ikeda discovered that the flavouring from these is the MSG and that it has an

attractive meat-like flavour. MSG is now manufactured on a large scale all over the world, and especially in Japan.

MSG is generally recognised as safe. However, it was reported some time back that MSG injected to young mice resulted in brain damage. Also, some individuals experience symptoms often comparable to those of heart attack, when served with food containing large amounts of MSG. The matter has now been thoroughly investigated and it has been concluded that there is no risk in its use. However, MSG which was being added to baby foods is now discontinued, as its benefits to babies are dubious.

Yeast extract has the same flavour enhancing property as MSG. It is found that, in this case, the flavour enhancing substances are the ribo-nucleotides. These are ten times more powerful than MSG.

Flour Improvers: These are bleaching and maturing agents; usually, they both bleach and "mature" the flour. These are important in the flour milling and bread-baking industries. Freshly milled flour has a yellowish tint and yields a weak dough that produces poor bread. Both the colour and baking properties improve by storing the flour for several months before making bread.

During storage, atmospheric oxygen oxidizes the carotenoid pigments responsible for the colour of the flour, converting them to colourless compounds. They also oxidize some of the proteins which form dough to give the latter increased strength and elasticity. These improvements can be obtained more rapidly with the use of chemical agents.

Chemical agents used as flour improvers are oxidizing agents, which may participate in bleaching only, in both bleaching and dough improvement, or in dough improvement only. The agent that is used only for flour bleaching is benzoyl peroxide $(C_6H_5CO)_2O_2$. This does not influence the quality of dough. Materials used both for bleaching and improving are chlorine gas, (Cl_2); Chlorine dioxide, (ClO_2); nitrosyl chloride,

(NOCl); and nitrogen di-and tetroxides, (NO_2 and N_2O_4). Oxidizing agents used only for dough improvement are potassium bromate, ($KBrO_3$); potassium iodate, (KIO_3), calcium iodate, [($Ca(IO_3)_2$] and calcium peroxide, (CaO_2).

Benzoyl peroxide oxidation takes several hours. The gaseous agents used for bleaching and improvement act immediately upon contact with flour. Oxidizing agents used for dough improvement only remain inactive until yeast fermentation lowers the pH of the dough sufficiently to activate them. As a result of late action they cause increase of loaf volume, improved loaf symmetry and improved crumb and texture characteristics. The dough improving oxidizing agents oxidize sulphydryl groups (-SH) in the procems of gluten to yield an increased number of intermolecular disulphide bonds (-S-S-), resulting in a tougher, drier, more extensible dough that gives rise to improved characteristics in finished products.

Humecants and Anticaking Agents: Humecants are moisture retention agents. Their functions in foods include control of viscosity and texture, bulking, retention of moisture, reduction of water activity, control of crystallisation and improvement or retention of softness. They also help improve the rehydration of dehydrated food and solubilisation of flavour compounds.

Polyhydroxy alcohols are water soluble, hygroscopic materials which exhibit moderate viscosities at high concentrations in water and are used as humecants in foods. Some of them are propylene glycol ($CH_3.CHOH.CH_2OH$), glycerol, and sorbitol and mannitol [CH_2OH $(CHOH)_4OH_2OH$]. Polyhydric alcohols are sugar derivatives and most of them, except propylene glycol, occur naturally.

Anticaking agents help prevent particles from adhering to each other and turning into a solid chunk during damp weather. The help free flowing of salts and other powders. These materials function by readily absorbing excess moisture, by coating particles to impart a degree of water repellency, and/

or by imparting an insoluble particulate diluent to the mixture. Calcium silicate (Ca SiO_3 X H_2O) can absorb liquids in amounts two and a half times its weight and still remain free flowing. In addition to absorbing water, some anticaking agents effectively absorb oils and other non-polar organic compounds.

Calcium silicate is used to prevent caking in baking powder, table salt and other foods and food ingredients. Because it can absorb oils, calcium silicate is a useful anticaking agent in complex powdered mixes and certain spices which contain free essential oils. Calcium and magnesium salts of long-chain fatty acids (e.g., calcium stearate) are used as conditioning agents for dehydrated vegetable products, salt and other food ingredients in powdered form. Other anticaking agents used in food industry are sodium silicoaluminate, tricalcium phosphate, magnesium silicate and magnesium carbonate.

Leavening Agents: Leavening agents produce light fluffy baked goods. Originally, yeast was used almost exclusively to leaven baked products. It is still an important leavening agent in bread making. When yeast is used, ammonium salts are added to dough to provide a ready source of nitrogen for yeast growth. Phosphate salts (sodium phosphate, calcium phosphate) are added to aid in control of pH.

To make light cakes, biscuits, waffles, muffins, and many other baked products, chemical leavening agents are used. Baking powders generate carbon dioxide for leavening purposes.

Nutrient Supplements: Nutrient supplements restore values lost in processing or storage, or ensure higher nutritional value than what nature may have provided. When foods are processed, there may be loss of some nutrients and additives may be added to restore the original value. For example, to produce white flour, wheat is milled in such a way as to remove the brown coloured part of the grain which is rich in vitamins and minerals. To restore the nutritive value, thiamine, nicotinic acid, iron and calcium, are added to the flour. Similarly,

vitamin C is added to canned citrus fruits to make up the loss of the vitamin during processing.

When manufactured foods are used as substitutes for natural ones, nutrients are to be added to the former to ensure that their nutritional value is at least equal to the natural product. For example, margarine is used as a substitute for butter on account of its cheapness. To ensure that the nutritional status of those who use margarine does not suffer addition of vitamins A and D to it, at least equal to that of the natural product, is necessary.

Some foods are to be fortified by adding specified nutrients in excess of what nature provides. Milk, for instance, which is a nutritious food is low in vitamin D content. Addition of this vitamin to milk has helped some countries to reduce the incidence of rickets. Similarly, cereals, baby foods and fruit juices are fortified with vitamins to improve the nutritional benefits. Some proteins of foods are deficient in essential amino acids and such foods are fortified with lysine and methionine. Similarly, foods are fortified with essential fatty acids, if they are deficient in them.

Iodine deficiency causes goitre. Sea foods are a source of iodine and where these do not form part of the diet, goitre may be endemic. Iodine in the form of potassium iodide added to common salt in controlled amount (iodized salt) is a safeguard against this disease. Similarly, fluorine in small quantities in food and water is required for normal tooth development. When there is this deficiency, controlled addition of fluoride to drinking water ensures effective protection against dental decay.

Non-nutritive Sweeteners: In many ways sucrose is an ideal sweetener; it is colourless, soluble in water and has a "pure" taste, not mixed with overtones of bitterness or saltiness. But it is rich in calories. The diabetics and overweights who must restrict their intake of sugar must have an alternative to sucrose. Thus, synthetic non-nutritive sweeteners having less

than two per cent of the calorific value of sucrose for equivalent unit of sweetening capacity came into use.

The first synthetic sweetening agent used was saccharin (sodium ortho benzenesulphonamide or the calcium salt), which is about 300 times sweeter than sucrose in concentrations up to the equivalent of a 10 per cent sucrose solution.

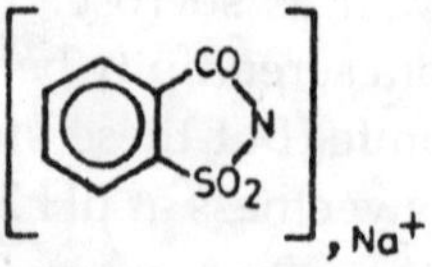

Sodium saccharin

Use of saccharin often leaves a bitter and unpleasant after-taste. Attempts to find better substitutes resulted in the accidental discovery of cyclamates (sodium and calcium salts of cyclamic acid—cydohexane sulpha-mate), which are about 30 times sweeter than sucrose but have little of the after-taste of saccharine. So, cyclamates were widely used as sweetening agents in the manufacture of soft drinks, other low-calories liquid foods and dietetic forms of foods. However, the use of cyclamates has been banned after high dosages were found to produce bladder cancer in rats, probably from the formation of cyclohexylamine, a known carcinogen.

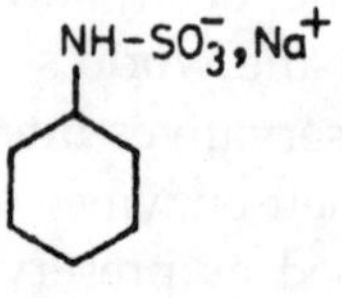

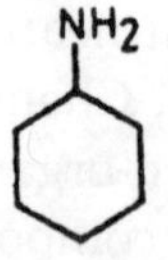

Sodium cyclamate **Cyclohexylamine**

Newer non-nutritive sweetening agents, ranging in sweetness from 10 to 3,000 times of sucrose have been discovered. Among them is glycyrrhiza acid, obtained from the roots of a European leguminous plant *Glycyrrhiza glabra* (licorice). The sweet taste of glycyrrhizic acid is detectable at one-fiftieth the threshold taste level of sucrose. It is used in

tobacco products, confectioneries and beverages. Neohespiridine dihydrochalcone isolated from citrus peels is about 1,000-2,000 times as sweet as sucrose.

The tropical African fruits, kutemfe and serendipity berry, contain low-calorie sweeteners. Kutemfe contains two proteins thaumatin I and II. On a molar basis these substances are about 10^5 times as sweet as sucrose. The protein substance, monellin, obtained from serendipity berries, is as sweet as the sweeteners from kutemfe. But these substances are unstable to heat and lose their sweetness at pH 2 at room temperature.

A potentially useful low-calorie sweetener is the diester of L-aspartic acid and L-phenylalanine. The methyl ester of L-aspartyl-L-phenylalanine, is reported to be 100-200 times sweeter than sucrose, with taste characteristics very similar to those of sucrose.

pH Control Agents: These include acids, alkalis and buffers. They not only control the pH of foods but also affect a number of food properties such as flavour, texture, cooking qualities, etc.

Preservatives: A preservative is defined as any substance which is capable of inhibiting, retarding or arresting the growth of micro-organisms, of any deterioration of food due to micro-organisms, or of masking the evidence of any such deterioration. It is estimated that nearly 1/5 of the world's food is lost by microbial spoilage. Chemical preservatives interfere with cell membrane of micro-organisms, their enzymes or their genetic mechanisms. The compounds used as preservatives include natural preservatives such as sugar, salt, acids etc., as well as synthetic preservatives. Chemical preservatives are generally added after the foods are processed. The role of some preservatives is considered in this section.

Sodium Chloride: This has been used as a food preservative from early times. Salt stops the growth of micro-organisms and interferes with the action of proteolytic enzymes. Salt also causes food dehydration by drawing out water from the tissue

cells. Salt is employed to control microbial population in foods, such as butter, cheese, cabbage, olives, cucumbers, meats, fish and bread. The amount of salt added determines the extent of protection afforded to the food. The term "brine" is used to denote the percentage of NaCl in the water phase of a food. In the preservative action of NaCl there is synefgistic action with other intrinsic factors such as pH, or extrinsic factors such as temperature, partial pressure of oxygen, etc.

Sugar: Sugar aids in the preservation of products in which it is used. The high osmotic pressure of sugar creates conditions that are unfavourable for the growth and reproduction of most species of bacteria, yeasts and molds. The preservative action of moderate strength of sugars can be improved if invertase is used to increase the concentration of glucose relative tc sucrose. Foods in which sugars aid preservation include syrups and confectionery products, fondant fillings in chocolate, honey, jams, jellies, marmalades, conserves, and fruits such as dates, sultanas and currants.

Sulphur Dioxide: Sulphur dioxide has been used in foods for long as a general preservative. It is used in the treatment of fruits and vegetables before and after dehydration to extend the storage life of fresh grapes, prevent the growth of undesirable micro-organisms during wine making and in the manufacture of fruit juices. Sulphur dioxide is also the most useful agent for the prevention of browning reactions in dried fruits. Most cut fruits are treated with sulphur dioxide to prevent enzymic browning.

Forms in which sulphur dioxide is employed as a preservative include the gas (SO_2), the sodium or potassium bisulphites ($NaHSO_3$ or $KHSO_3$), sulphites (Na_2SO_3 or K_2SO_3), and metabisulphite ($Na_2S_2O_5$; or $K_2S_2O_5$). In aqueous solutions, sulphur dioxide and the sulphite salts form sulphurous acid (H_2SO_3) and ions of bisulphite (HSO^-_3) and sulphite (SO^{2-}_3). At low pH values (lower than 4.5), the undissociated sulphurous acid predominates and inhibits the growth of yeasts and molds. At high values, the $HSO^{2'}$ion is effective against

bacteria but not against yeast. The sulphurous acid antimicrobial activity may be due to the reaction of bisulphite with acetaldehyde in the cell, the reduction of essential disulphide linkages in enzymes and the formation of bisulphite addition compounds which interfere with respiration.

One of the defects of use of sulphur dioxide is it leaves an unmistakable taste in the mouth. It also causes the breakdown of vitamin B-l, so that foods containing sulphur dioxide may not be good sources of this vitamin.

Sorbic Acid: Straight chain monocarboxylic acids and α-unsaturated fatty acid analogues have antimicrobial activity. Sorbic acid (CH_3-CH=CH-CH=CH-COOH) and its sodium and potassium salts inhibit molds and yeasts, in foods such as cheese, baked products, fruit juices, wines and pickles. The antimycotic action of sorbic acid is due to the inability of molds to metabolise the conjugated unsaturated structure.

Acetic Acid: Acetic acid (CH_3COOH) in the form of vinegar (4 per cent acetic acid) has been used to preserve pickled vegetables from antiquity. Acetates of sodium, potassium and calcium are used in bread and other baked goods to prevent ropiness and the growth of molds, but they do not interfere with yeasts. The acid is also used in foods, such as catsup, mayonnaise and pickles, primarily for flavour, but these products also benefit from the concurrent antimicrobial action. The antimicrobial activity of acetic acid increases as the pH decreases.

Propionic Add: Propionic acid ($CH_3.CH_2.COOH$) and its sodium and calcium salts exert antimicrobial activity against molds and some bacteria. The acid finds extensive use in the bakery field, where it not only inhibits molds effectively but is also active against the ropy bread organism

Bacillus Mesentericus: The toxicity of propionic acid to molds and certain bacteria is related to the inability of the organisms to metabolise the three-carbon unit.

Benzoic Acid: Benzoic acid is widely used as an

antimicrobial agent. Its sodium salt is more soluble in water than the free acid and hence it is generally used. However, once in the product some of sodium benzoate is converted to the acid form. The undissociated acid is the form with antimicrobial activity. The acid is most active against yeasts and bacteria, and least active against molds. It exhibits optimum activity in the pH range 2.5-4.0 and thus is well suited for use in acid foods, such as fruit juices, carbonated beverages, pickles and sauerkraut.

Parabens: These are alkyl esters of p-hydroxybenzoic acid. The methyl, ethyl, propyl and heptyl esters are generally used. These are effective inhibitors of molds and yeasts, but are relatively ineffective against bacteria. They are active at pH 7 and higher and have little effect on flavour.

Epoxides: Epoxides are cyclic ethers that destroy all forms of micro-organisms including spores and even viruses. The epoxides used as preservatives are ethylene oxide and propylene oxide.

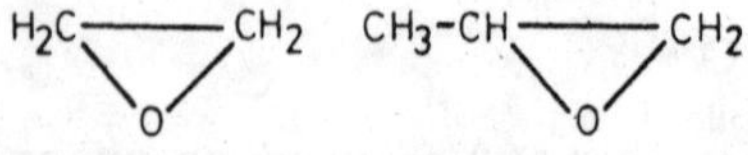

Ethylene oxide **Propylene oxide**

In order to achieve intimate contact with micro-organisms the epoxides are used in the gaseous form and, after adequate exposure, the residual epoxide is removed by flushing and evacuation. Their use is limited to dry items, such as nuts and spices. Spices often contain a high microbial load and they cannot be sterilised by heat because of the instability to heat of volatile flavour compounds. Thus, treatment with epoxides is a suitable method of reducing microbial load of spices.

Antibiotics: Antibiotics are antimicrobial agents produced naturally by a variety of micro-organisms. Antibiotics are of great chemotherapeutic value in controlling pathogenic micro-

organisms in living animals. Their use in food preservation could lead to the development of resistant strains of organisms, thus making their medicinal use ineffective. Therefore, the use of antibiotics as food preservatives is not permitted in certain countries, while some countries allow the limited use of a relatively few antibiotics. These include nisin, pimaracin, chlorotetracycline (aureomycin) and oxytetracycline (terramycin).

Nisin is used to control the growth of sporeforming bacteria of dairy products, such as cheese and condensed milk. Nisin is nontoxic to human beings and does not lead to cross-resistance with medical antibiotics. Pimaracin is an antifungal substance and used to control the growth of fungi in cheese and sausages. Its toxicity in human beings is low and it is active in low concentrations of 10-100 parts per minion. Chlorotetracycline and oxytetracycline are used to control the growth of bacteria in fish and poultry. The residual antibiotics are destroyed by usual cooking methods.

Diethyl Pyrocarbonate: Diethyl pyrocarbonate $(H_5C_2O-\overset{O}{\overset{\|}{C}}-O\overset{O}{\overset{\|}{C}}-OCl_2H_5)$ is used as an antimicrobial food additive for fruit juices, wines and carbonated beverages. It acts as a "cold sterilising" agent for aqueous solutions. Its advantage is that following its action it is readily hydrolysed to ethanol and carbon dioxide. It is active in low concentrations (120-300 parts per million). In concentrated solutions, it is an irritant. It is likely to react with ammonia, which is ubiquitous in plant and animal tissues, forming urethane (ethyl carbamate, $C_2H_5-O-\overset{O}{\overset{\|}{C}}-NH_2$) a carcinogen. Because of this, it is of limited use today. Dimethyl pyrocarbbnate, a lower member of this group of compounds, may be widely used in future as a cold sterilising agent.

Stabilisers and Thickeners: These compounds function to improve and stabilise the texture of foods, inhibit crystallisation (sugar, ice), stabilise emulsions and foams, reduce the stickiness of icings on baked products and encapsulate flavours. Substances used as stabilisers and thickeners are polysaccharides, such as gum arabic, guar gum, carrageenan, agar-agar, alginic acids, starch and its derivatives, carboxymethylcelluloses and pectin. Gelatin is one noncarbohydrate material used extensively for this purpose. Stabilisers and thickeners are hydro-philic and are dispersed in solution as colloids. These swell in hot or even cold water and help thicken food. Gravies, pie fillings, cake toppings, chocolate milk drinks, jellies, puddings and salad dressings are some among the many foods that contain stabilisers and thickeners.

Other Additives: There are a number of food additives that provide functions other than those indicated above. Clarifying agents like bentonite, gelatin, synthetic resins (polyamides and polyvinylpyrrolidone) are used to remove haziness or sediments and oxidative deterioration products in fruit juices, beers and wines. Enzymes are added to bring about desirable changes; rennin for producing curd and cheese, papain for tenderising meat and pectinase for clarifying beverages. Firming agents like aluminium sulphates and calcium salts are used to keep the tissues of fruits and vegetables crisp. Freezing agents like liquid nitrogen and dichlorofluoromethane, which are extremely volatile and rapidly evaporate at ordinary temperatures, are used to chill foods. Solvents like alcohol, propylene glycol and glycerine are used to dissolve suspended flavours, colours and many other ingredients. Packing gases, such as inert gases, are added to packets of instant foods to prevent oxidative and many other changes.

The use of additives, as well as their numbers, is increasing with years. Use of additives has helped not only to produce more food but also to protect what is produced until it is harvested, processed, marketed, bought and eaten. Use of

additives has made it possible to feed an increased population and in a better way than before. In the advanced countries of the world, the use of food additives has resulted in an increase in the number of food items available from about 900 in 1941 to 7,500 in 1975.

Sugar and salt, together with other nutritive sweeteners, corn syrup and dextrose, account for about 90 per cent of 1,800 intentional additives consumed. There are some fifty five additives which are moderately used, half of these are used as leavening agents or to control pH. Use of all other additives amounts to about 10 g per person per year. The consumption of several flavours and micronutrients may be less than 0.01 mg per person per year.

Role in Preservation: Although most food additives are apparently harmless, the rapidly increasing number and types of chemicals added to our food has increased concern regarding their harmfulness. Their use has provoked many emotional responses from consumers. There are those who strongly believe that essentially all chemicals in foods are bad and we should eat only natural foods. There are others who feel that there is nothing to worry about food additives since there is no absolute proof that any chemical has harmed a human being. It thus becomes necessary to consider if food additives are safe and how to protect the consumer about unsafe additives.

There are many types of food hazards. Millions of people die every year due to microbiological hazards in the form of food-borne infections caused by unsatisfactory handling practices. There is the nutritional hazard due to sub-optimal intake of several key nutrients, due to poverty, ignorance, indifference and misinformation. Then, there are the environmental pollutants due to pollution in water. Some natural foods contain toxic substances (natural toxicants). They cause harm when consumed in sufficient quantity. Pesticides applied to food crops result in pesticide residues in food, and this is a hazard. Finally, there is the hazard of the food additives. Of all food hazards, microbiological and nutritional hazards

are far greater than others; but are widely ignored. The hazards from food additives and others are remote and small; yet they receive the most attention of consumers, administrators and others.

The presence of synthetic additives does not necessarily mean that a food is harmful and the fact that a food is completely "natural" is no guarantee of its safety. Eating halibut liver can cause vitamin-A poisoning. Cabbage, lettuce, spinach, tea and a number of other foods contain very small amounts of benzopyrene, a known carcinogenic chemical. Lima beans, sweet potato and tapioca contain compounds which give deadly hydrogen cyanide in the human intestine. Some vegetables (cabbage, cauliflower, turnip brussel sprouts, etc.) can cause goitre in some susceptible individuals. Spinach and rhubarb have high oxalic acid content which can precipitate kidney stones. On the other hand, many synthetic food additives, such as vitamins, citric acid, etc., are identical to safe chemicals found in natural foods. So it is not a question of natural versus synthetic additives. The issue is the need and safety of the additive. The need for food additives has already been discussed at the beginning of this chapter and tests for the safety of food additives are to be considered now.

Evaluation of Safety: In the evaluation of safety of additives it has to be proved that harmful effects have not occurred with the addition of the chemicals. As it is difficult to carry out toxicity studies with human beings, animal experiments are carried out. The simplest way to express toxicity is by the value LD_{50}, i.e., the dosage (lethal dosage) necessary to kill 50 per cent of a population of test animals under precisely stated conditions. The four following types of toxicity studies are employed.

1. Acute toxicity, or single dose experiment
2. Subacute toxicity (daily ingestion for days or weeks)
3. Chronic toxicity (daily ingestion for weeks to months)
4. Studies of carcinogenicity, mutagenicity, teratogenicity and reproduction.

Acute Toxicity Tests: In acute toxicity studies, groups of animals (two species) are given a wide range of single doses of the additives orally or by injection into the blood stream. The LD_{50} values in a specified period of time (24 hours, 7 days) and abnormalities seen on autopsy may provide clues to the inherent toxicity of the additive.

Subacute Toxicity Tests: In a subacute toxicity test, two or more species of animals (usually rats and dogs) are used. The duration of test is 90 days. The study is made at different dose levels. They vary from near zero to those sufficiently high to produce some type of adverse effect, after the study has continued for several days. Regular evaluation of the state of health of the animals is made and the physiological and biological tests are carried out. Complete autopsy and histological examination of all organ systems of these animals given high doses of additives are carried out. These studies help determine the acceptable daily intake, in milligrams per kilogram of the body weight, allowed for human being.

Chronic Toxicity Tests: Chronic toxicity tests are conducted in a manner similar to the subacute test but the duration of the test extends for 1 to 2 years. The number of animals used at each dose and the number of dose level will be more than in the subacute test. Additives are administered at 10 to 100 times human allowable levels calculated from subacute tests. This type of test also helps determine the carcinogenicity (cancer causing) and mutagenicity (heritable changes in genetic material) of the additive. Long-term tests also help study the effect of additives on the permanent structural or functional changes induced in embryo (teratogenicity) and in reproduction, i.e., male or female fertility, litter size, litter weight and number of the surviving young.

An additive which according to the above evaluation procedure is not harmful is considered safe in some countries and permitted as a food additive. The control of additives in foods is being gradually improved and tightened and testing procedures are becoming more rigorous in different

economically advanced countries of the world. The laws regulating the use the additives differ in different countries. In order to try to formulate food standards on an international basis, the FAO and WHO have set up a permanent commission designated as Codex Alimentarius, to develop international and regional food standards, which include standards of food additives and contaminants, with the hope of improving food standards all over the world. The proposals of the Codex Alimentarius is highly valuable to developing nations which as yet have few standards of their own.

Limitations of Tests: Tests for safety do not solve all the problems associated with the use of food additives. Results of animal experiments, though very useful, have their own limitations. Not all biological species respond in the same way to the same compound. Further, metabolic pathways of several animals are not identical to those of man. This raises the question of which results with animals are really applicable to human beings? In addition, a chemical that has been thoroughly tested and found to be harmless may react synergistically with another chemical to produce harmful (or beneficial) effects greater than with each of the chemicals used. It is financially and scientifically not feasible to test for all synergistic reactions with the many thousands, even millions, of the natural and synthetic chemicals. Then there is the question of what is to be done when the test for safety reveals a low order of risk or leaves doubts about toxicity.

Safety Versus Hazards: The benefits involved in the use of food additives are many, such as increased shelf-life, quality, nutritive value and economic saving. Even if there is some risk involved in the use of a food additive it becomes necessary to accept it if the benefits are sufficiently great. The benefit of saccharin as a sweetener, in the absence of an alternative, to diabetics and obese persons outweighs the risks. Nitrite used for meat preservation also prevents the growth of the micro-organism *Clostridium botulinum.* The banning of the use of nitrite for its suspected adverse effect greatly increases the

toxic effects of the microbes in meat. There is no such thing as an absolute safety in food additives. Even common salt, which is essential for life in small amounts, can probably kill a person if taken at one time in large quantity. Similarly, metals like copper, chromium and zinc are essential for life in trace amounts. These can also be toxic and mortal if ingested in large amounts. The toxic effects observed with some food additives are at levels far higher than are present in food. Factors like age, sex, weight and nutritional status, are also to be taken into consideration in considering the hazards of an additive. Fortunately, our bodies possess a multitude of mechanisms for dealing safely with small amounts of toxic substances from any source. Thus, from a rational standpoint, the use of additives in foods should be accepted when they serve some useful purpose. As one cannot live in modern society without them, we must ensure that their use is controlled so as to gain benefits while eliminating hazards and abuses.

Unwanted Additives: The unintentional incorporation of chemicals into food is as widespread as intentional addition and may present health hazards. The sources of contamination are radioactive fallout, thousands of chemicals used in agricultural production, animal food additives and accidental contaminants during food processing.

Radioactive Fallout: Radioactive fallout through nuclear explosions is a serious modern problem. Nuclear explosions inject into the stratosphere as well as into the atmosphere considerable amounts of smaller particles called fission products. These contain unstable atoms, called radio-isotopes, which spontaneously break down emitting radiations and particles that are highly injurious to living tissues. The fission products finally reach the ground through rain, snow or wind. This is known as fallout. The fallout matter reaches man directly in drinking water, fruits and vegetables or indirectly through animals, which eat contaminated feed or graze on contaminated pastures.

Many of the fission-produced radioisotopes have half-lives (length of time that has to elapse before the amount of radioactivity measured has dropped to half the original value) of less than one day. Some have a half-life of a few days (8 days for ^{131}iodine) while others are radioactive for many years. The half-life of ^{90}strontium is 28 years and that of ^{137}cesium is 30 years. Even ^{131}iodine, with a half-life of only 8 days, is hazardous for the normal functioning of thyroid. ^{90}Strontium is considered to be the most hazardous radioactive isotope. It is closely related to calcium and the body treats both strontium and calcium in exactly the same way. ^{90}Strontium induces bone cancer and leukaemia. ^{137}Cesium passes through the human body quickly but it emits radiations which, through blood, reach all tissues and can be a potential genetic danger.

Agricultural Contaminants: Chemicals in the form of insecticides, fungicides, herbicides (in general biocides), growth promoting substances and protectants, etc., are extensively used in large numbers in agricultural production. Without mem much food would be lost. Commercial production of some crops would be impossible if chemicals are not used. It is estimated that 23 per cent of the commercial cabbage crop and 37 per cent of the potato crop would be lost if chemicals were not used. Small quantities of chemical residues often remain in such crops.

The residues of the pesticide DDT has been observed in small amounts in soil, water, vegetables and animal tissues in all parts of the world. From these sources, the pesticide residues reach man. The widespread use of this chemical as an insecticide has contaminated even the air we breathe. The presence of pesticide and other residues in food is a serious international problem. WHO has attempted to control the extent of contamination by prescribing the limits for the amount of many pesticides that may be present in foods. Some countries have banned the use of chemicals like DDT.

Animal Food Additives: The use of some antibiotics as

food additives has been considered in Subsection 29.15.11. They are also used as plant and animal additives. In some countries (particularly the USA), about 80 per cent of animal feed it treated with small quantities of antibiotics for enhancing growth, improved feed utilisation and the checking of intestinal flora of animals. This has helped to produce less expensive meat and poultry. In all the cases where antibiotics have been used residues may remain in meat. As already indicated, the presence of antibiotics in foods may result in the development of strains resistant to antibiotic drugs.

The synthetic female hormone, diethylstilbesterol (DES), is used on chicken, cattle, sheep as implants and as a daily additive to the feed. This helps the conversion of foodstuffs into meat more efficiently in such animals. On an average the growth is one-fourth faster than in untreated animals; and the increase is more in muscle than fat. Residues of DES when present in food are potential cancer hazards.

Contamination of food in any way is bad. Just as in the case of food additives, food contamination must be considered in the light of any benefits that ensue. Use of pesticides is hazardous; but without pesticides much food would be lost. Some type of packing may result in contamination but the protection of food requiring long-distance transport from dirt and infection requires packing. Again, contamination of any type in very small quantities may not be harmful and in many cases the reported hazard has not been proved conclusively.

Man's basic drive is for food to satisfy his hunger. Food is intimately woven into the physical, economic, psychological, intellectual and social life of man. It is a part of his culture and is filled with many different meanings and symbolism for all individuals at various ages and stages of their maturity.

Agricultural produce, such as cereals, pulses, fruits and vegetables, and reared animals for slaughter, milk, eggs, etc., are foods or food raw materials. When consumed, foods undergo digestive and other changes to supply the body its

requirements. After production and before consumption, foods are subjected to numerous adverse physical, chemical, microbial or parasitic factors which may cause their spoilage or cause disease when consumed. To prevent these and prepare food for immediate or future use requires processing, preservation and storage. Food for consumption should have the proper appearance, colour, juiciness, texture, odour and taste. During the past few decades great advances have been made in the study of the whole field of the properties, preservation and processing of raw food and of the behaviour of finished food products.

There is a tremendous growth of world population. In spite of increased food production there is a danger of shortfall in supply. Consequently, there must be greater and more efficient utilisation of the existing food sources and development of entirely new ones. Also, artificial foods, which have been controlled with respect to the attributes of eating quality, nutritional content and ease of assimilation, have to be developed to meet the requirements of man.

Main Sources of Food

Food is a more basic need of man than shelter and clothing. It provides adequately for the body's growth, maintenance, repair and reproduction. Food furnishes the body with the energy required for all human activities—it provides materials required for the building and renewal of body tissues and the substances that act to regulate body processes. An individual food, such as milk, may fulfil all these functions or, as in the case of sugar, any one function. However, all the above functions of food must be served by the diet. All through history food has played important roles besides that of nutrition. Wars have been fought for food. People have been known to steal, rob and kill for it. Explorers have searched the world for new foods. Sciences have been built on food and focd related discoveries have been rewarded. Fortunes have been made and lost in foodstuffs. Food has helped in creating history!

Food is also a Source of Power: People have been made to starve for submission. Wars have been won by blockading food supplies to the enemy. Power of food may be exerted in society and in the family, through its use as a reward or punishment. A child may be sent to bed without food because of bad behaviour or may be given something special as a reward for some in order to maintain the body in good health. Most foods fulfil more than one function as they are complex mixtures of a number of chemical substances.

Nutrients Foods are composed of dozens or even hundreds of different kinds of substances—the nutrients which when consumed in adequate amounts, fulfil all the functions of the body. Six general classes or kinds of nutrients found in all foods are carbohydrates, fats, proteins, vitamins, minerals and water.

Carbohydrates make up the bulk of our diet. They are our chief source of energy. About 70 per cent of the energy requirements for all body functions is obtained from carbohydrates. Energy is produced by the oxidation or 'internal burning' (cellular respiration) of carbohydrates in the animal cells using oxygen. Carbohydrates also help in the utilisation of proteins and fats. Carbohydrates when consumed in excess are converted into fats (some glycogen also) to be used when needed.

The main sources of carbohydrates in the diet are starch and sugar. The sources of the former are mainly cereal grains (wheat, rice, etc.) or tubers (potato, sweet potato, cassova) and those of the latter are sugarcane and fruits.

Fats or lipids are the most concentrated form of energy in the food. They furnish more than twice the number of calories per gram furnished by carbohydrates or proteins. When compared to carbohydrates, fats contain a less percentage of oxygen and more of hydrogen, and consequently on oxidation yield more energy. Generally about 30 per cent of human energy requirements are met by fats. When excess energy is supplied to the body, it is stored as fat.

Fats are abundant in both plant and animal materials. In plants they may be confined to the cytoplasmic membrane or may also be present as reserve material. The fat content of fruits (except avacado and olive) is poor. Fat up to about 15 per cent is present in the germ of cereals. Nuts, such as groundnuts, are rich sources of fats. Butter from milk is an important source of fat. The adipose tissue of animals consists mainly of fats.

Proteins are the major source of building material for the body. They play an important role as structural constituents of cellular membranes and function in the maintenance and repair of body tissues. Proteins also function as biocatalysts. The food value of the protein depends upon the nature and content of its amino acids, which are its structural units. The excess of protein not required for building may be used as a source of energy.

Proteins are found in both animal and plant tissues. Meat, fish, poultry, eggs, milk and cheese are good sources of protein foods from animal sources. Pulses and cereals contain considerable amounts of storage proteins. Soyabean contains over 40per cent protein on dry weight basis. Nuts and seeds are also good sources of proteins. Starchy vegetables contain up to 2 per cent protein. Other vegetables and fruits are poor sources of proteins.

Vitamins are "accessory nutrients". They are required for the proper utilisation of the bulk food of the diet—carbohydrates, fats and proteins, and for the maintenance of good health. Vitamins together with minerals are involved in small quantities in the regulation of body processes. They are constituents of enzymes, which function as catalysts for many biological reactions within the body.

Vitamins are found in plant and animal tissues. Their content in plant tissues varies widely depending upon the growing condition, stage of maturity, handling, processing and storage. They are not uniformly distributed in plant tissues.

Vegetables and fruits are good sources of vitamins. Wheat is an excellent source of B vitamins, but the bran and germ, containing the bulk of these nutrients, are usually removed during processing. Since vitamins are simple organic substances they are easily synthesised and the synthetic vitamins are added to enrich or supplement those found in food products.

Minerals also act as catalysts for many biological reactions within the body. Their other functions include the building of bones and other structural parts of the body, muscular contraction, transmission of messages through the nervous system, and the digestion and utilisation of nutrients in food. Some minerals like calcium, phosphorus, iron, magnesium and sulphur are required in large quantities, while others like zinc, copper, iodine, manganese, cobalt, etc. are required in small quantities.

Minerals are found in foods from animal and plant sources. The mineral content of plant foods varies depending upon the mineral elements present in the medium of cultivation. The distribution of a particular mineral element varies in different tissues. Minerals, as vitamins, are added to food to enrich it.

Water is second only to oxygen in importance for the body. It is an ideal medium for transporting dissolved nutrients and wastes throughout the body.

Apart from the consumption of water as such, body needs of water are supplied by the foods we consume. Some foods contain a high percentage of water. Apart from this, oxidation of carbohydrates, fats and proteins in the body yields water.

Characteristics of Food

In addition to the above nutrients foods contain enzymes which function as catalysts in chemical reaction, colouring material and flavour compounds achievement. Food is now being used as an experience of revolt against "the establishment".

Food is a source of security. An infant learns security from the way his mother feeds him. His behavioural patterns will be influenced by the extent to which he feels secure as regards his food supply. Similarly a growing child gains confidence and a feeling of belonging when he knows there is food in the house and he will be fed. People feel reasonably secure when they have enough food stored up to take care of them during periods of scarcity. Familiar foods give a sense of security when one has to eat away from home.

Food is a status symbol. The well-to-do eat foods the common man cannot afford, even though they may not be nutritious. It is prestigious to use polished white rice instead of brown rice which protects one against beriberi. The same is true of white flour and white bread which have replaced "black bread". The status factor associated with certain foods used by the so-called upper class makes others prefer them, to easily available, less expensive and often more nutritious foods. We serve certain foods for family meals and different foods when we have guests.

Food is a symbol of hospitality and friendship throughout the world. We express our hospitality to a guest through an offer of food or a drink. In times of disaster or sorrow we take food to the affected persons. In our country offering a cup of coffee or tea is a symbol of friendship.

Food is an outlet for emotion. As a relief from tension one may eat or overeat. For some people loneliness and boredom are relieved by continuous nibbling at food. Anger and frustration may turn one against food. Specific foods are associated with unhappy experiences. Foods consumed by some people are unacceptable and even revolting for others. Such concepts have no rhyme or reason, nor are they related to nutritive value; they are just emotional reactions.

Thus, for an average man, food is much more than a substance supplying nutrients for health. It is the sum of his culture and traditions, emotional outlet, gratification of

pleasure and a relief from stress, a means of communication, security, status—all of these interwoven in the fabric of life and unconsciously expressed in food likes and dislikes. The psychological and emotional reaction to food do not yield easily to reasoning or scientific facts about nutrition.

Food Intake and its Regulations: All animals must eat to live and all have mechanisms that direct them to take food. In almost every case some control is exerted over the amount and kind of food that is taken. When hungry the animal responds by locating and ingesting foods and eating stops when hunger is satisfied.

Hunger, Appetite and Satiety: Hunger is usually an unpleasant sensation that compels a person to seek food and eat it, it is a physiological, condition which is associated with the contraction of the stomach. The contractions are forceful and occur for a period and then die away as the stomach passes into a resting stage. Hunger conditions that subside without eating will reappear later with greater intensity. In addition to stomach contractions resulting in tenseness, rumbling and a feeling of emptiness, a hungry person may experience certain general sensations, e.g. weakness, irritability, occasional headache or even nausea.

Appetite in most people is a pleasant sensation that causes a person to desire and anticipate food. An appetite can be for a certain kind of food. Appetite also has physiological components but is basically a psychological state. It is less easily localised than hunger, and is usually felt in the mouth or palate. It appears to depend more on the odour and memory of pleasant food. Appetite is clearly distinct from hunger, as a person may express a desire for some food at the end of dinner when he is comfortably replete.

Satiety is the sensation accompanying the satisfaction of the desire for food that comes after eating. It is not just the opposite of hunger and has far fewer sensations. While hunger, builds up slowly, satiety occurs rapidly. Absence of the desire

for food, when it is needed, is an abnormal condition and is known as Anorexia.

Regulation of Hunger: Hypothalamus is a small portion of the brain located at its base near the brain stem. This has been identified as the area which evokes the sensation of hunger and satiety. In the central portion of the hypothalamus (the ventromedial hypothalamus), the destruction of a small area causes animals to eat voraciously and become obese, because of destruction of the satiety centre. Again, the destruction of a small centre in the side region of the hypothalamus (the lateral hypothalamus) results in the opposite effect, that is, the animals refuse to eat because of the destruction of feeding centre and become anorexic. A stimulation of the satiety centre causes organisms to stop eating and that of the feeding centre causes them to eat.

The satiety and the feeding centres of the hypothalamus are connected by nerve fibres. When stimulated, the satiety centre sends signals to the feeding centre to inhibit feeding activity. That is, the feeding centre controls the feeding behaviour and satiety centre regulates it.

A number of factors stimulate the satiety centre so that it can regulate the feeding centre. There are sensory, metabolic and hormonal influences controlling the food intake.

Sensory Influences: Stomach contraction results in hunger. But hunger contraction is not the only feature of food intake regulation, as hunger contraction continues even when the main nerve to the stomach is severed. However, there are stretch and chemo receptors in the gastro-intestinal system that record the distensions of the stomach and the presence of food after a meal, which relay information to the brain resulting in short-term regulation.

Several other stimuli involved in the gastro-intestinal system contribute to the sensation of eating. Sensory stimuli coming from the taste, smell and texture of food are relayed to the cortex of the brain, which are then transmitted to the

satiety centre in the hypothalamus which results in signals to stop or to continue eating. Further, some psychological factors also have an influence on this centre.

Metabolic Factors: There are glucose receptors in the satiety centre which are receptive to glucose utilisation. When food is taken, blood sugar level increases and the rate of utilisation of glucose in the tissues rises. Then the receptors are stimulated, and the satiety centre signals responses to stop eating. Some hours after eating, the blood glucose level falls resulting in a low utilisation rate. The satiety centre then stops sending signals to the feeding centre. This mechanism of food intake regulation is known as glucostatic regulation.

The regulation of food intake over a long-term is regulated by the amount of fat stored in the adipose tissues (lipostatic regulation). According to this theory precise information on the fat stores is relayed from the adipose tissues to the nervous control centre. If the stores are filled, signals are given to stop eating and when the fat level is reduced, eating continues till the satiety centre is activated. The manner in which the information on the fat content of adipose tissue is transmitted is not clear.

The plasma concentration of amino acid patterns produced by the diet have an effect on the amount of food intake. Since many brain neuro-transmitters are affected by the supply of amino acids, it is definitely possible that the feeding behaviour could also be so influenced (aminostatic regulation).

Hormonal Centre: A number of hormones, such as insulin, glucogen, etc. which are involved directly or indirectly in the regulation of the utilisation of glucose which can be sensed by the hypothalamus glucose receptors, regulate the intake of food.

Various Trends in Food

Man eats what his forefathers ate, if possible, and what his environment offers. A young child does not form fixed food habits, but is patterned by adults, who eat certain foods

and not others. The foods which his father does not like and which his mother therefore does not serve, do not become familiar to the child and would not be eaten by him. The environment—the physical, psychological and social setting, which relates to the culture of a group—also determines the food patterns.

Historical: Some of our knowledge of what early man ate comes from archeological studies of cave drawings of food getting and preparation activities. Other evidence of what primitive people ate includes the study of remnants of discarded food found by archeologists in mounds, pits, bogs, lake beds and tombs. The remnants of human faces (coproliter) also tell the story of early human food. From such remains of prehistoric man a picture of the food eaten by him has been reconstituted with some degree of certainty from the study of anthropologists and other scientists.

Primitive men lived as hunters and gatherers. They collected their food from wild animals and plants. They depended upon fruits, nuts, roots and other plant foods, meat from animals and fish catch in seas, lakes and rivers. They were forced to spend their days and nights in search of food. They roamed from place to place to correspond to the changing season, the coming and going of various fruits and leaves or the migration and movement of game animals. They lived this way till about 10,000 years ago in a few places and 5000 years ago inmost of the world.

Gradually, food gatherers learned to domesticate plants and animals. They were no longer roaming about eating what was available. They settled down, built shelters and raised plants and animals to provide food. Plant domestication began in China about 10,000 B.C. followed by India, the eastern Mediterranean area and Africa. The first crops to be grown were wheat and barley from wild grasses. Simultaneously, livestock were domesticated. Cows were domesticated first and then sheep and goats. Milk was probably the first food to be extracted from animals. In the early stages, food

production did not develop beyond an inefficient subsistence level and small farming communities developed around the areas of farming. The development of agricultural skills over the last two centuries and consequent supply of a sufficient amount of food, its preservation and storage, resulted in the emergence of cities and urban civilization.

One of the first great changes which occurred in man's food pattern must have been when he learnt to use fire to cook. It is not known when he first used fire for this purpose. Man eats most of his food cooked and this is one of the many characteristics which separate him from other animals. The type and amount of fuel available determines the pattern of food of a society. In areas where fuel is scarce, quick cooking foods are prepared. The use of chapatis, thin unleavened bread, and stew-like mixtures has developed in our country, and in other parts of Asia and Africa because their preparation requires meagre equipment and little fire.

Many changes took place in food patterns prior to middle ages. As man moved from one place to another, he found new foods growing in his new settlement. Since he usually carried the seeds of foods grown where he lived to his new place of residence, there was a migration of foods. Thus, sugar went from India to other parts of the world. Sweet oranges from China and sour ones from India went to Europe with the early overland traders. The dispersion of foods indigenous to one country into another has always increased the variety of man's diet from ancient times. Today, with the rapid means of communication, the differences in the diets of different countries are becoming less marked.

Population Effect

Population Growth: The population of the entire world when agriculture or settled cultivation was started over 10,000 years ago was about 15 million. The estimate of the world population about 2000 years ago was 250 million. This was doubled in 1650. Since then there has been a threatening

increase in the population growth reaching a billion in 1830,2 billion in 1940 and 4.42 billion in 1980, with more than half of the world population living in Asia. The regions of fastest population growth are Africa and South and Central America, although the world's largest growth and largest absolute increase are in South Asia. It is estimated that at the turn of the century the world population will be 6.13 billion and 90 per cent of this would be in developing countries. At present the world population is increasing by about 80 million per year.

The population of India at the time of partition of the country (1947) which was about 350 million jumped to 683 million in 1981 and is expected to exceed one billion in 2001. Now, the annual addition to the population of India is over 15 million.

The world is already facing hunger. According to recent (1985) Food and Agricultural Organisation (FAO) report, abnormal food shortages have been reported in 30 countries of Africa, Asia and Latin America. The present growth rate of the world population tends to further aggravate the food situation. Also, about one-third of humanity consumes two-thirds of the world's food supply. About one quarter of the people in developing countries are undernourished. India is one of the countries which suffers from undernourishment on a vast scale.

The people of developing countries consume food corresponding to 1800-2000 kilocalories (Kcal) per day (Indian consumption is 1969 kilo-calories). The consumption in developed countries is in the range of 3000-3500 Kcal. A utilisation of under 2000 Kcal per person per day can be definitely said to be on the low side. Within any country, the distribution of food is very much in favour of the better-off people.

In developing countries, the inadequacies of food are not only in the quantity of food, but also in its quality, leading

to malnutrition as well as undernutrition. The diet of people in these countries is unbalanced, because the bulk of their diet consist mainly of cereals—rice, maize, millet and starchy root. Over 80 per cent of their calories are derived from these food sources. The lack of variety in their food and particularly the shortage of protective foods leads to deficiency diseases. Protein-calorie malnutrition, particularly among children, is extremely common in these countries. In 1975, in Africa, Asia and Latin America, 20.9 million out of 34.6 million children of the age group 1-4 suffered from protein-calorie malnutrition.

Food Production: From 1950 to 1971 there was an unprecedented growth in world food production. Yields on existing crop land were raised dramatically by energy intensive agriculture. The introduction of high yielding wheat and rice strains in parts of Asia and Africa resulted in a large increase in yield and this is referred to as the "Green Revolution". During this period the world grain production increased from 631 million tonnes to 1237 million tonnes; resulting in an increase in grain availability per head, per year from 251 to 330 kg.

Even after 1971, there has been a gradual increase in the world grain production. It stood at 1432 million tonnes in 1980 as against 1237 million tonnes in 1971. But the output per head barely kept pace with population growth. During this period the availability of grain per head per year came down from 330 to 324 kg.

The position in developing countries was even worse. There was a decline in their total grain production. These countries face a widening food gap between their production and consumption. Their food import needs to be increased dramatically. The total food import of developing countries in 1984 was of the order of 100 million tonnes.

In India, the progress in agricultural production during the last 35 years has been very impressive. There has been a consistent upward trend in foodgrain production except during

the years marked by acute weather aberrations. The foodgrain production has risen from 51 million tonnes in 1950-51 to a record production of 151.5 million tonnes in 1983-84 and further increases are anticipated in the coming years. Our requirement in 2000 A. D. is of the order of about 190 million tonnes.

The increase in agricultural production during the last three decades is a result of the increase in the application of science and technology in agriculture. Great strides have been made in agricultural research, education and extension. However, we cannot be complacent and ignore major problems concerned with food production. Farm management efficiency has to be improved. Input factors like water, seeds and fertilizers are tc be increased. The diminishing size of an average land-holding and its adverse effect on the net income of the farmer has to be avoided. The consistent damage to the ecosystem, whose maintenance is essential for sustained agricultural advance should be prevented. Our dry farming has to be improved.

Prospects for Future: The shooting population growth, rising energy costs, doubts about the adequacy of agricultural resource base, disproportionate share of available food by the well-to-do, lagging behind of new agricultural knowledge and unfavourable weather conditions have given rise to pessimism about the ability of the earth to feed its people. This assessment is also explicit or implicit in the reports of a number of international institutions concerned with world food production and demand.

It appears as though one need not be frightened about the world food situation. Actually, hunger is not new; it is as old as history. The world lives from hand to mouth, never more than one crop away from hunger. Actually, the world food problem is diminishing. The average life-span is lengthening even in the poorest countries. The life expectancy at birth is now said to be 58 years, which was 42 years in 1960. The child death rate is dropping. In the low-income countries the death rate among children in the age group 1-4 declined from 30 per

thousand in 1960 to 20 in 1978. These recently born children were getting the food and health care that increased their survival rate by 14 per cent in eighteen years.

There is a decline in birth-rate. In low-income countries, the birth rate in 1960 was 48 per thousand; by 1978 it had fallen to 39. The decline in birth rate has decreased the world population growth rate from 2.0 per cent to 1.8 per cent.

The average food production in developing countries has crept upwards though it is not ahead of population growth. More land could be brought under cultivation though such development would require enormous investment and much time. There is great scope for increasing food production in the developing countries; the yield per hectare in some of them is less than half of what it is in developed countries. The poorer countries are arranging for proper distribution of available food through fair-price shops and food-for-work programmes.

The discoveries in agriculture are not lagging behind. International research has resulted in the production of high-yielding varieties of wheat and rice which has helped increase the world food supply. The number of agricultural research institutions with good financial resources is increasing. It may be hoped that the work in such institutions will result in technological breakthroughs in fields, such as gene-splicing, textured vegetable proteins, nitrogen fixation from non-leguminous crops, salt tolerant crops, wild crosses of plants and animals, improved photosynthesis, long-range weather forecasting or computer-aided agriculture management.

Food products are being used for non-food purposes, e.g., because of fuel scarcity, some countries are using petrol mixed with alcohol. If this is practised on a world scale, produce from 5 per cent of farmland would be required for this purpose. We can ill afford this when hunger is threatening the world.

A large amount of vegetable material like grasses are available. They are not eaten because they consist largely of

cellulose which man is unable to digest. Proteins are extracted from grasses and other vegetable matter. Such protein concentrates are not used as foods now as they lack both taste and smell. But they could be used as foods after improved processing.

After the oils are extracted from nuts, a residue rich in proteins is left. This residue at present is mostly used as animal food or fertilizer. It is estimated that the amount of protein contained in such oilseed residues is sufficient to provide twice the world's deficit of protein. Oilseed cake can be processed and used as food. It is already being used to a certain extent in some poor countries. This could be expanded.

New foods can be obtained in future. We are now mostly depending on foods derived from plants grown on land. Unicellular organisms, such as yeasts, algae and fungi may be cultivated on liquids. These are rich sources of high-quality proteins and could be developed as food. New foods may also be developed by chemical synthesis.

Thus, it should be possible to have sufficient food for future generations if we make the needed investment to develop and distribute new agricultural knowledge, the developed countries share their knowledge and experience with the less developed ones and the developing countries are diligent enough to improve their agriculture and reduce the rate of growth of their population number. Action should also be taken to prevent the conversion of a high share of our food into non-food products, and to develop food from non-traditional sources. Scientific and technological improvements will help meet the food needs of the future.

Bibliography

Achaya, K.T. : *Indian Dairy Products,* Asia Publishing House, Bangalore, 1974.

Adam, H. : *Food Science,* John Wiley and Sons, New York, 1982.

Austin, A. and Ram, A. : *Studies on Chapati-Making Quality of Wheat,* Indian Council of Agricultural Research, New Delhi, 1971.

Bachman, W. : *Food Chemistry,* Maclaren and Sons Ltd., London, 1955.

Balsdon, J.P. : *Life and Leisure in Ancient Rome,* Bodley Head, London, 1966.

Beersteeker, E. : *The Biochemistry of Vitamins,* Reinhold, New York, 1950.

Bharathi, V.V. : *Anxiety Behaviour in Children,* Discovery Publishing House, New Delhi, 1992.

Board, R.G. : *A Modern Introduction to Food Microbiology,* Blackwell Scientific Publications, London, 1983.

Burton, T.L. : *Recreation Research and Planning,* Allen and Unwin, London, 1970.

Cambell, A.M., Penfield, M.P. and R.M. Griswold : *Experimental Study of Foods,* Houghton, London, 1990.

Cameron, Spencer : *Food Science,* Pargamon Press, New York, 1984.

Challoway, D.H. : *Nutrition and Physical Fitness,* W.B. Saunders Co., Philadelphia, 1979.

Cherry, G.E. : *Town Planning and its Social Context*, MacGraw-Hill, London, 1970.

Cherry, J.P. : *Protein Functionality in Foods*, American Chemical Society, Washington D.C., 1981.

Chichester, C.O. : *The Chemistry of Plant Pigments*, Academic Press, New York, 1972.

Dale, E. : *Management Theory and Practice*, McGraw-Hill, New York, 1973.

Dharamraj, E. : *Food and Hotel Legislations and Policies*, New Age Publishers, New Delhi, 1998.

Dumazedier, J. : *Towards a Society of Leisure*, Free Press, New York, 1967.

Dyke, S .F. : *The Chemistry of Vitamins*, Interscience Publishers, New York, 1965.

Edmunds, F. : *Environmental Administration*, McGraw-Hill, New York, 1973.

Eley, G. : *Wild Fruits and Nuts*, E.P. Publishing Ltd., Yorkshire, 1976.

Feiffer, M. : *Giving Places*, MacMillan, London, 1985.

Furia, T.E. : *Hand Book of Food Additives*, C.R.C. Press, Ohio, 1977.

Geiss, I. : *Proteins : Structure, Function and Evolution*, Menlo Park, California, 1983.

Gell, D. L. : *International Hotel Industry Policy*, Van Nostrand Reinhold, New York, 1990.

Gyorgy, P. and Pearson, W.N. : *The Vitamin*, Academic Press, New York, 1967.

Haschemeyer, A.H. : *Proteins : A Guide to Study of Physical and Chemical Methods*, John Wiley & Sons, New York, 1973.

Hudson, F. : *Development in Food Proteins*, Applied Science Publisher, London, 1982.

Inglett, G. : *The Quality of Foods and Beverages : Chemistry and Technology*, Academic Press, New York, 1981.

Jackson, R. : *The Geography of Recreation and Leisure*, Hutchinson, London, 1972.

Jasscler, D.K. and M.A. Joslyn : *Fruit and Vegetables Juice*, The AVI Publishing Co., Connecticut, 1971.

Johnson, A.H. : *Encyclopedia of Food Science*, The AVI Publishing, Connecticut, 1978.

Kacharoo, P. : *Pulse Crops of India*, ICAR, New Delhi, 1970.

Kadans, J.M. : *Encyclopedia of Fruits, Vegetables, Nuts and Seeds*, Parker Publishing Co., New York, 1973.

Kalyani, T. : *Home and School Environment*, Discovery Publishing House, New Delhi, 2003.

Kinton, R. : *Practical Cookery*, Edward Arnold, London, 1981.

Krause, M.V. and L.K. Mahann : *Food Nutrition and Diet Therapy*, W.B. Saunders Co., Philadelphia, 1979.

Kumari, P. : *Role of Anganwadi Workers in Integrated Child Development Services Programme*, Discovery Publishing House, New Delhi, 1994.

Lal, H. : *Total Quality Management : A Practical Approach*, New Age Publishers, New Delhi, 2000.

Law, C. : *Urban Tourism : Attracting Visitors to Large Cities*, Mansell, London, 1993.

Less, R. and E.B. Jackson : *Sugar Confectionery and Chocolate Manufacture*, Leonard Hill Books, Burks, 1973.

Lickorish, L. J. : *Hotel Industry and International Balance of Payments*, International Institute of Scientific Travel Research, Geneva, 1953.

Lloydryall, A. and W.T. Pendzer : *Handling Transportation and Storage of Fruits and Vegetables : Fruits and Peanuts*, The AVI Publishing Co., Connecticut, 1974.

Mackinney, G. and A.C. Little : *Colour of Foods*, The AVI Publishing Company, Connecticut, 1972.

Malhotra, R. K. : *Food Service and Catering Management*, Anmol Publications, New Delhi, 1998.

Marton, I.D. and A.J. Macbod : *Food Flavours*, Elesvier Scientific Publishing Co., New York, 1982.

Masefield, G.B. : *The Oxford Book of Food Plants*, Oxford University Press, London, 1975.

Mate, S.A. : *Bakery Technology and Engineering,* The AVI Publishing Co., Connecticut, 1972.

————— : *Cereal as Foods and Feeds,* The AVI Publishing Co., Connecticut, 1959.

Mayo, J. : *The Psychology of Leisure Travel,* CBI Publishing Company, Boston, 1981.

Mcwilliams, M. : *Food Fundamentals,* John Wiley & Sons, New York, 1979.

Medlik, S. : *Hotel Management: Past, Present and Future,* Heinemann, London, 1976.

Mellor, J. D. : *Fundamentals of Freete-Drying,* Academic Press, London, 1978.

Meyer, L.H. : *Food Chemistry,* Reinhold Publishing Corporation, New York, 1961.

Michael Eskin, N.A. : *Plant Pigments, Flavours and Textures,* Academic Press, New York, 1979.

Murthy, R.B. : *Breeding Procedures in Pearl Millet,* ICAR, New Delhi, 1977.

Nath, P. : *Vegetables for the Tropical Region,* Indian Council of Agricultural Research, New Delhi, 1976.

Oberleas, D. : *Trace Elements in Health and Disease,* Academic Press, New York, 1976.

Palmer, T. : *Understanding Enzymes,* John Wiley & Sons, New York, 1981.

Pamela, V.P. : *Directory of Wines and Spirits,* North Wood Publication, London, 1974.

Paul, P.C. and Palmer, H.H. : *Food Theory and Application,* John Wiley & Sons, New York, 1972.

Peckham, G.G. : *Foundations of Food Preparation,* MacMillan Publishing Co., New York, 1979.

Pigman, W.W. and Horton, D. : *The Carbohydrates : Chemistry and Biochemistry Plants,* C.S.I.R, New Delhi, 1969.

Pigram, J. : *Outdoor Recreation and Resource Management,* Croom Helm, London, 1993.

Potter, N.M. : *Food Science*, The AVI Publishing Co., Connecticut, 1978.

Pudney, John : *The Thomas Cook Story*, Michael Joseph, London, 1953.

Puruthi, J.S. : *Spices and Condiments*, National Book Trust of India, New Delhi, 1976.

Ranganna, S. : *Manual of Analysis of Fruit and Vegetable Products*, Tata McGraw-Hill Publishing Co. Ltd., New Delhi, 1984.

Rani, Usha D. : *Nutrition and Religion*, Discovery Publishing House, New Delhi, 2003.

Rao, D. V. : *Community and School Nutrition Education*, Discovery Publishing House, New Delhi, 1998.

Roberts, D.V. : *Enzyme Kinetics*, Cambridge University Press, 1977.

Rogers, C.A. : *Profitable Poultry Keeping in India and the East*, Treasure House of Books, Bombay, 1966.

Ross, G. F. : *The Psychology of Hotel Management*, Hospitality Press, Melbourne, 1994.

Schapira, J. : *The Book of Coffee and Tea*, St. Martin's Press, New York, 1975.

Schultz, G.E. and R.H. Schirmer : *Principles of Protein Structure*, Springer-Verlage, New York, 1979.

Shreedevi, V. : *Nutrition Education*, Discovery Publishing House, New Delhi, 1997.

Smartt, J. : *Tropical Pulses*, Longman Group Ltd., London, 1976.

Stadlmare, W.J. and. Cotterill, O.Y. : *Egg Science and Technology*, The AVI Publishing Co., Connecticut, 1973.

Stumbo, C.R. : *Thermobacteriology in Food Processing*, Academic Press,, New York, 2000.

Swern, D. : *Bailey's Industrial Oil and Fat Products*, John Wiley & Sons, New York, 1964.

Tandon, G.L. : *Preservation of Fruits and Vegetables*, Indian Council of Agricultural Research, New Delhi, 1960.

Tandon, H.P. : *Egg and its Care,* Ministry of Agriculture, Govt. of India, New Delhi, 1973.

Thangam, P.E. : *Modern Cookery for Teaching and Trade,* Orient Longmans Ltd., Bombay, 1983.

Thankamma, J. : *Food Adulteration,* S.G. Wasani, New Delhi, 1976.

Tressler, D.K. : *Food Technology the World Over,* The AVI Publishing Co., Connecticut, 1963.

Uma, K. : *Poverty, House Hold Fold Security and Nutrition in Rural Areas,* Discovery Publishing House, New Delhi, 2000.

Underwood, E.J : *Trace Elements in Human and Animal Nutrition,* Academic Press, New York, 1977.

Veldhnis, M.K. : *Food Science and Technology,* The AVI Publishing Co., Connecticut, 1977.

Vidya, D.B. : *A Textbook of Nutrition,* Discovery Publishing House, New Delhi, 1996.

Weiss, T. J. : *Food Oils and their Uses,* The AVI Publishing, Connecticut, 1970.

Whitaker, J.R. : *Principles of Enzymology for Food Science,* Marcel Dekker, New York, 1977.

Williams, P.N. : *The Chemical Constitution of Natural Fats,* Chapman and Hall, London, 1964.

Winitz, M. : *Chemistry of Amino Acids,* John Wiley & Sons, New York, 1961.

Yasoda, P. : *Determinants of Nutritional Status of Pre-School Children,* Discovery Publishing House, New Delhi, 1998.

Yukic, T.S. : *Fundamentals of Recreation,* Harper Row, London, 1963.

Index

E

F

N

O

P

S

T

V

W

Z

❑❑❑

carbohydrates, vitamins, mineral salts, fats, and proteins should provide about three hundred calories per day for a normal adult. It comes under Food Chemistry, which has been dealt with, at length in this multi-volume.

Encyclopaedia of Food Chemistry, is an admirable step in this direction. These books throw abundant light on all aspects of food chemistry. It must prove to be a proud possession of students, scholars, academics, intellectuals and of course the general readers, who may like it, equally.

Wise comments and positive remarks for the betterment of these books in future, are solicited.

— Editor

Preface

Food Chemistry is a newer sub-discipline, under Food Science. This branch of knowledge enriches us about the composition and quality of food stuffs, we intend to consume. In a scholar's view, we are what we think, while we think what we eat and drink. This is an undeniable fact that our thoughts are determined by the kind of food we take. Food not only serves as a source of energy, it also plays a significant role in the making of one's personality.

Another scholar opines that for a healthy life, a healthy diet is a must. Our body mechanism is controlled by a number of elements, called vitamins, proteins, enzymes, carbohydrates and other simple and complex chemical elements.

All the abovesaid elements find their origin in various food articles, such as, milk, meat, egg, vegetables, cereals etc. Each has its own characteristics. For a healthy diet, different nutrients have to be taken in a proper proportion, because excess of any of these may result in disturbing the system of our body and affecting the health.

A proportionate or a balanced diet is one, which consists of all the necessary nutrients in a proper manner, required for the development and maintenance of the body. It depends upon age, health and profession. For instance, a diet, rich in

Contents

ANMOL PUBLICATIONS PVT. LTD.
H.O.: 4374/4B, Ansari Road, Darya Ganj,
New Delhi-110 002 (India)
Ph.: 23278000, 23261597
B.O.: No. 1015, Ist Main Road, BSK IIIrd Stage
IIIrd Phase, IIIrd Block,
Bangalore - 560 085 (India)
Visit us at: www.anmolpublications.com

Encyclopaedia of Food Chemistry

First Published, 2007

ISBN 978-81-261-3399-4 (Set)

PRINTED IN INDIA.

Printed at Amit Enterprises, Delhi.

Encyclopaedia of
FOOD CHEMISTRY

Volume 2

Manpal Singh

ANMOL PUBLICATIONS PVT. LTD.
NEW DELHI - 110 002 (INDIA)

Encyclopaedia of
FOOD CHEMISTRY

Encyclopaedia of
FOOD CHEMISTRY

Encyclopaedia of
FOOD CHEMISTRY

Volume 3

Manpal Singh

ANMOL PUBLICATIONS PVT. LTD.
NEW DELHI - 110 002 (INDIA)

ANMOL PUBLICATIONS PVT. LTD.
H.O.: 4374/4B, Ansari Road, Darya Ganj,
New Delhi-110 002 (India)
Ph.: 23278000, 23261597
B.O.: No. 1015, Ist Main Road, BSK IIIrd Stage
IIIrd Phase, IIIrd Block,
Bangalore - 560 085 (India)
Visit us at: www.anmolpublications.com

Encyclopaedia of Food Chemistry

First Published, 2007

ISBN 978-81-261-3399-4 (Set)

PRINTED IN INDIA

Printed at Amit Enterprises Delhi.

Contents

Preface

Food Chemistry is a newer sub-discipline, under Food Science. This branch of knowledge enriches us about the composition and quality of food stuffs, we intend to consume. In a scholar's view, we are what we think, while we think what we eat and drink. This is an undeniable fact that our thoughts are determined by the kind of food we take. Food not only serves as a source of energy, it also plays a significant role in the making of one's personality.

Another scholar opines that for a healthy life, a healthy diet is a must. Our body mechanism is controlled by a number of elements, called vitamins, proteins, enzymes, carbohydrates and other simple and complex chemical elements.

All the abovesaid elements find their origin in various food articles, such as, milk, meat, egg, vegetables, cereals etc. Each has its own characteristics. For a healthy diet, different nutrients have to be taken in a proper proportion, because excess of any of these may result in disturbing the system of our body and affecting the health.

A proportionate or a balanced diet is one, which consists of all the necessary nutrients in a proper manner, required for the development and maintenance of the body. It depends upon age, health and profession. For instance, a diet, rich in

carbohydrates, vitamins, mineral salts, fats, and proteins should provide about three hundred calories per day for a normal adult. It comes under Food Chemistry, which has been dealt with, at length in this multi-volume.

Encyclopaedia of Food Chemistry, is an admirable step in this direction. These books throw abundant light on all aspects of food chemistry. It must prove to be a proud possession of students, scholars, academics, intellectuals and of course the general readers, who may like it, equally.

Wise comments and positive remarks for the betterment of these books in future, are solicited.

— Editor

1

Significance of Cuisines

Significance of Italian Cuisine

Italian cuisine is a collection of regional cuisines. The culinary tradition dates back to 2000 years. When people in other parts of Europe were still living on a staple diet of millet and oatmeal porridge, in Sicily the Greek conquerors were feasting lavishly. A few centuries later derived a typical Roman feast which began with peachen's eggs, truffles, oysters, asparagus, lobsters, fried sausages, olives and plums, followed by the main course—stuffed geese, chicken, pigs, stewed flamingoes, tongue and peacock's brain, goose liver pates game. The art of cooking declined with the fall of Roman Empire and it was not until 15th century that it revived in the merchant cities of Genoa, Venice, and Florence. Catherine De' Medici, on her marriage in 1338 to the future Henri II of France, took Florentine cooks with her and introduced the French to the skills of cooking. Italians introduced artichokes, haricot beans, peas and other vegetables outside Italy. Best Italian food is, as a rule, still eaten in its own region—beef in Tuscany, artichokes in Rome, peas in the Lazio and Marches. Pasta is consumed all over Italy. The risotto in Venice

must always be served all'onda, rippling, a reference to its liquid consistency and appropriate to this city of the sea. North East Italy is famous for ham—prosciutto de pum. Parmesan cheese its full name is Grana Parmigiano Reggieno from Parma. Herbs are much used in Tusan cooking—they are thyme, oregano, mint, basil, parsley, tarragon. Some of the famous Italian dishes are as follows:

Zuppa di fagioli (bean soup), Minestrone, Spaghetti Bolognese, Zuppa alla pavese (consomme with egg and cheese), Obsobuco alla milanaise, Scallopine de vitello alla bolognese (Veal escalope Bologna style), Risotto alla milanese, Cassata alla siciliana, Pizza alla Andrea (Black-olives, tomato, anchovies), Pizza Margherita (tomato sauce, garlic, oregano).

With meal starters serve a dry white or light rose wine.

Rice and pasta light red or robust rose wine.

Pizza a rough red wine

Fish rose or medium dry wine

Meat full bodied red wine with Game, Roast, Chicken, Turkey light red wine.

Cheese—serve dry or medium white or light rose, with Gorgonzola full bodied red wine.

Desert wines—Aleatico de Paglia, Marsah, Mocoato, Vino Santo.

Dry white wines—Copri bianco, Fruscati (secco), Orvieto (secco), Soava, Valtellina.

Medium dry white—Lacrima Christi del vesuvio, verdicchio di Jesi

Semi Sweet—Albana di Romagna, Frascati, Orviato

Sparkling—Asti spumante, Lacrima Christi

Light Red—Bardolino, Chianti, Grignolino, Grumello, Nebiolo, Sassela

Generous red—Bar bareseo, Bar bera, Barola, Inferno, Santa Modena

Pasta-aselutta—Dry pasta with sauce, Pasta secca-eggless pesta

Pasta all' uenva—egg pasta, Maccheroni—Generic term for Pasta.

Significance of Greek Cuisine

Greek and Roman customs had an influence of their cuisine on the cuisine of Syrians, Lydians, Armenians, Babylonians, Cilicians and Mesopotamians. King Khusroo I and II had spent enormous sum on fantastic banquets lavishly laid, dazzling with luxuiy on promotion of food.

Islamic Empire stretched across North Africa, Spain and Cicily. Nowadays many Middle East countries such as Syria, Lebanon, Egypt, Iran, Turkey, Greece, Iraq, Saudi Arabia, Yemen, Sudan, Algeria, Tunisia, Morocco all have Islamic culinary linkage.

The religion of Islam is the most important part of Middle East culture. The code of religion is derived mainly from Quran. Prophet Muhammad repeated injunctions about kind of food permitted and not permitted. The following are forbidden: (1) Animals dead before they are slaughtered and those killed for other reasons than that of food. The animal should only be slaughtered as per the Halal technique which consists of severing the neck, the wind pipe, two jugular veins without lifting the knife off until the severance is complete. (2) Animal slaughtered as an offering to pagan deity or in the name of deity. (3) Alcoholic or fermented liquids and all inebriated liquors. (4) Pig's flesh. (5) Blood.

A large variety of spices and dry nuts were extensively used in the culinary preparation as these were brought from China and India by Arabians. In the medieval recipes some spices like cumin, coriander, cinnamon, ginger and masticwere used. A Lebanese spice mixture is made up of one part of hot pepper (Cayenne), two parts of sweet pepper (paprika) and

two parts of cinnamon. Arab favourite, called taklia is ground coriander fried with crushed garlic. All spice or Jamaican pepper was also used as it has the flavour of 4 spices—cinnamon, nutmeg, clove and ginger.

Fresh herbs like coriander, dill, mint, chervil, rigani, oregano and marjoram were used in salads and stews. Saffron was used in Morocco, Persia, Saudi Arab for delicate aroma and magnificent yellow colour. Turmeric (oriental saffron) was used with chicken, veal, potatoes and chick peas.

Cooking was generally done over a type of primus called a "Fatayel". It was a long, slow procedure and pans were sometimes left to simmer overnight, nowadays necessity have passed and cooking is done in oven or tandoor. Large degs were used for dum cooking by sealing the pot with dough. The utensils were made of copper, brass, silver, beaten or engraved, sometimes inlaid with silver or other metals, sometimes they were having oriental decorative patterns. Traditionally fresh yoghurt is served with the food. It is also used in cooking dishes in Persia, Turkey, Lebanon and the Arab Emirates.

Persians like chees delicately flavoured, sweet and sour, cooked with vinegar, lemon, sugar and sour juice of pomegranate.

Cooking was done in "Alya", fat obtained by rendering lamb's tail. Today cooking is done in clarified butter called "Samna" or corn or nut oil. Jews use corn or nut oil in their cooking while copts prefer seasame oil.

Bismillah is uttered by all diners at the table before the meal begins. Twelve rules of etiquette are observed at the meal. The first four are: (1) To know that God is the provider. (2) To be satisfied with what he has provided. (3) To say in the name of God at the beginning of the meal. (4) God to be thanked at the end of the meal. Next four are customary: (l) Wash hands before eating. (2) To sit at the left of the table. (3) To eat with 3 fingers. (4) To lick the fingers after eating.

The last four rules are (1) To eat out dish that is immediately in front of you. (2) To take small pieces. (3) To chew the food well. (4) Not to gaze at others. At the end of the meal coffee is served.

The following are a few food specialities of Middle East:

Gebna Beida: It is a fresh cottage cheese from buffalo milk. It is served as an Hors d'oeuvre *i.e.,* (Mezze).

Taramasalata: Creamy mixture made by cod's roe pounded with soaked bread, garlic, onion, lemon juice and olive oil. It is garnished with black olives and served with thin toast. It is a Greek dish.

Cacik Cucumber and Yoghurt salad: It is a cucumber salad flavoured with garlic, mint, dressed with yoghurt. It is a popular Turkish dish.

Salata Meshwiya: This salad is very popular in Tunisia (North Africa). It is almost similar to salad nisgoise with lemon dressing.

Sanbusak: A savoury pastry from Syria, Lebanon and Egypt, prepared by short crust paste, half moon shape, filled with cheese, egg washed, baked until golden and crispy.

Kotopitta: A Greek chicken pie wrapped in fila pastry or flaky pastry.

Hamud: A chicken soup with calery, lemon juice and garlic.

Beid bi Lamoun: It's a Greek Avgolemono soup.

Melokhia: Egypt's national soup, has chicken stock to which chopped melbkhia (herb), garlic paste (taklia) often meat and chicken balls are added.

Fata: Prepared on Bakrid day in Egypt, by the sacrificed lamb bones and flesh, to which rice is added, flavoured with fried garlic, vinegar and garnished with chopped parsley.

Chakehouka: Tunisian omelette with capsicum strips, sliced onions, tomatoes, whole egg is dropped without stirring.

Blehat-Samak: Poached fish, flaked mixed with chopped onion, garlic, parsley, coriander, bread crumbs, cumin, egg for binding and seasoning shaped in small finger shape, rolled in flour, deep fried. Simmer in fish stock, cool sprinkle with parsley, serve as Horsd'oeuvre.

Ferique: Egyptian chicken dish with whole wheat, hard boiled egg and turmeric.

Shish Kebab: Turkish dish, prepared from diced leg of lamb or beef fillet, marinated in olive oil, lemon juice, rigani, tomato pulp, quartered tomatoes, pieces of onion and sweet pepper are put on a skewer in between the cubes of meat and then roasted.

Egyptian Lamb Stew: The leg of lamb is first roasted, then stewed in stock turmeric, lemon juice, seasoning. Served with rice.

Mefarka: Fried minced lamb, mixed with chopped broad beans, beaten egg, set in mould, serve with lemon juice.

Mishmishiya: It is a lamb stew, thickened with aprieot and almond paste, coloured with saffron and flavoured with coriander, cumin, mastic and cinnamon powder.

Rutabiya: Rutab is Arabic for dates. Minced lamb boil stuffed with almonds and garnished with dates.

Rishta (Fresh Noodles): In Persian rishta means thread. Fresh egg noodles, cooked and tossed in butter, served hot.

Calsons: Similar to cheese raviote to which the mixture or parmesan, grayere cheese and egg yolk is stuffed.

Couscous: National dish of North African countries.

Bamia: An Egyptian dish, basically lamb and okra stew with tomato condasse and tomato puree.

Seleq: A Saudi Arabian dish, made of lamb, rice cooked in milk. Served with cucumber, lettuce and tomatoes. Pure honey is also stirred in each portion.

Khubz (Eish Shami): Common bread Middle of East.

Iflagun: It's an Armenian bread made of yeast fermented dough, topped with powdered spices, grated cheese and pista.

Baklava: Greek pastry dessert, prepared from very thin sheets (Fila or Phyllo) stuffed with chopped walnuts, almond, sugar, cut diagonally, baked, pour sugar syrup and serve cold.

Ataif: Arabian pancakes, prepared by yeast fermented batter, shallow fried, dipped in sugar syrup, spread with thick cream (eishta). Sprinkled with chopped pista or almonds on the cream.

Significance of German Cuisine

The cooking of the coastal region its dominated by the sea. In Hamburg the city's elegant restaurants serve plaice, turbot and sole frequently on the menus. Excellent shellfish—oyster, mussels, bright red Heligoland lobsters and herrings are delivered fresh to the shops everyday. Chunks of eel, dried fruit and fresh vegetables are cooked in a strong ham stock flavoured with herbs and frequently garnished with dumplings. From the fertile coastal plain comes some of the finest meat and poultry. Smoked ham and satisages come from the northern and central regions. Mettwurst, a smoked minced pork sausage, Leberwurst, liver sausage are sold throughout Germany. The smoked ham of Westphalia is famous all over the world. The taste of sweet and savoury combinations are enjoyed in Germany with Sauerbraten. Baden, however, is the one region of Germany where the cooking can be ranked as haute cuisine. Hiedelberg boasts of the finest asparagus in Germany. Spatzle are served whenever there is a rich gravy to be mopped up—with stews, fricassees or a roast saddle of Black Forest venison served in a sour cream sauce. German cooking is wholesome, unpretentious family cooking. Bayerische leberknodetsuppe (Bavarian liver dumpling soup) Sauerbraten, Rotkhol (Red cabbage and Apple), Roast saddle of vension—Baden-Baden style.

Significance of Spanish and Portuguese Cuisines

It has a unique cuisine which can be described as a blend of East and West. Spanish and Portuguese cooking is simple,

sober and conservative. The natural flavour of the ingredients are undisguised by sauces or dressings. Basically a peasant cuisine, it relies on the simple excellence of the natural produce in the most favoured regions. In Spain every province has its own imaginative cocido, stew. These cocido have been developed to fine degree, around the central plateau of Spain—to derive a simple but fine meal from a single pot *e.g.* cocido madrileno. Chorizo, a rich ruddy brown sausage, with a warmly spiced and smoky flavour, is the essence of Spanish cooking. Andalusia, the region which is known for gazpacho, a cold tomato-based soup. Paella a la valenciana—the classic finished dish is a colourful, fragrant melange of saffron tinted rice, lobster, shrimp, chicken, clams or mussels and vegetables.

Spanish omelette—may have peas, spinach, mushrooms, garlic, parsley, chopped ham, anchoives and chorizo sausage pieces. Cocido madrileno—traditionally Spanish stew which has dried and fresh vegetables or a selection of different meats.

Cheese

Hard		Semi Hard	
Parmesan	Cheddar	Fontina	Danish blue
Reggiano	Cheshire	Bel paese	Roquefort
Romano	Tillamook	Port salut	Gorgonzola
Sardo	Swiss	Monterey Jack	Stilton
	Gruyre		
	Edam		
	Gouda		

After Dinner Cheeses for a Chees

Hard Cheeses Sliceable	Semi Hard Cheeses	Blue Veined	Fermented Soft
Tillamook	Bel paese	Stilton	Fromage de Brie
Swiss	Fontina	Roquefort	Camembert
Gouda	Port salut		Coulommiers

Significance of French Cuisine

France has great climatic variation to the south. The heat of the Mediterranean adds flavour and richness to fruits,

vegetables and herbs including the aromatic delights of basil and garlic which are the flavours of Sun. In the east, south-west and centre mountain massifs dictate a tougher, more solid peasant fare, based on chestnuts, cabbage and hard cheeses which can be stored through the winter with grace—notes of trout, crayfish and woodier mushrooms. In the north there is the finest of fish, oyster, lobster which are cooked to perfection in the fine cream and butter.

French cooking has two groups of basic ingredients:

First there are the fats—olive oil, goose fats and butter. Second are the basic flavourings of garlic, shallot and onion. The regions where olive oil is used are most firmly determined by the climate. To the west it overlaps with goose fat in the great foie gras regions. To the north it overlaps with butter, as one goes up the Rhone into the Lyonnais and Burgundy. Garlic and olive oil match as they grow abundantly in the Mediterranean region. The shallot is the great flavouring of Boredeaux.

Cooking of France is only fifty years old. It was discovered and recorded in the 1920s after the First World War. It dramatically changed European life. Montaigne who, in the 16th century, travelled in his own country and through Switzerland, Germany and Italy to record the things he ate as well as the things he saw and experienced. In France cooking is influenced greatly by the commodities available locally, from town to town, even from village to village, local feeling and taste will ensure a variation of cheeses, wine charcuterie cakes and even of bread. At small roadside cafes the menu will vary according to what is available locally. To travel through this country is a gastronomic adventure. The following are some important regions and their traditional dishes:

Brittany, Touraine, Angou and Maine: Brittany is for fish, Oysters lobsters and langoustines are expensive but spider crabs, clams, mussels and scallops are abundant *e.g.* Mackerel a l'americaine, Matelote, Rillettes, Beurre Blanc, etc.

Normandy: Filets de sole normande, Canard a la bigarde, chick with cream and calvados.

Champagne, Picardy, Flanders, Artois: Filet de soles au champagne. He de France, Paris, Orleanais: Potage crecy, Tournedo Rossini. Alsac, Lorraine, Franche-Comte: Quiche lorraine, Le poule, au vinaigre.

Bvirgundy, Nivernais, Berry, Bourbonnais: Boeuf a la bourguignonne poularde derby.

Lyonnaise, Savoy, Auvergne, Dauphine: Soupe de cantal, gratin Savoyard.

Charentes, Poitou, Limousin, Marche: La Rochelle fish chowder, cuisse de grenouilles a la maraichine.

Guyenne, Gascony, Bearn: Poulè au pot (Boiled chicken Bearn Style), Haricots Blanc au beurre de Gascogne (use goose fat).

Langue doc: Cassculet.

Provence and Corsica: Bouillabaise (fish stew with saffron).

Significance of British Cuisine

Britain has been producing and exporting the best of venison, pork, beef, lamb, lobsters and scallops to France. Britain's own native seakale, the most delicious vegetable as well as globe and Jerusalem artichokes, courgettes, aubergines and peppers are grown in abundance in addition to delicious fruits like pears and grapes. The real pleasure of English, Welsh, Irish and Scottish eating is a pleasure on a Sunday lunch for the family, a weekend high tea or a dinner party with friends. Salmon, plaice or scampi are available in abundance. Puddings, for centuries the national theme are good examples. One can try the art of cooking in steak, kidney and oyster pudding, sussex pond pudding, yorkshire pudding.

To end the dinner one can try an apple pie, gooseberry fool. Following are some famous dishes:

Cockie-leekie: A Scottish dish, a beef stew with prunes.

Smoked Finnan-Haddock with egg sauce: "Finnan" correctly, Findon, is a village near Aberdeen, famous for its cured haddock, by oak smoke.

Potted Crab: Preserving meat and fish in butter has always been a peculiar English habit, served as a first course with toast, or as a lunch dish to be followed with a green salad. Salmon in pastry with herb sauce.

Steak Kidney and Oyster Pudding: A classic English dish from Wales.

Shepherds pie, Lancashire hot pot.

Irish stew: a stew from Ireland with meat and potato.

Roast beef, yorkshire pudding: best roasted meat with a delicious crisp better pudding prepared by the housewives in the north of England. Cheddar Cheese from Wales.

Haggis: Stuffed sheep's stomach a famous dish from Scotland. Bakewell tarts: from England.

Mince Meat and Pies: Everybody in England eats mince pies at Christmas. There are slight variations in flavour from household to household. The best mince meat includes hard liquor, apart from the steak, is the one used by most people.

Original English drink is London Dry Gin. Scotland's contribution is Whiskey Irelands' contribution is stout beer.

Significance of Mexican Cuisine

Mexican cuisine is a delicious experience and worth trying. It is one of the world's oldest cuisines, its true origins are with Mayan Indians, whose civilization rose and fell long before the Spaniards visited Mexico. The best historical reference to Mexican cuisine is records kept by the Spanish conquerors of what they found in the New World. The Spaniards were welcomed into the courts of the emperor Montezuma and found great riches as well as wealth of culture. Great feasts

were served each night, composed of many dishes. The dishes were kept warm on charcoal braziers for the guests to sample. Tortillas were made from cornflour in many different shapes and sizes. Dishes featuring wild game and elaborate sauces were also served. Beans and chillies were used extensively and many kinds were cultivated. Chocolate and fruit drinks were also served toquench the thirst. Tropical fruits, papaya, banana, avocado and exotic guavas were served.

The short reign of the French emperor Maximilian and the empress Carlota brought new dimensions to Mexican cuisine. Maximilian brought chefs of Austria, France and Italy to cook for his court. The native population borrowed from the continental cuisine and added vermicelli, noodles, white continental bread and other delicacies to the native dishes. During the World War II Germans introduced them to beer, nowadays the beer of Mexico is recognised as one of the finest in the world. A dish, for example, tamales may be cooked in different ways through the country and still be called Mexican. The basic of Aztec times still remains in the recipes, corn beans and chillies are still the life of Mexican cuisine. Tortillas are still made from corn in the traditional way, patted and shaped by hand and cooked on a hot comal (griddle) before being fried as totopos, filled with beans, meat, fish, vegetables or cheese or just served hot as an accompaniment to every meal. The same dough of tortillas are also vised for making tamales—to thicken sauces and as the base of a gruel like beverage known as Atole. Beans are cooked in earthenware pots and the eaten sparked with chilli. Chillies are grown all over Mexico in 61 different varieties. Chicken, pork, turkey, eggs and cheese are used frequently in the diet. Beef is used in Northern Mexico and fish in the coastal areas. The sap from the Maguey Cactus is fermented to form the local beverage Pulque. Tequila is a famous drink from Mexico. There are 5 recognised meal times each day, few people can afford or desire to eat all 5 meals.

Desayuno—early breakfast, consists of rolls, coffee or chocolate.

Almuerzb—brunch consisting of fruits, bread or tortillas, eggs or meat and coffee.

Comida—main meal, consists of 6 courses, appetizers followed by soup, a rice dish, chicken or fish, the entree beans served as a separate course, dessert and coffee.

Merienda—served at 6 p.m., sort of high tea, consisting of coffee or chocolate and cookies.

Cena—a supper served late and is very light.

Enchiladas—chicken breast rolled in Tortilla. Arroz Roso—Red Mexican rice. Cocada—Coconut and milk pudding. Crepas de cajeta—Burnt milk crepes.

Significance of Items for Drink

Wine gives a luxurious quality to many dishes. In general an inexpensive red wine or a white wine that is not too dry will be suitable for all dishes.

It is always safe to use red wine for meat and game recipes. White wine can be used for fish or any pale or delicately coloured dish. White wine may be added to stock to enhance the flavour. Red wines make a dish richer and gives a dark delicious gravy.

The Golden rule is "It's better to use too little than too much." Twice the amount won't make the dish twice as good, on the contrary may be nasty.

Fortified and Sweet Wines in Cooking: Traditionally Port is used in jugged hare and Madeira or Marsala in cooking of kidneys. Fortified wines should be used discretely as their alcoholic content is higher, to reduce their alcoholic content the dishes should be cooked longer.

The addition of fortified or sweet wines to fruits are very pleasant. These can be used for baked or stewed fruits or can be added to fruit salads or pour over trifles. Gin can be used in chicken dishes. Rum is used with fruits and raisin. Brandy is also frequently used for cooking game, chicken and birds.

Brandy is added to pates, terrines, it can be used to preserve fruits *e.g.* cherries, apricots, peaches and plums, used for moistening and enriching the mixture, it can be used in cakes, mousses etc. Just pierce the surface and inject some—this is called spiking. Liqueurs can be used for spiking, can be used for flambe, it goes well with fruits and plain ice-cream.

Bitters—normally a cocktail ingredient have a little role in cooking. Angostura bitter imparts a slightly spicy herb like flavour to the dish, it can be added to sandwich spreads and savoury butter, to salad dressing, to mayonnaise and in stewed fruits or compotes.

Cider can be substituted for white wine in many dishes especially white wine dishes with fish. Do not mix cider with cooking wine.

Traditional Character of French Menu

Hors D' Oeuvres: It is the first course of the meal. This course is composed usually of tangy, salty, sour, hot-natured items which help stimulate the appetite. The hors d'oeuvre should have eye appeal, should be decorative. The term hors d'oeuvre usually applies to a variety of side dishes offered as appetisers such as Potato salad, Anchiovies, Prawns, Olives, Russian salad, Herring, Sardines, Cold Egg dishes, Mushrooms, Artichoke, Asparagus etc. and also to single items served as a preliminary appetiser course before the soup. For *e.g.* dishes like melon, caviar, oysters, smoked salmon, salami, sausage. Smoked ham and pate de foie gras, when served before a rather heavy menu consisting of number of courses, should be light and very delicate. With a light lunch hors d' oeuvre may be somewhat more substantial and nourishing.

This course originated in Russia where the guests helped themselves to small dishes prior to sitting down at a table for their meals. Now it is popular all over the world and appears even on a simple menu. The function of this course is to stimulate the appetite by producing more saliva and other digestive juices in our system. This course stimulates the pallet so the dishes are attractive colourwise. A good hors d'oeuvres

service can increase the turnover of the restaurant by producing a psychological effect.

France has also adopted a custom imported from Russia of serving hors d' oeuvre as Zakuski. Zakuski is kind of a meal before meal, made up of hors d' oeuvres washed down with liberal draughts of wine and liqueur. Formerly in Russia, this was served in the ante-chamber adjoining the dinning room.

Hors d' oeuvres are divided into two main categories (1) Cold and (2) Hot. The cold hors d' oeuvre are further divided into (a) Simple and (b) Deluxe. Formerly hot hors'd oeuvre were called Entrees Volantes or Petit Entrees.

Hors d' oeuvres are served cold excluding meat balls, cheese, balls, pakora, kababs etc. It can be put in a 8 compartment or 5 compartment hors'd oeuvre tray with a dessert spoon and dessert fork. The cutlery required is a fish knife and fish fork and is served on a cold half plate. In case of hot hors doeuvre, a hot half plate is given along with a small knife and small fork. Some special items are served in special dishes.

Examples of hors' d oeuvres are as follows:

Salad: Dishes made up of herbs, plants, vegetables, eggs, meat, and fish seasoned with salt, vinegar and pepper with or without other ingredients. According to Brillat Savrin, salads *"Freshen without enfeebling and fortifies without irritating"*. Salads are further sub-divided into two categories (a) Simple and (b) Compound.

(a) *Simple Salads:* Plain salads could be green salads (raw) or salads of cooked vegetables consisting of one kind of vegetable.

Raw Salads (A Servir Crues)

1. Chopped Celery with vinaigrette dressing.
2. Green Cabbage (the tender parts cut in julienne) with mustard cream.

3. Turnip tops (Green) vinaigrette.
4. Leeks (The green part) vinaigrette.

Plain Salads made from Cooked Vegetables

1. Artichokes (Bottoms) vinaigrette.
2. Aubergine (Round egg plants blanched in salt water)—Mustard Cream (2: 4 ratio of mustard and cream seasoning and few drops of lemon juice.)
3. Beetroot roundals—vinaigrette dressing.
4. Lettuce (Blanched and drained) vinaigrette.
5. Leeks (White part cooked in water)—Mayonnaise.

Compound Salads (Salades Composes)

1. *Salade Americaine:* Cucumber cut in thin slices, previously soaked in salted water then rinsed and seasoned with vinegrette, surrounded with lettuce hearts with slices of tomatoes and quarters of hard-boiled eggs.
2. *Salade Andalouse:* Boiled rice, seasoned with oil, vinegar, salt and paprika, chopped onion and parsley and a little grated garlic. Arranged alternate heaps of julienne of sweet pimentoes and quarters of tomatoes sprinkled with chopped chervil (herb).
3. *Salade Favourite:* Asparagus tips, crayfish, sliced white truffles seasoned with oil, lemon juice, salt and pepper. Sprinkled with celery and herbs.
4. *Salade Japonaise:* Pineapple, tomatoes cut into square pieces, orange segments mixed with acidulated cream.
5. *Salade Mikado:* Patatoes cut into dices, mixed with shrimps, bound together with mayonnaise, flavoured with soya sauce.
6. *Salade Nicoise:* Dices of potatoes, french beans bound with vinaigrette dressing, decorated with fillets of anchovy, olives and capers. Garnished with quarters

of tomatoes, sprinkled with chopped chervil and tarragon.

7. *Salade Orientale:* Boiled rice, finely chopped onion, quartered tomatoes, black olives, mixed with oil, vinegar, salt and paprika. It is surrounded with sweet peppers sprinkled with chopped chervils.
8. *Salade Russe:* Macedoine (dices) of boiled seasonal vegetables like peas, carrots, turnips, potatoes etc. bound with mayonnaise.
9. *Salade de pomme de terre:* Cubes of boiled potatoes with chopped shallots, dressed with vinaigrette.
10. *Salade Astoria:* Grapefruit and orange segments with sliced red and green pepper served on a bed of lettuce leaf with french dressing.
11. *Salade Diplomate:* Juliennes of pineapple, celery, apple and chervil dressed with creamed mayonnaise.
12. *Salade de pomme:* Dices of apples bound with mayonnaise garnished with chopped parsley.
13. *Salade Waldorf:* Dices of apple, celery and walnut bounded with mayonnaise sauce.
14. *Salade Viande:* Brunoise of meat with chopped shallots, hard-boiled egg, dressed with vinegrette and garnished with finely chopped parsley.
15. *Salade Imperial:* Lettuce leaves, filled with asparagus tips and truffles.

Shell Fish: The shell fish served are Oysters, Snails, Prawns, Crayfish, Shrimps, Scamp, Mussels etc. These could be served assorted from an hors d' oeuvre tray. Cutlery is given accordingly. For *e.g.* Prawn cocktail is served in a cocktail cup along with a teaspoon and quarter plate as an underliner. Lobster is served in a half plate with lobster cracker and lobster pick. Snails are served in a snail dish with a snail fork and snail tongs. Oysters are served live with an oyster fork.

Eggs: Eggs could be served cold as an hors d'oeuvre. Some examples are:

1. *Oeuf dur Mayonnaise:* Hard-boiled eggs cut into halves coated with mayonniase.
2. *Oeuf Farcis:* Hard-boiled eggs, take out yolk, mix with the desired stuffing and pipe out into white of an egg.
3. *Plover Eggs:* These eggs are roughly the same size as those of a pigeon, have a light green coloured shell with black spots. When cooked, the white of the eggs takes a milky tint and is not quite as firm as the white of other eggs.

Plover eggs are usually served hard-boiled and cold. The merits of this hors d'oeuvre lies mainly in the rarity of the eggs and their high price. Other eggs which could be served are that of pigeon, quail, partridge, seagull and others.

American Relishes

1. Various types of fruits and vegetables with vinegar. Seasoned with cinnamon and cayenne pepper.
2. Tinned small finger-sized sardines are served filleted or whole decorated with lemon or parsley.
3. Anchovy fillets in oil or with hard-boiled eggs accompanied with a salad (Lettuce chiffonade, dices of boiled potatoes, celery cut into fine julienne).

Oignon (Onions)

1. *Oignon Escoffier:* It consists of button onions, sauteed in oil, cooked with garlic-flavoured tomato puree and reduced vinegar, flavoured with bay leaf, peppercorn and garnished with sultanas and served chilled.
2. *Oignon a La Greque:* Boiled in water with lemon juice, a pinch of salt, coriander seeds, peppercorns and bouquet garni, drained and served cold. Bouquet Garni is made up of thyme, parsley, bay leaf.

Olives: Olives are served cold on an hors d'oeuvre tray without dressing. Olives which can be served are, green, black and stuffed. Stuffing is made of red pepper or almonds.

Tomate Farcie (Stuffed Tomatoes)

Tomate a l' antiboise: Tomato pulp stuffed with tuna fish, hard-boiled eggs, capers, parsley, chervil and tarragon blended with strongly flavoured mayonnaise to which anchovy essence has been added and arranged on an hors d'oeuvre tray.

Spring Vegetables (Printaniere): Dices of french beans, carrots and turnips cut into very small squares, peas and asparagus tips cooked in water, drained, seasoned with oil, vinegar and salt and pepper.

Black radishes with vinegrette, dressing or sour cream and pickled gherkins can also be served.

Fruits: Various types of fruit served are:

Avocado Pear: It is an exotic sub-tropical pear-shaped fruit with a shiny green skin and has a soft watery pulp. It is cut into two halves and served. It has a nutty flavour and is thus served, dressed with other fruits and savoury dressings like french dressing, lemon, peas, mayonnaise, tomatoes and Thousand Island dressing. It is presented on a lettuce leaf with lemon slices and is served on a cold half plate with a teaspoon, a dessert spoon and fork.

Melon: Various types of melon are served:

(a) *Cantaloupe:* It is known by its rough rigid skin, green lines, soft and yellow pulp.

(b) *Honey Dew:* It has a firm, shiny, wrinkled skin with a sweet pulp.

(c) *Water Melon:* It has a red sweet pulp and black seeds.

Large melons are cut into small portions and served ice cold on a cold half plate with a dessert spoon and a dessert fork. Accompaniments are castor sugar and ground ginger which are served in a sugar dredger. Honey dew melon is served garnished with glazed cherries or a slice of lemon. Now ginger sugar is available commercially also.

Grapefruit: It is served with or without skin in a cocktail cup. The fruit is cut into two halves and the segments are

detached with a grapefruit knife and placed in the cup on a half plate and a doily paper, with a teaspoon. It is topped with a cherry and a dash of cherry brandy.

Fruit Juices: Assorted fruit juices are served in a pony tumbler underlined with a quarter plate, a daily paper and a teaspoon. Juices of tomato, carrots, orange, mango grapefruit, pomegranate, pineapple, sweet lime etc. are served.

Some mixed drinks like Kanji, Jaljeera, Panna are also served and are tangy and spicy. Various fruit juice cocktail are also served in pony tumblers castor sugar is usually passed in a sugar dredger as an accompaniment.

Mixed Fruit Juice Cocktail: Juices and segments of assorted fruits are served sprinkled with castor sugar and garnished with a cherry.

Grapefruit Cocktail: Juice and segments of grapefruit garnished with a cherry and castor sugar is added as an accompaniment.

Tomato Juice Cocktail: Tomato juice is served with three accompaniments:

1. Worcestershire sauce
2. Lemon wedges
3. Celery Salt
4. Tabasco Sauce (optional)

Sea Food Cocktail (Fruits De Mer): These cocktails are made of Prawns (crevette rouge), Cray fish, Shrimps, Lobster, Scampi, and other. It is presented in a stemmed glass, shell or a special silver dish which is lined with glass. The sea food is arranged on the lettuce, is masked with cocktail sauce, sprinkled with chopped parsely and garnished with a slice of tomato or olive and lemon slice. The underliner for the cocktail cup is a quarter plate and the cutlery is a teaspoon.

Canapes: Canapes are shaped pieces of bread approximately 6 mm (1/4 in thick) brushed over with melted butter and placed under a salamander and coloured on both sides. Could be shallow fried also.

Canape Baron: Slices of bread garnished with slices of fried mushrooms, grilled bacon and poached bone marrow.

Canape Ritchie: Creamed Haddock garnished with slices of hard-boiled egg.

Canape Quovadis: Grilled Soles garnished with small mushroom heads.

Canape Charlemagne: Garnished with shrimps bound with a curry sauce.

Angles on Horseback: Poached oysters wrapped in streaky bacon and grilled on skewers.

Devils on Horseback: Stone-cooked prune the inside stuffed with chutney and sprinkled with cayenne pepper. Wraped in streaky bacon and grilled on skewer.

Canape Nina: Half a small grilled tomato, garnished with mushroom heads and pickled walnut.

Galantine: A dish made from boned poultry or meat, stuffed and pressed into symmetrical shape. Galantine are cooked in a gelatine stock.

Vegetables

Asparagus: It is served on a large plate which is placed on an inverted large fork so as to tilt the plate towards one side, where the sauce which is vinaigrette or mayonnaise, collects at the tips of the asparagus. It is eaten with fingers. But a fork and knife is kept ready in case a customers asks for it.

Artichokes: It is served cold on a cold half plate with sauce vinaigrette or mayonnaise. When served hot, it is served on a hot half plate with melted butter cutlery given only when asked for. Eaten with fingers.

Note: A finger bowl with warm water and lemon slice must be provided for the above mentioned two dishes.

Savoury Jelly Or Aspic: It is obtained either from concentrated beef stock or by dissolving gelation in hot consomme. Generally, some meat, poultry, fish vegetables or

fruits are placed in small fancy bowls. Then hot aspic is poured over them and allowed tso set in the refrigerator. Then it is demoulded and served cold on a cold half plate. Cutlery required is a fish knife and fish fork.

Hot Hors D'oeuvre: These are quite popular and generally preferred in winter. Some examples are as follows:

1. *Beignets or Fritters:* A general term used to designate any fruit or mixture of food which has been dipped in batter and then deep fried.

   For *e.g.*

   (a) Pakoras (Indian)

   (b) Beignets de Cerevelle—Poached brain in vinegar and court bouillon cut into slices, marinated with oil, lemon juice. Dipped in butter and deep fried.

2. *Bouchees or Patties:* Small meat pieces, vegetables, minced poultry stuffed into the puff paste *e.g.* chicken patty.
3. *Brochettes or Skewers:* Small pieces of meat, vegetable, poultry pierced through a small thin iron rod and grilled on an open fire. *e.g.* Seekh kabab, Tikka etc.
4. *Croquette or Rissoles:* Cylindrical-shaped rissoles dipped in egg, bread crumbed and deep fried. Items used are potatoes, mixed vegetables, kidney, brain, minced meat, fish etc.
5. *Nalesnikis (Russian):* Pressed curd, cheese and butter rolled into pancake and batter fried.

Deluxe Hors D'oeuvre

Caviar: It is the roe of fish belonging to the Sturgeon family which has been freshly caught and sieved to clear it of the skin, veins and fat. This roe is then sauteed and packed in tins or glass jars or tubes. Caviar is also made from the roe of:

(i) Beluga

(ii) Sevruga

(iii) Sterlet

(iv) Starka

(v) Osetrova

The best quality of caviar comes from the Black Caspian Sea and the fish are caught while spawing in winter and spring when they swim up the river. The port of Astrakha exports the majority of the world's supply.

Beluga is of light grain and light colour. Severuga is of small grain and a little in darker in colour. Osetrova is of small grain and a little darker colour. Generally the best quality roes are lightly sauted and therefore do not preserve well. Although it is generally true that the lighter coloured Caviar is the best, the preparation and storage ultimately determine the quality. The best Caviar is obtained from Russia and Persia. Caviar being expensive should be stored at approximately 1°-2°C and should be handled carefully, using a horn, wooden or plastic spoon to transfer it to a smaller jar for service. Caviar should look bright and wholesome. It is packed in air tight jars. Pressed Caviar is the one which has been prepared by blanching and pressing and then packed. It is of inferior quality. It is also packed in air tight jars. It is served very cold.

The best quality is served directly from the tin or jars on a bed of crushed ice or surrounded by crushed ice. A portion is approximately 15 gms or 1/2 ounce. Accompaniments with Caviar are:

1. Hot thick toast and butter.
2. Blinis (50 paise sized Russian pancake).
3. Chopped parsley.
4. Chopped shallots.
5. Lemon wedges tied in a muslin cloth.
6. Special cruet set consisting of cayenne pepper and pepper mill.
7. The white of hard-boiled eggs and sieved yolk of an egg.
8. Caviar is served on a cold half plate with a gold-plated teaspoon and the cutlery given is a caviar knife (gold

plated). In the absence of a caviar knife, a fish knife may be given.

Caviars and their types

. ***Votariga—Boutargue:*** Red caviar is prepared from Salmon, Pike, Grey Mullet and other fish. The grain is much larger than that of real caviar and is pink to red in colour. It doesn't have the flavour of real caviar.

Dark Caviar/Substitute: A type made from various fish roe's and sold as cheap caviar, but very inferior in flavour.

Vesiga—Visigha: Vesiga is the dried marrow from the backbone of Sturgeon. When soaked it resembles sago cooked in water.

Pate De Foie Gras (Goose Liver paste): This expensive delicacy is obtained from the liver of specially flattened goose and is produced mainly in Strasberg. It is available either plain or with truffles in tins of various size and in brick shape, with a layer of truffles in the centre. It is also available in round pastry cases (Fois grass en coute) and in earthenware. Fois gras is the classic first course for any lunch, dinner or supper menu. It is also used as a garnish for some dishes like Tournedos, Rossini and is included in the stuffings for certain chicken dishes. A puree or mousse of fois gras is available and is suitable for sandwiches and certain stuffings. Accompaniments are:

(i) Hot toast and butter

(ii) *Truffles:* It is served in its original container on a bed of ice. Hot water in a jug is served with Pate.

Oysters (Les Huitres): An oyster is a shell fish enclosed in two shells, the top shell being flat. These shells are hinged together at a point by a muscle attached at the centre of each shell by means of which it opens and closes them.

Care should be taken to ensure that oysters are alive when purchased. If the shell is open, they should be discarded since the Oysters deteriorates rapidly.

If the Oysters contains a milky substance, this is known

as white sickness. If oysters are full of greyish liquid, it is grey sickness and if it is heavily flooding with black spots, it is black sickness. It means that the Oysters are breeding. Oysters under these conditions are not edible and should be returned.

It is important that the Oysters should be kept in wooden containers, deep shell downwards, covered with a wooden lid which must be weighed on to keep the oysters from gaping. It is stored in a cool damp room, preferably not in a refrigerator.

English Oysters are in season from September to April and are not in their best in other months as they are breeding. British Oysters are called Native. The best Caviar are the Royal Native or Royal Whistable Perfect, Pooles, Falumouth, Helford, Shanmov.

The best known continental oysters are French Marennes Verts, Marennes Blanche, Belons Caucale, Britavni, Dutch Belguan, Victoria, Portuguese.

The Oyster is opened with special care to avoid damage. Accompaniments with oysters are:

(i) Brown bread and butter
(ii) Cayenne Pepper
(iii) Pepper mill
(iv) Chilli vinegar
(v) Tabasco Sauce
(vi) Lemon Wedge

It is served in a soup plate on a bed of crushed ice with a large plate underliner. 6 to 8 oysters are served and the cutlery provided are oyster's fork placed at right side of the cover at an angle of 45°. Instead of the oyster's fork, a fish fork may be provided when not available.

Smoked Salmon (Saumon Fume): It is a famous Scottish delicacy. It is a sea water fish measuring about 5-6 feet. It is pink in colour. The fillets are paper thin. Smoked salmon is produced in Scotland, Denmark, Holland, Norway and other

countries. It is smoked by hanging in a room where cedar wood or saw dust is allowed to smoulder the fish.

It is served in a cold half plate with the fillets rolled on a fork and unrolled on the plate. Accompaniments are: (i) Brown bread and butter (ii) Half lemon tied in a muslin cloth (iii) Pepper mill and cayenne pepper (iv) Chilli vinegar

Snails (Escargot): Snail shell is opened and the meat is removed. It is cooked with butter and seasoned. The meat is ripped back into the shell. Six snails are served in a shell dish and hot garlic is poured on the top. The cutlery provided are snail tongs and snail fork. Button onions soaked in red wine.

Diagram of Tray Containing Special Equipments: Snail dish, butter dish, pepper mill, sugar dredger, nutmeg grater, oil & vinegar stand, carving knife & fork, lobster pick, snail fork, snail holder. (Clockwise from left)

Grapefruit spoon, lobster cracker. (In the centre)

Potage Or Soup: It is an extract of meat, vegetables, fish or poultry in stock or water. It is probable that soup in its earliest form was a complete meal, because it was found hearty, nourishing, economical and wholesome. The function of the soup is to supply immediate requirements of the nutrients in the body. Being in the liquid form it is absorbed quite easily and is utilised very quickly in our body. In the absence of hors d' oeuvre, it also helps to stimulate the appetite. Generally there are two types of soups on the menu, they are: Thick and Clear. When a menu contains two soups, one must be clear, the other thick. If only one is to be served, it may be either clear or thick.

Usually soups are served piping hot, but many soups can also be served cold. Clear soups in particular can be served in jellied form. The cutlery required for clear soups is a dessert spoon. It is served in a consomme cup with a half plate and consomme sauces. The cutlery for thick soups is a soup spoon and is served in a soup plate with a large plate as an underliner. It is served from a soup tureen poured into the soup plate

away from the guest. Nowadays a common soup bowl with underliner is used for both types of soups.

Major Accompaniments: Fried bread croutons and sippets are served as an accompaniments with tomato soup. Vegetable garnishes are added to clear, cream and veloute soups. Parmasean cheese is served with some bouillons and with many Italian soups.

Cheese straws are served as accompaniments with turtle soup and Potage Germaine. Slices of poached beef are served as garnish in some bouillons.

Wine and Fortified Wines in Soup: Fortified wines may be used to finish certain soups. They are added at the time of service.

Dry sherry is added to turtle soup, at time of service.

White wine, Red wine and Maderia may be added to soups while cooking.

Brandy is generally used in making Shellfish soups.

Classification of Soups

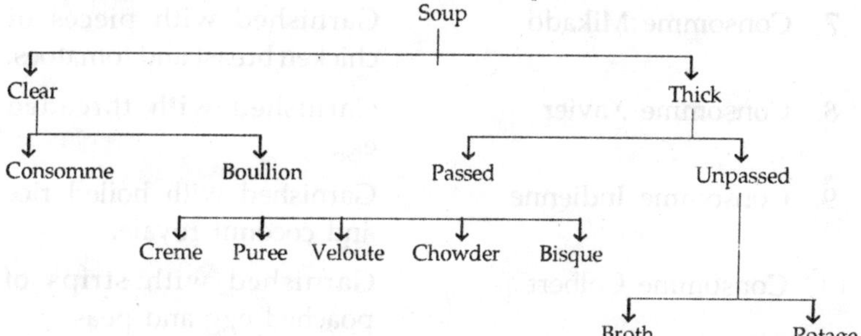

Clear Soups

(i) *Consomme:* It is prepared by clarifying stock of fish, poultry, beef or mutton etc. It is flavoured with bouquet garni. The clarifying agent is the white of an egg. It is amber-coloured, transparent and crystal clear in appearance. It is completed at the point of service with a garnish that may consist of cuts of vegetables, savoury egg custard, small meat or fish quenelles (small balls), shredded savoury pancakes and many other items. Some consommes are thickened with tapioca.

Hot Consomme

1. Consomme Brunoise	Garnished with dices of vegetables, like carrots and turnips.
2. Consomme Julienne	Garnished with juliennes of vegetables, like carrot and turnip.
3. Consomme Celestine	Garnished with strips of pancakes.
4. Consomme Royale	Garnished with small cubes of savoury egg custard.
5. Consomme Andalouse	Garnished with ham juliennes, boiled rice, Royale custard, strips of tomatoes and vermiceilli.
6. Consomme Carmen	Garnished with tomato juliennes, capsicium and boiled rice.
7. Consomme Mikado	Garnished with pieces of chicken breast and tomatoes.
8. Consomme Xavier	Garnished with threaded egg.
9. Consomme Indienne	Garnished with boiled rice and coconut royale.
10. Consomme Colbert	Garnished with strips of poached egg and peas.
11. Consomme Crecy	Garnished with juliennes of carrots.
12. Consomme Favourete	Garnished with julienes of artichoke, potato balls, mushroom and thickened with tapioca.
13. Consomme Waldez	Garnished with natural royale, peas royale and carrot royale.

14. Consomme Florentine — Garnished with strips of blanched spinach.

15. Consomme St. Germain — Garnished with fresh green peas.

16. Consomme Dubary — Garnished with flowerets of cauliflower.

17. Consomme A'L' Ambassador — Chicken consomme thickened with tapioca, garnished with chicken quenelles (small bells of meat/fish) and profite roles filled with puree of foie gras and chervil.

Cold Consomme

1. Consomme A'L' Essence: — Cold consomme flavoured with chopped celery. de Celeri
2. Consomme: — Cold consomme flavoured with tomato, garnished with strips of tomatoes, vermicilli and red pimentoes and sauted sorrel.
3. Gellide Volaille: — Chicken flavoured cold consomme.
4. Vichy Choise: (Spanish): — Cold potato (cold) soup.

Bouillons: A bouillon is an unclarified meat or poultry stock or a combination of two. It should be amber in colour and clear in appearance with small pieces of vegetables, beef and chicken served as a part of the complete dish and added at different stages in a cooking process. For Example:

Petite Marmite: The consomme simmered with beef, celery, carrots, cabbage, leeks, winglets of chicken and turnips garnished with beef, served in marmites (earthenware pot)—

country soup made in France, are served in the container in which it is cooked, called Marmite.

Thick Soups

Passed Soups

Puree: It is a thick passed soup with milk and butter. Puree soups are made from dry pulses or fresh vegetables cooked in water then passed through a sieve, soup strainer or liquidiser. Vegetables which contain a high level of starch will thicken by themselves, those with a low level of starch require additional thickening agent, such as potatoes, or rice. Puree soups are generally garnished with bread croutons or bread snippets are as follows:

Puree Egyptianna: Made out of yellow Egyptian peas with a liaison of egg yolk and cream.

Puree Solferino: Half parmentiere (potato), half tomato soup garnished with carrot balls and potato balls.

Puree Parmentier: Made of potatoes finished with milk and butter, garnished with croutons.

Puree Malakoff: Puree of tomatoes and potatoes, finished with milk and butter, garnished with croutons and julienne of spinach.

Puree Pompadour: Tomato puree, garnished with sago and juliennes of lettuce.

Puree Saint Germian: Puree of green peas finished with milk, butter and cream, garnished with croutons.

Cremes: Thick passed soup finished with cream. Examples are:

Creme d' asparagis: Blanched asparagus cooked with Bechamel and finished with cream.

Creme De Crecy: Carrot soup thickened with bechamel and finished with cream.

Creme Danoise: Soup made from cucumber and roux, finished with cream, garnished with cubes of cucumber.

Creme De Volaille: Roux cooked with chicken stock, garnished with small cooked pieces of chicken and finished with cream.

Creme De Volaille Princess: Roux cooked with chicken stock, garnished with chicken and asparagus tips.

Creme Longchamp: Purbe of fresh peas garnished with vermicelli, shredded sorrel and chervil, finished with cream.

Veloute: It is a mixture of vegetables puree meat and veloute sauce. *E.g.* Chicken veloute, Celery veloute, Almond veloute, Veloute Carmen (tomato flavoured veloute garnished with strips of pimentoes).

Chowder: Originated in America. These are thick heavy soups owing to their consistency due to potatoes. Examples are:

Clam Chowder: It is made from diced salted pork, onions, potatoes, tomatoes, celery and green pepper and clams. Finished with thyme, parsley and butter.

Seafood Chowder: Soup made from diced salted pork, onions, potatoes, tomatoes, celery, green pepper and seafood. Finished with thyme, parsley and butter.

Oyster Chowder: Soup made from diced salted pork, onions, potatoes, tomatoes, celery, green pepper and oysters, finished with thyme, parsley and butter.

Bisque: Soup prepared from shellfish like lobster, prawns, oysters etc. It is thickened with liaison of egg yolk and cream for e.g.: Bisque d' Homard.

Lobster Bisque (Bisque De Homard): It is made up of fish stock lobster and fish veloute and flavoured with root vegetables and tomato puree, finished with a liason of egg yolk and cream.

Convenience Soups: A wide variety of soups are available in convenience form. They may be canned, frozen, or powder soups. They are classified exactly in the same way as the fresh soups, though the finished product may well have been

produced by a totally different method. Any instructions given by the manufacturer for the production of these soups, however, should be followed. Additional garnish can be put at the time of service, such as cooked juliennes of vegetables, cooked chicken or cooked rice or barley. All this adds variety, texture and colour.

Unpassed Soups

Broth: A broth soup is a brunoise of various kinds of vegetables cooked in, chicken or gamestock, with the addition of cereal and garnish with meat or poultry according to the type of broth. The type of stock used is indicated by the name of the broth being made. Chicken broth is made of white chicken stock and mutton broth is made with white mutton stock. All vegetables for broth soups are cut into brunoise. Any additional cereal thickening such as barley or rice is cooked and added to the broth at the completion stage for *e.g.* Scotchbroth, Game broth, Mutton broth, Chicken broth.

Potage: This unpassed soup is generally farinacious in nature and does not require any thickening agent.

Potage Minestrone: It is an Italian soup made of chopped onions, white of leeks, all sorts of vegetables like tomatoes, dices of french beans, tossed in pork fat, moistened with water or consomme. Spaghetti, rice, peas, garlic, parsley and chopped basil Parmason cheese is added at the time of service.

Potage Paysanne: Paysanne of vegetable like onions, carrot, turnips, leeks, celery, french beans and peas tossed in butter, cooked in stock with bonquet garni, garnished with chopped parsley.

INTERNATIONAL SOUPS (Potages Grangers)

Minestrone: (Italian) Made of chopped onions, leeks and vegetables tossed in pork fat with water and cooked consomme, or with tomatoes, dices of beans, spaghetti, rice, peas, garlic, parsley and chopped basil.

Turtle Soup: (England) Consomme of beef, chicken and turtle, flavoured with basil, marjoram, sago, rosemary, thyme, coriander, peppercorns, bay leaf and thickened with arrowroot, garnished with warm cherry.

French Onion Soup: Consomme garnished with fried onions served with melba toast with cheese gratinated under the salamander (Soupe A l'Oignon)

Cock-A-Leekie: (Scotland) Chicken and veal consomme garnished with juliennes of leeks, prune and chicken.

Mulligatawany: Broiled spices cooked in water with tomatoes, onion, served with rice, sliced onions and lemon wedges. In case of non-vegetarian, it is garnished with small pieces of mutton which is cooked with the soup.

Bortsch Polanaise: (Poland) Duck flavoured consomme, strongly flavoured with parsley, roots vegetables, celery, fennel, marjoram, peppercorns. Garnished with juliennes of leeks, carrots, beetroot, cubes of beef and duck. Accompaniments are:

(i) Warm Beetroot Juice

(ii) Sour cream

(iii) Bouchees stuffed with duck force meat.

Bortsch Koop: (Russia) Julienes of leeks, carrots, onions, celery, beef, moistened with consomme. Garnished with minced beetroots and cereal with bouchees filled with chicken, forced meat and beetroot juice.

Hongroise: (Hungarian) Dices of beef seasoned with paprika, tossed in butter with onions, and flour, moistened with consomme and garnished with/dices of potatoes, pinch of fennel, crushed garlic and fried croutons.

Biere: (German) Veloute with beer, salt, sugar, pepper, cinnamon, garnished with toast.

Gazpacho: (Spain) It is a cold vegetable soup of cucumber, tomato and capsicum, garnished with dices of cucumber, tomato and capsicum.

Olla Podrida: (Spain) Thick soup made out of black beans.

Poison: Most fish are edible and the fish world is an enormous source of food. The most nourishing fish are river water fish such as Heel, Lamprey, Salmon, Trout, Mackerel, Turbot, fresh Herring. Among the least nourishing fish are Brain, Sole, Lemon Sole.

Fish have some proportion of protein as meat but the nutritional value of 500 gms of fish is generally equivalent to that of 300 g of meat. It contains significant amount of vitamin A and vitamin D. It is rich in phosphorus and iodine, and contains less fat than meat.

The water content of all fish is 75%, albuminides constitute about 18%, the fat content varies from 25-26%. The flesh of fish contains less nitrogenous substances than other meat and is easily digestable, therefore good for patients and children. One should be careful as far as freshness is concerned. The predominant smell of fish can be diminished by adding a little vinegar or white wine to poaching stock or court boullion. Fish is stored at a temperature of -18°C and thawed at 70°C.

	Different	***Types of Fish Available***
	Oily Fish	
	English Name	French Name
1.	Anchovy	Anchois
2.	Eel	Anguille
3.	Herring	Hareng
4.	Mackerel	Maquereau
5.	Salmon	Saumon
6.	Tuna	Thon
7.	Trout	Truite
8.	Sardines	Sardines
9.	Sprats	Sprats
10.	Pilchard	Royan

White Flat Fish

	English Name	French Name
1.	Brill	Barbue
2.	Sole	Sole
3.	Pomfret	Stromatee
4.	Turbot	Turbot
5.	Common Skate	Raie
6.	Hake	Collin
7.	Plaice	Carrelet

White Round Fish

	English Name	French Name
1.	Cod	Carbillard
2.	Hake	Collin Merluche
3.	Lamprey	Lamproie
4.	Gudgeon	Goujen
5.	Balluga	Beluga

Shellfish

	English Name	French Name
1.	Lobster	Homard
2.	Prawn	Crevette rouge
3.	Shrimps	Crevette gris
4.	Oyster	Huitre (Huitre)
5.	Mussels	Moule
6.	Crab	Crabe
7.	Scampi	Langoustine
8.	Snails	Escargots

Classes of Fish: Fish may be classified into two major categories:

1. *Vertibrial:* It is covered by scales.
2. *Invertibrial:* Shellfish covered with shell. Further classified into two categories: (a) Molluses (b) Crustaceans
   (a) *Molluses:* Have soft structure, partially or fully enclosed in shell which is largely made up of mineral composition *e.g.* Oysters, Clams, Mussels, Scallop.
   (b) *Crustaceans:* Have brush-like shells and have a covered body *e.g.* Lobster, Crab fish, Cray, Shrimps etc.

Composition and Nutritive Value: Lean fish has less than 2% fat that is edible. Medium fat fish has 2-5% fat contents only. Oily fish has more than 5% fat. Fish is usually more pigmented and has very little connective tissues present in the tail.

Calorie Count of a Fish Dish

Food energy	140 K. Cal.
Protein	17 gms.
Fat	5 gms.
Iron	10 mg.
Thiamine	0.03 mg.
Niacin	2.7 mg.

Whole fish is an excellent concentrated source of high quality protein.

Fat: While spawning, fish fat is transferred from the fish to roe in the spring season.

Carbohydrates: Shellfish have carbohydrates in the form of glycogen. The sweet taste of various shellfish is also due to glucose production by the enzymes action on glycogen.

Mineral: It is higher than in meat. Oysters are rich in iron and copper.

Fish Roe: Roe is a mass of fish egg, consisting of sacs of connective tissue.

Cured Fish: Fish is cured to preserve it by smoking.

Canned Fish: Generally fish like Salmon, Sardine, Shrimp, Tuna are canned.

Cuts of Fish

Filet (French)/Fillet (English): Refers to the flesh of fish. Flat fish yields four fillets. *e.g.* Soles, Pomfret, Brill. Round fish yield two fillets *e.g.* Cod, Whiting, Salmon, Herring. They are suitable for poaching.

Darne: A cut/slice from the middle of a round fish.

Supreme: This term generally applies to fillets of large fish cut in small pieces. They are suitable for poaching, shallow frying, deep frying, grilling or baking.

Goujon and Goujonettes: This term refers to fillets cut into small strips of 6.8 cm x 1 cm.

As the name implies Gojuonettes are smaller versions of Goujon and are cut into strips of 3.4 cm x 5 cm.

Troncon: It refers to a slice weighing 180-250 g cut to the bone from a large flat fish *e.g.* Halibut, Turbot, Brill. They are suitable for brilling and grilling.

Delice: Fillet of flat fish; the quarter of the head end and quarter of the tail end of the fillet are folded under the skin, the side being folded inwards. Suitable for planning.

Paupiette: Paper-thin fillet of fish, stuffed and rolled.

Fish Menu Examples Poached Fish

Filet De Sole Au Vin Blanc Glace: (Poached fillet of Sole glazed with white wine sauce) Poached fillet of Sole in fish stock flavoured with chopped shallots, and finished in the

oven. The sauce is prepared from poaching liquor, white wine, butter and cream. The dish is glazed under the salamander.

Delice de Sole Veronique: Poached fillet of Sole in white wine sauce, garnished with grapes.

Supreme de Cabillard Duglere: (Supreme of Cod Duglere) Fish is poached with chopped shallots, parsley, and coated with bechamel sauce, sprintled with cheese and dotted with butter and gratinated. Then garnished with tomato concuss.

Filet de Stromatee Bonne Femme: Fillet or pomfret with fish veloute and mornay sauce, gratinated under the salamander and garnished with sliced mushrooms.

Supreme de Carbillard Mornay: Poached fish flavoured with chopped shallots coated with mornay sauce and gratinated under the salamander.

Cold Fish Mousse: Minced preparation made by using cold pre-poached Salmon or boiled Lobster covered with aspic jelly made up of fish stock.

Deep Fried Fish

Filet de Stromatee A L'Orly: Fillet of fish marinated with lemon juice and seasoning dipped in yeast batter, deep fried, garnished with lemon wedges, pickled and served with tomato sauce.

Filet de Stromatee A L'Anglaise: Fillet of Pomfret marinated with lemon juice and seasoning, dipped in egg, rolled in bread crumbs and deep fried to a golden brown colour. Garnished with lemon wedges and parsley. Served with tartar sauce.

Filet de Sole Frit A La Francaise: Fillet of fish dipped in milk, passed through seasoned flour, deep fried to a golden brown colour, garnished with lemon wedges and parsley.

Shallow Fried

Filet de Stromatee Meuniere: Fillet of Pomfret marinated with lemon juice passed through seasoned flour and shallow fried, served with melted butter (Beurre noisette.)

Truite Meuniere Au Amande: Shallow fried Troutes with almonds, garnished with almonds strips, added to the nut brown butter.

Steamed Fish: All fish dishes that are prepared by shallow or deep frying may alternatively be steamed. The fish should be covered with buttered grease proof paper to avoid discolouration and to prevent moisture from the steam diluting the cooking liquid.

Grilled Fish: Grilling is cooking of fish in the griller with fat or oil to prevent sticking or burning. When cooked, grilled fish should be served with a garnish of lemon wedges, parsley, butter and parsley.

Examples of Grilled Fish

Sole Grillee: The fish is passed through seasoned flour and grilled. Then garnished with sliced lemon and served with parsley butter.

Filet de Sole St. Germain: Fillet of Sole, passed through seasoned flour, melted butter and bread crumbs, The fillets grilled until light golden in colour. Garnished with buerre noisette, potato and sprig of parsley. The sauce Bearnaise is served separately in a sauce boat. (Bearnaise is a derivative of Hollandaise with chopped tarragon and chopped chervil).

Filet de Sole Caprice: Fillet of Sole passed through seasoned flour and butter, grilled, garnished with banana halves, served with sauce Robert.

Shellfish Preparations

Homard Americane: The flesh of lobster is cut into pieces, seasoned with salt and cayenne pepper and fried in butter with chopped onions and shallots. Moistened with burnt brandy and white wine and fish stock. Chopped tomatoes and puree of tomatoes are added. Cooked for 20 minutes, the cooking sauce is reduced and poured over the Lobster pieces.

Homard Cardinal: The lobster is split, the flesh removed and cut into slices and cooked with butter and seasoning. Then

the cooked slices are put into the shell, mixed together with American sauce, sprinkled with grated cheese and glazed in the salamander.

Homard Newburg: Cooked in court bouilion, meat removed from shell and cut into regular slices. The slices are cooked in butter, moistened with brandy and Madiera and coated with liasion of cream, egg yolk and brandy, and gratinated.

Homard Thermidor: The cooked lobster are dressed and coated with mornay sauce, sprinkled cheese and glazed in the salamander.

Indian Fish Preparations

Tandoori Lobster: Lobster marinated with ginger, lemon juice, ajwain, garam masala, garlic paste and seasoning. The lobster is skewered and roasted in a clay oven, served with mint sauce and lacha onions and lemon.

Tandoori Pomfret: Three deep incisions are made on each side and marinated with egg, curd, ginger-garlic paste and seasoning. Then skewered and cooked in a moderately hot clay oven, served with lemon wedges, mint sauce and lacha onions.

Kadai Jhinga: Ajwain flavoured prawns with tomato chutney, finished with garam masala, kasuri methi and lemon juice.

Goan Prawn Masala: Prawn cooked in a gravy of onions, cumin seeds, cloves, cinnamon, ginger-garlic paste and vinegar.

Goan Fish Curry: Fish cooked in a gravy of onion, tomato, ginger-garlic paste, tamarind and coconut milk.

Amritsari Machli: Deep fried fish, marinated with lemon juice, seasoning, ajwain, ginger-garlic paste, orange colour and deep fired at the time of service and served with, lemon wedges, lacha onion and split green chillies. (Chaat masala sprinkled in the end.)

Service of Poisson Course: It is served in a hot half plate and the cutlery provided is a fish knife and fish fork on the

cover. If fish is served as a main course, then it is served on a large plate with accompaniments and the cutlery provided is fish knife and fish fork.

Entree: The word entree literally means 'Entrance'. This course was initially served before the roast course or Roti, but now it is served before the main course or Releve.

Food items that may be served as entree are:

1. All Hot Hors D' Oeuvres can be served as entree.
2. Cereals can also be served as entree *e.g.* Riz a 1' Italianne. It is made of boiled rice with mushrooms, chopped onions and tomato puree.

Different Kinds of Spoons Used in an 11 Course Meal

3. *Eggs:* Different preparations of egg can be served as an entree. For example:
   (a) *Spanish Omelette:* A fluffy omelette with red and green pepper (capsicum) and french beans.
   (b) *Creamed Eggs:* Fried eggs, put in a dish, covered with cheese, topped with cream and served in the same dish.
   (c) *Scotch Eggs:* Hard-boiled eggs, covered with a layer of minced meat and fried.
   (d) *Bird's Nest:* Hard-boiled egg, covered with boiled mashed potatoes, dipped in beaten egg and rolled in vermecilli, deep fried to a golden brown colour, served with tomato sauce.
4. *Organ Meats:* Organ meats like kidneys, liver, heart, pancreas, brain etc. are served as entree. They are sauted in butter and then served.
5. Sausages are served with stewed apples, veal and pork pies.
6. *Italian Pasta:* (Farinadous Products)

Pasta was invented by Marco Polo in Italy. Pasta is made of egg, flour, colouring agents like spinach or beetroot juice. The colouring agents of pasta also adds taste to it.

Examples of Pasta Products:

1. *Fettucini:* Flat strips of pasta, approx 1/2 cm broad.
2. *Spaghetti:* Solid, cylindrical long noodles.
3. *Macaroni Hollow:* Cylindrical small noodles of 1/2 mm thickness.
4. *Ravioli:* Two sheets of pasta stuffed and cut into fancy shapes.
5. *Lasange:* Thick strips of pasta about 1" in thickness.
6. *Canneloni:* Stuffed, cylindrical pasta product.
7. *Gnocchi:* Round dumplings of pasta.
8. *Capellini:* It is the finest ribbon pasta. It can be homemade or bought ready-made in a shop. It is of the thinnest variety. The word 'capellini' comes from capelli d'ngels which means angel hair.
9. Lasagnette Flat Ribbon pasta, 18 mm wide with a roughened edge. It is a smaller version of lasange, which are strips of pasta 1' thick.
10. *Taglia Telli:* It is the flat pasta, 16 mm wide made from wheat. Like all pasta, it is served with a variety of sauces.
11. *Fedelli:* Fine cylindrical pasta, similar to vermicelli, available fresh or dried.
12. *Lasareccia:* Pasta curled into shape, with a twist at one end.
13. *Festonati:* Pasta which is garland shaped. The word 'festonate' means festoon or garland.
14. *Cavatappi:* Rich pasta in a twisted shape.
15. *Fusille Bucate:* Pasta in shape of tiny sprigs.
16. *Penne:* Small/large size ribbed version of pasta.
17. *Fucilli:* Pasta with a spiral shape of cockscew.
18. *Pizza:* Fermented, round paste product.

Preparation of Pasta Products: The Pasta is plunged into hot boiling water with oil and salt, and then moved to cold running water.

Sicilliene	Pasta tossed in butter with cheese, minced beef and red pimentoes, chopped shallots and tomatoes.
Italienne	Pasta tossed in butter to which grated cheese is added.
Milanaise	Pasta tossed in butter to which grated cheese is added with tomato sauce.
Napolitaine	Pasta tossed in butter with tomato sauce, juliennes of ham and calf's tongue.
Vongole	Pasta tossed in butter with seafood and tomato sauce.

Some other Entree Preparations

Cervelle Beurre Noir Ou Noisette: Poached and sliced brain is seasoned and chopped parsley, drops of vinegar and butter are added and heated until slightly brown and poured over the grains.

Cervelle Frit A L' Anglaise: Brains dipped in egg, rolled in breadcrumbs deep fried and served with tomato sauce.

Foie D' Agneau A L' Anglaise: Lamb liver is cut in slices, grilled and dressed with bacon rashers.

Foie D'Agneau Espagnole: Lamb liver is cut in slices, grilled, garnished with grilled tomato, onion roundels fried in oil with parsley.

Foe D'Agneau Lyonnaise: Lamb liver is cut into slices, fried, garnished with sliced onion and cooked in butter, a drop of vinegar added.

Rognons D'Agneau Americaine: Grilled kidneys, dressed with half-grilled tomato, surrounded with grilled bacon rashers and some nut brown butter added on top.

Rognon D'Agneau Champignon: Lamb kidneys are cut into slices, tossed in butter, mixed with mushrooms and chopped shallots.

Rognon D'Agneau Bercy: The Lamb kidney is cut in slices, tossed in butter, and coated with Bercy sauce.

Service of Entree: It is served on a hot half plate, and the cutlery provided is a small knife and small fork. A dessert spoon is kept on the sideboard, in case the guest demands it.

Releve: It is known as the Main Course in English. Also known as Releve or Piece de Resistance. Big joints of veal, lamb, meat, poultry etc. are served with heavy accompaniments and garnish. The accompaniments of Releve are vegetables and potatoes. Potatoes are served in the 2 o'clock position while vegetables are served at 10 o'clock position, in a hot large plate and the cutlery provided is a large knife and large fork. Main course can be:

1. Continental : Accompaniments are vegetables and potatoes.
2. Chinese : Accompaniments are a Chinese cruet set comprising of chilli vinegar, chilli sauce and soya sauce.
3. Indian : Accompaniments are PPC/APC—Pickle/Achar, Papad, Chutney and Green Salad.

Type of Menu

Chicken

Poulet Saute Arlesienne: Chicken sauted in oil to which white wine and garlic is added, garnished with roundels of fried egg plant, fried onions and a heap of tomato concuss.

Poulet Saute Chasseur: Sauted chicken with white wine, brandy, tomato concasse, demi-glazed, sliced mushrooms and chopped shallots.

Poulet Saute Hongroise: Sauted chicken with paprika, cream, chopped onions and tomato concasse, served decorated with a border of rice.

Poulet Saute Marango: Sauted chicken to which white wine, tomatoes and garlic is added, garnished with fried eggs, croutons, Cray fish, Craw fish and chopped parsley.

Poulet Saute Laperouse: Sauted chicken topped with Hollandaise Sauce, garnished with artichoke bottom filled with mushroom puree.

Poulet Saute Mexicaine: Sauted chicken with white wine garnished with grilled capsicum, filled with chopped mushroom and tomato concasse.

Poulet Maryland: Breast of chicken dipped in egg, rolled in breadcrumbs and shallow fried. Its accompaniments are:

(i) Banana fritters
(ii) Grilled tomato
(iii) Grilled bacon rasher
(iv) Pommes crouquette
(v) Sweetcorn pancakes.

Poulet Cordon Bleu: Breast of chicken stuffed with ham and cheese, dipped in egg, rolled in bread crumbs, deep fried to a golden brown colour.

Poulet A L'Kiev: Breast of chicken stuffed with garlic butter, dipped in egg, rolled in breadcrumbs and deep fried. Served with garlic butter and tomato sauce.

Lamb: Mutton chops, Seekh kababs, Mutton cutlets can be served. Some examples are:

Gigot de Mouton Roti: Roast leg of mutton on a bed of root vegetables served with roast gravy.

Navarin de Mouton: Mutton stew made from boneless mutton, passed through seasoned flour and browned. Cooked in a demi-glaze along with sauted turned potatoes and carrots.

Lancashire Hot Pot: White stew made from boneless meat. White stock is added and then alternate layers of potato and onions are added. Cooking is finished in the oven and served from the hot pot dish.

Noisette D'Agneau: Lamb chops seasoned, grilled, and served with melted butter.

Blanquette De Mouton A L' Ancienne: White mutton stew made from boneless pieces of lamb and veloute, garnished with button onions and mushrooms.

Lamb Shaslik: Boneless lamb, skewered along with slice of onions, tomato and capsicum, marinated with lemon juice, cooked on an open spit fire, served on a bed of pulao, rice and tomato sauce.

Beef

Fillet de Boeuf A L'Arlesienne: Roasted fillet of beef garnished with roundels of fried egg plant, roundels of fried onions and tomato concasse.

Boeuf Steak A L'Americaine: Minced beef made into round shapes like a cake with a depression in the centre, filled with egg yolk and grilled. Accompaniments are:

(i) Capers

(ii) Chopped Onions

(iii) Parsley

Tournedos A L' Andalouse: Tournedos shallow fried, put on a base of croutons, garnished with fried egg plant, chipolatas and stuffed pimentoes.

Boeuf Strogn off: Russian stew with capsicum, celery, mushrooms and onions and served on a bed of pulao rice.

Veal

Escalope De Veau A L'Anglaise: Escalope (boneless fillet) of veal, dipped in egg, rolled in bread crumbs, deep fried to a golden brown colour. Accompaniments are:

(i) Grilled Bacon

(ii) Parsley Potatoes

Veau Cordon Bleu: Escalope of veal stuffed with ham and cheese, dipped in egg, rolled in bread crumbs, deep fried to

golden brown colour. It is served with roundels of lemon and beurre noisette.

Weiner Schnitzel (Austria): Escalope of veal, marinated with lemon juice and seasoning, shallow fried, garnished with anchovy, hard-boiled egg, lemon and parsley.

Pork: Ham, bacon, sausages, salami, roast leg of pork are served. For e.g.:

Suckling Pig: The pig is cleaned, and the from inside is filled with stuffing. It is braized in the oven with onions and aromats (bed of spring vegetables). If served hot, its accompaniments are roast gravy, small apples, puree of spinach and roast potatoes. If served cold; the cold dish is coated with brown aspic jelly and decorated with coloured butter. The whole apple is placed in the mouth of the pig.

Other Meats

The Jugged Hare: The Hare is cut into eight positions, marinated in red wine, sauted in butter with onions, bacon and carrots. Sauce is made from the blood of hare and onions and mushrooms. Accompaniments are heart-shaped croutons, red currant jelly and chestnut.

W=ild Bear: The skin and internal organs are removed. It is filled with stuffing and braized in the oven along with onions and aromats.

Horse Meat: This meat became popular in Germany and Italy during World War II. It is consumed in the form of steaks and sausages.

Squirrel: Belongs to the rodent family. It is cut into eight portions, marinated with red wine, sauted in butter with onions, bacon and carrots.

Whale: It gained popularity during World War II. It is mostly consumed by poor sections of society. It is eaten in the form of steaks alongwith mushrooms, tomatoes and onions.

Service of Releve: It is served on a hot plate and the cutlery provided is a large knife and large fork.

Sorbet: It is the sixth course of the menu. It is therefore a rest between the courses *i.e.* main course and Roti. It helps in digestion. 10-12 minutes are allotted for the Sorbet service. Russian and Egyptian cigarettes are passed with Sorbet. It originated in Arabia and China. It was introduced in 1660 by Francisco Porcopio.

It is iced water flavoured with champagne or any other liqueur. It is sweetened and coloured with fruit and fruit juices.

Sorbet A La Sicilienne: It is iced-water with melon pulp flavoured with Champagne, served chilled in a parfait glass, with a parfait spoon as cutlery and a quarter plate with doiley paper as an underline.

Granite: It is a light Sorbet made from fruit syrups.

Marquise: It is fresh pineapple or strawberry juice combined with whipped sweetened cream. Flavoured with Kirsch (Cherry Brandy) and served in a sten med glass.

Spoom: Iced-water flavoured with champagne to which double the quantity of meringue is added.

Sorbet Alexandra: Juices of fruits combined with desired liqueur, filled with strawberry water and served in a parfait glass.

Sorbet A L'Americain: Lemon or orange juice flavoured with Champagne, served chilled in a stemmed glass.

Service of Sorbet: It is served in a parfait or stemmed glass, with a quarter plate and doiley paper as underlines.

After the clearance of the main course, the dessert cutlery stays on the cover. An ashtray is placed on the cover in the centre. Cigarettes are passed on a quarter plate with a doily paper and match box.

Roti: In this course roasted game birds are served. For example chicken, pigeon, turkey, quail, pheasant, patridge, snipe, duck, goose, etc. may be served in this course. It is

accompanied by a salad served on a half moon-shaped plate which fits on top of the plate in which the roast is being served.

Menu Examples

Faisen A La Normande: Brown pheasant roasted in the oven with butter to which fried apples, cream and avocados are added.

Caille En Cocotte: Quail tossed in butter, wrapped in vine leaves or bacon rashers, cooked in cocotte dish and served with croutons fried in butter.

Dindonneau A La Chipolata: Trussed turkey cooked in cooking stock and Madeira and garnished with chipolatas.

Service of Roti: It is served on a half plate and the cutlery required is a small knife and small fork. A side plate or quarter plate is put on the sideboard along with a side knife. If the guest asks for bread, the quarter plate and knife are put on the cover. When serving roast of large birds, the joints may be served on the larger plate. The cutlery required is a large knife and large fork. If the roast is part of the main course, it may be served as main course.

Legume: Different kinds of vegetables like mushroom, artichokes, asparagus, carrots are served in this course.

Artichaut Provencale: Artichokes boiled ten minutes prior to being cooked, and tossed in butter, cooked in a cocotte dish, along with fresh peas and lettuce.

Artichaut Lyonnaise: Artichokes cut into quarters, tossed in oil with fried onions and garnished with chopped parsley.

Asperges Flamande: Asparagus boiled, coated with sauce, made from butter and egg yolk.

Service of Legume

It is served on a cold or hot half plate depending upon the dish and the cutlery provided is a small knife and small fork.

Vegetables like artichoke and asparagus are eaten with fingers, therefore a finger bowl is passed in the end of meal.

If served cold, accompaniments are vinaigrette or mayonnaise.

If served hot, accompaniments are Hollandaise and Beurre Noisette.

When serving asparagus tips, an inverted fork is placed under the plate, so that the plate tilts towards one side, where the appropriate amount of sauce collects, which is vinaigrette or mayonnaise.

Entrement: Entrement is a sweet course. Hot and cold puddings, cakes and Indian sweets are served in this course. These days, this course is merged with the dessert course.

Indian Sweets: Various items such as Rasgulla, Gajar ka halwa, Kheer, Rasmalai, Phirni, Jalebi are served.

Creme Caramel: It is prepared with a mixture of egg, milk, sugar, which is poured into a jelly mould, lined with carmelised sugar and baked in an oven on a double boiler and served cold or hot.

Apple Strudel: Thinly rolled out dough made of flour, egg, baking powder, stuffed with cooked apples, nuts and flavoured with cinnamon and clove, with castor sugar. Baked in the oven and served with Melba sauce.

Bread and Butter Pudding: Triangles of bread coated with butter, arranged in a pre-dish, topped with egg custard mixture, garnished with sultanas, and baked.

Diplomat: Triangles of cake or cubes of cake arranged in a pre-dish, topped with custard sauce and baked.

Queen's of Pudding: Triangles of bread, coated with jam, topped with custard sauce and baked in an oven.

Riz A L' Impertrice: Rice pudding in which rice is cooked with sugar, milk and liason of egg yolks and gelatine, enriched with cream. The mixture is spread on a sweet salver in which the bottom has a layer of lemon jelly and the top a layer of strawberry jelly. Decorated with fresh cream and fruits.

Bavaroise Rubane: Tricoloured Bavarois layers of vanilla, strawberry and chocolate.

Savarin Aux Fruit: Yeast fermented cake soaked in sugar syrup which is flavoured with rum and decorated with cream, fruits and almonds.

Souffle: It is prepared from milk, sugar, egg yolk, egg white, beaten to a stiff point and set in the refrigerator.

Mousse: It is made from milk, sugar, gelatin and chocolate.

Vanilla Blamanch: It is made from milk, sugar, milk powder and cream.

Charlotte Royale: Charlotte mould is lined with swiss roll slices. Strawberry jelly is poured in the mould to form a thin layer. Then the mould is filled with vanilla bavarois and set in the refrigerator. It is demoulded in a sweet salver which has a thin layer of lemon jelly.

Baked Alaska: Layers of cake arranged alternately with layers of icecream, topped with meringue and baked in the oven for a second.

Examples of Cakes/Gateaux are:

Gateaux Ananas: Pineapple cake.

Gateaux Moka: Coffee cake.

Black Forest Cake/Gateaux Foret Noir: Cake covered with sweetened cream and decorated with shredded chocolate.

Gateaux Religious: A base of short crust filled with bavarois mixture. On the side six to seven eclairs are coated with chocolate or coffee and a choclate-coated profit roll is put in the top.

Profit roll Suchard: Balls of choix paste filled in with whipped cream, topped with chocolate sauce, garnished with nuts. It is served in an ice-cream cup with a quarter plate with doiley paper as an underline and a teaspoon as cutlery.

Gateaux Milk Feville: Thousand layer cake which is topped with icing and jam.

Ice-creams: Various ice-creams which can be served are Casatta, Butter Scotch, Vanilla, Strawberry, Chocolate, Rum and Raisin, Mango, Pistacho etc.

Fried Ice-cream: Balls of ice-cream coated with a thin layer of cornflour and fried in very hot oil till it attains a golden brown colour.

Service of Enterment: Enterment is served in a cold half plate and the cutlery provided is a dessert spoon and fork. Ice-creams served in an ice-cream cup or cocktail cup, with a quarter plate with doily paper as an underliner with a teaspoon or ice-cream spoon.

Bonne Bouche: The Bonne Bouche course comes after the Enterment course. It includes savoury dishes.

Anchovies on Toast: Toast with fillets of anchovies, decorated with sieved hard-boiled egg yolk and chopped parsley.

Roe on Toast: Shallow fried roe, juice lemon, coated with beurre noisette and served on a toast.

Haddock on Toast: Poached haddock with cream sauce, spread on the toast.

Mushrooms on Toast: Sauted mushrooms topped with mornay sauce and gratinated under the salamander.

Welsh Rarebit: Grated cheddar mixed with beer, salt, pepper, cayenne pepper and worcesterchire sauce bound with a liasion of egg yolk and creams. Spread on a toast and glazed in the salamander.

Angles on Horseback: Poached oysters wrapped in streaky bacon and grilled.

Devils on Horseback: Prunes rolled in bacon and grilled.

Scotch Woodcock: Scrambled eggs on toast garnished with anchovy and studded with capers.

Quiche Loraine: Short crust pastry filled with bacon and cheese, covered with egg custard and baked.

Canapes: All types of canapes can be served. Both hot and cold.

Cheeses

English Cheeses: They are hard cheeses, creamy in colour.

Cheddar

Cheshire: Hard cheese, crumbly in texture, comes in red, white and blue colour. The blue variety is rarely available.

Lancashire: Hard cheese, creamy white in colour.

Caerphily: It is a semi-hard cheese, creamy white in colour with a delicate flavour and smooth texture.

Derby: Hard cheese, pale honey in colour.

Dorset Blue: It is a hard-pressed cheese with blue veins, crumbly in texture.

Double Gloucester: It is a hard straw-coloured cheese.

Leicester: Hard cheese, crumbly texture and russet coloured.

Wensley Dale: Hard cheese, white in colour, blue veined.

Stilton: Blue veined cheese.

French Cheeses

Brie: Soft or cream cheese made from cow's milk.

Camembert: Soft or cream cheese yellow in colour with a thin rind.

Carre De L'est: Soft or cream cheese.

Demi Sel: Soft or cream cheese with/or sour milk cheese.

Point L'Eveque: Semi-hard cheese with thin rind.

Port Salut: Semi-hard cheese with hard rind.

Roquefort: Blue veined cheese, creamy in texture and crumbly. It is matured in the rocks/caves of Roquefort.

Swiss Cheese

Gruyere: Hard cheese, pale yellow in colour.

Emmentale: Hard cheese having cavities or holes.

Dutch Cheese

Edam: Hard cheese with yellow or white rind and a waxy texture.

Gouda: Hard cheese, yellow in colour with yellow or red rind, flat in shape with rounded edges.

Italian Cheese

Gorgonzola: Blue cheese with semi-solid texture, soft to touch.

Parmasan: Hard, dry cheese having pinpricks.

Danish Cheese

Danish Blue: Blue veined cheese made from cow's milk.

Accompaniments with Cheese:

1. Special cruet set consisting of salt, pepper, mustard.
2. Celery is served in a glass bowl on a bed of crushed ice.
3. Radish are served when in season, in a glass bowl.
4. Castor sugar passed in sugar dredges for cream cheese.
5. Assorted cheese biscuits like cream crackers.

Service of Bonne Bouche: It is served in a half plate with a small knife and a small fork.

Service of Cheese: Cheese is presented on a cheese board or displayed on a trolley. Cheese is served on a quarter plate. It is cut into pieces with the help of a cheese knife. It is picked with the beak of a cheese knife and the cutlery provided is a small knife and a small fork.

Dessert: In the dessert course, fresh and dry fruits are served. Fresh fruits are arranged in a fruit basket which is called Corbielle des Fruits.

Fresh fruits may be divided into four main categories:

Soft Fruits: Fruits like banana, papaya, grapes etc. come under this category.

Hard Fruits: Fruits like mangoes, pears, apple etc. come under this category.

Dry Fruits: Include walnuts, cashewnuts, almonds, pistachio etc.

Citrus Fruits: Fruits like grapefruit, oranges, sweet lime may be included in this category.

Service of Fruits: Fruits may be served in a fruit basket or a halfplate. Cutlery provided is a fruit knife and a fruit fork. Grapes are served along with grape scissors and a bowl of fresh water to wash the grapes. A nutcracker is passed for nuts. A finger bowl and fresh napkins are passed after the service.

Cafe: Coffee is not a course but it is customary to serve coffee at the end of a meal. Coffee is served in a coffee cup with sauces and coffee spoon. Coffee is served or without milk and white or brown sugar.

2

Important Stuff

In an adult 4 per cent of the body weight is from minerals. There are 24 minerals in the body. Mineral substances are important in maintaining good health because they are part of our tissues, skeletal structure and the body fluids. Hormones, enzymes and vitamins contain minerals. Vital organs like heart and nervous tissues cannot function normally in its absence. Water balance, regulation of fluids, and the acid base balance in the body depend to a great extent on certain mineral balance in the body. Neuromuscular activities, coagulation of blood and fluid movement in the tissues are under the control of certain inorganic substances in the body.

Minerals are inorganic substance that are present in the body tissues like phosphomin, iron, sulphur, zinc or copper or soluble minerals in the body fluids like sodium chloride or mineral components in the bones and teeth and hormonal secretions. Water balance in the body is regulated by minerals like sodium and potassium. Our body has calcium, phosphorus, sodium, potassium, chlorine, iron, magnesium, copper, iodine, cobalt, zinc, aluminium, arsenic, bromine, fluorine, nickel,

chromium, cadmium, selenium, silicon, vanadium, and molybdenum as minerals. Each one of this has some role in maintaining the structure and function of vital organs of our body. Few milligrams of a mineral element decide the skeletal growth or the mental ability of a person.

Optimum intake of minerals and favourable environment for its utilisation are equally important to maintain mineral balance. During the growth period and also during pregnancy, lactation and old age, certain minerals are very essential to maintain health and growth.

Calcium: The most abundant mineral element in the body is calcium. Nearly 99 per cent of the calcium in the body is present in the skeleton. The remaining part of calcium performs a variety of functions in the body. About 850-950 gms of calcium is present in a healthy man.

Functions: The most important function of calcium is the ossification of bones. If calcium content in the blood is below normal, bone calcium is depleted and serum level is brought to normal. Parathyroid hormone regulates calcium level in blood and its metabolism in the bone. The bones constitute a mineral reserve which may be drawn upon in need. Calcium is also essential for the formation of teeth and for the contraction of the heart and muscle. The regulation of muscle irritability is dependent upon the serum calcium level. Calcium is also required for normal nerve impulse transmission and excitability of nerve fibres and nerve centres. Calcium forms part of acelytcholine which is necessary for the transmission of nerve impulses. Calcium helps in the coagulation of blood and is needed for permeability of capillary walls. Calcium exerts its influence in the absorption of vitamin B_{12} from the intestine. The mineral has been shown to activate certain enzymes, especially lipase and certain proteolytic enzymes, adenosine triphosphate and rennin.

Factors Affecting Calcium Absorption: The most important factor promoting calcium absorption is vitamin D content in

the body. Utilisation of calcium is more when it is needed by the body. For example, during growth the absorption of calcium is more than in adults. Calcium content in the diet increases the absorption of calcium. Phosphates and calcium ratio in the diet also influences the absorption. If the ratio is 1:1, absorption rate is high. If phosphate ratio is high, calcium absorption is low. Phytic acid which is present in cereals combines with phosphorus and forms insoluble calcium salts and calcium phytate and inhibits calcium absorption. Abnormal intake of fats and fatty acids reduces the calcium level in blood and its metabolism in the bone. The bones constitute a mineral reserve which may be drawn upon in need. Minerals like iron which forms insoluble phosphates interfere with phosphorus absorption which indirectly affects calcium absorption. Absorption of calcium is facilitated by low intestinal pH. Normal acidity of stomach due to the presence of hydrochloric acid is essential to stimulate the absorption of calcium. Parathyroid hormone also aids in the absorption of calcium from the stomach. If cereals with husk are used in the diet, calcium absorption is reduced. Whole flours, oatmeal, and preparation with these, if not fermented, increases the inhibition of calcium utilisation. On fermentation of the enzyme phytase splits up calcium phytate and then increases its absorption.

Like phytic acid, oxalic acid is present abundantly in leafy vegetables and certain fruits. This forms an insoluble compound calcium oxalate and thus diminishes the absorption of calcium. Magnesium, if taken in abundant quantities, often interferes with calcium absorption and results in poor calcium absorption. During sprue, steatorrohea or in biliary fistula fat is absorbed poorly and thus accumulated fats help the formation of calcium soaps. Good quality protein content in the diet helps calcium absorption. During rickets citric acid increases calcium absorption.

There are many factors which hinder the absorption of calcium. Phytic acid, oxalic acid, magnesium and fat reduce the absorption of calcium. Cereals contain phytic acid which

combines with calcium to form insoluble calcium phytates. Phytic acid is present in the outer parts of certain proteolytic enzymes.

Deficiency: Calcium deficiency symptoms are usually accompanied by a lack of phosphorus and vitamin D. Any of the factors which affect calcium absorption can lead to its ineffective utilisation. Vitamin D deficiency alone can produce calcium deficiency symptoms. Pure calcium deficiency is osteoperosis. In osteoperosis serum calcium level is normal but body store is reduced. During old age the capacity of the body to absorb and utilise calcium is diminished. In women hormonal imbalance during menopause or removal of ovaries at earlier stages results in osteoperosis. Corticosteroid injection for any purpose may result in osteoperosis. In osteoperosis pores are seen in the bones due to decalcification. Back pain, loss of weight, fracture of bones for no reason and loss of teeth are some of the clinical symptoms. Calcium and vitamin D deficiency results in rickets and osteomalacia among children and adults, respectively.

Tetany is another symptom of calcium deficiency. In tetany there is hyperexcitability of the nervous system. A series of spasmodic movements take place in the muscles. Depending upon the level of serum calcium symptoms of tetany vary. Irritation, confusion, pain, involuntary muscle spasm, bronchial spasm, spasm in the gastro-intestinal tract, pain and convulsions are again symptoms of tetany. If parathyroid gland is in hypoactivity, calcium is not properly assimilated and results in chronic tetany. In this condition nervous changes occur and impairment of memory takes place. Depression, irritation, loss of hair and nails, roughness of the skin, pitting of teeth and epileptic fits are other symptoms of this condition. Cataract is common in such patients.

Hypercalcemia: Hyper-vitaminosis-D or Hyperparathyroidism produces hypercalcemia. Normally infants of 5-8 months get this problem. Loss of appetite, vomiting, constipation and flabby muscles are the symptoms.

Sources of Calcium: Milk and milk products are the best sources of calcium. Calcium is absorbed maximum from milk and milk products because of the presence of lysine and lactose in it. Ragi is the cheapest source of calcium; therefore it is known as 'poor man's milk'. Sesame seeds and leafy vegetables are also good sources of calcium. But the availability from these sources is not insured as in sesame seeds calcium is present in its husk and on dehusking it is removed. In leafy vegetables the presence of oxalates inhibits its absorption. Molasses and vegetables like peas, beans, potatoes, pulses, and small fish and dried fruits are rich in calcium. Unabsorbed calcium is excreted through faecal matters.

Requirements: The FAO/WHO expert group (1984) has recommended 400-500 mg of calcium per day for an adult man and woman. During pregnancy and lactation the requirement increases to 1,000 mg/day. For infants 500-600 mg and for children of between 1 and 9 years 400 to 500 mgs and for age group 10 to 15 years 600-700 mg are recommended.

Phosphorus: No other mineral takes part in a number of functions as phosphorus in different parts of our body. An adult human body contains about 400-700 mgs of phosphorus. About eighty per cent of the body's phosphorus is deposited in the bones and teeth as calcium phosphate. The remaining portion is located as sodium phosphate and potassium phosphate in the body fluids and in cells. Phosphorus is also present in many enzymes and co-enzymes and co-carboxylase. Thus it takes part in metabolism.

Functions: Phosphorus takes an important role in the calcification of bones. It also plays an essential part in carbohydrate metabolism in phosphorylation of glycogen. Phosphorus is an essential constituent of co-enzyme and co-carboxylase enzyme system in the oxidation of carbohydrate, proteins and fat metabolism. Phosphorus compounds control the storage and release of energy through ATP, ADP system. Phosphorus is present in the nucleoproteins which are an integral part of the cells. Phospholipids, which are present in

various cells of our body, contain phosphorus. Inorganic phosphates in the body fluids act as a buffer system in the body.

About seventy per cent of ingested phosphorus is absorbed. If iron, aluminium and magnesium are present in excess amount in the diet they interfere with the absorption of phosphorus. The proportion of calcium and phosphorus is an important factor which regulates its absorption.

Requirements: The daily requirement is about 1 gm. Phosphorus requirement is similar to calcium for various groups.

Sources: Animal foods like meat, fish, poultry and eggs are excellent sources of phosphorus. Milk and milk products are good source of phosphorus. Though cereals and legumes are good sources, phosphorus is less available from them as it is present in the form of phytic acid which is not utilised well.

Iron: Iron is a component of the complex protein haemoglobin, a haemoprotein component of the red blood cells. It is the pigment of red blood cells. Of the 3 to 5 gms of iron present in our body more than half of it is present in haemoglobin. The rest of it is stored in the liver, spleen, bone marrow and muscles. In the muscle it is present in the form of myoglobin. Iron is also present in some enzymes like cytochromes, catalase and peroxidase.

Functions: The most important function of iron is to form a constituent of haemoglobin which takes part in the transportation of oxygen from the lungs to the tissues. This oxygen is used by the body for the processes of metabolism. In its deficiency the oxygen-carrying capacity of the blood is reduced. In red muscles myoglobin, a pigment substance, is formed with the participation of iron. This also has the same oxygen-transporting function like the haemoglobin. During strenuous activity this muscle is involved. In our body there are certain cytochrome protein pigments with iron as an essential component. They act as respiratory enzymes. Certain flavin enzymes also contain iron in them.

Absorption of Iron: Dietary iron is absorbed primarily from the duodenum by the mucosa. Depending upon the need of iron in the body its absorption varies. If the haemoglobin concentration of the blood is satisfactory, little or no iron is absorbed and it is excreted in the faeces. Ferrous iron is absorbed more efficiently than ferric iron. In the intestine most of the iron is reduced to ferrous state. Absorption of iron is enhanced in an acid medium. Vitamin C also enhances the absorption of iron, in the intestinal tract, because of its strong reducing capacity. Protein content in the diet helps the absorption of iron to some extent Sulphur containing amino acids reduce iron which helps its absorption. Excess of calcium phosphate, phytates and oxalic acid interfere with iron absorption by forming insoluble iron salts or oxalates. Minute traces of copper help the formation of haemoglobin. The rate of absorption of iron is affected by the magnitude of iron store in the body and the rate of erythropoiesis. Malabsorption syndromes also reduce iron absorption.

Deficiency of Iron: Anaemia is a condition where the haemoglobin level is lowered in the blood. Shortage of iron in the diet is the most common cause of anaemia. Iron, folic acid, proteins, B_{12}, copper, cobalt, vitamin C and pyridoxine are the nutrients required for erythropoiesis. Folic acid deficiency produces megaloblastic anaemia.

Symptoms of Anaemia: When the haemoglobin level falls below 12.5 gm/100 ml of blood, it results in the diminished oxygen-carrying power of the blood. Iron deficiency anaemia is otherwise known as hypochromic anaemia. In this condition the number of RBC is not much reduced but the quantity of haemoglobin is less. The pallor of skin and tissues, fatigue, weakness, giddiness, blurred vision, anorexia, heartburn, palpitation, dysproea, oedema over feet, headache and a feeling of tiredness are some of the symptoms. The nails show koilonychia and tongue and papillary atrophy are seen. Among women menstrual disturbances like menorrhagia or phases of amenorrhoea take place.

Nutritional megaloblastic anaemia occurs due to dietary deficiency during old age, increased requirement during pregnancy, lactation, prematurity, rapid growth and in endocrine disturbances. Haemolytic anaemias occur during malignancy, rheumatoid arthritis, pulmonary tuberculosis and in skin diseases. Alcoholism and side effects of drugs also result in anaemia. In diseases like idiopathic steatorrhoea, tropical sprue and in partial gastractomy anaemia takes place. During pregnancy this type of anaemia usually occurs in the second trimester. Weakness, palpitation and fatigue are followed by fainting bouts.

Anaemia must be treated with ferrous sulphate tablets and iron-rich diet.

Hemosiderosis: It is a disorder of iron metabolism. In this condition large deposits of iron are made in the liver and in the recticulo endothelial system. Iron with carrier protein which is known as transferrin of circulation becomes saturated with iron. When transferrin is unable to bind all the absorbed iron this metabolic disorder of iron takes place. If iron vessels are used for almost all cooking; much more iron is ingested in the body, which may result in hemosiderosis. The Bantu population of Africa is an example of this condition.

In hemolytic anaemia a large number of red blood cells are destroyed and this can lead to hemosiderosis. Prolonged prophylaxis with massive doses of iron can often produce hemosiderosis.

Sources of Iron: Liver, kidney, heart, lean meat, egg yolk and shellfish are the best sources of iron. Dried beans, legumes, dried fruits, nuts, green leafy vegetables, whole cereals, enriched grains and molasses are good sources. Milk is a poor source of iron.

Requirement of Iron: Recommended intake of iron is calculated, based on its loss and absorption. For an adult man weighing 55 kg the iron requirement is 24 mg/day and for a

woman of 45 kg weight 32 mg is the requirement. During pregnancy 40 mg and during lactation 32 mg of iron is required.

For infants till one year 1 mg/kg body weight and for children of 1 to 9 years of age 20 to 25 mg of iron are required. For boys and girls between 10 and 12 years of age 30 to 25 mg and for boys 13 to 15 years of age 25 mg and for girls from 13 to 15 years 35 mg of iron are required. During later adolescents for boys (16-18 years) 25 mg and for girls 35 mg is the recommended allowance.

Magnesium: The human body contains minute quantities of magnesium in the bones along with calcium and phosphorus and the rest in soft tissue. An adult human body has about 25 gm of magnesium.

Functions: Magnesium plays an important role in normal calcium and phosphorus metabolism in man. It is also present in certain enzymes like co-carboxylase which decarboxylase pyruvic acid. Magnesium also takes part in the phosphorylation in metabolic pathways.

Magnesium is easily absorbed from the intestine. Excess magnesium is excreted in the faeces. Normal human serum contains about 2 to 3 mg/100 ml.

Deficiency: Prolonged starvation brings magnesium deficiency, vomiting and intestinal fistula. Diabetic acidosis is associated with raised serum level of magnesium. Neuromuscular irritability, tremors, cramps, twitching and convulsions occur in its deficiency. Confusion, disorientation and visual hallucinations are some other complaints with magnesium deficiency.

Sodium: Sodium is essential for the normal functioning of the body. Sodium plays an important role in the regulation of acid-base balance and water metabolism in the body. Sodium in the form of sodium chloride is ingested directly through food and many food materials contain this mineral. Almost all the body sodium is found in the extracellular fluids of our body. Through sodium the osmotic pressure is also regulated

by the body. A healthy kidney regulates the excretion of sodium according to the requirement. During kidney diseases sodium is retained in the body and oedema takes place. Under normal conditions 90 per cent of ingested sodium is excreted in the urine in the form of sodium chloride and sodium phosphate. Sweat contains some amount of sodium. Sodium helps to contribute alkalibility of the gastrointestinal tract.

Sodium plays a special role in controlling the heartbeat by helping in its origin and maintenance. Contraction of muscles and regulating the permeability of cell membranes are the other functions of sodium.

The adult human body contains about 100 gm of sodium. On an average diet about 3 to 5 gins of sodium is excreted in the urine. In profuse perspiration sodium chloride is depleted from the body. In sodium depletion muscular cramps, weakness, headache and vascular collapse take place. In decreased water intake the serum level of sodium is increased and this condition is known as hypernatraemia. Blood urea is increased and cellular water is depleted during dehydration and in hyponatraemia. In infants shrinkage of brain causes rupture of cerebral veins and further cerebral thrombosis. Retention of sodium produces oedema and ascites. In renal diseases, cardiac failure, hepatic cirrhosis with ascites, toxaemia of pregnancy, prolonged administration of corticosteroids sodium must be restricted through diet. Normally 4-5 gm sodium is injested through food and it is restricted to 250 mg, 500 mg, or 750 mg depending upon the condition.

Requirements: For an adult 1-2 gms of sodium and for children 350 mg-1 gm is recommended. For infants 150-400 mg is enough.

Sources: Animal foods like meat, meat extractives, fish, poultry, milk, milk products, nuts, bakery items, salted biscuits, egg-white, dried fruits, roots like beetroot, carrot, radish and leafy vegetables like spinach are excellent sources of sodium.

Abnormalities of Sodium Level in Blood: Hyponatraemia and hypernatremia are two conditions which manifest the sodium level in the blood.

Hyponatraemia: When serum sodium level falls below normal the condition is known as hyponatraemia; various conditions lead to this symptom. Severe dehydration, decrease in blood volume, decrease in blood pressure, circulatory failure are some of them, prolonged vomiting, diarrhoea, chronic renal diseases with acidosis, Addison's disease due to adrenocestical deficiency are the common causes for hyponatraemia.

Hypernatraemia: In this condition the plasma sodium iron content is higher than normal. Hyperactivity of adrenal cortex as in cushing's syndrome, prolonged treatment with cortisone, ACTH and sex hormones are some of the causes of hypernatraemia. Deficient water intake due to vomiting, nausea, difficulty in swallowing with sore throat, oesophageal disease, cerebral impairment, increased water loss through respiration diarrhoea, excessive urination in diabetes mellitus, chronic renal diseases, increased sodium intake and deficient sodium excretion brings hypernatraemia.

Potassium: In contrast to sodium, potassium is present in the intracellular fluids of the body. It is an important constituent of cells. In adult body 250 gm of potassium is present. Most of it, that is, 97 per cent is found in the intracellular fluids and the remaining 3 per cent in extracellular fluids. The other important function of potassium is that it influences the cellular excitability of smooth, skeletal, cardiac and nervous tissues. If potassium content is high muscular irritability is increased. Diminished calcium and magnesium also result in such a condition. Involuntary muscular function is controlled by potassium content in the body. When the potassium level in the myocardium is changed it is reflected in the electrocardiogram. Potassium is also necessary for the acid base balance in the body.

Potassium deficiency shows weakness in muscular functions, mental disorientation and cardiac irregularities. Hyperpotassemia results from kidney diseases when renal excretion is less. In dehydration these symptoms occur. Deficiency of potassium is not common unless as a complication of diseases. Deficiency condition shows hypokalaemia during operation when food cannot be ingested. In diabetic acidosis this condition occurs. Deficient intake due to starvation and excessive loss from gastro-intestinal tract leads to deficiency.

Hypokalaemia occurs due to deficient intake of it while a surgical patient is exposed to prolonged parent therapy, excessive loss from gastro-intestional tract through persistent vomiting, fistulae or diarrhoea before and after surgery, the patient may be on a poor diet which may enhance the excretion of potassium. Surgical cases due to low intake produces hypokalaemia.

In diabetic acidosis potassium is lost through dehydration on insulin therapy, sugar is converted to glycogen and starch. This glycogen formation depletes some potassium from extracellular fluids and serum potassium level falls and it is dangerous.

While diuretics are administered considerable loss of potassium through urine leads to hypokalaemia. Nephritis and other kidney diseases, tumours of large intestine, treatment with steroids, deficiency disease like kwashiorkor, megaloblastic anaemia, heatstroke and violent exercise are some of the other causes which produce hypokalaemia. Prolonged hypokalaemia causes injury to myocardium and kidney.

Hyperkalaemia: When the potassium level in the serum is high the condition is known as hyperkalaemia. This occurs due to diminished urine volume, intake of potassium salts of penicillin, iodide, acidosis, endogenous tissue breakdown, Addison's disease, kidney failure due to kidney disease or chronic renal failure and rapid intravenous administration of potassium are some of the reasons for hyperkalaemia. Cardiac

and the central nervous system are affected by this condition. Anorexia, weakness of the muscles, diminished tendon jerks, low heart sounds followed by peripheral vascular collapse and general apathy are some of the symptoms.

Trace Elements: Elements which occur in our body in small quantities are known as trace elements. Copper, iodine, zinc, manganese, molybdenum, fluorine, selenium, nickel, chromium, cadmium, silicon, vanadium, strontium, cobalt and chlorine are some of the trace elements. Though these trace elements are present in traces they are very essential to maintain the structure of various organs in good condition. RBC formation, thyroxine production, formation of healthy bones, haemoglobin formation and proper enzymatic actions are some of the important functions of trace elements in our body.

Iodine: The significance of iodine as an essential trace element lies in its role in thyroxine production. Thyroxine is the hormone produced by the thyroid gland. Iodine is the/ main component of thyroxine and a normal -adult has only about 10 mg of iodine in the thyroid. The absorbed iodine combines with a thyroxine binding protein. Thyroxine plays an important role in carbohydrate metabolism. Deficiency of thyroxine interferes with the normal physical and skeletal development of the child during the growing period. Iodine also helps the tissues in its oxygen consumption. Conversion of glycogen to glucose occurs only in its presence. Thyroxine increases the heart rate and depletes bone calcium.

Deficiency: If the iodine intake is deficient the thyroid gland enlarges and this condition is known as simple or endemic goitre. The normal thyroid gland weighs about 25 gin. In goitre it enlarges from 200 to 500 gm. Iodine deficiency goitre results in hypofunctioning of the gland. If diagnosis is done early and iodine supplements are given the thyroid may become normal. If timely treatment is not given the enlargement will become permanent. During the growing period if severe deficiency occurs it results in retarded growth and the condition is known as cretinism. People living mainly in submountain zones are

affected. In India residents of sub-Himalayan regions and Kashmir and parts of adjoining plains are the victims. The most affected are among girls between the ages of 12 and 18 years and boys from 9 to 13 years. The areas where goitre is common is known as the goitre belt. According to the latest data available, about 40 million people are affected with goitre. In these areas air, water, soil and foods grown are poor in iodine which results in its high incidence. Goitre can be prevented by the regular use of iodized salt.

Goitrogenic Substances in Foods: Certain foods of brassica genus like cabbage, cauliflower, turnip, soyaflour and oil contain substances which react with iodine present in the food and make it unavailable. These substances are known as goitrogenic substances. Consumption of these foods in large quantities makes the iodine non-available to the body.

Requirement of Iodine: An adult requires about 0.15 to 0.2 mg of iodine and for infants and children 0.05 to 0.1 mg daily is enough.

Sources: Iodine is present in minute quantities in all foods. The iodine content of foods depends on the iodine content of the soil. In certain parts the soil contains less iodine and the foods grown in such places are iodine poor. The soil of mountainous regions is usually iodine-poor soil. The best sources of iodine are seafoods, common salt and vegetables; cereals and legumes are poor sources of iodine. Most of the iodine in the vegetables is in their leaves and flowers.

Copper: Copper is found in all tissues of our body. It is an essential element in the formation of haemoglobin. About 100 to 150 mg of copper is present in the human body. The brain, liver, heart and kidney contain the highest amounts. In the blood, copper is present in the plasma and RBC. Copper is absorbed rapidly into the blood.

Copper is a component to certain enzymes in our body. It is part of the enzyme cytochrome oxidase which is essential for glucose metabolism. It is also part of tyrosinase, an enzyme

necessary for converting the amino acid tyrosine to melanin, the dark pigment of the skin. Copper stimulates the absorption of iron. Copper is a constituent of the elastic connective tissue protein elastin.

Deficiency: Copper deficiency is not common in human beings. In malnourished persons iron is not absorbed properly and anaemia occurs. Premature babies fed on cow's milk develop anaemia due to copper and iron deficiency.

Requirement: An adult person requires about 2 mg/day. During pregnancy and lactation the copper requirement is increased to 3 mg/day. Infants from 0 to 12 months require 0.5 mg to 1 gm and children 2 gm and adolescents 3 mg of copper.

Sources: Millets like cholam, bajra, raw rice, whole grams, cow gram, dried peas, red gram dal, vegetables like drumstick are good sources of copper. Liver is the richest source of copper. Organ meat like kidney is also a good source of copper.

Zinc: In an adult 2 to 3 gms of zinc is present. It is present in certain parts of the eye, certain glands and in their secretion. Liver, muscles, bone and hair contain some of the remaining zinc. Blood, especially the RBC, has some zinc. In small amounts it is present in tissues and bone cells.

The important function of zinc is its role in enzymatic action. At least in 25 enzymes of digestive and metabolic action zinc takes part. It is also present in an enzyme carbonic anhydrase which is essential for the transport and elimination of carbon dioxide. Co-carboxypeptidase and alkaline phosphatase contain zinc in them. Zinc plays an essential role in the formation of DNA and RNA. It is a constituent of insulin and it is necessary for the formation of connective tissues. Zinc also aids in the healing of burns and wounds.

Deficiency: Zinc deficiency in animals results in growth failure, loss of appetite, degeneration of the male sex glands, testes, skin lesions, alopaecia or loss of hair, changes in the epithelial and cutaneous tissues, dermatitis and scaling and

cracking of the paws. Abnormal bone formation, thickening and metabolism also occur during zinc deficiency. In human beings zinc deficiency produces anaemia, growth retardation and delayed genital maturation.

Requirement: Infants require 3 to 5 mg. For children 10 to 15 mg and during pregnancy and lactation 20 to 25 mg of zinc are required.

Sources: Seafoods, meat poultry and eggs are good sources of zinc. Cereals, legumes, and nuts contain considerable amounts. Fruits and vegetables are poor sources. On milling eighty per cent of zinc is lost

Cobalt: In an adult body about 1.1 mg of cobalt is present. It is found in the blood, liver, kidney, spleen and pancreas. It is also present in minute quantities in all tissues.

The most important function of cobalt is that it forms an integral part of vitamin B_{12}. In animals, if the feed is deficient in cobalt severe anaemia occurs. Addition of cobalt to feed or injection of vitamin B_{12} cures this condition. In human beings vitamin B_{12} synthesis occurs in the colon. At that phase absorption is not taking place. So preformed B_{12} must be ingested for the cobalt requirement. But deficiency of cobalt is not common among human beings. Cobalt also takes part in the enzymatic action and also in thyroid function.

Requirement: Since deficiency is not known, thc requirement is not studied. In pernicious anaemia B_{12} deficiency is present and on administration of B_{12} it disappears.

Sources: Cereals, beans, peas and organ meats like kidney and liver are good sources.

Manganese: About 20 mg of manganese is present in an adult. Bones, liver, pancreas, kidney and pituitary gland contain manganese.

The manganese functions in many enzyme systems. It activates certain enzymes which take part in the digestion and metabolism of carbohydrates, proteins and lipids. In the

synthesis of ATP cholesterol, fatty acids and RNA, the manganese plays an important role.

Manganese deficiency in human beings is not known. Experimental deficiency symptoms are slowing of growth, abnormalities in bone, depressed reproductive functions, neurological weakness of the limbs, ataxia and lack of balance. Among birds a condition known as perosis occurs on manganese deficiency. It is characterised by bone deformities and dislocation of bones.

Sources: Cereals, bran, dried beans, peas, green vegetables and nuts are good sources. Tea and coffee are high in manganese content. But animal foods are relatively poor in it.

Chromium: An adult body contains about 6 mg of chromium. Chromium is present in skin, hair, muscles, brain, tissues, adrenal glands and in body fats.

An important function of chromium is its role in glucose tolerance factor formation in the body. Chromium niacin complex factor helps the action of insulin on glucose. Chromium is also essential for the activating enzymes involved in carbohydrate, protein and fat digestion and metabolism. The serum cholesterol level was raised on experimental animals in chromium deficiency.

Requirement: The exact requirements is not known. A normal balanced diet supplies enough chromium.

Sources: Organ meats like liver, meats, wholegrain cereals and brewer's yeast are good sources of chromium.

Fluorine: Fluorine is present in the body as fluoride. It is present in teeth, bones, thyroid gland, skin and liver.

Fluoride is necessary for the formation of caries-resistant teeth. Small amounts of fluorine prevent dental cariesin children. It promotes calcification.

Fluoride and Tooth Structure: For healthy tooth formation some amount of fluorine is essential. Though the exact mechanism by which it prevents dental caries is not fully

understood, it is assumed that fluorine is incorporated into the tooth enamel. This reduces the solubility of enamel in the acids produced by bacteria. In its absence the enamel is soluble in the acid and tooth decays. In experimental animals fluoride seems to help normal growth during the growing period, but in human beings its role is not yet proved.

Toxicity or Fluorosis: If water contains more fluorine, dental fluorosis occurs. The enamel of the teeth loses its natural lustre and becomes rough. Opaque white Patches with yellow or brown stains are found irregularly on the teeth. Ridges or diffused spots and pitting occur over the teeth. The surface of the teeth gives a corroded appearance.

Skeletal Fluorosis: The bones of vertebral column, especially the cervical part of the spine, Pelvis and bones of the lower extremities, are affected in this condition. Hypercalcification of the bones and the collagen take place due to excessive fluorine deposition. Stiffness and pain are the earlier tymptoms. Weakness and spasticity of lower limbs, wasting of the distal muscles of hands, deep jerks on lower extremities and numbness or girdle pain are the other symptoms of skeletal fluorosis. Gradually the person becomes crippled as he cannot bend or squat for become stiff.

Requirements: Water containing 1 to 2 ppm prevents caries and so it is considered as the requirement.

Molybdenum: The adult human body contains only 9 mg of molybdenum. It is present in the liver, kidney, adrenal glands, bones, skin and tissues. Molybdenum is essential for the normal enzymatic system and its, function. Enzymes like xanthine oxidase and aldehyde oxidase make use of molybdenum. These enzymes are essential for the proper nucleic acid and iron metabolism. Molybdenum seems to aid in preventing teeth decay. The requirement of molybdenum is not known. Whole-grain cereals, dried beans, peas and dark green leafy vegetables are good sources. Meat is also an excellent source.

Special Use of Oils and Fats

Fats and oils are used in cookery to enhance flavour and taste of food. Moreover, fat gives much more energy to body; from every gramme of fat 9 calories are given. In India most of our fats and oils are extracted from vegetable oilseeds. A solid fat-like ghee has a decisive place in Indian cuisine and it is often substituted by vanaspati. Liquid vegetable oils are hydrogenated and converted to solid fats. Hydrogenated fat is incorporated with vitamins A and D.

The Oils: Common oils used in cooking are groundnut oil, sesame oil, mustard oil, safflower oil, coconut oil, soyabean oil, cottonseed oil, palm oil and sunflower oil.

Choice of fat for cooking depends on various factors. Flavour, taste, saturation of fatty acids and presence of double bonds, melting points, smoking temperature and nutritive value are some of them.

Liquid oils rich in linoleic acid like safflower, cottonseed, and groundnut are good in suppressing the formation of scales on the walls of blood vessel. The characterisitc flavours of various oils are due to the presence of very small quantities of certain substances which have no nutritional qualities. Cooking in fat gives fuliness of satiety along with taste to the dishes. Now most of the oils are refined to make them colourless and odourless.

Composition of Fats: Fats and oils are classified into two groups based on their origin as vegetable and animal sources.

Vegetable fats and oils are again broadly divided into three groups based on their composition. They are: Group I which has fatty acids with 16 or 18 carbon atoms; Group II has erucic acid; and Group III has short chain fatty acids. Groundnut oil, cottonseed oil, sesame oil, sunflower seed oil, sunflower seed oil and soyabean oil are with 16 or 18 carbon atoms and they are termed as Group I. Mustard oil and rapeseed oil have erucic acid in it and they are in Group II. Coconut oil and palm oil are with short chain fatty acids in Group III.

Oil	Linoleic acid content (%)
Coconut oil	2
Mustard oil	15
Ghee	2
Palmoil	10
Palmolein	12
Safflower oil	70
Cottonseed oil	50
Sunflower oil	50
Groundnut oil	40

Animal fats are also classified into three groups based on their composition. Group I is milk fat which contains butyric acid and short chain fatty acids. Group II comprises animal body fat with fatty acids of 16 to 18 carbon atoms. Ox fat, pig fat and egg fat are examples of this group. In Group III fish oils with large amounts of polyunsaturated fatty acids are present herring, cod, halibut and sardine oils fall in this group.

Melting points of fats depend on the saturation of fatty acids. Solid fats like lard with high saturated fatty acids have high melting point and liquid oils have a low melting point.

Smoking Temperature of Fats: When fats and oils are heated to a high temperature decomposition of fats occurs and at a point visible fumes are given off. This temperature is called the smoking temperature of fats. The smoking temperature of fat is low when amounts of free fatty acids are high.

Smoking Temperature of Some Fats and Oils

Oil or Fat		Smoking Temperature
Butter	...	208° C.
Cottonseed oil	...	322° C.
Coconut oil	...	138° C.

Contd...

Oil or fat		*Smoking temperature*
Lard	...	194° C.
Olive oil refined	...	234° C.
Groundnut oil	...	162° C.

Fat absorption is high if foods are cooked in fat with low smoking points.

Smoking temperature of fat varies depending on the surface area of oil exposed while heating. The penetrability of the produce to be fried also determines the fat absorption while frying.

Shortenings: Fat is used as a shortening agent due to its ability to give a soft crisp texture to fried or baked products. In bakery items fat plays a significant role.

In wheat products fat brings shortening of gluten strands and in other dishes shortening is through gelatinisation of starch. To bring shortening effect one fat or a mixture of fats is used. In biscuit-making plastic shortenings are used which contribute a white, opaque and creamy appearance to the shortening. Butter and ghee are generally preferred as shortening agents. In certain preparations fat blends homogeneously with non-fat substance to form an emulsion and examples of such preparations are sweets like halwa, mysore pak or cake.

Changes in Fats: Changes in fat during heating are related to its smoking points. Free fatty acid is increased on heating. This is due to the change brought about by the moisture content of food materials. It hydrolyses the cooking fat and increases the free fatty acid content. Therefore, deep frying foods in fats with high smoking temperature is preferred. Ghee has a lower smoking point and therefore other oils or hydrogenated fat are suggested for deep frying.

When fat is used for deep frying, temperature control is essential. If high temperature is used to heat fat, it affects its

nutritive value and in experimental cases it produced a toxic effect. Certain changes take place in fat during heating and this brings down the unsaturated fatty acid content. Repeated heating of fats brings chemical changes and results in poor absorption. It also increased blood cholesterol level in experimental animals. If the fat is heated too high, the food will char outside and remain raw inside. If the temperature of fat is too low the food will get soaked in it. In correct temperature food will be crisp and uniformly cooked. To get the best result fat must be heated to its smoking point. When fumes appear insert food material, fry and lower.

Fats and Oils: The heat after frying the first batch allow the fat to heat to required temperature and add the second batch to fry. Absorption of fats by food materials varies depending on its smoking temperature. If fat with low smoking point is used to fry, more fat is absorbed and fat with high smoking point is required less to fry items. Food items with milk, sugar, or eggs require low temperature to fry them. If food items are fried at high temperature more fat is absorbed. If the food item has greater surface area fat absorbed is more and if surface area is small less fat is absorbed. The moisture content of food item also affects fat absorption. With more moisture content fat absorption also increases. Hard materials also take more fats.

Salad Dressing: Fat is used as an emulsion in salad dressing. Along with fat, egg yolk, vinegar, salt, spices and starch are used in salad dressing. Mayonnaise, french salads and cooked salad dressing are the common methods used. Winterised oils are used in all these methods.

Mayonnaise: It is a semi-solid emulsion of vegetable oils, egg yolk or whole egg, vinegar and seasoning. Egg yolk functions as a seasoning agent.

French Dressing: Vinegar, oils (soyabeans, cornoil, sesame oil or sunflower oil) and salt and spices are added. It is mixed vigorously. Mustard powder is also added.

Characteristics of Some Common Fats

Fat source fatty acid	Poly unsaturated fatty acid	Saturated fatty acid	Mono-unsaturated
Groundnut	31	19	50
Sesame	43	14	43
Safflower seed	67	7	26
Cottonseed	47	23	30
Coconut oil	2	92	6
Mustard oil	63	5	32
Butter	3	70	27
Soyabean	64	18	18
Lard	10	96	40

Lipids: Lipids are heterogeneous groups of compounds with same properties. Fat is the common household name given to lipids.

Like carbohydrates, fats are organic compounds composed of carbon, hydrogen and oxygen. But they differ from carbohydrates in that they contain much less oxygen and much greater proportions of carbon and hydrogen. Seventy-six per cent carbon, twelve per cent oxygen and twelve per cent hydrogen are present in fats. True fat has one molecule of organic ester of glycerol and three molecules of fatty acids. Fats vary in their structure due to the presence of different fatty acids with glycerol molecules. If the glycerol in a fatty acid is combined with three fatty acid units such fats are called triglycerides. Fats are insoluble in water and soluble in organic solvents like ether, benzene or chloroform. Their cooking properties depend upon the type of fatty acids present in them.

Classification: Lipids are classified into (a) simple lipids, (b) compound lipids, (c) waxes, and (d) derived lipids.

Simple Lipids: These are esters of fatty acids with various alcohols. Fats are esters of fatty acids with glycerol Oils are liquid at 20°C while fats are solid at 20°C. Combination of fatty acids and sterol alcohol form cholesterol. Food fats contain neutral fats. Fat is stored in the body as adipose tissue. This is also neutral fat.

Compound Lipids: When fatty acids and glycerol combine with other organic compounds they are known as compound lipids. Examples of compound lipids are phospholipids and glycolipids.

(i) *Phospholipids* (phosphatides) are substituted fats in which phosphoric acid and a nitrogenous base take the place of one fatty acid radical. These are present in vegetable oils and in the fats of animal tissues in the blood and in egg yolk. Phospholipids help the absorption of fat in the body, and also help the transfer

of fat from one place to another. They have prominent role in the functions of the nervous tissue. Lecithin and cephalin are the two important forms of phospholipids in our body.

Lecithin is composed of glycerol, two fatty acids in ester linkage, phosphoric acid and choline. This is the most common form of phospholipid.

Cephalin contains nitrogen compounds other than choline, phosphoric acid, glycerol and fatty acid esters. Phosphatidyl ethanolamine and phosphatidylserine are the common forms of cephalin.

(ii) *Glycolipids* are compounds which contain a carbohydrate molecule with fatty acids and a nitrogenous base. Cerebrosides are the best example of glycolipids. Cerebrosides occur largely in the brain. Galactolipids, glycolipids, and sulfolipids are examples of cerebrosides.

Waxes: These are esters of higher fatty acids and alcohol other than glycerol. Cholesterol esters, sterol esters, esters of fat-soluble vitamins are the common waxes.

Derived Lipids: Substances derived from one fat on hydrolysis or by enzyme activities are called derived fats. Examples of derived lipids are fatty acids like stearic acid, linoleic acid, butyric acid, alcohols like glycerol, cetanol and lanol. Sterol alcohols like cholesterol, ergosterol and sistosterol are also derived fats.

Sterols are nutritionally important. They are classified as animal sterols, plant sterols and mycosterols.

Cholesterol is the precursor of steroid hormones and bile salts. Hormones like progesterone, oestrogen and adrenal corticoids contain cholesterol. Sodium glycocholate and sodium taurocholate and the bile salts have cholesterol in them. Cholesterol esters produced by the liver regulate the cholesterol content in the blood. It is present in all animal and human

tissues. The white matter of the brain has 1-14 per cent cholesterol while the grey matter has 60 per cent on dry basis. Sebum secreted by sebaceous gland also has large amounts of cholesterol.

Properties of Fat: The properties of fat are based on its composition and structure. Melting point of fat is the most important physical quality. Fats which have a high proportion of saturated fatty acids are solid at room temperature and unsaturated fatty acids are liquid at room temperature; If the melting point of a fat is above 50°C or 122° F it is poorly utilised by the body.

The most important chemical properties of fats are hydrogenation, saponification, emulsification and rancidity. All these reactions are related to their chemical structure. The double bond linkages in the unsaturated fats make them readily available for all chemical reactions.

Iodine Value (number): Iodine value gives the number of grammes of iodine which combine with 100 grammes of fat Iodine value depends on the number of bonds it makes at the site of double bonds. This gives the degree of unsaturation of fats.

Saturated and Unsaturated Fatty Acids: Saturated fatty acids have all the carbon atoms in the chain saturated with hydrogen atoms. When a double bond is present between two carbon atoms the fatty acid is termed as unsaturated fatty acid. The degree of unsaturation varies according to the presence of double bonds. If one or more double bonds are present in a fat it is called unsaturated fatty acid. If there are more double bonds in a fat it is called polyunsaturated fatty acid.

Palmitic, stearic and butyric acids are saturated fatty acids. Oleic, linoleic, linolenic and arachidonic acids are unsaturated oils. They have one, two, three and four double bonds, respectively. The unsaturated fatty acids are often described as essential fatty acids. Vegetable oils are rich in linoleic acid and fish oils and animal fats are rich in arachidonic acid.

Saturated and Unsaturated Fatty Acids

Common name (saturated fatty acids)	*Molecular formula*	*Unsaturated fatty acids*	*Molecular formula*	*Double bonds (Nos)*
N-Butyric	C4H8 02	Palmitoleic	C16H30 02	1
Caproic	C6H12 02	Oleic	C18H34 02	1
Caprylic	C8H16 02	Elaidic	C18H34 02	1
Capric	CIOH20 02	Vaccenic	C18H34 02	1
Laurie	C12H24 02	Linoleic	C18H32 02	2
Myristk	C14H28 02	Linolenic	C18H30 02	3
Palmitic	C16H32 02	Ylinolenic	C18H30 02	3
Stearic	C18H36 02	Eleostearic	C18H30 02	3
Arachidic	C20H40 02	Arachidonic	C20H32 02	4

Percentage Composition of Some Fatty Acids in Common Oil and Fats

Fat	*Oleic acid (C18.1)*	*Linoleic acid (C18.2)*	*Polyunsaturated fatty acids*
Butter fat	3.3	4	2
Coconut oil	8	1.6	-
Cottonseed oil	30	56	2
Egg fact	52	9	3
Groundnut oil	65	17	Trace
Olive oil	65	15	Trace
Pig fat	47	4	3
Rapeseed oil	28	15	59
Red palmoil	52	5	Trace
Safflower seed oil	31	51	-
Sesame oil	45	41	Trace
Soyabean oil	29	51	6
Sunflower seed oil	14	73	1
Sardine oil	17.3	2.5	30.1

Some of the chemical reactions of fats are related to its unsaturated status.

Hydrogenation: Unsaturated fats have a tendency to combine with hydrogen at the site of its double bonds. Hydrogenation is such a process. Vegetable fats are exposed to hydrogen at high temperature in the presence of an alkali. Nickel or cobalt is used as a catalyst. For example, groundnut, coconut, cottonseed and other oils are usually used for hydrogenation. On hydrogenation liquid fats turn into solids. Vanaspati is an example of hydrogenated fat from groundnut oil.

Rancidity: On keeping, fats develop an unpleasant odour and this is known as rancidity.

There are two types of rancidity: hydrolytic and oxidative rancidity. One of the common reactions which cause the spoilage of fat is hydrolytic rancidity. In hydrolytic rancidity lipase hydrolyses fats into free fatty acids. Fatty acids with low molecular weight give an unpleasant odour to fats which is easily spoiled.

In oxidative variety unsaturated fatty acids are oxidised. The addition of oxygen may give rise to peroxides which on decomposition yield aldehydes and ketones. They have a foul odour. These peroxides are oxidising agents. They destroy important vitamins present in fats. This oxidation is prevented by the presence of antioxidants in the fat. Vegetables oils have tocopherols or vitamin E as a natural antioxidant in them. There are α β γ (alpha, beta, gama) tocopherols. Among these alpha tocopherol is the most powerful antioxidant.

There are some other substances which can be used as antioxidants which give a prolonged shelf life to fats. Ascorbic acid or vitamin C, gum guaiac acid and propylgilate are some of them.

Saponification of Fats: The ester linkage of fatty acids with glycerol brings this process of saponification. Neutral fats are rich in ester linkage. When neutral fats are heated with

sodium or potassium hydroxide the fat molecules may readily rupture at the ester linkage. It is thus hydrolysed into glycerol and sodium or potassium salts of fatty acids and this is known as soap. Thus, the formation of soap from the hydrolysis of fat with heat and alkali is termed as saponification. Saponification of fats, when performed under standard conditions, is used as an index to find out the molecular weight of fatty acids in a fat.

Emulsification: Fats and oils are insoluble in water because they are lighter than water.

When fat forms a homogeneous mixture with water it is called emulsification. This property is made use of in cookery.

Essential Fatty Acids: Nutritionally important fatty acids are known as essential fatty acids (EFA). Since the body cannot synthesise them the EFA must be supplied to the body through foods. Linoleic acids, linolenic acids and arachidonic acids are the essential fatty acids.

Functions of Fat in the Diet: The most important function of fat is due to its highest calorie density in foodstuffs. One gramme of fat supplies 9.3 kcal while 1 gramme of carbohydrate and protein supply 4.3 and 5.6 kcal respectively. Fats are the carriers of fat-soluble vitamins A, D, E and K. Fats also exert a thiamine-sparing action. Fats have a protein-sparing action also. Essential fatty acids help to maintain tissues, their normal structure and efficiency. Fats, especially phosphatides, form an essential part of nervous tissues. Adipose tissue, where fat is stored, serves as an insulation material in injury. Vital organs of the body are thus protected. In growth period it promotes growth. It also enables proper sexual maturity, especially in female experimental animals. Fat provides flavour and palatability to food.

Digestion and Metabolism of Fats: There is no digestive change for fat in the mouth. In the stomach a little fat is mechanically separated from other foodstuffs. Since carbohydrate and proteins are partially digested in the stomach

fat is set free. In the stomach fat forms a coarse emulsion and is eventually released into the duodenum as small portions. Even though there is not much digestive action in the stomach for fat, digestion is delayed in the stomach by fat. The emulsified fat, known as chyme, takes three to four hours to transfer it from the stomach into the intestine. Pancreatic lipase is the most powerful enzyme in fat digestion. This lipase breaks the alpha position of triglycerides and makes them free fatty acids and 1,2, diglycerides. 1,2 diglycerides are hydrolysed to 2, diglycerides and fatty acids. They combine with bile salts and emulsify fats. They reduce the surface tension of fat and make it available for enzyme activity. Bile salts promote emulsification and help the fat into partial solution of fats and fatty acids. The water-soluble glycerol is quickly absorbed. Short chain fatty acids which are also soluble in water are transported into the portal blood. The products of fat digestion which reach the intestinal wall are resynthesised into new 'triglycerides by the intestine. In the jejunum triglycerides are absorbed. These triglycerides pass into the lymphatics in the form of small scattering particles called chylomicrons. Thorasic duct carries them into the systemic circulation.

Fats which are quickly absorbed are readily available for the tissues. If the fats remain in the gastrointestinal tract for a longer period for utilisation they add satiety value to the meal.

Absorption of fat is based on its digestibility, melting point, the size of the fat molecules and its quantity. The age of a person is another factor which affects the absorption. Old people utilise fat slowly.

Lipids in Blood: In the post-absorptive state plasma contains about 500 mgs of total lipids/100 ml of blood. In this 180 mg/ 100 ml is cholesterol. Neutral fat occurs in blood plasma largely in the form of chylomicrons. In the liver triglycerides are resynthesised and they combine mainly with cholesterol and form beta lipoproteins and are passed into the circulation. Cholesterol esters are converted to bile acids in the liver, re-

excreted into the small intestine and reabsorbed. The concentration of bile may affect cholesterol synthesis in the intestine. Only forty percent of cholesterol in the circulating blood is from food. The remaining portion is synthesised in the liver and intestine. Cholesterol synthesis in the body is controlled by various factors like calorie intake, cholesterol intake, fat intake and essential fatty acid content of the fat. Liver, skin, intestine, adrenal cortex, kidney and lungs are the organs which synthesise cholesterol. Acetate is the precursor of cholesterol and so all substances which yield acetate become precursor of cholesterol. Apart from fatty acids, carbohydrates, ethyl alcohol, alanine, and certain amino acids yield acetate or pyruvic acid or glycogen or glucose which can in turn form acetate. Thus every foodstuff forms acetate and cholesterol while calorie restriction and essential fatty acid deficiency inhibit hepatic cholesterol synthesis.

The adult human body synthesises about 2,000 mgs of cholesterol daily. The dietary intake of cholesterol ranges from 500 mg to 1,200 mg depending upon the consumption of milk, butter, eggs, meat and fish. Apart from these foods coconut oil and hydrogenated fat increase the blood cholesterol level. Unsaturated fats or essential fatty acids reduce the blood cholesterol level. Safflower oil, sesame oil, sunflower oil and soyabean oil are the unsaturated fats.

Metabolism: The metabolism of fatty acids and glycerol are separate. Glycerol enters the glycolytic pathway through the formation of 2 glycerophosphate. Fatty acids are oxidised to acetyl-CoA and it is again oxidized to Co_2 and water through Kreb's cycle of carbohydrate metabolism. Energy is the end product in the metabolism of fat also. In the stepwise reactions two carbon atoms are added or subtracted using co-enzyme A. For this reaction ATP or Adenosine triphosphate is required. The enzyme which catalyses the reaction is thiokinase. In the next step dehydrogenation of fatty acids takes place. Fatty acids are dehydrogenated into alpha beta unsaturated fatty acid. Alpha beta unsaturated fatty acids are converted or

hydrated to hydroxyacyl-CoA. Hydroxy acid is oxidised to keto acid. The last step in the fatty acid oxidation is the thiolytic cleavage of Keteacyl-CoA ester to yield a fatty acid-CoA derivative. This is repeated and Acetyl-CoA molecules are released. Different fatty acids yield varying numbers of acetyl-CoA. For example, one molecule of steraic acid gives 9 molecules of acetyl-CoA and 1 molecule of palmitic acid gives 8 molecules of acetyl-CoA and 1 molecule of lauric acid gives 6 molecules of acetyl-CoA.

Acetyl-CoA thus formed is oxidised through Kreb's cycle and glycerol is converted into glyceraldehyde - 3 - phosphate and it enters into the glycolysis pathway.

Summary of Fat Metabolism

(1) Fatty acid - α β unsaturated fatty acid

+ H2O

↓

(Beta) hydroxy fatty acid

Keto acid

↓

Acetyl CoA

(2) Glycerol————→ Glycerophosphate

↓

Glyceraldehyde - 3- phosphate.

Ketone Bodies: Normally acetyl-CoA combines with oxaloacetic acid which is oxidised in the TCA cycle. When large amounts of fat are metabolised in the absence of carbohydrate more acetyl-CoA is formed. When there is not enough oxaloacetic acid from carbohydrate, two molecules of acetyl-CoA condense to form acetoacetyl-CoA. This is hydrolysed into acetic acid and later converted into P-hydroxybutric acid and acetone. These metabolic end products of fats are known as ketone bodies. When the concentration of ketone bodies in the blood increases the condition is known as ketosis. When it is excreted through urine it is called ketonuria. In normal condition acetoacetate is either oxidised directly to carbon dioxide and water or converted to two

molecules of acetate which are later oxidised. In diabetes and during fasting ketone bodies are formed in excessive amounts. Aoart from fats, amino acids like leuine also produce ketone bodies.

Lipo Proteins and Lipotropic Factors: Lipo proteins are lipids containing protein in blood and other parts of the body. The insoluble lipids when attached to proteins form soluble lipo proteins. Cholesterol, neutral fats and phospholipids bind them with protein to form lipo proteins.

Lipo protein in the liver along with choline enhances the oxidation of fatty acids. On diets which are high in fats with saturated fatty acids and which are low in protein or in choline there is an increased deposition of fat in liver. A high cholesterol diet also leads to the production of fatty liver. An increased choline intake will bring down the fat in the liver. This effect of choline on fat release is known as lipotropic action. Methionine likewise exerts a marked lipotropic action owing to its ability to promote the synthesis of choline by the transfer of methyl groups to suitable precursors. Inositol (B-complex vitamin), tryptophan and glutamic acid exert this lipotropic action. Essential fatty acid deficiency can also produce fatty liver, pantothenic acid or riboflavin deficiency may also lead to it

Requirements: The requirement of fats is about 50 to 60 gms for a normal person. The dietary fat is used for energy and it is also a source of essential fatty acids. At least fifty per cent of the fat must be from vegetable oils which are rich in essential fatty acids.

An average Indian adult derives about 20 per cent of energy from dietary fat.

Sources: Vegetable oils, like groundnut oil, coconut oil, mustard oil or gingelly oil, hydrogenated oils, butter and ghee and lard are the common sources of fats.

Deficiency of Fats: During infancy a deficiency of essential fatty acids results in perianal irritation and skin changes like

dry skin and itching. In adults phrynoderma is easily cured by essential fatty acids with vitamins B2 and A.

Kinds of Matters

Spices and Condiments: Spices and condiments are accessory foodstuffs mainly used for flavouring food. Some condiments are good to fair sources of carotene. Red chillies and coriander are examples of this. Green chillies are rich in vitamin C. Turmeric and tamarind are fair sources of iron. Garlic promotes intestinal synthesis of certain vitamins. Among the spices and condiments fenugreek seeds have the highest amount of protein, that is, 26.2 per cent. Cuminseeds, dry chillies and coriander seeds have a protein content of 18.7,15.9 and 14.1 per cent, respectively. Caraway has a protein content of 17.1 and fat content of 21.8 per cent in it. Coriander, cuminseeds and dry pepper have a fat content of 16.1,15.0 and 6.8 per cent, respectively. Cuminseed is also rich in calcium. About 1.080 gm of calcium is present in 100 gms. Asafoetida and coriander are also good sources of calcium. Asafoetida, cuminseeds, cloves, fenugreek seeds, turmeric, tamarind and omum are good sources of iron. Cardamon, green chillies, coriander, cuminseeds, caraway and green pepper are rich sources of carotene. Dry chillies, nutmeg, cuminseeds, coriander and fenugreek seeds are good sources of thiamioe. Even though these condiments and spices are rich in vitamins and minerals they are used only in minute quantities. Black pepper, coriander, cumin, ferugreek, mustard seeds, chillies, nutmeg, cloves, cinnamon and cardamom are used as spices in our country.

The popularity of spices is due to their stimulating effect on appetite next to its flavour. Certain spices and condiments have medicinal effect. Turmeric is believed to be anti-inflammatory. In anti-allergic drugs turmeric is used. Turmeric also reduces blood cholesterol level. Consumption of spices and condiments in large quantities produces irritation in the mucous membranes of the alimentary canal. Peptic ulcer is common among people who use condiments and spices liberally.

Herbs like curry leaves, coriander leaves, mint leaves and leaves of citrus fruits are used for preparing chutney powders and they are also used in many curries.

Gelatin Dishes: When the protein in connective tissues and bones—the collagen—is heated with water gelatin is obtained. To bring this change in collagen it has to be heated to 95°C. Some peptide bonds of collagen are hydrolysed to gel.

Commercially it is available in granulated, flake or sheet forms. It easily dissolves in hot water and on cooling it sets to a gel. Concentration of gelatin, cooling temperature, the presence of minerals like calcium and concentration of sugar affects the quality of the product. With a 2 to 3 per cent of gel concentration and at temperatures below 15°C a good gel is formed. Sugar in 6 to 10 per cent concentration enhances the speed of gel formation. If the gel is heated it melts and on cooling it sets again. Now vegetable gels are also available in the market.

Gelatin is used for decorating various types of desserts. For ice, souffles, baked ice-cream, or other creams, jelly trifles, caramel puddings, flans, milk puddings, sponge puddings, with stewed fruits, milk jellies, purees of fruits, junkets and with baked fruits gelatin can be used to make them more attractive. Gelatin acts as a stabiliser and reduces crystal formation of ice-creams.

Emulsifying and Stabilising Agents: Substances which, when added to food, gives a uniform dispersion of oils and fats in aqueous media are called emulsifiers or stabilising agents.

In food industry it is often liberally used and they are harmful. Agar, edible-gums, brominated vegetable oils, gelatin, albumin, lecithin, starches, hydrolysed proteins and stearyl lastaric are some of the examples.

Sweetening Agents: Sugar, sugarcane juice, jaggery, molasses, honey, saccharin and cyclamates are the common agents used as sweetening.

Sugar (sucrose): Sugar is almost pure sucrose made up of glucose (dextrose) and fructose (laevulose). One gramme of sugar supplies 4 kcals and a teaspoonful (5 gms) provides 20 kcals. Brown sugar and white sugar are similar in nutritive value. Sweets made from sugar produce a sense of satiety and so they must be taken after a meal. If sweets are consumed at the beginning of a meal they reduce appetite. Sugar syrup is used as a preservative for fruits, jams and preserves.

Excessive consumption of sugar and sweets affects the teeth adversely.

Sugarcane Juice: The juice obtained by crushing sugarcane is a refreshing drink. Addition of sugar and lemon juice improves its taste and flavour. Sugarcane is good for hepatitis patients. Jaggery is produced by heating sugarcane juice. It has been reputed to be good for the heart. Molasses is a byproduct in the manufacture of sugar. Fermentation of molasses produce alcohol.

Honey: Honey is a golden-coloured syrup made by bees from the nectar of flowers. It is a mixture of glucose and fructose. The quality of honey depends on the flower blossom. Honey is a soothing drink for patients with pharyngitis and tracheitis. Honey contains 17.7 per cent water; 36.5 per cent dextrose, 403 per cent laevulose and 1.9 per cent sucrose.

Gur or Jaggery is obtained from sugarcane juice after processing. Palmyra palm, dale palm, or coconut palm is also used for it.

Saccharine and Cyclamates: Saccharine and cyclamates are used as sweetening agents for calorie-conscious people. Saccharine was discovered in 1879. It is 350 times sweeter than sugar and has a bitter taste in concentrated form.

Cyclamate was discovered in 1937. It is 30 times sweeter than sucrose. Cyclamates are harmful to the body if consumed in large quantities. Malignant bladder tumours were produced in experimental animals. Excessive use of soft drinks with cyclamates produce renal tubular acidosis.

Leavening Agents: Leavening agents are substances which produce carbon dioxide in the product. This carbon dioxide contributes a porous appearance to the product as in bread, cake or in biscuits. Baker's yeast and baking powders are usually used as leavening agents. Baker's yeast when added to the dough, ferments sugar and produces carbon dioxide and alcohol. This gives the characteristic structure to the product. Baking powders are leavening agents produced by mixing an acid-reacting material and sodium bicarbonate with or without starch. There are three types of baking powders—fast acting, slow acting and double acting based on their speed of action.

Flavour Components in Foods: Food industry is a fast growing field. Much progress has been made in understanding the chemical nature of flavours in foods. Many artificial flavouring materials are now available in the market. The compounds which give the flavour characteristic in one food is given below:

Food7		*Main flavour compound*
Banana	—	Isopentyl acetate
Lemon	—	Citral
Grape-Concord	—	Methyl anthranitrate
Potato	—	2-Isopropyo-3 methosy pyrazne
Beetroot	—	Geosmin
Apple	—	Ethyl -2- methyl butyrate

Energy and Heat

Carbohydrates are compounds which contain carbon, hydrogen and oxygen. Oxygen and hydrogen are present in carbohydrates in the same proportion as in water. They are the main sources of energy for our body and the only source of energy for our nervous tissues.

Carbohydrates are mainly distributed among plant foods with few exceptions, like glycogen, lactose and ribose which

are present in muscles or in liver, human milk and animal cells, respectively.

In various parts of the plants like leaves, stems, fruits, seeds and roots, carbohydrate is stored as starches and sugars. And the plant sources supply about 45 to 80 per cent of the energy requirement of the people throughout the world. Plants through the process of photosynthesis manufacture carbohydrate and after its use store them as polysaccharides.

Importance: Carbohydrates are chemically known as saccharides, as carbon, hydrogen and oxygen form saccharide groups. Carbohydrates are classified, based on the number of saccharide groups in them:

a. Monosaccharides

b. Disaccharides, and

c. Polysaccharides.

Monosaccharides: On hydrolysis or by digestion, monosaccharides cannot be simplified as they are the simple form of carbohydrates. This is the absorbable form of carbohydrate in our body. Based on the number of carbon atoms present in them, monosaccharides are grouped into triose (3 carbons), tetrose (4 carbons), pentose (5 carbons), and hexose (6 carbons).

Monosaccharides

Biose,	Triose	Tetrose	Pentose	Hexose
($C_2H_4O_2$)	($C_3H_6O_3$)	($C_4H_8O_4$)	($C_5H_{10}O_5$)	($C_6H_{12}O_6$)
(e.g. Glycolical-dehyde)	(e.g. Glyceral dehyde)	(e.g. Erythrose)	(e.g. Xylose, ribose)	(e.g. Glucose, fructose)

Biosetriose and tetrose are not nutritionally significant

Pentoses like ribose, xylose and arabinose are of little dietary significance, xylose and arabinose are widely distributed in many root vegetables and in many vegetables. Ribose is part of riboflavin and DNA and RNA Body can synthesise it and so it is not a dietary essential.

Xylose and Arabinose are not present in free state in nature. Both of them are present in gums of various origins like woodgum, cherry gum or gum arabic.

In human nutrition only hexoses are of importance. Commonly found hexoses are aldose and ketose. The three monosaccharides of hexose group are glucose or dextrose, fructose or levulose and galactose and mannose.

Glucose: Also known as grape sugar, glucose is an aldose sugar. It is a white crystalline substance easily soluble in water with a sweet taste. Glucose is readily absorbed from die stomach. Glucose is used to detect the disorders in carbohydrate metabolism because it is in the form of carbohydrate circulating in blood. A normal healthy person has 80-100 mgs of glucose/ 100 ml of blood. Glucose is present in fruits and honey.

Fructose known as fruit sugar is a ketosugar. It is sweeter than glucose. Fructose enters the cells without the aid of insulin and so it is recommended for diabetics. Honey is the richest source of fructose (30-40%). Fructose on reduction yields a mixture of sorbitol and mannitol.

Galactose is not present in nature. It is present only in milk sugar lactose; on hydrolysis lactose yields galactose and glucose. Galactose is present in cerebrosides in brain and nervous tissues.

Disaccharides: They are complex sugars with two saccharide groups and are formed by the condensation of two monosaccharides. On hydrolysis or toy digestion they are split into monosaccharides. Disaccharides have a structure of $C_{12}H_{22}O_{U}$. They are water soluble, diffusible and crystallisable. They vary in their sweetness.

Commonly found disaccharides are sucrose, maltose and lactose.

Sucrose, otherwise known as invert sugar, is the commonest form of sugar in the diet. Either by hydrolysis or by enzyme action sucrose is simplified into glucose and fructose. It is present in sugarcane, beetroot and in many fruits and

vegetables, and honey. Sucrose is formed by the condensation of one molecule of glucose and one molecule of fructose.

Maltose is a disaccharide which contains two molecules of glucose. It is otherwise known as malt sugar. Starch is converted to maltose before breaking down into glucose. It is present in sprouted grains. Cereal grains are rich in maltose.

Lactose is also known as milk sugar and contains glucose and galactose. Cow's milk contains about five per cent of lactose and human milk has about 6.8 per cent It is easily digestible but it is not as sweet as cane sugar. Lactose is synthesised in the mammary gland. It is less in sweetness compared to other sugars.

Trisaccharides: They are with three saccharide groups but are not of any nutritional significance. The fibrous parts of certain beans and root vegetables which are not digestible contain trisaccharides, e.g., Raffinose is present in beetroots and cottonseed meal.

Tetrasaccharides: They contain four monosaccharides and like trisaccharides they are of no nutritional significance, e.g., stachyose and scorodose.

Polysaccharides: They are large complex molecules of monosaccharides linked together. Several hundreds of glucose units are linked together in polysaccharides. They are insoluble in water. Starch, dextrins, glycogen, pectin and cellulose are the common forms of polysaccharides.

Starch: Plants store carbohydrates in the form of starch and it is the main source of nourishment for human race. Cereal grains, seeds, roots like potato, tapioca, yam and plantain contain considerable amount of starch. On cooking, starch absorbs water and it swells and ruptures. This thickening quality of starch is used in cookery to produce a variety of dishes. Different sources of starch behave differently. Maize starch and cornflour are better 'thickening agents' than rice or wheat starch. All starches are broken down into glucose in the digestive system.

The characteristics of starch molecule depend upon the way in which 2,000 or more glucose units are linked. The number of glucose molecules in starch vary from 2,000 to 15,000. The glucose chain in starch are of two types. They are amylose and amylopectin.

(i) *Amylose:* Amylose is a long chain of glucose and starch has about 10-20 per cent of amylose molecule in it One molecule of amylose has 500-5,000 glucose molecules. Amylose is easily soluble in water and it is mainly responsible for the stiffening of cooked rice on keeping.

Amylose chain - G-G-G-G-G-G-G-G-G

(ii) *Amylopection:* Starch contains about 80-90 per cent of amylopectin in them. Amylopectin molecule has more glucose units. Usually one amylopectin molecule has 50,000-5,00,000 glucose molecule. Glucose molecules are arranged in branches.

```
                G
                G
                |
          G-G-G-G-G-G G
                G
                 -G-G-G-G
                 -G-G-G-G
           -G-G-G-G-G-G G
G
G
G
G
G
```

Amylopectin molecule

Small branches of amylopectin has 20-30 glucose units and it has a colloidal property and this is responsible for thickening of starch water on heating.

Dextrins: When starch is partially broken into fragments either by digestion or by acids they are called dextrins. They are like starch itself. Dextrin is broken down into maltose.

Starch is broken down into-amylodoxtrin, erythrodextrin, achrodextrin and maltose.

Pectin is a polysaccharide with no nutritional significance. But it is useful in the preparation of jam and jelly and thus it contributes to the palatability of foods.

Glycogen can be described as the animal starch as it is in this form animals store carbohydrate in the body. Glycogen is stored in the liver and muscles. This is the form of immediate energy for the body. About 350 gms carbohydrate is stored as glycogen in the body.

Cellulose is an insoluble, indigestible polysaccharide. More than 3,000 glucose units are there in cellulose but it is not of human utilisation. Cattle can digest cellulose. Even though it is not of much food value it provides bulk to the diet and thus helps movements in the large intestine. It prevents constipation and to an extent cancer of the bowel. It reduces the cholesterol level in blood as well as body weight. A high fibre diet can help in the treatment of obesity as it delays digestion and contributes satiety to obese people.

Sugar Alcohols: There are three sugar alcohols available commercially. They are sorbitol, mannitol and dulcitol. Sorbitol is used by diabetic patients. D-sorbitol or D-glucitol is an alcohol made from glucose by hydrogenation. By this process the aldehyde (CHO) group of glucose is reduced to alcoholic groups (CH,OH). Mannitol is produced from mannose, a monosaccharide which does not occur free in nature.

Functions in Body

Energy Supply: The most important function of carbohydrate is to supply energy or fuel for the body. Energy from carbohydrate is utilised by the body for immediate use. Nervous tissues and erythrocytes use glucose for their energy whereas other tissues can make use of fatty acids for energy.

Special Functions of Carbohydrates in the Liver: Carbohydrates have protective and detoxifying action in liver,

liver is protected against bacterial toxins by producing glycuronic acid or acetify groups from carbohydrates. They are excreted through kidneys. Since the defences of the liver against toxic agents are of great importance to the body, this function of the liver is very valuable. For this purpose glycogen is made use of and thus carbohydrates play a significant role in removing poisonous substances from the liver. Carbohydrates, especially lactose, help the growth of desirable bacteria in the small intestine and help the synthesis of some B-complex vitamins in the intestinal tract. Lactose also helps in the absorption of calcium.

Carbohydrates and the Heart: Carbohydrate is used as the most efficient fuel for muscular exercise. Glycogen stored in the heart muscle is used for this purpose, especially in an emergency.

Carbohydrates and the Central Nervous System: Continuous supply of carbohydrate is essential for nervous tissues to function normally. Glucose as such has a specific influence in maintaining the functional integrity of nervous tissues.

Carbohydrates and Protein-sparing Effect: If carbohydrate supply is enough in the diet, protein is spared for important functions of it. This effect exerted by carbohydrate is called protein-sparing effect Insufficient intake of carbohydrate forces amino acids to be de-aminated and converted to fatty acids for energy purpose.

Carbohydrates and Non-essential Amino Acids: Carbon skeleton of carbohydrate is used by the body for the synthesis of non-essential amino acids.

Carbohydrates and Fats: Carbohydrate is essential for the oxidation of fat. In extreme restriction of carbohydrate, fat may be metabolised rapidly.

Carbohydrates in Diet: Carbohydrate provides flavour and variety to the diet Cellulose and other indigestible polysaccharides add bulk to the diet Carbohydrate also retains water content of colon which helps intestinal motility.

Rich Sources: The best sources of carbohydrates are cereals like rice, wheat, millets like ragi, maize, roots and tubers like potato, tapioca, sweet potato, yam and chalcosis, pulses, sugar and jaggery; honey, fruit and vegetables are the other sources of carbohydrates. Banana, apple, plantain and dried fruit are also good sources of carbohydrates.

In an average Indian diet carbohydrates are the cheap sources of calories. One gram of carbohydrate gives 4 kcals and an average Indian diet contains about 300 gms of carbohydrate.

Consumption and Digestion: In the mouth salivary digestion takes place. Carbohydrates are broken down by mastication and the saliva prepares the broken pieces of food for swallowing. Ptyalin, the enzyme present in saliva, acts on starch splitting it into dextrin and maltose. This action continues in the stomach till the acidity of the gastric glands interferes. There is not any digestive action on carbohydrates in the stomach as there is no enzyme in the stomach to act on carbohydrates.

The main digestive action on carbohydrate takes place in the small intestine. Amylase of the pancreatic juice hydrolyses the remaining starch into maltose. The intestinal/ mucosa cells contain sucrose or invertose, maltose and lactose. These enzymes convert sucrose into glucose and fructose, maltose into glucose and glucose and lactose into glucose and galactode, respectively.

The end products of carbohydrate digestion—glucose, fructose and galactose—are absorbed in the intestine. Bulk of these sugars are absorbed in the duodenum and jejunum and small amounts are absorbed in the ileum. Glucose and galactose are readily absorbed while fructose is absorbed slowly. These monosaccharides are absorbed by the capillaries in the walls of the intestines and they are carried by the bloodstream to the liver and the tissues.

The absorption of carbohydrate from the intestine is

controlled by certain factors like condition of the intestinal tract and muscle tone, endocrine glands like anterior pituitary, thyroid and adrenal cortex and their functions and vitamin B-complex content in the diet.

Carbohydrate Metabolism: Through portal circulation the absorbed monosaccharides are carried to the liver. There galactose and fructose are converted into glucose. The liver cells release some glucose into the bloodstream and blood carries it to the tissues. In the tissues this glucose is metabolised to release energy. Excess of glucose is polymerised in the liver into glycogen and stored in the liver and muscles. This is again converted into glucose when energy is required. This process helps the bloodstream to maintain the normal blood sugar level.

Glycogen is formed from glucose, fructose and galactose and it is shown below:

Glycogen Galactose
↑ ↓
←
Glucose-1 – Phosphate → Galactose-l-Phosphate
Glucose → Glucose-6-Phosphate
Fructose—Fructose-6-Phosphate.

Formation of Glucose from other Monosaccharides

The carbohydrate metabolism in the body involves a series of chemical reactions. Glycogen formation, breakdown of glycogen and glucose to supply energy, formation of glucose from amino acid and glycerols of fat and formation of fat from carbohydrate are the main changes related to carbohydrate metabolism.

Breakdown of glycogen and glucose to yield energy involves a series of gradual reactions. Some of these chemical reactions take place in the absence of oxygen and it is known as the analrobic phase of carbohydrate metabolism. In the aerobic phase of metabolism oxygen is used. In total 38 ATP are yielded by each molecule of glucose: 8 ATP through

glucolysis, 6 from pyruvic acid to Acetyl-CoA, 24 from TCA cycle.

Anaerobic Phase of Carbohydrate Metabolism (Glycolysis)

In glycolysis glycogen is degraded to glucose and it is converted to lactic acid.

The energy release in the process of glycolysis is calculated from the breakdown of ATP (Adenosine Triphosphate) and regeneration of ATP. Two molecules of ATP are used and four molecules of ATP are formed. Glycolysis takes place in muscular tissues.

Glycolysis is also known as Embden-Meyerh of-Parnas scheme of carbohydrate metabolism. The basic steps involved in glycolysis are given in Fig.

Aerobic phase of carbohydrate metabolism involves oxidation of lactic acid into pyruvic acid and pyruvic acid to Acetyl-CoA. Acetyl-CoA undergoes a series of a chemical reactions whereby carbon dioxide, water and CoA are formed.

Glycogen

Glucose-1-Phosphate

Glucose Glucose-6—Phosphate

Fructose—-6—Phosphate

Fructose-l-6-diphosphate

Glyceraldehyde-3-phosphate →Dihydroxy acetone phosphate

1,3 Diphosphoglyceric acid

3 Phosphoglyceric acid

2 Phosphoglyceric acid

Pyruvic acid

Kreb cycle———————— Lactic acid

The conversion of Acetyl-CoA is otherwise known as Kreb's Tricarboxylic acid or T.CA. cycle.

Oxalacetic acid reacts with Acetyl-CoA and this is repeated.

Most of the absorbed glucose is oxidized releasing energy. During this process carbon dioxide and water are also formed. Some portion of glucose is converted to fatty acids and some amount combines with free amino group to form amino acids.

B Vitamins like thiamine, niacin, riboflavin, pantothenic acid and B6 and minerals like magnesium and phosphorus are very essential for the proper metabolism of carbohydrates.

Certain hormones like epinephrine, thyroxine and glupxine and glucagon and islands of Langerhans are essential to maintain normal blood sugar level in a person. A healthy person's blood shows the fasting sugar level as 80-100 mgV 100 ml. of blood. After a meal it rises to 130-150 mgV 100 ml. of blood. If the blood sugar level is above this the person is termed as a diabetic patient

If the carbohydrate intake in the diet is inadequate, it leads to malnutrition and other metabolic disorders. Tissue proteins and fats will be utilised for energy purpose. If excess carbohydrate is taken, it will contribute towards obesity and diabetes.

Regulation of Blood Sugar: Blood circulates glucose continuously to each and every cell of the body as a source of energy and for the synthesis of a number of substances. From the blood if glucose is taken, the liver immediately replaces the blood glucose.

In the fasting state blood sugar level is 60-85 mgm per 100 ml of blood. Soon after a meal it rises to 140-150 mgm per 100 ml. If the body is in a good carbohydrate metabolic pathway this concentration will fall down to a normal level of 60-85 mgm. The liver very efficiently maintains the normal blood sugar level. If the blood sugar concentration is high this condition is known as Hyperglycaemia. Hypoglycaemia occurs when the blood sugar level is below normal level. This occurs in certain abnormalities of liver function or when insulin is produced in excessive amounts by the pancreas.

The liver is the only organ in our body which can either supply glucose to the circulation or to remove the excess sugar from blood. Glucose is made available to blood from absorption, from intestine glucogenesis, i.e., glycogen is produced from glucose and reconverts it; glucoseogenis, i.e., de-aminated amino acids are used for the synthesis of glucose, glycogenolysis is reconversion of stored glycogen to glucose and lipogenesis, i.e., excess glucose is converted to fatty acids and the use of glycerol to form glucose.

3

Eatable Items

Fish

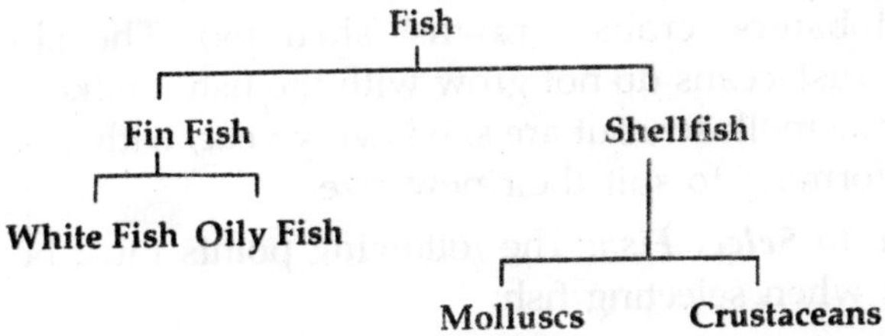

Fin fish are vertebrates and have skin and scales which cover the body. They move with the help of fins. They are subdivided into:

(1) *White fish* which are mainly flat fish, and contain oil only in the liver. Most of these are deep sea fish. Common local examples are pomfret, sole etc.

(2) *Oily fish* which are mainly round fish and contain fat all over the body. The amount of fat varies from 1.5% to 20% in different varieties. These fish are often pigmented and tend to be surface fish. Example: mackerels, sardines etc.

Shellfish, as the name denotes, have a shell covering the body. They are invertebrates. They are subdivided into:

(1) *Molluscs*: (a) Bivalves (e.g. oysters, scallops, mussels, clams and cockles) which have two distinctly separate shells joined by a hinge-like membrane. The movements of the shell are controlled by a strong muscle. When the muscle relaxes, the two halves of the shell fall open. The shell also opens when the organism dies, thus exposing the contents of the shell to contamination from outside, resulting in quick putrefaction. (b) *Univalves:* (e.g. whelks and winkles). These are recognised by the characteristic spiral formation of their shells, which unlike those of bivalves are not divided into halves.

The shells of molluscs increase at the rate of one ring per year to allow for the growth of the organism. The age of the mollusc can thus be roughly estimated by the number of rings on the shell.

(2) Crustaceans have a segmented, crust-like shell (e.g. lobsters, crabs, prawns, shrimps). The shells of crustaceans do not grow with the fish, unlike those of the molluscs, but are shed every year, with a new one forming to suit their new size.

How to Select Fish: The following points must be borne in mind when selecting fish:

(i) Eyes should be bright and not sunken.
(ii) Gills should be red.
(iii) The tail should be stiff.
(iv) The flesh should be firm and not flabby.
(v) The scales if any should be plentiful.
(vi) There should be no unpleasant odour.
(vii) To test a cut piece, press down with a finger, and if an impression is left then the fish is stale.
(viii) Any tendency for the raw flesh to come away from the bones is a dangerous sign.

Scaling and Cleaning:

(i) To scale, use the blunt side of the knife. Holding the fish by the tail, scrape firmly from the tail towards the head. Wash well to remove any loose scales,

(ii) Remove fins and trim tail,

(iii) To clean fish such as herrings, mackerel, etc., slit the belly, using a pair of kitchen scissors, from the head to almost half way down. Remove the head if desired,

(iv) In the case of flat fish such as pomfret, etc., cut away the gills, and make a small opening in the belly. Remove the entrails and wash well. If the head is to be removed, do this with a semicircular cut at the base of the head.

To fillet the fish: Slit the fish down the backbone with a sharp knife and separate the flesh from the backbone on each side with short sharp strokes. Then slit from the side and take out each fillet.

To bone herrings After slitting and cleaning, open gently. Turn over and place the inside on hard board. Press firmly along the backbone to loosen the bone. Turn over and starting at the head raise the backbone with the thumb and forefinger. Pull the bone steadily away from the fish. After pulling out the bone, rinse the fish well and drain.

Cooking of Fish: The flesh of fish is very delicate and, therefore, great care must be taken to prevent the flesh from breaking during cooking. When boiling or cooking fish, the liquid should be at simmering point before fish is added as the connective tissue in fish is softer and the quantity is less than in meat. When preparing fish stew or fish curry, fish is put into cold liquid to flavour the gravy as well. Whole fish may also be covered with a cold liquid and brought to boil. This is done to prevent shrinkage of the skin. Steaming is a better method of cooking fish as it is a gentle method. Fish cookery is not a special method of cooking. All methods of cooking can be applied bearing in mind that the delicacy of the flesh of the fish and the fat content of the fish determine the cooking

Cleaning of Fish

What to do	*What to do*
Do not work without having a basin prepared to hold the fish bones, and another one for the offal.	Wash and dry fish with a clean swab. Cut off all the fins and trim the tail; re move the eyes if necessary.
Do not work directly on a table; use a chopping board.	Remove the scales, if any, by scraping them with a knife from tail to head, then remove the gills.
Do not throw the fish scales on to the table; remove from the knife by warping the knife against the side of the basin to collect them.	Remove all the entrails, roe, liver etc., from the gill-slits t h e m by pulling the gut out with a fork, or the hook of a small ladle handle. Make a small opening, to 5 cm. long, from the vent to the belly of fish such as herring, whiting, trout, and salmon, to remove the roe and clean the fish properly. For turbot, brill, etc. make the opening behind the head.
Don't mix different fish together; always begin with one kind and place them on one side to be washed with the others, when they have all been cleaned from inside.	
Do not have your table covered with scales or offal of fish. Keep your chopping-board and table clean by swabbing often.	After removing roe, such as herring roe, place them imm-ediately in a basin of cold water.
Do not forget, when removing the entails of a fish, to clear the bones of the blood vessels; and ensure that there is nothing left inside the fish by feeling with one finger.	If the head has to be removed, cut it from the gill in a V-shape.
Do not cut the ligament between the head and the body when removing the gills from small fish.	Wash the inside of the fish well under running cold water. With a clean or clothe swab or cloth dry inside as well. Then inspect the interior of the fish carefully.
Do not burst open or cut off the belly of a small fish when cleaning the inside	
Do not place a fish on a trav unless it has been well washed, dried and caiefully examined.	

method. The time taken is always much less than for cooking meat. An average timetable is as follows, but much depends on the size and shape of the fish.

Boiling and baking		10 minutes per 500 gm and 10 minutes over
Steaming		15 minutes per 500 gm and 15 minutes over.
Stewing		Half to three quarter of an hour formes time the stew begins to simmer.
Frying (for small fish or thin pieces)		5 minutes
Grilling		5 to 7 minute's
Lean fish	Steamed or	Poached, deep-fried or baked in a
Oily fish	Baked	Sauce. Baked, grilled or pan-fried.

Boiling of fish: Fish is generally boiled in a court bouillon. This is made of salt, water, milk and lemon juice for sea fish with white flesh. For salmon, trout and shellfish and various types of river fish the following court bouillon is used.

Court Bouillon

Water	1 litre
Salt	15 gm
Vinegar	75 ml
Carrots (sliced)	60 gm
Onion (sliced)	60 gm
Bay leaf	1
Parsley stalks	2 to 3
Peppercorns	6
Thyme	1

Simmer all ingredients for 30 to 40 minutes.

Strain and use as required.

The stock after fish is boiled can be used as a base for different types of sauces. Fish is cooked when the flesh separates easily from the bone. Test at the thickest part nearer the bone. It there is no bone then the fish is considered done when a creamy substance begins to run from the fish.

Steaming of Fish: Place the fish in a steamer. Sprinkle with salt and pepper. The water in the steamer must be fast boiling all the time the fish is cooking, if large, the fish can be turned once. Test as for boiling.

Grilling of Fish: Heat the grill and grease the rack. Clean fish. Sprinkle over with salt, pepper and lime juice. Gashes should be made on the fish to allow heat to penetrate, otherwise the outside becomes dry before the inside is cooked. For large fish, baste frequently with melted fat. For oily fish such as herrings and sardines, no extra fat is required.

Frying of Fish: Fish can be fried whole, if small, or cut into steaks or fillets or pieces if large. For cut pieces, it is preferable to coat the fish with egg and bread crumbs or seasoned flour and milk.

Baking of Fish: This method is most suitable for medium sized fish or the middle portion of a large fish. Scale and clean, but leave the head and tail on. Stuff the belly with a savoury forcemeat (11 gm fresh bread crumbs, 60 gm fat, grated rind of half a lime, 2 tsp chopped parsley, ½ tsp mixed herbs, salt and pepper and egg or milk to bind). Sew up with a needle and coarse thread. Do not fill the fish too tightly or it will burst, since the stuffing swells when it is cooked. Place the fish in a baking tray with a little fat, in a moderately hot oven, and cook till the flesh leaves the bones. To serve, remove the thread. Garnish with parsley and lemon. Serve with a sharp sauce.

Soused Herring or Mackerel: Soused herrings and mackerel are served as an hors d'oeuvre or for breakfast. Scale and clean the fish. Cut off the head and tail. Remove entrails. Wash well. Put back roe if any. Place on a plate. Sprinkle over with a little

salt and leave for a couple of hours. Drain the fish. Roll up and place in a casserole dish. To 500 gm of fish add 12 peppercorns, 1 bay leaf and a large onion thinly sliced. Pour over equal proportions of vinegar and water, the liquid being just sufficient to cover the fish. Bring quickly to boil and simmer for a few minutes before removing from the casserole. Serve garnished with parsley.

Shellfish

Choosing Shellfish: They should be bought in season as far as possible. They should be medium- sized and of good weight. It is best to buy them' alive, particularly oysters, crabs, and lobsters. If they are dead, the following precautions must be taken: (a) The claws of crabs should be springy and not hanging down, (b) The eyes should be bright, (c) The tail of lobsters should spring back when stretched out. (d) Shrimps and prawns must be crisp, (e) Oyster shells should be tightly closed.

Preparing and Serving Shellfish: Shellfish can be served in a variety of ways. Lobsters and crabs are boiled alive. The meat is removed after cooking, flaked and replaced in their own shells with sauces or served in salads or other cold dishes. The meat is mixed with sauce and served in patties, or in scallop shells or on toast and canapes. Smaller shellfish such as prawns and shrimps must be shelled first, the intestines (the black thread- like substance found at the back) removed, then washed and cooked. They are served whole, boiled or steamed, stewed or in curries, fried with masala or in batter, devilled and in soups. Shrimps and oysters can be covered with sauces. Lemon and vinegar are served with shellfish to aid in digestion.

What to Serve with Fish:

When fish is served as a main dish, it is usually served with potatoes, a salad or a cooked vegetable and a sauce. Here are some suggestions.

FISH	POTATOES	VEGETABLES	ACCOMPANIMENTS
Fried or grilled	Fried, boiled or sauteed	Green salad, cole slaw, green peas or beans, carrots, tomatoes parsnips, cauliflower,	Hollandaise, tartare, caper, tomato, mustard or cucumber sauces, lemon, parsley, butter, chutney, pickled beetroot or red cabbage.
Boiled or steamed	Boiled in jacket or mashed	Green peas or beans, carrots, tomatoes, spinach, celery, green salad, cucumber salad, parsnips, hot beetroot, sprouts, cauliflower, onions.	Hollandaise, cheese, horseradish, chutney, parsley, egg, fennel, anchovy or lemon sauces.
Baked	Baked, boiled or sauteed	Green peas or beans, carrots, green salad, tomatoes, cucumber salad, spinach, celery, parsnips, hot beetroot, sprouts, cauliflower, Onions.	Hollandaise, mustard, anchovy, cucumber, tomato, tartare, caper or brown sauces, lemon, parsley, butter.

Meat: The term meat is generally applied only to butcher's meat—beef, mutton, lamb, pork and kid.

The tenderness of the meat depends on the age and feeding of the animal, on the hanging and preparation after killing. Muscle structures have an effect on the tenderness and the texture of the meat. The skeletal muscles are made up of fibres which are composed of proteins, salts and nitrogenous extractives. The fibres are grouped parallel to each other in bundles held together by connective tissue and surrounded by a sheath of heavier connective tissue. The size of the bundles varies with different muscles and determines to some extent the texture of the meat. There are two kinds of connective tissue, collagenous or white connective tissue and elastic or yellow connective tissue.

The flavour of meat is determined by various factors such as feeding, age, sex and maturity. The older the flesh the greater the flavour in most cases; young flesh is not so highly flavoured but is sweeter and more tender. Flesh of male animals is stronger in flavour than that of females unless the animal has been castrated. Well developed muscles are usually more strongly flavoured, e.g., chicken leg as compared with breast. The longer the meat is hung the stronger the flavour, owing to certain chemical changes. Meat should not be eaten immediately after killing.

It has to be hung till tender. Shortly after death rigor mortis sets in and the muscles stiffen. Meat must be hung at 1° to 3°C (about 34° to 38°F) at least, till this passes off and until the acids which then develop have had a chance to soften the connective tissue. The enzyme present in the tissues also helps to make the meat render and more juicy. In cold countries this process takes at least 3 days but it is much quicker in warmer temperatures, varying between 12 to 24 hours. If so desired meat can be hung for a longer period. Not all meats should be hung like this and veal and pork particularly should be eaten fresh. They should also be thoroughly cooked before use.

Generally speaking, old meat and meat with considerable muscular development will be darker in colour and stronger in flavour. The flavour of the meat varies depending on the saltiness and sweetness of the blood, the species, the age and sex of the animal and the degree of maturity when eaten. Another factor is fat. Pork and mutton for example, have individual flavours depending on the fat. The third factor is the amount of sugar in the meat which on cooking tends to caramelise. Wherever possible select meat which is marbled. This is meat where the fat is interspersed between the muscle fibres and connective tissue. This shows good feeding. Marbling helps to keep the meat moist when cooked by dry methods. Marbled meat is also more nutritious.

There are no special methods of cooking meat and the fundamental methods of cooking can be deftly adapted to suit the particular joint. The amount of connective tissue present in the meat determines to a great extent the method of cooking. When connective tissue is present in large amounts, the meat, being tough, is cooked by moist methods. For those cuts which have only a small amount of connective tissue, dry heat cooking may be used. Long-fired cuts of meat require longer cooking than short-fired ones. They should be stewed or braised to make them tender. Shorter-fired meat is cooked by dry heat. The most developed muscle fibres are usually the longest and these coarser fibres are found in the legs, neck and tail. The shorter fibres are found in the fillet, rump and loin.

Dry heat cooking (roasting, frying and grilling or broiling) develops the characteristic meat flavour, coagulates the proteins, thus making fibres firm, hardens the connective tissue, and changes the colour. This change in colour is caused by a change in the composition of haemoglobin. Shrinkage is observed in meat cooked by the dry heat method. Moist cooking (braising, boiling, stewing, etc.) besides developing flavour and coagulating the meat, softens the collagenous tissue, thus rendering meat tender. Elastic connective tissue is softened only slightly by moist heat. There is no shrinking in meat cooked by moist methods.

When cooking meat it is desirable to retain the juices in the fibres. Steaks, roasts, etc., are cooked either at a constant temperature of 149°C (about 300°F) for beef, veal and lamb and 177°C (about 350°F) for pork, or by the searing method, where the meat is cooked at a high temperature for the first 10 minutes and then cooked at a lowered temperature. This method was supposed to help in retention of juices but this has been disproved although the aroma, flavour and colour that accompany searing are preferred by some. Steaks are usually grilled at 204.5°C (about 400°F). Studies have illustrated that a lower oven temperature 171°-177°C (about 300°-350°F) as compared to a higher temperature 218.3°-232.2°C (about 425°-450°F) results in less drip loss, less shrinkage, increased juiciness and more uniform colour throughout a cut. Overcooking or extended cooking at high temperature denatures the protein of meat, eggs, fish and birds or they toughen or become stringy or all three.

If a roast is cooked at a high temperature, the time of cooking is decreased, but dripping and evaporation losses are higher, and the meat is not so evenly cooked. The degree of 'doneness' also affects this loss, the steak done 'rare' or 'underdone' showing' less than that done 'medium' or Veil done'. A meat thermometer helps in determining the degree of 'doneness'. The thermometer should be inserted so that the bulb is at the centre of the cut but not in contact with the bone. For thin cuts the thermometer is not very satisfactory.

Meat meant for boiling should be dipped into salted boiling water and then allowed to simmer till tender. A pressure saucepan may be used to save time when cooking less tender cuts of meat. Care should be taken to prevent cooking of meat in a pressure cooker or at a high temperature longer than necessary, otherwise it becomes hard, dry and flavorless.

Frozen meat may be prepared by the same methods as fresh meat. It need not be thawed before cooking. If it is not thawed, however, the cooking time is considerably increased, to as much as two to three times the time required for cooking

thawed or fresh meat. Thawing can be done in the refrigerator or at room temperature. When meat is not thawed, it does not retain a coating such as egg and bread crumbs.

Tenderisers such as papain (taken from raw papaya) may be used to soften the fibres before cooking. These must be applied to the meat, which should then be allowed to stand for one hour for a cut of 2.5 cm. (1") thickness. The acidity of vinegar, lemon juice, curd, cider and wine which can be added before or during cooking also helps to tenderise meat.

Sign of Quality: Mutton and lamb lamb is meat from a lamb under one year old):

(i) A good quality animal should be compact and evenly fleshed,

(ii) The lean flesh should be firm, of a pleasing dull red colour and of a fine texture or grain,

(iii) There should be an even distribution of surface fat, which should be hard, brittle and flaky in structure, and of a clear white colour,

(iv) In a young animal the bone is pink and porous, so that when it is cut, a small amount of blood can be seen. With progressive aging the bones become hard, dense, white and inclined to splinter when chopped.

Order of Dissection:

(i) Remove the shoulders,

(ii) Remove the breasts,

(iii) Remove the middle neck and scrag end.

(iv) Remove the legs.

(v) Divide the saddle from the best end.

Approximate weight of the carcass of an Indian animal:
Lamb: 10 kg. (22 1b.)

Mutton: 16.8 kg. (37 lb.)

Whereas the approximate weight of an English animal is:

Lamb: 14 kg. (32 lb.)

Mutton: 23 kg. (50 lb.)

Pork

Fresh pork is at its best from September to April (the months with an 'r' in them). At 5 to 6 weeks it is known as suckling pig.

Sign of quality

(i) Lean flesh should be pale pink, firm and of a fine texture,

(ii) The fat should be white, firm, smooth and not excessive,

(iii) Bones should be small, fine and pinkish,

(iv) The skin or rind should be smooth.

Pork: Fresh pork is at its best from September to April (the months with an 'r' in them). At 5 to 6 weeks it is known as suckling pig.

Sign of Quality:

(i) Lean flesh should be pale pink, firm and of a fine texture.

(ii) The fat should be white, firm, smooth and not excessive.

(iii) Bones should be small, fine and pinkish.

(iv) The skin or rind should be smooth.

1. Leg
2. Loin
3. Spare-rib
4. Belly
5. Shoulder
6. Head (whole)

   Trotters

   Kidneys

   Liver

Order of dissection:

(i) Remove the head.

(ii) Remove the trotters,

(iii) Remove the leg.

(iv) Remove the shoulder.

(v) Remove spare ribs.

(vi) Divide loin from the belly.

Joints and their Uses

	JOINTS	USES		APPROX. WEIGHT (Lamb)	Mutton)	
			French	(Lamb)	Mutton)	
(i)	Shoulder	(L'epaule)	Roasting Stewing	2.00 kg	3.5 kg.	
(ii)	Leg	(Le gigot)	Roasting Boiling	2.75 kg.	3.75 kg.	(iii) Breast
(La poitrine)		Roasting	0.9 kg Stewing	1.75 kg.		
(iv)	Middle neck	(Le collet)	Stewing	1.15 kg.	1.7 kg.	
(v)	Scrag end	(Le cote decourverte)	Stewing Broths	450 gm.	0.9 kg.	
(vi)	Best end	(Le carre)	Roasting Grilling Frying	575 gm.	20 kg.	
(vii)	Saddle	(La selle)	Roasting Grilling Frying	1 kg.	3.5 kg.	

(viii)	Kidney	(Le rognon)	Grilling Saute Braising	1.5 kg.	2.25 kg.
(ix)	Heart	(Le coeur)	Braising	115 gm	60 gm.
(x)	Liver	(Le foie)	Frying		350 gm.
(xi)	Sweetbread	(Le ris de veau)	Braising Frying		
(xii)	Tongue	(La langue)	Boiling Braising		

Joints and their Uses

	JOINTS	USES	APPROX.
	French		*WEIGHT*
(i) Thigh	(La cuisse)	Roasting Boiling	4.55 kg.
(ii) Loin	(La longe)	Roasting Frying Grilling	5.45 kg.
(iii) Spare rib	(Le cote decou- verte de pore)	Roasting Pies	1.35 kg.
(iv) Belly (Breast)	(La poitrine)	Pickling Boiling	1.8 kg.
(v) Shoulder	(L'epaule)	Roasting Sausages Pies	2.7 kg.
(vi) Head (whole)	(La tete)	Brawn	3.6
(vii) Trotters	(Le pied)	Grilling Boiling	
(viii) Kidney	(Le rognon)	Saute Grilling	
(ix) Liver	(Le foie)	Pate	

Bacon and Ham: Bacon is made by preserving the meat of carefully bred pigs. The preservative used is salt. This can be applied in two ways:

(i) by covering the meat with dry salt and allowing at least 4 weeks penetration;

(ii) by using a salt solution (brine).

This is injected into the flesh and then the meat is immersed in a salt solution for 5 days. Sometimes a combination of the two methods is used—brine is injected into the meat and the sides are then stacked and covered with dry salt for 2 to 3 weeks. It is then matured for about 3 weeks under controlled temperature. Desirable bacteria acting on the meat juices and salt produce the characteristics associated with well matured bacon.

This bacon is now ready for sale as 'green' or 'pale' bacon. In certain parts of the world the bacon is preferred smoked. So it is then hung over smouldering wood dust for 2 days. This extracts excessive moisture and gives the bacon a distinctive flavour. Smoking also improves the preservation as it acts as an antiseptic. This is particularly suitable in hot climates.

Good quality, well cured bacon should have a pleasant smell; the rind should be thin and smooth, the fat firm and free from any yellow marks, the lean part a good deep pink colour and the flavour mild and mellow.

1. End Collar
2. Prime Collar
3. Fore Hock
4. Top Back
5. Top Streaky
6. Back and Ribs
7. Prime Streaky
8. Short Back
9. Thin Streaky

10. Long Back
11. Flank
12. Corner Gammon
13. Slipper
14. Middle Gammon
15. Gammon Flock

Ham and Gammon: Ham and gammon come from the hind leg of a pig Gammon is cut off the side of a cured porker. Ham, on the other hand is cut off at the fresh pork stage and includes a substantial part of the loin. It is cured after cutting, usually by the long process of dry salting.

Cuts and their Uses

CUT	USE
(i) End collar	Boiling; Braising
(ii) Fore hock can be bought boned and rolled	Boiling
(iii) Prime collar	Boiling (either whole or cut into smaller pieces)
(iv) Top streaky used for frying and grilling. Can be boned and boiled.	Thinly sliced rashers can be
(v) Top back (good lean cut) in thin rashers and grilled.	Boiled or braised whole or cut
(vi) Back and ribs	Cut into regathers and grilled.
(vii) Prime streaky (the best cut of streaky bacon).	Boiled in one piece or cut into rashers.
(viii) Short back (one of the best cuts for breakfast frying)	Frying
(ix) Thin streaky Frying (Crisp and well cooked)	Grilling
(x) Long back or grilled quickly.	Sliced thinly and fried gently

Contd...

	CUT	USE
(xi)	Flank (good cut to be served with liver or other meats and in a minced bacon recipe)	Boiling Frying
(xii)	Corner gammon (lean and well flavoured) Boiling whole (serve hot or cut into	rashers for grilling.
(xiii)	Slipper (gammon lean small cut)	Boiled whole or cut into rashers.
(xiv)	Middle gammon or lean rashers for frying.	Boiled whole (serve hot or cold)
(xv)	Gammon hock	Should be partly boiled and then baked (serve hot or cold.)

Beef

Signs of Good Quality:

(i) The lean meat should be bright red with small flecks of white fat.

(ii) The fat should be firm, brittle in texture, creamy white in colour and odourless. Older animals and dairy breeds have fat which is usually deeper yellow in colour,

(iii) Beef in good condition should have streaks of fat running through the lean section and the flesh should rise again quickly after being pressed with the fingers. Any cut surface should present a slightly moist appearance to the touch.

Order of Dissection: A whole side is divided into the forequarter and hindquarter, the division being between the wing ribs and fore ribs.

Hindquarter:

(i) Remove the rump, suet and kidney.

(ii) Remove the thin flank.

(iii) Divide the loin and rump from the leg. (The leg will consist of topside, silverside, thick flank and shin)

(iv) Remove the fillet.

(v) Divide the rump from the sirloin.

(vi) Remove the wing ribs.

(vii) Remove the shin.

(viii) Bone out the aitchbone.

(ix) Divide the leg into topside, silverside and thick flank.

Forequarter:

(i) Remove the shank.

(ii) Divide in half down the centres.

(iii) Take off the fore ribs.

(iv) Divide into joints.

Joints and their Uses

	Joints French	*USES*	*Approx Weight*
(i) Shin	Jambe de derriere (Jambe de devant)	Consomme beef tea, stewing	4.00 kg.
(ii) Topside	Culotte de boeuf	Braising, stewing, second-class roasting	7.5 kg.
(iii) Silverside	Gite a la noix, semelle	Salt beef	10.00 kg.
(iv) Thick flank	Gite a la noix	Braising and stewing	6.7 kg.
(v) Rump	Rump steak, colutte de	Grilling and frying as steaks or braised boeuf	7.2 kg.
(vi) Sirloin	Aloyau	Roasting, grilling and frying	10.00 kg.
(vii) Wing ribs	Cote d'aloyau	Roasting, grilling and frying steaks	5.9 kg.
(viii) Thin flank	Bavette	Stewing, boiling, sausages	4.00 kg.
(ix) Fillet	Filet	Roasting, grilling and frying	2.72 kg.
(x) Fore rib	Cote premiere	Roasting and braising	6.5 kg.
(xi) Middle rib.	Cote deactivate	Roasting and braising	8.00 kg.

Contd...

	Joints French	*USES*	*Approx Weight*
(xii) Chuck rib	Cote du collier	Stewing and braising	4.08 kg.
(xiii) Sticking piece	Cou (collier)	Stewing and sausages	10.00 kg.
(xiv) Brisket	Poitrine	Salted and boiled	6.8 kg.
(xv) Plate	Plat de cot	Stewing and sausages	5.9 kg.
(xvi) Leg of	Gigot mutton	Braising and stewing	10.5 kg.
(xviii) Shank fat,	Jarret bone of kidney	Consomme, beef tea (trumeau)	4.08 kg. 7.2 kg.

N.B.: Weight of aitchbone x 10 = total weight of bones in hind quarter.

Minute Steak: Sirloin cut into 1.25 cm. (½")thick slices, flattened with a cutlet bat dipped in water to make the steak as thin as possible, trimmed and grilled or fried.

Sirloin Steaks: Sirloin cut into 1.25 cm. (½")thick slices, trimmed and fried or grilled.

Double Sirloin Steaks: Sirloin cut into 2.5 cm. (1") slices, trimmed and fried or grilled.

Porterhouse and T-bone steak: Complete slices of sirloin including the bone and fillet.

Chateaubriand: This is a double fillet steak taken from the head of a fillet of beef, 3.5 to 10 cm. ("-4) thick, with an average weight of 340 gm. to 1 kg. It is generally served grilled at a temperature of 204.5°C (about 400°F), done to the taste of the customer.

Fillet Steak: These are approximately 115 gm. to 170 gm. in weight cut into 2.5 cm. (1") thick slices trimmed and grilled or fried.

Tournedos: Taken from the middle portion of the fillet. Remove all fat.

Fillet Mignon: Taken from the tail end of the fillet. All fat and sinews are removed and it is minced or sliced as required.

Entrecote Steak: A slice of steak cut between two bones of a rib of beef, or cut from the contrefilet (bone sirloin or undercut), and weighing about 650 gm.; sufficient for four persons.

Entrecote Minute: Cut as above from the sirloin or contrefilet as thinly as possible, about 1 cm. (2/5") in thickness, well flattened, and served grilled or sauteed.

Chicken

Choosing Poultry: Only young birds are suitable for roasting; older birds may be boiled , steamed, braised or stewed. A roasting chicken should be about nine months to a year old; geese and other birds for roasting should not also be older than one year.

Signs of a Young Bird

(i) The feathers, especially the quills on the wings, should be easy to pull out.

(ii) There should be down all over the body, but especially under the wings.

(iii) No long hair on the body.

(iv) The skin should be white or clear and smooth.

(v) The feet should be supple, with smooth, even, overlapping scales,

(vi) The comb and wattle should be small and not well developed.

Signs which Show that a Bird is Fresh:

(i) Feathers - light and fluffy,

(ii) Eyes - prominent and clear.

(iii) No marked or unpleasant smell,

(iv) Feet - moist and not stiff and dry

(v) Skin - clear with no dark or greenish tinge.

Points which Denote Good Condition in Ducks: The breast should be plump. The bill and feet are yellow in young birds, but they darken with age. The webbing of the feet should be smooth, tender and easily broken, and the underbill soft and pliable.

Preparation

Storing poultry: Fowls and ducks are best used fresh and should not be stored longer than 1 or 2 days before cooking. Protect them from flies and hang by the feet in a dry airy place.

Plucking: If poultry is plucked while still warm the feathers are easier to remove.

(i) When plucking, do not sit in a draught or the feathers will blow about.

(ii) Put the bird on a large sheet of strong paper.

(iii) Take care not to tear the skin, especially when plucking game birds.

(iv) Hold the bird firmly with the left hand and pluck with the right.

(v) Pluck out small handfuls at a time, starting from under one wing then continue plucking until one side is completely clear.

(vi) Remove all the pin feathers with a knife.

Singeing: A bird can be singed in either of the two following ways. Care should be taken not to scorch or blacken the skin.

(i) Hold the bird by its head and feet, and quickly and carefully singe off the hairs over a small gas jet.

(ii) Hold the bird in the left hand and singe with a lighted taper or paper spill.

The Giblets: These can be used for making stock for gravy to serve with the roast bird, for enriching the stock when the bird is boiled, or for making giblet pie or giblet stew. The liver and kidneys are not boiled with the fowl, as they tend to

darken the meat. They may be half-cooked, chopped and used as part of the stuffing or cooked in other ways. Instructions for cleaning giblets are given below.

Liver: Carefully cut away the gall bladder, taking great care not to break it.

Stomach: Remove any fat and reserve it. Scrape the gizzard lining well.

Heart: Cut free of tubes, cut open, and scrape away all congealed blood.

Kidney: With a large bird, such as a turkey or goose, these are large enough to be used. Cut them open and cut out the core.

Neck: Scrape away all congealed blood.

Wash the liver, gizzard, heart, kidney and neck in salt water and rinse well.

Drawing the Tendons: If a bird is not young and tender, it is advisable to draw the tendons from the legs before roasting.

(i) Carefully cut through the skin of the leg at the knee joint, taking care not to cut the tendons at the same time,

(ii) Break the joint and swing the leg free to expose the tendons,

(iii) Pull out the tendons one at a time with a strong metal skewer. There will be 7 of them in each leg.

Stuffing: Birds can be roasted, stuffed or unstuffed, but they are usually stuffed. Stuffing can be put inside the bird only, or both inside and at the crop, covered by the skin of the neck.

Veal forcemeat may be used for the inside and sausage meat for the crop or vice versa. Pack the veal forcemeat, or any similar stuffing, fairly loosely to allow for swelling.

Trussing:

Method (A)

A trussing needle and some thin, clean string will be necessary. A trussing needle is a strong steel needle about 23 cm. (9") long with a large eye.

(i) Fold the skin of the neck over the back of the bird, and cross the tips of the wings over, to keep it securely in position.

(ii) Turn the bird breast upwards and put the tail through the vent.

(iii) Thread the trussing needle and pass it just under the top joint of the right wing through the bird and out through the same joint in the left wing. Run the needle back through the bird so that it enters and comes out just above the bottom joint of the wings.

(iv) Tie the ends of the string securely in a bow, taking care not to drag the flesh.

(v) Re-thread the needle, press the legs well into the sides towards the wings, plumping out the breast, and run the needle through the back of the bird, just under the thigh joint, and return through a point near the ends of the legs. Tie both ends of the string firmly together.

Method (B)

(i) Fold the skin of the neck over the stuffing and secure firmly with a small skewer.

(ii) Turn the bird breast upwards and put the tail through the vent.

(iii) Pass a large skewer through the wing pinions to hold them securely in position,

(iv) Tuck the legs well into the sides of the bird to plump out the breast, then tie the legs firmly together with a string; pass the string underneath the bird, cross it, then bring it up again and twist it round the wing. Skewer and tie firmly.

Trussing a Fowl for Boiling:

(i) Cut the skin of the legs all round at the knee joint, crack the joints and pull the shanks off with as many of the sinews as possible.

(ii) Thread the trussing needle and secure the wings in the same way as for the roasting bird, or fix them securely with a skewer,

(iii) Carefully loosen the skin round the legs, and slip them into the body of the bird. Secure in position with the trussing needle and string or with a skewer.

Vegetables

Test for Freshness: All vegetables, particularly green ones, are best when taken straight from the garden. But this is not always possible so it is essential that one should know how to judge a well-grown vegetable. A well-grown vegetable is usually tender and free from pests.

Vegetables are classified into three divisions:

(i) Root vegetables, (ii) Leafy vegetables, (iii) Other vegetables grown above the soil.

Root Vegetables: These should be firm and heavy for their size. They should be free from earth and grubs. Carrots should be heavy, smooth skinned and orange in colour.

Leafy Vegetables: These should be crisp and give a crunchy sound when squeezed. Break the vein of a leaf; it should give a sharp snap. It the leaves are limp, then the vegetable is not fresh.

Other Vegetables: Cabbage and cauliflower should be free from greenfly. The cabbage should have a firm heart and the outer leaves should stand the same test for freshness as green leafy vegetables. Cauliflower should have a close, white flower. Tomatoes should be of a rusty red colour and should not be soft. Runner beams and French beans should not be stringy. Peas should be full in the pod, and green. Ladies fingers should

snap if the ends are broken. Brinjals should have a smooth skin and should not be heavy for their size.

Importance in Diet

Vegetables play an important role in the human diet. Properly chosen and eaten, cooked or raw, they make an invaluable contribution toward the supply of vitamins and minerals. Root vegetables are good sources of carbohydrates, and pulses (peas, beans, lentils, etc.) of vegetable proteins. A good balance of ingredients essential for growth and maintenance in human beings is more easily attained by a diet made up of both plant and animal life. Vegetables also provide variety in a meal, help to make a meal attractive by introducing colour, and furnish roughage.

Mushrooms have been found to contain one or more antibiotics, one of which is effective against tuberculosis and germs which cause boils. Mushrooms which have always been a food favourite are rich in vitamins and low in calories. Seven average sized mushrooms have only 1-2 calories.

Whenever possible at least two vegetables should be served in addition to potatoes. Salads should be made popular and should be served at both meals.

Salads are dishes made up of meat, poultry, game, fish, shellfish, eggs, vegetables, fruits and milk products—normally served cold but can be served hot or frozen. They can be made out of a single ingredient or a combination of ingredients. Salads are normally served as an accompaniment to the main course, a meal by itself or a sweet course (dessert). There should be special stress on hygiene as most of the ingredients are served raw. The ingredients used and the colour, flavour, texture and presentation may be varied according to the purpose.

There are four basic parts in a salad:

The Underliner: These are generally greens either shredded or in large pieces. Tear the greens instead of cutting with a

knife. The greens must be crisp and chilled. To keep the greens crisp, it is necessary to store them in a manner that retains the maximum amount of water in the cells. This could be achieved by storing them at refrigeration temperature in the hydrator drawer wrapped in a moist cloth or in a covered container. The loss of water in cells can be reversed. To do this reduce the temperature to 2-3°C (about 34°F), and moisten the greens. When the cells are once again filled with water the greens will be crisp.

The Body or Heart of the Salad: This will form the major component and can be made of one ingredient or a combination of ingredients. Ingredients should look sparkling fresh with each individual piece intact. Fruits are particularly fragile and should be handled as little as possible. Dressings These make the salad more appetising although diet conscious people today eat their salads without a dressing.

Dressings: Dressing must be carefully selected. Depending upon the salad, the dressing could be either basic French or an emulsion. Pleasing flavours can be added to and variations made in the dressings by adding ingredients like cayenne pepper, tabasco sauce, diced hard boiled eggs, assorted herbs, a variety of cheeses, chutney, etc. Dressings should enhance the salad and not obscure it.

Garnish: The garnish should be the focal point and should be carefully selected for its simplicity and to enhance eye appeal and flavour. Edible garnish is more appropriate than inedible ones. Sometimes the dressing is also used as a garnish by piping it in a decorative form on the salad.

Special stress should be given to the quality of ingredients, eye appeal, simplicity, neatness and contrast or harmony in colour and in texture. All ingredients should be edible, clean, fresh and free from defects, such as bruises, rotten spots, insects, dirt, sand and insecticide. Raw fruits and vegetables should be at their optimum stage of maturity. All inedible Portions should be removed. Remove any excess moisture from ingredients as it thins the dressing, making it less effective.

The aesthetic qualities are determined by the form and shape of the food as well as by the colour. Skilfulness and artistic cutting and arranging of salad ingredients can create an appealing effect.

Try for an uncluttered appearance.

Always prepare salads as close to time of serving as possible.

Making of Vegetables

They should be washed before peeling. A knife or a peeler may be used.

The purpose of cutting is to ensure uniformity of size and even cooking.

The types of cutting used for vegetables are dicing, chopping, grating, slicing, shredding, cutting into rings, mincing, etc. The object of preparation is to remove spoilt, soiled and indigestible parts.

Cooking: Fruits and vegetables are cooked for many reasons:

(i) To preserve them,

(ii) To soften the product,

(iii) To improve the flavour.

(iv) To increase the digestibility.

Whatever the methods of cooking applied, the objective is to lose as little as possible of the natural flavours, colour and nutrients.

Cooking affects the palatability of the vegetables, i.e., their texture, flavour and colour. While palatability may not affect the nutritive value of vegetables, it affects their desirability.

All the different methods of cooking can be effectively employed in the preparation of vegetables.

The effect of cooking on starch, protein and cellulose brings about the change in texture observed in vegetables when they are cooked. The predominant cause for change in texture will

depend upon the composition of the vegetable. For example, with leafy vegetables which contain a high percentage of fibre, the change in texture is primarily due to the effect of cooking on the hemicellulose, whereas in potatoes the gelatinisation of starch is the chief factor. While this is true, it has also been noted that the change in any one ingredient is not the only cause that makes vegetables tender. For example, gelatinisation of the starch in a potato is completed before the potato is considered done.

As the amount of protein in vegetables is small, the effect of coagulation during cooking is not of great significance.

Alkalies such as baking soda aid in softening of vegetables as they speed up the solution of pectic substances and dissolve the hemicellulose. Acids, on the contrary, harden the vegetables, as acid precipitates pectin.

The texture of the cooked vegetables can be controlled to a great extent by the length of the cooking time. The longer the cooking period, the softer the product.

Flavour: The flavour of vegetables depends upon many substances—sugar, acids, tannings and a variety of volatile substances. There is a predominance of sweetness in beetroot, peas, corn and carrots, but each vegetable has other ingredients which contribute to its characteristic flavour and aroma. The pronounced bitterness or astringency of certain vegetables is due to the presence of tannin. The characteristic odour of the onion family—chives, shallots, onions, garlic and leek—is due to a readily volatile substance—allyl propyl disulfide. As this substance is soluble in water the longer the vegetables are cooked in water, the milder is the flavour.

The colour of the cooked product will depend on the kind and amount of colour pigment, the method of cooking and the acidity of the cooking medium. The colour pigments of fruits and vegetables are chlorophyll, carotenoids, flavones and anthocyanins. Tannings, though not colour pigments, undergo

chemical changes during cooking and thus affect the colour of many cooked vegetables.

Chlorophyll: Chlorophyll is made up of two compounds, chlorophyll 'a' and chlorophyll 'b'. Both are organic esters and both contain magnesium. The reactions taking place during cooking affect these radicals. Chlorophyll is slightly soluble in water but dissolves easily in many organic solvents. Like other esters it is subject to hydrolysis, the reaction taking place in the presence of alkali, forming chlorophyll liens—salts of chlorophyll which are intensely green in colour.

In the presence of acids, chlorophyll undergoes another type of decomposition with the loss of magnesium from the molecule. The resulting compounds are called phaeophytins. They are yellow to brown in colour. Both reactions, with acids and with alkalis, are irreversible in the sense that chlorophyll cannot be formed again by treating the decomposed products with a neutralising agent.

Any cooking process which involves the use of acids or alkalies will affect the colour of green vegetables. During the life of the plant the chlorophyll is held in the cytoplasm of the plant cells in such a way that it is protected from the regions of changing acidity. When cell life is destroyed, the constituents of the cells intermingle and the green colour fades, becoming yellow and brown.

When a green vegetable is cooked the cell structure gets slowly destroyed and the acid of the cells gradually comes in contact with the chlorophyll. Hence green vegetables cooked for long periods lose their greenness. If the cooking pan is kept open during cooking, some of the acids of the vegetables may evaporate, preserving some of the green colour. When soda is added to the water in which green vegetables are cooked there is no loss in colour. The alkali neutralises the acidity of the vegetable and also speeds up the cooking process. Loss of colour also depends on the temperature of cooking. At high

temperatures there is greater hydrolysis and, therefore, greater loss of colour.

Carotenoids: Carotenoids include two types of colour pigments—the hydrocarbons, e.g. carotene as in carrots, and the oxygenated hydrocarbons, e.g. lutein, the colouring matter in egg yolk.

These pigments like all other hydrocarbons do not dissolve readily in water. They are also not affected by acids or alkalies. The oxygenated carotenoids are also chemically resistant to acids and alkalies as well as insoluble in water. Other carotenoids are xanthophyll, associated with carotene, lycopene (the red pigment in tomatoes and watermelon), zeaxanthin (the yellow pigment in egg yolks and corn), capsanthin (in red pepper) and fucoxanthin (in brown seaweed).

Anthocyanins: Anthocyanins are colour pigments which are found more in fruits and flowers than in vegetables. The colour varies from red to blue. Red or purple cabbage, purple broccoli, beets and skins of aubergine and radish contain large amounts of anthocyanins. There are 28 known anthocyanins. They differ in their degree of solubility and sensitiveness to conditions which produce colour changes. They are inclined to be red in acidic and blue in basic solutions. The colour changes of anthocyanins, unlike those of chlorophyll, are reversible.

While cooking vegetables containing anthocyanins, the loss of colour as well as change of colour must be considered. Beets lose their colour or 'bleed' during the preparation. This is due to the great solubility of anthocyanins. High temperatures bleach this pigment, the colour returning partly when it is cooled.

Natural acids in fruits increase the redness of the cooked fruit. The colour of cooked red cabbage depends on the acidity of the cooking water. Tap water gives it a purple colour. Red cabbage is produced by adding acid to the water. The addition of cooking soda gives a bluish product.

Flavones occur in all plants, but particularly in white onions, white cabbage, cauliflower, etc. In an acid medium, they are colourless, while in a basic medium they are yellow. The reaction like that of arithocyanin, is reversible. Vegetables such as cauliflower can be made to change colour from white to yellow and back by holding them in fumes of acid or ammonia. Flavones are readily soluble in water as can be observed from the colour of the alkaline vegetable water, which is yellow. During cooking it is important to see that white vegetables do not turn yellow. The addition of acids such as lime juice or vinegar will keep them white.

Boiling: Vegetables are usually put into boiling liquid. This is to prevent the nourishment escaping into the liquid and also to curtail the action of plant enzymes which have harmful effects on vitamins. The exceptions to this are dried vegetables and root vegetables which should be started in cold water. Salt is generally added to the water, except when cooking dry vegetables, which do not cook well if salt is added at the commencement. The lid is kept on generally, except when cooking green vegetables, where the volatile acids present in chlorophyll must be allowed to escape or else they will discolour the vegetables. Lime juice brings out the whiteness of cabbage and cauliflower if added before cooking.

Potatoes: Scrub and scrape new potatoes, peel and eye old ones. Keep under water to avoid discolouration. To boil, put in salted water and cook till tender, drain and use as required.

Roast potatoes: Parboil, drain and cook in dripping around the meat for at least 45 minutes.

Saute potatoes: Slice cold boiled potatoes and fry in very little fat till brown.

Potato chips: Two methods are used:

(i) Peel and cut them in to finger lengths. Parboil, drain and put into hot fat. Remove when half done. Put into smoking fat or boiling fat and brown just before serving.

(ii) Cut the peeled potatoes into finger lengths. Dry on a towel. Deep fry in a pan till a faint blue smoke rises. Fry chips until golden brown. Drain on absorbent paper, add salt.

Spinach: Remove the stalk and damaged leaves. Wash very thoroughly in several batches of water to remove grit and mud. Put into a pan. Add salt. Place over gentle heat, when the juice begins to run over spinach, raise temperature and boil for 15 minutes. Drain and chop or sieve as required. Add pepper and a small knob of butter to taste.

Before cooking spinach and other leafy vegetables, soak them for some time in water to which a little sugar has been added. This improves the flavour.

Green peas: Shell peas and place in boiling water with teaspoon of sugar, salt and a sprig of mint. Boil till tender. Drain. Take out the mint and add a little butter.

French beans: String the beans. Slice them diagonally and cook in boiling salted water till tender (approx. 20 minutes). Cook with the lid off. Drain well and add a knob of butter.

Brussels sprouts: Found in a temperate climate. Cooked the same way as cabbage but kept whole. Slit stalk end into four before cooking.

Cabbage: Wash well, remove coarse leaves but use dark outer leaves. Cut in half. Remove the hard centre stalk, shred. Cook in a little boiling and salted water till tender (10 to 15 minutes approx.). Drain well. A few bacon rinds can be added to the water. Serve plain or tossed in melted butter.

Cauliflower: Wash thoroughly, remove outermost leaves if necessary but leave some on. Turn over and cut across the stalk. Put the cauliflower into boiling salted water and cook for 20 to 30 minutes.

Broccoli: Three types—white, green and purple.

Purple and green: To prepare, remove coarse stalks and

wash the leaves and sprigs. Boil. Serve same way as cabbage or cauliflower. Particularly good with white saxice.

White: Prepare and cook as for cauliflower.

Turnips: Peel rather thickly. Cook in boiling salted water till tender; this takes 20 minutes to one hour. Drain, add white pepper and salt and butter. The cooked turnips can also be mashed up.

Vegetable Marrow: Very young vegetable marrow may be boiled whole and unpeeled. Cook in salted water for 15 minutes. Drain and serve with melted butter. Large marrows must be peeled and the seeds removed. Drain well and serve sprinkled with brown bread crumbs or with cheese sauce. Marrow may also be baked whole, (peeled or not) and stuffed with any savoury filling. Bake in a moderate oven at 177°C (350°F) for one hour. Cover the marrow with greased grease-proof paper.

Onion: Remove the root and the outer skin. Apart from being used as a flavouring agent, onions can be boiled and served as a vegetable or used stuffed.

Tomatoes: Grilled, fried, stewed, baked or raw.

Carrots: Scrub and scrape lightly. Peel old ones. Cut into strips or into roundels. Add just enough water, salt and a little butter. Cook till the carrots are tender and the water has evaporated. Serve sprinkled with chopped parsley. Young carrots can be boiled whole.

Corn on the Cob: Remove the sheath and silk threads. Cook in salted, boiling water till tender. Serve with melted butter.

Sweet Pepper: Capsicum, red or green. This can be served cooked or raw. To prepare the pepper, wash, cut out the stalks and remove the seeds. Red and green peppers may be cooked together to make a colourful vegetable dish. Slice and cook in a little salted water. Add a knob of butter. They may also be served stuffed, baked, boiled or stewed.

Celery: Wash celery thoroughly, using a brush to remove all soil. Separate the stalks. Tie into bundles. Cook in boiling salted water till tender. Drain and serve with white sauce.

Beetroot: Cut out the stalks about 2.5 cm. (1") from the root, taking care not to damage the skin. Wash and boil in salted water till tender (about 2 hours). Peel off the skin. Cut into cubes or slices and either serve hot, coated with white sauce, or cold in vinegar.

Leeks: Remove the coarse outer leaves. Cut out the roots and green top. Split the leaf end so that it can be washed thoroughly. Tie in bundles and boil in salted water. Drain. They are usually served with white or cheese sauce.

Broad Beans: Shell the beans. Cook them in boiling salted water and drain. Serve with parsley sauce. Skin older beans before serving.

Parsnips Peel thinly, using a potato peeler. Cut as required. Boil or bake.

Asparagus: Cut the woody portion, scrape the white part lightly. Remove any coarse spine. Tie in bundles. Place upright in a pan of salted water and boil for 10 minutes. Then lay the asparagus flat and cook till tender. Drain well, remove the string and serve with melted butter.

Artichokes: Only the lower parts of the leaves and lower sections are edible. Wash well. Cut off the stems and trim lower leaves if necessary. Place head downwards in boiling salted water and boil gently until tender, for about 30 minutes with the lid off. When cooked, the leaves should be easy to pull off. They may also be dipped in a thin batter and fried. Serve hot with melted butter or margarine or hollandaise sauce. Do not pour the sauce over the artichokes. The lower sections without the leaves may also be used, either raw or cooked, in salads.

Artichokes (Jerusalem): Scrub well, peel, and to keep them white until ready to be cooked, place in a bowl of water to

which a little vinegar has been added. Boil or bake. Serve plain boiled in the same way as potatoes or with white sauce. A very good au gratin can be made with these artichokes. Can be served in soups, separately or with another vegetable such as celery.

Colour Reaction of Vegetable Pigments

Pigment	*Vegetable in acid*	*Colour in alkali*	*Colour reaction reaction*	*Colour to metals*
Flavone	Cauliflower	Colourless (white)	Yellow	All-yellow Iron-brown
Anthocyanins	Red cabbage Beetroot	red	blue to green	Iron-blue Tin-purple
Carotenoids	Carrots	orange	orange	
Chlorophyll	French beans	olive green	bright green	Copper-bright
	Spinach			Iron-green

Pulse

Nutritive Function: Pulses are a good source of vegetable proteins and vitamin B complex. Being substances with very little taste, they should be combined with spicy and strong foods, e.g. bacon, ham, etc.

In Western preparations pulses are combined with onion, tomatoes, leeks, celery and bacon rind.

In Indian preparations pulses are combined with spices, coconut milk, tomatoes, leafy vegetables and herbs.

Cooking:

(i) If with skin, soak overnight in boiling water to soften skin, e.g. haricot beans, thick peas, etc.

(ii) Wash first before soaking so that the pulses can be cooked in the same liquid in which they are soaked. Those with no skins like lentils can be soaked in cold water. Bring to boil and cook with lid on.

(iii) Add salt after cooking, as salt inhibits the normal cooking process of lentils and pulses. Those with skins can be made into a puree, as it is the skin that causes indigestion.

In Western preparations these foods are rarely served plain boiled. They are generally mixed with other foods and browned by baking or frying. Use liquids for soups and gravies.

Cheese: Cheese, one of the oldest known food products, is universally popular. The chemistry involved in changing semifluid milk into semisolid cheese is complex and many factors enter into the final curing of cheese to produce the different varieties but they all begin with the action of rennet or lactic acid which performs the same basic function, that of coagulating the milk proteins. Simply defined, cheese is the solid or casein portion of milk separated from the whey. The curdling of the casein is brought about by the action of rennet or lactic acid. The many different cheeses are the result of variation in the making and curing of each variety after initial curdling has taken place.

Many experiments and inventions in many countries of the world have produced cheeses ranging in texture from very soft to very hard and in flavour from extremely mild to very sharp. These cheeses are known by thousands of names, generally with names taken from the localities where they were first made.

Classification: Cheeses are classified into large general groups according to whether they are started by the rennet or lactic acid method. They may also be classified as unripened, mould ripened and bacteria ripened or as soft cheese, hard pressed cheese or blue veined cheese.

Cheese is one of the most highly concentrated of all protein foods. It is also readily digested. Experiments have shown that from 90 to 99% of all cheese is digested. It is also a complete protein. Since approximately 10 litres of fluid milk are required to make 1 kg. of cheese, cheese contains many of the nutrients

of milk in highly concentrated form—milk proteins, fat, fat-soluble vitamins and minerals.

Wholemilk cheese contains the same properties as milk. Certain cheeses such as blue veined are made out of skimmed milk and, therefore, are less nutritious.

Normally cow's milk is used to make cheese, but certain well known cheeses are made from goat's milk. The texture of the cheese from goat's milk differs slightly from that of cow's milk. It is more crumbly. Sheep's milk can also be used. The quality of the cheese depends to a great extent on the breed and the condition of the animal and the fodder given to it. Cheshire cheese is said to owe its fine flavour to the wild radish on which the cows feed, and its special nature is due to the mineral in the soil. Cheshire cheese, therefore, cannot be made in any other place as cheddar cheese can.

Certain cheeses develop a blue vein in maturing. This is sometimes a purely natural development, often sporadic and unpredictable. Sometimes fresh cheese is inoculated with pieces of blue cheese to catch the mould by contact. In some cases special bacilli are introduced. The mould is sometimes strengthened by brushing the cheese clean while the skin is soft, dipping it in whey and then rubbing it slightly with butter. This is done once a day for 10 to 20 days. Sometimes cheese is pierced with a copper wire. The main classes of cheese are:

- (i) Unripened soft cheese, e.g. Cottage cheese, Cream cheese, Neufchatel.
- (ii) Ripened in moulds by bacteria, e.g. Brie and Camembert.
- (iii) Ripened by bacteria, e.g. Limbourger.
- (iv) Semi-hard, ripened by bacteria in moulds, e.g. Gorgonzola, Roquefort and Stilton.
- (v) Semi-hard, ripened by bacteria, e.g. Brick cheese, Munster.

(vi) Very hard cheese without gas holes, e.g. Cheddar, Edam, Gouda and Cheshire.

(vii) Very hard cheese with gas holes, e.g. Gruyere, Swiss Cheese, Parmesan.

Glossary:

Acid, acidity a description of a pleasant tang; it can be a defect if too pronounced.

Ammoniated a term describing cheeses that smell or taste of ammonia, a condition that affects the rinds of overripe cheese, primarily those with bloomy rinds such as Brie and Camembert. A hint of ammonia is not necessarily objectionable.

Annatto a yellow-orange dye extracted from the seeds of a South American plant and used to colour such cheeses as Cheddar, Momolette, Double Gloucester, Edam and many others.

Bloomy rind the white fleecy rind that develops on certain surface-ripened cheeses like Brie, Camembert, Double or Triple Creams and some Chevres. It is formed by spraying the surface of the cheese with spores of penicillium candidum while it is curing.

Chevres the French term for goat cheeses.

Gummy a negative term used to describe an overly plastic texture, as well as overripe rinds that have become sticky or gooey. Gumminess is undesirable in any context.

Salty most cheeses have some degree of saltiness; those lacking in salt are said to be dull or flat. Pronounced saltiness is characteristic of some cheeses, but over saltiness is a defect.

Springy a descriptive term for cheese with a resilient texture that springs back when gently pressed. Ripe or nearly ripe soft-ripened varieties should be springy.

Characteristics, Mode of Serving and Place of Origin of the Commonly Used Cheeses

	Cheese	Characteristics	Mode Of Serving	Place Of Origin
(i)	Cottage	Simplest of all cheese varieties. Uncured cheese made of cow's milk. Now made commercially from pasteurised skimmed milk to which lactic acid culture is added. Cream may be added if desired.	Salads, sandwiches, appetizers, cheesecakes and as paneer in Indian curries.	Uncertain,
(ii)	Cream Cheese	A soft uncured cheese made of cow's milk with cream added. The special richness an d smoothness comes from the whole milk with added cream. One of the few world cheeses made by the lactic acid rather than the rennet method of coagulating the curd. It is extremely mild in flavour.	Salads, sandwiches and appetizers.	Uncertain.
(iii)	Neufchatel	A soft rennet cheese made of cow's milk. Prepared in the same way as Cream Cheese with as lightly higher moisture content and lower fat content than Cream Cheese. Sometimes flavoured with pimento and other condiments.	Same as above.	France.
(iv)	Brie	Soft ripened cheese in small discs, made from cow's milk. One of the most famous and a great favourite in Europe.	Served after a meal and as a buffet cheese.	France.
(v)	Camembert	Soft ripened rennet cheese, made from cow's	Served at the end of a	France.

Contd...

Cheese	Characteristics	Mode Of Serving	Place Of Origin
	milk. Strong flavour like that of ammonia. It is a world favourite.	meal with crackers, fruit, etc.	
(vi) Limburger	Soft, highly flavoured cheese made from cow smilk. One of the most delicate and difficult cheeses to make.	Served at the end of a meal.	Luttich, Belgium.
(vii) Gorgonzola	Hard blue-veined cheese made of cow's milk.	Served after meals, also used fors alads and b uffets.	Gorgonzloa, near Milan, Italy, France.
(viii) Roquefort	Blue-veined, semi-hard. Has a sweet, piquant flavour. Only world-famous variety made from sheep's milk.	To be served at the end of a meal. Also used for salads and buffet serv-	England.
(ix) Stilton	Semi-hard, blue-veined cheese. Made of cow's milk (whole with addea cream), —spicy flavour. Surface rind is crinkled and brown with the interior creamy white and marbled by blue veins.	ices. Served at the end of a meal and with salads.	
(x) Brick	Distinctly American cheese developed in Sand wiches and buffets. the mid-nineteenth century. Texture ranges from very firm to fairly soft and elastic. Flavours vary from extremely mild to sharp, depending on degree of curing.	Sandwiches and buffets.	America.
(xi) Munster	Semi-hard, whole milk cheese. European Munster is highly flavoured and sharp. American Munster is mild in flavour, the curing time being shorter.	Sandwiches and buffets.	Munster (originally in Germany, now in france.)

Contd...

Cheese	*Characteristics*	*Mode Of Serving*	*Place Of Origin*
(xii) Cheddar	One of the most popular and well-known cheeses of the world; made from cow's milk (whole or partly defatted). Hard and smooth. The English Cheddar is one of the oldest and best known of English cheeses. It is white in colour, close in texture and clean and mellow in flavour. American Cheddar, known as American Cheese, varies in colour from white to yellow. It is factory-produced, using lactic acid as a starter, with rennet added later.	Served after meals. Used in sandwiches and in preparation of foods.	Cheddar, England.
(xiii) Edam	Round, cannonball-shaped, bright red cheese. Produced by dipping ripened Holland. cheese in red-coloured paraffin. It is hard and rubbery in texture. In flavour it is like Cheddar, but milder and sweeter. It is made out of partly defatted cow's milk.	Served after meals of buffets.	Northern Holland.
(xiv) Gouda	Semi-hard. It has the same flavour as Edam and is also dipped in red paraffin after ripening. It varies in size from Edam, weighing between 5-25 kg. (10-45 lbs.). Baby Goudas weighing about 500 gm. (1 1b) are now available. Goudas are made from partly defatted cow's milk.	Served after meals.	Southern Holland.
(xv) Cheshire	One of the oldest of English cheeses made of cow's milk. A hard cheese characterised by its loose, flaky crumbly texture ad solid in rich yellow colour or in its natural white colour.	Serve after meals, or in such dishes as Welsh rare bit. fondues etc.	Cheshire County, England.

Contd...

Cheese	*Characteristics*	*Mode Of Serving*	*Place Of Origin*
(xvi) Gruyere	Hard, with gas holes, and a nut-like, salty flavour. Made of cow's milk, which is usually partly defatted.	Served after meals.	Gruyere Switzerland
(xvii)Swiss cheese (known as switzerland.)	One of the world's most famous cheeses. It is hard, with gas holes, and a nutlike, sweet flavour. Made from partly defatted cow's milk (known) with gas holes, and a nutlike, sweet flavour. Made from partly defatted cow's milk. as Emmenthal in Switzerland	Served after meals, in sandwiches and with salads.	Emme Valley, italy.Emmenthal in
(xviii) Parmesan	Hard, granular texture, sharp flavour. Made of partly defatted cow's milk.	Served after meals and for flavouring foods.	Parma and Lodi. Utakt.
(xix) Port-Salut	Semi-hard with rubbery texture and a flavour between Limbourger and Cheddar. It is made of whole slightly acid cow's milk. Oka cheese made in Canada belongs to the same family.	Served after meals	trappist Monasteries, France.
(i) Derby	Close buttery texture with a pale honey colour. It is mild when young but develops a fuller flavour as it matures.	served after meals and for flavouring foods.	England.
(ii) Lancashire	Crumbly texture. Has a mild flavour when young but develops a full and rather pungent flavour as it matures.	Ideal for crumbling over soups, hot-pos, or etc., Also well known for tosting quality	England.
(iii) Wensleydale	Flaky texture with subtle flavour. Clean, mild and slightly salty. It has an after taste of honey and is pale parchment in colour.	Used in preparation of foods.	Yrokshire, England.
(iv) Double Gloucester	Velvet texture like butter, resembles Cheddar in flavour and has a pale straw colour.	Served after meals or with fresh fruit salad and cream.	England.

Processed Cheese: Pasteurised processed cheese is the food prepared by mixing together, with the aid of heat and the addition of a small amount of emulsifying agent (not exceeding 3% of the total weight of the finished product), one or more cheeses of the same variety or two or more varieties. The emulsifying agents commonly used are sodium phosphate and sodium citrate.

Many cheeses are available in pasteurised processed form. Among the most common ones are American Cheddar, Swiss, Brick, Limburger, etc. Processed cheeses are widely served after meals. For cooking they have several advantages. They melt smoothly and quickly without separation or stringiness. They are also uniform in flavour and texture.

Cheese Cooking: Cheese cookery is an old art. Classic favourites like fondues, Welsh rare bit, macaroni and cheese, cheese pies and cheesecakes are well known. There are countless other dishes incorporating cheese which are of more recent origin. Cheese blends happily with many other food flavours and hence is a popular ingredient.

Cheese is a protein food and like all other proteins it is toughened by high heat. All cheese dishes should be cooked at low temperatures whatever the dish is. Whenever possible cheese should be melted in a double boiler or chafing dish rather than over direct heat.

Grate or chop cheese finely and dilute with some kind of starchy food such as flour, bread crumbs, macaroni, etc.

Add a small pinch of bicarbonate of soda tsp to a kg. tsp to a 1b). This softens the cheese, prevents stringiness and makes it more digestible.

Cook by moist heat whenever possible, or at least see that there is some moisture included in the dish. If and when possible, add the cheese only at the last moment after the rest of the food has been cooked. This is to prevent long cooking of cheese.

Up keep: The keeping quality of cheese varies greatly. Some of the hard varieties such as Swiss and Parmesan have a high keeping quality. Soft cheeses such as Camembert, Cream Cheese, etc. are highly perishable.

All cheeses require special care in handling once they are cut, for once cheese is cut it tends to dry rapidly. Mould may also form on any natural bulk cheese. This is in no way harmful, and can be cut off before being served without detriment to the quality of the remaining cheese.

Pasteurised processed cheese keeps well when leftover portions are rewrapped in the original paper and placed in the refrigerator. Bulk cheese or any hard cheese variety may be wrapped in wax paper, aluminium foil or plastic material. Tightly covered refrigerator dishes of plastic or glass are excellent for keeping cheese of all varieties.

All cheeses should normally be kept under refrigeration but before serving, they should be left out for a few hours. No cheese should be served chilled. The flavours of all cheeses are best when the cheese is neither cold nor over-warm.

Dishing and Garnishing: A good cook not only takes pride in preparing a dish correctly but also in presenting it properly to appeal to the different senses. This is done by suitable decoration and the way it is placed on the serving dish. One of the chief points that must be borne in mind when both dishing and garnishing is neatness. Decoration that is done without restraint or badly carried out is worse than no decoration at all. Decoration done on hot dishes must be done quickly to prevent cooling of a dish which should be served hot. The colours of the garnish and the flavour should blend with the main dish. Proper garnishing has a three-fold role in food service. It helps "dress" the plate, "style" the product and adds colour, flavour and eye appeal of its own.

Accompaniments: Accompaniments are dishes such as sauces, salads or vegetables which are used to make a dish complete. They provide variety and improve the nutritive

value of the meal. What accompaniments should go with what, is usually determined by custom, based on a desire to enhance the flavour, to increase the digestibility and to improve a dish, e.g. roast beef with Yorkshire pudding, roast pork with apple sauce, roast duck with sage and onion stuffing and orange sauce, fried fish and chips with tartare sauce or tomato sauce, roast chicken with roast potatoes, buttered peas and bread sauce.

Forcemeat or stuffing are used to add flavour and piquancy to meat. They also help to increase bulk.

Sage and Onion Stuffing

Ingredients	*Quantity*
Onions	225 gm
Fresh bread crumbs	50 gm
Salt and Pepper	to taste
Butter	30 gm
Powdered sage or	
Fresh sage (chopped)	1 tsp
Method	

1. Boil chopped onions till tender, in just enough water. 2. Add butter and star till melted. 3. Add remaining ingredients, mix well and use as required.

Forcemeat Stuffing

lime	
arsley	2 tsp
Mixed herbs	tsp
Salt and Pepper	to taste
Egg or milk to bind	

Method

1. Mix dry ingredients. 2. Add binding. 3. Mixture should be crumbly should not be pressed heavily and not mixed too wet.

BREAD SAUCE

Ingredients	*Quantity*
Milk	300 ml
Onion	10 gm
Margarine	30 gm
Fresh bread crumbs	50 gm
Cloves	2
Mace	a pinch
Lemon rind or a strip of lemon	Seasoning

1. Boil the milk. Infuse the hot milk with onion, cloves, mace and lemon rind for 30 minutes. 2. Strain over fresh bread crumbs, add seasoning and margarine; stir well to a smooth consistency. 3. Stand in a warm place to thicken. Reheat and beat well; add one tablespoon cream if desired.

Sauces: Sauce is a liquid adjunct to a dish such as meat, poultry and vegetables used to moisten the food, to enhance the flavour, to provide contrast in taste or colour and sometimes to improve the digestibility (e.g. Roast pork with apple sauce); generally thickened by means of a liaison such as roux egg, blood, etc.

"A correct sauce is that wonderful production of the culinary art which forms a pleasant and exquisite accompaniment to all kinds of fish, meat, poultry, game and vegetables." The skill and knowledge of a cook is shown in no other part of culinary art so clearly and prominently as in the way the sauces are prepared. To make a perfect sauce is the height of the art of cooking.

The most simple dishes can be made appetising by the addition of a good plain sauce while the most recherche dishes can be improved and made still more palatable by a well-made sauce. Sauces in cookery are the essence of elegance. Every

sauce, whether plain or rich, must possess a decidedly distinct flavour and character. Plain sauces should be simple and pure so that they taste of the materials employed, from which they take their name. Richer sauces always require a longer and slower process for their preparation.

Sauces were hardly made use of in English cookery till the beginning of the nineteenth century. Today there are at least 650 different types of sauces and gravies. A clever cook can devise as many varieties as he needs in sauces as an artist can paint pictures.

The art of sauce-making consists in preparing liquids from various materials by cleverly extracting and combining certain flavours into the liquid. Besides this, a good palate as well as the experience and skill of the most accomplished cooks and a thorough knowledge of the taste of those for whom the dish is being prepared, are necessary.

All sauces should be smooth, glossy in appearance, definite in taste, and light in texture.

Gravy is not a sauce. It is generally the juice of meat or vegetables seasoned but not thickened like a sauce. According to textbooks on cooking, a gravy may be called a sauce but a sauce cannot be called a gravy. Many of the more elaborate sauces have gravies as their foundation ingredient. It is easier to call a thinner liquid a gravy and a thicker one a sauce.

Liaison

The various processes of thickening sauces as well as soups are called "Liaison'. There are six distinct methods known for thickening sauces:

(i) Liaison with roux.

(ii) Liaison with egg.

(iii) Liaison with butter and cream.

(iv) Liaison with kneaded butter and flour (manied).

(v) Liaison with blood.

(vi) Liaison with farinaceous products.

Roux: The most popular and the most widely used thickening is roux. Roux is a mixture of flour and butter cooked to a definite degree - white, fawn or brown colour. Flour and butter are generally used in the same proportion. Roux can be prepared beforehand and stored, but it should be kept in airtight containers.

Approximately 60 gm of roux is required for 60 ml of liquid. If the roux is used cold, the stock used can be hot or cold. If mixed hot, the liquid, which should also be hot, must be gradually added to the roux away from the fire. This is then stirred over the fire till it boils. Roux should be allowed to cool slightly before the liquid is added to prevent the formation of lumps. All roux must be stirred constantly during the process of cooking.

White roux or Roux Blanc': This is a mixture of flour and butter cooked in a pan over a moderate fire without allowing it to get discoloured.

Blond roux or 'Roux Blonde': This is prepared by melting a certain quantity of butter and stirring in the same quantity or a little less of flour and by cooking it over a slow fire or in the oven till it gets a fawn colour.

Brown roux or 'Roux Brun': This is generally known as stock roux and can be prepared in bulk and stored. It is prepared more or less in the same way as 'Roux Blonde' but is cooked for a longer period to get a deeper brown colour. It is better to finish a brown roux in a slow oven so as to make it more brown without burning. This will also improve the aroma. With brown and blond roux, a mirepoix is added to give the-necessary flavouring to the sauce. Mirepoix is an essence or extract of meat and vegetables. The following ingredients are used: 225 gm bacon, 1 carrot, 1 bay leaf, 1 sprig thyme, 2 small onions, and 2 cloves garlic. Chop all ingredients and fry without browning.

Egg Liaison: This is a thickening composed of egg yolks beaten up and diluted with a small quantity of cream, milk or

cold white stock. Cream is used more often than stock. When the soup or sauce is boiling it is removed to the side of the fire and a spoonful of the boiling liquid is added to the egg mixture gradually and then the whole is poured slowly into the sauce and stirred over the fire without allowing it to boil. It is passed through a tummy before it is served. This liaison is used largely for banquettes, white ragouts (usually of vegetables) and white fricassees.

Butter and Cream Liaison: Butter and cream are incorporated in sauces and soups just before they are served. After the liaison has been added the sauce or soup is stirred vigorously without reheating. The flavour of the sauce alters if butter and cream are added too soon or if the sauce is allowed to boil after the addition of the liaison. The same applies to a plain butter liaison as well. Butter is added in small amounts the moment the sauce is taken off the fire. It is then stirred with a whisk and served without being reheated.

Kneaded Butter and Flour Liaison: Knead as much flour into the butter as it will absorb, to form a soft paste. Mix in small portions into a hot thin sauce. Stir constantly till the butter melts.

Blood Liaison: This is mostly used with hare or other game. It is made by preserving the blood of hare or game to which is added a little vinegar to prevent clotting. It is then strained through a fine sieve and generally added to the sauces.

Farinaceous Liaison: Arrowroot, corn-flour, potato flour, rice flour or other farinaceous products are commonly used for thickening sauces. Dilute with a little cold milk, cold stock or water. Pour through a strainer into boiling liquid. Stir continuously until it boils. Simmer gently for 10 minutes.

4

General Items

For Child-bearing Mother: Pregnancy is a normal physiological phase where rapid growth of foetus takes place in the mother's womb. The foetus in mother's uterus grows more rapidly than after birth. The zygote develops into a seven-pound baby within 9 months. At the time of birth, the infant is generally 9 months old. Optimum development of the infant is necessarily a function of parental diet. Inadequate maternal nutrition results in low birth weight of the infant and high depletion of mother's body reserves of nutrients. Premature death, maternal death and low vitality of the infants are due to poor nutritional status of pregnant mothers.

Before pregnancy a woman needs nutrients for growth and maintenance of her body. Good nutrition keeps her healthy. During pregnancy additional requirement for all nutrients occurs to enable the foetus to grow normally in the uterus. A full-term infant has about 300 ml of blood, 500 gms of proteins, 30 gms calcium, 15 gms phosphorus and about 300 to 400 gms of iron. Apart from these foetal reserves in the body, the pregnant mother needs nutrients for the development of the uterus, breasts, placenta, amniotic fluids and reserves for

parturition. A full-term pregnant mother gains about 10 to 12 kg body weight and only an adequate diet can provide the nutrients for the normal weight gain. If the pregnant mother is underweight the infant born would be small in size and low in birth weight. Infectious diseases and infant mortality rate are common among such infants. Like underweight, overweight among pregnant mothers is not desirable. Overweight at the beginning of pregnancy or an excessive rate of weight gain during the second and third trimester results in pre-eclampsia and eclampsia. Correct weight gain during pregnancy is 1.5 kg in the first three months and 1.5 kg and a little more during the last two trimesters. Sudden changes in weight gain or weight loss are harmful.

Precaution

Nausea and Vomiting: Morning sickness like nausea and vomiting are common in early pregnancy. They may continue in some women till later stages of pregnancy. Even though the aetiology is not clear mental stress or emotional disturbance aggravate the situation. Diminished secretion of hydrochloride acid brings heartburn and gastric distress. Mild morning sickness can be controlled by consuming high carbohydrate foods such as biscuits or crackers. Small meals must be taken to help digestion. Fatty foods like sweets, fried items and strongly flavoured vegetables must be reduced, besides minimising coffee or tea.

Constipation: In the second and third trimesters constipation occurs among pregnant mothers. The amount of pressure exerted by the developing foetus on the digestive tract, lack of enough exercise and thiamine deficiency are the causes of constipation among pregnant mothers. Lack of fluid, fibre and bulk or cellulose also lead to constipation. If purgatives are used, excessive potassium is lost and this can lead to constipation. Excessive use of irritating foods, fried items and emotional distress affect the normal bowel evacuation and result in constipation. Liberal intake of fruits, wholegrain

cereals, leafy vegetables, other vegetables and fluids and a happy mental attitude are suggested to avoid constipation.

Toxemia: Toxemia occurs most commonly among patients who are overweight at the time of conception and during pregnancy. Pregnant mothers with underweight at conception and who fail to gain enough weight during pregnancy also manifest it. Excessive protein, salt and vitamin C metabolism in the body cause toxemia. Poor nutritional status also contributes to toxemia of varying degrees. During toxemia oedema, hypertension, albuminuria and convulsions occur. Kidney or heart diseases can manifest symptoms of toxemia. Sodium is restricted to 200 to 800 gms depending upon the condition and just enough protein is suggested.

Anaemia: About half the number of pregnant women in our country are anaemic. One out of every four maternal deaths are due to anaemia.

Complications of pregnancy, both foetal and maternal, can take place in an anaemic mother. The maximum rate of growth in a baby occurs before birth, in the fourth month of foetal life. Maternal malnutrition, especially anaemic condition, affects the growth of the foetus. Foetus grows from 15 gms of its weight at 12th week to 3,200 gms around the 40th week. Infants of anaemic mothers are frequently born with a subnormal iron store. Still birth, premature birth, toxemia of pregnant mother and various infections and diseases of the newborn infants occur if the mother is anaemic. A pregnant woman is termed as anaemic if from 28th week onwards the haemoglobin level is less than 10 gms per 100 ml of blood. The normal haemoglobin level is 15 gms per 100 ml of blood. Even if it is above 12.5 gms per 100 ml of blood the person is not considered anaemic. Poor loss due to any cause can also result in anaemia. Apart from rich sources of proteins, iron, vitamin C, folic acid and B_{12}, copper and molybdenum, supplements of iron, that is, 30 mgs of elemental iron per day, after the first trimester of pregnancy prevent anaemia.

Diabetes in Pregnancy: A diabetic woman who is pregnant requires additional nutrients. So she has to adjust the insulin dosage. Glycogen depletion, hypoglycaemia, acidosis and frequent infections are the complications that occur in her. Milk secretion also fails in a diabetic mother.

Dietary Recommendations in Pregnancy: During pregnancy energy allowance should be increased so as to support the growth of the foetus, placenta, maternal tissues and for the increase in basal metabolic rate. During the first trimester 5 per cent BMR and during the second and third trimesters 12 per cent BMR increase takes place. For an Indian woman with 45 kgs weight the total energy lost of pregnancy is about 62,500 kcals. Some energy is deposited as fat during pregnancy for lactating period. Additional intake of 300 kcal/ day during the second and third trimesters is the recommendation of the ICMR Committee in 1981. Besides, 35 gms of cereals and 10 gms of sugar or jaggery must be added to the normal balanced diet to derive the extra energy.

Protein intake must be increased for the development of the foetus and placenta and for necessary maternal tissue formation; 14 gms of protein along with the normal requirement will satisfy the demands of protein in the second and third trimesters of pregnancy. An additional quantity of 15 gms of pulses and 100 ml of milk along with the normal requirement will furnish the extra demands of protein.

Additional requirement of fat is not necessary during pregnancy. Since the foetal organs, especially the liver and brain, contain phospholipids rich in essential fatty acids, a normal balanced diet satisfies this demand.

During pregnancy minerals like calcium, phosphorus and iron are required more. A full grown foetus has about 25 to 30 gms of calcium. Most of the calcium deposition occurs during the last two months of pregnancy. The first set of teeth begins to form in the foetus in the eighth week. But its calcification starts just before birth—0.5 gm of calcium is

recommended during the second half of pregnancy. Phosphorus allowance must be equal to that of calcium.

During pregnancy about 400 mgs of iron for haemoglobin formation and more than 240 mgs of iron for storage in the foetus are required. Loss of iron through the placenta and blood during parturition are about 90 mgs. To meet all this iron requirement during pregnancy 8 mgs of iron per day is additionally recommended. Iodine, an essential constituent of thyroxine, must be supplied in enough quantities to prevent goitre in infants and pregnant mothers.

All vitamins are required in additional quantities during the gestation period. But for most of the vitamins the exact amount needed in additional quantities is not known. Intake of 400 I.U. of vitamin D as against 200 I.U. of normal amount is recommended. If a well balanced diet is taken, enough vitamin C and riboflavin will be supplied. Thiamine requirement is related to energy consumption and 0.5 mg per 1000 kcal is enough during pregnancy also. Based on the observation that urinary excretion of metabolites of tryptophan is higher in pregnant mothers it is assumed that conversion of the amino acid into niacin is more and 2 mgs more of niacin is recommended during pregnancy to about 200 mgs and it is difficult to supply it through food sources alone. So supplements with folate is essential. About 0.5 mg of B 12 a day is recommended during pregnancy.

The requirement of various nutrients during pregnancy is given along with balanced diets. Five or six small feeds are better than three or four meals so as to avoid fullness in the stomach. Leafy vegetables, fruits, other vegetables and fluids tend to maintain normal bowel evacuation. Rich, highly spiced and fried foods, heavy desserts, rich gravies and excessive amounts of salts must be avoided not to gain excessive weight. It is better to use jaggery instead of sugar as it contains more iron.

Due to tradition and habits many food fads are prevailing in our country. In the light of nutritional knowledge many of

these food fads were proved to be prevalent without any scientific basis. Many mistaken beliefs still prevent pregnant mothers from consuming very useful foodstuffs. For example, in South India papaya, which is an excellent source of vitamin A, is avoided for fear of abortion. In the same way pumpkin and jackfruit which are good sources of carotene (vitamin A) are avoided, being classified as hot foods. In certain places, the quantity of food eaten during pregnancy is cut down stating that delivery will be difficult. Wheat, jaggery, meat, fish, eggs, mangoes, tea and coffee are also tabooed as hot foods. There is not scientific basis in such beliefs and a pregnant mother can consume all foods like a normal woman. The composition of a balanced diet during pregnancy includes additional quantities such as 35 gms of cereals, 15 gms of pulses, 100 ml of milk or milk products and 10 gms of sugar or jaggery. For non-vegetarians some modification is normal.

A Sample Diet during Pregnancy

Time		*Meal*		*Menu*
Time 6.30	-	Bed coffee	-	Coffee
8 a.m.	-	Breakfast	-	Chapati, Dal curry, Soyabean milk, Guava
10 a.m.	-	Mid-morning	-	Lime Juice
12.30 Noon	-	Lunch		Vegetable pulav, Sprouted green gram salad, Curry-leaves chutney, Lassi
2.30 p.m.	-	Mid-time		Tender coconut water
4 p.m.	-	Tea	-	Tea, Moong dal laddoos
6 p.m.	-			Amaranth soup
8 p.m.	-			Chapati, Liver-tomato curry, Cucumber salad, Fruit cup.

Contd...

Time		*Meal*		*Menu*
9 p.m.	-	Milk	-	
This diet provides		Calories		2,600 kcals
		Proteins		62 gms.
		Calcium		1.4 gms.
		Iron		45 mgms.

Composition of Balanced Diet during Pregnancy

Items		*Quantity*
Cereals	—	475 gms
Pulses	—	55 gms
Leafy vegetables	—	100 gms
Other vegetables	—	40 gms
Root and tubers	—	50 gms
Milk and milk products	—	250 ml
Oils and fats	—	40 gms
Fruits	—	60 gms
Sugar and jaggery	—	40 gms

For Lactating Mother: The nutritional link between the mother and the child continues even after birth. The newborn baby depends for some period solely on breast milk for its sustenance. Unfortunately, most of the modern mothers do not want to breast-feed their children because they fear that they will lose their shape and charm. But breast-feed is not only the birthright of the baby, but proper emptying of breasts reduces the chances of mastitis and even of breast cancer. Studies have shown that mothers who have never fed their children have higher rate of malignancy. Nutritionists are of the opinion that there is no food equivalent to breast milk for a newborn baby. Nature has designed it to be a complete food for the first few months of a baby's life. Breast milk immunises

the baby against infection also. To secrete enough milk for the baby the mother should have nutritious foods. Nutritional needs of a lactating mother are higher than that of a pregnant mother. The quality and quantity of breast milk depends on maternal diet. In an inadequate diet the quality of mother's milk is maintained by drawing the nutrients from her body reserves and from tissues and bones. That is why the milk secreted by poor women is equal in its nutritive content to that of lactating mothers from developed countries of the world. The diets consumed by many lactating mothers in our country are very poor. A lactating mother requires more calories so as to secrete enough milk and to meet the high BMR during this period. The ICMR recommends about 550 kcal/day additionally during the first 6 months and 400 kcal/day from 6 to 12 months for a lactating mother. An average of 600 ml of milk is secreted by a lactating mother in India. About 420 kcals are supplied through this milk. The efficiency of conversion of diet calories to human milk calories is only 60 per cent.

The human milk contains 7.2 gms of milk protein and 14.4 gms of protein is required to produce so much milk protein. Therefore, 25 gms of additional protein per day is recommended for a lactating mother during the first six months of lactation. The calcium content of breast milk of Indian mothers are about 30 to 40 mgs/100 ml and to meet the additional needs 500 mgs of calcium is recommended along with the normal requirements of 500 mgs. Thus 1 gm of calcium is recommended for a lactating mother.

The concentration of iron in breast milk is about 0.72 mg/600 ml. During lactation menstruation is not common, thus about 1 mg of iron is saved. So iron is not recommended by the ICMR committee for a lactating mother. But care must be taken to provide a balanced diet to meet the normal iron requirement. Since our Indian diets are predominantly based upon cereals, the high phytate content interferes with iron absorption. Good calcium and vitamin C ratio can minimise this phytate interference but our diets are again deficient in these nutrients.

The quantity of vitamin A in 600 ml human milk is 300 mgs and to meet this requirement 300 μ gm (Microgram) of retinol or 1600 μgm carotene is recommended by the committee.

The quantity of ascorbic acid in human milk is only 15-30 mgs. ICMR committee recommended an additional 40 mgs of vitamin C per day. The thiamine content in breast milk is 60 μgms per 600 ml. Depending upon the energy requirements thiamine demand varies. The normal requirement of 0.5 mgm per 1000 kcals is applicable to lactating mother also. For riboflavin an additional recommendation of 0.3 mg/ day is enough to meet the needs of a lactating mother. The niacin requirements during lactation is 4 mgms additional to normal requirements. Increased fluid intake is recommended during lactation period to enable the normal functions in the body. There are no foods which require restriction as it is customary in various parts of India based on food fads and fallacies. A lactating mother requires adequate nutrition for the proper development of the baby.

The composition of a balanced diet during lactation is based on the characteristic of a normal person. Additional requirements must be met by making slight variations in the intake of food groups. The ICMR committee of 1981 recommended an additional intake of 60 gms of cereals, 30 mgs of pulses, 100 ml of milk, 10 gms of fat and 10 gms of sugar. These additional quantities of various foodstuffs supply about 521 kcals besides 25 gms of protein and other requirements by the body during lactation. If such a diet is consumed, a lactating mother can produce enough milk for the baby without affecting her health.

Composition of Balanced Diet for a Sedentary Lactating Mother

Foodstuff		*Quantities*
Cereals	-	500 gms
Pulses	-	70 gms

Contd...

Foodstuff		*Quantities*
Leafy vegetables	-	100 gms
Other vegetables	-	40 gms
Roots and tubers	-	50 gms
Milk and milk products	-	250 ml
Oils and fats	-	50 gms
Fruits	-	60 gms
Sugar and jaggery	-	40 gms

For Old Age Persons: Good nutrition throughout the life serves as a sound insurance for health for the years of old age. The process of aging brings about marked physiological changes in the body. Inadequate dentition, diminished sensitivity to taste and smell, diminished secretion of hydrochloric acid in the stomach and digestive enzymes, biliary impairment, if any, which interferes with fat digestion, irregular bowel evacuation, general ill-health, economic or emotional insecurity and unwanted feelings are some of the problems common among old people. Diseases of old age like diabetes, hypertension and other heart diseases also pose more problems to them. In modern society, with nuclear families, problems of old age are varied. In a joint family, old people are wanted and respected and they enjoy their position. But in our modern society old people often feel neglected and psychological problems created by this reflects as many ailments. Loss of appetite is a common complaint of old people.

Nutrition requirements of old change from normal adult requirements. The lowered metabolic rate reduces the caloric requirement by about 25 per cent compared with normal adults. Physical activity is also less compared to a normal person. If an old person has normal body weight adequate calories can be given. For obese people the calorie intake should be adjusted to reduce body weight whereas for underweight persons the

calorie intake should be enough to bring them to normal weight. For a normal sedentary worker 2,100 kcals for males and 1,700 kcals for females are enough to maintain normal condition.

Protein deficiency is common among old people. Protein-splitting enzymes like pepsin, trypsin and erepsin are mild in their action and so protein is not digested properly. Absorption of nutrients is poor among old people due to changes in intestinal wall. Loss of appetite and difficulty in chewing and masticating food must be supplied to old people to provide 1.5 gms of protein per kg of body weight. Though pulses are rich sources of protein, they produce flatulence among old people. Cooking methods adopted for geriatric people must be like steaming or boiling, making food easily digestible. Fat digestion is difficult and delayed in old age. Moreover, cholesterol level could be high among old people and so it is better to avoid saturated fat from animal sources, coconut and palm oil. Vegetable oils reduce the blood cholesterol level and 40 to 50 gms of such fats or oils can be used.

Assimilation of minerals is poor in old people compared to a normal person. Osteoporosis and anaemia are common among elderly people. Poor absorption of minerals and hormonal imbalance, especially of androgen and oestrogen, produce osteoporosis in old people. About 0.8 to 1.00 gm of calcium and 30 to 40 mgms of iron are recommended for them. Chronic arthritis and hypoactivity of the thyroid diminishes the activity of bone marrow which results in anaemia. Generally vegetables, raw vegetables or fruits are consumed in less amounts by old people which produces signs and symptoms of various vitamin deficiencies. B complex vitamin deficiency is common and along with a good diet a multi-vitamin supplement is essential.

Fluid intake must be liberal so as to form 1 to 1.5 litres of urine daily. Since many renal diseases are common among old people proper urine output must be checked. Only proper urine formation eliminates the urea, uric acid and other metabolic by products.

Since constipation is a common complaint of this age bulk or roughage must be included in the diet. Too much of refined carbohydrate foods, sweets and juices lack bulk and produce constipation. Muscle tone of the intestine is reduced and physical activities are also less, so only roughage in the diet can stimulate the intestinal movements.

For non-vegetarians the amount of milk is reduced to 400 ml and pulses to 55 gms for males and 45 gms for females. One multi-vitamin mineral tablet is recommended for them. Individual variation is common among old people depending on their health status.

Nutrition of Workers: Requirement of various nutrients for labourers and industrial workers differ from normal adults. Calories, thiamine and riboflavin requirements are high. Working efficiency of a person depends to a great extent on the pattern of diet consumed. To get the maximum output a labourer must have adequate diet. Manual work demands more energy and to assimilate energy sources properly B vitamins are essential. Energy requirement varies according to the climatic condition and pattern of work, that is, whether under the sun or inside a factory.

Composition of Balanced Diet for Geriatrics Over 60 Years

Food Stuffs	*Males*	*Females*
Cereals	320 gms	220 gms
Pulses	70 gms	55 gms
Green leafy vegetables	100 gms	125 gms
Other vegetables	75 gms	75 gms
Roots and tubers	75 gms	50 gms
Fruits	75 gms	50 gms
Milk	600 ml	600 ml
Fats and oils	30 gms	30 gms
Sugar and jaggery	30 gms	30 gms

Protein requirement is not increased for labourers, but if the energy requirement is not met from energy sources proteins are utilised for energy purpose. Climatic conditions again affect the protein requirements, for, instance, about 50 per cent more proteins are required in cold climate.

Vitamins must be present in adequate amount in the diet. Thiamine is essential to maintain normal appetite because deficiency of it interferes with metabolism of carbohydrate. Therefore, 0.5 mg of thiamine per 1,000 kcals is the required amount. Riboflavin and niacin are also required in good amount to enable proper utilisation of energy.

Minerals like calcium, phosphorus and iron must be the same as in the diet of a normal adult. Adequate nutrition ensures working efficiency. Undernourished persons reduce their voluntary activity. An anaemic person cannot work hard or concentrate on anything. His output is low and this yields low income. Purchasing capacity of such people is low and as a result they buy foods insufficient for the family and consume inadequate diet. Inadequate nourishment brings ill-health and this cycle is continued. Low-cost foods can be substituted for high-cost ones. But often people are not aware of the cheap nutritious sources which can nourish them well. Nutrition education is essential to select low-cost foods like roots and millets like ragi or jowar, cowgram, horsegram, soyabeans, cluster beans, groundnuts, leafy vegetables, amla, papaya and locally available vegetables and fruits.

Composition of Balanced Diet for Labourers or Industrial Workers

Food Stuffs	*Males*	*Females*
Cereals	630 gms	575 gms
Pulses	80 gms	60 gms
Green leafy vegetables	125 gms	100 gms
Other vegetables	100 gms	100 gms
Roots and tubers	100 gms	100 gms

Contd...

Food Stuffs	*Males*	*Females*
Fruits	60 gms	60 gms
Milk	400 ml	400 ml
Fats and oils	50 gms	40 gms
Groundnuts	50 gms	50 gms

For non-vegetarians 60 gms of meat or fish or one egg can be included and pulses reduced to 55 gms and 50 gms for males and females, respectively.

For Youths: Adolescence is a period of rapid growth after infancy. The rate of growth reaches its peak between eleven and fourteen years for the girl and between thirteen and sixteen years for the boy. Internal activities like secretion, hormonal reactions, basal metabolism and biochemical reactions are more during this stage. Pubertal growth demands more body-building substances and basal metabolic rate is increased which demands more energy. Since the period of adolescence is accompanied with considerable stress due to physiological and psychological changes attitude towards diet is often very unhealthy. Boys are usually well-fed in adolescence as they prefer to be tall and well-built with strong muscles. Therefore, an adolescent boy is more receptive to form good dietary habits. Girls are often self-conscious of their figure and they avoid many foods labelling them as fattening. Complexion, pimples and other marks are often associated with certain foods and their consumption. Withdrawal attitude, a common problem of adolescents, is often taken on food. Weight control is another important problem with adolescent girls and they eliminate essential nutrients in this effort. Skipped meals, poor lunches, snacking in-between, munching in-between, consuming large quantities of soft drinks and salty tit-bits which reduce the appetite are the common unhealthy dietary habits observed among adolescents.

Nutritional requirements for boys and girls differ at adolescence.

Requirement of Various Nutrients

Group	*Age (years)*	*Calories (kacls)*	*Protein feW*	*Iron (mgm)*	*Retinol (gm)*	*β Carotene (mgm)*	*Thiamine (mg)*	*Riboflavin (Mg)*	*Niacin (mgm)*	*Vitamin C(mg)*	*Calcium (mgm)*
Boys	10-12	2420	42.5	30.25	600	2400	1.2	1.5	16	40	.0.4-0.5
Girls	10-12	2260	42.1	30.25		"	1.1	1.4	15	40	0.4-0.5
Boys	13-15	2660	51.7	25	750	3000	1.3	1.6	18	40	0.6-0.7
Girls	13-15	2360	43.3	35	"	"	1.2	1.4	15	40	0.6-0.7
Boys	16-18	2820	53.1	25	750	3000	1.4	1.7	19	40	0.05-0.06
Girls	16-18	2200	44.0	35	750	3000	1.4	1.3	15	40	0.05-0.06

Composition of Balanced Diet for Adolescents

Age	*Cereals tables*	*Pulses tables*	*vege- (Ml)*	*Leafy vege-*	*Other tubers fat*	*Roots & Milk*	*&*	*Oil jaggery*	*Sugar and (years)*	*Group*
Boys	10-12	420 gm	45 gm	50 gm	50 gm	30 gm	250	40 gm	45 gm	
Girls	10-12	380 gm	45 gm	50 gm	50 gm	30 gm	250	35 gm	45 gm	
Boys	13-18	420 gm	70 gm	50 gm	50 gm	30 gm	250	40 gm	45 gm	
Girls	13-18	380 gm	70 gm	100 gm	50 gm	30 gm	250	40 gm	45 gm	

On the contrary, an adolescent girl requires all nutrients in good quantities not just for the rapid growth but also to obtain optimal storage for later requirements during pregnancy and lactation. Underweight is undesirable for them because it helps further susceptibility to infection. Studies on obstetric performance of adolescents have shown that high incidence of eclampsia, toxemia, miscarriage and premature deliveries occur among undernourished pregnant adolescent mothers.

The ICMR expert group suggested the following composition of diet to derive the above recommended allowance for adolescent groups.

For School Children: Majority of our school children consume inadequate diet and so they are malnourished. School children usually skip their meals due to various reasons. Poverty, ignorance and disturbed emotional status due to maladjustment in schools are some of the factors which produce malnutrition among school children. Western countries have shown a general trend towards an increase in the average national height of the people with each succeeding generation. Better socio-economic condition and better nutrition contribute to this. But this is not true in our case. Diet surveys carried out in our country have shown that diet consumed by school children is deficient in calories, proteins, vitamin A, riboflavin, folic acid and iron. Deficiencies in the diet are both qualitative and quantitative. More than 50 per cent of our school children are anaemic. The growth rate is poor and they gain low body weight and height. Their capacity to put maximum effort for work is poor. They are unable to concentrate, their power of grasping is reduced and their learning ability is poor. Thus an undernourished child is hurdled in its physical and intellectual development. Periodical check-up of height and weight manifest if there is retarded growth.

Nutritional requirements of boys and girls are more or less the same till the first 9 years. After that there is variation in some nutrients. Recommendation of allowances of various nutrients for school children suggested by the ICMR is given below:

Recommended allowances of these nutrients can be derived from a balanced diet with the following composition:

Recommended Dietary Intake of Nutrients

Age (Years)	*Calories (kcals)*	*Proteins (gm)*	*Calcium (gm)*	*Iron (mgm)*	*Retinol (mgm)*	*β Carotene*	*B1 (mgm)*	*B2 (mg)*	*Vitamin C (mg)*
1-3 year	1220	22.0	0.4-0.5	20-25	250	1000	0.6	0.7	40 mg
4-6 "	1720	29.4	0.4-0.5	20-25	300	1200	0.9	1.0	40 mg
7-9 "	2050	35.6	0 4-0.5	20-25	400	1600	1.0	1.2	40 mg

Foods	***Age Group*** ***1-3 years***	***4-6 years***
Cereals	175 gms	270 gms
Pulses	35 gms	35 gms
Leafy vegetables	40 gms	50 gms
Other vegetables	20 gms	30 gms
Roots and tubers	10 gms	20 gms
Milk	300 ml	250 ml
Oils and fats	15 gms	25 gms
Sugar and jaggery	30 gms	40 gms

Among non-vegetarians 50 per cent of pulses can be deleted and instead 1 egg or 30 gms of meat or fish can be added.

For Pre-school Children: The growth rate declines after the child is one year old, but the foundation of good health is laid during the pre-school age. In India about 20 per cent of the total deaths occur among toddlers in the age group of 1 to 4 years. A child who has failed to grow during this crucial period may not make up the loss in growth even with an excellent diet in later life. Deficiency of vitamin A which leads even to blindness and anaemia are common disorders found in children in the age group of 1 to 5 years. About 1 to 2 per cent of pre-school children suffer from severe deficiency diseases

like kwashiorkor and marasmus. Studies in India have shown that the performance of children who had earlier suffered from malnutrition, was clearly inferior to that of children who had not gone through malnutrition. Their diets in general consist of cereals, roots, tubers and vegetables. Important items like pulses, leafy vegetables, yellow vegetables, milk and milk products and other protein sources and fruits, are consumed much below their requirement. Deficiency of protein calories, vitamin A and iron are very common among this group due to their inadequate dietary habits. Non-availability of protective foods, low purchasing capacity, illiteracy and ignorance about the importance of nutrition during this period, traditional habits, food fads and fallacies, insanitary living conditions and prevalence of infectious diseases are the main causes of malnutrition. Physical and mental retardation set in and high mortality takes place among infants. It has also been shown by studies that the measurement of head circumference usually indicates that the brain volume is less among malnourished children. Malnutrition reduces memory and hearing ability and impairs intellectual functioning.

Other Requirements: Energy requirement varies according to age, activities, climate and the growth pattern. Proportionately, a child requires more calories per kilogram of body weight compared to an adult person. This is mainly due to the high basal metabolic activities, extra physical activity of the child and extra energy needed for growth. The ICMR Committee (1981) suggested that the ideal weight chart must be consulted before recommending energy requirements.

Height and Weight of Indian Pre-school Children

Age Yrs	*Boy's Height cm.*	*Girl's Height cm.*	*Boy's Weight kg.*	*Girl's Weight kg.*
1-2 yrs	82.61	79.89	10.94	10.21
2-3 yrs	91.14	89.63	12.79	12.11
3-4 yrs	98.36	96.21	14.78	13.79
4-5 yrs	104.70	104.19	16.12	15.85

Energy Allowance for Children

Age Group	*Body Weight*	*Energy Requirement*
1-3 yrs	12.03 kg	1220 kcals or 5.1 MJ
4-6 yrs	18.87 kg	1720 kcals or 7.2 MJ

Protein allowances for children of 1 to 3 years is 1.83 gms/kgs or 22 gms. For 4 to 6 year old, 1.56 gms/kgs or 29 gms to ideal weight is recommended.

Balanced Diet for Pre-school Children

Age group	*Cereals*	*Pulses*	*Leafy vege-tables*	*Other vege-tables*	*Roots & tubers*	*Milk*	*Fats & oil*	*Sugar or jaggery*	*Fruits*	*Egg*	*Non-vegetarian Meat or fish*
1-3 years	175gm	35gm	40gm	20gm	10gm	300ml	15gm	30gm	60gm	1	30gm
3-6 years	270gm	35gm	50gm	30gm	20gm	250ml	25gm	40gm	60gm	1	30gm

From 1 to 6 years calcium requirement is 0.4-0.5 gms and iron requirement is 20-25 mgs. Retinol during 1 to 3 years 250 μgm and 3 to 6 years 300 μgm is enough, β Carotene is required 1,000 μgm and 1,200 μgm respectively. Also 0.6 mgm of thiamine and 0.7 mg of riboflavin and 8 mgs of niacin are recommended for 1 to 3 years. For both groups 40 mgms of vitamin C and 100 mgms of folic acid and 200 I.U. of vitamin D are recommended. All these nutrients are required for maintaining health and well-being of the growing child. Since childhood is the age of rapid growth, proteins of high biological value must be included in the diet. Iron deficiency is common among this group, food rich in iron must be made available in the diet. Low intake of iron was found in a high percentage of 1 to 5 year group which leads to iron deficiency anaemia.

If such a balanced diet is provided to the pre-school child, it will be reflected in normal growth pattern. An average of 12.4 cms height and 2.5 kgs weight is gained by a pre-school child during 1 to 2 years. The growth pattern of a pre-school child of 2 to 3 years is 8.9 cm height and 2.1 kgs weight per year. During 3 to 4 years there is a slight decline in the growth

pattern, that is, 7.3 cms height and 2 kgs weight per year. Height gain of a pre-schooler of 4 to 5 years per year is only 5.6 cms and the weight is 1.8 kgs. Even though there are different factors which affect the height and weight gain of a child the proportionate rate has a pattern and marked variation in it must be taken seriously.

For Infants: Nutrition during infancy lays the foundation for health. Growth is rapid and changes in body composition take place at this age. Chemical maturation of the body is accomplished and internal activities occur at a high speed. Basal metabolic rate is also high. Compared to an adult in terms of body weight, an infant needs all nutrients in more quantities. A healthy newborn baby doubles its birth weight by the fifth month and triples by one year. The average birth weight of Indian infants is less than the average European infant. But the growth rate of Indian infants is on par with the developed countries in the first few months when it depends entirely on breast milk. A diet survey conducted in various parts of India shows that a lactating mother of lower socio-economic class consumes only about 1800 kcals and about 40 gms of protein which is much below to her requirement. Despite this, she is able to breast-feed her baby successfully during the first six months. Qualitatively her milk is not inferior to a well-nourished mother. But beyond three to four months, breast milk alone is not able to supply the needs of the infant. Supplementary foods are needed after three months. Sometimes breast milk is inadequate to meet the nutritional needs of the infants even up to six months. Continuous stress on the mother may result in her ill-health and reduction in the quantity of her milk. The nutrients from mother's body are withdrawn for milk production which leads to severe deficiency condition in her which in turn is reflected on the baby. Malnutrition impairs growth and development of the baby. Such babies have poor resistance to infection. Intellectual potentiality is impaired besides reduction in working efficiency in adult life. Studies

conducted at the National Institute of Nutrition (1981) indicate that children from well-fed, well-to-do sections of the community are taller and heavier and their resistance power against diseases is far better.

Nutritional Requirement of the Infant: Infants in the age group of 0 to 3 months are mostly breast-fed and lactating mothers of poorer section also secrete enough milk for their baby in the first three months. In our country, breast-feeding is traditionally prolonged and continues for one or two years. A well-fed mother secretes 850 ml of milk up to three months. Mothers from poorer sections secrete to an average of 600 ml/day. The energy requirements for infants of 0 to 3 months are 120 kcals/kg of body weight. If the mother secretes 600 ml of milk the infant gets 420 kcals per day.

Protein intake of healthy infants is found to be 2 gms/kg of body weight Six hundred milliliter of breast milk supplies about 7.2 gms of protein for the baby. Fat content of breast milk is 3.8 per cent and about 24.8 gms of fat is supplied by the breast milk. Fat in breast milk supplies 50 to 60 per cent of energy; 100 mgs of iron and 0.5 mgm to 0.6 mgm of calcium per kilogram of body weight is required by the baby. Calcium builds the skeletal structure of the baby's body.

Vitamins are essential for the rapid development of the infant. Breast milk supplies 140 gms of vitamin A during the six months of life. This is not enough when the requirement is 400 mgms of retinol during 0 to 6 months and 300 kcals gms from 6 to 12 months. They utilise the reserves in the body. Vitamin A deficiency is very common in our children. In India there is plenty of sunshine which is a rich source of vitamin D and 200 I.U. of this vitamin D is sufficient to meet the requirements and 20 mgs of vitamin C is recommended for infants. In 100 ml of breast milk 15 mgms to 20 gms of thiamine is present Thus about 0.17 gms of thiamine is ingested through breast milk and the requirement is 59 mgs/kg of body weight,

requirement of riboflavin is 0.25 gms/1000 kcals. Niacin requirement for infants is not studied and 780 mgms/kg of body weight is recommended by the committee. For infants recommended requirement of folic acid is 25 mgms and B 12 is 0.72 gm.

Growth Rate of Infants

Infancy is a period of rapid growth and the growth rate is given below:

Age	*Body Length*		*Weight*	
(Months)	*Male (cm)*	*Female*	*Male (kg)*	*Female*
At birth	47.5	47	3.1	3
1	50.6	50	3.8	3.7
2	53.7	53	4.8	4.7
3	56.8	56	5.2	5.1
4	58.8	57.9	5.8	5.2
5	60.8	59.8	6.4	6.1
6	62.8	61.7	7.0	6.6
7	64.5	63.4	7.4	7.1
8	66.2	65.1	7.8	7.5
9	68.9	66.8	8.0	7.7
10	70.2	68.1	8.2	7.9
11	71.5	69.4	8.4	8.1
12	72.8	70.7	8.6	8.3

Breast-feeding: Scientific studies have shown that the watery human milk is what the human infant needs. Cow's milk or buffalo's milk has the composition fit to match the rapid growth of its calves and not for a comparatively slow growing human infant. The nutritive value of breast milk is better than buffalo's or cow's milk. During the first two days after childbirth the yellowish fluid that is secreted is known as colostrum. This is very nutritious and good for the baby as

it is a good source of vitamin A and contains substances which protect the baby from diseases. But many mothers do not feed the colostrum to the baby with the belief that it is not good for the child. The average quantity of milk secreted by an Indian woman is 600 ml or equal to four glasses of milk.

From the table, it can be seen that the protein content of cow's, buffalo's and goat's milk is three times that of human milk. The casein content is also more. In the stomach, hard curd is formed in children fed on animal milk. Fat content is almost the same except in the case of buffalo's milk where it is about twice as much as that in other milks. Minerals and vitamins are almost the same in all milks.

Composition of Human, Cow's, Buffalo's and Goat's Milk (100 ml)

Nutrients	*Human*	*Cow*	*Buffalo*	*Goat*
Proteins	1.2 gm	3.30 gm	3.8 gm	3.3 gm
Fat	3.8 gm	3.7 gm	8.5 gm	4.1 gm
Calories	71 kcals	69 kcals	100 kcals	76 kcals
Lactose	7.0 gm	4.8 gm	4.4 gm	4.7 gm
Calcium	33 mg	125 mgm	210 mgm	130 mgm
Iron	0.15 µmg	0.10 µmg	0.20 µgm	0.05 µgm
Vitamin A	48 µmg	47 µgm	60 µgm	36 µgm
Thiamine	0.02 mgm	0.04 mgm	0.05 mgm	0.05 mgm
Riboflavin	0.04 mg	0.18 gm	0.10 mg	0.12 mg
Vitamin C	4 mgm	2 mgm	2.5 mgm	2 mgm

Advantages of using breast milk as an infant's food are many. No substitute has ever been developed that matches the numerous advantages of this specific baby food. It contains various elements in the correct proportion required by the baby. Human milk is a dilute fluid which is easily digestible

and it is the ideal starting food for the baby. Breast milk is readily available and it is very economical. By giving any food, breast milk can be produced in a lactating mother. No time is required for its preparation and the chances of contamination are nil. Artificial feeding requires complete cleanliness and in a country like ours where sanitary conditions are poor infections are more among artificially fed babies. Colic pain and respiratory diseases are common among such children. Poliomyelitis, influenza, mumps and infectious diseases are also common among artificially fed children because breast milk contains high concentration of antibodies which give resistance against infections. Milk allergy, constipation and fullness of the stomach, gas formation and gastro-intestinal discomforts are less among breast-fed infants. A mother who is not feeding her baby due to some reason subconsciously feels guilty.

The infant gets close physical relationship with the mother which provides emotional security and well being to the baby. Apart from these advantages, emptying the breasts reduces the chances of mastitis and cancer. Milk secretion is reduced if the secreted milk is not drained by suckling the baby. Only by emptying it further secretion is stimulated. Worry, tension, anxiety and emotional disturbances reduce the secretion of milk. Cleanliness of breast and nipple, care of inverted, fissured and cracked nipples are essential to ensure proper feeding. If the child is sick and is not feeding then the mother must empty the breast either manually or with the help of a breast pump. An infant must be fed at a breast for twenty minutes. An adequate diet, relaxed attitude, enough rest and willingness to breast-feed serve as the key factors to secrete enough milk by a lactating mother. Suckling position is also important. The mother should offer her whole breast to the baby rather than just the nipple, because the baby will not suck the whole milk secreted by the breast by sucking at the nipple. The feeding of the baby must be developed by every mother. Individual variations are there in this matter. For the first few days most

of the babies are reluctant to suck as they are not very hungry. From the third day onwards the infant may want to feed as many times as it is fed. This may be to get the closeness and warmth of the womb from the physical contact in the new environment. From the second week onwards most babies can be trained to follow a routine of feeding.

In diseases like tuberculosis, typhoid, severe neurosis, psychosis, septicaemia and in eclampsia breast-feeding is not advocated.

Artificial Feeding: Industrial revolution, urbanisation and life-style of modern women, fment level is added. Gradually the quantity of water is reduced as the infant grows.

The number of feeds and dilution of milk vary according to the age of the infants. Seven feeds are enough for the first month. From second to seven months six feeds, and eight months to one year five feeds are recommended. From the third month onwards fruit juices and from the fourth month onwards cereals cooked in different forms must be introduced. From the sixth month onwards egg yolk, mashed vegetables, dals, vegetable soups and other forms of food can be introduced. This pattern is essential even for children who are breast-fed.

For the first month dilution of milk with equal quantity of water may be done with the addition of 5 per cent cane-sugar. For instance, for the feed 400 ml milk, 200 ml water and 20 gms of sugar can be added. In the second month, 500 ml milk, 300 ml water and 30 gms of sugar are enough. From three months to 1 year 100 ml of milk is added and 100 ml of water is reduced and 10 gms of sugar is also added.

Various processed infant foods are available in the market. The nutritive value of these infant foods is given on the container. Certain nutrients like vitamin D, iron, and other minerals are fortified in some foods. Dilution of these formulas must be done. Otherwise indigestion and other gastro-intestinal disturbances may occur. If the dilution is more enough nutrients

will not be present in the formula and the baby will not show the normal growth pattern. Since vitamin C is absent in processed foods it must be supplemented.

For low-income groups, cheap nutritious substitutes can be prepared for an infant from groundnut or soyabeans. Instead of diluting the expensive artificial foods into watery milk these substitutes can be prepared at home.

Groundnut Milk: Select fresh groundnut, shrunken or shrivelled nuts are harmful. Shell and roast gently for 5 to 10 minutes. Rub off the pink skins and soak one cup of white nuts in clean water for 2 hours. Drain and grind the soaked nuts to a fine paste sprinkling with water. When the paste is ready dilute it with 5 cups of water and filter through a fine cotton cloth. Boil the milk for 10 minutes and keep it aside for 8-10 hours in a closed vessel. Afterwards remove the layer of fat from the top and the liquid can be used as a substitute for milk.

Soya Bean Milk: Select fresh beans and discard discoloured or spoiled ones. Wash three times with water. For 1 cup of beans use 3 to 4 cups of water for soaking. Soak for 6 to 8 hours. Drain and throw away the water and rinse the beans with clean water. Grind them to a fine paste using 2 cups of boiling water and a pinch of baking powder of soda bicarbonate. Strain through a muslin cloth and add one more cup of boiling water. Boil this milk stirring constantly. Add 2 cardamoms to improve the flavour.

Children who are unable to tolerate animal food can be given these two types of milk. Artificial feeds must be given at body-temperature. The hole in the nipple should be just enough to avoid rapid flow of milk. If a large gulp is taken, air is swallowed along with it, which causes discomfort and regurgitation. A small hole in the nipple provides very little milk and the baby may become exhausted by sucking. Along with the feed some air is swallowed, therefore the baby must be made to burp.

Sterilisation of bottles, nipples, bottle-caps, strainer, spoons and other equipment used is essential.

Other Requirements: Liquid and solid supplements, vitamins and mineral supplements are essential to meet the nutritional requirements of the baby. Juices of fresh fruits such as oranges, tomatoes, grapes, or soup of drumstick leaves can be introduced to meet the vitamin C requirement. One teaspoon of fruit juice can be given to the baby from the third week onwards. At this stage the fruit juice must be diluted with equal quantities of boiled water. This can be increased to about three teaspoons (without dilution) by 2 to 3 months. Boiled leafy vegetables, carrots and tomatoes can be introduced from the third month. After boiling, mash and strain the vegetables. Add salt and lime juice. Mashed banana, pumpkin, egg yolk, meal soup without seasoning, porridge, double-cooked cereals and pulses, roasted pulse-cereal powders sweetened with jaggery are the other items that can be introduced by the sixth month. Ripe fruits are also good for them. Only one food must be introduced at a time, though a variety of items can be given to familiarise with new tastes. Only a small quantity of food must be given to the baby in the beginning. Do not force the baby to take more, otherwise it will reject it in course of time. If the baby vomits or shows dislike for a food do not force him to eat it. After an interval start again and if the dislike persists substitute it with another food. Provide variety in supplementary foods because infants like older people prefer it. If proper supplementation is not provided to the infants its growth is retarded. Foods rich in proteins, calories and other nutrients must be supplemented to prevent malnutrition because the baby is growing vigorously at this stage.

Deficiency of protein leads to physical and intellectual dwarfism in the baby. Lack of adequate vitamin A in the diet results in poor vision even in the case of an infant because mother's milk and baby's store of vitamin A may not be enough. ICMR reports that about 12,000 to 14,000 children become blind every year. Deficiency of vitamin A is observed as one

of the important causes for it. Deficiency diseases are not the only consequence of malnutrition but they again impair the power of resistance of an infant and expose them to a number of diseases. Respiratory diseases, gastro-intestinal disturbances and fevers are the common ailments during infancy.

Some recipes and their nutritive value are given in this chapter. These can be included in the diet of an infant to improve the nutrient intake. Green gram dal, kheer, rice wheat or ragi porridge, ragi milk, halwa, chandan kheer, idli, groundnut halwa, pongal, sweet kichiri, vegetable soup and fruit juices are some of the foods that can be given to a pre-school child.

Ragi, Wheat or Rice Porridge

Ragi flour	-	30 gms (2 tbsp) or any other flour.
Groundnuts	-	30 gms
Jaggery	-	50 gms
Water	-	2 glasses

Method: Mix ragi flour in a small amount of water and make it into a paste. Roast and grind groundnuts into a paste. Add ragi flour paste to boiling water and stir continuously. Crush the jaggery and dissolve in half cup water. Strain and add it to the ragi mixture. Add groundnut paste and allow the mixture to cook for five minutes and remove from the fire.

Nutritive Value: This porridge supplies about 11.5 gms proteins and 225 kcals.

Ragi Milk Halwa

Ragi (grains)	-	100 gms
Milk	-	half cup
Jaggery	-	50 gms
Water	-	2 glasses

Method: Clean and soak ragi for 10-12 hours and grind into a paste. Dilute with water and strain the mixture through

a thin cloth. Add milk and cook it on low heat. Stir to avoid lump formation. Add jaggery to the cooked milk.

Groundnut Halwa

Shelled groundnuts	-	100 gms
Jaggery	-	50 gms

Method: Soak groundnuts for 6 to 8 hours and grind into a paste. Add 20 to 25 ml of water to the paste and keep on fire. Add jaggery, allow to cook for 5 minutes and remove from the fire.

Pongal

Rice	-	20 gms (one fistful)
Dal	-	20 gms (2 tbsp)

Method: Cook dal, and a Hide salt, and mash it well with a ladle. Cook rice, add mashed dal. Add a little ghee to it Serve this with a soup prepared from greens.

Nutritive Value: 6.2 gm protein, 120 kcals (without ghee).

Kheer Pongal

Rice	-	50 gms (4 tbsp, heaped)
Amaranth	-	50 gms (one medium size bunch)
or any other greens		
Green gram dal	-	50 gms (4 tbsp)
Groundnut oil	-	10 gms
Salt	-	to taste
Cumin	-	a pinch

Method: Roast rice and green gram dal and powder them. Boil amaranth, mash and strain. Mix amaranth puree with rice and dal powders and make into a paste. Add this mixture to the boiling water. Fry cumin in oil and season the pongal.

Nutritive Value: 4.0 gm protein, 100 kcals. Low Cost Recipes

Chandan Kheer

Roasted bengal gram flour	-	25 gms (2 tbsp)
Milk	-	200 ml (1 glass)
Sugar	-	15 gms

Method: Add a little milk to the bengal gram flour to form a paste. Boil the remaining milk and add bengal gram flour paste to it. Stir continuously to avoid lumps. Allow it to cook for five minutes. Add sugar and remove from the fire.

Nutritive Value: 4.0 gms protein, 110 kcals.

Sweet Khichiri

Wheat	-	45 gms (4 tbsp)
Roasted bengal gram dal	-	45 gms (3 tbsp)
Jaggery	-	50 gms (4 tbsp)

Method: Lightly roast wheat and bengal gram dal and powder them. Make into a thin batter. Add water to jaggery and make it into a syrup. Add jaggery syrup to the batter and allow to cook till there is no raw flavour.

Nutritive Value: 5-0 gms protein, 150 kcals.

Drumstick Leaf or Spinach Chapati

Drumstick leaves	-	25 gms
Onion (big)	-	15 gms
Salt and water	-	enough
Wheat flour	-	50 gms
Chillies (green)	-	1

Method: Chop onion and chilli. Clean and cut leaves into pieces and cook. Mix all three to wheat flour and make it into a soft dough. Divide into balls and make chapathi and cook it.

Nutritive Value: 8.0 gm proteins, 196 kcals.

Benefits of Food

Food is basic for survival of life. It is needed for the body's growth, repair and reproduction. Food provides energy required for all life activities. Satisfaction of hunger is usually the primary criterion for adequate food intake. But, satisfaction

of hunger itself is not a safe guide for selection of proper food. Also, when the types and variety of foods available for consumption are many and constantly change, a scientific knowledge of nutrients in food becomes essential. Hence, there is a need for an introduction to nutrition in a book on foods.

Good health is a major factor in our happiness and for an active life. Food is vital for building strong bodies and promoting good health. The science of nutrition helps in planning a diet, which improves growth rate, longevity and health. This is only one aspect of nutrition. In addition to material aspects, nutrition should help to promote emotional, energetic and spiritual quality, resulting in holistic health.

Finally, the role of food is not just for providing for life and health. The dynamic force of food should interact with the dynamic force of one's total being. Thus, this chapter deals with food in respect of nutrition, health and consciousness.

Qualitative Food: There are different concepts of nutrition. The most commonly studied aspect is the one based on laboratory research on how the various food components of our diet are digested, absorbed and metabolised to carry out various activities of the body. It also helps to understand the various diseases which occur due to malnutrition and deficiency. The wealth of information encompassing these aspects of nutrition is many times more than that of food.

Diet: The three basic foods which constitute our diet are carbohydrates, fats and proteins. Carbohydrates and fats are considered to be energy providing foods whereas proteins are growth-promoting substances. Mineral salts and vitamins, which are required for many metabolic activities, form the minor components of the diet. For a diet to be adequate and balanced, these foodstuffs must be ingested in the correct proportions.

Carbohydrate is the fuel on which the body functions are based. Energy produced through oxidation of carbohydrates forms the primary source of energy for the body. In terms of

quantity carbohydrates make up the bulk of the diet and consist of simple sugars like glucose, fructose and sucrose and starch from food grains or tubers like potato. The sugar should form only a small portion of carbohydrate intake as large amounts lead to obesity and serious health problems.

Fat is the most concentrated source of food energy, providing on a dry weight basis, more than twice the number of calories derived from protein or carbohydrate. Fat is taken in lesser quantity than carbohydrates. Food fat serves as a carrier for fat-soluble vitamins A, D, E and K. Fat is stored in the body to afford energy for later use, its high caloric density and low solubility providing superior qualities for purposes of storage. Dietary fats may be from plant or animal sources, but, healthwise, vegetable oils containing poly-unsaturated fatty acids are preferable. The deposits of fat in adipose tissue provide both insulation and protection for the body. With the exception of central nervous system, virtually all tissues of the body utilise fatty acids directly as a source of energy.

Protein in the diet is necessary to provide sources of nitrogen and amino acids to be utilized in the synthesis of body proteins and other nitrogen containing substances. The proteins of the body contribute to tissue structure and are involved in a variety of important metabolic functions. Proteins in excess of these needs serve as a source of energy. The nutritional value of a protein depends upon the composition of its amino acids. Vegetable foods generally contain small quantities of protein and the amino acids present are rarely in the proportions required by animal tissues. Therefore there is a danger of malnutrition if only one vegetable food forms the major component of the diet. However, a good vegetarian diet can be worked out which provides the complete range of protein requirements by using a wide variety of protein-containing vegetable foods. These include cereals, legumes, nuts, fruits and other vegetables. Most of the animal proteins contain a high proportion of essential amino acids in balanced amounts and are termed biologically good proteins.

The minor components of food contribute about 2 per cent of the dry weight of our diet. Of these vitamins form the accessory food factors. These are complex organic components present in very small quantities in natural food and absorbed into the body from the small intestine. They possess no energy but are essential for good health and maintain the body's normal metabolic activity. If the diet is deficient in a particular vitamin, metabolic activity is impaired. This produces a disorder symptomatic to that particular vitamin deficiency. This is known as deficiency disease. This deficiency disease can be avoided by supplementing the diet with the missing vitamin.

Mineral salts are the second important minor components of our diet. Some minerals in the form of their salts (sodium, calcium and potassium) are found in the body in large amounts while others like zinc, copper, cobalt, manganese and iron are present in small amounts. Some of these metals are part of several enzymes and complex organic molecules, which play a crucial metabolic role. Our requirement of minerals is met by the foods we eat, particularly vegetables and leafy stuff, which constitute our diet. If need be, there could be an external supplement of minerals.

Digestion: The process by which our food is broken down from complex insoluble substances into simple soluble ones is called digestion.

Digestion starts in the mouth; when food is properly chewed, there is thorough mixing of salivary gland secretion with food. The salivary enzyme— α-amylase—acts on the starch component of the food breaking it into shorter components - dextrins. When food enters the stomach its contents become acidic due to gastric secretion and starch digestion due to enzyme ceases. Further digestion of starch into sugar takes place in the small intestine. The mucosa of the stomach possesses many tubular glands, which are lined with different types of cells such as chief cells and parietal cells. Chief cells secrete the inactive enzyme pepsinogen and

prorennin while parietal cells secrete a 0.05% solution of HCl which makes the pH of the stomach contents 1-2.5, ideal for the optimum activity of the stomach enzymes. The acid promotes the conversion of pepsinogen to pepsin which hydrolyses protein into smaller polypeptides. Hydrochloric acid converts prorennin to rennin which in turn coagulates caseinogen, the soluble protein of milk, into the insoluble calcium salt of casein. The calcium salt is then digested by pepsin.

Relatively little digestion of fats occurs in the stomach because the pH of the stomach is too low for gastric lipase to be active. In the lower regions of the stomach, triacylglycerols (fats) are mixed with proteins, carbohydrates, gastric juice and other substances. Partial degradation of this mixture and churning action of the stomach result in the formation of chyme. This results in the elevation of pH, which allows some gastric lipase activity to occur. Chyme then enters the duodenum where it is mixed with pancreatic juice, which contains bile salts, pancreatic lipase and esterase. Pancreatic lipase ultimately breaks down triacylgrycerols to fatty acids and glycerol.

The pancreas is a large gland whose cells called acini produce a variety of digestive enzymes that are poured into the duodenum via the pancreatic duct. They include amylase to convert amylose to maltose, lipase to convert fats to fatty acids and glycerol, trypsinogen, which when converted to trypsin by enterokinase, digests proteins into smaller polypeptides, chymotrypsinogen which is converted to chymotrypsin to digest proteins to amino acids, carboxypeptidases to convert peptides to amino acids, and nucleases to convert nucleic acids to nucleotides.

Absorption: Digestion reduces food to a solution of fatty acids, glycerol, glucose, fructose, nucleotides and amino acids. If these small, soluble molecules are to be of any use to the cells of the body, they must be moved out of the intestine into the blood stream. The process of moving nutrients and water from the intestine to the blood constitutes absorption.

Since the small intestine is the place where nutrient and most water absorption takes place, we should expect it to have a large exposed surface area. This is exactly the case. The small intestine is quite long and is packed into the abdominal cavity as a series of coils. Besides, the surface of the small intestine also contains folds and enormous numbers of fingerlike process called Villi. Additionally, each villus is covered with epithelial cells, and the plasma membrane of fingerlike projections called microvilli, it further enhances the absorptive surface area. The residues of food that cannot be digested or absorbed leave the small intestine and pass through the ileo-colonic valve into the large intestine. There, water and vitamins are absorbed, and waste materials are compacted into faces.

The absorbed nutrients do not directly go into systematic circulation. Instead, everything is channeled into the portal vein, which goes to the liver where it breaks up into tiny capillaries. The liver cells then process the absorbed food for proper use when they pass into the blood stream through another vein, venacava. The liver also stores absorbed and processed food for release into blood stream when required.

The study of nutrition based on the biochemical make-up of the diet, the action of digestive juices on it, the role of the pancreas, liver, etc., has given us valuable information on various aspects of nutrition. The information available is voluminous. Foods, in addition to nutritionally important components, contain a lot of other substances in small quantities, which have significant pharmacological activity. Also cooking and food additives affect the nutritional value of food. The composition of food also varies depending on conditions of production and processing. Their effect varies from person to person who eats it. Although, the information available on nutrition is voluminous, it is responsible for some contradictions one observes in the literature.

Health Care: Knowledge of nutrition helps man to know the types and quantities of different foods to be taken in our

diet to maintain good health. According to ancient Indian concepts, health based on just the diet is *Arogya*—a state of no disease with which the, physician is primarily concerned. According to the World Health Organisation (WHO), the term "health" means more than the presence or absence of symptoms and has a more comprehensive concept of a multidimensional, global well being.

Dimensions of Health: Growth of psychological and social sciences has shown that illness is conditioned by psychological, social and cultural factors. Epidemiological diseases are due to social and cultural factors, while psychosomatic diseases are due to conditions of mind. While the psychosocial concepts of health are important, they are not enough for the total well-being of man. For, as Dr. Brisht points out, "if the components of health are restricted only to physical, mental and social parameters, then a pack of wolves who are physically strong, mentally alert and socially well-knit would be ideally healthy". If this is so, a group of human beings would be no better than a pack of wolves.

Man is a multidimensional being and he has a spiritual perspective too that transcends other modalities of life. In his case the definition of health should be enlarged to cover spiritual well-being in addition to physical, psychological and social well-being. The Sanskrit word for health is *"Swasth"*, which means "footed in the self true health exists only when man's consciousness is fixed in spiritual self, the *"swa"*.

Most persons believe that philosophy, religion, ethics and morality constitute spiritual health. These are only approaches to spiritual truth and not the spirit itself. When we go beyond these misconceptions and understand the spiritual reality by identification, we discover that there is an unity of experience that cuts across the barriers of time and space. Therefore, spiritual health conveys a special concept of a holistic and integral viewpoint of health. The WHO, in its programme of health for all, has recognised the spiritual dimensions of health.

Spiritual Health: Spiritual health evokes different images to different persons. To the materialistic Western mind it is paranormal healing, to the Eastern mind it is a boon or miracle of grace that heals our suffering and pain. To the materialist spiritual health is either a hoax or faith healing, while to — a time rationalist it is moral and ethical values in life which help man become a better race. To some others the word "spiritual" is a supreme something or someone whom none has known or can know.

According to Vedas and Upanishads (Indian scriptures), the spiritual state is a reality, which cannot be described but experienced. Seekers of consciousness have come in contact with spiritual reality and their life has been moulded by its influence. They have discovered in it something quite other than our customary notions of moral, ethical and religious values. Spiritual knowledge is a state of knowledge of that, the ultimate reality; it is revealed in a few, veiled in others, and present in all.

The seers who have experienced and lived spiritual life affirm that "The spirit is other than the mind". It is unity and oneness while the mind is duality and division. It is peace and bliss, harmony and truth, while the mind dwells in pain, pleasure and indifference. The mind also fumbles through error and ignorance. The spiritual state of health cannot be defined. It is best communicated and understood in the silence and stillness of our being. It is the highest perfection man is capable of through self-evolution.

This higher and greater consciousness is not tangible to the mass of humanity as yet. The next cycle of humanity will see the evolutionary emergence of this spiritual dimension. The resurgence of interest in Yoga, the debate over the possibility of consciousness independent of material reality, the turn towards global life, the liberalisation of spiritual pursuits, the growing need for man for freedom, love and oneness, inclusion of the spiritual dimensions in the field of psychology, education,

health and the findings in modern physics, are some areas indicative of such phenomena.

Those who have realised spiritual heights affirm that it is possible for man to enter and live in it and with a greater consciousness and power. If we attain that state, the disorders and conflicts that afflict humanity can be solved and it will release us from the yoke of suffering and death, and impart harmony and peace we so much lack now as a species. It is this that justifies and even makes it necessary to include the spiritual dimension as a factor with regard to health. This factor, being a source and origin of all, will also include lesser terms of our existence - mental, social and even physical. The spiritual dimension is all comprehensive, the simplest and the most fundamental. It should be sought after by man who is missing peace.

Science and Spirituality: Science is concerned with matter. In its quest to understand matter science has acquired incredible powers and harnessed it for the purpose of improving man's outer life. In the process, life has been reduced to a mechanical formula and man has been reduced to a machine among other machines.

Spirituality is the study of self and consciousness based on tapping the hidden powers of life and mind. Spirit or consciousness runs as a common thread through all phenomena. It is not an exclusive dimension that exists in isolation without any hold on the creature that inhabits it. The spiritual dimension includes the material and can intervene subtly to alter the laws and processes of the material universe and change the course of purely material and biological processes. Thus spirituality not only helps in attaining spiritual dimension but also it subtly intervenes to alter the laws and processes of the material universe.

Science seeks the truth of matter reality. Scientific methods and equipment can measure gross events. Scientific experiments can be seen, verified and reproduced. Scientific observations and inferences can be codified, resulting in physical laws.

Spirituality seeks after the truth of spirit, which is quite an other reality. Spiritual dimensions are subtler truth, which cannot be measured by conventional equipment. Subtler levels can only be experienced. One must reach a level of experience, then go to the next level and on and on till reaching an awakened state. Here, life and mind are open systems and keep growing and evolving, which upset old experiences.

Thus, science and spirituality are proceeding on their own lines of enquiry in search of a common goal. In the sincerity of their search, they will meet and when they meet they will realise that spiritual and material reality are two phases of one reality. The scientist would learn that "One Force" gives birth to the many. The spiritual scientist would discover that matter is created by a condensation of the spirit of One Reality that gives birth to various forms, phenomena and processes. This will result in the integration of human consciousness from a fragmented one, when one experiences bliss.

Evolution of Consciousness: Man's life is riddled with disease, death and infirmity. His attempts to conquer them will remain precarious at his own level of evolution. At best, he may arrive at a healthy equilibrium with his environment. All the stress and strain that besieges him is a call to evolve to a state of consciousness more perfect than his.

Man's problem of want, greed and lust and desire are due to his sense of self or individuality. His false sense of ego-self (I-ness) has to be shed off and replaced by true "I", the individual soul in him. This evolution of consciousness results in his feeling of peace, fullness and joy, with their immediate results of progress, health and fitness in the physical, quietness and goodwill in the vital, clear understanding and a general feeling of security and satisfaction. It will open the heart to a deep, pure and calm capacity to love without possessiveness and turbulent attachment. These will help one to be calm and balanced, free from fear, things which help in healthy living and even cure one of diseases.

It is possible for consciousness to evolve further, when man will be immune from all types of diseases and even death. But, this type of consciousness is possible only for a few. Contact with such evolved persons will help others to awaken, the sleeping soul in them, better than any other means.

Basic Needs of Food

Quality is the ultimate criterion of the desirability of any food product. The overall quality of a food depends on the nutritional and other hidden attributes, and sensory quality as assessed by means of human sensory organs. The absence of nutritional qualities and possible presence of hazardous microbes, environmental contaminants, food toxins and chemical additives cannot be easily judged by the consumer. Governments in many countries protect the interest of the consumer regarding nutritional and hidden attributes by stringent controls to assure good food quality and enactment of food laws regarding inspection, grading, packaging and labelling of foods. Qualities of foods are evaluated by sensory organs—eye, nose and mouth—or by the use of instruments.

Quality Judgement: Sensory evaluation consists of judging the quality of food by a panel of judges. The evaluation deals with measuring, analysing and interpreting the qualities of food as they are perceived by the senses of sight, taste, touch and hearing. By the sense of sight, the size, shape and colour of foods and other characteristics, such as transparency, opaqueness, turbidity, dullness and gloss, could be perceived. Colours of foods contribute immeasurably to one's appreciation of them. In addition, colour is associated with other attributes, e.g., the ripeness of fruits is judged by colour. The strength of coffee and tea is judged, in part, by the colour of the beverage.

Other sensory organs, the nose and mouth, are utilised to obtain information on flavour. Flavour embraces the senses of taste, smell and a composite sensation known as mouth-feel.

Sensory evaluation may be designed to reflect common preference, to maintain the quality of food at a given standard,

for the assessment of process variation, cost reduction, product improvement, new market development and market analysis.

Selection of Panel of Judges: Actually, one extremely discriminating, painstaking and unbiased individual would suffice for tasting. However, human judgement is individual and is not always consistent. Physical conditions of the individual, psychological factors and environmental factors may affect one's judgement. Further, one individual may not be able to discriminate different aspects of the food quality. For these reasons, for sensory evaluation, a panel of judges is used. Members of the panel should be carefully selected and trained to find out difference in specific quality characteristics between different stimuli and also direction and/or intensity of difference. The panelists should be of sound health and average sensitivity and conscientious individuals. The number of members in the trained panel should be small, varying from 5 to 10. The semi-trained panel should consist of 25 to 30 members. The findings in the case of the untrained panel should be based at least on 100 independent judgements.

Preparation of Samples: Careful sampling of the food is necessary for sensory evaluation. A well-mixed homogeneous sample, if possible, is desirable. With non-homogeneous materials, samples should be taken from the same area. Samples to be tested should be prepared by identical methods. All samples in a series should be at the same temperature. The number of samples used in any one session depends on the sensory nature of the test product and the evaluation method used.

Types of Tests: Different sensory tests are employed for food evaluation. The tests are grouped into four types; they are difference (discrimination) tests, rating (quantitative differences) tests, sensitivity tests and descriptive tests. The selection of a particular test method will depend on the defined objective of the test, accuracy desired and personnel available for conducting evaluation.

Difference Tests: Difference tests are used to determine if there is any difference between or among samples. The three

basic types of differences tested are simple difference, directional and quantitative difference, and quality preference difference. The methods of difference testing commonly used are the paired comparison, duo-trio and triangle (triad) tests.: In paired comparison test two samples, one standard and the other experimental, are presented to the panelists to determine if the samples are different and the directional difference in a specific characteristic. The duo-trio test employs three samples, two identical and one different. The panelists are first given the reference sample and then the other two samples are given successively in random order and asked to match one of them with the first. The triangle test also employs three samples, two identical and one different. All the three samples are presented simultaneously to the panelists who are asked to determine which of the three is the odd sample.

Rating Tests: These tests are difference tests with a quantitative aspect through direction and degree of judgements using suitable defined scales or scores. A number of methods are employed for this purpose. In a ranking test, judges are asked to rank a series of samples in the increasing or decreasing order of a specified characteristic, such as flavour, odour, colour or texture. A single sample (monadic) test is used for testing foods that have an after-taste or flavour carryover which precludes the testing of a second sample in the same session.

In two sample difference tests, the panelist is served with four pairs of samples. In two pairs, the test sample is a duplicate of the reference sample. In the other two pairs, the test sample is the test variable. The panelist is asked to judge each pair independently as to the degree of difference between the test sample and standard on a scale indicating no difference to large difference.

When more than one test variable is to be evaluated, multiple sample difference test is employed. Each panelist is served 3-6 samples depending upon the number of test variables. One sample is a known standard. The panelist

compares the test sample with the known standard. One test sample is a duplicate of the standard. The difference of the test sample from the standard according to the scale—none, slight, moderate and large—is evaluated.

A hedonic rating test is used to measure the degree of pleasurable and unpleasurable experience of the food product on a scale of 9 points from "like extremely" to "dislike extremely".

The numerical scoring test is used to evaluate a particular characteristic of one or more samples indicating the rating as excellent, good, fair and poor.

Sensitivity-threshold Tests: These tests measure the ability of the individual to smell, taste or feel specific characteristics in food or beverages. The tests are most commonly used in selecting panel members for evaluating, and with materials such as spices for assessing the intensity of odour or flavour.

Descriptive Tests: These tests help identify the perceptual characteristics of a product and express them on an agreed scale. The panelists record the impact of all quality attributes in a total perspective but not a single attribute judgement with precision, for statistical analysis. The tests include flavour profile tests to describe the aroma and flavour intensity characteristics of food and textural profile tests to describe the sensory manifestation of structure of inner make-up of foods, comprising concepts of texture, body and consistency of foods.

Judging: The environment is an important factor in judging food. There should not be any distraction so that the judges can concentrate. Judgements should be made independently and it is preferable to have panel booths and no communication between panel members should be allowed. The best- time of the day for sensory testing is an hour after any normal meal, i.e., mid-morning or mid-afternoon, when the judges are neither too well fed nor too hungry. The panel members should not smoke, chew pan or supari or take intoxicant for at least half

an hour before the test. Use of odoriferous substances (cosmetics) should be avoided. The panelists should be provided with room temperature water for rinsing between samples.

Results: The results of evaluation are recorded in suitable evaluation cards prepared for different-type methods. The data are tabulated, averaged and analysed for the answer to the question posed in the experiment. This method may not be sufficiently trustworthy in many cases for it to be stated with confidence that the results obtained are significant. For a correct interpretation of results, their statistical analysis is necessary.

Development and Growth: The evaluation of food quality does not rely mainly on human sensory organs. Objective methods have also been developed. These methods use instruments or a standard method to evaluate food quality by physical and chemical methods. The results obtained by these methods are reproducible, can be permanently recorded and are less susceptible to errors than the sensory methods of evaluation. Results obtained from the objective methods of evaluation should correlate with those obtained from sensory evaluation to ensure consumer acceptance. Objective evaluation includes chemical, physicochemical, microbial and physical methods of analysis.

Chemical methods include the determination of nutritive value of foods before and after cooking, products of food decomposition and adulterants in food. For details of the method, one has to refer to manuals on food analysis, such as the publication of the Association of Official Analytical Chemists (AOAC). Physicochemical methods include determinations, such as the hydrogen ion concentration by the use of a pH meter, sugar concentration and the degree of hydrogenation of fats by a refractometer, and a qualitative analysis of sugar solution with a polarimeter. Some properties of foods which depend on the structure or the physical arrangements of their components, such as whipped cream, or meat, vegetable and fruit tissues, are examined by microscopic methods.

The most widely employed objective methods are the measurements of physical properties by the use of instruments. Measurements of the appearance and volume of foods are of importance, particularly in the evaluation of baked products. A record of size, shape and grain (cell structure and distribution) of such foods can be recorded by photography and the volume determined by suitable methods.

Measurement of Colour: The importance of colour in food, the natural food colours, the changes they undergo during cooking, processing and storage are worth consideration. Natural and synthetic colours are also added to foods to bring the produce to the generally accepted standard. Therefore, the colour of food is measured to identify its properties.

A number of instruments are used for colour measurement. A simple method is to match the colour of the food with coloured chips or coloured glass. Disc colourimetry is another method used. Different coloured discs' are spun on a stage so that the colours merge into one colour without flickering. The test sample is placed adjacent to the spinning disc and the colours are matched. Tintometers are simple instruments used to determine the colour and its depth in foods.

For a more reliable measurement of colour, spectrophotometers are used. In the case of clear or transparent solutions, spectrophotometric measurements give quantitative results. As most foods are opaque in nature, reflectance spectrophotometry or, more commonly, tristimulus colourimetry is used. In a tristimulus system, colour is specified by three attributes: dominant wavelength, brightness and purity, according to CIE system (Commission Internationale de L' Eclairage). These three attributes refer respectively to the actual colour, the luminosity and the strength of colour. Several colourimeters are available for the evaluation of these attributes or complete specification of colour.

Measurement of Texture: The texture of food is an important factor in its acceptance. The textural properties include the

mechanical properties of hardness, cohesiveness, adhesiveness, chewiness, crispness, gumminess, viscosity and elasticity (springiness). A number of instrumental methods are available to evaluate the texture of various foods.

The tenderness of meat is an important quality for consumer acceptability. One aspect of tenderness of meat is tested by measurement of structure of meat with a penetrometer. This measurement gives an idea of how easy or difficult it is for the teeth to bite into a piece of meat. A more commonly used device to measure tenderness is an apparatus (Warner-Bratzler shear) which measures the force needed to cut the meat in simple shear, usually across the fibres, and estimate the force necessary to chew the meat.

The textures of fruits and vegetables are evaluated by a number of methods. A puncture testing is used to evaluate the firmness of fruits. It measures the amount of force required to penetrate the sample to a specific depth. A shear press is used to study the tenderness of fruits and vegetables. This consists of a rectangular box with evenly spaced slits in the bottom. A series of blades is moved through a sample of food. As the blades move the food is compressed, sheared and extruded through the openings in the box.

Some characteristics of foods depend upon their rheological properties. Rheology is the science of flow and deformation of materials, both liquids and solids, and has three aspects: viscosity, elasticity and plasticity.

Viscosity or consistency is an important factor in influencing the quality of a large number of food products. The acceptability of foods like salad cream, tomato products, jams, jellies, mayonnaise, syrups and fruit pulps, depends upon their consistency. Measurement of viscosity is also important in the processing of foods. A number of instruments are available for the measurement of viscosity and the type of instrument used depends on the nature of food.

Solids do not Flow: However, some solids can be deformed

by force and they recover when the force is removed. This is elasticity. Gels like those of pectin, gelatin and starch, baked custard and rennet curd are examples of elastic solids. There are a number of objective tests to determine the firmness of gels. A material that is plastic resists flow until a force is applied; but unlike elastic flow, plastic flow is not reversible.

Texture of dough and batters determines the quality of the finished product. The consistency of batters is determined by a line-spread apparatus which indicates the nature of dispersion of incorporated air. The consistency and stability of doughs are measured with a farinograph which measures the force required to turn mixer blades at a constant speed during mixing of the dough. A micrograph also gives information similar to that of farinograph.

Textures of baked products such as pastries, cookies and crackers are evaluated by shortometer. The instrument measures the force required to break the product. A compressimeter is used to evaluate firmness of bread crumb or softness of a baked product. The force required to break through (shearability) a sample of a baked product is also obtained using Warner-Bratzler shear.

Texture as perceived by human beings is a composite of characteristics. To arrive at a textural profile, different instruments have to be used. New instruments have been developed which could evaluate more than one characteristic constituting the texture of a food.

Non-nutritional Constituents and Food Safety: A diverse group of Non-nutritional chemicals find their way into foods. Their presence in foods represents real or potential health risks of different characters and magnitudes to persons consuming them. The undesirable constituents that affect the safety of foods include toxicants naturally occurring in foods, toxins resulting from microbial growth, environmental contaminants—such as those arising from processing and accidental contaminants—and chemical additives.

Naturally Occurring Toxicants: Certain plant and animal products contain natural constituents that are toxic. When foods containing them are consumed in sufficient quantities, the toxins present may prove to be hazardous.

Toxicants in Plant Foods: Plants are capable of synthesising a multitude of chemicals that cause toxic reactions when consumed by men or animals. Pulses contain a number of toxic substances, such as protease inhibitors, lathyrogens and flavism agents, cyanogens, hemoagglutinins, and saponins. Some of these toxins are also present in other foods, e.g., protease inhibitors in cereals and potatoes. Saponins are present in spinach and asparagus. Goitrogens (which cause hypothyroidism and thyroid enlargement due to thioglucosides) are present in cabbage and related species, in rapeseed and mustard. Practically all foods contain proteins which may act as allergens in sensitive individuals. Some varieties of mushrooms are poisonous. Oxalic acid, a constituent of rhubarb, spinach and beet, may cause oxalic poisoning in certain individuals.

Toxicants in Animal Foods: Tissues of marine forms of animals contain toxic chemicals and cause adverse responses when eaten. About 500 species of marine fish are known to be poisonous (ichthyotoxism) when eaten and many of these are among edible varieties. The toxic agents arise from blue-green algae and are then passed directly to herbivorous fish and indirectly to carnivorous species. Poisonous syndromes from their ingestion are variable in character and are usually designated by the kind of fish involved.

Shellfish have concentrated toxins from plankton constituting their food supply. The ingestion of shellfish (clams and mussels) results in paralytic shellfish poisoning. The toxic agent has been isolated. It is not destroyed by heating.

Microbial Toxins: Some micro-organisms promote desirable changes in foods under controlled conditions. Use of yeasts in

making breads and alcoholic beverages, and molds and bacteria in making cheese and fermented dairy products, are examples of beneficial effects of micro-organisms. However, they also cause harmful effects and are involved in most cases of food spoilage.

Mycotoxins: Mold growth on foods, especially if they are moist, is a common observation. Some molds, during their growth, produce toxic substances of various kinds. These substances are referred to as mycotoxins and the poisonous effect on men or animals as mycotoxicoses. Mycotoxins remain in the food long after the mold producing them has died and can therefore be present in foods that are not visibly moldy. Further, many mycotoxins, but not all, are stable and survive the usual conditions of cooking or processing. Mycotoxin contaminated foods when eaten are thus hazardous to man. Foodgrains, especially bajra, rye and jowar, get infected with the parasitic ergot fungus, *Claviceps purpurea.* Ergotism, a toxicosis, results from eating grain contaminated with this fungus. Eating moldy grains also results in alimentary toxic aleukia (ATA), a mycotoxicosis.

Aflatoxins: Mycotoxins produced by some molds, *Aspergillus flavis* or a *parasiticus,* are known as aflatoxins. A *spergillus* develop in many foods, particularly groundnuts and cottonseed, and their cakes and flour. Aflatoxins have been much studied. There are 14 chemically related toxins and one of them, aflatoxin B, is most frequently found in food and is the most potent carcinogen known. Liver cancer due to aflatoxin ranks high in our country.

Bacterial Toxins: Just as the growth of spoilage molds can introduce toxic substances into foods, the growth of certain bacteria can also result in production of toxic substances. These include *Clostridium botulinum, Staphylococcus aureus* and *Bacillus cereus.* These produce sufficient toxin in the food to cause illness in a susceptible host. In addition to these, *Bacillus lichiniformis* and *Bacillus subtilis* are organisms for which evidence for toxin production is beginning to accumulate.

The disease caused by the toxins of *C. botulinum* is known as botulism. The toxins are produced after an incubation period of 18-36 hours. They act on the nervous system and are potent poisons. A concentration of less than 1.0 μg is fatal for man; death occurs in 1 or 2 days. If one survives the effect of the toxin, convalescence is slow and takes 6-8 months. The toxins produce in human beings disturbances of vision and difficulties with speaking and swallowing as the mucous membranes of mouth, tongue and pharynx usually become dry. Progressive weakness and respiratory failure set in.

The botulism toxins are proteins. Six serologically distinct toxins designated as A, B, C, D, E and F are known. The protein is produced as a single polypeptide chain and when released from the bacterial cell is cleaved by a protease producing two fragments—∝ and β. The toxin is probably associated with a hemoagglutinin and thereby protected from proteolysis and denaturation in the gut. The toxin acts by inhibiting the release of acetylcholine from cholinergic nerve endings. The neurotoxin absorbed from the gut binds to nerve endings through fragment β while the a-fragment is responsible for neurotoxic effects.

The toxins causing botulism are thermolabile; they lose their activity on heating for 30 min at 80°C. The cause of botulism in man is inadequately heated or cured foods. Canned foods, low in acid, could be hazardous if they have not been heated sufficiently to kill the spores of the organism which might be present.

Staphylococcal food poisoning is perhaps the most commonly experienced form of food borne disease, but it is much less serious than botulism. To be pathogenic to man, the organism must be present in sufficient quantity, the food must support toxin production and an incubation period of 2-6 hours is required. The toxic period will last for 12-24 hours and results in nausea, vomiting, diarrhoea and abdominal pain. Severe cases will result in collapse and dehydration. However, there will be no fever.

S. aureus toxins are also proteins, single-chained and globular. At least seven serologically distinct toxins are recognised. Although termed entero-toxins, they are probably neurotoxins. They are absorbed from the gut and activate receptors on the abdominal viscera, the stimulus reaching the vomiting centre.

The toxin of *Bacillus cereus* is produced after an incubation period of 8-16 hours and it is active for 12-24 hours. Abdominal pain, diarrhoea, vomiting and nausea are the features of the toxin. Two or more toxins are involved in this food poisoning, one causing diarrhoea and another inducing vomiting.

Bacterial Food Poisoning: Over 50 genera of bacteria have been associated with food spoilage, food poisoning, food-borne diseases or general contamination. In general, Gram-negative, non-aciduric, nutritionally non-fastidious bacteria are commonly associated with the spoilage of moist proteinaceous foods, related aciduric genera with acid fruit products and lactic acid bacteria with foods rich in nutrients and fermentable carbohydrates.

Food poisoning bacteria grow in a food and infect the persons eating it. They form populations sufficiently large to colonise the gut of a susceptible individual. This situation obtains with *Salmonella* spp, *Vibrio parahaemolyticus, Escherichia coli, Yersinia enterocolitica, Clostridium perfringens* and *Compylobacter jejuni*. With these organisms, a casual relationship between a specific micro-organism and a specific disease exists.

Salmonella are ubiquitous, but they are more numerous in foods, such as pork, poultry and eggs, and products containing them. The illness caused by the organism is known as salmonellosis. As the multiplication of the organism in the intestine after ingestion of the food takes time, the illness starts 12-36 hours after ingestion and continues from 1-7 days. Diarrhoea, abdominal pain, vomiting and fever are caused by this type of poisoning. Infants are more susceptible to this

infection than adults. Acids and temperature tend to destroy the micro-organism.

Clostridium perfringens grows in the alimentary canal producing the poison, some 8-12 hours after the ingestion of contaminated food. The symptoms of illness are diarrhoea, abdominal pain, nausea but rarely vomiting, and no fever. The toxin is a protein which is the structural component of the spore and acts on the membranes of the cells of the gut disturbing the flow of Na^+ and Cl^- ions. The spores of the organism can withstand heating at 100°C for one hour, but there are marked variations between strains.

Contaminants arising from Processing: These include residues that become part of food as a result of processing, handling and distribution of food. Some of them are the following.

Fumigants are used to sterilise food under conditions in which steam heat is impractical. Ethylene oxide is a commonly used fumigant. This and other epoxides react with food components to produce toxic products or destroy essential nutrients. Ethylene oxide reacts with inorganic chlorides to form ethylene chlorohydrin which is toxic.

Solvents are used for the extraction of oil from oilseeds. Solvents like trichloroethylene that was formerly, used as a solvent reacts with the substance being processed, with the production of a toxic product, although the chemical used in processing it is itself nontoxic.

During processing of foods a number of changes can be induced in their lipids. On prolonged heating, oxidative and polymerisation reactions take place and these bring down the nutritional efficiency of the processed product.

Smoking of meat and fish for preservation and flavouring is an old practice. Such foods get contaminated with polycyclic aromatic hydrocarbons (benzopyrine, for example), many of which are carcinogenic. In areas where smoked fish is consumed in plenty the incidence of stomach cancer is high. Smoke from

woods also contain many other classes of compounds such as phenols, acids, carbonyls and alcohols most of which have toxicological activity.

Metal Contaminants: Metals are one of the many unintentional contaminants of foods. They find their way into foods through air, water, soil, industrial pollution and many other routes. Metals, when present beyond small quantities, are toxic. Mercury, cadmium, lead, tin, etc., are the toxic metals present in foods.

Mercury: Toxicological effects of mercury depend on the chemical form involved. Elemental mercury and inorganic salts of mercury are much less toxic than organic mercurials. Methyl mercury is very poisonous and practically all the methyl mercury in the diet comes through fish taken from water contaminated by the industrial use of mercury. Mass poisoning due to mercury has been reported in Japan in areas where the fish consumed were caught from mercury-contaminated waters.

Cadmium: Cadmium is becoming a major environmental pollutant as the metal is extensively used in industries. The metal is readily absorbed from air, water and food, and is built up in the body. Ingested cadmium is stored in the kidney in the form of a metal-protein complex. Long-term exposure results in renal tubular damage, liver dysfunction and testicular damage.

Lead: The human system absorbs lead not only from foods but also from water and leaded dust in the air. Lead contamination of food is due to water conveyed through lead pipes, contact with machines, equipment, packing, etc., containing lead, and lead containing sprays and dusts used as pesticides of fruits and vegetables. Lead, absorbed by the human system, brings about changes in the kidney and arteries that tend to shorten life.

Other Metals: Metals may enter foods from certain utensils. Enamel-ware of poor quality contributes antimony and galvanised utensils zinc, which are toxic. A major source of tin contamination of foods is tin plate. A variety of containers

with tin plate are made for storing all types of processed foods. Canned foods, if acidic, and foods stored in tins after opening, change in colour or may develop a metallic flavour that is unpalatable. Small quantities of metal are dissolved when foods are cooked in aluminium utensils but this does not impart harmful properties to foods. Copper is an essential trace element required by the human body but copper-contaminated foods are toxic.

Other Contaminants from Food Processing: These include lubricants, boiling water additives, packing material, etc. Mineral oils are used as lubricants for the extrusion of foods and rapid economic production of bread by permitting it to come clearly out of the baking pan. Chemicals are added to boiler feed water to prevent scale formation as a result of hardness of water. When steam from the boiler is used in food processing, small amounts of added chemicals may be carried into the food. To strengthen paper used for food packaging, resins are used, which might migrate to the food.

In addition to the contaminants discussed above, food will contain a number of other contaminants. Further, a number of chemicals are intentionally added to foods to accomplish improvements in nutritional value, maintenance of freshness, creation of desirable properties or aid in processing.

Food Adulteration: Adulteration of food consists of substituting it wholly or in part by any cheaper or inferior substance or of removing any of its constituents, wholly or in part, which affects adversely the nature, substance or quality of the food. According to the Indian Prevention of Food Adulteration Act (PFA) 1954, any ingredient which, when present in food, is injurious to health is an adulterant. Thus, according to the PFA definition, many of the food contaminants discussed in the previous section are adulterants and come under the purview of food adulteration.

Adulteration of foods was practised even in ancient times. It was not a serious problem then because business was on a

small scale and the transactions involved a large measure of personal accountability. Later, with the increased centralisation of food processing and distribution, and corresponding decline in personal accountability, intentional adulteration of food increased. Even in the early nineteenth century, many adulterants were used in articles like pepper, essential oils, vinegar, beverages (coffee, tea, beer, wines), butter, bread, etc. In recent years adulteration of foods has become a serious problem the world over. In western countries there is a rising concern over the safety of the food supply and many remedial actions, both voluntary and enforced, have been taken.

The percentage of adulterated food sold in various parts of India is 30 to 35 per cent. Unfortunately, in our country, there are still no strong consumer associations to give proper education to the public at large about the various methods of food adulteration and ways to detect it. Food adulteration, apart from cheating the consumers, often results in disorders or diseases. Some of the foods commonly adulterated in India and the adulterants encountered in them are the following.

Pulses are an important component of the Indian diet. Pulses like masoor, black gram and chana, other as gram, split pulses (dals) or gram flour, are mixed with the corresponding forms of khesari pulse. Consumption of khesari dal for a longtime produces lathyrism, which results in paralysis of the lower limbs.

Some seeds, barks, leaves and other matter are dressed up to look like genuine foodstuffs for marketing as genuine products or to adulterate the pure ones. For example, roasted tamarind and date seeds are ground into coffee powder, and exhausted tea leaves or coloured sawdust are mixed into fresh tea. Exhausted bark or seeds of coffee, spices, etc., are dried, suitably coloured and substituted for the fresh commodity. Powdered bran and sawdust may be present in wheat flour and ground spices. Strips of jute or foreign plant material may be suitably dyed to be sold as saffron. Easily obtainable seeds

are substituted for cumin, cardamom, black pepper, mustard seeds, etc.

Edible oils and fats are adulterated with cheap edible and non-edible oils. Argemone seeds are obtained from *Argemone mexicana* which grows widely in the country. The black seeds resemble mustard and are used to mix with mustard seeds. Argemone oil itself, extracted from the seeds, is used to adulterate oils such as coconut, sesame and groundnut. Argemone oil is poisonous and its use results in dropsy in human beings. Orthotricresyl phosphate is a colourless industrial chemical freely soluble in oils but if soluble in water. This is used to adulterate some oils. The chemical can cause permanent damage to the nervous system. Oils and fats are also adulterated with petroleum products causing gastrointestinal disturbances. Other common adulterants are animal fats in vanaspati and vanaspati in ghee.

Stones, sand, marble and earth are usually found mixed in foodgrains and pulses and their flours. Talc and chalk powder are used to adulterate wheat flour, arrowroot powder, panir and confectionery. Starch is used as a filler in milk, milk products and confectionery. Starch is also found in ground turmeric, coriander, mixed spices, etc. Addition of excess common salt to powdered spices, such as curry powder, is a common practice. Water is a common diluent of milk, curd, etc. It is also added to non-alcoholic beverages in more than the recommended quantities. Methyl alcohol is an adulterant of alcoholic beverages. This has resulted in a number of serious liquor tragedies in this country.

The most frequently used food adulterant in India is colouring matter. Colour is used in many foods such as milk products, confectionery (including Indian sweetmeats), soft drinks, alcoholic beverages, tea and spices. Colours are also added to foods such as egg preparations, bakery products, fruit products and others. More than 70 per cent of the colour containing marketed samples of foods are found to contain non-permitted colourants like the coal-tar dyes and mineral

pigments like lead chromate and red or yellow earth. Use of foods containing non-permitted colourants results in various health hazards.

Measures to Control: The governments in many countries play an important role in ensuring food quality. National official standards are set to safeguard the consumer's health and ensure fair food trade practices. In 1963, the FAO and WHO established a Commission for setting of international food standards.

Codex Alimentarius: The Codex Alimentarius are international standards set by FAO/WHO for all the principal foods, whether processed, semi-processed or raw, for distribution to the common man. It also includes standards in respect of food hygiene, food additives, pesticide residues, contaminants, labelling and presentation, and methods of analysis and sampling. A Codex, Standard may be accepted by a country in its entirety or with more stringent requirements for trade and distribution of the food within its territory. A country which accepts Codex Standards cannot hinder the distribution in its territory of imported foods which conform to the standards.

Indian Standards: In India, food standards have been prescribed based on the International Codex Alimentarius with suitable modifications to suit the country's conditions. The most important standards are those set by the Indian Standards Institution (ISI).

The ISI is responsible for preparing and promoting general adoption of standards at national and international levels. The institution has an Agriculture and Food Products Division Council and a number of sectional committees dealing with various commodities. The standards are formulated and revised from time to time after eliciting the opinion of interested parties and based on the recommendations of a technical committee. Standards are available for a number of foods and food products like foodgrains and foodgrain products, bakery and confectionery products, processed and fresh fruits and

vegetables, protein-rich food products, spices and stimulant foods like tea, coffee and cocoa. The institution has also published specifications on alcoholic and carbonated beverages, meat products and fish and fisheries products.

Chemicals used in the manufacture of food products to maintain their nutritional quality, enhance keeping quality and stability, etc., are subject to ISI standards. Before a chemical is permitted to be added, it is ensured that the substance is toxicologically safe, is added in safe dose and is of the required purity.

Hygienic quality of food is very important. Unless the factory producing the food is governed by the strict hygienic code in respect of layout, plant and personnel, the food produced cannot be considered safe, however nutritious it may be. Therefore, ISI has prepared codes for sanitary conditions in the food processing, bakery, soft drinks, ice cream, fruit and vegetable canning industries.

ISI has also worked out standards for sensory evaluation of foods. It safeguards the food trade interest of India, by maintaining effective liaison with organisations like the International Organisation for Standardisation and Codex Alimentarius Commission.

The Indian Standards Institution operates a certification marking scheme under which manufacturers are licensed to use the ISI mark on goods produced by them in conformity with the relevant standards. The ISI mark on products has been very helpful to consumers, organised purchasers, exporters and export inspection authorities. Food products covered under the ISI Certificate mark scheme are bakery and confectionery products, cereals, pulses and their products, dairy and allied products, beverages and food additives.

In addition to ISI standards, a number of other steps have been taken by the Government of India to control the quality of foods. Some of them are the following.

The Agricultural Produce (Grading and Marketing) Act provides for the grading and marketing of agricultural and

other products. The Agmark Standards, incorporate grades, 1, 2, 3 and 4, or special, good, fair and ordinary, based on the physical and chemical characteristics, intrinsic as well as acquired, during processing or otherwise. The Agmark gives the consumer an assurance of quality according to the standards laid down. The grading system makes better buying and selling of agricultural produce.

The Preventing of Food Adulteration Act, with its various amendments, provides for the prevention of adulteration of food. The Director General of Health Services on behalf of the Central Government or the Chief Officer in charge of Health Administration in a state operates PFA rules under the Act.

The Vegetable Oil Control Order stipulates that any vegetable oil product, unless it conforms to the standards of quality and offers requirements for vanaspati or bakery shortening or margarine, shall not be manufactured, stocked or sold. There is also an order to regulate the production, distribution and quality of solvent extracted vegetable oils and vegetable oil products.

The Fruit Products Order (FPO) lays down minimum standards relating to the quality of various fruits and vegetables. The order specifies standards of sanitation and hygiene to be followed in the factories and gives directions regarding packing, marking and labelling of containers.

The Meat Products Order provisions are meant to control production, quality and distribution of raw and processed meat.

There are thus numerous organisations in the country to control the quality of food, its production and distribution. Sometimes confusion occurs among consumers and manufacturers regarding the multiplicity of food standards, such as ISI standard, PFA standard and Agmark standard for the same product. Hence, it is felt that there is a need to have a single organisation to ensure food quality control, whether it is checking for food adulteration or ensuring quality control for home consumption or export.

5

Items Used as Food

World of Drinks

About sixty to seventy per cent of the total body weight of a person is water. Water is present inside the cells of tissues (the intracellular water) and outside the tissue cells (the extracellular water). The extracellular water is present in the blood plasma, water in the space between the tissues is known as interstitial water and in lymph. Water and electrolytes are essential constituents of cellular metabolism. It is not only a component of our body but also acts as a vehicle for chemical transport. Water and electrolytes must be supplied to the body regularly to maintain the cellular function and to regulate the excretion through kidneys, lungs and skin.

Necessity of Water:

Water intake and loss must be balanced. Water is in jested through food and also as drinking water. Apart from this, a considerable amount of water is formed in the tissues through the oxidation of carbohydrates, fats and protein in their metabolism.

Water Intake and Loss from the Body

Water intake	(Ml)	Daily excretion	(Ml)
As fluids	1,500-1,750	Urine	1,200-1,500
Water in solid food	600-900	Perspiration	700-900
Water from oxidation of carbohydrate, protein, fat metabolism	300-350	Faeces	100-200
Total	2,400-3,000		2,400-3,000

Water is lost through urine, skin, lungs and faecal matter. In a normal healthy person the- water intake and water loss is more or less equal. The excretion through the kidneys and skin changes according to the climatic conditions. In a dry climate more perspiration and in the cold weather more urination takes place. If the fluid intake is less thirst occurs. When two per cent of the body weight is lost dehydration starts.

Water is absorbed quickly and it is circulated to the tissues. From the tissues water is passed on to the capillaries. The capillary walls and cell membranes separate the water compartments. Water can move in and out of all cells by osmosis. The osmotic pressure in the plasma proteins regulates the exchange of water from tissues to capillaries. Antidiuretic hormone of posterior pituitary, hypothalamus, electrolytes like Na and K and adrenal cortex hormone aldosterone regulate the water metabolism in the body.

Antidiuretic hormone of the posterior pituitary increases the reabsorption of water in kidney tubules by increasing the permeability of distal tubules and collecting tubules. This mechanism enables them to retain more water in the body.

Hypothalamus is having the thirst centre. In dehydration plasma is concentrated and osmoconcentration stiumulates the thirst centre of hypothalamus. This induces the person to be very thirsty. By taking water the water balance in the body is regulated.

Various solutes present in the body fluids also help to regulate water metabolism. Plasma proteins exert low osmotic pressure and this helps to maintain the water in the blood vessels. Aldosterone increases reabsorption of Na by the kidney.

About 170 litres of fluid pass through the glomeruli of the kidneys daily. A healthy person's kidneys excrete about 1.5 litres of urine from the above fluids. The urine formation is reduced if the intake of fluids is less. Then the urine becomes more concentrated with urea which forms the main bulk in urine. The normal output of urea is 30 mg for which 750 ml of urine must be formed. If urine formation is less urea is retained in the body which is harmful to the body; when less urine is formed in summer season more sweat is formed. Excessive sweating occurs with increased external temperature and humidity, fever and exercise. Water is also lost through insensible perspiration.

Apart from maintaining the intracellular and extracellular water metabolism, water also takes part in a number of activities. All the vital functions of the body depend on the presence of proper water content. Digestion depends on its juices and all digestive juices are watery in nature. Urine, the excretory medium of our body, contains 97 per cent water. Water also helps in regulating body temperature. By evaporation of water, by respiration and through sweating the body temperature is regulated.

Requirement of Water: Water requirement of a person varies with climate, age, activity, dietetic habits and body build. Activity is an important factor which affects the water requirements. When activity is more intense more heat is produced and more water is needed to dissipate the heat produced. The type of diet also influences the water requirement. A high mineral content increases the water requirement. Consumption of sweets demands extra water for their dilution in the stomach. Beverages like tea, coffee and cocoa are diuretics and they stimulate excretion of water. If

excess water is injested the kidney responds to it after 13-50 minutes and the extra water is excreted within three hours. The water balance will be maintained by the excess urine output. The sensation of thirst is one means of meeting the body's need for water.

Dehydration: Water depletion occurs when water loss is not met by water intake. On water depletion changes occur in the body fluids. Thirst, weakness, loss of weight, dryness of the skin, mucous membranes, tachycardia, disorientation and delerium are the symptoms manifested by dehydration. If not corrected coma and death occur. The urine changes in its specific gravity and concentration. The salt level is diminished in the urine and serum levels of sodium, urea and chlorides are elevated. Intercellular water is depleted to compensate the water content of the secretions. This results in cellular dehydration and vital organs fail to function. Plasma volume and blood volume are reduced which leads to reduced output and further cardiac failure.

Acid-base Balance: Acid-base balance refers to the regulation of the hydrogen ion concentration of body fluids. From metabolic processes acids are continuously produced and have to be eliminated. Lungs and kidneys are the major agents in carrying out this function in our body. Normally the body is maintained in neutrality. In some pathological conditions like diabetes mellitus or fasting for weight reduction and in renal failure the acid-base balance is affected. Acidosis occurs in this condition.

Carbonic acid is the main form of acid formed in the body from metabolism. Carbohydrates, protein and fats yields Co_2 and water during their metabolism. The hydrated form of carbon dioxide is carbonic acid. Intermediary products like lactic acid, pyruvic acid, keto acids, uric acid, sulphuric acid, phosphoric acid from phosphoprotein are acidic in nature.

Certain foods in the body produce alkaline ash as the

cations (Na^+. K^+) will allow the body to retain more bicarbonate ions thus producing alkaline effect. The neutrality of the body is maintained through the body fluids and their circulation, the acid buffer system like bicarbonate carbonic acid through haemoglobin and its capacity to form carbamale, regulation in respiration and the final effort from kidney on emergencies.

Acidosis: It is a condition in which the hydrogen ion concentration is increased or there is an excessive loss of base. The ratio of bicarbonate to carbonic acid is less than 20:1 and the pH is below 7.35.

Alkaksis: This is a condition in which the hydrogen ion concentration is decreased or the base is increased. The ratio of bicarbonate to carbonic acid is greater than 20:1 and the pH is 7.45.

Necessity of Milk: Milk is commonly considered as a complete food as it contains all six of the essential foodstuffs. Milk protein is of excellent quality and it promotes growth and maintenance of body tissues. Therefore, milk is an ideal food for infants and children. But milk is very low in iron and ascorbic acid content Calcium and phosphorus levels in milk are very high. The ratio of calcium and phosphorus and the presence of vitamin A help its maximum utilisation. Milk is an excellent source of vitamin A. In it the cream is rich in vitamin A and if the same is removed as in slammed milk, its vitamin A content becomes low. Riboflavin, a B vitamin, is present in significant quantities in milk. If the milk is exposed, to light riboflavin is lost. Fats present in milk are readily digested. The carbohydrate of milk is lactose and it is hydrolysed into glucose and galactose. Whole milk contains 4.7 per cent carbohydrate, 4 per cent fat, 3.3 per cent protein and 88 per cent water. Fats in the form of oleic, palmitic and stearic acids is present in milk. Milk protein is casein and whey protein has lactalbumin and lactoglobulin.

The Average Composition of Milk and Milk Products (gm\100 ml)

Milk product	*Protein (gm)*	*Fat (gm)*	*Calcium (mg)*	*VitA. (I.U.)*	*Vit.C. (mg.)*	*Calories (kcal.)*
Buffalo's milk	4.3	8.8	210	160	3	118
Cow's milk	3.2	4.1	149	150	2	67
Buttermilk	0.8	1.1	30	...	...	15
Skimmed milk	2.5	0.1	120	...	...	29
Condensed milk	9.1	8.4	280	430	...	328
Cheese	24.1	25.1	790	273	...	348
Goat's milk	3.7	5.6	170	60	2	84
Human milk	1.0	3.9	20	70	3	67
Curd	2.9	2.9	120	40	4	51

In developing countries, the per capita availability of milk is poor whereas in developed countries it is more.

Among these buffalo's milk has a higher proportion of total solids and fats than cow's milk. Casein coagulation is also high.

Goat's milk is rich in cream and poor in protein compared to buffalo's milk. Human milk is easily digestible because it becomes solid on reaching the stomach. Casein is the chief protein of milk. It is a phosphoprotein which contains protein and phosphoric acid. An enzyme rennin aggregates and precipitates casein and it is coagulated or curdled. Bacterial action also curdles milk. Casein is present in milk in the form of calcium caseinate. Rennin converts calcium caseinate into calcium and paracasein. Again it forms a smooth, soft clot of calcium paracaseinate which is non-dispersible. If milk is overheated it forms a weaker clot. Dried milk also forms a weak clot. Milk also contains lactalbumin, a simple protein of the albumin class. Prolonged and direct heating of milk also forms a weak clot. Milk also contains lactalbumin, a simple protein of the albumin class. Prolonged and direct heating of milk helps it to settle down at the bottom and sides of the

container in which milk is heated. Use of double cooker rectifies this. The souring of milk is due to the presence of lactose, the carbohydrate. Lactic acid-forming bacteria use lactose as food to produce lactic acid in milk. Fat is present in emulsified form. Creamy layer of milk consists of fats. Milk contains potassium, sodium, calcium and magnesium chlorides and phosphates. On heating milk calcium precipitates on the sides of the container. Of all food materials milk is the best source of calcium and careless heating reduces its availability. Even though iron is present in minute quantities it is highly nutritional.

Processing of Milk

Pasteurisation of Milk: Milk is a favourable medium for bacterial growth. Louis Pasteur, the French scientist developed the heat treatment of milk for inactivating disease-causing micro-organisms.

Milk-borne diseases are tuberculosis, typhoid, hoof and mouth diseases, undulant fever, scarlet fever, septic, sore throat gastroenteritis and diphtheria. But pasteurisation destroys certain advantageous bacteria also. There are two methods of pasteurisation—the flash method and the holding method. In the flash method the milk is heated at a temperature of 160°F for 30 to 60 seconds and in the holding method the milk is heated to temperatures of 140° to 145° F and held there for at least 30 minutes. In both the methods, after the heat treatment, the milk is cooled quickly to about 43° F. Pasteurisation reduces the thiamine and ascorbic acid content whereas the nutritive content of other vitamins is not much affected.

Evaporated Milk: Evaporated milk is produced by evaporating 60 per cent of water from milk and the product is then sterilised and sealed in cans. To evaporate, the milk is heated at a high vacuum from 130° to 150* F. The concentrated product is homogenised to reduce the fat globules and then placed in cans, heated and sterilised at temperatures from 240° to 245° F for IS minutes. Evaporated milk is fortified with vitamin D.

Condensed Milk: Evaporated milk is incorporated with 40 per cent carbohydrate to form condensed milk. Sucrose or dextrose or a mixture of these two is added.

Dry Milk: Whole milk or defatted whole milk is used in its preparation. From milk most of the water is removed. There are two methods applied to form dry milk-spray process and roller process.

In the spray process fresh milk is sprayed under high pressure into a drying chamber where dry air is passed. The current of dry air evaporates water and the dried solids settle to the bottom of the chamber.

In roller process fresh milk is spread over the surface of hot revolving metal rolls. The water evaporates and milk solids remain in thin sheets. If the fat is removed from whole milk the keeping quality is more. Dried whole milk does not keep so long because the fats tend to become rancid.

Various Milk Products: Cheese, yoghurt or curd, buttermilk, low-fat milk, non-fat milk cream and butter are the common milk products used.

Cheese is mainly from casein of milk with fat and a little whey. To precipitate cheese lactic acid or rennin are used. For commercial cheese making rennin is used and for homemade cheese lactic acid is used.

Cheese is rich in protein and fat and a considerable amount of lactose is also present. In homemade cheese calcium is less but other minerals like phosphorus are there in both. Compared to milk vitamin A content is the same but thiamine and riboflavin are dissolved in the whey. Cheese is used to spread in items like butter as it contributes flavour. In vegetables, fish and meat preparations and in custards and souffles cheese can be used. Low temperature is suggested for cheese cookery. Buttermilk is prepared from curd; it is churned and the butter removed. Yoghurt or curd is a custard-like product prepared by fermenting milk. In commercial production streptococcus

thermophilus is added to milk to ferment it. Low-fat milk is commercially prepared for people who prefer low calories from milk with a fat flavour. Non-fat milk is skimmed milk where the fat is removed entirely. Vitamin A and D are fortified in it. Creams with different levels of fats are commercially produced. The fat content ranges from 18 to 40 per cent. Cream combined with sugar, stabilisers and emulsifiers is often sold in cans. Commercially, cream is prepared by using a cream separator which enables the separation of fat at different levels. Whipped cream is an air-in-water foam in which fat globules are piled up in the film and it is used in salad dressings and in desserts.

Milk is used in many preparations for its solvent capacity and its nutritive contribution. Milk is used as a dispersing medium, especially in puddings and for scum formation with its coagulating property. Vegetables with tannin or acid cause milk to curdle. Scum formation and curdling occur when the protein in milk becomes insoluble in heat or in the pH change. Use of low temperature and short heating periods prevent scum formation. Acidity of the content produces curdling. Adding acid food towards the end and mixing it in low temperature reduce curdling.

Types of Milks in the Market: Homogenised milk, separated milk, toned milk, sterilised milk, separated milk (skimmed milk), recombined milk, filled milk and flavoured milk are the common fluid milks available in the market.

Homogenised Milk: In this, fat globules are broken up mechanically to less than one microne in diameter. It is easily digested but cream is not obtained on separation.

Separated Milk (Skimmed Milk): This is prepared by removing fat from milk in a cream separator. Very little fat is in it.

Toned Milk: Toned milk is prepared by mixing milk reconstituted from skimmed milk powder with buffalo milk which has 7.0 per cent fat.

Recombined Milk: It is a homogenised product prepared from milk fat, non-fat milk solids and water. It should be pasteurised.

Filled Milk: Filled milk is the homogenised product prepared from refined vegetable oil and non-fat milk solids and water.

Sterilised Milk: Cow or buffalo milk is sterilised in bottles by fixating to l5 minutes.

Flavoured Milk: Chocolate, coffee, cardamom or any other edible flavours, colour and to flavour milk.

Other Beverages: Beverages are materials used as drinks for the purpose of relieving thirst and introducing fluid in the body, nourishing the body and stimulating or soothing the individual. Beverages are classified into four groups according to the reactions they produce.

(1) *Refreshing* to relieve thirst

(a) Plain or carbonated water.

(b) Ginger ale and other bottled beverages.

(c) Fruit juices and fruit-ades.

(d) Iced tea or coffee.

(2) *Nourishing*

(a) Milk pasteurised, skimmed, evaporated, dried, malted, buttermilk, and chocolate and cocoa.

(b) Eggnog made with rum, whisky, brandy, fruit juices, coffee and chocolate.

(c) Fruit juices with egg-white or whole egg.

(d) Glucose lemonade or orangeade.

(e) Tube feeding feeds.

(3) *Stimulating*

(a) Eggnogs made with whisky, rum, brandy, tea or coffee, other alcoholic beverages.

(b) Tea or coffee.

(4) *Soothing*

(a) Warm milk.

(b) Hot tea.

Coffee and tea have no food value unless served with milk and sugar. Cocoa in itself will increase the food value of the milk which in used its preparation. Coffee, tea and cocoa, all have a certain amount of stimulating effect. Iced coffee, tea and cocoa are used all the year round between meals as a refreshing drink. Milk, eggnog and other cold beverages made with milk or milk and egg are valuable for serving between meals when the protein content of the diet needs to be increased. Wines and other liquors which are used occasionally to provoke appetite, or to stimulate or flavour furnish additional calories. Cold beverages may be used with meals or between meals. They are more appetising when chilled. The choice of the beverage will depend largely upon whether it is taken for its food value or merely as a refreshment. In any case a cold drink served between meals adds interest to the monotonous hours in the daily routine of a person.

Tea: Tea is obtained from the leaves and flowers of teabush, which grows in sub-tropical countries, such as China, India and Japan. The top tiny leaves and buds or flowers at the end of the shoot furnish the choicest and the most expensive brands designated as flowery pekoe. Orange pekoe is the most commonly marked tea which consists of the first and second leaves just below the buds.

The kind of tea is determined by its treatment. Subsequent to picking, the green tea undergoes no fermentation process and is light in colour and rich in tannin. Black tea is fermented to a dark colour and loses some of its tannin.

A good cup of tea will contain a high proportion of volatile oils and a minimum amount of tannin. Tannin is a stimulant and has approximately the same psychological effect as caffeine. Tea, which is allowed to stand too long in the container with

the leaves, contains a considerable amount of tannin which causes the finished beverage to be bitter and unpalatable.

Coffee: Coffee is the beans of coffee plants which grow in tropical countries. It is tasteless in its green form. Roasting develops the volatile oils, specially caffenol which give the characteristic aroma and flavour to coffee. The typical individual brands depend upon the various types of coffee used and the length of the time of roasting. Bitterness is increased when combined with substances like chicory.

Coffee alone has no food value, but it does furnish a certain amount of stimulation of caffeine which raises the blood pressure and stimulates renal activity and momentarily masks fatigue. As a rule a cup of coffee contains from 1.5 to 2.5 grains of caffeine depending upon the strength , of the brew. Tannin, another alcoholite present, interferes with digestion. The amount of tannin in coffee depends upon the method of preparation. Since it is soluble in hot water, the longer the coffee is brewed, the greater will be the tannin content.

Coffee beans are ground to increase the surface area for maximum extraction of the flavour. However, coffeole is rapidly lost, the loss being greater in freshly ground coffee. Ground coffee should be kept in a tightly covered tin in a cool place and should be used promptly after purchase.

Drip and vacuum coffee contains the least amount of tannin, caffeine and coffeole. Percolated coffee contains slightly more caffeine and less coffeole than drip and vacuum coffee. Boiled coffee should be strained. It contains more caffeine and tannin than either the drip or percolated coffee.

Instant coffee is made from pure coffee which has been pulverised. The extraction of caffeine and tannin is high, but the amount in the finished beverage will depend upon the amount used. Some instant coffees are mixed with malted cereals.

In decaffeinated coffee 75 per cent of caffeine is removed. It does not possess the stimulating property of coffee. Coffee

substitutes are made from a combination of roasted grains and cereals.

Cocoa: Cocoa is the ground product of cocoa beans and often some of the fat has been removed. The cocoa beans are in fleshy pods. The well ripened pods are feathered and dried for 24 hours after which they are cut open and the beans are removed. The beans then undergo a series of processes like fermenting or sweating, removing the pulp, washing, drying and roasting. Chocolate and cocoa, unlike tea and coffee, have a high nutritive value. The roasted beans, likewise the bitter chocolate, contain approximately 50 per cent of fat, about 80 per cent protein, 10 per cent starch, and 7 per cent sugar. Like tea and coffee, they also contain a stimulant and tannin. The stimulant known as theobromine has approximately the same physiological effect, though not to such a marked degree. There is only a small amount of stimulant in one cup and the chief nutritive value of the drink lies not in the cocoa but in the amount of milk and sugar used in it.

Coconut Water: Tender coconut water is high in sugar and electrolytes like Na and Potassium.

Fruit Juices: Fruit juices in themselves are not only pleasantly refreshing but they are easily digested and especially useful in increasing the fluid intake. The citrus juices and tomatoes are excellent for the ascorbic acid content Fresh, frozen and canned juices may be served plain or flavoured. They may be reinforced with egg white or whole egg. The following fruits include commonly used juices: (1) citrus—orange, lemon, grape fruit, lime; (2) Berries—blackberries, loganberries, raspberries, strawberries; (3) tree fruits—apple, apricot, peach, pear, plum, prune; (4) other fruits—grapes, pineapple; (5) vegetables—tomato; (6) Sugarcane juice, sweet toddy and coconut milk are some of them.

Milk: Milk is most of all liquid food since it has more nutritive value than any other beverage. Milk beverages add protein, calories, calcium and other nutrients to the diet. The milk flavour of food permits its use in a great number of ways.

It may be plain, malted, or acidulated. It may be reinforced with egg and sugar depending upon the type of nutrient needed and it may be flavoured with chocolate and other flavouring extracts.

Egg: Eggs are specially useful for increasing the protein content of beverages. Eggnog in which whisky or other alcoholic beverages are used is the most important stimulating beverage. Eggnog made with strong coffee is only mildly stimulating but lends variety to the general liquid diets. Eggnogs contain egg, cream, and possibly milk and sugar in addition to fruit juice.

Sugar: Sugarcane, fructose and lactose may be used to sweeten beverages and to increase the caloric value.

Bask Rules in the Preparation of Tea and Coffee: The best tea produces a pale coloured infusion. Freshly infused tea has good flavour and it is harmless for digestion. Reheating or overheating of tea results in continued infusion and it extracts more tannin which gives a bitter taste Id tea. Moreover, this is often harmful to the digestive system.

The quality of tea is an important factor in giving good taste to the product. Since tea leaves absorb water, they must be stored in airtight containers.

Water used for the preparation of tea also has the key role in producing good result. Water should be freshly drawn before it is used for tea. Previously drawn water and previously boiled water should not be used for preparing tea as it makes the tea insipid. To get the best flavour water should not be boiled before or after tea leaves are added. As soon as the tea leaves are added to boiling water it circulates from the bottom to the top and from the top to the bottom.

To prepare good tea, one teaspoonful of tea leaves for 1 cup is ideal. For one litre of water 15 gms of tea leaves are enough. It is always better to measure tea leaves to get good result. The vessel in which tea is being prepared must be warmed before making tea. Instead of adding tea leaves to water, pour boiling water into a wanned pot in which tea

leaves are added. Do not disturb this for five minutes. Let the leaves infuse. If it is kept for more than five minutes, it gives a bitter taste and less then five minutes yields a weak tea.

Milk and sugar can be added according to individual taste. To get a good cup of tea, it must be served hot.

Coffee: There are different methods of preparing coffee. The pot method, the filter method, the urn brewing method and the vacuum brewing method are the common methods.

Coffee powder must be fresh to get the best flavour and taste. The quantity of coffee powder used must be just enough. For one cup of coffee 8.5 gms or 1/2 teaspoon of brewing coffee is required. If instant coffee is used one teaspoon is enough to make 1 cup of coffee.

Freshly drawn water must be used to prepare coffee.

Coffee is generally prepared by any one of the following:

1. The pot method.
2. Filter method.
3. The percolator method.
4. Instant coffee preparation.

The Pot Method: In this method an earthenware pot or jug is used to prepare coffee. Warm the earthenware pot or jug, put in 3 level spoons or 11/2 heaped teaspoons of fresh coffee powder (coarse grind) for an 8-ounce cup. Pour boiled water over the powder and stir. Cover the vessel and let it stand for 5 minutes. Strain the coffee through a fine meshed sieve or cloth. Add milk and sugar to taste.

Filter Method: Stainless steel or brass filters can be used to filter coffee in this method. If brass filter is used it should be tinned properly to get good results. Put 2 to 3 level teaspoons coffee powder for 8 ounces water. Press the plunger down lightly over the powder and pour boiling water over the plunger in a circular motion. Allow to stand for 5 to 7 minutes. Pour out coffee from the lower vessel and add milk and sugar to taste.

The Percolator Method: Keep coffee powder in the middle of the percolator and pour boiling water through it to the upper section. Coffee is collected from the bottom section. It takes thirty minutes to percolate down. Vacuum coffee makers are also available in some places.

Instant Coffee: Strong brew is made by percolation of roasted and ground coffee which is sprayed into fine atomised particles from the top of a chamber into a mass of very hot air. The flavour is retained through highly developed technology. It is quickly packed into airtight bottles. Varying degrees of chicory is mixed with coffee powder to give slight bitterness and colour to brew. Now mixtures of coffee soluble with solids and sugar in the form of thick pastes are also marketed. By adding hot water it can be used instantly.

Alcohols: Beer, ale, toddy, wine whisky, gin, brandy, rum, vodka, arrack and cider are some of the alcoholic beverages. Beer is made by fermenting barley malt extract with yeast and ale is the same thing with more stimulation. Toddy is from fermenting sap of palm, coconut and palmyra or date palm.

Wine: Wine is prepared by the fermentation of grapes with yeast and cider is apple wine.

Whisky: It is an alcoholic distillate from a fermented mash of grains.

Gin: It is an artificial drink from diluted pure ethyl alcohol.

Brandy: This is made from fruit juices like grapes and apples.

Value of Non-vegetarian Food

Flesh of all edible animals including mammals and poultry is termed as meat. Meats of cattle origin, beef and veal, of hog origin, pork, sheep, lamb, mutton, flesh of partridge, duck and game birds are used commonly as food.

Meat: Muscle of meat is composed of 75 per cent water, 20 per cent protein and 5 per cent fat, carbohydrate and

minerals. The composition of meat varies with its sources, season of the year and pH of the meat. Protein comprises about four-fifths of the solid material. True fats, fat-like substances called phospholipids, inorganic substances and vitamins are present in fats.

Veal is the meat from cattle slaughtered three to fourteen weeks after birth. Calf is from fourteen to fifty-two weeks old animal. Beef is from cattle over one year old. A heifer is a female cattle that has never borne a calf and a cow is a female cattle that has borne a calf. Meat has skeletal or muscular cuts and organ cuts. Skeletal or muscular cut has muscle, tissue, connective tissue, bone fats and nerve tissues. Muscle tissue and connective tissue supply the qualities of meat in cookery. Muscle structure is made up of bundles of muscle fibres called fasciculi. Connective tissues are scattered throughout the muscles where they play a connecting and supporting role. White fibrous and yellow elastic fibres with proteins like collagen and elastin are present in all muscles. Collagen contributes strength and elastin supplies elastic quality to meat Collagen reacts with water in hydrolysis to produce gelatin and brings about a tenderising effect to the meat. But no sack change is brought in elastin and if elastin is more in muscle fibres meat will be hard even after cooking. This can be remedied by chopping or grinding the elastin fibres.

Fats in meats are distributed between fasciculi. Fat is laid as a protective layer around the organs. The fat deposited with the muscle structure is known as marbling. Muscle haemoglobin is present in small amounts. Muscle cells of living animals contain glycogen which is otherwise known as animal starch. But immediately after the animal is dead glycogen is converted into lactic acid. The vitamin content of different meats varies. The B vitamins—thiamine, riboflavin and niacin—occur in significant amounts in all meats. The meat of hog origin is rich in thiamine liver is rich in vitamin A. Iron-containing substances and B vitamins are more in organ meats.

Rigor Mortis: After the animal is slaughtered certain changes occur in the meat and it stiffens. This process is known as rigor mortis. On slaughtering the animal adenosine triphosphate is synthesised from glycogen and lactic acid, ammonia and other products are formed from it. Myosin and actomyosin are found in contracted muscles and stiffening occurs to it. If the slaughtered animal has a high glycogen level, it produces a low pH and the quality of meat will be good with less rigor mortis. Fasting, exercise, nervous exhaustion or insulin injection cause low glycogen level and the carcass has a high pH. Undesirable dark colour and poor flavour and texture of meat result from this. Slow and long slaughter produce more adrenaline in the animals and this results in sticky, gummy and dark coloured meat. Rigor mortis reaches its peak within 24 hours and then the muscle softens.

Quality and Grading: Appearance, tenderness, juiciness and flavour of the meat decide the quality of meat. Good quality flesh must be firm, the fasciculi should be small and velvety in texture, with characteristic colour of the meat-bright cherry red for beef, pinkish-red for lamb, greyish pink for veal and pork; bone must be small, red and spongy in appearance and the fat must be white and firm. To grade the meat quality grades are given and they are prime, choice, good, commercial, utility and cutter and conner. For lambs and veal cull is the term used instead of utility. For pork grades of 1,2, and 3 are used. Meat of top quality is more tender than lower quality grade. Tenderness of meat increases palatability and it varies according to the cuts of meat. The most tender cuts are from backbone and ribs. Less tender cuts are from the shank, shoulder, neck and under parts of the body. Under part cuts of beef are flank, plate and brisket. The same cuts from lamb and veal are known as breast.

Meat Cookery: Palatability, appearance and flavour are enhanced by cooking meat. Cooking also makes meat more tender and digestible. There are two methods generally applied to meat cookery—dry heat and moist heat methods. In all

methods of cooking the interior of the meat changes. Heat penetration, substances added to meat and composition of the meat affect these changes. Mainly the method adopted for cooking depends upon the cuts of meat. Less tender cuts contain more connective tissues than tender cuts and so moist heat methods convert collagen to gelatin and it has a tenderising effect upon tissues. Roasting and broiling are the methods used to cook by dry heat methods and these methods can be applied to tender cuts of meat. For less tender cuts moist heat methods are used. Moderate temperature is recommended for this method. Water, juices and fluids are used for most heat methods. The addition of sodium chloride or sodium and potassium phosphate has been reported to increase the tenderness of meat.

Dry Heat Methods: *Roasting:* Roasting is cooking by means of radiant heat. Meat is roasted either in the open or in a closed tin in an oven. In covered roasting evaporation is diminished, and it cooks faster too. The water runs out of the meat instead of evaporating from the surface. An enclosed roast loses more weight and water than the open roast. Cooking to rare stage only increases the juiciness and flavour. If the pieces of meat lack fat, a thin layer of fat on top will be very effective. Water should not be added to oven roast.

Broiling and Pan Broiling: Slices from tender cuts of meat can be cooked rapidly without adding moisture or fat by radiant heat from a broiler or in an uncovered pan. The time required depends on the stage of cooking required and the size and shape of the pieces. This method gives a uniform result. The temperature should be higher than in roasting. It can be regulated either by changing the heat input or by changing the distance of meat from the heat source.

This method is suitable for cooking tender cuts of beef and lamb or ground meat on the pattern of ham, bacon and sausages.

Frying: Meat is fried when it is partly or completely immersed in hot fat during cooking. When the meat is

completely immersed the method is called deep-fat-frying. When a small amount of fat is added it is called pan-frying or sauteing. The fat should be kept below smoking point. A temperature of 350° F for pre-cooked meat and 300° F for raw meat is usually satisfactory. This method is suited to lean meats such as liver or cooked steak.

Cooking by Moist Heat: The cooking medium is steam of water.

Braising and Pressure Cooking: Meat may be cooked in steam with or without pressure. When the meat is first browned in its own fat or a small amount of added fat and allowed to cook in its own steam without pressure the method is known as braising. Meat braised as a large piece is called a pot roast and that as small pieces, a fricassee. Braising is preferable to boiling for fresh pork, because it produces less hardening and drying of fibres when meat is cooked to a safe stage of cooking.

Stewing or Simmering: Meat may be cooked by immersing it in water. To minimise the hardening effect of high temperature simmering is preferable to boiling. Vegetables are added to stews to enhance the flavour.

Generally shrinkage of meat cooked by moist heat method is greater than that in dry heat methods.

Tenderising Meat: Papain, an enzyme derived from papaya, tenderises beef. Solution of papain is often injected into the beef shortly before slaughter. Meat of old animals is tough and stringy and usually this meat requires tenderising. Tougher cuts of meat require a long, slow cooking period. Prolonged cooking changes collagen into gelatin in moist heating. Dry heat hardens it. Acid hastens this conversion of collagen into gelatin; tomato juice and vinegar tenderise meat. High temperature gives a tasteless dry-product-like quality to tough meat. Cutting the tough meat into smaller pieces enhances the flavour of meat. Mining, grinding or pounding of meat helps to break the muscle fibres.

Vitamin retention is more in dry heat methods. Roasting or broiling retains 65-70 per cent of thiamine whereas moist heat retains only 25 per cent of thiamine.

Gelatin: Gelatin is a protein obtained from collagen on hydrolysis. Connective tissues of cartilage, tendons, bones, skins and hoofs are used to prepare gelatin. After preliminary treatments a dispersion of collagen in water is obtained. And this dispersion is boiled for several hours. Water is evaporated and gelatin is obtained from collagen.

From the nutritional point of view gelatin is inferior because it is an incomplete protein. The property of gel formation is used in cookery. In desserts gelatin is used to provide a spongy appearance to the product.

Processing of Meat: Canned meat, cured meat, smoked meat, dehydrated meat and sausages are the common processed form of meat.

Canned-meat: The meat is cooked and filled in the can along with the gravy, leaving a little head space.

Cured-meat: Meat is pickled by the use of curing agents like sodium chloride, sodium nitrate and vinegar.

Smoked-meat: Smoking helps to preserve the meat and develop flavours in it The temperature and period of smoking vary with the type of meat

Dehydrated-meat: The meat is cut into pieces and cooked in steam for 30 minutes at 10 lb pressure. It is passed through a meat chopper and the chopped meat is dried in a drier.

Sausages: Cooked chopped meat is ground to a paste with seasonings and packed casings. It is available as fresh, cooked, smoked, cooked and smoked, dry, and semi-dry.

Fish: Fish is a very good source of calcium, protein, vitamins A, D and iodine. There are two main classes into which fish may be grouped as vertebrate fish or fin fish and shell fish. Cod, mackerel and salmon have backbones and fins, whereas a shellfish is more or less completely covered with a shell. Oysters, lobsters and crabs are examples of shellfish.

Protein is the main component in fish and its content ranges from 15 to 20 per cent by weight of flesh. The fat content in fish varies from 1 per cent to 20 per cent or more. Hilsa has a fat content of about 19.4 per cent. The fat content of shellfish is less than 5 per cent. Since the fats in fish are unsaturated fats even in therapeutic diets this can be recommended. The flesh of fish is devoid of carbohydrates except in the case of shellfish. The mineral content of fish is variable; 1 to 1.5 per cent of the flesh of fish is with mineral elements. Small fish which can be consumed with their bones are rich in calcium. Oysters and shrimps are rich in calcium. Most fishes are not very good sources of iron except oysters. Sea fish are high in iodine content. Oysters are also rich in copper. The flesh of most fish contains fair amounts of thiamine and riboflavin and some amounts of niacin. Fatty fish are rich in vitamins A and D. Fish livers are rich sources of these vitamins.

Freshness of Fish: Fish deteriorates quickly and so requires careful handling. The quality of freshness of fish is indicated by bright red gills, clear eyes, bright and slightly bulging but firm flesh when pressed with a finger, and elasticity. The fresh odour of fish suggests its freshness. Suggestion of putrefaction is easy to detect from its foul smell. The tail of a fresh fish is stiff.

Spoilage of Fish: Fish is highly perishable. The spoilage of fish is due to microbiological, physiological and biochemical changes.

Processing of Fish: Fish protein concentrates, fish meat, smoked and salted fish are the common forms of processed fish. Fish protein concentrate is used for supplementing diets.

Fish meals is not suitable for human consumption. It is used for poultry and for animal ration.

Preservation of Fish: Cold storage, freezing, canning and drying are the common methods of preservation of fish. Salting before drying is essential.

The Eggs: Eggs are used in many preparations due to

the variety they provide to food. Eggs are important ingredients of recipes for breakfast, lunch or dinner for any person. Nutritive importance, flavour, pleasing colour, thickening power, foam-forming ability, emulsifying capacity, stabilising capacity and the binding property of eggs are utilised in menu planning.

An egg has two parts—outer egg-white and inner egg yolk. The egg is held in position in the shell by two membranes which lie close together. About 63 per cent by weight of an egg is made of egg-white. It has three layers—an outer portion of thin egg-white, a thick jelly-like layer and a thin portion which surrounds the yolk. The central part of the egg is the egg yolk, and the yolk is surrounded by a thin covering called vitellin membrane.

The egg-white is rich in protein. Albumin, globulin and mucin are the proteins in egg-white. It is easily coagulated by heat and it provides the jelly-like character of thick egg-white. Minute quantities of sodium, potassium, magnesium, calcium, clorides and phosphates and riboflavin are present in the egg-white.

The egg yolk contains lipovitellin as the proteins and it is phosphoprotein, and it resembles the casein of milk. The egg yolk is rich in fat, that is, about one-third of the yolk is fat, especially of cholesterol and lecithin variety. The fats in egg are finely dispersed and are in emulsified forms. The mineral content is more in egg yolk compared to the white. Egg yolk is an important source of iron and it is also rich in sodium, potassium, calcium and magnesium. Lipovitellin and lipovitellinin are the lipo proteins in egg yolk which act as emulsifying agents. Egg yolk is rich in vitamins, especially thiamine riboflavin, vitamins A and D and their content in egg depends upon the hen's diet.

The whole egg is a rich source of all nutrients except vitamin C. The protein in egg is considered as a complete protein because of its high biological value and digestibility. It is often mentioned as reference protein. But a raw egg-

white contains a protein known as avidin which renders the vitamins biotin unavailable to the body. Duck egg-white contains another substance known as trypsin-inhibitor which inhibits the action of trypsin on protein. Heating destroys avidin and trypsin-inhibitors.

Role of Egg: Egg protein coagulates on heating. The coagulum formed is a smooth and jelly-like mass. It supplies a thickening effect to the products. If the temperature is high, the coagulum is toughened and it shrinks and water comes out of the product. This is called syneresis and it occurs in custards and scrambled egg preparations.

Egg is cooked in its shell, scrambled, poached, fried or used in the preparations of deserts, and cakes and other confectioneries. Egg is used in food mixtures to contribute flavour and colour to the products. Cakes and puddings owe their colour and flavour to eggs. In the preparation of mayonnaise the protein of egg stabilises the oil in water dispersion. The protein of egg stabilises the air in liquid dispersion. Egg is also used for thickening various desserts and creams and patties. Since egg is highly perishable the quality has to be judged by internal appearance flavour and external shell appearance.

Candling: Candling is used as a method to detect the interior quality of egg. In a closed box with a small opening a lighted candle has to be placed. The depth of air cell can be observed by placing the egg in front of the opening. In a fresh egg the white is stationary and slight movements of yolk are permitted by the white. Yolk is in the centre and it does not move nearer to the shell. In a poor quality egg the yolk moves. The shell of good quality egg is firm and not stained or cracked.

Formation of Dark Greenish Discolouration: When egg is boiled for more than 15 minutes a dark greenish colour may be formed on the surface of the egg yolk. This green colour is due to the formation of FeS on the surface of the yolk. Yolk contains more iron and the white has more sulphur in its amino acids. On boiling it yields H_2S. The H_2S combines with

the iron of the yolk and it gives the dark green colour in the middle. In high temperature this change takes place immediately.

Sample of Vegetarian Diet

A fruit is the edible fleshy seed-bearing part of a plant or tree. Fruits are favoured by all due to their attractive appearance, appealing flavour and pleasing odour. Apart from their nutritive value they are also appreciated for their appetising effect. Fruits contain starch. On ripening the starch is converted into fructose and glucose and these sugars contribute to the sweetness of fruits. Water is the chief constituent of fruits. About 75 to 95 per cent of the total weight is from water. Fruits as a whole are rich sources of vitamin C and vitamin A and minerals like sodium, potassium, magnesium, fibre and cellulose. Characteristic flavours of fruits are due to the presence of organic acids in them. Citric acid and malic acid are the acids present in fruits. Grapes contain tartaric acid. Acid content of various fruits varies. Citrus fruits are the best sources of vitamin C. Dried fruits like raisins, dates and figs are rich in iron. Calcium is present in citrus fruits. Banana contributes a very high amount of carbohydrates and about 23 per cent of it is sugar. Jackfruits, figs and cherries are also very good sources of carbohydrates. For vitamin C guava is a cheap source. About 300 mg vitamin C is present in 100 gms of guava. Cashew fruit is also an excellent source of vitamin C. Lime, mango, papaya and orange are other sources of vitamin C. Fruits have alkaline reaction in the body.

Fruits: Fruits with different colours have carotenoids, anthocyanins and flavanols in them. Fruits also contain enzymes like ascorbic oxidase and phenol oxidase. The vitamin A value of fruits is in the carotene in yellow fruits. Organic acids and salts are present in fruits. Dried fruits like dates and raisins are sources of iron. Potassium is the most important salt in fruits. Potassium is present in citrus fruits as citrates and in grapes as tartarate.

Pleasing colours of the fruits are due to their pigment content. In ripe fruits carotenokl pigments are present in different forms. Red, purple and blue shades of fruits are due to the presence of anthocyanine pigments in them.

Tannins and Browning Reaction: Tannins are present in almost all fruits. They contribute bitter taste and astringent quality and discolouration of cut surfaces of fruits on exposure to air. When certain fruits are pared a brown colour appears on their surface which becomes darker as the fruit is allowed to stand. Apple, banana, plum, guava and peach show this change. Tannins and flavone pigments give brown coloured pigments on oxidation. Oxidising agents present in fruit cells catalyse the reaction. If the fruits are cut with an iron knife, tannins and flavones react with iron salts forming a greenish purple compound which discolours the fruits. Acidic juices of fruits like lime, grape or pineapple are sprinkled on the fruits to prevent browning action.

Enzymatic browning occurs only when the fruits or vegetables are cut into pieces. The natural phenolic compounds present in tissues of plants do not come into contact with the phenoloxidase which is also present in some tissues. Practical methods for preventing enzymatic browning are to change the pH and the use of anti-oxidants which retard oxidation. By using the acidic juices of fruits the pH can be brought down.

Non-enzymatic browning occurs due to the formation of brown condensation products from reducing sugars and amino acids on heating. To prevent non-enzymatic browning storing the food materials at a low temperature, between 0° and 5°C, and addition of sulphur dioxide in the case of dehydrated vegetables are recommended.

Browning reaction affects the nutritive value of food, especially the lysine content of it. Loss of nutrients of fruits on cooking is negligible.

Prevention of Browning: Thermal inactivation of oxidase enzymes, elimination of O_2 from the systems, change of pH to

prevent enzyme action, and use of antioxidants are the measures to prevent enzymatic browning.

For non-enzymatic browning low temperature (0º-5°C), addition of SO_2 and keeping the moisture content low are some measures.

Vegetables: Edible parts of plants like bulbs, tubes, roots, seeds, fruits, stems, leaves or flowers are considered as vegetables. Vegetables as a whole are important sources of minerals and vitamins and certain vegetables have starch, cellulose and water. In meal planning and for dietary calculation vegetables occupy a prominent place. To check their sufficiency and also for convenience, vegetables are grouped into green leafy vegetables, and roots and tubers and other vegetables. Legumes like double beans, french beans, cluster beans and peas are also included into other vegetables. Vegetables are the best sources for calcium, iron, sodium, chlorine, cobalt, copper, magnesium, manganese, phosphorus and potassium. Carotenes, the precursors of vitamin A and ascorbic acid are present in abundant quantities in leafy and yellow vegetables. Vegetables are poor sources of calories except in the case of roots and tubers. Raw legumes and beans are good sources of proteins. Vegetables as a group supply a high amount of water and the mineral elements present in them contribute the alkaline substances in the body. This enables to maintain the acid-base balance of the body.

Cabbage, cauliflower, coriander leaves, lettuce, spinach, amaranth and drumstick leaves are the common leafy vegetables. Of all the vegetables green leafy vegetables are the richest in carotene and the carotene content depends upon the green colour. If the colour is deep the carotene is more in it. Green leafy vegetables are fair sources of proteins and good sources of folk; acid, ascorbic acid and calcium. Some leafy vegetables contain oxalic acid which interferes with the absorption of calcium in the diet. Leafy vegetables act as a buffer and maintain the Proper alkalinity of the blood by balancing the acidity of acid-producing foods like meat.

Roots and Tubers: Roots and tubers are good sources of starch and fair sources of protein, vitamin B and vitamin C. Roots and tubers include tapioca, potato, sweet potato, carrots and beetroots. Carrots are rich sources of Vitamin A. The starch in roots is easily digestible. Potato, a root vegetable, is a fair source of proteins, ascorbic acid, and B vitamins. Sweet potato is a poor source of proteins but it is a fair source of B vitamins and ascorbic acid. The yellow variety of sweet potato is a rich source of pro-vitamin A or carotene.

Other Vegetables: Other vegetables are good sources of minerals, cellulose and water. They are also good sources of vitamins. Common vegetables like ladyfingers, gourds, brinjals, pumpkins, green peas and all other vegetables fall in group Tomato is also considered as another vegetable. Green peas and yellow pumpkin are fair sources of carotene.

Cooking of Vegetables: Cooking affects the colour, texture and nutritive value of vegetables. Cooking methods can destroy the good qualities if not handled properly. Common methods of cooking are boiling, steaming, baking, frying and pressure cooking. Before applying these methods the chemical structure of vegetables must be considered. Vegetables contain different chemical compounds like pigments, tannin, enzymes and flavouring substances. Common pigments in vegetables are chlorophyll, carotinoid, flavcnes and anthocyanins. On heating these pigments undergo certain chemical changes which affect the colour of the finished product. Vegetables also lose organic acids and nutrients in cooking water. Acid or alkali content, and the cooking media also affect the appearance and quality of the product

Chlorophyll: Chlorophyll is the green pigment in leaves which is not readily soluble in water. In an acid medium chlorophyll turns an olive green and in alkali it becomes bright green. But the use of alkali to retain the green colour in salad vegetables destroys vitamin C and thiamine. Cooking in copper vessels turns chlorophyll into deep bluish green colour. Addition of sodium carbonate affects the texture and the

cellulose becomes mushy. Cooking with a lid and in minimum time retains the colour . The Carotenoid is yellow orange in colour and insoluble in water. It is soluble in ether and acetone. Carotenoids are present in different forms like carotenes, Ly copence, xanthophylls, cryptoxanthin and citroxanthin. Carotenoids are present in yellow vegetables like carrots, sweet potatoes, tomatoes, and xanthophyll in all vegetables. Cryptoxanthin is present in papaya, orange peel and yellow maize. Carotenoids are unchanged in acid or alkali and so vegetables with carotenoid pigments may be cooked with a cover. Such vegetables can be cooked by any method as the carotenoids remain unaffected.

Flavanoids: Flavones and flavonols are together known as flavanoids. Anthocyanins, anthoxanthins, leucoanthox-anthins and catechins are together known as flavonoids.

Anthocyanins are reddish water soluble pigments occurring in fruits and vegetables. They are destroyed by heat and oxidation. Passion fruit, apple, cherry, plum, grapes, blackberries are some of the fruits with anthocyanin in them. Beetroot has anthocyanin as its pigment Anthocyanin is water soluble and in acid medium it turns blue. In some vegetables flavones and anthocyanin are combined together which gives a green colour on cooking in alkaline medium.

Another chemical substance present in plants is tannin. Tannin and ferric salts form a greenish purple compound. On oxidation this turns brown which imparts a brown colour. Certain vegetables show discolouration if an iron knife is used to cut it. If vegetables rich in tannin are cooked in milk, the milk curdles.

Effect of Cooking: At a high temperature the cellulose is disintegrated and protopectin, a non-dispensable substance in water, undergoes hydrolysis and makes it dispersible. This changes the structural framework of the vegetables and makes it soft. If it is overcooked the vegetable becomes mushy and unpalatable. Some vegetables lose their natural flavour on

cooking. Strong flavoured vegetables like onion, cabbage and cauliflower must be cooked by suitable method either to retain or to lose the flavour. If a mild flavour is desired the vegetables can be cut into small pieces and cooked in a small amount of boiling water in a covered pan. If vegetables are in large pieces, large quantity of water can be used in an open pan. The plant acids are dissolved in water and the open lid allows the acids to escape.

Green vegetables turn brown because the plant acids and cholorophyll are disintegrated on cooking. Cooking time must be reduced so as to prevent their contact. Use of pressure cooker, large quantity of boiling water to cook the vegetables and use of an uncovered part are some of the measures that can be used to reduce the colour change.

Retention of Nutrients: Vegetables are rich in mineral salts and vitamins and most of them are water soluble. Some of them which are not soluble in water are destroyed by heat or by oxidation. If cooking water is thrown away considerable loss of mineral salts, especially of sodium, potassium and chlorine due to leaching occurs. To avoid this either use minimum , amount of water or the excess water after cooking can be used for other preparations. Cut the vegetables as big as possible to reduce the cooking lose bring the cooking water to boiling before inserting vegetables. Cook just enough. Washing vegetables after cutting enables considerable loss of minerals in the water. About 40 per cent minerals and thiamine are lost if the rice is washed twice or thrice -before cooking. Washing once and utilising the rice water are better methods for retaining nutrients. Absorption method of cooking can be adopted for rice cooking.

Vitamin A or carotene is not affected when food is cooked in water. But shallow frying or roasting brings considerable loss of these vitamins. If sodium bicarbonate is added especially for dals or pulses most of the thiamine is destroyed. Roots like potato, sweet potato and other roots which can be cooked with their skin must be cooked so as to prevent leaching out of

nutrients. They can be boiled first and later cut into pieces. Vitamin C is the most unstable vitamin among water soluble vitamins. Washing, cutting with an iron knife, cutting into smaller pieces and prolonged cooking lead to oxidation of vitamin C content of food items. If tamarind or lime, which is high in acidity, is added to cooking water it has a preservative effect on this vitamin. Since vitamin C is leached the least possible amount of water must be used and the cooking done in a covered vessel.

Cooking just before consumption is another method to preserve vitamin C content in the foods. Use of pressure cooker minimises the loss of nutrients. Consumption of raw vegetables provides all nutrients without any loss. If boiling is the method used to cook vegetables put them only into boiling water and bring the cooking water again to boiling temperature. For strong flavoured vegetables use little boiling water and let them remain uncovered. For other vegetables smaller amounts of water and tightly covered untensils are recommended. Avoid the use of baking soda.

Cereals

Dried edible seeds of certain plants are called pulses. Split pulses are known as dals. Pulses are good sources of proteins and B vitamins. Even though vitamin C is not present in pulses, sprouting of pulses makes it a very good source of vitamin C. Pulses are the best supplements for cereals. Generally pulses contain about 20 to 28 per cent of protein but they are inferior in quality compared to animal foods. Methionine, the limiting amino acid in cereal proteins, is present in good amount in pulses. Thus cereal pulse combination improves the nutritional quality of the diet. This combination is the most economic way to provide protein. Pulses contain fair amounts of minerals like calcium and iron. About 8 to 10 mg of iron and about 100-200 mg of calcium are present in 100 gms of pulses. Pulses contain fair amounts of thiamine, riboflavin and niacin. Sprouting of pulses improves the carotene content and vitamin C.

The Losses of Vitamins in Some Vegetables during Cooking

Name of vegetable	*Method of cooking*	*Loss of* Tiamine	*Loss of* Riboflavin	*Loss of* Vitamin C
Beans	Boiling	29%	20%	42%
Cabbage	„	32%	18%	45%
Carrots	„	20%	10%	30%
Potato	„	25%	9%	23%
Peas	„	28%	17%	40%
Spinach	„	15%	10%	28%
Sweet potato	„	20%	13%	24%
Beans	Steaming	22%	14%	32%
Cabbage	„	18%	9%	38%
Carrots	„	15%	10%	25%
Peas	„	24%	11%	30%
Potato	„	20%	12%	25%
Spinach	„	16%	12%	25%

Pulses: Green gram, red gram, bengal gram, black gram, cow gram, horse gram, lentils, peas and soyabeans are the common pulses used in our cookery. All these pulses are good in protein, carbohydrate and vitamin B and fair in mineral content. But vitamin C is not present in them. On sprouting this deficiency can be rectified. To sprout, cleaned pulse is soaked for 24 hours and then it is spread on a muslin cloth, sprinkling water occasionally on it. After 24 hours tiny sprouts appear on the surface. These sprouts are very rich in vitamin C and carotene. Sprouted Pulses are very tender and cooking is not necessary. They can be used as raw salad by sprinkling lime juice and salt or as raitas by adding curd and salt to it. As cooking destroys vitamin C it is better to use them as a raw item.

Vitamin 'C' Content of Sprouted Legumes

Period of Germination	*Vitamin C*
Dry legumes	Trace
After 24 hrs germination	7-12 mgms/100 gm
After 48 hrs germination	11-18 mgm/100 gm
After 72 hrs germination	13-20 mgm

If pulses are not cooked properly they contain much undigestible materials. But if pulses are used as flours they are completely digested and absorbed. In raw pulses trypsin inhibiter is present and it is anti-digestive in nature. Among the various pulses green gram is digested best. If more than 100 gms of pulse is consumed in a day it is not properly digested and gas is produced. If diarrhoea or other stomach complaints are present gas is produced by consuming small quantities of pulses. Bengal gram has the efficiency to reduce serum cholesterol level by increasing faecal excretion of total bile acids.

Lathyrism: Lathyrism is a paralytic disease seen in India and Spain where a type of pulse called khesari dal. — Lathyrus

sativus — is used liberally in food. In India Khesari is extensively cultivated in Madhya Pradesh, Bihar and Orissa. In Uttar Pradesh, West Bengal, Maharashtra, Gujarat, Karnataka and Andhra Pradesh it is cultivated in limited extent. Kesari dal is very cheap and highly nutritious. The flour prepared from this dal is used as a staple food in these areas. In these areas most of the agricultural labourers get khesari dal as their wage and about 250-500 gm of it is consumed by a person in a day. As a result about 4 per cent of population in these areas are crippled. Surveys in two districts of % Madhya Pradesh (Rewa and Satna) showed 32,000 paralytic victims with lathyrism. The cause of lathyrism is the production of a toxin known as BOA A-B oxalyl amino L alamine which affects the nervous system. The toxin is produced more by the busk and its consumption causes high incidence of disease. The Indian Council of Medical Research has developed certain methods to destroy the toxin.

The disease occurs in different stages. The onset is sudden. But in the first stage itself it can be detected. On exposure to cold the calf muscles contract and backache occurs. An awkward manner of walking with short jerky steps and crossed gait are common. Muscular stiffness necessitates the use of a stick to maintain balance. Later spastic paralysis of the lower limbs, and exaggerated knee and ankle jerks occur. And at this stage the patient needs crutches for support. Finally correct posture becomes impossible as the knee joints cannot support the weight of the body. The patient crawls by throwing his weight on his hands. The serum globulin and urinary glutamic acid are increased in this condition.

The National Institute of Nutrition identified the toxic factor and a simple feasible solution was suggested by them to destroy the toxin. The household procedure is to steep the dehusked grain is boiling water. The seeds are drained of water and sundried and thus the toxin is eliminated. About 90-95 per cent of the toxin is leached out by this method.

Soyabean: Soyabean is otherwise known as the miracle

bean because it is a storehouse of proteins, fat, vitamin A, thiamine, riboflavin and nicotinic acid. The protein content in soyabean is about 43 per cent and fat content is 19.5 per cent. The proteins of soyabean consists of all the essential amino acids in proper proportion except methinine and cystine. It has a good amount of lysine and it can be used for supplementing rice which is deficient in lysine. Apart from the nutritive importance it is also made use of in the production of penicillin and other antibiotics for industrial purposes. Enamel paints, varnishes and printing inks are made out of soyabean.

Even though it is rich in nutrients, unless it is processed properly, proteins and carbohydrates of soyabeans are poorly digested. Soyabean contains a factor which inhibits the protein digesting enzyme, trypsin. Heat destroys it and after cooking if it is used the protein is properly digested. Soaking, steaming, germinating and fermenting are other processes which improve the nutritive value. Soyamilk is very wantious, but soyabean seems to produce goitre at times because it interferes with iodine metabolism. Soyamilk has about 4.2 gm of 3.4 gm of fat and 0.75 gm of salts. Soyabean can be parched and used like parched bengal gram. It can be used as a substitute for Dal gram flour in recipes.

The Cooking: Soaking, use of sodium bicarbonate, acids, alkali or calcium salts quantity of cooked pulses. Dals absorb water before they can be cooked. If dal is soaked in cold water overnight or in warm water 1 4 to 5 hours it can be cooked within 30 to 40 minutes. But if hard water is used for cooking it takes more time. (a) If sodium bicarbonate is added to cook pulses cooking time is reduced. (b) vitamin like thiamine destroyed. Addition of acid helps the cooking of pulses whereas as retard the cooking. Addition of formic acid of the pulse also influences the cooking quality. The exact causes for these variations are not found out.

Cereals: Cereals and millets form the staple food of human race. Wheat, rice, maize (corn), barley, oats, jowar, ragi and bajra are the common cereals and millets used. Cereals as a

whole are rich sources of starch or carbohydrate, and good sources of protein in an Indian diet. Millets like ragi are a rich source of mineral calcium. Cereals are also good sources of B group vitamins.

Wheat: Wheat is one of our important food crops. Next to rice wheat is the most extensively grown foodgrain.

Wheat varies in its nutritive value, especially its protein content depending upon the seeds, variety in soil and water supply. The protein content varies from 7 to 20 per cent. In the 'green revolution' new varieties of superior quality wheat were produced. Sonalika, safed lerma, kalyan sona, sharbati sonora are some of these new varieties of wheat.

The wheat grain consists of three main parts. The germ or embryo is rich in protein and oil, the endosperm is rich in starch and fair sources of proteins and the outer coverings such as pericarp.

Various wheat products available are broken wheat, semolina, whole wheat flour (atta) and white flour. These flours are made from different kinds of wheat. Hard wheat, durum wheat and soft wheat are the different types of wheat from which flours are made. Flours of different percentages of extraction are obtained. Depending on the extent of milling, flour made from hard wheat has an extraction of 83 per cent and it is used for chapati making. With 75 per cent extraction white flour or soft wheat products are obtained. Biscuits, cakes, crackers and breakfast items are prepared from this flour. Semolina is prepared from hard wheat and durum wheat. It is granular wheat and it is used for the manufacture of macaroni products like noodles.

Composition of Wheat Flour: Wheat is a very good source of carbohydrates and proteins; 100 gms. of wheat provides 350 calories. Wheat contains carbohydrate in the form of starch; 66 to 68 per cent of flour is carbohydrate. Wheat flour has cellulose and hemicellulose and its quantity enhances the quality of dough prepared from it

Protein: Wheat flour contains glutenin and glidin as proteins which are commonly known as gluten. Wheat flour contains about 12 to 12.5 gms of protein per 100 gm. The strength of wheat flour is based on die quality of gluten in it Gluten is comparatively poor in amino acid lysine. Pulses, soya bean and groundnut are good in lysine content and they can be supplemented with wheat.

Lipids: Wheat flour contains 1.88 per cent fat; its kernel has about 3 per cent and bran contains about 4.8 per cent fat. Fat is present in the gluten of the flour.

Minerals: Wheat flour is poor in calcium content. The iron content of whole wheat flour is high but refined flour has less iron. Phosphorus content in whole flour is high compared to refined flour. But this high phosphorus content interferes with the absorption of calcium and iron.

Vitamins: Wheat is comparatively a good source of thiamine, riboflavin and nicotinic acid. But cooking results in loss of these B complex vitamins. During chapati making about 33 per cent of thiamine is destroyed and during fermentation much more thiamine is lost. In the preparation of puri 40 to 90 per cent of thiamine is lost. During bread making about 10 to 30 per cent lysine is destroyed. If the bread is toasted again this vitamin is lost Bulgar wheat is much more nutritious. Wheat is parboiled in the preparation of bulgar wheat. During this process vitamins and other nutrients in the outer layers of the seeds move to the centre. Thus bulgar wheat is better in its auditive value.

Processed Wheat Products: Bread, biscuits, toast and breakfast cereals like wheat flakes and shredded wheat are processed wheat products. The autrivtive value of bread, especially fresh bread, is similar to two-thirds of wheat flour in its quantity. In the preparation of biscuits thiamine and protein is lost. On baking, apart of thiamine is lost and lysine is also lost. The average protein content in biscuit is 5 to 8 per cent as against 12 per cent in wheat flour. In the preparation of toast heat is applied on bread slices and part of thiamine

is again destroyed. Lysine is also lost considerably. Readymade breakfast items are also inferior in nutritive value.

Rice: Rice is the staple food not only in India but also in many of the South-East Asian countries like Ceylon, Burma, China and Japan. Rice with its husk is called paddy. Husk is the loose covering of rice. It has no nutritive value. About 20 per cent by weight of paddy is busk. Removing the husk is called milling. Bran, aleurone layer, embryo or the germ and endosperm are the other layers. Bran is die outermost layer which contains nutrients and cellulose. Aleurone layer which lies beneath the bran is rich in vitamins, minerals, proteins and fats. Embryo or the germ is at the base of the grain which is rich in all nutrients and vitamin E. Endosperm forms 75 per cent of the grain and it comprises mainly starch. On milling vitamin B lost considerably. Parboiling improves the vitamin content of rice.

Parboiling of Paddy: In the conventional method of parboiling paddy is steeped in cold water for two to three days. Then it is steamed for about 5 to 10 minutes and men dried under the sun. During soaking paddy is fermented and at times gives out a foul smell. B complex vitamins in the pericarp and germ penetrate into the endosperm and get fixed into the grain. If parboiling is not done with care certain moulds present in the grain grow and produce toxins like aflatoxin. This can be rectified by using modern techniques in food processing. Through hot soaking process developed by CFTRI (Central Food Technological Research Institute) in Mysore is can be rectified. In this method paddy is soaked in water at 65° to 70°C for about 3 to 4 hours. Then water is drained and the soaked paddy is steamed in the same vessel for 5 to 10 minutes. This paddy is dried in sun or in mechanical driers.

The nutritive value, especially of thiamine and niacin, is improved by parboiling.

Nutritive Value of Rice: Rice contains carbohydrate in the form of starch. Though it is a source of protein, it is a poor source of fat, iron and calcium, provides about 335 calories per

100 gm of dry weight. The protein content is about 7 per cent which is not an appreciable amount since it is consumed in large quantities, rice supplies a good amount to proteins. In poor Indian diets cereals or millets are the main source of proteins. Though the protein content is low in rice compared to wheat the quality of rice proteins is superior to wheat proteins. Rice protein deficient in lysine which can be improved by supplementing the with pulses. High yielding varieties of rice are cultivated in our country. The protein count of hybrid variety rice like IR-20 is 9 per cent against 7 per cent of ordinary rice.

Rice is poor in its mineral content especially calcium. Iron content is also negligible, therefore rice-eaters must depend on other sources of minerals. Green leafy vegetables can be used for miner supplementation.

Rice is a poor source of carotene or vitamin A, and vitamin B present in good amount. Since it is present mainly in the outer lay polishing reduces the vitamin content. As polishing is increased content of B vitamins decreases. In 20 per cent polishing thiamine (B. content is only 40 mg/gms whereas in parboiled rice at 20 per a polishing thiamine is 200 mg/100 gm. Niacin content is 0.6 mg/100 gm, in 20 per cent polished rice whereas in parboiled rice it is 2.8 mg/l gm. Among people who depend entirely on polished rice, a burn sensation in the feet, tingling and numbness occur. This is due to deficiency of thiamine in the diet. Even though rice is a good source of B complex vitamins. Its high solubility in water makes it non-available after various processes of cooking. About 60 per cent of it is lost through washing, if rice is washed twice or thrice. In the cooking water , remaining B complex gets dissolved. If the "conjee" water is thrown i this amount is also lost. Cooking rice in just enough water and utilise the excess water, if any, are the useful tips to make use of this vitamin.

Gelatinisation: Starch is present in cereals. This is insoluble in water, but in hot water it dissolves and forms colloidal dispersions. It swells and forms a sticky substance. In cereals

about 65 to 80 per cent starch is present. Next to cereals roots like tapioca, potato and sweet potato are with starch granules. These roots and tubers contain about 22 to 39 per cent of starch. In starch there are two components—amylose and amylopectin. Amylose is about 15 to 30 per cent and amylopectin about 70 to 85 per cent. In different sources of starch it is present in different proportions. Amylose is soluble in water and amylopectin is insoluble in water. On 5 per cent concentration starch forms gels. Among the various sources of starch corn starch is the finest. This property of starch to gel is exploited in cookery. In many preparations starch is used as a thickening agent. To get a good gel, gelatinisation temperature of different starches must be known. Usually it starts to gel at 65° to 72°C. The enzymes in the starch are the other factors which affect the gelatinisation. Commercial products like sago, custard powder, corn syrup, glucose or dextrose and dextrimaltose are prepared from tapicao starch or from corn starch.

Millets: The common millets used in our country are jowar, ragi and bajra. Though millets are comparatively cheap they are superior in nutritive value. Usually people prefer rice or wheat as their cereals. If these millets are used more many deficiency symptoms among various groups can be minimised as they are excellent sources of vitamins and minerals.

Ragi: In India ragi is used mainly in Karnataka and Tamil Nadu as a staple food. In Kerala, ragi is used as a weaning food. Actually ragi is considered as a 'poor man's milk' by nutritionists because of its high calcium content. It can be easily substituted as a breakfast item. If we consider its cost it is cheap and at the same time rich in minerals that are lacking in rice or wheat. Like wheat or rice, ragi also contains 7 to 10 per cent protein and is equivalent to the above cereals in its carbohydrate content. It is considered superior because of high calcium content. Ragi is also a good source of iodine. No other cereal contains iodine. Another special feature of ragi is that the carbohydrate in ragi is digested slowly and so it can be given to diabetic patients because the sudden rise of blood

sugar can be avoided by this. In certain hybrid varieties of ragi protein and iron contents are more and thus it is superior in its qualities. Ragi can be incorporated with pulse instead of rice in all breakfast items.

Bajra: Bajra is used in certain parts of the country as a staple food. It consists about 12 per cent of protein and calcium and iron content is compared to rice. Bajra is a better source of vitamin B, especially thiamine and riboflavin.

Jowar: Jowar is used in Maharashtra, Andhra Pradesh, Karnataka, Tamil Nadu and Madhya Pradesh. The protein content in jowar is 10 per cent and the amino acid proportion is better. It has more B complex vitamins compared to rice. But certain varieties of jowar inhibit the availability; of niacin and cause the deficiency symptoms of pellagra.

Effect of varying concentration of sugar syrup and crystallisation

Boiling temperature° C	*Sugar percentage (Syrup)*	*Cold water test*
110-112° C	78-80	Does not form a ball
113-115° C	83	Soft ball, eg., fondant
118-120° C	86	Firmer ball, eg., caramels.
128-130° C	90	Hard ball, but still plastic, eg., candy
138-140° C	94	Very hard ball, eg, toffees
149-154° C	98	Very hard and brittle, eg., brittles.

To prepare the syrup add a small amount of sugar in water and heat, when the whole sugar is dissolved examine the consistency.

Beverages

Enjoying the Tea

The most popular non-alcoholic beverage, tea consists of the dried leaves of the tea plant grown in India, Ceylon, China

and Japan. Chinese tea contains less tannin than the other varieties. Tea leaves have a characteristic oval shape and serrated edge. The younger the leaves the better the quality of the tea. The treatment of the leaves after they are picked varies according to the tea required.

For black tea the leaves are withered, then rolled till they become soft and mushy. This is done to break up the fibre and cells of the leaf to liberate the constituents so that extraction is easy. The leaves are then fermented. During the process of fermentation some of the tannic acid in the leaves oxidizes and is converted into less soluble forms while more essential oils develop. After fermentation is complete, the leaves are fired in a drying machine. For green tea the leaves are withered in hot pans at a temperature of 71°C (about 160°F) or steamed, then rolled to break them up and liberate their juices, and then fired.

The chief difference between black tea and green tea is that the former, is fermented while the latter is not. Since the purpose of fermentation is to make the tannin less soluble, an infusion of green tea has more tannin in it.

The best tea generally produces a pale-coloured infusion and the depth of colour is not necessarily a sign of strength. Freshly infused tea is harmless to normal digestion; continued infusion extracts the tannin, a bitter substance with a decidedly harmful effect upon the digestive organs.

Storage of Dry Tea: Tea very quickly absorbs odours and moisture. So it should be kept in airtight containers, whether in the store or on the counter.

If the tea is in chests, they should be kept off the ground and at least 16 cm. or 6 inches away from the walls. This allows a current of air to circulate round the tea chest and averts dampness. Once a chest is opened it is advisable to put a close-fitting lid over it. When issuing tea from it, be sure that it is sent out in a tin fitted with a lid that can be firmly replaced after use, so that the natural aroma of the tea is retained.

Next in importance to the tea leaf is water, which is used for the preparation of tea. Water varies in different parts of the country. In some places water is hard and at other places it is soft. Tea which gives good results in soft water may not necessarily do so in hard water. For this reason samples of water drawn from different areas of the country are used for blending teas.

Water should not be too hard or too soft for making good tea. Artificially softened water is unsuitable for making tea and should be avoided.

Water should be freshly drawn before it is put on to boil.

Water which has been lying in the kettle for several hours or which has been boiled previously makes tea flat and insipid, as also, water which is allowed to continue boiling for some time before it is used. When the water used is off the boil, approximately half the leaves float and remain floating throughout the infusion. The rest remain at the bottom from the time the water is added. The movement of the leaves is very slight and tends to be from the bottom upwards. The flavour of the tea is poor.

With freshly boiled water approximately two-thirds of the leaves float from the start. There is a marked circulation of the leaves from the bottom to the top and from the top to the bottom. The liquor has good body and good flavour.

When water is over-boiled all the leaves remain clotted at the bottom from the start of the infusion to the end. The liquor is thin and lacks flavour and character.

Preparation of Tea

(i) *Use Good Tea and Remember the Recipe:* one teaspoonful per person and one for the pot. The ideal is 60 gm of good tea to 4 litres of boiling water. There should be no guesswork; weigh or measure tea for each brew.

(ii) *Use Freshly Drawn, Freshly Boiling Water:* Take water from a cold water tap, use it when it bubbles fiercely.

An automatic boiler provides this, but if bulk boilers are used, two small ones are better and more economical than one large. Measlired water may be boiled in metal pots (degchis) and transferred to another pot (already heated and containing tea leaves), or tea leaves can be put directly into the pot in which water has reached boiling temperature. Remove the pot from the fire and allow the tea to infuse for 5 minutes.

(iii) *Heat the Pot:* Tea pots or urns in which tea is to be prepared should be heated before making tea. Unless this is done the water will go off the boil rapidly, thus preventing correct infusion of the tea leaves.

(iv) *Take the Teapot to the Boiling Water:* The water must be at boiling point to enable the leaves to infuse properly, so the teapot should be brought near the boiling water outlet to ensure the shortest length of pour possible. To a measured quantity of water add a measured quantity of tea leaves and remove the pot immediately from the fire. Allow it to infuse for 5 minutes and strain. If electric or other boilers are used for boiling water, their outlet taps should be such that the maximum water can flow through them in the minimum of time.

(v) *Brew for 5 Minutes:* Never more, never less. Tea should be made of good average strength to allow being watered down if necessary, as nothing can be done about strengthening tea which is too weak to start with. Removable infusers enable the leaves to be with draw easily and prevent stewing. Stir well before pouring.

Approximately 190 to 200 cups of tea can be obtained from 500 gm of tea leaves.

Adding of milk first or last does not make any significant difference in the taste of tea—but let people have their choice—some like to add milk first and some afterwards. Sugar must

be added last. Tea is also taken hot with sugar and slices of lemon. This is known as Russian Tea.

Iced Tea: Prepare strong tea. Pour over crushed ice on which is placed a sprig of mint. Top with lemon slices.

Tea contains the stimulating principle caffeine, tannin and some volatile oils.

Service with Coffee

Coffee is the berry of a plant grown in Kenya, Ceylon, Brazil, West Indies, India and other places.

Coffee has become one of the important beverages in India. There has been a steady increase in the consumption of coffee in recent years. It is also one of the major plantation crops of India, particularly South India. The two main varieties of coffee grown in India are, Coffee Arabica and Coffee Robusta. Coffee Arabica originated in Yemen in Arabia and was brought to India in the seventeenth century by a Muslim called Bababudan Sahib on his return from his visit to the Holy Land, Mecca. Coffee Robusta was brought to India from Java in 1910.

After the fully ripe berries are harvested, they are processed in two ways, by the 'wet' method or by the 'dry' method.

The 'wet' process is very elaborate. Ripe berries called cherries are first passed through pulpers where their outer skin is removed. The pulped beans are then allowed to ferment for about 12 to 36 hours. They are then washed in running water. This leaves the beans encased in a soft parchment covering. They are then dried in the drying yard. Great care is taken at this stage to see that only a little moisture remains and the parchment covering becomes brittle. This produces what is known as Plantation or Parchment Coffee.

The 'dry' method is a simpler process. Immediately after harvesting, the coffee berries are dried in the sun until the three layers—the outer skin, the sticky layer and the inner parchment covering—dry into a husk, leaving the beans loose

inside. The beans are separated from the husk by being pounded in a mortar or passed through a huller. This is then graded and is called Native Coffee or Cherry Coffee.

In the days of sailing ships when a voyage to Europe took about six months or so, Native (Cherry) Coffee stowed away in damp holds lost its colour and became white. Later, when steamships went speedily through the Suez Canal, coffee drinkers in Europe complained that the coffee had lost its flavour. To overcome this, a new process known as 'monsooning coffee' was evolved. This is done in Mangalore and Tellicherry on the west coast of India.

When the monsoons break in early May or June, Native (Cherry) Coffee is evenly spread in layers from four to six inches thick, in airy godowns which are open on all sides. Every four or five days the beans are raked. They are then packed in gunny bags and stacked in piles so that the monsoon breeze can blow continuously on each bag. Once a week the beans are repacked or poured from one bag to another, so that they do not develop moulds and also to ensure uniform monsooning. After six weeks of this treatment the coffee becomes silvery white in colour and is considered fully matured and ready for export.

After the beans have been treated to make either 'parchment' or 'cherry types, they are sent to the curing houses for the final process required to prepare the coffee for the market. It is here that the parchment is peeled off Parchment Coffee.

The stimulating principle in coffee is caffeine. It also contains tannin and some volatile and aromatic oils. As long as coffee is in good condition the distinctive aroma remains. It should always be stored in airtight containers. Refrigerated storage is best for keeping coffee.

Roasting and grinding: The aroma and flavour associated with coffee is brought about by roasting. Roasting must be done evenly without any scorching. It should be done on a

slow fire and the beans should be uniformly roasted. When the beans stop swelling and obtain a golden-brown colour, the roasting is complete. If roasting goes beyond this point the essential oils exude and the beans turn shiny. When it reaches this stage the coffee has no value. Roasting can be done in a charcoal roaster or in an electric roaster. A medium roast and grind is considered to give the best results.

Roasted beans keep fresh much longer than powder and so grinding must be done just before use. Too finely ground powder clogs up the filter if used, and deteriorates rapidly when stored. The coarser the powder, the less the extraction. As a standard for brewing coffee, 450 gms. (16 oz.) gives 53 cups [187.5 ml (6 fl.oz.) coffee, 62.5 ml (2 fl.oz.) milk per cup].

Coffee can be prepared by many methods. The pot method, the filter method, the urn brewing method and the vacuum brewing method are commonly used.

Just one cup of coffee does some amazing things to the body. The temperature of the stomach jumps up by 10-15%. Salivary glands double their flow; the heart beats 15% faster; the lungs work 13% harder; blood vessels dilate in the brain and near the heart; the metabolism rate goes up 25%; and the work-load of the kidney doubles.

Rules for Making Coffee

(i) Fresh coffee is the best; so buy quantities to last not more than a week.
(ii) Use the quantity required for making coffee,
(iii) Use freshly drawn water,
(iv) The coffee-maker must be rinsed with hot water before use and thoroughly washed and dried before being put away. Never brew less than three-fourths of the coffee-maker's capacity; use a smaller one instead.

Pot Method

(i) Warm an earthenware pot or jug. Put in 3 level teaspoons of fresh coffee powder (coarse grind) for each 250 ml (8 oz.) cup.
(ii) Pour water which has started to boil over the powder

and stir. Cover the pot and let it stand near the fire for 5 to 7 minutes.

(iii) Pour the coffee through a fine meshed sieve or cloth. Add milk and sugar to taste.

Filter Method: Several types of filters are available. Stainless steel or brass filters are the best but the latter should be properly tinned, or else the coffee will be spoilt. Glass or china containers are good but are fragile. Copper should not be used.

(i) Put in 3 level teaspoons of coffee powder (fine or medium grind) to each 250 ml or 8 oz. cup. Press the plunger down lightly over the powder.

(ii) Pour water which has just come to boil over the plunger in a circular motion. Let it stand for 5 to 7 minutes.

(iii) Coffee can be poured out straight from the lower vessel and milk and sugar added as required.

Percolator Method

(i) Place the ground beans in the centre section of a clean warm percolator on a fine strainer fitted inside and resting on a paper filter (a pinch of salt is mixed with the coffee).

(ii) Pour fresh boiling water slowly through the top section.

(iii) The water passes through the coffee, is strained and collects in the bottom section of the apparatus.

Cona coffee machine (a patent percolator) consists of two flameproof glass bowls joined by a glass drainer tube which also acts as a filter. A measured amount of ground coffee is placed in the top receptacle and water in the lower one. Heat is applied, causing the water to rise, meet and mix with the coffee. The whole operation is automatic, and the beverage returns to the lower bowl and is served immediately or placed on a warming unit to be held at 85°-88°C (about 185°-190°F) (never allow coffee to boil).

Coffee is made in large quantities in the urns using the filter or percolator method. It is then held at 85°-88°C (about 185°-190°F) temperature and should be served within the hour.

All equipment used for the preparation of coffee should be kept scrupulously clean. When using a Cona or other mechanical percolator be sure to:

(i) Scrub upper tube regularly with a stiff brush.

(ii) Remove rubber collars and wash all parts thoroughly,

(iii) Replace rubber collars when they become heat softened.

If a coffee urn is used

(i) Clean inside of urn after each batch,

(ii) Clean faucets at close of each day.

(iii) Leave plenty of water in urn when not in use.

(iv) Clean gauge glasses twice a week,

(v) Give urn a thorough cleaning twice a week.

Turkish Coffee: Turkish coffee is rich and syrupy and has a peculiar aroma of its own, caused by the preliminary method of drying coffee: the 'wet' method. The ground coffee is first mixed with sugar in a small china or copper pot fitted with a long handle. Water is added, heat applied and the coffee is flavoured with rose water. It is served piping hot.

Cocoa: Cocoa, besides being a stimulant, is also a food. It is prepared from the seeds of a tree grown in South and Central America, the West and East Indies and along the Gold Coast and adjacent areas in Africa.

The pods are gathered in heaps and cut open with sharp rounded knives. The cocoa beans which are covered with a moist glistening white pulp are scooped out. Oxidation begins almost at once causing the beans to become brown. Therefore, as soon as possible, they must be placed on fermenting heaps or spread in the sun to remove the moisture so that they are not spoilt. However, fermentation is necessary to get the finest possible flavour. This is accompanied by a rise in temperature and the transformation of natural sugars to acetic and other acids. After several days, this operation is complete and the beans are allowed to dry. They are then ready to be packed and shipped.

In the manufacture of cocoa and chocolate the first step is thorough cleaning to remove any foreign substance which may be present. The beans are then roasted in revolving cylinders by currents of air heated to a temperature above 204.4°C (about 400°F). At exactly the right moment they pass from the roaster and are quickly cooled to prevent further change.

The beans are then conveyed into chambers where they are broken into fragments called nibs. At the same time the shells are lifted away from the nibs by air currents and removed from the chambers. The nibs are quite dry. They are then milled to reduce their size. In this process they pass between three sets of steel-encased grinding stones. Here the cocoa butter, the natural fat of the cocoa bean averaging about 55 per, cent, is released from the cells in the form of a free-flowing liquor of rich, dark colour and heavy, distinctive aroma. This product is the basic ingredient of all forms of chocolate and milk chocolate.

Cocoa is the second derivative of chocolate liquor. After undergoing heat treatment for flavour development, the liquor is pumped into powerful hydraulic presses which expel the free-flowing cocoa butter until the original 55 per cent is reduced to less than half that quantity. The solid cakes of cocoa remaining in the presses are then ejected and reduced to a fine dry powder in crushers and pulverisers. The powder is then cooled and wind-sifted to get the pleasing brown colour and extreme fitness essential for high quality.

Cocoa contains the bromine and caffeine as well as starch, fats, nitrogenous bodies and salts, so it is both a stimulating drink, and a food.

Cocoa can be prepared in milk only, or in milk and water, mixed to suit the taste of the individual. The powder is mixed with sugar to avoid lumping. A little cold milk is added and a thick, even paste made. Either hot milk or hot milk and water are poured over. The preparation is then allowed to boil for

a few minutes which improves the flavour. A pinch of salt added at the beginning enhances the flavour.

Rechauffe Cooking: Although freshly cooked food is better than reheated food, in the interests of economy and food management, reheating of food forms an important part of cooking. A clever cook should take pride in seeing that the second appearance is appetizing and that the maximum amount of nourishment is retained. Great care must be taken in reheating food, particularly meat. Since it is already cooked, further cooking will toughen the fibres and harden the proteins, thus making the food indigestible. Therefore, the first rule is never recook, only reheat.

Food may be warmed in a moderate oven if carefully covered. If the vessel containing food can be placed in a tin containing water, over-drying is prevented. Steaming is one of the best methods of reheating as the heat is gentle and does not dry the food. A covering is necessary to prevent the condensed steam from falling on the food. Frequently it may be necessary to do more than just reheat and, therefore, an ingenious cook must find ways and means of using up cooked food to make new dishes, e.g., cutlets, minced coilops, shepherd's pie, hash, and curry can be made without having to recook.

Rules for Reheating Food:

(i) Never recook, only reheat.

(ii) Make use of all scraps of food, e.g., vegetables, sauces, gravy, etc., but be sure that they are absolutely fresh,

(iii) Remove gristle, bone and skin but not fat unless in excess because fat helps to keep the food moist,

(iv) Divide finely so that flavourings and seasonings can penetrate quickly and the food can be reheated quickly to avoid recooking.

(v) Cook any ingredient such as vegetables before adding it to the food that has to be reheated, e.g., if a cutlet has to be made from roast, the potatoes should be

boiled first and then mixed for binding. The short, quick reheating does not allow time for the cooking of raw ingredients,

(vi) Additional moisture is always necessary in the form of sauce or gravy as most of the original moisture will have been removed.

(vii) A binding is essential for croquettes, rissoles, etc., but see that the binding is properly cooked before being added, e.g., as with Panada.

(viii) Cooked meat and fish become insipid when cold. So they must be carefully seasoned and flavoured in reheating. Meat is improved by the addition of ham, tongue, salted meat, onions, tomatoes, mushrooms, curry paste and fresh vegetables, while fish is improved by the addition of lemon juice, parsley and other mixed herbs.

(ix) It is often necessary to coat food to protect it from direct heat and, therefore, merely reheat: e.g.

(a) Covering with potato as in shepherd's pie.

(b) Coating with egg and bread crumbs.

(c) Enclosing in pastry as in mutton patties.

(d) Enclosing in batter as in pan rolls,

(x) Dish daintily and garnish neatly.

(xi) Serve with a good gravy or a sharp sauce as an accompaniment,

(xii) Serve hot. Half-warmed, reheated food is far from appetizing.

Bibliography

Achaya, K.T. : *Indian Dairy Products,* Asia Publishing House, Bangalore, 1974.

Adam, H. : *Food Science,* John Wiley and Sons, New York, 1982.

Austin, A. and Ram, A. : *Studies on Chapati-Making Quality of Wheat,* Indian Council of Agricultural Research, New Delhi, 1971.

Bachman, W. : *Food Chemistry,* Maclaren and Sons Ltd., London, 1955.

Balsdon, J.P. : *Life and Leisure in Ancient Rome,* Bodley Head, London, 1966.

Beersteeker, E. : *The Biochemistry of Vitamins,* Reinhold, New York, 1950.

Bharathi, V.V. : *Anxiety Behaviour in Children,* Discovery Publishing House, New Delhi, 1992.

Board, R.G. : *A Modern Introduction to Food Microbiology,* Blackwell Scientific Publications, London, 1983.

Burton, T.L. : *Recreation Research and Planning,* Allen and Unwin, London, 1970.

Cambell, A.M., Penfield, M.P. and R.M. Griswold : *Experimental Study of Foods,* Houghton, London, 1990.

Cameron, Spencer : *Food Science,* Pargamon Press, New York, 1984.

Challoway, D.H. : *Nutrition and Physical Fitness,* W.B. Saunders Co., Philadelphia, 1979.

Cherry, G.E. : *Town Planning and its Social Context*, MacGraw-Hill, London, 1970.

Cherry, J.P. : *Protein Functionality in Foods*, American Chemical Society, Washington D.C., 1981.

Chichester, C.O. : *The Chemistry of Plant Pigments*, Academic Press, New York, 1972.

Dale, E. : *Management Theory and Practice*, McGraw-Hill, New York, 1973.

Dharamraj, E. : *Food and Hotel Legislations and Policies*, New Age Publishers, New Delhi, 1998.

Dumazedier, J. : *Towards a Society of Leisure*, Free Press, New York, 1967.

Dyke, S .F. : *The Chemistry of Vitamins*, Interscience Publishers, New York, 1965.

Edmunds, F. : *Environmental Administration*, McGraw-Hill, New York, 1973.

Eley, G. : *Wild Fruits and Nuts*, E.P. Publishing Ltd., Yorkshire, 1976.

Feiffer, M. : *Giving Places*, MacMillan, London, 1985.

Furia, T.E. : *Hand Book of Food Additives*, C.R.C. Press, Ohio, 1977.

Geiss, I. : *Proteins : Structure, Function and Evolution*, Menlo Park, California, 1983.

Gell, D. L. : *International Hotel Industry Policy*, Van Nostrand Reinhold, New York, 1990.

Gyorgy, P. and Pearson, W.N. : *The Vitamin*, Academic Press, New York, 1967.

Haschemeyer, A.H. : *Proteins : A Guide to Study of Physical and Chemical Methods*, John Wiley & Sons, New York, 1973.

Hudson, F. : *Development in Food Proteins*, Applied Science Publisher, London, 1982.

Inglett, G. : *The Quality of Foods and Beverages : Chemistry and Technology*, Academic Press, New York, 1981.

Jackson, R. : *The Geography of Recreation and Leisure*, Hutchinson, London, 1972.

Jasscler, D.K. and M.A. Joslyn : *Fruit and Vegetables Juice*, The AVI Publishing Co., Connecticut, 1971.

Johnson, A.H. : *Encyclopedia of Food Science*, The AVI Publishing, Connecticut, 1978.

Kacharoo, P. : *Pulse Crops of India*, ICAR, New Delhi, 1970.

Kadans, J.M. : *Encyclopedia of Fruits, Vegetables, Nuts and Seeds*, Parker Publishing Co., New York, 1973.

Kalyani, T. : *Home and School Environment*, Discovery Publishing House, New Delhi, 2003.

Kinton, R. : *Practical Cookery*, Edward Arnold, London, 1981.

Krause, M.V. and L.K. Mahann : *Food Nutrition and Diet Therapy*, W.B. Saunders Co., Philadelphia, 1979.

Kumari, P. : *Role of Anganwadi Workers in Integrated Child Development Services Programme*, Discovery Publishing House, New Delhi, 1994.

Lal, H. : *Total Quality Management : A Practical Approach*, New Age Publishers, New Delhi, 2000.

Law, C. : *Urban Tourism : Attracting Visitors to Large Cities*, Mansell, London, 1993.

Less, R. and E.B. Jackson : *Sugar Confectionery and Chocolate Manufacture*, Leonard Hill Books, Burks, 1973.

Lickorish, L. J. : *Hotel Industry and International Balance of Payments*, International Institute of Scientific Travel Research, Geneva, 1953.

Lloydryall, A. and W.T. Pendzer : *Handling Transportation and Storage of Fruits and Vegetables : Fruits and Peanuts*, The AVI Publishing Co., Connecticut, 1974.

Mackinney, G. and A.C. Little : *Colour of Foods*, The AVI Publishing Company, Connecticut, 1972.

Malhotra, R. K. : *Food Service and Catering Management*, Anmol Publications, New Delhi, 1998.

Marton, I.D. and A.J. Macbod : *Food Flavours*, Elesvier Scientific Publishing Co., New York, 1982.

Masefield, G.B. : *The Oxford Book of Food Plants*, Oxford University Press, London, 1975.

Mate, S.A. : *Bakery Technology and Engineering,* The AVI Publishing Co., Connecticut, 1972.

——————— : *Cereal as Foods and Feeds,* The AVI Publishing Co., Connecticut, 1959.

Mayo, J. : *The Psychology of Leisure Travel,* CBI Publishing Company, Boston, 1981.

Mcwilliams, M. : *Food Fundamentals,* John Wiley & Sons, New York, 1979.

Medlik, S. : *Hotel Management: Past, Present and Future,* Heinemann, London, 1976.

Mellor, J. D. : *Fundamentals of Freete-Drying,* Academic Press, London, 1978.

Meyer, L.H. : *Food Chemistry,* Reinhold Publishing Corporation, New York, 1961.

Michael Eskin, N.A. : *Plant Pigments, Flavours and Textures,* Academic Press, New York, 1979.

Murthy, R.B. : *Breeding Procedures in Pearl Millet,* ICAR, New Delhi, 1977.

Nath, P. : *Vegetables for the Tropical Region,* Indian Council of Agricultural Research, New Delhi, 1976.

Oberleas, D. : *Trace Elements in Health and Disease,* Academic Press, New York, 1976.

Palmer, T. : *Understanding Enzymes,* John Wiley & Sons, New York, 1981.

Pamela, V.P. : *Directory of Wines and Spirits,* North Wood Publication, London, 1974.

Paul, P.C. and Palmer, H.H. : *Food Theory and Application,* John Wiley & Sons, New York, 1972.

Peckham, G.G. : *Foundations of Food Preparation,* MacMillan Publishing Co., New York, 1979.

Pigman, W.W. and Horton, D. : *The Carbohydrates : Chemistry and Biochemistry Plants,* C.S.I.R, New Delhi, 1969.

Pigram, J. : *Outdoor Recreation and Resource Management,* Croom Helm, London, 1993.

Potter, N.M. : *Food Science,* The AVI Publishing Co., Connecticut, 1978.

Pudney, John : *The Thomas Cook Story,* Michael Joseph, London, 1953.

Puruthi, J.S. : *Spices and Condiments,* National Book Trust of India, New Delhi, 1976.

Ranganna, S. : *Manual of Analysis of Fruit and Vegetable Products,* Tata McGraw-Hill Publishing Co. Ltd., New Delhi, 1984.

Rani, Usha D. : *Nutrition and Religion,* Discovery Publishing House, New Delhi, 2003.

Rao, D. V. : *Community and School Nutrition Education,* Discovery Publishing House, New Delhi, 1998.

Roberts, D.V. : *Enzyme Kinetics,* Cambridge University Press, 1977.

Rogers, C.A. : *Profitable Poultry Keeping in India and the East,* Treasure House of Books, Bombay, 1966.

Ross, G. F. : *The Psychology of Hotel Management,* Hospitality Press, Melbourne, 1994.

Schapira, J. : *The Book of Coffee and Tea,* St. Martin's Press, New York, 1975.

Schultz, G.E. and R.H. Schirmer : *Principles of Protein Structure,* Springer-Verlage, New York, 1979.

Shreedevi, V. : *Nutrition Education,* Discovery Publishing House, New Delhi, 1997.

Smartt, J. : *Tropical Pulses,* Longman Group Ltd., London, 1976.

Stadlmare, W.J. and. Cotterill, O.Y. : *Egg Science and Technology,* The AVI Publishing Co., Connecticut, 1973.

Stumbo, C.R. : *Thermobacteriology in Food Processing,* Academic Press,, New York, 2000.

Swern, D. : *Bailey's Industrial Oil and Fat Products,* John Wiley & Sons, New York, 1964.

Tandon, G.L. : *Preservation of Fruits and Vegetables,* Indian Council of Agricultural Research, New Delhi, 1960.

Tandon, H.P. : *Egg and its Care,* Ministry of Agriculture, Govt. of India, New Delhi, 1973.

Thangam, P.E. : *Modern Cookery for Teaching and Trade,* Orient Longmans Ltd., Bombay, 1983.

Thankamma, J. : *Food Adulteration,* S.G. Wasani, New Delhi, 1976.

Tressler, D.K. : *Food Technology the World Over,* The AVI Publishing Co.; Connecticut, 1963.

Uma, K. : *Poverty, House Hold Fold Security and Nutrition in Rural Areas,* Discovery Publishing House, New Delhi, 2000.

Underwood, E.J : *Trace Elements in Human and Animal Nutrition,* Academic Press, New York, 1977.

Veldhnis, M.K. : *Food Science and Technology,* The AVI Publishing Co., Connecticut, 1977.

Vidya, D.B. : *A Textbook of Nutrition,* Discovery Publishing House, New Delhi, 1996.

Weiss, T. J. : *Food Oils and their Uses,* The AVI Publishing, Connecticut, 1970.

Whitaker, J.R. : *Principles of Enzymology for Food Science,* Marcel Dekker, New York, 1977.

Williams, P.N. : *The Chemical Constitution of Natural Fats,* Chapman and Hall, London, 1964.

Winitz, M. : *Chemistry of Amino Acids,* John Wiley & Sons, New York, 1961.

Yasoda, P. : *Determinants of Nutritional Status of Pre-School Children,* Discovery Publishing House, New Delhi, 1998.

Yukic, T.S. : *Fundamentals of Recreation,* Harper Row, London, 1963.

Index

F

G

H

I

J

L

M

N

O

P

Q

R

S

T

V

W

Z

□□□

119A

Encyclopaedia of
FOOD CHEMISTRY

Encyclopaedia of
FOOD CHEMISTRY

Volume 4

Manpal Singh

ANMOL PUBLICATIONS PVT. LTD.
NEW DELHI - 110 002 (INDIA)

ANMOL PUBLICATIONS PVT. LTD.
H.O.: 4374/4B, Ansari Road, Darya Ganj,
New Delhi-110 002 (India)
Ph.: 23278000, 23261597
B.O.: No. 1015, Ist Main Road, BSK IIIrd Stage
IIIrd Phase, IIIrd Block,
Bangalore - 560 085 (India)
Visit us at: www.anmolpublications.com

Encyclopaedia of Food Chemistry

First Published, 2007

ISBN 978-81-261-3399-4 (Set)

PRINTED IN INDIA

Printed at Amit Enterprises, Delhi.

Contents

Preface

Food Chemistry is a newer sub-discipline, under Food Science. This branch of knowledge enriches us about the composition and quality of food stuffs, we intend to consume. In a scholar's view, we are what we think, while we think what we eat and drink. This is an undeniable fact that our thoughts are determined by the kind of food we take. Food not only serves as a source of energy, it also plays a significant role in the making of one's personality.

Another scholar opines that for a healthy life, a healthy diet is a must. Our body mechanism is controlled by a number of elements, called vitamins, proteins, enzymes, carbohydrates and other simple and complex chemical elements.

All the abovesaid elements find their origin in various food articles, such as, milk, meat, egg, vegetables, cereals etc. Each has its own characteristics. For a healthy diet, different nutrients have to be taken in a proper proportion, because excess of any of these may result in disturbing the system of our body and affecting the health.

A proportionate or a balanced diet is one, which consists of all the necessary nutrients in a proper manner, required for the development and maintenance of the body. It depends upon age, health and profession. For instance, a diet, rich in

carbohydrates, vitamins, mineral salts, fats, and proteins should provide about three hundred calories per day for a normal adult. It comes under Food Chemistry, which has been dealt with, at length in this multi-volume.

Encyclopaedia of Food Chemistry, is an admirable step in this direction. These books throw abundant light on all aspects of food chemistry. It must prove to be a proud possession of students, scholars, academics, intellectuals and of course the general readers, who may like it, equally.

Wise comments and positive remarks for the betterment of these books in future, are solicited.

— Editor

1

Basic Items

Types of Food Production

Cooking Fishes : Fish provides many attractive and satisfying dishes. Fish is a high protein food, supplying on an average a little more than 5 gms. of protein per edible ounce.

Fish could be got from lakes, rivers of seas. Freshwater fish are considered to be superior in flavour to salt water fish. A fish that has fat is superior in flavour to lean fish. The season also has a bearing on the fish for better taste.

The fat content helps to determine the method of preparation. Fish that has fat will produce superior eating qualities, if baked, or boiled because of their natural fat which keeps them from drying during cooking.

Lean fish are best poached, boiled or steamed, the flesh is firm and holds together during the cooking period. Both lean and fish with fat can be sauteed or fried which give excellent results.

Fumet de poisson is concentrated fish stock.

Classification of Fish

(1)	White fish (Round)	(a)	Round, cod, haddock, hake, whiting, Mumbai duck, chamli, pholi and turai.
	Lean fish (Flat) White fish has oil in the liver.	(b)	Flat, pomfret, brill, lemon sole plaice, turbot, sangara, halibut, Malabar sole.
(2)	Oily fish		Salmon, mackerel, herrings, Malabar sardines, hilsa, bindi, salmon trout, trout and skate.
	Shellfish are sub-divided into:	(1)	Molluscs—Clams, mussels, scallops and oysters. (Immobile)
		(2)	Crustaceans—Crayfish, lobster, prawns, shrimps, scampies, crabs and turtles. (Mobile)

Selection of Fish : When purchasing fresh whole fish, the following points should be observed:

(1) Quality in fresh fish is determined by the condition of the skin which should be shiny and of good colour.

(2) In white fish, the flesh should be really white, not faintly yellow.

(3) The body should be rather heavy in relation to the length, and the flesh plump and springy.

(4) Eyes should be bright and not sunken.

(5) Gills should be red.

(6) For freshness, the fish should have a pleasant, slightly salty smell, while stale fish has an ammoniac odour, which increases with its staleness.

(7) If the imprint of the fingers remain, after it has been pressed, then it is not fresh.

(8) If the scales come out themselves, then the fish is stale.

(9) Raw flesh should not come away from the bones, as it is the sign of fish deterioration

(10) The tail should be stiff. Fish should be stored at 1°C.

Shellfish : The food value of shellfish is similar to that of

fish, as it contains protein, calcium, magnesium, iron, phosphorus and iodine. Whilst shellfish is difficult to digest, it is a good body-building food. Because they are scavengers, shellfish are regarded as a dangerous food and a source of food poisoning. But this is largely due to poor buying and storage, or that they are misused during the cooking process. Debeard is to remove beard from shellfish such as oysters and scallops.

Storage : Where possible, shellfish should be purchased alive, so as to ensure absolute freshness and stored in a cold room, in a box covered with crushed ice and a damp sack and cooked as soon as possible after purchasing.

Choice of Shellfish

(1) Freshness in shellfish can be recognised by brightness of eyes and stiffness of flesh and tail.

(2) Claws, when pulled, should spring back and both the claws attached.

(3) They should smell fresh.

(4) They should be heavy in proportion to their size.

Shrimps (Crevettes Fr.) : There are several kinds and the best known are the brown and red shrimps. They are available all the year round in the coastal areas. They should have a fresh smell and not be sticky or slimy, which would indicate decay. They are mainly used for decorating fish dishes, making fish sauces, seafood cocktails, hors d'oeuvres, potted shrimps, omelettes, savouries, curries and pakodas.

Prawns (Crevettes Fr.) : There are two principal types, the sword prawn (Crevette rose) which is pink, and the common prawn (Crevette gris) which is pinkish grey and smaller. They are larger and more delicate in flavour than shrimps. Obtainable all the year round, usually cooked, with the same points of quality as for the shrimps. They are used in fish dishes, seafood cocktails, canape Moscovite, salad, hors d'oeuvres and curries.

Scampi (Langoustine Fr.) : Mainly a Mediterranean shellfish. The variety used here is the "Dublin Bay Prawn'.

They resemble a small lobster, about 8" long and only the tail part is used. Usually purchased in the frozen state as it is difficult to obtain fresh. Main uses are fried, e.g. meuniers, a l'Anglaise, a la Franchise, a l'Orly, or poached and finished, e.g. Newburg.

Lobster (Homard Fr.) : Probably the most highly rated of all shellfish.

(a) European...smooth shelled first three pairs of legs, terminating pincers.

(b) American/spiny or rock lobster.

Is of rich dark blue colour tinged with purple. Grows very slowly and when 5 years old will measure some five inches overall, having shed its tail some 20 times in the course of this period. Specimens of 12 to 14 inches long of 5 kg. have been known.

They should be alive when purchased and should have both the claws attached. Weight in relation to size is extremely important when purchasing. Parasites like barnacles clustered on the shell indicate age. The flesh of the cock lobster is thought superior to the flesh of the hen, Hen lobsters have a broader tail than cock lobsters.

Lobsters are served cold in cocktails, hors d'oeuvres, salads, sandwiches, and on buffets. Hot preparations include soup, grilled and served in numerous dishes with various sauces and also as garnish to fish dishes.

Crayfish (Ecrevisse Fr.) : A small freshwater lobster found in rivers. Average market length in 3" to 4". Of a dull green or brownish colour above and paler brown or yellowish below. They dig themselves into river banks by day and come out at night for food. Generally used for garnishing cold dishes for buffet work but any recipe given for lobster could be applied.

Crawfish (Langouste Fr.) : Greenish brown with yellow markings on tail. This turns to brick red after cooking. They have no claws but have long antennae. Owing to their size and

appearance, they are used mainly on cold buffets, but hot lobster recipes can be applied, points of quality and season would be the same as for lobster.

Crab (Crabe Fr.) : *A* shellfish of many varieties, fresh and salt water. Common varieties known as the edible crab. They should be alive when purchased and both the claws attached to the body. The claws should be large and heavy. There is usually more flesh on the hen crab, but it is considered of inferior quality to the cock. Used for hors d'oeuvres, cocktails, salads, dressed crab sandwiches and bounchees.

Scallop or Scollop (Coquille de St. Jacques Fr.) : *A* species of bivalve molluscs with ribbed rounded shells. When fresh, the shells are tightly closed and they should feel heavy for their size. Whilst they are normally very muddy, they should not smell unpleasant nor be covered with barnacles, which indicate age.

Method of cooking are usually poaching or frying, dressed back in the cleaned top half of the shell with a border of Duchesse potato piped around and finished with a sauce and garnish.

Mussels (Monies Fr.) : An edible mollusc, of which there are two main species, the common mussel from the cold regions and the large mussel from the Mediterranean. A distinction must also be made between wild mussels gathered in their natural beds and rocks, and cultivated mussels, which are tender and delicately flavoured, small but very plump.

Mussels can cause skin eruptions and digestive disturbances. However, as they are always cooked, dangerous microbes are destroyed.

When purchasing, shells should be tightly closed, large, and not having an excessive number of barnacles and no offensive smell. After purchase, they are best kept in a box of crushed ice with a damp sack, and placed in a cold room. Maybe served raw, hot, or cold, as hors d'oeuvres, cocktail or garnish.

In conclusion, it must be said that shellfish have a specific use and can be featured in luncheon menus also.

Cockles : These little shellfish are marine molluscs found mainly in the mouth of rivers. The white fish is enclosed in two concave shells of equal size, white in colour. Cockles are traditional delicacies.

They are raked up from sandbanks when the tide is out, and they are usually sold boiled and shelled. The flavour is not as delicate as other shellfish, but is interesting and savoury.

The type found on the Atlantic coast of France, called Bucardė-Sourdone, is edible. It is also known in France under the name of conque, sourdon and poor man's oyster.

Cockles are eaten raw and can also be prepared like mussels.

Winkle or Periwinkle (Bigorneau)

A name applied to many small gastropod molluscs. Winkle farms have been set up where winkle are cultivated.

Generally speaking winkles are eaten raw, but they can be cooked like cockles. To eat it, the winkle is extracted from its shell with a pin.

Clam : A name applied to many edible bivalve molluscs (hard-shelled animal with two halves) which live in slimy sand and are found at low tide. Some of those clams are cultivated in special beds along beaches. Auray and Roscoff are the principal French centres of the clam trade.

Clams are mainly eaten raw but they can be cooked in the same manner as mussels, oysters and many other shellfish. The North American people are very fond of this shellfish, and it is prepared in innumerable ways.

There are two types of clam—soft-shelled clams and hard-shelled clams.

Oyster : The oyster has been known to man from the earliest times. Nevertheless, there is no positive evidence that the Egyptians and Assyrians ever ate this delicious mollusc. It seems to have been unknown to the Jews, also there was no mention of them in the Bible. On the other hand, the Celts gathered oysters and fed on them abundantly.

In France, right up to the nineteenth century, it was thought that the coastal oyster-beds were inexhaustible. So much so that the Ordinance of 1681, which was enacted to protect mussels, left the oyster fisheries free from control.

The natural oyster-beds continued in all coastal areas and it became necessary to guard the oyster-beds in the bay.

Nutritive Value of Oysters : Raw oysters contain vitamins, phosphorus salts, chalk, iron, copper and manganese in a readily assimilable form. They also contain a very high proportion of iodine and are sometimes eaten to combat anaemia. The risks of infectious diseases being spread through the water contained in oysters is small.

Tradition has established the hygienic rule that oysters may not be eaten during the warm months of the year—that is to say, when there is no 'R' in the name of the month. Oysters are "in season" from September to May, although the idea that they are inedible during the summer months, the breeding period, is erroneous. Pacific coast oysters are sold all year long.

Various Types of Oyster : There are two types, the wild oyster, which has a very pronounced sea water flavour and the cultivated oyster. In the cultivation of oysters, seed-oysters are affixed to tiles and reared in farms some way out to sea. When they have reached a certain size, they are transferred to the fattening beds, which are always situated at the mouth of a river, the mixture of fresh and sea water being essential to induce overgrowth of the liver in which the fattening consists. Whitsable, Colchester and Helford are the chief British centres of oyster cultivation.

Storage : Oysters are stored in barrels or boxes, covered with crushed ice or damp sacks and kept in a cold room.

Fish should be stored for a long duration at -18°C to -24°C.

Quality and Purchasing Points

1. Oysters must be alive; this is indicated by the firmly closed shells.

2. They are graded in size and the price varies accordingly.
3. Oysters should smell fresh.
4. They should be purchased daily.
5. They are in season from September to May.

Indian Varieties : The production of fish in India as compared to other countries is rather low. The major fish of India are sardines, mackerel, prawns, tuna, etc.

Fish found along the Indian coasts, rivers, ponds, etc. are given below:

Name of Fish	*Found in*	*Remarks*
Sunera	Along Mumbai coast	The liver yields 67% oil.
Karanj	Mumbai and Thiruv-ananthapuram coast	Popular for its liver oil.
Nali	Coromandel coast	Oil is extracted from the liver, which is rich in Vitamin A.
Chamli	Cape Comorin,	The flesh is good food. An excellent food fish.
White or Rain-bow sardines	Chilka Lake, Mumbai	Small fish, tinned used for hors d'oeuvres and sandwiches.
Hilsa	West Bengal, Andhra, Tamil Nadu and West Mumbai region	Culinary excellence. It has a high fat content. Found in a coast smoked, canned and fresh condition.
Bhing Indian Mackerel	Inshore waters of India and all along the West coast	It is a scaly fish. It is much appreciated as food, and is in great demand. It is salted, cured or pickled or smoked.
White and Grey Pomfret	Along the coasts of India	They are much in demand and do not have a very fishy flavour. They are sold fresh, salted and cured.
Anchovies	Estuaries from India to Malaysia	They are small fish curried in oil or brine. Used as garnish.
Tunnies	Deep waters around Maldives and Lakshadweep, Kerala coast.	Cured and canned in oil or brine, exported as tuna fish. Its flesh tastes like meat and it is a very large fish.
Soles, Malabar soles	South coast	They are poor in fat content. Dried soles are sold in northern markets.
Cod	Off Indian coasts and in the Indian Ocean	It is round fish. Small cods are called codling. It is a big fish
Begti or Bhetki	Seas from India to China. Also in estuaries of large rivers	It is a popular fish in North India. It is an expensive fish.

contd....

Name of Fish	*Found in*	*Remarks*
Rohu	West Bengal.	Eaten in many parts of India, particularly in West Bengal. Sold in fresh form. Suitable for pond culture, it belongs to the carp family and is a valuable food fish.
Malli	Northern region rivers	It is a big fish, and is expensive. Good for curries.
Surmai	All Indian seas	Used for curries, pakodas, etc.
Indian Salmon	East and West coast	They are mostly consumed fresh. They are sometimes sun-dried, smoked or tinned.
Mumbai duck	Abundance around Mumbai seas and estuaries from India to China	It is a cheap fish and is eaten by the poor. Sold in a dried form. Has a very strong flavour.
Milk Fish	Seas of Indo-Pacific regions	Used for fish dishes.
White mullet Salmon Trouts and Trouts	Kashmir and Nilgiris.	It is a river water fish and cultivated on trout farms Average weight 120 gms. to 240 gms. Trouts are grilled, shallow fried or smoked.
Pholi	Fresh and brackish waters of India	The fish is relished both in fresh and dried state. Soup made from it is reported to be given in measles.
Shellfish Prawns (Jinga)	West coast, Malabar coast. Estuaries and backwaters in Kerala state, Chilka lake, etc.	Prawns are of different species, e.g. chapola jinga, bagda jinga, dhanbone jinga. Sizes vary from 13 cm. to 30 cm. Prawns are available in a fresh form. They may be shelled and sun-dried, smoked, canned, frozen, pickled etc.
Shrimps	Same as above. Inshore waters.	Smaller in size and fished along with the prawns, seldom exceed the length of 2.5 cm. They serve as a valuable food fish.
Rock lobsters or spiny lobsters (Crayfish)	On rocky bottoms from shallow waters to deep waters. Mumbai coasts. Tamil Nadu state.	Used for garnishes, decorating fish dishes. Attains a length of 30 cm. Sold fresh or in a frozen form.
Crabs (Kenkra)	Mumbai, Chennai, West Bengal coasts.	Crab soup is popular and other dishes are also made.
Molluscs		
Edible oysters,	Places round the Indian	Nutritious and delicious. Used

contd....

Name of Fish	*Found in*	*Remarks*
Clams	coasts which have brackish waters attached. Same as above.	for curries.
Sea mussels, Brown mussels	South of Quilon to cape comorin on the West coast and on the East coast.	Considered a delicacy by the coastal people.

Pre-preparation

Descaling and Cleaning

(1) Soaking the fish in cold water for a few minutes before descaling, helps in removing scales more easily.

(2) The blunt side of the knife should be used.

(3) The head of the fish is held with the left hand and holding the knife vertical, scraping is done starting from the tail, working towards the head, the scales are scraped off. The fish is then washed to remove any loose scales.

(4). Cut off the fins, remove the head. The entrails should be removed by cutting the length of the fish from the vent end to the head on the belly side.

Filleting : Cut the flesh along the line of the backbone and raise the fillet from the middle of the back, to the sides, first working towards the head, then the tall.

Skinning : Hold the tail end of the fish in the left hand, first sprinkling salt on the fingers for a good grip. Skin the flesh (skin side down) from tail to head, with quick short sawing strokes of a sharp knife. Point the knife blade towards the skin so that no flesh is wasted.

Cuts of Fish : The common cuts of fish are:

Fillets	Deboned, long fiat pieces of fish, without the skin. Flat-fish gives 4 fillets.
Paupiette	The fillet of fish is laid flat on the table, stuffing is spread on it, then it is rolled up, tied with a string, to keep the shape. Can be poached.

Supreme	Large fillets of fish cut on a slant of a large round or flat fish e.g. supreme of turbot.
Meuniere	Pan frying of fish, finished with butter noisette, chopped parsley and lemon juice.
Goujons and Goujonettes	Strips approx. 8 cm long by 1 cm. wide cut from fillets of fish. Goujonettes are cut smaller and are mainly used as a garnish. Can be crumbed and deep fried.
Tron, con	It is thick piece of fish 4-5 cm. thick, cut on the bone from a flat fish like the turbot. It is a fish cutlet or steak with bone.
Darne	A piece of fish cut across and through the bone of a large, whole round fish such as cod, salmon which is 2-3 cm. thick.
Mignon	Fillet of fish folded as a cornet (triangular fold as for piping bag).
Phe	Fillet flattened and folded in two.
Medallion	A medallion shaped portion from large fillets.
Fish Farce	Puree of fish, bound with whole eggs panada and cream.

Significant Liquors Used

Court Bouillon : To preserve and enhance the delicate flavours of fish, the liquors in which they are poached are prepared with care. The poaching liquor is called court bouillon. Different types of court bouillon are :

White Court Bouillon : Sliced onion, bay leaf, parsley stalk, peppercorns, juice of lemon, salt, milk and water.

Ordinary Court Bouillon : Ingredients—Vinegar, water, peppercorns, sliced carrots, onions, salt, parsley stalks, thyme and bay leaf.

White Wine Court Bouillon : Ingredients—White wine, water, minced onions, bouquet garni, salt, peppercorns, sliced carrots.

Red Wine Court Bouillon : Ingredients—Red wine, water, minced onions, bouquet garni, parsley stalks, salt, peppercorns and sliced carrots.

Method : For all the different court bouillons, the ingredients are simmered, strained and used for poaching, boiling, stewing, etc.

Particular Concerns of Application

The method of cooking depends upon the characteristics of the fish, its texture, size, form, percentage of fat and its strength of flavour. Over cooking of fish should be avoided. Oily fish are best for baking and grilling and lean fish are good for steaming, poaching, etc. Either lean or fish with fat are suitable for all types of frying,

Boiling : The method is suitable for the whole fish, e.g. Malli, Surmai, Indian Salmon and Trouts, etc. Thick cuts of fish could also be boiled. Whole fish cleaned could be put in court bouillon and boiled. The other method is to wrap the whole fish with cheese cloth (to prevent it from breaking), and to retain the shape, could be tied to a wooden plank. This is put into a pan, large enough for the fish and court bouillon used, as the cooking liquor. It should be sufficient to cover the fish and should be brought to the boil and then the temperature should be lowered and allowed to simmer. If the fish is cooked, then a skewer will easily pierce. 10-15 minutes per 500 gms of fish is enough to cook the fish.

Bisque : It is a rich classical soup made from shellfish, shells, tomato, brandy and cream.

Steaming : For steaming, the fish should be well seasoned and sprinkled with lemon juice to keep the flesh white. Greased, grease-proof paper should be put at the bottom of the steamer before cooking. This helps in lifting the flesh out and does not allow it to stick to the steamer, before cooking. The fish does not stick or break, and can be easily lifted. Cooking time varies from 10 minutes for fillets to 20 minutes for cutlets, and larger pieces. Steamed fish must be well drained before service and should be accompanied by a good, sharp, well seasoned sauce.

Poaching : This method of cooking is suitable for small

whole fish, cuts or fillets. Court bouillon should be used for poaching as it enhances the flavour of the fish.

Cooking liquor barely covers the fish, covered with grease-proof paper and poached in the oven. In a moderate oven, it takes 15-20 minutes and the cooking liquor could be used for making sauces.

Baking : Round white fish is suitable for baking. The fish should be cleaned, and the belly is stuffed with force meat (breadcrumbs, chopped onions, nutmeg, lemon rind, egg yolk, a little butter, seasoning) or any other stuffing and the edges are sewed. Butter is applied to the fish and then it is baked in a baking tray. Fish augratin and fish souffle are also baked. Baking time is 10 minutes per 500 gms.

Frying : Small whole round of fiat fish, goujons and fillets are suitable for deep frying. As the flesh is very delicate, the fish must be coated before frying. The following are the coatings normally used:

(1) Flour, egg white and breadcrumbs

(2) Milk and flour

(3) Flour batter

(4) Flour and beaten egg

Fillets of fish are always coated with flour and then dipped into any of the above protective coating and the flour helps the coat to stick. The cooking time is about 5 minutes.

Shallow Frying : Fillets, tron, cons, supremes, steaks or small fish are shallow fried. The term 'meuniere' is used when fish is shallow fried. The fish is coated with seasoned flour and shallow fried in butter. The cooked fish is masked with melted brown butter and garnished with fried or fresh parsley and lemon slices.

Grilling : It is a suitable method for various cuts and types of fish. They are seasoned, oil applied and cooked over heat on the grill or in a salamander. The grill should be greased, so that the fish does not stick. Only fish do not need extra

basting. Grilled fish is garnished with lemon and parsley and accompanied with a suitable sauce or butter. The time taken or grilling is about 5 to 7 minutes.

Stewing : Bouillabaise and Matelote are French regional stews in which a mixture of different fish are used. Pieces of different fish and shellfish with vegetables, herbs and fish stock are cooked and the cooking liquid is not thickened.

Fish Kettle : Long narrow, deep pan used for poaching.

Dishes of Fish

Pomfret

Pakodas : Dip pieces in garlic flavoured besan batter and deep fry, serve with pudina (mint) chutney.

Diable : Spread with mustard, sprinkle with breadcrumbs and grill. Serve with Sauce Ravigote.

Maitre d'Hotel : Serve parsley butter (*Maitre d'hotel Fr.)* with fried, poached or grilled fillets.

Flamande : Fillets poached in white wine, mushrooms, parsley, chopped shallots. Reduce the sauce, garnish with slices of lemon.

Portugaise : Poach in white wine, chopped onions, garlic, parsley and tomato concasse. Dish and coat with reduced sauce.

Trout (Fr. Truite)

Bretonne : Cook the fillets in butter and serve with finely minced mushrooms and salted shrimps.

Frite : Paner al'Anglaise, small fillets, deep fry and garnish with lemon slices.

Grenobloise : Coat the fillets in seasoned flour, cook in butter, pour brown melted butter (beurre noisette Fr.) garnish with lemon slices and capers.

Meuniere : Coat the fillets in seasoned flour, cook in butter, garnish with lemon slices, chopped parsley and melted brown butter poured on top.

***Aigulette Murat* :** Fillets cut into strips coated with seasoned flour and sauteed in butter, served with strips of artichoke bottoms and Parmentiere potatoes.

***Tandoori* :** Marinade in yoghurt, garlic, ginger, green chilli paste, orange colour, garam masala and salt. Skewer the fish and place in tandoor (mud coated oven and wood is the fuel used) and roast it. Apply hot butter and serve with onion rings and lemon slices.

Indian Salmon

***Coquille Mornay* :** Poached and placed in shells, coated with mornay sauce and bordered with duchesse potatoes and gratinated.

***Espagnole* :** Fillets cooked in butter placed on sauteed tomato concasse and served with fried onions and sauteed juliennes of capsicum.

***Ouglerce* :** Fillets poached in white wine, fish fumet, chopped onions, parsley and tomato concuss and remove the fillets, reduce the sauce, cook and mix with veloute sauce, coat the fillets with the sauce and serve cold.

***Bercy* :** Poach the fillets with chopped onions, parsley, white wine and fish stock. Reduce, add butter and flour (*Beurre manie Fr.*) cook, coat the fish and gratinate.

***Veronique* :** Poach fillets in fish stock, reduce, add butter, garnish with destoned muscat grapes coated with sauce and gratinated.

***Nori* :** Glossy purple black seaweed used in sushi and cold fish dishes. Indian Fish Dishes

Amritsari Machchi (Fish)

Fish covered with masala, i.e., amchoor, besan, red chillies powder, jeera, ginger and garlic seasonings.

The average serve of whole fish on a table d'hote menu is 200-250 gm.

Dishes of Eggs : The most popular and widely used eggs are of hens in India, although eggs of turkeys, guinea fowls, ducks and geese are also used.

The colour of egg shells and egg yolks varies with the breed of hen and the food they eat; but this makes no difference to the food value. They are rich in proteins and contain valuable amounts of iron and vitamins A & B. Eggs are nourishing and easily digestible. As such, they are given to invalids and infants. The ideal storage temperature for eggs is 2°C-5°C.

Eggs are essential to all kinds of cooking, not only in the preparation but as a food in their own right. They produce meals that are economical in price, and with a minimum of waste and time. By the term eggs, we mean those of the domestic hen, but these are not the only edible eggs. There are those of the ducks, geese, turkeys, also of wild birds such as plovers and gulls. Other eggs eaten as food are a variety of fish, i.e. (roe) herrings and sturgeon (caviars). Eggs have been described as a "Complete food" because they are highly nutritious and used in a range of recipes, yolks and eggs are used separately.

A Valuable Food : An egg is a compact and almost complete food. It is equal to seventy calories of energy and contains carbohydrates, fats, proteins, mineral salts, vitamins and water.

Vitamins A, B, D are all necessary for health and growth, Iron, phosphorus and calcium are essential for building and maintaining healthy bodies. These furnish all the essential amino acids for building and repairing body tissues. From the culinary point of view, an egg comprises two separate parts. The white or albumen and the yolk. The average egg is 10% shell, 30% yolk, 60% white; 73% water is contained in its constituents.

To test the freshness, place it in water. A bad egg will float, a good one will sink and varying degrees of freshness would be indicated between these two positions.

Uses : In all culinary preparations eggs are used for binding, colouring, enriching, increasing volume, coating, etc.

Strengthening an Egg White : This is done by adding a pinch of salt/Corn-flour/or arrowroot to the whites just before reaching their peak to stiffen the final result.

Storing : The shell of an egg is very porous and for this reason, eggs must be stored away from contaminating odours. Keep in a cool dry place. Ideal temperature: approx. 2°C-5°C.

Fresh eggs may be preserved and stored in water glass, a gelatinous odourless substance.

Note: The dirty grey rings which appear when hard-boiled eggs are sliced, are caused by the iron in the yolk and the sulphur in the white combining under certain conditions to form a black chemical compound called iron sulphide. This can easily be overcome if the right methods are followed.

Separating Whites and Yolks : If a little of the yolk gets into the white when separating, remove with a piece of the broken shell. If the whites are for whipping this is important, as the white will not whip, the slightest smear of grease on any of the utensils will also prevent the white from whipping to a stiff froth. Use only copper, stainless steel, or china bowls for whisking egg whites.

Cooking of Eggs: The basic principle of egg cooking centres around the ability of the proteins in both the yolk and the white to coagulate when heated. The rate of coagulation and the firmness of the gel depends on the intensity of heat, cooking period and the presence of other Ingredients such as milk, water, sugar and salt. Sugar increases the temperature at which coagulation takes place, and acid and salt speed up the process of coagulation. Extended cooking or high heat toughens eggs. While boiling eggs, if high temperature and long cooking is applied, the iron and sulphur compounds in the white are released to form ferrous sulphide. It is a blackish or purple ring around the yolk. To avoid this, as it spoils the appearance of the eggs, fresh eggs should be used and simmered. When the eggs are boiled, they should be placed in cold water, to release the pressure inside that discolours. The time required for soft-

boiled eggs is 3 to 5 minutes and hard-boiled eggs about 10 minutes.

The coagulation quality of eggs makes them usable for various purposes and thus they form an important raw material for cooking. They can also be combined with other ingredients to produce a variety of dishes.

Forms of Using : Various forms of using eggs in food preparation are:

Binding : Addition of eggs to minced meat and mashed vegetables, etc. helps to bind the mixture. As the heat coagulates, the proteins are bound into a cohesive mass. It helps to retain the shape of mutton croquettes, meat loaf, medallions, hamburgers, koftas, kababs, etc.

Coating : The egg or egg batter help to give a coat to the food items and prevent them from disintegrating and give them a protective coating. Many of the food items such as fish fillets, croquettes, cutlets, etc. are dipped into the batter before crumbing and then fried. Eggs are also used for preparing pancake batters (eggs, flour and milk).

Leavening : By beating the egg whites, a foam is made up of air bubbles, surrounded by a thin elastic film of egg white. This mixture, when added to products such as sponge cakes, meringues, souffles, etc. increases the volume and the egg white film hardens. Addition of sugar to egg whites, makes it stable, smooth and the foam does not collapse easily. Egg yolk has a less foaming power because of its fat content. Egg is used as the principle ingredient for chou paste from which eclairs, beignets, fritters and protleroles are made.

Emulsifying : Eggs form stable emulsions. For example, mayonnaise, oil and vinegar separate out unless oil droplets are coated with substance, that keeps them from separating. Eggs are the emulsifiers that give a smooth mayonnaise sauce. It is also used as an emulsifier in icecreams, cakes, cream puffs.

Eggs enhance colour and sheen.

An egg wash gives a shiny appearance to pastry and baked dough products.

Eggs also enrich and enhance the flavour and taste.

Cakes and farinaceous dishes are enriched by adding eggs, which improve the flavour and taste.

***Thickening*:** Eggs help to improve the consistency of gravies, curries, sauces and soups. Egg liaisons used in soups and sauces help to thicken and Improve the consistency. When used in custards, the heat coagulates the eggs and makes the custard firm.

***Garnishing and Decoration* :** Slices, sieved or quarters of boiled eggs are used to decorate or garnish dishes such as— salads, biryanis, curries, Vienna steaks, etc. For Consomme Xavier, threaded eggs are added as a garnish, and for prawn cocktails eggs are the garnish.

***Clarifying* :** Consommes are clarified with egg whites.

Various Courses in which Eggs are used:

***Hors d'oeuvre* :** Used for various hors d'oeuvres, such as egg mayonnaise, egg stuffed, egg ducklings, etc.

***Soups* :** In the preparation of consommes, thickening of veloutes, for garnishes such as slices of savoury custard (royal), threaded eggs (Xaviere).

***Eggs* :** They are placed as the third course before the fish. They are featured only on luncheon menus and not dinner. The egg dishes served are scrambled, Encocotte, Egg Portugaise, etc.

***Farinaceous* :** For raviolis, cannellonies, and noodles, eggs are used to prepare the paste.

***Fish* :** Batters are prepared out of eggs for the coating prior to crumbing.

***Meat and Poultry* :** Chicken cutlets, hamburgers, meat, beef, Vienna steaks, get their binding from eggs.

***Salads* :** Hard-boiled eggs are used in salads as main ingredient, border or as a garnish.

Sweets : Eggs bind and enrich the pudding and sweet dishes.

Savouries : Scotch woodcock, Welsh rare bit, cheese souffle, devilled eggs, are all made with eggs.

Cooking Method

Boiling : Lower the egg in boiling water without cracking shells— four minutes for soft boiled eggs and ten to fifteen minutes for hard boiled eggs. Cool quickly if it is to be served cold. It prevents green rim appearing around the set yolk and stops egg cooking.

Poaching : The egg is cooked in simmering water to which a little salt or vinegar has been added to help set the yolk more quickly. Break the egg and lower carefully in the water and use an egg ring to keep shape.

Scrambling : Beaten egg and milk is cooked over a low heat in a little melted butter until set, but not too firm. Stir the mixture constantly to distribute the white and yolk throughout. Add salt after cooking to prevent drying and toughening of the protein.

Frying : Preheat the pan to a moderate temperature and cook the egg gently to avoid the bottom browning or catching while the yolk remain unset. Never add eggs to a hot pan or the whites will cook quickly.

Omelette : Stir mixture to facilitate even cooking so that the eggs in centre and edge cook evenly.

Baking : Place egg in a ramekin dish, cover with cream to prevent drying. Cook in moderate oven for 6 to 8 mins.

Cooking Cereals and Dals : Rice and pasta feature in many cuisines as either a staple food or because of their bland taste, as a traditional accompaniment to spicy or strongly flavoured dishes. As both are economical items to either purchase or make, they can be included in various recipes at a reasonable cost, which is helpful to the caterer. As pasta is available in

many forms it can be used with some flexibility in a range of sweet and savoury food.

Rice : It is a cereal and is available in the following forms:-

1. *White Rice :* It is a popular variety and is purchased in both grain for desserts and long grain for main dishes. It has a bland and soft texture.
2. *Brown Rice :* The total grain, inclusive of the outer husk which makes it more nutritious alternative to white rice- It is available in both short and long grain and requires a slightly longer cooking time to soften the husk, although it will never achieve the same softness as white rice. It has a nutty taste and slightly chewy on the palate.

Cooking Rice: The objective when cooking rice is to keep the grains separate, thereby producing a light fluffy result. When the rice cooks, it absorbs water and expands, while at the same time releasing starch into the water. As a general rule, rice multiplies three times during the cooking which implies that careful calculation should be used to achieve the appropriate quantity of cooked rice.

Boiling Rice: The objective when cooking is to keep grains separate and getting a light fluffy result. When rice cooks, it absorbs water and expands, while at the same time releasing starch into the water.

Two methods can be used for boiling rice, one method is to throw away the excess water by straining the rice. The second method is the absorption method by using less water, a ratio of two to one.

Different dishes are made of rice, example: rice soup, fried rice, rice pilaff, mexican rice, etc.

Using Rice in Recipes : Rice can also be used in some of the following preparations:-

Puddings: Uncooked rice is placed in a casserole and an egg custard mixture is poured over the top. It is cooked slowly

for around one and a half to two hours in a water bath until the custard is set.

Rice Salad: Cool boiled rice is mixed with other ingredients and in a well flavoured dressing. The salad could be made in advance or made with leftovers.

Binding: For meat loaves, meat balls or croquettes, kofta etc, cooked rice can be added to the rice to thicken, bind or to give substance. Rice flour is used to thicken or bind dishes.

Filling for Vegetables: E.g. Stuffed peppers, tomatoes, zucchini (turai) dolmades or as an ingredient for stuffings of roast meat.

Cooking Pasta : Allow plenty of water in a large saucepan about eight times the amount of pasta being cooked. Pasta tends to double during the cooking.

Add a good pinch of salt to the water along with a little oil, and bring to a rolling boil. Add the pasta and cook for fifteen to twenty minutes until, 'aldente' that is it is firm when bitten through the centre. Lasagna sheets should be popped into the boiling water individually and allowed to cook separately for a minute to prevent sticking. Once the pasta is cooked, it is drained and rinsed in either hot or cold water, depending on its intended use.

Avoid leaving the pasta in a colander or it will tend to stick together on cooling. It should be served immediately.

Pasta can be reheated in the same way as rice, that is by steaming or in the microwave. It will last for one or two days once cooked and can be frozen as part of a prepared recipe for several months.

There are many sauces that are traditionally served with pasta, ranging from the popular Bolognese (rich meat sauce) or marinara (with seafood) to carbonara (with beaten egg, bacon and cream).

The variety of sauces used is almost endless, like the pastas available. Sauces are used to escalate the nutritional value of the dish.

Pasta, like rice is a high carbohydrate food and the addition of a protein based sauce consisting of meat, fish, chicken, or eggs will instantly boost its balance of nutrients.

The predominant flavour in many pasta dishes is Italian in influence and garlic, tomatoes and herbs, such as oregano or basil feature regularly.

Catering Considerations : Pasta can be used as a substitute for potato in a recipe and it is a filling dish, like rice. Serve only one course of pasta. Avoid serving large quantities. Pasta can be used as an ingredient in soups, salads, entrees, and casseroles as well as the traditional recipes. Allow 40 gms of uncooked pasta per person for main course 20 gms per person for a first course.

Catering Considerations : When served as an accompaniment to casseroles, curries or other foods containing a sauce it is used to mop up the sauce that would otherwise run across the plate.

When planning menus it is wise to remember that rice can be used as substitute for potatoes, should never be served together in the same meal. Avoid repeating rice in courses. It is ideal in all seasons and can be served as a salad in summer or as a pilaff in winter.

Pasta : It is a versatile food that can be an ingredient in many courses, although it is usually associated with savoury dishes. Home made pasta is easily prepared from a simple dough and if handled correctly, can provide a flavour some alternative to the commercial varieties. Each type of pasta differs in size, shape and sometimes colour. The addition of various ingredients like finely chopped spinach (green) tomato paste (red) and carrot juice (orange) will produce different effects. Own pasta can be made in a pasta machine.

The main ingredient of pasta is flour, which should be a strong variety, usually made from durum wheat. Wholemeal flour is proving a popular alternative to the plain dough although it should be remembered that extra water may be

required to produce a soft dough and kneaded lightly. Eggs are sometimes added to a smooth dough. Water and oil are to be added to the flour which is then mixed to a smooth dough and kneaded lightly. Eggs are sometimes added to the basic dough for flavour and colour.

Once the pasta has been prepared, it is rolled out thinly. Cut into the desired shapes and left to dry out before cooking. When it is completely dried out, it can be stored in an airtight container.

Varieties of Pasta

Bows : Squares of paste cut with a crimped edge and pinched in the centre.

Shells : Resembles a seashell with a wide, curved side.

Macaroni: Small thin tubes with a slight curve.

Twists : Two thin strands of pasta that are twisted together.

Spirals : Single, wider strands of pasta that are twisted.

Tagliatellel/Fettuccine: Long, flat, ribbon shapes of varying widths.

Rigatoni : Wider tubes than macaroni and have a ridged edge.

Wheels : Circles that have a spoked pattern through them.

Ravioli : Small squares of pasta that have a savoury filling between two layers of pasta.

Spaghetti: Long thin strands.

Vermicelli: Thin strands that are wound together into nests.

Varieties of Meat : In order to know how the animals are slaughtered and to buy meat sold in a sanitary condition, it is important to know how the animals are slaughtered in hygienic conditions. Meat bought must be from a slaughter house that has professionals working.

Slaughtering and Butchery : In Europe slaughter houses are called abattoirs. The slaughter houses have space for meat

storage and deal with large amounts of offals. A slaughter house is functional, compact in space, has satisfactory working conditions, proper equipment and has provision for the highest sanitary standards.

The animals are slaughtered by professionals in a systematic way.

Pre-slaughter Steps

Inspection : Animals are inspected by medical inspectors to assess whether they are fit for human consumption; then only, they are deemed fit for slaughtering.

Resting of Animals : The animals are kept in clean dry enclosures (pens) to rest for 24 hours. The animals are given plenty of water during resting,

(a) It helps to maintain normal heartbeat of the animals.

(b) It helps in the drainage of blood.

(c) It helps to have normal body temperature as it increases the keeping quality of the carcass and minimises the bacterial growth.

(d) It speeds up the glycogen stored in the muscle to turn to lactic acid, which lower the pH of the muscles and reduce the growth of bacteria and act as a natural preservative.

(e) It shortens the rigour mortis period. (The carcass is hung in cool temperature during the period, the muscles become stiff. Certain enzymes/acids are produced that soften the meat.) Technically It is called ripening or ageing. The meat is tenderised because of the action of the acids. In a fatigued animal the rigour mortis period is longer, whereas in a rested animal it is short.

Fasting : Animals are fasted for at least 8 to 12 hours.

(a) It helps to reduce the amount of undigested food and faeces in the intestinal tract.

(b) Bacterial infection are minimised.

(c) It improves the quality of the meat. The blood left in the tissues is free from absorbed nutrients.

Washing : After resting and fasting, the animals are passed through a tunnel containing showers from top and bottom with lukewarm water and the animal is kept on a sliding platform and is washed from all sides and superficial dirt is removed.

Stunning : It is to render the animal unconscious so that it may bleed freely and it may easily be killed. It is also done to avoid pain to the animal, while it is being killed.

Different ways of stunning are:

(a) *Knocking:* The animal is hit with a heavy hammer (4 lbs.) on the head to make it unconscious. Not a very popular method nowadays.

(b) *Carbon Dioxide Chamber:* The animals are made unconscious by putting them in these chambers, but not popular as the weak animals may die.

(c) *Cartridge Pistols/Captive Bolt:* This is done by damaging the brain by placing the pistol between the eyes, this renders the animal unconscious.

(d) *Electric Tongs:* This is the most popular method. The ends of tongs are placed behind the ears and a current (60-70 volts) is passed. This renders the animal unconscious for about 15-20 minutes.

Sticking : Sticking is done with a knife. 6 inches long blade and it is spear pointed. It is done soon after stunning when the rate of the heartbeat and the blood pressure are high. This assists bleeding. In India, Muslims and some other communities take meat from animals slaughtered" by the Halal method in which throat is quickly cut with a sharp knife with short strokes and bled. The jugular veins and carotid arteries are cut. Sikhs only accept meat which is slaughtered by the Jhatka method, i.e. the head is severed from the body by a single blow from a long, heavy, butcher's knife.

Bleeding : The animals are bled, to the fullest possible

extent. The animals are hung on hooks with their heads downwards, as neither the animal nor the slaughterer comes in contact with the blood. Drains for the blood lead outside the building either to a blood collecting tank or to a processing room.

Flaying and Cleaning : Air is blown between the flesh and skin to create space and the skin is pulled off. The stomach and intestines are removed, inspected and sent to the tripery section. Offals are removed and sent to the offal section.

The carcass is now split into hindquarters and forequarters or left whole and sent to the butchery section.

Meat : The different types of meat are: Lamb, Mutton, Beef, Veal, Pork and Game meat.

In order to cook meat properly, it is important to know the composition and the structure of meat.

Muscle tissue is composed of very minute fibres bound together in bundles. They are tiny tubes filled with water, in which are held the mineral constituents, vitamins and emulsified fat, protein of meat. The muscle fibres are held by connective tissues which are of two kinds. The white connective tissue is collagen, and the yellow connective tissue is elastin.

Qualitative Elements

Fat : The fat present in meat indicates the quality, as it gives a good flavour. Fat in animals is found around the music fibres or between muscle fibres. It is interspersed between the muscle fibres. The fat seen as small flecks of fat within the muscle is known as marbling and enhances the flavour of meat. Fat contributes moistness to the meat, makes it tender, enhances flavour and makes it succulent.

Colour of the Meat : Older animals have darker coloured meat and a yellowish fat, whereas the younger animals have a lighter coloured meat with a firm, dry, creamy white fat.

Tenderness : The tenderness of meat depends upon the

food the animal is fed on. The finer the grains, the more tender is the meat. The amount of connective tissue is directly related to the tenderness of meat. Cuts of meat with much connective tissue are tougher than those containing little connective tissue. The least used muscles such as are found in the loin and rib cuts, are more tender than those that are fully developed such as those found in neck, chuck, breast, etc.

Factors that Make Meat Tender

Moist Heat: White connective tissue called collagen changes into gelatine when moist heat is applied to it. Yellow connective tissue is called elastin, and this does not change by cooking or marinading. It should be either discarded or finely minced or chopped, which breaks it down.

Tenderisers: When used, tenderisers make the meat tender. Commercial tenderisers contain proteolytic enzyme, such as papain, an enzyme found in raw papaya and its leaf. Raw papaya paste applied to meat helps to tenderise it. Acidic foods like vinegar, tomatoes, curds, lemon, tamarind help to tenderise the food also.

Ripening or Ageing of Meat: Meat should not be eaten immediately after killing as rigour mortis sets in and stiffens the muscles. The muscle is hung in cool conditions 1°-2°C (34°-36°), the time period varies depending upon the type of meat, i.e. beef, mutton etc. There is an improvement in tenderness, flavour moisture and colour in the meat. Veal and pork are not hung.

Marinating: Tough meats soaked in an acidic solution, i.e., wine, vinegar, curds make the meat tender and enhance the flavour.

Mechanical Pounding and Grinding: Mechanical pounding and grinding is done to meat to break down the connective tissue. Grinding breaks and cuts the muscle fibres and connective tissue making it possible for all ground meat to be prepared in a fashion similar to that used for tender cuts. Pounding is used to tenderise the meat; this process breaks and tears only the surface meat fibres and connective tissue.

Cooking Methods : The method of cooking which is most suitable for a particular piece of meat is decided by the kind of animal from which it comes, its condition, age, quality, its fat content, the cut of the animals and the size of the cut and its tenderness. The tenderness and the amount of fat in the tissues influence cooking method and eating quality. Developed muscle fibres usually take longer to cook. A suitable cooking method should be used for various cuts of meat, that preserve nutrients, develop tenderness, preserve juiciness, thus resulting in the least shrinkage.

Moist Heat : Braising, boiling or stewing are methods of cooking in which liquid is the medium of cooking. Collagen in the meat, when subjected to moist heat, changes into gelatine. For braising , the meat is browned and then placed in the oven between 250°-300°F for cooking. Medium heat makes the meat tender.

Steaming : Meat cooked in a steamer is less tender and juicy and does not have such a good flavour. The moisture in the tissues of the cuts of meat changes the collagen or connective tissue into gelatine and tenderises the meat.

Dry Heat Roasting : For roasting, a constant temperature of 250°-300°F should be used, till the internal temperature of the meat has reached 180°F, as it makes meat palatable and enhances flavour and there is less shrinkage. After it is cooked, the meat should be browned by increasing the oven temperature. A meat thermometer helps in determining the degree

Approximate Roasting Time for Meat

Beef	—	15 to 20 minutes per 455 gms. and 20 mins. over.
Mutton	—	25 mins. per 455 gms. and 20 mins. over.
Lamb	—	20 mins. per 455 gms. and 20 mins. over.
Veal	—	25 mins. per 455 gms. and 20 mins. over.
Pork	—	25 mins. per 455 gms. and 25 mins. over.

Internal Temperature Recommended for Roasting Meat

Beef	*Mutton*
Rare 60°C (140°F)	60°C (140°F)
Medium 71°C (160°F)	71°C (160°F)
Well done 82°-85°C	82°-85°C
(180°-185°F)	(180°-185°C)
All pork meat is well done.	

Grilling : Tender cuts of meat are used for grilling such as cutlets, boti kababs, seekh kababs, chops, beef, ham steaks and bacon. The temperature and distance of the grid on which meat is cooked should be varied according to the thickness of the meat. A thin piece can be grilled by turning only once and thick pieces, a number of times. Oiling aids cooking and browning the meat helps the appearance, by giving a sheen to the surface. Steaks and chops 2½ cm. to 4 cm. thick should be cooked in moderate heat. A steak 1½ cm. and 2 cm. thick should be exposed to higher temperature. To test whether it is cooked, the meat should be pressed. If it resists, then it is cooked. The pressure brings the juices to the surface and their colour is a guide.

Hamburgers are usually griddled, i.e. placed on a flat hot surface, with a little fat and cooked on both sides. The hamburgers are placed on one side and allowed to brown and when the juices rise to the top, they should be turned on the other side and cooked.

Only those meats are fried that have a protective coating and binding, e.g. meat croquettes, meat balls, cutlets, scotch eggs, etc.

Judging the Quality

Beef

(1) Fresh meat must be hung to allow it to become tender. The colour darkens after it has been hung.

(2) Lean meat should be bright red, with small flecks of fat interspersed in the muscle (marbled).

(3) The fat should be firm, brittle in texture, creamy white in colour and odourless.

Veal

(1) The flesh should be pale, pink, firm, not soft or flabby.

(2) Cut surfaces should be moist.

(3) Bones in young animals are pinkish, white, porous and with a very small amount of blood in their structure.

Lamb

(1) Carcass should be compact and evenly fleshed, having an even coat of fat.

(2) Lean flesh is firm and of a pleasing, dull red colour and of a fine texture of grain.

(3) Fat should be evenly distributed, hard, brittle, flaky and clear white in colour.

(4) The bones are porous in young animals; as age progresses, they are smooth, white and brittle.

Pork

(1) Lean flesh should be pale pink changing to rose as the animal matures.

(2) The fat is white, firm, smooth and not excessive.

(3) Bones must be small, slender and pinkish.

(4) The fat ought to be white, smooth and not excessive in proportion to the bacon.

(5) The lean meat should be deep pink in colour and firm.

Liver/Offal

(1) The liver of animals should be fresh and not dry.

(2) Should be smooth in texture and not contain tubers in the flesh.

(3) There should be pleasant odour.

Kidneys (1) Should be fresh.

(2) Certain amount of fat should be attached to the kidneys as it keeps them moist.

(3) The fat should be crisp and not sticky.

Tongue (1) Lamb or ox-tongue should be fresh.

(2) There must not be an excessive amount of waste at the root end.

Sweetbreads (1) They should be creamy, white in colour.

(2) They should have a pleasant smell.

(3) They should be fleshy and large in size.

Oxtail (1) They should be of good size and lean.

(2) There should be no signs of stickiness.

Head (1) Pig's and calfs head should not be sticky.

(2) They should be well fleshed and odourless.

Lamb/Mutton : Lamb is the meat of a young sheep, mutton the flesh of the mature sheep or goat. *Lamb* usually is the flesh of animals not more than 14 months of age. The average weight of an imported lamb is approx. 32

lbs. (16 kg.), and for mutton, it is upto 50 lbs. (25 kg.). The average weight of an Indian lamb is 10 kg. to 14 kg. For mutton, it is between 20 kg. to 22 kg.

Dissection of Joints

Lambs/Mutton

1. Remove the shoulder.
2. Remove the breasts.
3. Remove the middle neck.
4. Remove the scrag-end.
5. Remove the legs.
6. Divide the best end from loin. (2 loins attached is called saddle.)

The joint that has two legs and the saddle attached, and when it is cooked whole is called a baron of lamb.

Some of the dishes prepared with small cuts of lamb of mutton, are obtained from certain main joints as follows:

Lamb or Mutton Chop

1. The prepared loin is cut into chops approx. 4 cm. thick and trimmed. Grilled or fried.
2. The chops are taken from the leg end of the loin. Trimmed, usually are braised, roasted or fried.
3. It is obtained from the saddle and roasted. The dish, crown roast is very popular. It looks interesting with the paper frills placed on the ends of the bones.
4. The chump end of the loin is removed along with the fillets. The loin is then deboned and thick slices are cut from the meat slantwise. Flattened, trimmed, to cutlet shape and then roasted or grilled.

For Indian dishes, the cuts of meat used are not the same as above. No specific joints are used except for chops. Kashmiri chops, and chop covered with mashed potatoes are popular; for the Kashmiri dish "Kabargah" chops are also used. For kormas, roghanjosh, kaliyas, etc. (Indian curries) any part of mutton is used. Minced meat is used for koftas (meat balls), kababs, shami kababs, seekh kababs and mutton cutlets, etc. When deboned meat is to be used in the form of juliennes or dices, the legs or shoulders are used, e.g. jhalfrezi, mutton do peaza, etc. For biryanis and pulaos, any cut of the meat is used.

Pork : The flesh of a pig is called pork. Most of the pork meat comes from animals not more than a year old. It generally has more fat than other meats. A suckling pig is about 5 to 6 weeks old.

First quality joints of pig are the leg and loin of pork.

Order of Dissection

1. Remove the head.

2. Remove the trotters.
3. Remove the legs.
4. Remove the shoulder.
5. Remove the spare ribs.
6. Divide the loin from the belly.

Ham : It is taken from the hind leg of the pig, preserved by curing or pickling in brine, then dried and smoked. Ham is prepared from fresh pork meat. Boiling of ham is done by placing it in cold water first and soaking it. Then it is placed in a suitable container covered with cold water and brought to boil slowly, till it is cooked. (Cooking time is 18 to 20 minutes for 455 gms. of ham.) Hams after boiling can be decorated with chaud froid sauce. Gammon is taken from the leg of the pig that has been reared for bacon, and the meat is cured. Gammons are mild and do not keep as long as true hams.

Bacon : Bacon is obtained from the sides and back of a baconer (a pig reared and specially fed to yield bacon). The bacon is acquired by taking the meat and preserving the meat by salting. After the meat has matured, the bacon is sold as "green" bacon. Smoked bacon is very popular and this is done in smoke chambers, which gives it a distinctive flavour, and helps to preserve it for a longer time. The wood that gives good flavour are from the breech tree, or oak tree, etc.

Beef : Beef and veal are got from the bovine animals and they are classified by their ages-

Veal	—	Flesh of calf (less than three months of age) which lives on milk.
Calves	—	The animals are from 3 to 8 months old.
Beef	—	The meat is taken when the animal is above 8 months.

Beef is the most popular of all the edible meats in the Western countries. In India, beef is not very popular and buffaloes are slaughtered in some places and sold as beef.

Beef is the flesh of steers, heifers, cows, bulls and stags. The age and the sex has an influence on the taste and quality of meat.

Order of Dissection

Side: The carcass is divided into two lengthwise. For Hindquarters:

1. The whole side is divided between the wing ribs and the foreribs.
2. Remove the rump suet and kidney.
3. Divide the loin and rump from the leg (topside, silverside, thick flank and shin).
4. Remove the fillet.
5. Divide the rump from the sirloin.
6. Remove the wing ribs.
7. Remove the shin.
8. Bone out the pitch bone.
9. Divide the leg into three remaining joints (silverside, topside and thick flank).

For Forequarters:

1. Remove the shank.
2. Divide in half down the centre.
3. Take off the foreribs.
4. Divide into joins.

Average Weights of Joints and Preparation of Beef

Joints	*French Terms*	*Weight*	*Methods of Cooking*	*Preparation*	*Yield Net Weight of Cooked Meat*
1	2	3	4	5	6
Sticking piece	Collier	10½	Stewing	Used for preparing pies, and broths beef tea.	50%
Chuck ribs	Cotes	5 kg.	Braising and stewing	Beef tea, pies.	50%

Contd...

Joints	*French Terms*	*Weight*	*Methods of Cooking*	*Preparation*	*Yield Net Weight of Cooked Meat*
1	2	3	4	5	6
Mid ribs	Cotes	8 kg.	Roasting and braising	Could be roasted on the bone	50%
Foreribs	Cotes	6¼	Roasting and braising	Cutting 4 ribs from the ribs end removing.	50%
Wing ribs	Cotes de boeuf	6 kg.	Roasting	Roasted with the bone, gristle removed and tied with a string to keep shape.	
Sirloin or half saddle	Alloyau de boeuf	9½	Roasting and poeling	From the sirloin comes the whole fillet. The porter house steak, entrecote and the contre fillet. The fillet is removed, which lies along the bone on the inside of the meat. The sir-loin is then deboned, gristle removed and tied to retain shape.	50%
Boned sirloin (Undercut)	Contre		Roasting, grilling and poeling	The joint is completely deboned after removing the fillet. It can be larded and tied with a string.	50%
Fillet	Fillet de boeuf	3 kg.	Roasting, grilling and poeling	The fillet is the best part of the animal. The thin silvery skin should be removed. All sur-plus fat is trimmed and the sinew is cut away which lies along the top of the fillet. Pique with larding bacon allowing 66 gms. of fat for 455 gms. of lean meat.	60%
Rump	Culotte de boeuf	7 kg.	Roasting braising, steaks for grilling	A part of fillet lies on one side of the bone. Debone, trim and cut straight across.	50%
Thick flank	Tranche	6½ kg.	Braising, steaks prepared for braising, stewing.	The joint is cut into two equal joints, placing slices of beef fat over and tied with string. For steaks, cut into 1½ thick slices, approx. 130 gms. For Paupiettes, i.e. steak puddings, carbonades, goulashes the meat is used from this joint.	35%
Silverside	Gite a" la noix	10 kg.	Braising, pickling	Meat for pickling is placed in a brine tub and soaked for 5 to 6 days. It is washed before cooking. If the meat is thick, inject brine into the meat.	55% (Boiled)
Topside	Tranche tendre	7½ kg.	Braising stewing, roasting	Cut into pieces and remove excess fat layer. Lean meat should be larded. Meat can be minced. It is the leanest part of the buttock end. 1st class braising joint, 2nd class roasting joint.	55% (Roasting)
Thin flank	Bavette	4 kg.	Stewing	Fat is removed, cut into 2" cubes or as required. It is a cheap cut. Pressed beef is prepared.	35%

Contd...

Joints	*French Terms*	*Weight*	*Methods of Cooking*	*Preparation*	*Yield Net Weight of Cooked Meat*
1	2	3	4	5	6
Plate	Poitrine	5¼ kg.	Boiling, stewing	Used for the manufacture of sausages.	55%
Brisket	Poitrine	6½ kg.	Pickling and pressed beef boiling	Also used for the manufacture of sausages. It is cheap cut with plenty of fat. Used in the preparation of corned beef.	50%
Leg of Mutton	Tallon du Collier	11 kg.	Braising, stewing	Steaks are prepared and braised.	55%
Shin and Shank	Jambe	9 kg.	Boiling	Remove fat and gristle and cut into 3" pieces. Mince meat and use bone for stock. Beef tea is prepared. Beef Terms	

Beef Terms

Carcass — Whole animal after dressing from the slaughter house.

Side — A carcass that has been split down the centre of the spinal column.

Quarter — A side that has been sub-divided into two sections as—Forequarters—Hindquarters.

Crop — Forequarters minus the plate, brisket and shin.

Baron — A complete back (uncut) which is a pair of sirloins undivided.

Offals — Products derived from the process of slaughtering other than the carcass, the edible internal organs, e.g. liver, heart, sweet-bread, head, tongue, tail, brains, kidneys.

Steaks : Steak is a slice of beef varying in thickness, usually

taken from fillet, rump or contre fillet. Steaks are tender and have a good flavour and are very popular. They are usually cut into individual portions, that are thick enough to broil well and to develop a crisp well browned exterior, while retaining a juicy interior. Steaks are obtained from different parts of the animals such as:

Entrecote Steaks : These steaks are taken from the deboned sirloin without the fillet. This joint, known as contre fillet, undercut (faux fillet). From this contre, fillet are cut those famous steaks like Minute Steaks (Fr. Entrecote Minute), Sirloin (Fr. Entrecote), Double Sirloin Steaks (Fr. Entrecote Double).

One cm. slices are cut from the flesh of contre fillet. They are thin steaks, flattened and weigh approximately 115 gms. Sirloin steak (Fr. Entrecote). It is a slice of steak, often taken from contre fillet. It is a little thicker than minute steak, 1.25 cm. thick, approximate weight is 150 gms.

Double Sirloin Steaks (Fr. Entrecote Double) : It is a steak bigger and heavier than Entrecote steak. It is 2½ cm. thick and the approximate weight is 250-300 gms.

Porter House Steak and T-bone Steak : Like the Entrecote, it is cut from the sirloin of beef, usually from the best part of the loin, 3 or 4 cm. from where the fillet begins. It is cut or sawn right through including bone, loin and fillet. It is approximately 4 cm. thick and weighs about 650 gms. This is a nerous portion and is large enough for 2 portions. T-bone steak is an individual porter house steak.

Fillet Steaks : Special names have been given to the steaks cut from the different areas of the beef fillet.

Chateaubriand : It is a large steak, cut from the head of the fillet, which has been trimmed of nerves, etc. Approximate weight is 250 gms. to 1 kg. It is 3.5 cm. to 8 cm thick.

Fillet Steaks: It is cut from the upper middle part of the prepared fillet. The steaks are 2 cm. thick and the approximate weight is 100 gms. to 150 gms.

Tournedos : A round barrel shaped steak cut from the lower middle of the beef fillet. It is 2.3 cm. thick and weights approximately 100 gms.

Fillet Mignons : It is prepared from the tail of the fillet, weighs approximately 50 gms. It is small fillet. The ends of tail at one time were used to prepare carpet-bag steaks.

Steaks from other Parts of the Animal

Point Steak : It is approximately 200 gms. in weight and is cut from the triangular part of the rump which is generally 2 cm. thick.

Tartare Steak : Finely chopped fillet of beef, finely chopped onion, parsley and capers. This is the only steak that is served raw.

The above ingredients are mixed, well moulded into flat cakes, a depression is made in the centre of the cake with an egg yolk placed in it, and served garnished with neat bouquets of onion, parsley and capers all round the steak.

Cooking and Serving of Steaks : Steaks are grilled or sauteed. For thicker steaks, a lower temperature (325°F) should be used for cooking. Steaks could be cooked under done as rare and medium done. The cooking time of steaks varies according to the thickness and size of the steak.

Different garnishes could be served with the steaks. They should be accompanied by a suitable sauce, e.g. sauce bearnaise, sauce diable or chateaubrai and sauce.

A Popular Form : The origin of this popular form of meat food is difficult to trace. It was over two thousand years ago that the Romans enjoyed this delicacy and various forms were known as Tomacina, Circelli, etc. The word "sausage" can trace its ancestry to the Latin salsus (salted) which might suggest that the original type was prepared from cured meats, which could be preserved for a longer time. The ancient town of Salamis provided a place, association for salami and its fame evidently spread, as we now have many types; Italian, French

and German Salami, usually with a prefix indicating the locality of origin.

Crepinettes are flat sausages made from fresh pork and wrapped in caul. Sausages can be considered under three main sections, though there will be, of course, some overlapping between the sections.

Composition : Sausages consist of forcemeat prepared with minced pork, poultry, veal and lamb.

Casings : The intestines of sheep, hog and cattle according to the required diameter. The natural casings are available but require a great deal of labour to be processed satisfactorily.

Cellulosic or artificial intestines are made of edible or inedible cellulosic material. They are available in all sizes and are adapted to specific uses. The casings can stand boiling temperature and they are porous and let the smoke go through.

Forcemeat : Pork is used as a basic ingredient for forcemeat stuffing. Pork is minced, ground, chopped or diced into what is called "Forcemeat". Every forcemeat has a binding such as rice, potatoes, panada or fat which holds meat. It is used for the preparation of many speciality items and is also used as stuffings or garnishes for meat, eggs, fish, poultry, game or vegetables.

The composition of forcemeat is made up of 40% dominant flavour meat, i.e. chicken, meat, lambs; 30% pork meat—it gives flavour and moisture; 30% pork fat imparts tenderness, binds and contributes to flavour.

To this mixture herbs, spices, nuts and seasoning are added.

The forcemeat is used to prepare terrines, pates, sausages, galantines.

Fresh Sausages : As they are a perishable product, speed in production and distribution is essential. Pork, beef and veal meat is used. Some of the sausages have cereals added to the meat and seasoning, e.g. chipolatas, bokwurst.

Cooked Sausages : Beef, veal, ham, tongue and pork are

used to prepare the sausages. Sometimes a binder is used. The minced meat is cooked and then stuffed into casings. Liver sausages, black pudding, Frankfurters, etc.

Dry Sausages : Pork is used, which is an important constituent. The Food Regulations necessitated the pork to be heated to a temperature not lower than 140°F or refrigerated not below 48°F as a precaution. Dry curing is done after pickling and then smoked. The dry sausages are salami, mortadella.

Ingredients : Standard quality ingredients should be used, with proper seasoning and binder. Fresh beef and pork have greater binding qualities than the frozen variety. Good quality meat should be used, pork fat is added, if necessary. The internal fats have a higher melting point. The fats should be absorbed by the binder or the excess oil will collect under the skin of the sausage and be detrimental not only to the appearance but also to the taste.

A product containing no lactose or milk fat but containing 85% Protein was found and it is known as gervita. By using gervita in sausage, the fat contained therein is in a stable form and the loss on cooking is reduced. Another important aspect is that it swells and minimises shrinkage. The meat used must be finely minced.

Binders : Such as bread, rusk, rice, flour, farina (potato flour) are used in small quantities.

Flavouring and seasoning ingredients used are cayenne pepper powder, paprika, allspice, cinammon, cloves, pistachio nuts, mustard powder, sage, thyme, etc. Salt, vinegar, saltpetre, herbs act as preservatives. The mixing of the ingredients should be thorough to get good results.

Sausage casings are obtained from the intestines of sheep, cattle and pigs, which are carefully cleaned. Cellulose and synthetic casings are being widely used. Different diameter casings are available for the sausages.

The mixture is pumped into the casings. The sizes, different ingredients and processing give different names to the sausages.

Different Kinds

Frankfurt Sausages : It is a German speciality and is most often used to accompany sauerkraut. They are served hot as hors d'oeuvres with grated radish. They are made from a forcemeat composed of lean pork, trimmed of all fat, minced with bacon. Saltpetre is added to give a slightly pink colour. They are smoked, which gives them the typical yellowish outside colour. German Frankfurters are thus called if they come from Frankfurt of its immediate surroundings.

Strasbourg Sausages : These sausages are made like smoked Frankfurt sausages.

They are mainly used for garnishing sauerkraut.

Chipolata Sausages : They are long twisted sausages made in very small sized casings. These are divided into small sausages and can be grilled, sauteed in butter or poached in white wine. They are made of pork meat and rusk is the binder with pepper, salt, pimento, nutmeg, thyme and cayenne pepper. When cut, it is slightly pinkish colour and is served with roast turkey and Rognons saute turbigo dishes.

Mortadella : It is a larger oval shaped type of sausage of Italian origin. The Mortadella of Bologna has the highest reputation. Mortadella is also made in France. This is a smooth sausage prepared with pork and beef with pork fat, garlic, pistachios, cardamom, cloves, salt, pepper, etc. It is filled in beef bladders and smoked.

Salamis : They are smoked and usually served sliced. They are medium sized sausages of beef and pork with garlic and salt, peppercorns, etc. The texture varies from coarse to very finely cut meat.

The most renowned is the Bologna.

The different salamis are:

Belgian salami: It consists of beef and pork in a distinctive white rasing- without garlic.

Italian salami: It consists of pork and beef with rice like speckles f fat It is well seasoned with wine and garlic.

Danish salami: It is made up of pork and veal, bright pink in appearance with a distinctive salty flavour and slight garlic flavouring.

Black Pudding: One of the earliest cooked sausages consisting of blood, oatmeal, onions, pepper, rice flour, pimentos, salt, mustard powder, etc. They are black in colour.

Poultry is the name given to domestic birds, specially bred to be eaten. It includes the different varieties and sizes of chicken, turkey, ducks, geese and guinea fowls.

Classification of Poultry according to the Age and Weight

Names	*French Name*	*Approx. Weight*	*No. of Portions*	*Method of Cooking*
1	2	3	4	5
Single baby chicken	La pouissin	300 gms to 350 gms	1	Roasting and grilling
Double baby chicken	La pouissin double	450 gms	2	Roasting and grilling tandoor
Medium chicken	La Poulet reine	750 gms— lkg.	3-4	Poeling, roasting. encasserole. tandoor.
Large chicken	La poularde	1 ½ -2 kg.	6-8	Boiling, braising. galantines, stewing
Old hen	La poule	2½-3kg.	—	Boiling (used for stocks, sauces and soups)
Duck	Le canard	3-3 ½ kg.	6-8	Roasting, braising
Duckling	Le caneton	2-2½ kg.	3-4	Roasting, braising
Turkey	Dinde	9-12 kg.	250 gms. weight per portion	Roasting
Young turkey	Dindoneau	5-7 kg.	Same as above	Roasting
Goose	l Oie	5-7 kg.	10-14	Roasting
Guinea fowl	La pintarde	This bird is not common in India.		

Drawn poultry loses approximately 25% of its original weight. The word 'Volaille' in French means poultry, but when it appears in the menu, it means chicken.

When the size of the bird has been chosen, the next point is to recognise signs of young and fresh birds, as it helps in the preparation of chicken dishes.

Signs of a Fresh and Young Chicken

(1) Soft fluffy feathers should be all over the body, especially concentrated under the wings.

(2) The quills of the feathers should be easy to pull out.

(3) The skin should be clear and smooth with no bluish tinge spots.

(4) The feet should be supple with smooth overlapping scales, and if fresh, moist.

(5) Combs and wattle (the skin hanging near the beak) should be small and well developed. The comb being bright red.

(6) Conformation or the shape of the bird is important, and the bird should be free from deformity.

(7) The eyes should be prominent and clear.

(8) It should not have a marked unpleasant smell.

Conditions in Different Birds

Chicken and Fowl : The breast should be plump, firm flesh, pliable breastbone.

Turkey : The breast should be plump, the wattle red. Under four and half kg should not be chosen or the proportion of bone to flesh will be great.

Duck : The breast should be plump. The bill and feet are yellow in young birds, but they darken with age. The webbing of the feet should be smooth, tender and easily broken and the under bill soft and pliable.

Goose : The legs should be smooth and the breast plump.

Feet should be yellow and smooth, and webbing should be easily torn. After slaughtering the birds, they should be held at a cool temperature, but, not frozen, for 4 to 8 hours. Turkey should be held for 12 hours in cold conditions.

Food Value : The flesh of poultry is more easily digested than that of butcher's meat. It contains proteins and is therefore useful for building and repairing body tissues and providing heat and energy.

Cooking Methods : The methods used for poultry depend upon the size of the bird and the nature of the flesh. Some birds are young, tender and lacking in fat, others are tough and require moist heat for tenderising. Older fowls have richer flavour to impart to soup and creamed dishes than young chickens. The characteristics of the flesh have led to special methods of preparations for the specific classes of poultry. Young birds are cooked by grill, tandoor or roasted as they are tender. Chicken pies or chicken dumplings, and pakodas are made with fat fowls. Matured birds are cooked by moist cooking, spices are added when flavour is lacking and are enriched with fat such as fricassee, chasseur, kormas, curries, etc. Lean young birds are fried as it enhances Palatability.

Cleaning

(1) Remove any pin feathers with a small knife.

(2) Singe (place it over an open flame to burn the hairs) but do not scorch.

(3) Cut the head off and slit the neck skin lengthwise on the back.

(4) Cut off the neck as close to the body.

(5) Put the forefinger from top and loose the crop, intestines.

(6) Make a slit near the vent (rectum) and loosen the internal organs with the fingers.

(7) Draw out the innards being careful not to break the gallbladder.

(8) Clean the gizzard, keep the liver, neck and heart.

Cutting for Sauteing : This is the Jointing of a chicken at the raw state. The cuts are made mostly through the natural joints to produce two wings, two pieces of breast meat, two winglets, two drumsticks, and two thin pieces. The carcass can be used for stocks.

When cut this way, a chicken will lend itself to many preparations, i.e. fricassees, blanquette, saute, curries, pies, tandoor chicken pieces.

	English Cuts	*French Name*
1.	Drumsticks	Pilon de Cuisse
2.	Thighs	Gras de Cuisse
3.	Winglets	Aileron
4.	Wings	Aile
5.	Breast	Poitrine or Blanc
6.	Carcass	Carcasse

Trussing

For *Roast:* The legs are blanched in boiling water for a few seconds and the scales removed with the aid of a cloth. The claws are trimmed. To facilitate carving, the wishbone is removed. The bird is then trussed, using two strings. The trussing needle is a strong steel needle about 6" long with a large eye. Thread the trussing needle and pass it through both the thighs and then through the wing, the neck, which has the Skin covering the neck opening, and then through the other wing and tying both the ends of the string securely. The winglets should be crossed over the breast. A second string is passed over the legs and under the white breast meat and secured leaving the prepared legs protruding beyond the rear of the chicken.

Stuffing : Stuffing is put inside from the neck and between the skin of the breast and the breast of the chicken. This makes the chicken look plump. Stuffing for murgh mussallam is done in the hollow from where the internal organs have been removed and cleaned.

Preparation of Tandoori Chicken : The neck is cut off and the skin is removed, internal organs are removed. Slits are made in the flesh on the thighs and breast so that heat could penetrate easily. The chickens are marinaded in a mixture of beaten curd (yoghurt), lime juice, orange colour, garlic and ginger paste, Kasuri methi, salt, pepper and garam masala (mixed spices) powder.

For Grilling : Remove the wishbone and the legs from the first joint. Cut through the backbone, open out, remove back and rib bones, flatten slightly.

Supreme : This is the wing and half of the breast of a chicken, or the whole of the white meat from one side, with the trimmed wing bone attached. The white meat of one chicken yields two supremes.

Ballotine : The boned-out raw leg of chicken which can be stuffed, braised, stewed or sauteed according to the recipe.

Accompaniments and Garnishes : For roast chickens, game chips, watercress, roast gravy and bread sauces are served separately.

Grilled chicken— Fried straw potatoes, piquant or devilled sauce.

Tandoori chicken—Onion rings, lemon and green chillies.

Chicken korma— Chopped green dhaniya and fried cashewnut slices and silver leaf.

Chicken pakodas— Garam masala powder sprinkled on top, mint or sonth ki chutney.

Chicken curries—Chopped green dhaniya and juliennes of ginger

Dishes of Chicken

Chicken Maryland : 4 pieces are taken from the chicken and leg is deboned (2 breasts,, and 2 legs) coated with seasoned flour, dipped in eggs and bread crumbed and fried. Served with grilled bacon rashers, corn galettes, banana fritters, and fried parsley.

Chicken Ballotine Chasseur : Deboned chicken, stuffed and tied to maintain shape. Saute the chicken, add mushrooms, white wine and brown sauce tomatoes, and cook on top of the stove. Then add chopped tomatoes and sprinkle chopped parsley and fried heart-shaped bread croutons.

Chicken Mousseline : The word mousse means 'a finely pounded substance to which cream has been added'. This can be served hot or cold. It consists of finely minced chicken, cream, egg whites, seasoning and nutmeg moulded and poached. Served cold.

Chicken Salad (Salade de Volaille)	Roasted chicken, hard-boiled eggs, tomatoes, gherkins, pineapple, onions, lettuce, olives, vinaigrette or mayonnaise, lettuce and parsley. (Mix all ingredients except olives and parsley and lettuce with dressing.)
Chicken Galantine	Chicken stuffing—(Veal, pork, ham, pork fat, larding bacon, truffles, pistachios, spices, salt, brandy and eggs.). Remove the bones from the back of the chicken, by making a slit and working round the bony structure. On cloth, place the deboned chicken, place a layer of stuffing, placing strips of fat, bacon and nuts in between. Roll like a sausage and tie the ends securely and poach in chicken stock for 1¼ hrs. Remove, tighten the ends and again cook for 15 minutes. Drain, cool and unwrap the cloth, decorate with clear chicken aspic.
Chicken Chaud Froid	Spring chicken, supreme sauce, chicken aspic, truffles, court bouillon. Use for decoration: tomatoes, lettuce leaves, curry

	leaves, etc. Truss the chicken and poach slowly in court bouillon (15 minutes per 450-500 gms.) Cool in cooking liquor when done, remove skin, coat the chicken in supreme sauce + aspic and cool it. Decorate with truffles, tomato and curry leaves.
Chicken Saute Viennoise (Poulet saute Viennoise Fr.)	Egg and crumb the deboned chicken pieces, cook in butter and serve with lemon slices, chopped parsley on top, mask with brown butter and serve.
Chicken Florentine (Supreme de Volaille Florentine Fr.)	Poach dress on a layer of cooked spinach leaves, coat with mornay sauce and glaze.
Chicken Stanley (Poularde Poche Stanley Fr.)	Stuff chicken with rice, truffles, mushrooms, onions and butter. Truss and poach in chicken stock, remove chicken, reduce liquor, add cream and place chicken on rice bed (rice and mushrooms) and coat the chicken with sauce and serve.
Chicken Laperouse (Poulet saute Laperouse Fr.)	Saute chicken in the sauce pan, add brandy, cream, mushrooms puree and yolk of eggs, garnish with artichoke bottoms containing mushrooms puree and gratinated.

The other dishes that are made out of chicken are: Quenelles, medaillons, cutlets, chops, croquettes, salpicons, salads, bouchees, etc.

Various Cold Dishes

Pates — Forcemeat is put in various types of utensils and is not covered while cooking. Actually the word pate is a dish prepared of meat

or fish enclosed in a pastry or a dough and baked in the oven. A pate is a savoury pie and can be served hot or cold. Liver pates are made of liver. Fat seasoning finely minced and cooked and passed through strainer to get a smooth paste and moulded.

Terrines — The terrine is made up of forcemeat of any type of fleshy items such as pork, veal, beef or lamb, poultry, game and seafood. The coarse forcemeat is moulded in earthenware utensils from where it gets its name. Covered with a light fitting lid during the cooking process and served in the same mould. Terrines are always cold.

The Art of Carving : The art of carving in front of the guest is difficult to teach by theory or with the help of diagrams. However, an attempt has been made with the aid of diagrams.

Practice, good tools, careful preparation of the joint before cooking, and a knowledge of the bone structure is essential.

Selection of Tools : For most joints, a knife with a blade 10" to 12" long and about 1" wide is required.

For poultry or game, a knife with a blade 8" long is more suitable For ham, a long thin flexible knife is preferred.

A carving fork is needed to hold the meat and serve the slices.

A steel is necessary to set the edge of the knife. The manner of holding the knife to set it on the steel is important; it should be laid at a low angle (nearly flat), and it should be drawn downwards in a semicircular motion on both sides of the steel.

Preparation of Joints : The preparation of joints before cooking is most important in carving at the table, if the carving is to appear effortless as it should, and not a tug of war.

As bones cannot be cut through at the table, any bone

which makes carving difficult should be removed before cooking, or should be sawn through so that it can be removed easily after cooking. In this respect, the collaboration of the butcher is required. It will of course be necessary to inform the butcher of your requirements.

Hotels and restaurants, having a larder chef, prepare their joints to suit the requirements of their particular establishment, thereby ensuring the maximum possible economy.

Bone Structure : Knowing where to expect bones is most important in carving. This refers to those bones that are part of the internal structure of the joint, and that could not be removed without changing the shape of this joint, i.e. the difference between a shoulder of lamb with the bone in a boned and rolled shoulder.

A joint is useful to locate the bones, but it will be necessary to study it beforehand.

The following methods of carving do not represent the only possible way. They are those that are generally accepted and practised.

Chicken : With wishbone removed before cooking:

1. Remove leg by cutting around skin A, stick a fork at XXXX and pull outwards, severing joint. Cut leg in two at B.
2. Carve the wing off at C, cutting through the joint.
3. Repeat for other leg and wing.
4. Cut on either side of the breast- bone, in order to remove the meat D. in two long pieces.

Without wishbone being removed : 1,2,3 as above.

4. Lay bird on its side, cut through X-X, and pull whole breast off: Divide in two at D, or cut at D first, lift off the meat from the bone.

Duck

1. Remove leg as for chicken.

2. Carve the wing C off, as for chicken but shorter.
3. Cut the breast in thin slices, holding the knife slightly inclined outwards.

Ducklings, having little flesh on the breast: It will be easier, after removing leg to cut away wishbone (if not previously removed), make incision along breastbone and lift wing and breast in one piece, slicing in two or three afterwards.

A small turkey can be carved without removing the legs.

1. Cut leg first at A, then under drumstick at B, slicing inwards, next carve the flesh off the thigh at C.
2. Carve the breast diagonally, starting at D, proceeding towards the wing E.
3. Serve a slice of white meat and a slice of black meat.
4. When all the meat is carved off the left side of the turkey, turn it round with the leg towards the carver and proceed in the same manner.

A large turkey can be carved without removing the legs, but it will be found easier to remove and carve them separately.

Partridge : Cut along each side of the breastbone, push knife sideways and downwards to detach the meat from the carcass, serve through shoulder and thigh bone, and serve leg, or breast and leg wing in the piece.

Guinea Fowl, Grouse, Pheasant, etc. similar to chicken.

Pigeon, Plover, Woodcock, Snipe, Quail, etc. are small birds normally served whole.

Hold the knuckle bone with a serviette (or paper frill).

Make an incision above the knuckle at A.

With the knife slanting towards the knuckle, cut two or three slices, at B according to size of joint.

Then proceed to cut alternate slices from C.

Finish the carving by slicing the fillet D.

Saddle of Lamb—Mutton—Veal

There are two methods for cutting of lamb :

Method 1

Make a cut on side of backbone (A-B) lifting the knife slightly to clear bone at bottom of cut.

Cut slices lengthwise downwards and lift them out at C. Slices of fat cut on the slope at D are also served. The fillets F are removed and cut on the slant. Proceed in similar manner for other side of the saddle.

Method 2

Cut crosswise on the slant as shown at E. This gives a good slice of fat and lean.

Shoulder of Lamb—Mutton

By far the most difficult joint to carve. There are three bones to deal with:

The Knuckle A the middle bone B the blade bone C

There are also lean parts and fat parts.

Caring Methods

1. On the underside of the blade bone C, there is a lean part, which will give 3-4 slices, by cutting outwards on the slant. From the middle bone B part, 3-4 slices of fat meat can be cut at D. Alternate these slices and you have easily disposed of these parts.
2. Next the best part of the shoulder is cut into slices E, from the blade bone end towards the middle bone B, starting at X, shaping the cut so as to reach the curve created by the blade bone. This part has fat and lean flesh somewhat equally divided.
3. Turn shoulder round, so that blade bone faces the carver. On the blade bone, there is a projection F. From each side of this ridge cut a few slices, slanting the

knife outwards. One side will give more slices as the ridge is not in the centre. This meat is fairly lean and the last cut, as H, will give some fatty meat to balance the portion.

Near the joint of the blade bone C and middle bone B, there is a gland that cut H will reveal and which should be removed.

Ribs of Beef

Preparation

1. Detach meat from backbone and saw through rib, line A-B.
2. Backbone may be replaced in position and held with string for roasting
3. Remove bone before sending to the table.

Carving : Slice from C to D holding the knife slightly inclined outwards, so as to cut the meat away easily on reaching the bone (rib).

Sirloin of Beef

Preparation :

As for ribs.

Carving :

Slice fillet or undercut F, lengthwise.

Turn joint over and proceed as for ribs.

Loin of Lamb—Mutton—Pork

Saw Cuts

Roasted on the bone, can only be cut if the backbone has bee sawn through before cooking.

Slice in thick pieces as shown.

NB: The sides of pork will need to be scored accordingly. The pitch bone should be removed.

Begin by cutting wedge X.

Carve in thin slices.

Slice part A towards the knuckle.

After a few slices, turn the ham over.

Slice part B, again towards the knuckle end.

Serve alternate slices from A and B.

A different method is to start by removing a wedge at X and cutting thin slices from both sides of this wedge, till all the top meat is carved. Part A.

Turn ham over and slice part B, towards the knuckle, starting at Y.

Start by cutting both corners off. Slice alternately A and B, starting from corners, serving a slice from each side.

bone is exposed at part B, remove patella bone, settle gammon steady and continue carving thin slices from part A.

Vegetable Dishes : Vegetables play a very important part, as more and more people are becoming vegetarians and vegetables are widely used in cookery. They are served in various forms, as accompaniments, garnishes, salads, hors d'oeuvres, bhajees, sookhi sabzi, curries, raitas, etc. There is a lot of repetitive work entailed in the preparation and cooking of vegetables. Vegetarian cooking demands a good knowledge of the principles of cookery and a high degree of skill. During the preparation of all kinds of fresh vegetables, there can be excessive amounts of waste unless it is efficiently and economically carried out. All vegetables, particularly the green ones, are best when taken straight from the garden.

Vegetables could be classified in two ways for culinary purposes:

Parts of a plant—

(1) *Roots*—Carrots, horseradish, beetroot and turnips.

(2) *Bulbous roots*—Onions, shallots, leeks, garlic.

(3) *Tubers*—Potatoes, Jerusalem artichokes, kachaloo.

(4) *Flowers*—Cauliflower, broccoli, globe artichokes.

(5) *Leaves*—Cabbage, lettuce, spinach, watercress, sarson.

(6) *Fruit*—Cucumber, tomatoes, beans, peas, pulses, pumpkin, tin das, lauki.

(7) Stem—Celery, rhubarb, seakale, chicory, asparagus.

The other way is to broadly classify such as—

Root vegetables. Leafy vegetables. Other vegetables.

Life Giving Value

Except for the dried vegetables such as dried legumes, vegetables have a high water content, which ranges from approximately 70 to 95%. Of the other foodstuffs present in vegetables, carbohydrates rank first in quantity. These may be present in the form of starch, sugar, cellulose, pectic substances. The quantity depends upon the part of plant from which the vegetable comes and the degrees of maturity—ripe or unripe. Starch is the chief nutrient of roots and tubers; potatoes have the highest percentage. Sugar is highest in beetroot and it is also rich in carrots and turnips. Cellulose, the fibrous framework or roughage becomes coarser and tougher as the vegetables mature. In some cases it is too tough to be eaten raw compared, e.g. to lettuce and cabbage.

If the dishes are well chosen and eaten raw or cooked, they contribute vitamins and minerals to the diet. They are provided by vegetables, as they contain relatively large amount of vitamins and minerals as compared to other foods.

Besides nutrient components, vegetables have certain other materials (components) which have no food value, but are of great concern in vegetable cookery. They have an effect upon flavour, colour, and general palatability of the cooked product. These materials are— flavouring substances, plant pigments and tannings.

Flavouring Substances : Many volatile and non-volatile organic acids contribute to flavour. But volatile aromatic oils are chiefly responsible for characteristic flavour and odour.

Some are present in a raw state and others develop when food is cooked.

Pigments : The attractive and varied colours of vegetables are due to the presence of pigments distributed in plant tissues like chlorophyll (green) and carotenoid pigments (red and yellow).

Tannings : Complex organic compounds are widely distributed in plants. They have astringent properties. Tannings of tannic salt form greenish purple compound, which upon oxidation, turns brown.

Selection of Vegetables : The quality of root vegetables is determined by their firmness and absence of blemishes. The skin should be smooth and not wrinkled for carrots. The potatoes should be of regular size and free from earth. The vegetables should be heavy for their size. Onions should have thin shiny skin and no signs of sprouting. Green vegetables should be crisp, have a fresh, green colour appearance. If the ends of the stems are broken, they should snap off easily and the leaves should not be limp. Cabbage should be compact and be heavy for its size. Cauliflower should be white, the flowerettes well formed and the leaves crisp. Spinach should be crisp with slender stalk ends, and the leaves large in size. Tomatoes should be bright red, firm, regular shaped with shiny skin. Beans should not be stringy and should break crisply under pressure. Pea pods should be plump. Brinjals should be light for their size, having a shiny and smooth skin. Lady's finger should snap if the ends are broken. Carrots should have a shiny skin and regular shape.

Vegetable Preparation : The nutritive value which vegetables contribute in the diet should be conserved as completely as possible. Most of the vegetables grow near the ground and get contaminated by dirt and sprays, etc. They must be properly washed especially those which are tô be are peeled thickly to remove the inner skin, which is tough and tasteless. Green vegetables should be washed well before cutting so that the grit is thoroughly removed. Potatoes and carrots

are washed first and thinly peeled. Vegetables are cut into different shapes such as dices, batons, juliennes (matchstick) roundels, triangular shapes, and they may be chopped, minced, sliced and grated, etc.

Saving of nutrients during the preparation and cooking of vegetables is important, and following are the ways to preserve nutrients:

(1) Do not let vegetables soak in water unnecessarily.

(2) Cook or bake vegetables in their skins.

(3) Keep air-oxygen away from vitamin rich foods by covering foods while cooking; not stirring food unnecessarily as it mixes air; cutting vegetables just before cooking, and serving food as quickly as possible when it is cooked.

(4) Use as little water as possible in cooking, so as to preserve vitamins, minerals, flavour, colour and texture of food.

(5) Baking powder should not be added while cooking vegetables.

(6) Do not expose food, milk in particular, to light.

(7) When preparing frozen vegetables, place directly into boiling water.

Effect of Cooking on Vegetables : Vegetables are cooked for various reasons such as improving their palatability and digestibility. In the case of certain food, flavours are improved and preserved longer by cooking.

Certain changes occur during cooking as:

Cellulose : It disintegrates and becomes soft and mastication is facilitated.

Flavour : Flavour is affected during cooking, and over-cooking adversely affects flavour. A covered pan increases the intensity of flavour.

Pigments : The colour pigments of fruit and vegetables are chlorophyll. carotenoids, flavones and anthocyanins. Green

vegetables because of their chlorophyll when cooked, for a long period at high temperature, tend to lose their colour. The green coloured vegetables are cooked in an alkali medium, to preserve the loss of colour, but certain nutrients are lost—carotene of carrots while cooking is not lost: in fact, it brightens the colour. Anthocyanins, in red and purple coloured vegetables like red cabbages, beetroots which are highly soluble in water and high temperature, bleaches them. Flavones are found in white onions, cauliflower, etc. Their white colour could be improved by cooking them in water to which lime juice or vinegar has been added.

The changes in carbohydrates take place as the heat partially or completely gelatinises starch. Swelling of the starch grains occur during cooking of vegetables and, in some circumstances, the sugar caramelises.

Proteins are negligible in vegetables except for legumes: heat coagulates completely or partially.

Cooking losses occur through the solvent act of water or the chemical decomposition which may be caused by the reaction of the cooking water or by heat. Other common cooking losses are through oxidation, or by volatilisation.

Vitamin C is lost because of high temperature. Vitamin A, Vitamin B2 and Niacin are heat stable.

***Methods of Cooking*:** Common methods used for vegetable cookery are boiling, frying and stewing, though methods are also used such as steaming, braising, grilling and baking.

***Boiling*:** Green vegetables such as spinach, cabbage, peas and beans, etc. are placed in boiling salted water and rapidly boiled because it helps to conserve the colour and flavour. Root vegetables are first placed in cold salted water and brought to boil and simmered till tender. Care must be taken to ensure that they are not over-cooked. The cooking liquor should be used and not thrown away. Blanching green vegetables, retains a bright colour, prevents discolouration, retains texture, stops further cooking and retains nutrients.

Steaming : Usually potatoes are steamed. Other vegetables also can be steamed in a pressure cooker.

Braising : It is mainly for celery, leeks, cabbage, onions, and lettuce. They are blanched and placed on a bed of cut aromatic vegetables and moistened with stock and braised in the oven, e.g. celery, spinach, etc.

Baking : Usually the vegetables are baked in their skins, often on a bed of salt. Brinjals for bharta are baked. Sweet potatoes and potatoes, also can be baked.

Grilling : Tomatoes and mushrooms are grilled.

Frying : Vegetables cut into various shapes, can be fried and they can be dipped in a batter and fried.

Continental Vegetables

Artichoke : It is cultivated for its flowering head—used for highclass hors d'oeuvres, garnishes and a vegetable course. The fleshy base is the part usually eaten. Tips are also eaten.

Asparagus : Agenteuil asparagus is very popular. The part that is eaten is the young shoot. It is usually served with butter or sauce and is available in India. It is used as a vegetable course or as a garnish. It is delightful when cold and served with vinaigrette of cut and reheated in butter as a garnish.

Endive : Belgian chicory (endive) is the best. Green endive leaves are very bitter. Therefore they are blanched to pale yellow colour to make them palatable. Cold preparations are served as a hcrs d'oeuvre. It is also used as a vegetable and for salads.

Leeks : Leek resembles *hara pyaz* (spring onions), but are bigger and have a lighter shade of green. They are used as a hors d'oeuvre. It is also used as a vegetable and for salads.

Celery : A vegetable that has a good flavour and is rich in mineral salts and Vitamin C. It is used as a hors d'oeuvre, soup, in sauces, garnishes, salads and served with cheese course.

Kale : The leaves are purplish or light red in colour. It

resembles a wild cabbage. It is boiled and has a nutty, slightly bitter flavour.

Parsnip : Widely used in England and it is like turnips and yellowish in colour. It is used in soups, as garnish or as a vegetable dish.

Swede : Often called the Swedish Turnip. It has yellowish coloured flesh. Mainly used as a vegetable.

Cardoon : Plant common to the celery of which the stalk leaves are eaten as a vegetable. Stalks are crisp and tender. It is similar to celery from which it takes its culinary uses.

Indian Vegetables

Aubergine (Egg plant, Baigan or Brinjal) : It is available all round the year. The colour is from light purple to blackish purple. Used as a vegetable course and as a garnish. Used widely for Indian dishes, in the form of curries, barva baigan, kachadi {slices dusted with atta and shallow fried), in sambar, bhartas, etc.

Cabbage (Bandh Gobi) : They are rich in Vitamin C, widely used in Western menu for salads, as a vegetable and in soups. The Indian dishes are bhajees, foogaths, thorans and for pulaos, etc.

Cauliflower (Phool Gobi) : It is a high-class vegetable according to Western standards and is used in soups, as a garnish and as an hors d'oeuvre. The Indian dishes for cauliflowers are pakodas, tahari, cauliflower sabzi, bhujia etc.

Cucumber (Kheera) : It should be tender without seeds. Used for hors d'oeuvres, cold soups, indispensable with salmon and fish dishes as garnishes. Doria is a keyword here, also used for cooked vegetables and savouries. In India dishes, cucumber is used for kachumbers, raitas, etc.

Lettuce (Salad ka Patta) : There are different types of lettuce—Romaine or Cos lettuce. Widely used as hors d'oeuvres, soups, garnishes, hot vegetable, accompaniment and salads.

Marrow (Lauki) : Member of the gourd family. It has a

high water content and has little food value. Mainly used as vegetable, raitas, halwas, etc.

Peas (Matar) : It is a popular vegetable used in hors d'oeuvres, soups, garnishes, and as a vegetable dish. It is also used in salads. The Indian dishes are matar chawal, matar keema, matar paneer, sookha matar, etc.

Pumpkin (Kaddu) : If the pumpkin is green, then it is prepared with the skin and if it is yellow, the skin is removed. It contains 90% water. It is used as a vegetable, pumpkin pie or halwa.

Spinach (Palak) : It is a nutritious green leaf vegetable. Extensively used as a garnish and as a vegetable, mutton palak, pakodas, etc.

Tomatoes : Very popular vegetable and used in various forms—purees, sauces, or paste. It is extensively used in hors d'oeuvres, soups, fish dishes, sauces, vegetables, salads, savouries, curries, chutneys, pulaos, raitas, etc.

Capsicum (Shimla Mirch) : It has very good flavour and is used in hors d'oeuvres, garnishes, stuffed or sauteed as a vegetable, and salad. In the West, there is a variety of the capsicum family. The large ones are called pimentos, which could be green or red. When dried and powdered, it is called paprika powder. The smaller ones, the green and red peppers are more pungent and are called chillies. Used for sauces, stews, and pickles, curries, subzi, etc. When dried and ground, it is called cayenne pepper.

Rhubarb : It is a vegetable and consists of a leaf stalk with a pleasing flavour. Used mainly for stewed pies and puddings. In India, rhubarbs are Pickled.

Mustard Leaves (Sarson ka Sag) : Extensively used in the north of India. For the Punjabis, it is a speciality dish.

Horse-Radish (Mooli) : It has a pungent flavour, is used in hors d'oeuvres, decoration for cold buffets, salads. It is used in India for bhajees, parathas, etc.

Turnips (Shaljum) : It is a common root vegetable. Main use of the turnip is as a flavouring vegetable and for garnishing and as a subzi.

Beetroot (Chukander) : The sugar-beet goes to sugar manufacturers, the common beetroot to the kitchen. Leaves can be used to prepare vegetables and salad.

Carrots (Gajar) : Contains sugar, mineral salts and certain vitamins. It is used for hors d'oeuvres, soups, sauces, garnishes, hot vegetables and salads, gajar ka halwa, pulaos, etc.

Onions (Pyaz) : These are available all the year round. Probably the most used vegetable in the kitchen.

Potatoes (pommes de terre, fr.) : Potatoes play an important part in the general diet of the population today. No menu seems complete until potatoes have been used, in the form of hors d'oeuvres or as an accompaniment to the main dish. For Indian menus, potatoes are used in the form of curries, subzis, raitas and taharis, etc.

Chemically potatoes are largely composed of water and starch, it is estimated that 99 gms. of potatoes yield approximately 80 calories. The sugar content of potatoes ranges from slight traces to as much as 10% of the dry weight of the potato. The two chief factors influencing sugar content of potatoes during storage are variety and temperature.

There are two main seasons for potatoes, the so called "Earlies" or "Pahadi" and the "Main crop" or "Deshi" potatoes. Therefore the potatoes are available all the year round.

Pre-preparation of Potatoes : Those potatoes should be selected that do not have deep eyes, irregular shape and do not have any discolouration during peeling and trimming. The green colour frequently seen immediately beneath the skin indicates the presence of chlorophyll. Much greening in potatoes in undesirable because of its association with a bitter taste.

The potatoes with thin skin should be scrubbed, scraped and cooked in their skins, and the ones with thicker skin

should be peeled. To prevent the potatoes from discolouring after they have been peeled, they should be kept in water.

Raw potatoes are shaped or cut into different ways: big or small dices, potato balls, matchstick strips, quarters, roundels, drum shaped rectangular strips, etc. Some of the shapes and sizes are:

Methods of Cooking : Potatoes could be fried, baked, roasted, steamed, boiled and sauteed.

The method of cooking could be selected according to the potatoes, as there are two different kinds of potatoes. The mealy or floury potatoes when cooked burst immediately on application of a fork and break to pieces and crumble. Baking, steaming in jackets, boiling and mashing are the methods of cooking used for this type of potato.

The other variety is close and waxy and the flesh is firm, which could be cut into different shapes, as it will retain its shape when cooked. Frying and sauteeing are the common methods used for this variety.

Potato Dishes

Baked Potatoes : (Pommes au Four Fr.) Wash and wipe equal size potatoes, place on a bed of salt and bake in the oven. When cooked, make slits—cross and open by given pressure on either side. Place a dot of butter and sprig of parsley and serve hot.

Arlie Potatoes (pommes Arlie) : Bake the potatoes as above, and when baked, cut the potatoes into two, lengthwise and scoop out and mix with chopped onions, butter and cream and pipe the mixture into the jackets, sprinkle with grated cheese and gratinate.

Macaire Potatoes : Bake the potatoes and mash them, season, add butter, make the balls and flatten, dust with flour and saute in butter.

2nd Method : Put the mashed potatoes in Pommes Anna mould and cook with clarified butter.

Byron: Prepare as above, hollow the top, sprinkle with Chester cheese, cover with cream.

Robert: To Macaire potato mixture, add chopped onions and chopped hard-boiled eggs, and cook the same way.

Bhurta: Bake/roast potatoes on gas or charcoal. When cooked, remove the skin and mash the potatoes. Heat butter, and chopped onions, green chillies, tomatoes, saute and add seasoned potato mixture and saute.

Bhurta Raita: Mix the above mixture into beaten curd (yoghurt), correct seasoning and sprinkle zeera powder (cumin) on top. Steamed/Boiled

Steamed Potatoes: (Pommes all anglaise, Fr.) Shape the potatoes, cook in a steamer and season.

Menthe: Same as above, add leaves of mint.

Maitre d'hotel: Same as cream potatoes with the addition of chopped parsley.

Jacquettes Potatoes (Pommes Jacquette): Wash and cut the skin in a spiral method and cook in steamer.

Parsley (Pommes Persillies Fr.): Same as anglaise, season, saute in melted butter and chopped parsley.

Vapeur: Same as steamed potatoes.

Cream (Pommes Creime): Cook the potatoes in water, peel and slice, moisten with milk, add butter, season and bring to the boil, add cream before serving.

Shallow Fried/Saute

Allemande: Cook potatoes in their jackets, peel and cut into thick rounds and well buttered.

Brune: Cook potatoes, peel, chop, shallow fry in butter till crisp.

Provencale: As sautee with addition of chopped garlic to the butter at last minute.

Lyonnaise: Slice potatoes and saute them, and then saute

the sliced onions and mix together, season, place in a tray and finish in the oven.

Saute Acru: Saute 1/8" dices of potatoes, finish cooking in oven, sprinkle chopped parsley, serve hot.

Aloo ki Chat: Parboil potatoes and shallow fry, sprinkle chat masala and lime juice and toss.

Tikkies: Parboil potatoes, mash and grind, add seasoning, prepare small flat cakes and shallow fry till golden brown on both sides and serve with tamarind chutney.

Oven Cooked

Anna: Potatoes should be cut into cylindrical shape, sliced in rounds, wash and dry and place the slices in layers in an Anna mould (round pan with lid sliding in the slot with handles on either side) with clarified butter, till the top of the pan, compress and close the lid, cook in the oven. When half done, turn for colour on both sides.

Boulangere: Slice potatoes and onions and saute, season, add stock, cook in the oven.

Chateaux: Turn the potatoes to the shape of a large olive, blanch, saute in butter and place in the oven for finishing the cooking.

Cocotte: Same as chateau, but smaller in size.

Fondates: Turn the potatoes into drum shape (egg size), saute in butter and place in a tray, add stock and seasoning and finish the cooking in the oven.

Flamende: Cook button onions turned carrots cooked, mix with Potatoes.

Delmonica: Cut dices ½, place in a tray, add milk, sprinkle bread-crumbs, dot with butter and place in the oven for cooking.

Hongroise: Chop onions, saute in butter and sprinkle paprika, add tomato concuss and large roundels of potatoes, add stock, cook and then place in oven and finish off cooking. Sprinkle chopped parsley and serve hot.

Deep Fried/Shallow Fried

Straw Potatoes: (Alumette Fr.) Cut potatoes, matchstick shape, fry in deep fat, add salt just before serving.

Bataille: Cut in dices, deep fry.

Bordelaise: Deep fry ½ dices, add fried finely chopped garlic, add chopped parsley before serving.

Frites: cut potatoes 2" length and ½ square. Deep fry.

Chips: Cut slices of potatoes, very thin, deep fry till golden brown.

Mignonette: Cut potatoes double of matchsticks and deep fry.

Paille: Cut potatoes in large juliennes and deep fry.

Parmentier: Cut ½ dices and deep fry.

Pont Neuf: Cut potatoes 2½ length, 1" width (regular size) and deep fry (shaped like a ruler).

Souffles : Trim potatoes in a rectangle, cut one-eighth of an inch thick, blanch in a moderately hot fat and drain, then put it in hot fat to effect puffing. Dry and strain.

Puree/Mashed Potatoes

Berny: Mash potatoes and egg yolk, butter, nutmeg and seasoning, stir well till it leaves sides of the pan (appareil duchesse), add chopped truffles, shape like an apricot, egg and coat with sliced almonds and deep fry.

Brioche: Prepare appareil mixture as above, prepare flat rounds and place a small ball of mixture in the centre, egg wash and gratinate.

Croquettes: Prepare appareil duchesse (mashed potatoes, nutmeg, butter, egg yolk, seasoning). Make cylindrical shape (2") egg and bread crumb and deep fry,

Duchesse: Pipe the apparel mixture into rosettes, brush egg and gratinate.

Marquise: (1) Take Duchesse mixture and mix with thick tomato sauce and shape like flat cakes.

(2) Take Duchesse mixture and pipe two three rounds leaving a hollow in the centre, put sauteed tomato concasse in butter in the centre, brush with egg and gratinate.

Gratinees: Mash potatoes, add nutmeg, milk, butter and seasoning, heap in gratin dish, sprinkle cheese and dot with butter and gratinate.

Mousseline: Mash potatoes and add whipped cream and mix well. Serve hot.

Indian potato dishes are : Aloo dum, aloo dum Banares, aloo dum Bhojpuri, aloo ki chat, kofta, bhajee, raita, aloo ki tahari, aloo ki roti, etc.

Oven Cooked

Anna: Cut potatoes in cylindrical shape into slices 1/16" thick. Put butter in an Anna mould.

Arrange slices overlapping the bottom lid slides in sides. Season and fill the mould up to the lid. It should be compressed and put in the oven and cooked. When cooked, remove the lid, clear the top by putting it back in the oven. Demould and serve.

Voisin: Same as Anna after the mould has been lined, add grated Parmesan cheese. Cook and serve as for Pommes Anna.

Darphin: Proceed as for Pommes Anna using coarse julienne of potato in place of sliced potatoes.

Cheese : Cheese is one of the by-products of milk. It is prepared from either cream of milk, or cream and milk mixed together. This is soured and coagulated with rennet or acetic acid. The whey is run off and the curd is fermented and cooked. Different degrees of heat and pressure produce different

types of cheeses. The harder and longer the cheese is pressed, the harder will be the texture. The degree of pressure denotes the type of cheese required. Cream cheeses are subjected to light pressure. They are then left to mature and the timing of this produces mild or strong cheeses, as they ripen with age. Cheeses range in consistency from soft creamy substance to an almost rocklike hardness. Cheese is made from cow's, goat's, or ass' milk. Some cheeses are flavoured with herbs or spices.

There are four main types of cheeses used in the kitchen, Parmesan (Italy), Gruyere (Switzerland), Cheddar (England) and Amul (India). These have a strong flavour which is not lost in cooking.

When cooking cheese, it should not be over-cooked, as it is liable to become very indigestible and is liable to "split" (the fat separating from other ingredients). It should always be combined with starchy foods as this aids the digestion. Cheese also becomes rubbery after over-cooking. Low temperature and short cooking is recommended.

Cheese is a very valuable food. It is roughly $\frac{1}{3}$ protein, $\frac{1}{3}$ fat and $\frac{1}{3}$ water. It has a large calcium content, and also contains vitamins A and D.

Cheese is available in natural or processed forms.

Processed cheeses are made by grinding the cheese and then cooking in a vacuum cooker and pouring into moulds and then wrapping in tin foil.

Blue cheeses are made by pressing the cheese, then skewering with copper wires and introducing a type of Penicillin mould into the cheese.

When storing cheese, it should be kept in a cool dark place and should be wrapped in grease-proof paper to prevent drying. If it is kept at a temperature which is too cold, it will dehydrate and if put in a warm place, it will "sweat". It should be stored away from foods or it will absorb flavour.

Cheese is served in salads, sandwiches, canapes, hors d'oeuvres. Sometimes they are served at the end of a meal. They are also served as a garnish for soups. Paneer (cottage cheese) is used for curries, pakodas, etc.

Cheese	*Origin*	*Description*
Bel paese	Italy	Mild Italian cheese. Large round and flat, creamy colour, texture soft. Mild sweet flavour, ripens quickly.
Brie	France	Soft texture, colour is reddish, brown crust, yellow paste.
Petite Suisse	Swiss	White colour, creamy texture, prepared from unsalted fresh cream.
Camembert	France	Colour and texture soft and creamy.
Gorgonzola	Italy	Semi-hard blue mould cheese, prepared from renneted cow's milk.
Parmesan	Italy	A hard white cheese. A kitchen cheese. Takes 2 to 3 years to mature.
Emmenthal	Swiss	In colour, it is ivory to light yellow, waxy texture. Has holes approximately of the size of a cherry.
Gruyere	Swiss	As above, but firmer in texture, holes are smaller, the size of a pea. Takes longer to mature than Emmenthal.

Contd...

Cheese	*Origin*	*Description*
Gouda	Dutch (Netherlands)	Dutch cheese pale yellow in colour. Waxy texture. Flat.
Edam	Dutch	Dutch cheese red rind, yellow in colour. Texture waxy. Round like a cannon ball with a flat top and bottom.
Roquefort	France	Blue mould cheese similar to Gongonzola. Has a semi-hard texture.
Neufchatel	France	A soft creamy cheese.
Limburger	Belgium	Soft highly flavoured cheese.
Brick	America	Very firm to fairly soft and elastic texture. Mild to sharp flavour.
Port du salat	France	Semi-hard, rubbery texture.
English Cheese		
Cheddar	England	Colour yellow, texture close. Clean and mellow, nutty flavour, many uses in the kitchen.
Cheshire	England	White or orange in colour. Texture crumbly. A strong flavoured cheese.
Stilton	England	Finest English table cheese. It is a hard blue mould. Texture flaky. It is usually spooned out for service.

Contd...

Cheese	*Origin*	*Description*
Lancashire	England	It is of a mild flavour, crumbly texture and when fully mature, can be spread like butter.
Wensleydale	Yorkshire	It has a flaky texture and is pale in colour. It has a "honeyed" aftertaste.
Caerphilly	Wales	Creamy white colour, semi-smooth texture and delicate flavour.
Derby	England	Close textured, pale honey colour. Mild when young but when fully matured, has much fuller flavour. Sometimes flavoured with sage.
Double	England	Corn coloured, velvety textured.
Gloucester		Similar to Cheddar in taste.
Dorset Blue	England	Crumbly texture, strong rich flavour.
Leicestershire	England	Russet coloured, slightly flaky texture and a mild flavour.

Sandwiches : Sandwiches have always been a tradition where catering has been offered to the people. Many ideas have been put forward as to their origin, the most feasible one being that the name came from the Earl of Sandwich, a compulsive gambler, To save time from the table, he asked for

a piece of meat to be placed between two slices of bread which he ate whilst playing.

Sandwiches are one of the most varied types of food prepared. They are usually one or two slices of bread, buttered with a filling of meat, chicken, fish, cheese or vegetable, etc. Bread may be toasted or fresh, a variety of shapes with an almost endless assortment of fillings.

The services of sandwiches varies with the purpose for which it is used. Snackbar sandwiches are substantial and a meal unto themselves. With rapid industrialisation and the speed in tempo of work, the use of sandwich as a meal in itself is becoming popular. It is served for tea parties, barbecues, supper or with coffee, used for picnic baskets. Sometimes sandwiches are taken as a full meal with fruit and milk.

Accompaniments of sandwiches are wafers, chiffonade (shredded) lettuce, watercress, juliennes of cabbage or carrots. The different kinds of bread that are used for the preparation of sandwiches are: White bread, brown bread, French loaf and bread rolls, or boat shaped bread.

Approximate quantities for average portions :

Bread : Allow 2 slices of bread per serving. The thickness and number of slices will vary in different catering establishments.

Buns : One sandwich bun per serving.

Butter : *Softened* 500 gms. butter will spread 96 bread slices if one teaspoonful is used per slice.

Bread Slices : 2 lb. sandwich loaf white bread cuts 28 ½") slices or 36 ($3/8$") slices with end crust.

Approximate quantities for average portions for various fillings of sandwich are:

Cheese Sliced : 480 gms. average 16 slices (approx. 30 gms. each).

Chicken : 500 gms. average 12 slices or 16 slices.

Peanut Butter : 455 gms. average l3/4 cups.

Tomatoes : 455 gms. average 3-4 medium tomatoes.

Other fillings used are cucumber slices, tomato slices, tuna fish, hard-boiled eggs, meat, olives, etc.

Cutting of Sandwiches : It is done in a number of ways, the type of filling and the method of serving the sandwiches determine the way to cut it.

Method of cutting filled layers into sandwiches, if it is to be prepared in bulk.

Important points to be observed for the preparation of sandwiches:

(1) Use stale bread (one day old), because it cuts better and finer.

(2) The bread should have close fine texture.

(3) Thinly sliced bread appeals more than thick slices.

(4) Crust should be removed neatly.

(5) Use a sharp knife preferably with a saw-edge and dip occasionally in hot water.

(6) Store bread at 75°F to 85°F (Ideal). Refrigerator temperature accelerates staling, though in moist, warm climate, store in refrigerator to prevent mould growth.

(7) Sandwiches may be stored for 12-24 hours at temperature below 40°F.

(8) Cream fresh butter facilities easy and even spread, as the creamy consistency glides over the slice.

(9) Use only fresh ingredients, e.g. crisp lettuce, watercress. tomato, etc.

(10) The filling should consist of ingredients which do not clash in flavour, e.g. avoid mixing a sharp flavoured cheese with anchovies.

(11) If the filling contains moisture, apply it just before serving, otherwise the sandwiches become soggy and will appear unappetising.

(12) If sandwiches are prepared well in advance, they should be wrapped mostly in a foil or alternatively in a sterilised damp napkin; this keeps them supple and fresh.

(13) For serving, they may be dressed on a paper doiley surrounded with lettuce leaves, chiffonade. etc.

Savoury Butters, or Spreads

Lemon : 225 gms of butter, 1 teaspoon of lemon juice, and fish paste, used for fish sandwiches.

Onion : 225 gms. of butter. 2 teaspoons onion juice. 1 teaspoon of water—Cold meats and sandwiches.

Olive : 225 gms. butter, ¼ cup minced olives—for fish or chicken sandwiches.

Sardine : 225 gms. butter 1/3 cup mashed sardines. 2 teaspoons of lemon juice.

Paprika : For watercress or tomato sandwiches.

Mayonnaise : For egg and salmon sandwiches.

English Mustard : Butter and English mustard for ham and beef sandwiches.

French Mustard : Butter and French mustard for cheese and tinned meat sandwiches.

Types of Sandwiches'

(1) *Sandwich Cake*—These look like iced layer cakes and for these, round bread loaves are used. They are sliced in two or three round slices, soft icing containing butter, cheese, etc. It is cut into wedges like cakes.

(2) *Loaf—The* loaf bread is taken and sliced lengthwise, filled and iced in the same way as sandwich cake. It is sliced across the bread and served.

(3) *Pinwheel*—The top crust is taken out of the sandwich loaf. It is then sliced lengthwise, and butter and soft filling spread evenly, which is of a colour to contrast with the bread. Each slice is rolled and wrapped firmly

in a foil and chilled till it is set. It is cut into thin slices and served.

(4) *Ribbon*—They are prepared by using two coloured breads. Two slices of brown bread and one of white or vice versa. The sandwiches are made with three slices, and two different coloured fillings are used. They are cut into thin strips and served.

(5) *Double Decker*—Three slices are used, the filling placed on top. The next slice is buttered and placed on top, lettuce, tomato and cucumber placed on top of this slice. The final slice is placed in position. Pressed, trimmed and held together with a plastic cocktail stick with stuffed olive or cherry.

(6) *Club*—"fills sandwich is usually served hot and consists of two buttered slices of toast. One slice is the base on which sliced cooked chicken, egg, lettuce, tomato, grilled bacon, and mayonnaise or ham is placed. The other slice of toast is placed on top, pressed, trimmed and decorated. Served between the folds of a paper napkin,

(7) *Bookmaker*—This is buttered roll or French bread spread with French mustard and a grilled steak in between.

(8) *Broadway*—Two layers of toast with smoked salmon, shredded lettuce, sliced hard-boiled egg, served with mayonnaise.

(9) *Conventional Closed Sandwich*—Two slices of bread with any fillings. Cut into two without cutting crust.

(10) *Tea Sandwiches*—Thin slices of bread are used with any fillings cut into triangles and crust removed.

(11) *Buffet Sandwich*—Smaller than conventional, triangles, fingers or squares and fancy shapes.

(12) *French Sandwich*— Consists of thin slices cut from French loaf and fillings added between two slices.

(13) *Grilled Sandwich*—Two slices of bread with fillings, grilled and then served. They are hot sandwiches.

Open or Smorresbrod : This famous sandwich originated in Denmark and it is an open sandwich having a fancy shape, it is easy to make, appealing to the eye, delicious to eat, and the cost is low, as left-overs can be used. The choice of filling and decoration is endless. The arrangement on the serving dish is most vital. Doyleys are used and the sandwiches are arranged so that the sandwich's colour scheme and filling blend well.

Examples of Smorresbrod

Capenhagener : Danish blue cheese, with raw egg yolk in raw onion ring.

Louisiana : Danish ham, Camembert cheese, lettuce, glace cheery.

Harlequin : Danish luncheon meat, with potato salad, watercress and radish.

Sunshine Salami : Sliced salami, with raw egg yolk in raw onion ring.

Tivoli Tongue : Sliced tongue, ring green pepper, horseradish, cream, tomato.

Guardsman : Brisket of beef, horseradish, cream, onion, tomato and parsley.

Roskilde Special : Danish ham, topped with chicken mayonnaise, tomato and cucumber.

Kronberg Salad : Ham, topped with Russian Salad and apple, sliced onion.

Frinkadeller : Danish meat balls, red cabbage and fanned gherkin.

Bacon and Apple : Savoury apple puree, two rashers of cold bacon, lettuce and tomato.

Captain's Breakfast : Pickled herring, slice of egg, capers, on lettuce.

Roulette : Cod's roe, Russian Salad, tomato, cucumber, gherkin.

2

Significant Features

The Department of Rural Development (Ministry of Agriculture and Irrigation), Government of India has overall charge of the programme. The State Department of Community Development or its equivalent is in charge of the ANP in the State, the officer responsible is the State ANP coordinator. At the district level, the District Collector will have overall charge of the programme. He will be assisted by other district development officials.

At the block level, the BDO will have overall charge of the programme. He will be assisted by a coordination committee including the head of the Panchayat Samiti, the Block Agricultural Extension Officer, the Mukhya Sevika, the Doctor in charge of the PHC and other block officials.

At the village level, the gram sevika and the gram sevak will be chiefly responsible for the implementation of the programme.

Initiation of the Programme : As soon as possible, after the start of the financial years on first April, the BDO and other officials of the new ANP blocks will receive an orientation to

the ANP and training in the preparation of the Block Plan. The BDO will then begin the work as the block plan is completed. He will submit copies to the State ANP coordinator and to the UNICEF field representatives, if necessary.

The State Government may be asked by the Government of India to submit a modified plan or provide additional information. As soon as the block plan is approved by the Government of India, the State will ensure that all posts in the blocks including those of gram sevikas, are filled. Preparatory activities such as the digging of wells will begin and the training of block and village staff will proceed.

Block Plans : Experience has shown that the condition and needs of blocks all over India and even of villages in the same block can be quite different. It is important to the ANP to plan realistically in terms of the needs and resources of each block and of the villages of that block. The BDO and his staff will be responsible for preparing a realistic block plan based on their local knowledge of the block.

The BDO and other block officials will receive special training in the preparation of a block plan. The BDO and the medical officer under the supervision of the State Nutrition Officer will conduct a survey to identify local nutritional problems. Then after consulting block and village workers as well as the villagers, the BDO will select the villages for ANP activities that will be suitable in these villages. He will write a detailed plan of action for the block. The importance of making realistic plans before ANP activities cannot be overemphasised.

Selection of Villages : To make the programme effective, it is important to introduce as many ANP activities as practical in the village simultaneously. Villages with facilities that will improve the chances for successful implementation of the programme should be given preference. Such facilities may include water for drinking and agricultural purposes, a road, bus services, proximity of a health centre or sub centre,

availability of land for garden. Other government programmes or activities in the villages in a block should be selected for intensive integrated development under the ANP. However, if other villages in the block show interest, they can also be assisted to develop particular activities.

The ANP activities must be planned so that they continue even after the 5-year programme of assistance is finished. If activities are introduced in a village only in the third and fourth year of the programme, it is not likely that they will become established before the end of the programme. Therefore, it is better to choose all of the village in the first two years of the programme and start all activities in these villages as soon as possible

Suitability to Local Conditions : If it is essential the ANP should be tailored to the special needs and conditions of each block and each village. In the past the practice has been to introduce ANP activities according to a standard pattern and a fixed financial limit. This is not realistic and is usually not successful. Now the needs of each village and the activities to be carried out will be used to decide what assistance and how much will be given to each block.

Nutrition education should be carried on in every village selected for ANP. Production activities must emphasise the growing of nutritious dals, vegetables and fruits. However, these activities can be carried out in many ways, depending upon the local conditions and resources.

For instance, if a community garden is not possible because village houses are scattered or because land or water is not available, it may be possible to plan on funds to increase the number of home gardens in that village.

Activities such as poultry or fisheries should be started only if local conditions are suitable and such activities can be maintained. Local officials and villagers must consider carefully the local conditions, resources and needs and then decide on the best plan.

Contribution of People

Unless the community itself is concerned about the usefulness of the activities of the ANP and unless the community is involved in planning and carrying them out, the programme will not succeed.

In making the block plan, the BDO should consult the villagers. He will then be able to assess the problems and resources of the villages and, together with the villagers, decide the activities to be started. The village organisation such as the Mahila Mandals and Yuvak Mandals play an important role in the ANP. Members of these organisations will receive special training. They will then be able to organise such activities as common gardens, nutrition education and balwadi training.

The ANP will not succeed unless the block staff and the villagers understand the purpose of the programme and the rules in maintaining the work. The ANP training that they will receive should explain to them their work and also that of others so that all can work together to make the ANP a success. As soon as possible, after the selection of new blocks, the BDO agricultural officer, education officers, medical officers, Mukhya Sevikas and other important block officials will receive an orientation to the programme and training in the preparation of the Block Plan.

Later, training will be given to gram sevikas, bal sevikas, teachers, panchayat officials and members of Mahila and Yuvak Mandals. The training will emphasise nutrition education. Mukhya and Gram Sevikas will receive practical training in teaching and demonstration. School teachers will be given material on nutrition so that they can make nutrition a part of their regular lessons.

The Government of India contributed Rs. 30,000 per year for each block. The state government contributed Rs.51,000 per year for each block. The United Nations Children's Emergency Fund (UNICEF) will contribute supplies of equipment, cash grants and assistance for training to a maximum value of

Rs. 1,60,000 per block, spread over the operational period of the programme. The figure of Rs.20,000 is a ceiling the UNICEF supports and does not necessarily represent the amounts that will be provided to every block. Actual expenditure in each block will be determined by the needs and capacities of the individual blocks. The central and state funds will be placed in a common pool and disbursements will be accounted for by the state government on block by block basis.

Significance of Mass Level Efforts

The pre-school age is the most crucial period in a child's life. This is the period when the child increases the activity and is exposed to the environment outside the home. Providing a nutritious and wholesome diet and immunising the child against diseases common during childhood are of utmost importance. Equally important is providing education to this age group through balwadis. Mere distribution of food to the children will not have a substantial effect unless the economic status of the mother is also simultaneously improved. For the successful implementation of programmes aimed at the overall development of the child, associate organisations like Mahila Samajams have to be fully involved in such programmes. After taking all the above aspects into consideration the Composite Programme for Women and Pre-school (CPWP) children was evolved and put into operation as part of the Applied Nutrition Programme in all the blocks in the state from the year 1975-76 in a phased manner with the following objectives:

1. To improve the health of the children (both physical and mental).
2. To inculcate a sense of discipline and equality among the children.
3. To promote healthy living habits.
4. To create a desire among the children to go to school and acquire knowledge.
5. To impart nutrition and health education to mothers.

6. To provide additional income, for the mother for giving special attention to the needs of the children.

   The main features of the regular ANP are found in the CPWP also. The CPWP has the following important aspects additionally:

   (i) Education to the children.

   (ii) Health check-up and immunisation of pre-school and pregnant and nursing mothers.

7. The medical officer of the Public Health Centre is expected to visit the balwadi at least once a month and the Auxiliary Nurse- Midwife once a week, for health check-up and immunisation. Growth charts for the children are maintained in duplicate—one kept by the balwadi teacher and the other by the mother. Growth charts to record the health status of the children and one DDS (Diet Drug Supplement) set costing Rs.400 is given by the UNICEF to every balwadi.

8. The target group consists of economically, socially and nutritionally backward children within the age group of 0 to 5 years and pregnant and nursing mothers. In each regular ANP centre 70 children and 30 mothers are enrolled. In a CPWP centre 100 children and 30 mothers are enrolled. The beneficiaries are selected by a committee consisting of the Block Development Officer or his representative, the Medical Officer of the Public Health Centre or his representative, the president of the Panchayat and the president of the Mahila Samajam. The socio-economic and nutritional status of the family will be the basic criterion for selection.

9. The feeding programme under both ANP and CPWP is implemented through organised Mahila Samajams. In the case of CPWP, Samajams having at least five cents of land with a building are chosen as implementing agencies. While sanctioning centres it is ensured that there is no other feeding centre within a radius of one kilometer. In the regular ANP feeding

centre feeding is conducted only once a day. In the case of the CPWP, children attending balwadi classes are fed a second time. In specially recommended cases of severe malnutrition, there is provision for one more feeding. In the regular ANP centres feeding is done 25 days in a month. In the CPWP centres feeding is conducted 26 days a month.

10. The nutritive value of the feed is 14 grammes protein and about 280 calories. The composition of the diet is as follows:

    (i) Regular ANP

    (ii) CPWP first feeding:

grain	80 gms
grain	80 gms
oil	07 gms
milk powder	30 gms
Second feeding: grain	30 gms

11. The cost of the feed, at subsidised rate, works out to 40 paise and 50 paise on an average in the regular ANP and CPWP centres, respectively.

12. In the CPWP centres the balwadi teachers are given an allowance of rupees 100 per month. In all other centres the conveners are paid a monthly allowance of 25. No separate preparation charges are paid. The fuel requirements are met out of voluntary contributions in kind collected by the Mahila Samajams. In fact, the whole programme revolves around the Mahila Samajams.

13. The programme is guided and supervised by the personnel of the Development Department. At the village level, lady Village Extension Officers are in charge of the programme. Their work is supervised by the Block Development Officer and the Extension Officer (women's welfare) at the block level. At the

district level, the Assistant Development Commissioner and the District Women's Welfare Officer, under the guidance of the Collector, control and supervise the implementation of the programme.

14. Apart from feeding, nutrition education, health check-up and immunisation, there is also the production component of the programme consisting of financial assistance and technical advice for goat rearing, backyard poultry, kitchen gardening, a building for balwadi, environmental sanitation and training of balwadi teachers. This is a critical component of the ANP, in so far as the objective is to include and inculcate among the vulnerable sections in the rural areas, the habit of producing and consuming with very little additional strain, foods of high nutritional value. The basic concept behind the package of services and activities under the ANP is itself one of extension, that is, to train and educate the rural poor about the various aspects of nutrition and production of essential food items of nutritive value so that in course of time the people themselves would look after their own nutrition requirements balanced.

One-meal-a-day Programme

15. In 1979, the International Year of the Child, the government decided that supplementary nutrition programme should cover all undernourished pre-school children in the state. Towards the realisation of this objective, the one-meal-a-day programme was introduced in 1979-80. During that year the programme covered 4.13 lakhs of pre-school children through 4,825 feeding centres. The programme is being contributed with the same number of feeding centres. The target group under this programme is children under five years and feeding is conducted 300 days in a year, that is, an average of 25 days in a month.

16. The quantity of food materials supplied under this programme to each beneficiary is as follows:

1. CSM/Bulgar Wheat	40gms
2. Pulses	10 gms
3. Rice	40 gms
4. Oil	10 gms

This meal costs 12.5 paise at subsidised rates,

17. In rural areas, the District Collector is empowered to administer this scheme through the staff of the Development Department. In urban areas this task is entrusted to the Social Welfare Department. The feeding programme is actually carried out by voluntary organisations. In the rural areas the beneficiary children are selected by a committee consisting of the Block Development Officer, the Medical Officer of the Primary Health Centre, the president of the Panchayat and the president of the local Mahila Samaj. In urban areas the Committee consists of the municipal commissioner, the municipal health officer or medical officer and the president of the voluntary agency concerned.

Special Nutrition Programme (SNP)

18. The Special Nutrition Programme was launched as a nationwide programme in 1971, to combat the incidence of malnutrition among the vulnerable groups like children below 6 years, pregnant women and lactating mothers. It was first introduced in the urban slums and tribal areas. Subsequently, it was extended to other areas like poverty-stricken nutritionally-deficient pockets.

19. The objective of the scheme is to supplement the diet of selected beneficiaries with additional protein and calories to make up the deficiency in their daily food intake. Under this programme special sweet bread from Modern Bakeries, special milk bread from private bakeries, milk from Milkman or cooperative societies,

locally available food consisting of rice and pulses, are supplied. The quantity of food supplied contains an average of 5 to 6 gms of protein and 250 to 300 kcals. In addition Adexeline capsules containing vitamins A and D are also being supplied.

20. Originally the number of beneficaries in most of the feeding centres sanctioned during 1979 and 1980 was fixed at 100 per centre. Ten per cent of the beneficiaries in the SNP feeding centres will be pregnant women and lactating mothers and the rest will be children below the age of six years.

21. While introducing the scheme the per capita rate per beneficiary per day was fixed at 23.5 paise consisting of 18 paise for food, 2 paise for transportation of food materials and 3.5 paise for administrative charges. During 1974 the per capita unit cost of food was enhanced from 18 paise to 20 paise per beneficiary per day. Subsequently, the per capita rate was enhanced to 30 paise per head per day (i.e., food charges 24.5 paise, transportation 2 paise and administrative expenses 3.5 paise).

Integrated Child Development Services (ICDS)

22. The latest approach to nutrition is to provide a package of services comprising nutrition, medical check-up, immunisation, referral services, day-care services, education for mothers and functional literacy for adult women. These services are extended with the Anganwadi as the focal point. There will be one Anganwadi for every thousand population. The entire population in a block will be covered by setting up the requisite number of Anganwadis. This scheme is known as the ICDS and is evolved along the lines of the national pattern. As soon as the ICDS starts functioning in a block, the Special Nutrition Centres are merged into it.

23. The ICDS, as a centrally sponsored scheme, has been

introduced through blocks in states. Central assistance is obtained for administrative charges. Equipment is donated by the UNICEF. Only the nutrition component of the scheme is financed out of the state plan outlay. Besides the above, there are ICDS blocks sponsored and financed by the state governments.

24. The food usually being supplied to the beneficiaries under the ICDS are gruel and green gram. The diet costs 30 paise per child below 3 years and 50 paise per child above 3 years. The target group under the ICDS is the same as in the case of the Special Nutrition Programme. The number of feedings in a year is also the same. Of the beneficiaries 10 per cent will be pregnant women and lactating mothers and the rest will be pre-school children.

World Food Programme (WFP)

25. During the Fifth Plan, the Government of India introduced a new scheme, the "World Food Programme" (WFP), as a part of the Special Nutrition Programme. During this period 2,000 feeding centres started functioning under this scheme. During 1979-80 another 300 feeding centres were also started. In each centre, there are 100 beneficiaries out of which 12.5 per cent will be pregnant women and lactating mothers and the rest will be children below the age of 6 years. At present WFP covers nearly 2,30,000 beneficiaries spread over 2,300 feeding centres.

26. Uppuma made out of soya fortified Bulgar wheat and butter oil/salad oil are being distributed to the beneficiaries under WFP as detailed below:

    Soya-fortified Bulgar wheat is given to pre-school children (0-6 years) 80 gms per day per head and to pregnant women and lactating mothers 125 gms per day per head.

    Butter oil/salad oil is given 10 gms each to all beneficiaries.

Organisational Arrangement—Social Welfare Department

27. The feeding centres under SNP and WFP in the State are run by the local bodies like municipalities, corporations, panchayats and voluntary organisations like Bharat Sevak Samaj, Gandhi Smarak Nidhi and Manila Samajams. In the case of SNP feeding centres there is one organiser and one helper to distribute the food materials to the beneficiaries. An amount of Rs. 20 and Rs.10 per month respectively is being paid to them as honorarium. The transportation of food materials to the SNP feeding centres is being arranged by the Government through the contractors. Under the ICDS the Anganwadi worker gets an honorarium of Rs. 175 per month and the helper gets Rs.50 per month. In the case of WFP feeding centres, an allowance of Rs.40 per month is being paid to the voluntary organisations as grant. In addition to this the local bodies like panchayats are given financial assistance for the purchase of cooking vessels to the feeding centres. The transportation of food materials from the Warehousing Corporation Depot to the feeding centres is being done by the voluntary organisations themselves. They are also permitted to sell the empty containers and to utilise the amount for meeting the transportation cost.
28. The Health Services Department operates with this department to provide health services in the nutrition programme. Medical check- up and immunisation to the beneficiaries in the centres in urban areas are being conducted by the health staff of the corporations and municipalities and those in the rural and tribal areas by the Medical Officers in Primary Health Centres. Government has accorded sanction to meet the additional expenses towards petrol charges in connection with the visit of Medical Officers of the Primary Health Centre to the nearby feeding centres

of WFP for the conduct of Health Education, Nutrition Survey and Immunisation.

29. The organisers and helpers are the nominees of the voluntary organisation responsible for implementing the nutrition programme. In every centre, there is an organiser and a helper. The organisers are responsible for the proper maintenance of printed records such as admission register, attendance register, stock register and medical check-up register. They are also vested with the responsibility of receiving the food supplies, their proper preservation, storage and distribution to the beneficiaries. The helpers are responsible for the preparation of food articles for distribution. They also render assistance to the organiser for spot feeding of children.

30. There are two officers, one in the gazetted cadre (District Social Welfare Officer) and the other in the non-gazetted cadre (Inspector) for each district for the supervision of the feeding centres. Four inspectors in the non-gazetted cadre are appointed in the directorate for attending to the scrutiny of the bills of the SNP. The supervisors are entrusted with the work of supervision of feeding centres, arrangement of Nutrition Education Camps, selection of location of feeding centres, payment of honorarium to the organisers and helpers of feeding centres, allotment of food materials consistent with the number of beneficiaries in each feeding centre, allocation of cost of food materials and transportation charges.

31. One-day training camps at the rate of two camps in each district are being conducted by the department for giving nutrition education and training to the organisers of the feeding centres and to the mothers of the beneficiaries. The importance of nutrition, child care and family planning are being taught by the Medical Officer and Block Development Officer, in the

camp in the feeding centres given to the supervisor-inspectors by the professors of medical colleges.

32. Some feeding centres in the States have now become defunct for various reasons such as non-availability of special bread, practical difficulties in serving cooked food made out of rice and pulse and irregularities noticed in the supply of food articles.

    The recommendation of the District Committee is therefore necessary for allotting and cancelling and redistributing the centres to voluntary agencies. These committees are not being convened at regular intervals. Hence, much difficulty is being experienced in the matter of starting, cancelling and redistributing the centres.

Health-Based Nutrition Programme

33. The Department of Health Services implements mainly three nutrition programmes of which two are meant for prophylaxis against nutritional anaemia among mothers and children and against blindness in children caused by vitamin A deficiency. Under these two programmes only iron and vitamins are supplied to the beneficiaries. Feeding is conducted only under the Integrated Health Package Programme. A brief description of the above programme follows:

Prophylaxis against Nutritional Anaemia among Mothers and Children

34. This programme was started in the year 1970-71. The family welfare programme supplies the cost of drugs as grant to the state government. Expectant and breast-feeding mothers, acceptors of family planning methods like IUD, tubectomy and pre-school children receive the benefit of this scheme. Tablets containing iron and folic acid are distributed through AN clinics, PH clinics and paediatric clinics of hospitals, PHCs and FW centres under the guidance of Medical Officers.

35. Haemoglobin percentage is obtained before starting the programme and at 3 months interval wherever facilities are available. The drugs are issued for a week at a time. Each beneficiary gets 100 tablets.

Adult: 60 gms iron   0.5 mgs folic acid

Child: 20 gms iron   0.1 mgs folic acid.

Difficulties experienced are irregular supply and lack of motivation to come to the centre. The most deserving beneficiaries often do not come to the centre or do not come regularly.

Prophylaxis against Blindness in Children by Vitamin A Deficiency

36. Blindness due to vitamin A deficiency is more prevalent in the preschool age group. Studies have shown that vitamin A in oil given orally is readily absorbed and stored in the liver from where it is gradually released for utilisation in the tissues. Vitamin A is procured by the Department of Family Planning and distributed to the State Health Department.

37. The vitamin A capsules are administered to children in the age group of 1 to 5 years in homes, personally by peripheral workers like AIM and EPHA. Two lakhs of IU/60,000 microgrammes of vitamin A is given at an interval of 6 months till the children attain 5 years of age. In view of the limited financial resources the programme is restricted to areas where there is maximum prevalence of vitamin A deficiency.

3

Nutritional Schemes

In India, several dietary and clinical surveys carried out in different parts of our country have revealed that malnutrition is rampant. Nutrition education assumes special significance in the Indian context because the problem of malnutrition in India is mainly due to ignorance, poverty and lack of knowledge regarding the value of foods. Dietary practices, especially in children and pregnant and lactating mothers, are often governed by social taboos based on food fads. Nutrition education is the foundation for improvement in the dietary habits of his vulnerable set of population. Rigid dietary habits need correction and only systematic nutrition education programmes can bring changes in dietary habits. Nutrition awareness is essential to bring changes in dietary habits and creating nutrition awareness entirely depends on education and training.

Nutrition education requires identification of needs and collection of baseline data enables to identify the needs. It is a popular belief that young mothers or adolescents must be educated about the basic principles of health and nutrition. Studies on nutrition education in our country have proved that

even elderly women need education as they are the ones who mislead the youngsters. Various topics of nutrition can be selected for nutrition education. Selection of foods and preparation of foods, feeding practices of various groups, special diets during pregnancy and lactation, infancy, pre-school age, supplementary feeding, weaning foods, low cost recipes, food sanitation, food-borne infections, food adulteration, nutrition and health, therapeutic diets, food fads and fallacies and deficiency diseases and their prevention are some common topics used for nutrition education.

Extension education methods can be utilised for nutrition education. Collection of background data, analysis of problems, plan of action, calendar of activities, selection of audio-visual aids, methods of teaching and evaluation are the various steps involved in nutrition education programme.

The various approaches suitable for nutrition education are (1) Individual contacts through home visits, consulting individual training, (2) Group contacts through simple talks, demonstrations and discussions, (3) Group demonstrations, (4) Hospital visits, and (5) Nutrition rehabilitation training. These are the methods suitable for rural groups.

Nutrition education and training can be implemented through schools and colleges. Good nutrition education, if followed effectively, can be a useful tool to bring economic development in the country as it provides good health. Better health enables people to improve their productive efficiency and it enhances the purchasing capacity and the consumption patterns of the population. Food and nutrition is thus intimately linked with production and economic development of a country.

Training in nutrition at various levels—schools, colleges, adult education centres, village level workers—is essential. Housewives are ignorant about the principles of nutrition as they are not trained anywhere. Thus the researches on these lines are wasted as the messages are not passed on to the people. Various feeding programmes and nutrition programmes which are implemented through various

organisations can be used as nutrition education channels. Feeding programmes under the family and child welfare departments, school feeding programmes, the composite programme for women and pre-school children, the applied nutrition programmes, world food programmes and other state programmes can be utilised for education purposes.

Various Nutrition-programmes

A number of nutrition programmes were initiated by the Government of India. Most of them are functioning under the guided supervision of development department, social welfare department and voluntary agencies in various states.

Applied Nutrition Programmes : The applied nutrition programme (ANP) is evolved on the basis of the fundamental premise that though malnutrition is primarily due to mass poverty it is also due to ignorance and that appropriate choice of foods and feeding practices without involving extra cost from the family can improve the nutritional status of the people, especially young children.

Objectives and Inducements

It is well known that many young children in India are undernourished. This undernourishment is dangerous. If children do not receive adequate nourishment when they are young, they will never grow to be strong, healthy and productive adults. Furthermore, undernourished children are more likely to become ill. Undernourishment is directly or indirectly responsible for illness and death.

The goal of ANP is to prevent the bad effects of undernourishment before they can occur. The ANP works at the village and family levels to improve nutrition, particularly of young children. To bring about such improvement the programme supports the local production of good and nutritious foods, consumption of such foods, especially by young children, pregnant women and nursing mothers, and education about good foods and proper feeding of young

children and women who are pregnant or nursing (nutrition education).

Improvement in nutrition will not be effective unless there is also improvement in health. The ANP should be implemented with local health workers to provide health check-ups, immunisation, safe drinking water and proper sanitation. The ANP does not have enough funds to assist all the village in a block. Therefore, it is necessary to choose those villages which have a reasonable chance of success. As many ANP activities as possible should be started in the village selected.

The ANP also does not have enough funds to cover all the families in the village selected to participate in the programme. Care should be taken to reach the most disadvantaged families such as the families of the Harijans, tribal people, small cultivators and landless labourers, which have young children, pregnant women and nursing mothers.

By concentrating on these selected villages and families it should be possible to demonstrate clearly the results of the programme. This should lead to the voluntary use of beneficial practices in other villages of the blocks which is called the radiation effect

Education of Nutrition : People will not change their beliefs and habits concerning food unless they are convinced that changes will benefit them and their children. Therefore, the goal of ANP is to educate village families about the benefits of nutritious foods, good child feeding practices and improved preparation and preservation of foods.

Many traditional practices such as prolonged breast-feeding and eating roti and rice with dal are basically good. Nutrition education should emphasise such traditional practices and at the same time should teach new practices where necessary. By using nutritious local foods and good feeding practices it is possible to improve the health of undernourished children at little or no extra expense to the family.

Nutrition education should be provided in cooperation

with schools, Balwadis, PHCS, sub centres, Manila Mandals and Yuvak Mandals. It can be given by people such as school teachers, Balsevikas, Mukhya Sevikas, Gram Sevikas and medical staff to the PHC and sub-centres. Efforts should be made to bring this eduction to the poorest villagers, who may not ordinarily participate in village organisations. The education should stress on basic messages about nutritious foods, child feed and child care.

The method of practical demonstration with the participation of the villagers ought to be used as much as possible. In the demonstrations, only local food and easily available equipment should be used; charts, posters and film shows should also be used to make the information clear and interesting.

Main Messages : Breast-feed as long as possible. Introduce some solid food 5 to 6 times a day. Do not reduce food in illness. Use health services available. Get children immunised. Keep yourself and your surroundings clean. Drink clean water. Have no more than 2 or 3 children 2 to 3 years apart. Try to eat some green leafy vegetables every day.

Activities in Production : The ANP gives many kinds of assistance for production of good nutritious foods depending on local conditions. The programme now emphasises assistance to home gardens, school gardens and community gardens. These gardens are used to grow dals, leafy vegetables and other nutritious vegetables and fruits. The products are then used to demonstrate the preparation and use of the foods.

Gardens should become as self-sufficient as possible. After the first season, seeds should be obtained either from the garden itself or from the sale of surplus products. Local manure should be used as much as possible. When available, waste water from the village should be used for irrigation.

Home gardens should be emphasised where possible, because of their importance in making nutritious foods available to the individual.

School gardens will be supported if the following conditions are met—suitable land is available close to the school and water for irrigation is available at a reasonable cost. The garden is made part of the regular school programme of the students and the students must look after the garden themselves.

Community gardens can be used both to demonstrate the proper cultivation of nutritious food and to produce food for feeding purposes. Cash crops may be grown to make the gardens self-supporting but the entire net profit of the garden must be used for demonstration feeding projects at health centres or balwadis. The garden should be run by a village organisation such as a Yuvak Mandal or Mahila Mandal.

Foods of Animal Origin : The ANP now emphasises the cultivation of nutritious dals, vegetables and fruits in home, school and community gardens rather than the production of foods of animal origin, which are much more expensive. Moreover, if local conditions are suitable, activities such as poultry, fisheries, piggeries and goat keeping can be supported.

Consumption Proceedings

Feeding Projects : Feeding projects under ANP will be chiefly for the purpose of practical demonstrations for nutrition education.

Agents for Change : The ANP cannot succeed unless villagers themselves accept the need for new practices. Therefore, the cooperation of village leaders is essential in planning and running the programme in each village. The support of women and youth is also important.

4

The Backdrop

Types of Cooking

French Cuisine

It is recognised by the world as one of the finest cuisines. The French are known for their artistic temperament and so is their cuisine.

History of French Cuisine : At the end of the Merovigian period, the convents which had preserved all the traditions of the gourmands increased in France; it was the beginning of culinary art.

Violet le Duc, in the course of his studies, said that the kitchens were large and high spit roasting was common, dishes were cooked on the fire itself. Ovens were installed in the twelfth century for preparing roasts. From the fourteenth century, sauces were a much appreciated part of culinary art. Menu was varied which included game, fish, fowl, butcher's meat. Vegetables were featured less.

At the end of the sixteenth century, Italian cooks and

pastry cooks were employed. They taught and prepared a number of dishes, which since then, remained in the French repertoire, e.g. cannelloni, revioli, gnocchis (Farinaceous dishes).

In the seventeenth and eighteenth century, Royalty and high ranking people took great interest in food and the cooks in gratitude named dishes after them. The cooks during feasts or banquets were honoured if the food was good and appreciated. If something went wrong, they were flogged in front of the guests. Confectionery, foreign dishes such as Caviare, Punch, etc. became popular.

Since the end of the nineteenth century, French culinary art has reached the highest point of perfection. Chefs like Escoffier and Prosper Salles, etc. published books on cookery, which had far-reaching influence and are known throughout the world.

France's contribution to the international culinary repertoire is well known. She has bestowed her own names on the majority of dishes creating technical terminology of international usage.

Gastronomy (style or custom of cooking or eating) is an art with defined rules; the same is true of its presentation. For French Cuisine- cooking and law are quite similar. With both, there is the challenge of problem solving, logic and reasoning.

Style of Cooking : French cooking is less complicated while cooking the garnishes served are more complex. It has a delicate flavour , as spices are used sparingly and they create a suspicion in one's mind, if they are incorporated in the dish. Great emphasis is laid on sauces and it is even called sauce *cookery.*

Every ingredient speaks its own language, Wines used for cooking enhance the taste, give an un-usual flavour, this being a unique feature of French cuisine. The most important rule is to use only the best ingredients and the proper blend of colour and taste of ingredients.

Garnishes and Accompaniments : Garnishes and

accompaniments play a key role in French Cuisine and dishes are identified by them. Food items are cooked in sauces or they accompany a dish, e.g. roast beef is served with Yorkshire pudding and is accompanied with horseradish sauce.

Great emphasis is laid on the presentation of the dishes, and garnishes and accompaniments make the dish colourful.

Ingredients play an important role as it has climatic variations in the South of France, the Mediterranean heat adds flavour and richness to fruits, vegetables and herbs. In the northern region, finest fish is available. In certain regions, mushrooms and hard cheese are popular. The wines for cooking come from Bordeaux, Champagne and Brittany.

The cooking medium used is butter, olive oil or salad oil, thus the dishes are not greasy, but give a lightness to the taste.

Spices and herbs give a subtle flavour, taste and originality to the dishes. Some of the spices and herbs are—parsley, celery, basil, tarragon, thyme, rosemary, chives, shallots, allspice, bayleaf, paprika, pepper, nutmeg etc.

Popular continental vegetables are—asparagus, artichokes, carrots, turnips, aubergines, broccoli, swede, seakale, cardoon, Brussels sprouts, Topinambow, mushroom, etc.

Vegetables are served with main courses and serve as one of the accompaniments, they act as fillers to meat dishes. They add colour to the menu and make them colourful. Vegetables are arranged as a painter arranges colour. A meal is completed with potatoes as they give a satiety value. They also accompany the main dish.

The main dish consists of flesh foods such as lamb, beef, veal, pork, chicken, duck, turkey, fish, etc. The fleshy foods should be of the best quality.

Cheese is extensively used and can be used as a base of dish, garnish to soups, blended with sauces, or grated and sprinkled on top of the dishes and gratinated. Cheese in a French menu forms a separate course during the meals and

is called *"Fromage"* in French. Almost every region of France has its own cheese and a few popular cheeses are Camembert, Neufchatel, Roquefort, Brie, etc.

Red and white wines are used for cooking. Wine is much more than just a flavouring agent, if used with discretion. It gives an unmistakable fillip to the simplest dishes, regouts, and in marinades. It acts as tenderisers for the drier and tougher cuts of meat. Sauces have the quality of giving contrast or blend with the ingredients, enhance taste, flavour and give colour to the dishes.

Chinese Cuisine

Chinese culture is the oldest civilization in the world. In unrecorded history, they were the first discoverers of fire. Friendship and food were inseparably linked. A gathering without food was considered incomplete and improper.

One thousand years before the birth of Christ, they were already experimenting with fancy cooking, recording their recipes on silk and bamboo. They are considered the best in the eastern World.

Famines greatly influenced Chinese cooking. Adversity forced the Chinese always to seek new sources of food. They have been gastronomically courageous, setting no limits. They put everything edible to use, finding wonderful ways to utilise strange plants and roots, lily bulbs, chrysanthemum petals, bamboo shoots, jasmine flowers, etc.

The Chinese have a strong sense of artistry which finds impression in Chinese cooking. Every aspect of food is taken into consideration palatability, texture, fragrance and colour. Every dish is given proportion, balance and harmony.

A few factors are common in Chinese and French cuisine. The French and Chinese are skilful cooks. Both seek simplicity as their ultimate goal, the simplicity that comes with the art of diligence. They both recognise that second class ingredients can never produce first class dishes.

Chinese cooking calls for maximum preparations of raw ingredients and minimum cooking method. "Fan & Cai"—the dishes have balance, combination and blend of colour, flavour, texture, shape and size.

Techniques of Chinese Cooking : The Chinese saute, steam, deep fry and roast with a difference. Baking is rarely done. Stir frying is the most popular method of cooking. It is similar to sauteing but it is done on intense heat. It is because of the short cooking time that fresh foods have to be used as *they need less cooking time.* The heat seals the juices and preserves colour, texture. Ingredients are cut into beautiful shapes, yet flavour is never sacrificed to appearance as it brings joy to eating. The ingredients are combined and contrasted—cold, tiny ingredients against big ones, dry dishes against those with gravies, sweet taste is contrasted with sour taste. They set crunchiness against smoothness. Crisp tender vegetables surrounded with exquisitely smooth sauces, stimulate the senses with contrasting tastes.

Various Ingredients : Basic ingredients that give Chinese cooking its characteristic and subtle taste are soya sauce, peanut oil, sherry, garlic, ginger root, pepper, chilli sauce, vinegar, corn-flour, noodles and sodium glutamate, bamboo shoots, bean sprouts, dried shrimps, shark's fin, pork, ham, eggs, scallops, etc.

The ingredients are usually combined and seldom cooked well. The bite size pieces in the dish aids the use of chopsticks for eating.

Five regional cuisines predominate, i.e. the South-Eastern—Canton, Western—Szechwan, sometimes called Chungking, North Eastern— Peking or Shantung, Central—Honan, and East Coast—Fahien.

Cantonese (South-East) : The various regional cuisines were influenced by many factors, such as in Cantonese, it was influenced by Western travellers. When they migrated, they established Chinese restaurants outside China.

Shantung Style : Shanghai was influenced by the active trade between Peking and Shantung. Peking, now called Beijing, was the site of Imperial Palace. Concentration of wealth attracted the country's best chefs, which brought cooking to its highest level. It had a reputation for mammoth feasts and banquets.

Szechwan Style : The province is hot, almost tropical. The food is highly spiced, peppery and somewhat oily. Deep frying is popular. Chicken Chilli Fry, Szechwan Duck is a favourite dish of this region.

Hunan Style : The yellow river runs through the Hunan province which is famous for its Yellow River Crop. It's known for its spiced concoctions, sweet and sour dishes and rich seasonings.

Fukien Style : Fukien province on China's east coast is famous for seafood and for clear light soups. They are known for their quality. Wine is sometimes used for its cooking.

Jiangsu in the last represents the four major original cooking styles of China.

The Chinese divide the temperature of heat into 'Military' (high or fierce, and medium) and 'Civil' (low or gentle, and weak) heat. Proper control of temperature and cooking time is the key to success or failure.

High or Fierce Heat : It is usually used for quick cooking of crisp and tender foods. Different kinds of frying, steaming, instant boiling, etc. all call for a high heat.

Medium or Moderate Heat : It can be used for quick-braising, stewing and boiling.

Low or Gentle Heat : It is used for slow cooking, allowing the flavours to penetrate through all the ingredients such as roasting and simmering.

Weak Heat : It is used for long-cooking, turning hard ingredients to soft ones. It is used for simmering, braising and stewing.

Here are some commonly used methods in Chinese Cooking.

One dish may require one, two or three methods; each will produce a different effect.

Chuan—Quick or Rapid Boiling : This simple cooking method is often used for making soups. Bring the water or stock to boil over high heat, add the ingredients and seasonings, serve as soon as the soup reboils. No thickening agent is added and the vegetable will be crisp and fresh.

Shuan—Instant Boiling or Rinsing : Thinly sliced ingredients are dipped into boiling water for a second or two, occasionally dunking it as if rinsing. Then serve with a dip sauce. This cooking method keeps the ingredients fresh and tender.

Ao—Stewing or Braising : Flavour a little hot oil with spring onions and ginger root, then stir-fry the ingredients for a short time.

Now add stock or water and seasonings. Simmer over a low heat. The food should be soft and tender.

Hui—Braising or Assembling : It is a method of cooking a dish that consists of several different ingredients. Stir-fry the ingredients first, add stock or water and seasonings, boil over high heat for a short while, then thicken the gravy before serving. Alternatively, prepare the gravy first, then add the partly cooked ingredients (deep fried or steamed), cook over low heat, thicken the gravy and serve.

Ban—Mixing Salads : This method does not actually involve cooking, but simply calls for cutting the raw or cooked ingredients, and dressing it with seasonings.

Giang—Hot Salads: Here the raw ingredients are parboiled or blanched first, then dressed with seasonings.

The difference between cold salad's and hot salad's dressing is as follows:

Cold Salad Dressing : Soya sauce, vinegar and sesame seed oil.

Hot Salad Dressing : Ginger shreds, sickwan peppercorn, salt, sugar and sesame seed oil.

Yan—Pickling : Pickle the food with salt and sugar, or with salt and wine. Dishes prepared this way have a subtle fragrance and crispness.

Jian—Shallow Frying : A flat-bottomed pan is used with little oil and medium or low heat. Seasonings are added when food is half done. The pan should be turned from time to time during cooking so that heat is evenly distributed.

The Chinese restaurants serve dishes which have mixed the styles of cooking. Cantonese is the most popular all over the world with Szechwan being the second popular style.

Indian Cuisine

Indian cuisine can be said to have evolved along lines parallel to Indian history. India has ancient cultural heritage which is dependent on religion, geography and socio-economic conditions. Traditionally Indian cooking has been handed down through the generations by demonstrations and word of mouth.

Strong impact was made on Indian cuisine during the reign of Mughals in the sixteenth century, who were fond of good living, cooking and eating. Muslim cooking was based on meat. Their influence was strongest in North and Central India. Since then, well-known Mughlai dishes were evolved which have soon developed into an important culinary art and is part of Indian cuisine. Extreme South, the Mughal influence was less, where cooking is mainly vegetarian. The regional dishes from different parts of India are influenced by religion, availability of ingredients and old traditions and customs, many of them being influenced by the different invasions. Thus it is a combination of cookery of many nationalities.

The preparations of various dishes have been handed down from one generation to another, from mother to daughter in homes, father to son working in Maharajah's kitchen, or those working as *Khansamas* and *Halwais* in catering establishments.

Pictures depict the keen interest the kings took in the preparation of food as they would sit giving instructions and supervising the cooking. It is the knowledge of our ancestors that has been handed down from one generation to another.

The cooking habits of India vary not within the religious communities but from area to area.

Regional cooking is influenced by the staple food of the area—in the North, wheat grows, the food is drier and the gravy is thicker. In the South, where rice is the staple diet, gravies are thinner. Climate too plays its role and because of the heavy rainfall, vegetables grow in abundance and make vegetarian cuisine varied and exciting.

Various Ingredients : The heart of Indian cooking is *masala,* the combination of spices and herbs which gives each dish its individuality. It is a fresh mixture of spices, mild and delicate or sharp and strong that may bring tears to the eyes. Masala may be in paste or powder form; spices are ground with vinegar, water or curd to form a paste. Masalas are dried in the sun or roasted on tawa and pounded into a powder form and kept in airtight containers. The combination of spices is endless and each cook follows his own taste and regional preference. Garam *masala* usually contains black pepper, cardamom, cinnamon, cloves, cumin seeds and javitri. It is ground into powder and is sprinkled over a dish just before serving. Red chillies and green chillies are essential to most Indian dishes. The common spices are ginger, garlic, onions, haldi (turmeric), dhaniya powder, jeera, methi, rai, hing. Fresh herbs are dhaniya leaves, curry patta, pudina and tulsi.

Acidic ingredients used are tamarind, curd (yoghurt), anardana (pomegranate seeds dried), lemon juice, vinegar and amchur (raw mango dried and powdered).

Hindus never eat beef, because the cow is sacred for them. Pork, mutton, chicken and vegetables are widely used. Indian Muslims do not eat pork but eat beef. Game birds are a delicacy; teetur and batair is a special favourite dish. The people staying

in coastal areas eat fish because it is cheap, delicious and plentiful. Some of the popular fishes eaten are: pomfret, rahu, Mumbai duck, malli, carps, bhetki, surmai, Indian salmon, trout, shrimps, prawns, sole, tuna, crayfish, crab, etc.

Indian breads are prepared with different methods of cooking ; poories, kachories, lucchies, bhaturas are deep fried, parathas are shallow fried and bhakar khani, nan and tandoori rotis are baked (Tandoor).

No Indian meal is complete without dal (pulses) and there are a variety of dals such as moong, channa, urad, arhar, masoor, lobia, rajma, gram, etc.

Vegetarian Cookery : As vegetables and fruits are found in profusion, there is an accent on vegetarian cookery in India. Many of the vegetables that are available in the West are found in India. The Indian vegetables are — laisvai, karondas, plantain stem and flowers, kamal kakri, khol khol, snake gourd, amaranth leaves, sarson ka saag, ganth gobi, tindas, parval, kathal and amla etc.

The cooking mediums used in various regions of India are—pure ghee- (clarified butter), vanaspati (hydrogenated oil), mustard oil, sesame oil, coconut oil, peanut oil etc.

Popular methods of cooking are frying and boiling. Traditional method used is spit roasting. A tandoor is a small Indian oven made of mud. Bread is flattened with wet hands and stuck to the inner sides of the heated tandoor and cooked. Chicken boti kabab, seekh kababs are skewered on an iron rod and it is placed in the hot tandoor and cooked.

Every Indian meal is a feast in itself. Typical way of serving Indian meals is in a thali (round metal plate) with small katoris (small metal bowls) placed all round in it. There are no different courses, all the dishes are served at one time including the sweet dish.

Usually a complete meal consists of dal (bland), mutton or vegetable curry (spicy, rich) sookhi bhaji (dry vegetable), raita (curd with mint to potato, etc), pickle/chutney (sour and spicy),

papad (crisp and crunchy), rice/bread/(chapaties, parathas, etc.) which is the base of a meal. Sweet dishes—kheer, srikhand or halwa or sayvain or phirni, etc. The Indian dishes look exotic, colourful and rich.

Tandoor is a mud oven shaped like a drum, with the top side open which is narrow compared to the base. The inner sides are plastered with mud. Wood is the fuel used.

Some popular regional dishes come from different parts of India such as:

Bengal	Sukhto, Ponch, Phooraner, Charchari, Macherjhal, Sandesh, Rasgullas
Kashmir	Gushtaba, Rishta, Kamargah, Methi, Chaman, Muthanjan
Gujarat	Amti, Srikhand, Khaman Dokla, Patrail, Irani
Maharashtra	Samosas, Khandvi, Dhansak, Baffat
South India	Dosa, Pulio darai, Murukku, Avial, Pachadi, Bondas, Sambar, Bisi-bela—Huliyana, Appam, Mysore pak, etc.
Punjab	Mukandwadi, Bhatura, Chhole, Sarson ka saag, Makki ki Roti.

Middle Eastern Cuisine

The Middle East countries language is "Arabic" throughout the region in its numerous and varied dialects. The religion is that of Islam. The various countries are inextricably linked culinary wise. Middle Eastern cooking, though sometimes elaborate, is easy. Some of the dishes, such as stuffed vine leaves may take some time to prepare. Lamb and mince are the favourite meat eaten. Cheap cuts of meat are used extensively. Alcohol is not used owing to the Muslim dietary law. Vegetables are used in abundance. Vine leaves, pine nuts,

cabbage, baniva (okra) lentils, chick peas, bitter herbs, olives, figs, grapes, raisins, dates, almonds are extensively used. Fish are eaten widely along the coast line.

A dish for the Kind consisted of hot and cold meats, rice jelly, stuffed vine leaves, marinated chicken and a sweet date puree. Some of the popular dishes are : meat grilled on the spit, meat fried in butter with a sauce, pudding made of eggs, honey, milk, butter, rice and sugar. A Dehkan dish, consisted of slices of salted mutton with pomegranate juice served with eggs. Veal and beef are cooked with spinach and vinegar. All kinds of game and poultry are also performed. Meats are usually marinated in yoghurt and flavoured with spices. Almond pastry is popular. Jams are made with quinces, dates are stuffed with almonds and walnuts. Pastries such as baklavas and Konafa are prepared.

Couscous and Burghul (cracked wheat) were adopted in most Middle Eastern countries.

The activities of cooking and eating reflect many subtly intricate facets of Middle Eastern character and ways of life. They are social activities. The religion of Islam is the most important part of Middle Eastern culture and the main foundations of the customs and traditions of the region.

The following is forbidden:

1. Use of animals dead before they are slaughtered, or those killed for reasons other than that of food.
2. Blood
3. Pig's flesh
4. Forbidden is the use of alcoholic or fermented liquids.
5. An animal that is being killed, the butcher will have to say "in the name of God, God is most great! and then cut its throat, "halaal" and let it bleed.

Cooking fats used were "alya" rendered fat from sheep or lamb. Nowadays oil, ghee, Samna i.e. butter, are used. Oils used are olive, sesame and corn oil,

A large variety of spices are used. Favourite combinations of spices or herbs are packed together and sold. Commonly used are; cumin, coriander, cinnamon, ginger, hot pepper (ayene) aniseed, sesame seed (taklia), ground coriander fried and crushed garlic.

Another mixture is cinnamon, nutmeg, cloves and ginger (four spices). The herbs used are coriander leaves, dill, mint, chervil, oregano and marjoram. Saffron is highly prized. Middle Eastern countries are — Syria, Lebanon, Egypt, Iran, Turkey, Greece, Iraq, Saudi Arabia, Yemen, Sudan, Algeria, Tunisia and Morocco. All are inextricably linked culinary wise.

The early origins of Middle Eastern food can be found in Bedouin dishes. The ancient inhabitants of the Middle East were prosperous and highly civilised states have developed luxurious culinary traditions, undoubtedly influenced by Greek and Roman customs because of the invasions and conquests.

New Zealand Cuisine

New Zealand settlers unlike their Australian counterparts, managed to graft onto the dreary "home" cooking which they brought from less well fed houses of Britain, a little livelier tradition they found in the land they took over.

Their produce is superb when fresh (rather than shipped, chilled or frozen halfway round the world) especially the dairy produce and the lamb. As in Australia, it is at its best when simply prepared.

The Maori influence however produces lots of interest, marinated raw fish is common. Paua (Abalone) pipis (which tastes like cockles) and toheroas (like clams) are excellent shellfish. The white bait in batter, paties are good. Maori vegetables are unusual puha and rauriki (sow thistle) are a little like spinach. Kumara is the sweet potato.

The Maori way of cooking meat, fish and root vegetable is in an earth oven or hangi. A hole dug in the ground. They are wrapped in flax leaves and placed on fiery stone on top

mnuka or tea tree wood which gives a special flavour. Water is poured in to produce steam and then the oven is sealed with special mats and earth.

Several hours later the food is supposedly cooked to perfection. The climate guarantees an extensive range of fruits, the most exotic being the Kiwi fruit, tamrillos (tree tomatoes) and passion fruit.

Australian Cuisine

It's a fine cuisine, took centuries to establish and it requires an indolent aristocracy to refine it. Australians have superb produce—marvellous beef and lamb, wonderful fish, first class vegetables and exotic tropical fruit. Nevertheless, these democratic people have never been much concerned with food.

There was no native cuisine to build on, the aborigines being nomadic and given to half roasting whole kangaroos and eating them dripping with blood. The settlers themselves were too busy settling to bother with anything but solid food to give them energy. Many of the interesting terms which describe uniquely Australian cuisine, are inherited from early pioneers and Bushmen who moved from place to place carrying their swag and billycan. Since the war, there has been a wave of immigrants from Europe and things looked up.

Most Australian meals have as their main ingredients, meat, chicken or sea food. The quality of Australian produce is high so even a simple meal of grilled chops and vegetables is a culinary treat. Famous Australian dish carpet bag steak, a giant piece of thick rump steak slit down the middle and stuffed with few oysters seasoned with salt, pepper and lemon juice and baked dishes made from Sydney rock oysters, snapper and Baramundi fish are popular. Plain steaks, roast lamb are eaten in abundance by Australians. You will constantly encounter Peache Melba invented in England by Escoffier for Dame Nellie Melba, the great Australian soprano. You will also see the meringue and passion—fruit cake made for Anna Pavlova.

Good wines are produced in Australia.

Australian Food

Kangaroo Steak : Loin of Kangaroo filled are used and roasted.

Anzac Biscuits : They are made with rolled oats, coconut, flour, sugar, golden syrup, batter, bicarbonate of soda and water. They were made and sent in tins to Australian soldiers at the battlefields during World War I.

Meat Pie : Individual mince meat pies, small in size. They are traditionally sold at sporting and other community events, usually smothered with tomato sauce.

Balmain Bugs : Unusually shaped flat crustaceans found in Sydney harbour. They have a mild flavoured white flesh and are usually barbecued or boiled.

Billy Tea : Tea made in a metal can called a billy, over an open fire. The billy has a wire handle and a tight fitting lid. Water is brought to the boil, the lid removed and tea added. The billy is then removed from the fire and tea drunk from metal cups.

Cockys Joy : A tin of either golden syrup or treacle was one of the basics that swagness used to carry with them. It had many uses like sweetening billy tea, spreading on damper or pouring over puddings.

Cuppa : A cup of tea.

Damper : A basic bread originally, made simply with flour and water, kneaded, shaped into a round and cooked on the coals of a fire made on the ground. Raising agents are now used to make damper nowadays.

Some campers made the dough and moulded a small amount around one end of a thick stick. The stick is then held over a fire. When the damper is cooked, golden syrup is poured in the hole left by the stick. It is also known as bush tucker. Tucker is a colloquial term for food.

Fish : Australia's coastline is a breeding ground for large fish like snapper, garfish, leather jacket, jewfish and flat head

are some of the most common types available. John Dory and Barramundi, both flavoured fish are very popular and highly regarded.

Granny & Smith Apples : Greens Kinned variety with crisp juicy flesh. Extensively used in cooking and are also an excellent eating apples.

Jumbuck : Young sheep.

Lamingtons : Pieces of light sponge cake dipped in chocolate icing and coconut.

Rock Oysters : Popular edible molluscs found on Australian seashores. Eaten straight from the shell, but some like them grilled or poached.

Pavlova : Basically a large meringue served with fruit and cream. Australian claim it was invented by a chef in Western Australia for the touring Russian dancer—Pavlova.

Tropical Fruits : When in season, mangoes, paw paws and pineapples are available. Kiwis and passion fruits are very popular.

Nouvelle Cuisine

In the early 70's chefs wanted to bring about a change in French Cuisine. It was done by H. Gault and C. Millan, as they were fed up to eat the same classical food everyday.

P. and J. Troisgros (brothers), M. Guerard, A. Chapel, R. Verce and Paul Bocuse spoke of changes and the revolution was started by emphasising on serving original salads, vegetable terrines, fish just cooked on the bone; and the prices were high. It took about 20 years for people to take Nouvelle Cuisine dishes. It was easier to get the recognition as many restaurants were not offering the French classical dishes.

The Nouvelle Cuisine was influenced greatly by the monthly publication by Gault and Millan. They increased from 10,000 copies to 20 lakh copies per month.

The change from the classical cuisine was brought about by the following points:

(1) A variety of dishes were served on the menu but smaller portions were served.

(2) Presentation of the food was given a lot of importance by serving it on a plate, so that the presentation was not spoilt by the service personal. Individual portions were placed on the plate and served.

(3) The sauces were served thinner, flour was not added for thickening.

(4) The cooking time was reduced, by stir-frying, frying, grilling, etc.

(5) Fresh foodstuffs had to be used and emphasis was laid on seasonal vegetables, fishes, etc. The items should be bought from the best suppliers.

(6) Unusual flavours, combinations were prepared. The dishes served by Paul Bocuse spit roasted, chicken and items like veal chops, steaks, fish are cooked in its own juices. The dishes were influenced by Chinese cuisine by preparing dishes with plenty of good quality crisp vegetables, tender meat items and cooking by stir-frying and steaming. To give a different taste, some Indian spices were used such as ginger, coriander, jeera and saffron. The latest equipment such as microwave ovens, high pressure steamers, vacuum cooking and inferred grillers are used. The Nouvelle cuisine has created interest but it has not got the same popularity as French classical cuisine; in India, only few restaurants are serving Nouvelle cuisine. Abroad, some of the Indian restaurants are offering Nouvelle Indian cuisine, less spicy and greasy, etc.

The change from the classical cuisine was brought about by the following points:

(1) A variety of dishes were served on the menu but smaller portions were served.

(2) Presentation of the food was given a lot of importance by arranging it on a plate, so that the presentation was not spoilt by the service personnel. Individual portions were placed on the plate and served.

(3) The sauces were served thinner, flour was not added for thickening.

(4) The cooking time was reduced, by stir frying, frying, grilling, etc.

(5) Fresh foodstuffs had to be used and emphasis was laid on seasonal vegetables, fishes, etc. The items should be bought from the best suppliers.

(6) Unusual flavours, combinations were prepared. The dishes served by Paul Bocuse were roasted chicken and items like veal chops, steaks, fish are cooked in its own juices. The dishes were influenced by Chinese cuisine by preparing dishes with plenty of good quality crisp vegetables, brined meat items and cooking by stir frying and steaming. To give a different taste, some Indian spices were used such as ginger, coriander leaf and saffron. The latest equipment such as microwave ovens, microwave steamers, vacuum cooking and non-stick utensils were introduced. The Nouvelle cuisine has created interest but it has not got the same popularity as French classical cuisine. In India, only few restaurants are serving Nouvelle cuisine. Abroad, some of the Indian restaurants are offering Nouvelle Indian cuisine, less spicy and greasy, etc.

5

Fundamental Aspects

This is not a special method of cookery but we have to use our knowledge of baking, boiling, steaming, etc., and in addition follow some rules which apply specially to food prepared for the sick. An invalid usually has a weak appetite and needs tempting because strength can only be regained through taking suitable food. Wrong food may kill a patient, while badly prepared food will retard recovery. Good food, well cooked and served will do much towards saving the patient's life.

Convalescents vary in appetite according to the nature of their complaint. But whether they have good appetites or not, their food should be easily digestible as they cannot take as much exercise as a normal person.

Compliant Food

(i) Consult the doctor about the patient's diet and obey his instructions.

(ii) Choose food that will supply necessary nourishment and be suitable to the illness,

(iii) Serve food which can be easily digested because in sickness the digestive system is often impaired.

(iv) Serve well-balanced meats, but pay special attention to the constituents according to the nature of the illness. In some cases meat, starches and sugar may have to be taken sparingly or even eliminated.

(v) In a serious illness, food is usually given in liquid form and should be as varied as possible,

(vi) In convalescence see that the solid food is light, easily digestible and served in small quantities,

(vii) As convalescence proceeds, increase the food in amount and variety.

(viii) Serve the best quality possible. Food should be absolutely fresh.

(ix) Vary the food and the method of cooking as much as possible. This stimulates appetite.

Perfect Dishes

Liquid : Fresh fruit juice, barley water, beef tea, egg flip, lemonade or orangeade.

Soft Food : Made with milk, e.g., junket or milk jelly, custard puddings (baked or steamed), blancmange, broths, light soups, rice conjee, etc.

Solid Food : White fish, fowl, rabbit, sweet bread or pancreas, liver dishes (on the advice of a doctor), khitchdi, etc.

Storage of Food

The fundamental principle in the storage of food is to keep it clean, cool and covered. Periodical inspection is a must, and there should be a regular turnover of goods. Daily cleaning, weekly cleaning and spring-cleaning are necessary to keep food free from dust and dirt.

Access to food stores must be restricted. No food should be kept in the kitchen or larder, apart from what is required during the day. The storage room should be dry, well-lit, ventilated, vermin-proof and clean.

Cupboards and shelves should be of simple design without dust-collecting ornamentation, unnecessary ledges or panels. Doors should be so designed as to enable the whole shelf to be clearly visible and accessible when doors are opened. Shelves must be constructed for specific purposes. They should be so fixed that they could be easily taken down and re-erected to facilitate regular and thorough cleaning. They should be narrow enough to enable all the goods to be easily accessible. Wide wooden shelves should not be made of planks which are joined, tongued or grooved but of 5 cm. x 1.25 cm. (2" x ½") slats laid with a clearance of 3.75 cm. to 5 cm. (1 ½"to 2") between each slat.

The one nearest to the wall should have a return board to form a lip. The back of the lip should be at least 5 cm. (2") from the wall, to leave room for the cleaning of the wall. Corners must be rounded, particularly where shelves on two walls meet at a corner, to avoid an uncleanable dust trap. As far as possible avoid having shelves above windows. Special shelving covered with marble, hard stone or tiles should be set aside for such materials as fat. One part of the shelf should be set aside for glass receptacles.

No foodstuff, even when in bins, boxes or cartons, should be stored directly on the floor. The bottom shelving should be at least 75 cm (26") above the floor. Containers for food should be provided, the nature of the foodstuff determining the nature of the receptacle required. These should be smooth, impervious, easily cleanable, resistant to wear denting, buckling, pitting, chipping, cracking and should withstand penetration by vermin, and the corrosive action of foods and cleaning compounds. Stainless steel, model metal, vitrified china, glass or plastic should be used for moist unpackaged food or beverages, and galvanised steel for dry packaged and unpackaged foods.

Solder must be nontoxic, and welded areas must be as resistant to corrosion as the parent metal. Food containers should stand clear of walls, should be kept scrupulously clean

and unless the articles they contain are naturally wet, should be completely dry, both inside and outside.

Segregate food to be used promptly from the food that is to be held for some time. Use food in the order of priority—first in, first out.

Stocks should be inspected once a week. A stock list should be made to tick off inspection dates and to indicate conditions.

Keeping food clean not only means keeping it free from visible dirt, but also keeping it in such a way as to prevent the multiplication of germs High temperature kills them, and freezing preserves bacteria in suspended animation. Hence food should be kept cool and dry. Perishables should be kept under refrigeration or, when no refrigeration is available, in the larder in closed containers or adequately protected with muslin or wire covers. Windows and doors should have fly-proof meshing.

Guide to Storage of Certain Fruits and Vegetables (R.H. 85 to 90%)

Commodity	*Temperature* °C	°F	*Approximate Storage life (weeks)*	*Approximate post-storage life (days)"*
Apples	2°-3°	35°-38°	18	9
Apricots	-1°-0°	31°-32°	1-2	-
Bananas	11°-13°	52°-55°	3	1
Beans	0°-2°	32°-35°	2-3	-
Beetroot	0°-2°	32°-35°	6-8	-
Brinjal	8°-10°	47°-50°	3-4	-
Cashew apples	0°-2°	32°-35°	5	-
Cherries	-1°-0°	30°-32°	2-3	-
Carrots	0°-2°	32°-35°	13-17	-
Cabbage	0°-2°	32°-35°	9-13	-
Cauliflower	1°-2°	34°-35°	4-6	-
Guavas	8°-10°	47°-50°	4	-
Grapes	0°-2°	32°-35°	8	-
Grapefruit	5.5°-7°	42°-45°	8	3
Gooseberry (Indian)	0°-2°	32°-35°	8	3
Jackfruit	11°-13°	52°-55°	6	-
Lemons	5.5°-7°	42°-45°	6	3

contd....

Commodity	*Temperature* °C	°F	*Approximate Storage life (weeks)*	*Approximate post-storage life (days)"*
Limes	8°-10°	47°-50°	8	-
Litchi	0°-2°	32°-35°	11	
Mangoes:				
Alphonso	8°-10°	47°-50°	4	-
Badshapasand,				
Dasharj				
Dilpasand				
Khuddus				
Neelum				
RaspuriSafeda,				
Bangalore	5.5°-7°	42°-45°	4-6	3-4
Mangosteen	4°-5.5°	39°-42°	7	-
Oranges	5.5°-7°	42°-44°	18	-
Onions (R.H.70 to 75%)	0°-2°	32°-35°	17-26	-
Papaya	4°-5.5°	39°-42°	5	3
Passion fruit (purple)	5.5°-7°	42°-45°	3	-
Pears	-1°-0°	30°-32°	12	-
Persimmons	0°-2°	32°-35°	7	-
Peaches	-1°-0°	30°-32°	4-6	-
Peas	0°-2°	32°-35°	2-3	-
Pineapples	8°-10°	47°-50°	6	-
Plums	-1°-0°	30°-32°	2-4	-
Pomegranates	0°-2°	32°-35°	7-11	1
Pommelos	5.5°-7°	42°-45°	23	-
Potatoes	3°	37°-38°	26-35	-
Radishes	0°-2°	32°-35°	6-8	-
Sapota (Chikkoo)	0°-2°	32°-35°	10	5
Tomatoes				
(a) ripe	4.5°-7°	40°-45°	1-1½	-
(b) mature (turning to yellow)	13°-15.5°	55°-60°	3-4	-
Turnips	0°-2°	32°-35°	2-3	-

*at a temperature of 21°-34°C (70°-93°F)

Poultry : Dressed and ready to cook poultry deteriorates rapidly unless carefully handled and stored under properly controlled conditions. Freshly killed poultry, after dressing,

should be drawn quickly, and wrapped loosely in waxed paper or aluminium foil and refrigerated. Such poultry can be kept for 2 to 3 days in a refrigerator at 3°C (about 38°F) or less.

Frozen poultry should be stored at -18°C (about 0°F) or less, and cooked soon after it has been defrosted. Generally speaking, frozen poultry should be used within 6 months after it has been processed to obtain satisfactory results.

Eggs : Eggs are highly perishable products and care must be taken in handling and storage. Eggs should always be stored in a refrigerator. They may lose as much of their fresh quality in 3 days at room temperature as in 2 weeks in the refrigerator. Store with the large end up. At a range of temperatures and relative humidity varying from -2°C (about 28°F) and 85 per cent R.H. and 2°C (about 35° F) and 75 per cent R.H., eggs may be stored for as long as 9 to 10 months. Frozen eggs should be stored at -18° C (about 0° to 20°F).

Treating with oil or wax before refrigeration also extends shelf life for several months. A pure, colourless, odourless, tasteless mineral oil is used and the processing done under controlled conditions.

Milk Bottled pasteurised milk can be safely kept for at least 3 to 4 days at refrigerated temperatures preferably between -1° and 4.4° C (about 34° and 40° F).

Know-how of up Keep : Knowing how to store food properly, and for how long you can keep it is very important: the food will be safe to eat; it will retain its flavour and texture; a high level of nutrients will be maintained; and you will not waste money on spoiled items.

Safety First : The three types of bacteria responsible for most cases of food poisoning (Salmonella, Staphylococcus and Streptococcus) are present around us all the time. Normally, they are not trouble some, but if given the right (actually wrong) conditions, bacteria can multiply in a matter of hours to dangerous levels. The contributing factors to bacterial growth-temperatures between 1-52°C (about 50°-125°F) and

time To thwart growth, do not be casual about handling food; always work with well-scrubbed utensils, cutting boards-and hands; cook or serve food as soon as possible after removing it from storage; refrigerate food immediately after the meal is over; always take precautionary care. It is often not possible to tell by taste or smell if food is contaminated.

Foods that Need Special Care : Give special attention to preparing and storming poultry, fish and sea food, meat, creamed mixtures, mayonnaise puddings and stuffings. Bacteria find these foods a good media to grow in. In particular: do not stuff poultry the night before; the cold stuffing may not heat up to a safe temperature when you cook the bird Never refrigerate cooked poultry with the stuffing in-store it separately If you have bought cracked eggs, use them only where they will be thoroughly cooked (baked or hard-boiled); Salmonella, which may be present on shells, could contaminate cracked eggs.

Know-how of Frozen Foods : It is safe to refreeze virtually all partially thawed foods if they still have ice crystals in them and are still firm in the centre. However, many foods (ice cream and uncooked baked goods for example) will not maintain top quality. Meat, fish and poultry purposely thawed in the refrigerator and kept no more than one day may be refrozen. However, do not refreeze combination dishes-pies stews etc., that have been thawed. With the exception of fruit and juice concentrates, foods thawed accidentally in the freezer over a period of days (because of a power failure, etc.) should not be refrozen, unless of course, they still have ice crystals. If food is completely thawed (on purpose or by accident), warmed to room temperature and left for more than two hours, throw it out. The exceptions: fruit and juice concentrates, which ferment when spoiled. Throw away fruit if flavour is "off".

Pantry-shelf Storage Temperature: Store food in well ventilated cabinets—not over a cooking range or near refrigerator's exhaust. Use cool areas for storing large amounts of potatoes, onions, etc., and for long-term storage of canned foods.

Time: Though most staples and canned foods will keep indefinitely, buy no more than you expect to use in the recommended storage times given later. While foods will be safe beyond the recommended storage times, flavours will fade and textures will wilt. Date foods. Then check cabinets every six months and use up the oldest items.

Buying: Purchase the freshest-looking packages—messy or shop worn labels indicate old stock. Don't buy cans with swollen ends—the food has gone bad. Dented cans may be purchased, provided they are not punctured.

Home Canning: Use first-quality foods and the best techniques.

Temperature: From 1° to 4.5°C (about 34° to 40°F) is best. Above 4.5°C (40°F), food spoils rapidly. Check temperature with a refrigerator thermometer or an outdoor thermometer.

Time: Use foods quickly—do not depend on maximum storage time.

Wraps: Use foil, plastic wrap or bags, airtight containers. When meat, poultry or fish is bought in a plastic wrapped package, loosen ends of package to dry surface moisture—bacteria grow faster on moist surfaces.

General Care: Clean refrigerator regularly to cut down food odours. Remove spoiled foods immediately so that decay will not be passed on to other food.

Food	*Storage Time*	*Special Handling*
		STAPLES
Baking powder	18 months	Keep covered and dry.
Bouillon cubes	1 year	Keep covered and dry.
Bread crumbs, dried	6 months	Keep covered and dry.
Cereals	4 months	Keep covered and dry.
Chocolate, premelted	1 year	Keep cool.
semisweet	2 years	Keep cool.
unsweetened	18 months	Keep cool.
Coffee, cans (unopened)	1 months	Refrigerate after opening.
Coffee, instant (opened)	2 months	Keep lid tightly closed.
(unopened)	6 months	

contd....

Food	*Storage Time*	*Special Handling*
Condensed and evaporated milk	1 year	Refrigerate after opening.
Food	Storage Time	Special Handling
Flour (all types)	1 year	Put in airtight container.
Gelatine (all types)	18 months	Keep in original packets.
Honey, jams, syrups	1 year	Keep tightly covered.
Nonfat dry milk	6 months	Put in airtight container.
Pasta	2 years	Keep tightly closed.
Pudding mixes	1 year	Keep in original packets.
Rice, white	2 years +	Keep tightly closed.
Salad dressing (all types)	3 months	Refrigerate after opening.
Salad oil	1-3 months	
Shortening, solid	8 months	Refrigeration not needed.
Sugar, brown	4 months	Put in airtight container.
confectioners	4 months	Put in airtight container.
granulated, molasses	2 years +	Keep tightly covered.
Tea	2 years	Put in airtight container.
	CANNED AND DRIED FOODS	
Fruits, canned	1 year	Keep in cool spot.
dried	6 months	Put in airtight container.
Gravies, canned	1 year	
Meat, fish, poultry, canned	1 year	
Pickles, olives	1 year	Refrigerate after opening.
Soups, canned, dried	1 year	Keep cool.
Vegetables, canned, dried	1 year	Keep cool.
	HERBS, SPICES AND CONDIMENTS	
Ketchup (opened)	1 month	
Herbs and spices		Transfer from cartons to
whole spices	1 year	airtight containers, keep
ground spices	6 months	away from sunlight.
herbs	6 months	Check aroma; when it fades replace.
Tabasco, Worcestershire	2 years +	
	DAIRY PRODUCTS AT REFRIGERATED TEMPERATURES	
Butter, margarine	1-2 weeks	Keep tightly wrapped or covered. Hold only 2 day supply in butter keeper.
Buttermilk, sour cream or yogurt	5 days to 2 weeks	Keep tightly covered. Turn unopened yogurt and sour cream containers upside down to prevent surface from drying. Once opened, store right-side up.

contd....

Food	*Storage Time*	*Special Handling*
Cheese		Keep all cheese tightly pack
cottage	5 days to	aged in moisture-resistant
cream	2 weeks	wrap. If outside of hard
hard and wax-coated		cheese gets mouldy, just cut
cheeses-Cheddar, Edam,,		away mould—it would not
Gouda, Swiss etc.		affect flavour. For longer
large pieces		storage, see freezer storage.
(unopened)	3-6 months	
(opened)	3—4 weeks	
(sliced)	2 weeks	
Parmesan (grated)		See pantry-chart. Does not need refrigeration.
Processed (opened)	3-4 weeks	Unopened processed cheese
Cream-light, heavy,	3 days	need not be refrigerated
half-and-half		Keep tightly covered. Do not
Dips-sour cream, etc.		return unused cream to original containers. This would spread any bacteria present in leftover cream.
Commercial	2 weeks	Keep tightly covered.
homemade	2 days	Keep tightly covered.
Eggs	2 days	Keep small end of egg down,
in shells	3 weeks	to center yolks.
whites	3 days	Store in covered container.
yolks	3 days	Cover yolkes with water, cover container.
Milk		
evaporated (opened)		
pasteurised	4-5 days	Keep containers tightly closed Do not return unused milk
nonfat dry, skimmed	3-4 days	Remove entire lid of a can to make pouring easier Keep covered.
sweetened condensed	4-5 days.	remove entire lid of can to
	FRUITS AND VEGETABLES—FRESH	
Fruits		Sort fruit; discard bruised or
apples	1 week	decayed fruit. Do not wash
berries, cherries	1-2 days	before storing—moisture
citrus fruits	1 week	encourages spoilage. Store in
citrus juices		crisper or moisture resistant
bottled, reconstituted		bags or wrap. Keep fruit
frozen, canned	6 days	juices tightly covered.
melons	1 week	

contd....

Food	*Storage Time*	*Special Handling*
other fruit	3-5 days	It is not necessary to remove canned fruit from can. Wrap uncut cantaloupe, honeydew, etc., to prevent odour spreading to other refrigerated food.
Vegetables		
beets, carrots, radishes	2 weeks	Remove leafy tops, keep in crisper.
mushrooms	1-2 days	Do not wash before storing.
onions, potatoes, sweet potatoes		Refrigeration not needed. See pantry-shelf chart.
shredded salad greens, cabbage	1-2 days	Keep in moisture-resistant wrap or bags.
unshelled peas, corn on husks, etc.	3-5 days	Keep in crisper or moisture-resistant wrap or bags.
other vegetables	3-5 days	Keep in crisper or moisture-resistant wrap or bags.
		MEAT, FISH AND POULTRY-FRESH UNCOOKED
Meats		
beef, lamb, pork and veal chops	3—4 days	Store fresh meats loosely wrapped. Partial drying of surface increases keeping quality. If meat comes packaged in plastic wrap, loosen ends. Store in coldest part of refrigerator or in meat keeper.
ground meat	1-2 days	
roasts	5-6 days	
steaks	3-5 days	
stew meat	1-2 days	
variety meats (liver, heart, etc.)	1-2 days	
Fish and shellfish		
fresh cleaned fish including steaks and fillets	1 day	Store loosely wrapped. Keep in coldest part of refrigerator or in meat keeper.
Clams, crab, lobster in shell seafood including shucked clams, oysters, scallops, shrimps	2 days	Cook only live shellfish.
Poultry ready-to-cook chicken, duck or turkey	2 days	Store loosely wrapped. Keep fresh poultry in coldest part of refrigerator or in meat keeper.
	CURED AND SMOKED MEATS	
Bacon	5-7 days	Keep wrapped. Store in coldest part of refrigerator or in meat keeper. Times given are for opened packages of sliced sliced meats, Unopened vacuum packs keep about 2
Bologna	4—6 days	
Corned beef	5-7 days	
Dried beef	10-12 days	
Dry and semi-dry sausages (salami, etc.)	2-3 weeks	

contd....

Food	*Storage Time*	*Special Handling*
Frankfurters	4-5 days	
Hams (Whole, halves)	1 week	
Hams, canned (unopened)	6 months	
Liver sausages	4-5 days	
Luncheon meat	3 days	
Sausages, fresh	2-3 days	
Sausages, smoked	2-3 days	
	CANNED FOOD AFTER OPENING	
Baby food	2-3 days	Store baby food covered.
Fish and seafood	1 day	Don't feed baby from jar;
Fruit	1 week	saliva may liquefy food.
Gravy, broths Meats	2 days	Store all canned foods tightly
	2 days	covered. It is not necessary to
Pickles, olives	1 month	remove food from can.
Poultry	1 day	
Sauce, tomato based	5 days	
Vegetables	3 days	
	OTHER FOODS	
Coffee	2 weeks	Keep tightly covered after opening
Honey, jams, jellies	2 weeks	Refrigeration not needed but storage life is lengthened if refrigerated.
Nuts	9 months	Refrigerate nuts after opening cans or packages. For longer storage, freeze.
Onions, potatoes sweet potatoes	2 weeks at room temperature	For longer storage, keep below 10°C (about 50°F), but not refrigerated. Keep dry, out of sun. Plan short storage in spring when sprouting is a serious problem.
Parmesan cheese	2 months	Keep lid tightly closed.
Peanut butter		
unopened	9 months	Refrigeration not needed, but
after opening	2 months	it keeps longer if refrigerated.
Refrigerated biscuits, rolls pastries, refrigerated cookie dough	Expiry date on label	Products keep better if stored at the back of the refrigerator, where it is colder.
Soft drinks	3-6 months	
Syrups		Same as honey
Whipped topping mix	1 year	
Wines, table	2-3 days	Keep tightly closed.
cooking	2-3 months	Keep tightly closed.

Cold Storage

Temperature: -18°C (about 0°F) or below is best. The maximum temperature should be -15°C (about 5° F). Check temperature with freezer thermometer or outdoor thermometer, or use this rule of thumb: if the freezer does not keep ice cream brick solid, temperature is above the recommended level. If this is the case, do not plan to store food for more than a week or two.

Time: Date foods with an "expiry date" according to maximum storage time recommended below. Longer storage is not dangerous, but flavours and textures begin to deteriorate.

Wraps: Use foil, plastic bags and wraps, freezer wrap or freezer containers.

Commercially Frozen Foods: Do not buy them if they are battered, a sign that they may have been partially thawed. If foods have been partially thawed, then refrozen, use them within a few days.

Home-Frozen Foods: Use good-quality foods and proper techniques. Freeze foods quickly in coldest part of freezer, then store in another freezer area.

Food	*Storage Time*	*Special Handling*
	MEAT, FISH AND POULTRY	
Meat—home frozen and purchased frozen		
bacon	1 month	If meat is purchased fresh on
corned beef	2 weeks	trays and in plastic wrap,
frankfurters	1 month	check for holes. If none
minced beef, lamb and veal	4 months	freeze in this wrap for up to 1 month. For longer storage,
minced pork	3 months	wrap over with foil, plastic
ham slices	1 month	wrap or freezer wrap.
ham whole	2 months	Other meats should be
luncheon meats	1 month	wrapped as above. Make
Roasts		package as airtight as possible. Keep frankfurters, bacon, etc., Keep meat purchased frozen in original package, Thaw and cook according to sale instructions.

Contd...

Food	*Storage Time*	*Special Handling*
Steaks		
beef	1 year	
lamb	9 months	
pork	6 months	
veal	9 months	
Fish—home frozen and		
purchased frozen		
fillets and steaks from		To home-freeze fish, wrap in
"lean" fish—cod, flounder,		foil, plastic wrap or freezer
haddock, sole	6 months	wrap. Make package as air-
"fatty" fish—bluefish,		tight as possible. Freeze in
mackerel, perch, salmon	3 months	coldest part of freezer. For
breaded fish	3 months	best results, prepare fish as
clams	3 months	directed in authoritative
cooked fish or seafood	3 months	freezing guide.
king crab	10 months	Keep fish purchased frozen in
lobster tails	3 months	original wrapping. Thaw arid
oysters	4 months	cook according to label
scallops	3 months	directions.
shrimp, unbreaded	1 year	
shrimp, breaded	4 months	
Poultry—home frozen or		
purchased frozen		
chicken,		For home freezing, prepare
whole or cut-up parts	1 year	as directed for fish and
chicken livers	3 months	seafood (above). Thaw
cooked poultry	3 months	commercially frozen poultry
duck, turkey	6 months	according to label directions.
Cook all thawed poultry	within 1 day.	
	FRUITS AND VEGETABLES	
Fruit-home frozen or purchased		
frozen berries, cherries,		
peaches, pears,		
pineapple, etc.	1 year	
(contd.)		
Food		
Storage Time		
Special Handling		
citrus fruit and juice		
—frozen at home	6 months	
fruit juice concentrate	1 year	
Vegetables—home frozen or		
purchased frozen		
home frozen	10 months	Cabbage, celery, salad greens,
purchased frozen-cartons,		tomatoes do not freeze
plastic bags or boil-in-bag	8 months	successfully.
	COMMERCIAL FROZEN FOODS	

Contd...

Food	*Storage Time*	*Special Handling*
Also see: meat, fish, poultry; fruits, vegetables; dairy products Baked goods		
breads, baked	3 months	Pick up frozen foods
breads, unbaked	2 months	immediately before going to
cakes		check-out counter. Place in
cheesecake	3 months	home freezer as soon as
chocolate	4 months	possible. Purchase only foods
fruit cake	1 year	frozen solid. Cook or thaw
sponge cake	2 months	according to label instructions
yellow or pound	6 months	
pies		
cream, custard	8 months	
fruit	8 months	
Main dishes		
meat, fish and poultry pies	3 months	
meat, fish and poultry		
casseroles	3 months	
TV dinners	6 months	
	HOME FROZEN FOODS	
Also see: meat, fish, poultry; fruits, vegetables; dairy products		
Breads	3 months	Package foods tightly in foil,
Cakes	3 months	plastic wrap, freezer or
Casserole dishes:		watertight freezer containers
meat, fish, poultry	3 months	For caseroles, allow
Cookies, dough, baked	3 months	room for expansion.
Nuts	3 months	Freese in coldest part of
Pies, unbaked	8 months	freezer.
	DAIRY PRODUCTS	
Butter, margarine	9 months	Store in airtight freezer container or wrapped in freezer wrap.
Buttermilk sour cream or yogurt		Do not freeze.
Cheese		
Camembert	3 months	Thaw in refrigerator.
Cottage, farmers' cheese (dry curd only)	3 months	Thaw in refrigerator. Do not freeze creamed cottage cheese—it gets mushy.
Cream cheese, Neufchatel	1 month	Thaw in refrigerator.
Hard cheeses—Cheddar, Edam, Gouda, Swiss, Brick, etc.	3 months	Cut and wrap cheese in small pieces for freezer storage. When frozen, may show mottled colour due to surface moisture. Thaw in refrigerator.
Processed	3 months	

Contd...

Food	*Storage Time*	*Special Handling*
Roquefort, Blue	3 months	Becomes crumbly after thawing Still fine for salads and melting.
Cream—light, heavy. half-and-half	2 months	Heavy cream may not whip after thawing. Use cream for cooking. Thaw in refrigerator.
Cream, whipped	1 month	Make dollops of whipped cream; freeze firm. Place in plastic bag or carton; seal; store in freezer. Place on top of dessert to thaw.
Eggs		
in shells		
whites	1 year	Store in covered container. Freeze in amounts for favourite recipes.
yolks	1 year	For sweet dishes: Mix one cup yolks with 1 tablespoon corn syrup or sugar. For other cooking: substitute ½ teaspoon salt for sugar.
Ice cream, ice milk, sherbet	1 month	Cover surface with plastic wrap or foil after each use to keep from drying out.
Milk	3 months	Freezing affects flavour and appearance. Use in cooking and baking. Allow room for expansion in freezer container. Thaw in refrigerator.

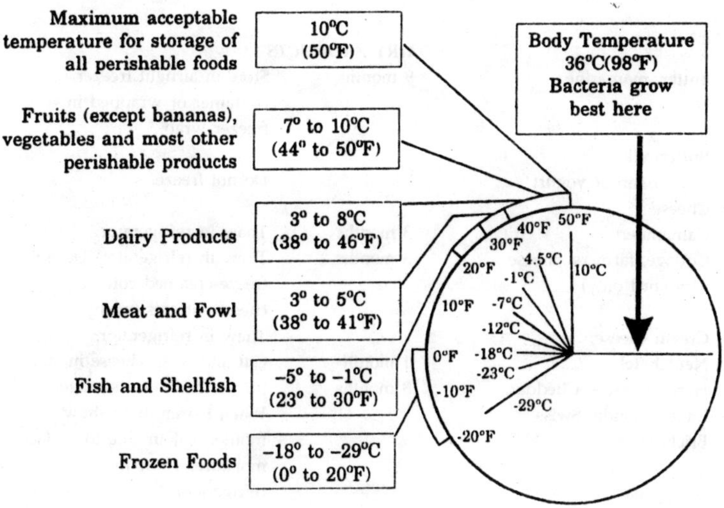

6

Dietry Plans

Preventing Adulteration : The practice of adulteration of foodstuffs and marketing of substandard or materials of poor quality is widely prevalent in our country. It has been roughly estimated that over 50 per cent of marketed foodstuffs available today are adulterated in one way or another. The percentage varies from region to region. It is an unpalatable fact for a nation to admit, but there is little doubt that the Indian consumer, unlike his Western counterpart, cannot take the purity of foodstuffs for granted. Despite the existence of legislation to prevent adulteration of foodstuffs, a very large number offered to the consumer today are injurious to health.

One of the causes of adulteration is that there is a wide gap between production and supply of food articles. The temptation for quick gain and easy profits amongst most merchants, slow action by the authorities, the poor enforcement of Prevention of Food Adulteration Act and leniency shown to traders in regard to the enforcement of the laws on food safety, lengthy procedural wrangles and delays, due to various loopholes in our legal system, contribute to the rampant adulteration of foodstuffs. The gullibility of the predominantly illiterate masses

coupled with ignorance, apathy and indifference make for easy adulteration. In a developing country, the cost factor is the most crucial point for a large segment of the population, the cheapest product is preferred to the best. There are a large number of mushroom industries to cater to this section.

Almost all foods, milk, cereals, dais, spices, ghee, oils and beverages are adulterated. The adulterant used generally mixes well with the major food article in colour, shape, size and appearance. The more highly priced foods and those foods which are in great demand are the ones more often widely adulterated. By adulterating the foodstuffs, the merchant is benefited in many ways. First more weight is added to the commodity. Secondly, addition of colour improves the appearance of the product and hence substandard product can fetch higher price. Thirdly, new product when combined with the old one, lowers the quality and at the same time substandard commodity is sold at a higher rate. The ingenuity, imagination and initiative shown in adulteration make it difficult for an average buyer to detect it at first sight. Sometimes, it can be detected only in well-equipped laboratories. Besides the relatively harmless forms of adulteration like addition of water to milk, moisture to butter, mixing of edible oils with cheaper edible oils, many types of adulteration are positively injurious to health.

Rice and pulses are often polished with talc, a silicate containing asbestos fibre, which is not removed even after washing, cleaning and cooking. The Japanese have traced cases of stomach cancer to this practice. The other common adulterants used in cereals such as rice and wheat are clay particles, grit, soap stone and weed seeds. Most yellow pulses are adulterated either with kesari dal or by polishing with metanil yellow, a cheap soluble coal tar dye that makes the pulses deceptively brighter and cleaner looking. Metanil yellow is a highly toxic substance with accumulative tendencies, which over the years build up enough toxicity in the body to cause testicular degeneration in the male. When khesari dal is

consumed continuously for a few months, it results in the development of lathyrism, a permanent paralysis of the lower limbs.

Spices, particularly because of their high cost, are the most adulterated item in food commodities. Turmeric is coated with lead chromate or metanil yellow to give a lustrous yellow colour. Black pepper is sometimes mixed with papaya berries. Foodgrains and pulses are often indiscriminately sprayed with insecticides to prevent infestation or to destroy pests; chillies are soaked in soluble coal tar dyes to give them a deeper hue, while sawdust, blackgram husk, and used tea leaves are utilised for adulterating tea dust/leaves. These are a few of the many common practices that are used today. Many cheap sweets, sherbets in variety of colours and ice fruits are adulterated with many non-permitted coal tar dyes.

These undesirable colours can impair the liver and can be the cause of cancers. Similarly, lead chromate when added to turmeric powder can cause lead poisoning resulting in stiffness of limbs and paralysis. Mineral oils and other non-edible oils when mixed with edible oils can induce nausea, purging, impaired liver functions and cancer. Continuous use of mustard oil with more than 11 per cent argemone seed oil causes the disease 'epidemic dropsy' in people. Further, many mould-infested grains and oil seeds are further processed into flours and oils. Here the main danger is due to the mould growth which produces many hidden toxic substances called mycotoxins. These can cause various diseases and they act as nervous and neurotoxins impairing the general health. Some also have been found to be carcinogenic.

The menace of food adulteration exists all over the world. In developed countries, due to the implementation of food laws for the last few decades, there has been a significant improvement in the quality of food. But due to tremendous improvement in food industry and technology, various chemical substances like artificial colours, preservatives, antioxidants, emulsifiers, flavour and taste improvers are increasingly being

used. Some of these pose a great threat to the health of the people. As against this, in developing countries adulteration is found to be rampant even in primary articles of food like milk, ghee, edible oils and foodgrains. Shortage of food and spiralling prices have aggravated the problem. Thus all countries in the world have to grapple in their own way with problems raised by food adulteration.

Prescribed Aspects

In India, as early as in 1860, adulteration of food was dealt with in certain sections of the Indian Penal Code (Sec. 272,273). Later it was included under the provisions of the Municipal Act. But as a comprehensive act, it was first promulgated in Bombay State in 1899. Subsequently similar acts were enacted by other states also. In the erstwhile Travancore State an Act—The Travancore Prevention of Adulteration Act—was enacted in 1931. Besides food adulteration, adulteration of other articles was also made an offence under this act But then acts were not comprehensive in many ways. The punishment prescribed was not deterrent enough to prevent adulteration of food efficiently. There was no uniformity of standards prescribed for various food articles.

The necessity of a comprehensive act applicable to all the states was felt. In 1938 the Central Advisory Board of Health was established. This Board in its report stressed the necessity for efficiently tackling the problem by all states. When the Constitution of India was framed the topic of prevention of adulteration of food was brought under the concurrent list and the P.F.A. Act was formulated in 1954. It was enacted on June 1,1955. After about ten years of implementation of the Act, the position was reviewed and some amendments were made in the Act in 1964 to plug loopholes and to give more protection to the honest traders. Again in 1976 some other important amendments were made to the present act to make it more effective and purposeful. These amendments came into force from April 1,1976.

Object of the Act : What the consumer wants is genuine, wholesome food, of nutritive value, free from any health hazard. He also expects that labelling must be truthful and there shall not be fraudulent claims of any kind by any statement, device or design. The interest of the trader is not only to sell the food articles to the consumers but also to make the maximum profit out of it. This conflicting interest of the consumer on the one hand and the traders on the other, lead to unhealthy practices in trade. It is to prevent such practices that the Act has been made. Though the consumer's interest is the foremost in the P.FA. Act, that of the traders have also been accommodated reasonably as far as scientific grounds and consumers' interest permit.

Important Provisions of the Act : Food has been defined under the Act as articles other than drugs and water, which are ordinarily used as food on drink and all articles which enter into and are used in the preparation of human food. Another definition of 'primary food' has also been introduced by the amendment act of 1976. Primary foods are products of agriculture or horticulture in its natural form.

The term adulterated has been defined under Section 2 of the Act: An article of food will be deemed to be adulterated if it is not of the nature, substance or quality demanded by the purchaser and is to his prejudice or not of the nature, substance or quality which it purports to be; if an article is partly or wholly substituted by the cheaper substance; if any constituent of the article is extracted from it, if it is prepared or exposed for sale under insanitary and unhygienic surrounding; if it contains prohibited colouring matter or preservative or permitted colouring matters and preservatives in excess of the permissible limits; if it does not conform to the standards under all these conditions it will be deemed to be adulterated. There is a provision that a primary food will not be deemed to be adulterated if it does not conform to standards due to natural causes beyond the control of human agency. In short, anything which lowers the quality of a food article has been brought wider the definition of adulteration.

False Brands

Sale of food articles without proper labelling, imitation of other food articles and coating food articles to give improved appearance are brought wider the definition 'misbranding'.

Prohibitory Section : Section 7 of the P.F.A. Act prohibits the sale of (a) any adulterated food, (b) any misbranded food, (c) any article of food for the sale of which a licence is prescribed except in accordance with the conditions of the licence, (d) any article of food and sale of which has been prohibited by the Food (Health) Authority in the interest of public health, (e) any article of food in contravention of the provision of the Act and Rules, and (f) any adulterant.

Crime and Punishment : Infringement of the provision of Section 7 is punishable under Section 16. Various offences and punishments for each have been detailed under these sections. This section has been amended by the amendment act of 1976. Non-graded punishment, depending on the nature of the offence has been laid down. For the sale of adulterated food the minimum punishment is six month's imprisonment and a fine of rupees one thousand.

If the adulterant is of such a nature as to cause death or grievous hurt when consumed by a person, the punishment is imprisonment for a term of three years which may extend to life imprisonment and a fine which shall not be less than rupees live thousand. Punishment has been prescribed for sale of misbranded food and the sale of adulterant. Preventing a Food Inspector from exercising his powers and duties under the Act, using a report of the Public Analyst of Director of Central Food Laboratory for advertising purposes, not issuing warranty, issuing false warranty, tampering of seized food articles are all punishable under the law.

Protection of the Traders : Section 7 of the Act prohibits the sale of adulterated food or any consumable article, by any person either by himself or by any person on his behalf. Prima facie it may seem that an innocent person who buys a food

article from a wholesaler or manufacturer and sells it without knowing that it is adulterated is punishable under the Act. Sections 14, 14 A, 19(2) and 20A are meant to protect honest vendors of food articles. Section 14 lays down that no manufacturer, distributor or dealer of any article of food shall sell such article to any vendor unless he also gives a warranty in writing about the nature of substance and quality of such food to the vendor. Insistence of warranty at the time of sale by the manufacturer, distributor or dealer would make it easy to pin down the responsibility with the offender. The amendment of 1976 makes this easier by incorporating a provision that a cash bill can also be considered as a warranty.

According to Section 14A, the vendor has to disclose to the Food Inspector the name, address and other particulars of the person from whom he purchased the food article. This helps the Food Inspector to proceed against the warrantor also. According to Section 19(2) of die Act, a vendor shall not be deemed to have committed an offence if he proves that he purchased the food article with a proper warranty, that he stored the article properly and that he sold it in the same condition in which it was purchased. Section 19(3) confers a right on the warrantor to appear in court and give evidence. The law thus protects the warrantor also against spurious warranties. When warranty is proved, the vendor can be exonerated and the warrantor can be proceeded against. As per Section 20A during the trial, when warranty is proved the court can proceed against the warrantor as though a prosecution has been launched against him.

Keeping Quality

There are two standards set up by the government to help the manufacturers, the traders and the consumers to ensure quality. They are 'Agmark' standard and 'Indian Standards Institution mark.

Agmark Standard : The Directorate of Marketing and Inspection of the Government of India have set up this standard.

This grades goods into (1) Special (Grade I), (2) Good (Grade II), (3) Fair (Grade III) and Ordinary (Grade IV). These standards also specify the type of packaging to be used for different rent products. An Agmark certificate is issued to traders and the mark on commodities provides assurance of the quality to the consumer.

Indian Standards Institution (ISI) : The ISI mark on food is a guarantee of good quality. This institution prescribes standards for various food items and issues certificates, to traders.

Common Food Adulterants

Foodstuffs	*Common Adulterants*
Milk and milk products	Water, removal of butter and addition of refined oil.
Milk, liquid	Fat addition of skimmed milk reconstituted from skimmed milk powder.
Milk powder	Starch, dextrins.
Cream	Other fats.
Ice-cream	Non-permitted colour, artificial sweetness, other fats and gelling agents.
Butter	Other fats.
Ghee	Hydrogenated fat.
Vegetables oils and fats	
Vanaspati	Animal fat and other high melting fats.
Vegetable oil	Argemone oil, mineral oil, orthotricresyl phosphate, cheap non-edible oils.
Spices and condiments	
Whole turmeric	Coating of lead chromaie or coaltar dyes.

contd....

Foodstuffs	***Common Adulterants***
Turmeric powder	Coaltar colour, yellow earth, starch or talc coloured yellow by coaltar dye.
Curry powder	Starch coloured red by coaltar dye.
Coriander seeds	Other seeds coloured green.
Coriander seed powder	Powdered bran or sawdust coloured green with dye.
Chilli powder	Starch coloured brown by coaltar dyes.
Mustard	Argemone seeds.
Cumin	Artificial jeera like product.
Black pepper	Dried papaya seeds.
Asafoetida	Resins and other plant gums.
Cereals	
Wheat and rice	Stones.
Wheat flour	Tapioca flour, talc
Semolina	Tapioca semolina.
Pulses	
Bengal gram dal	Khesari dal (Lathynis sativus)
Red gram dal	Coloured yellow with coaltar dye.
Bengal gram flour	Tapioca flour or starch coloured yellow with dye.
Sweetening agents and soft drinks	
Honey	Coloured canesugar syrup.

contd.....

Foodstuffs	***Common Adulterants***
Soft drinks	Artificial sweeteners (Saccharin).
Beverages	
Coffee powder	Exhausted coffee powder, roasted husk or date seed or tamarind seed powder.
Tea	Other leaves with added colour, exhausted tea leaves.
Miscellaneous	
Processed arecanut (Supari)	Other seeds or nuts broken and coloured.

Consumer Protection

For consumer protectic.. there are a number of laboratories in central and state level. T'-ey are set up to collect samples and to analyse it. Some of the Government agencies are:

(1) Municipal Laboratories in big cities.

(2) Food and Drug Administration Laboratories of States.

(3) Central Food Testing Laboratory of the Government of India, and

(4) Laboratories of the Export Inspection Council.

Municipal Laboratories

In municipalities of big cities like Mumbai, Delhi, Chennai and Kolkata they have set up their own testing laboratories. According to the Prevention of Food Adulteration Act the Municipal analyst qualification must be scrutinised. Chemical and microbiological analysis of all kinds of foods are essential in the laboratory. Nutrient analysis like fat content of milk,

contamination of water or pesticide residue in parts per million in foods etc. are some examples of the test conducted there.

These laboratories are under the Health Officer of the state. Health inspectors under him make periodical rounds of the various wards and collect samples of food suspected to be adulterated. Samples are sealed in the presence of the vendor and witness and sent to the municipal laboratory for analysis. The Health Officer can take action against the manufacturer of the food if he is found to be guilty. The manufacturer can defend himself in the court of law. Depending on the severity of crime he can be punished by a fine or imprisonment or both.

Food and Drug Administration : Some state Governments have laboratories throughout states. The Director of the Administration is empowered through his staff to collect samples which are suspected to be adulterated and have them analysed by the Health Officer of a municipality. The Food and Drug Administration has jurisdiction all over the state as compared to municipal laboratories which work only within city limits.

The Central Food Testing Laboratory : The Government of India has established a Central Food Testing Laboratory in Kolkata for carrying out analysis of all foods. These laboratories are very well equipped and carry out even sophisticated and sensitive analyses. These laboratories are the ultimate authority in determining whether a food sample is adulterated or not in cases of conflicting reports of analyses from two laboratories, the authorities concerned usually refer the case to this laboratory. Usually analysis carried out by these laboratories are accepted as the last word in analysis of food.

Export Inspection Council Laboratory : The Government of India has made it mandatory for all exporters to have their products analysed so as to ensure the quality of foods exported. Frozen food items like seafood, canned fruits and vegetables are also analysed before export. It also sees to it that all the foods exported conform to the minimum requirements as laid down by the council. The council has laboratories in all the

major parts and the samples analysed. Without the council's certificate no food can be exported.

Quality Checking Agencies

Quality Control Laboratories of Companies : Most of the companies which manufacture food products have well equipped quality control laboratories to check the quality of their products. The ingredients used for manufacturing and the products are examined physically, chemically and microbiologically. The taste, weight, volume, colour, keeping quality, packaging all these factors are checked.

In the dairy industry microbiological centre of the milk, are taken and the fat content determined, the former gives an idea of the sanitary condition under which the milk was collected, the latter whether any cream had been removed. Proper pasteurisation is also checked before supply.

In canning fruits and vegetables the fresh materials are visually inspected and all the bruised and spoiled ones removed. Oxidity of fruits like mangoes, cherries, tenderness of vegetables like green peas are determined to estimate duration of processing time, accelerated storage tests at higher temperature one conducted to ensure the keeping quality of the canned food.

In freezing, quality control checks at various stages of processing are carried out to see that the final frozen product conforms to the prescribed microbiological standards, weights after thawing and drying are taken to make sure that the customer gets the right weight. Biscuit, dairy, fruit and vegetable canning industries set up their quality.

Quality Control Laboratories of Consumer Cooperation: Apna Bazar in Mumbai, Super Bazar in Delhi, Chinthamani in Coimbatore have criteria for selection of food items which they sell in their shop. If the qualities are not fulfilled they will not accept materials.

Testing Laboratories : There are a number of private testing

laboratories which carry out tests of all food materials on payment. Government has recognised certain private laboratories and certificates given by them are valid for legal purposes.

Consumer Guidance Society : A consumer guidance society has been formed in India with Bombay as its headquarters and with branches in major cities. They create consumer awareness of the various forms of adulteration and develop consumer resistance to such adulterated food products through various communications and mass media.

Importance of Dietary Food : Food is essential for human existence just like the air we breathe or the water we drink. The food that we eat is utilised in the body and the assimilated substances are used for the growth and maintenance of tissues. A living organism is the product of nutrition. The human being requires more than 45 different nutrients for its well-being. Food materials ingested by the body are digested, absorbed and metabolised. Useful chemical substances derived from food by the body are called nutrients. A number of foodstuffs have to be selected to get all the nutrients. The health of a person depends on the type and quantity of foodstuff he chooses to make his diet For sustaining healthy and vigorous life, diet should be planned according to the principles of nutrition. Extensive research work carried out on human beings and on experimental animals throughout the world has provided us with sufficient knowledge on nutrition and health. World Health Organisation defines health as "the state of complete physical, mental and social well being and not merely the absence of disease or infirmity".

Importance of Food

Nutrition is the science of food values. It is relatively a new science which was evolved from chemistry and physiology. Nutrition is often mentioned as a branch of chemical science or biochemistry. The effect of food on our body is explained in nutrition. In other words, nutrition is defined as food at

work in the body. In a broader sense nutrition is defined as the combination of process by which the living organism receives and utilises the materials necessary for the maintenance of its functions and for the growth and renewal of its components. Nutrients are defined as the constituents of food which help us to maintain our body functions, to grow and to protect our organs. There are six major nutrients in our body. They are carbohydrates, proteins, lipids, vitamins and minerals. The human body requires 17 vitamins and 24 mineral elements for various day-to-day activities. The composition of human body is 60-62 per cent water, 17 per cent proteins, 14 per cent fat, 6 per cent minerals and 1 per cent carbohydrates. In infants the percentage of water is more as compared to an adult. In women water content is slightly lower whereas fat content is more than in men. Fat deposition in the body increases with age.

Percentage Composition of Human Body

Nutrients	*Man*	*Woman*
Water	60-62	54
Protein	17	15
Fat	14	25
Minerals	6	5
Carbohydrates	1	1

Vitamins are present in negligible amounts.

Relationships of food to health have been made from the research conducted by chemists, microbiologists, pathologists and nutritionist from the past two centuries.

Human nutrition is governed by many factors like food habits and behaviour, food beliefs, ethnic influences, geographic influences, religious and sociological factors, psychological factors, food and production, income, national and international food policies, food technology, processing, fisheries,

transportation, marketing, educational status and other mass media facilities.

The benefits of good nutrition are health, happiness, efficiency and longevity.

Wholeness of Nutritional Value

When all the essential nutrients are present in a correct proportion as required by our body, it is called optimum nutrition or adequate nutrition. Optimum nutrition is required to maintain good health.

There are certain signs of good nutrition. They are height and weight for the age, clear complexion, fresh and lively skin and hair, healthy pink nails, correct posture and gait, inquisitive and alert eyes, good appetite and bowel evacuation, emotional maturity and confident deeds and pleasing personality and optimism in life and overall health.

Undernourishment : When almost all nutritions are below the requirement, the condition is known as undernourishment Under nourishment may be defined as a state of partial starvation. An undernourished person manifests symptoms of deficiencies and feels unwell. Poor body weight, poor resistance to infection, weakness, apathy and general ill-health are symptoms of undernourishment.

Malnutrition : Malnutrition is an impairment of health either from a deficiency or excess or imbalance of nutrients. Malnutrition is a condition when one or two nutrients are less or are in excess in the body. This again manifests in disorders and discomforts. Severe malnutrition in certain phases of life can do irreparable damage to the body. Physical, mental and intellectual well-being in a person is affected due to malnourishment. Emotional upset and intellectual dwarfism create personality problems in later life. Impaired functional ability and deficient structural integrity result from malnutrition. Marasmus, kwashiorkor, xerophthalmia, scurvy, rickets, osteomalacia, beri-beri, pellagra and anaemia result

from the deficiencies of protein, calories, vitamin A, vitamin C, vitamin D, B-complex vitamins and minerals, respectively. Malnutrition creates lasting effect on the growth and development of a person. Permanent retardation of the central nervous system occurs due to protein calorie malnutrition in early life. Thus nutrients from food sources enable one to keep fit and maintain health. These substances include energy which gives the capacity to work, proteins which form our body muscles, bones, blood, body fluids, enzymes, hormones, antibodies, organs, skin, hair and nervous tissues. Food also supplies minerals and vitamins which protect our organs and regulate their functions and other physiological processes.

Classification of Foods : Based on these functions, foods are grouped into energy-yielding foods, body-building foods and protective foods. Nutrients which engage in these activities are known as energy-yielding nutrients, body-building nutrients and protective nutrients. Carbohydrase, fats and proteins release energy on metabolism in our body.

1 gm of carbohydrates provides - 4 kcals

1 gm of proteins provides - 4 kcals

1 gm of fat provides - 9 kcals

Cereal grains like rice, wheat, ragi and maize, roots and tubers like potato, sweet potato and tapioca are good sources of carbohydrates. Fats are considered the concentrated source of energy as they supply more than double the quantity of calories compared to carbohydrates and proteins. Though proteins yield energy, normally they are considered body-building foods. Nutrients like proteins, mineral salts like calcium, phosphorus, iron and water are body-building nutrients. Their sources are body-building foods. Just like a house is built from wood, brick, cement, sand, iron rods, nails and other materials our organs in the body are built from the substances derived from food.

Protein foods like milk, meat, fish, eggs, pulses, grams and nuts are essential to build our tissues and to form blood.

Our body functions are regulated by water, certain minerals and vitamins. They are essential for the well-being and working of the body. They are called protective foods. Foods rich in protein, minerals, vitamins and water are termed as protective foods. Water is necessary for various body processes.

Minerals like calcium help in controlling blood clotting, muscular contraction and for the efficiency of heart muscles. Iron is essential for blood formation and iodine is necessary for regulating body function through the thyroid gland.

Vitamins are essential for regulating the body process such as growth, muscular coordination of various organs and functions of several organs like eyes, ears, nose and skin.

Thus foods play a prominent role in providing physical, mental and social well-being which is otherwise known as health to the people. Health is reflected in a person's nutritional status. Nutritional status is the condition of the individual as influenced by the utilisation of the nutrients. Dietary history, physical examination and laboratory examinations reveal it

Healthy Diet

Food industry has brought out some novel protein-rich foods which can be used for combating malnutrition. Some of them are from unusual food sources which were never considered as consumable by our forefathers. Leaf and grass proteins, food yeast, algae and microbial synthesis of proteins from hydrocarbons are some of the novel food sources now available.

Leaf Protein Concentrates : Proteins from fibrous materials of leaf are separated and leaf protein concentrates with 30 to 35 per cent of the original nitrogen can be extracted from leaves. Outstanding work in this field has come from Pine et al in 1961. Leaf protein concentrates can be incorporated with milk or in common flours or dal powders for children's foods.

Food Yeast : Yeast is capable of converting carbohydrate to protein. About 16 to 33 per cent of this type of conversion

is possible by yeast. Thus it can be incorporated at 5 to 10 per cent level in children's foods. Food yeast is a pale yellow product. It has a characteristic nut-like taste. This can be incorporated in low cost foods for infants, it can also be used for overcoming protein malnutrition among children.

Algae **:** A unicellular green algae chlorella contains a protein content of about 50 to 60 per cent. This can be incorporated with soups, noodles and in baked items. Even though it is nutritionally good, its acceptability is very poor.

Microbial Synthesis of Proteins from Hydrocarbons **:** Certain petroleum hydrocarbons can be converted to proteins through microbial use. Its utilisation for human beings is not established even though as an animal feed it is used.

Stored Foods

Modern technology has made many food sources into processed items. Food industry is developing in accordance with modern living. Almost all food groups are now processed into readymade or ready for table items. Pre-cooked or half cooked items are also available. Examples of some of these items are given below.

Cereals: Starch is a polysaccharide and it is used in cookery. It is manufactured from corn, potato, sweet potato and tapioca. In the market it is available as tapioca starch, arrowroot starch and starch products like sago or custard powder where cornstarch is flavoured with vanilla and edible colour and corn syrup.

Rice is available in processed form as instant rice or quick cooking rice. It is prepared by cooking rice and dehydrating the cooked rice so as to retain a porous structure.

Puffed rice from paddy and from parboiled rice is available in the market.

Cornflakes from corn can be used as a breakfast item.

Barley is processed into barley flour and pearl barley. Malted barley is breakfast item.

Oats is available as oatflakes in processed forms.

Breakfast cereals and infant foods are often prepared from cereals. Flaked items from all cereals, puffed wheat, rice, barley and corn are available in the market. Shredded breakfast cereals like shredded wheat is a pre-cooked item.

Your Baby and her Diet

Most of the infant foods contain both cereals and milk and these pre-cooked cereals are eaten as gruels with water or milk. Rice noodles are popular in some of the Far East countries. Sago or sabudana is made from rice, or other cereals. Biscuits and breads are also processed cereal products.

Pulses are processed into puffed grams. Now commercial instant mixes of cereal, dal, fermented products are also available. Ready mixes of South Indian dishes like idli and dosa are prepared by using baker's yeast or baking soda and acidifying agents like citric acid. Puffed chickpeas and peas and canned dry peas are available as processed items.

Roasted groundnut and protein isolates from soyabean or peanut can be used in the production of toned milk and protein-rich biscuits or bread. Soyabean milk powder and butter are available in the market.

Milk powders, malted milk powders, various weaning foods, ice-cream mix, cream and condensed milk, cheese, khoa-based sweets, channa or paneer-based sweets like rasogulla and sandesh are some of the processed milk products. New milk-derived products like Miltone and Chai-sathi are available, especially in North India.

Fruits are processed into fruit juices. Today even bananas and guavas are used to prepare juice. Fruit juice concentrates and fruit juice powders are made from juices along with syrups or squashes. Cordials, ketchups, sauces, jellies, jams, marmalades, artificial syrups, dried fruit and salad, preserves, candies, glaced and crystallised fruits and fruit-cereal flakes are available in the market as processed foods.

Vegetables are processed into soups, purees, canned or frozen and dehydrated forms.

Egg powder, egg albumin powder, custard or caramelised custard powder and readymade puddings are some of the processed forms of egg available in the market.

Meat is processed into canned meat, cured meat, smoked meat, sausages, cutlets, soup powder and as dehydrated meat.

Strained baby foods with meat are being manufactured in some places.

Fish is processed into frozen, canned and dried forms. Fish protein concentrates and fishmeal and strained baby foods with fish are also available as processed items in developed countries.

All food items are processed in very many ways and they are marketed in various forms. Pickled and fermented items of innumerable sources are also available in the market.

Everyone knows that good diet is needed to sustain good health. By health we mean the well-being of an individual in physical, emotional and social conditions. Just the presence or absence of a disease cannot label a person as healthy. Sound emotional and mental condition of an individual is contributed by good nutrition. A healthy person will have a positive attitude towards life. A good natured person full of life reflects his better health standards. A good or adequate diet is known as 'balanced diet'. A balanced diet yields daily nutrients in the proper amounts and proportion required by the body.

Necessity of Nutritional Diet : Nutritional requirements vary according to age, sex, physical activities and other physiological conditions. The National Institute of Nutrition is engaged in research work on the dietary pattern of Indians and their requirements of various nutrients. The Nutrition Advisory Committee of the Indian Council of Medical Research made recommendations in 1944 for all nutrients. They were revised in 1958 and 1968. In 1980 further recommendations were made by the Nutrition Advisory Committee and the

latest requirements are available now for formulating balanced diets for various groups. A balanced diet must supply enough food to the body for deriving energy. Energy is simply expressed as the capacity to do work. For basic activities like heart beat, breathing and excretion energy is required.

Energy is stored as carbohydrate in foods. Cereals like rice, wheat, millets like ragi or bajra contain much carbohydrates in the form of starch. Starchyuits like banana and roots like tapioca, potato, yam, colocasia and sweet potato are rich in carbohydrates. Energy requirements for various age groups are given elsewhere and for an adult sedentary man 2,400 kcals are required and for a sedentary woman 1,900 kcals are recommended. In a balanced diet energy from cereals should not exceed more than 75 per cent of total requirements. Energy derived from fat or oil should not exceed 15 per cent of the total calories and from refined carbohydrates (sugar and jaggery) has been recommended around 5 per cent of the total calories.

A balanced diet must supply enough protein for building up our tissues in various parts of our body. When some tissues get worn out or used up proteins repair and replace them. Proteins are essential for the secretion of digestive juices and synthesis of enzymes and hormone production. Dietary proteins are used again for the synthesis of body ins. Protein requirements must be met by dietary proteins. Usually in requirement is in terms of body weight, that is, 1 gm of protein to body weight is the recommendation. Even though animal ins are of superior quality it is not a practical suggestion to depend on them for dietary proteins. In an average Indian diet pulses are a very important source of proteins. When nutritional scores are given to jus proteins egg has a high score of 90 per cent whereas vegetable proteins score only 45 to 50 per cent in the case of pulses and 45 per cent to cereals. But it can be improved by incorporating pulse and cereal proteins; 40 to 60 gms of pulses and 150 to 250 ml of milk can provide enough proteins in a vegetarian diet For non-vegetarians 20 to 30 gms of pulses

and one egg or 30 gms of meat or fish and 100 to 150 ml of k provide the protein requirement If no pulse is used non-elarians can take two eggs or 50 gms of meat or fish or 30 gms of fish neat and one egg.

Fats and oils are other major nutrients in a balanced diet. Fats are concentrated forms of energy. Fat soluble vitamins are utilised only in presence of fat For males 40 to 65 gms of fat and for females 20 to gms of fat must be included in a balanced diet. Vitamins are essential the proper functioning of our body. Small quantities of vitamins own dramatic effect on our body. Organs like eyes, alimentary canal, in and mucous membraneous structures depend on vitamins for their formal structure and functions. Leafy vegetables, other vegetables, yellow vegetables and roots and tubers provide almost all the vitamins required by our body. About a quarter kilogramme of vegetables is required in our daily diet to ensure vitamin supply. Raw vegetables or sprouted grams or fruits are essential to supply vitamin C. Minerals like calcium, phosphorus, iron and trace elements and salts are essential to maintain good health. Sodium chloride is essential to regulate water metabolism in our body. A balanced diet provides enough vitamins and minerals from natural sources. Tonics or supplements are not required or people who consume balanced diets. The Indian Council of Medical Research has designed Recommended Dietary Allowances for various nutrients. They have also formulated composition of balanced diet for various groups. These allowances are designed to serve as a guideline for planning diets. Suggestions are given for substituting foods for-non-vegetarians and additional allowances during special conditions.

To include the above nutrients in daily diet certain guidelines are suggested by the ICMR. Food items which contribute the same nutrients are grouped together and certain food groups are evolved. Foods are grouped into 11 groups based on their nutritive value. They are cereals and millets, pulses, nuts and oilseeds, vegetables, fruits, milk and milk

products, meat, fish, eggs and poultry, fats and oils, sugar and jaggery and spices and condiments.

Cereals and millets constitute 70 to 80 per cent of calories and proteins of low income group. Except ragi all cereals are poor in calcium. Dried pulses are rich in proteins and a cereal-pulse combination is an excellent substitute for animal proteins—pulses are also good sources of many B vitamins and sprouting enhances the vitamin C content of pulses. Nuts and oilseeds are good sources of proteins, certain B vitamins, vitamin E and minerals like phosphorus andiron.

Vegetables are the storehouse of carotene, riboflavin, folic acid, vitamin C and calcium. Vegetables also supply water and roughage to the body. Leafy vegetables, other vegetables and roots and tubers are the different types of vegetables. Fruits in general are a good source of vitamin C. Milk is nature's best food as it is almost complete except for iron and vitamin 'C. Like milk, egg also contains proteins of high biological value. Egg is a rich source of vitamin A and some B vitamins. Animal foods like meat, fish and poultry are rich sources of proteins and B vitamins. Fatty fish are good in fat, vitamin A and D and small fish with bones are good sources of calcium. Liver and other organ meats are good sources of proteins, vitamin A, B complex, B_{12} and iron.

Sugar and jaggery are good sources of energy and honey and jaggery contain minerals like iron. Condiments and spices are mainly used to enhance palatability.

For non-vegetarians pulses can be reduced to 50 per cent and instead of that, one egg or 30 gms of meat or fish can be included. If no pulse is taken two eggs or 50 gms of meat or fish can be included.

Significance of Balanced Diet : For planning a balanced diet for labourers, low cost nutritious foods must be included. Ragi, tapioca, sweet potato, clusterbeans, bittergourd, soyabeans, drumstick leaves, various types of leafy vegetables, cowgram, horsegram, groundnut, Indian gooseberry, guava,

papaya, mango, cashew fruit and other seasonal fruits can be used liberally to make a low-cost balanced diet Sprouted pulses and cheap vegetables that can be used as raw salads also must be included to meet the vitamin C requirement.

The ICMR Advisory Committee (1981) recommended the composition of balanced diet and it is given below:

Food Item	*Adult Man* Seden-tary	Moderate Work	Heavy Work	*Adult Woman* Seden-tary	Moderate Work	Heavy Work
Cereals	460 gm	520 gm	670 gm	410 gm	440 gm	575 gm
Pulses	40 gm	50 gm	60 gm	40 gm	45 gm	50 gm
leafy vegetables	40 gm	40 gm	40 gm	100 gm	100 gm	100 gm
Other vegetables	60 gm	70 gm	80 gm	40 gm	40 gm	100 gm
Roots and tubers	50 gm	60 gm	80 gm	50 gm	50 gm	100 gm
Milk and milk products	150 ml	200 ml	250 ml	100 ml	150 ml	200 ml
Oil and fat	40 gm	45 gm	65 gm	20 gm	25gm	40 gm
Fruits	60 gm	60 gm	60 gm	60 gm	60 gm	60 gm
Sugar and jaggery	30 gm	35 gm	55 gm	20 gm	20 gm	40 gm

For non-vegetarians pulses 50 per cent less + one egg or 30 gms fish or meat, if no pulse, 2 eggs or 50 gm fish or meat.

A Sample Diet for a Sedentary Woman

Time		*Meal*		*Menu*
6.30 a.m.	-	Bed coffee		
8 a.m.	-	Breakfast	-	Chapathi, Groundnuts-Tomato curry, Coffee
12.30 Noon	-	Lunch	-	Rice or Chapathi, Brinjal and horsegram curry Amaranth or spinach pugath, Lassi, Carrot salad.
4 p.m.	-	Tea	-	Vegetable vermicelli uppuma, Mint leaf chutney, Tea.
8 p.m.	-	Dinner	-	Wheat dosa or phulkas, Sprouted green gram curry, Fruit cup.

Quantities of various food groups can be distributed according to the composition given by ICMR. This above diet provides:

Calories - 2420 kcals

Proteins - 53 gms

The idea that the body's needs are definite and can be measured is a very recent one. We are so accustomed to eating to satisfy hunger that for centuries this criterion accepted by most people without question. Plenty was, in general, associated with health and want with starvation.

And yet, throughout the ages there have been those few who have seen a little further into the meaning of nutrition. We know from the book of Daniel of how he asked that he and his companions might eat a simple diet of pulses and water and how, after ten days during which this was permitted, 'their countenances appeared fairer and fatter in flesh than all the children which did eat the portion of the king's meat.' This was probably the first dietary experiment ever to be recorded.

We also know how, in the Middle Ages, illness and loss of life from scurvy led to the recognition that food could prevent and cure certain diseases but the connection between man's daily food and his health was only dimly acknowledged and in a very limited field.

Fortunately, the pleasures of a mixed diet led those who could afford it to eat a variety of foods. The greater the variety, the greater chance there was that all the vital and necessary factors would be included. Hence disease was most prevalent in institutions where the food was restricted and monotonous and among the poor who could not afford variety.

It was Sanctorius (1561-1636) who showed, by sitting on a chair suspended from the arm of a balance while he ate a meal, that afterwards the increase in weight which had taken place was gradually lost again. He described the loss as due to 'insensible perspiration.'

This proved the first basic fact that body-weight was connected with food eaten and that living, even without any conscious exertion, causes loss of weight.

This led in time to the acknowledgment that all the activities of the body derived their energy from the slow combustion of the foods eaten or (in starvation) of the body itself.

An adequate diet is one of that provides all the essential nutrients in sufficient quantities to meet the needs of the body. Thus far, the energy and nutrient needs, including carbohydrates, fats, proteins, minerals, vitamins, and water. The information concerning each of the nutrients in terms of the day's food is now brought together. Easy rule-of-thumb methods for knowing which foods and how much of each one are needed to provide an adequate diet have been developed on the basis of *food groups;* for this purpose, foods similar in nutritive value are considered together as a group. The pattern for the adequate diet is developed by recommending sufficient food servings from the various groups.

Various Human Requirements : The food shortages during and after the 1914-18 war and the resulting ill-health of children, especially in Eastern Europe, made the general public realise for the first time that starvation was not the only consequence of a poor diet. Enough food to satisfy hunger sometimes failed to protect from diseases of malnutrition. Food had to be of the right and, to be fully protective, had to contain the essential vitamins as well as the basic nutrients : proteins, fats, carbohydrates and mineral salts.

Information to have extracted from old records about diets known to have been associated with deficiency diseases. Beriberi had occurred most commonly in countries where rice formed the bulk of the food eaten and had been known to disappear when some of the rice was replaced by vegetables. A diet provided for the U.S.A. garrison of Porto Rico consisted of rice, beans, potatoes, onions and meat with some extra fruit and vegetables, and on this, cases of beriberi occurred. Using recent values for the vitamins B_1 content of these foods, we find that

the daily diet contained about 0.5 mg. of vitamin B_1 per man. This represents an average intake, and implies the some men ate less and some more, the former being without doubt those who succumbed to beriberi. The rations were later improved. Less rice and onions were given and more beans, potatoes and meat, and with the change, beriberi disappeared from the camp. The vitamin B_1 of the improved ration calculated by modern methods was just over 0.8 mg. per man — an increase of only 0.3 mg. but it made the difference between beriberi and protection. This difference is about two-millions of the total weight of a man's daily food, so it is not surprising that vitamins, active in such minute amounts, remained undiscovered for so long.

During the 19th century, scurvy was common in prisons and in other institutions, not to be punitive. In 1822 the meals in Millbank Penitentiary are described as:

For females — morning 9 oz. bread — pint gruel (made with barley)

noon 9 oz. bread — pint soup evening — pint soup.

The bread would have been made from stone-ground flour and would contain plenty of vitamin B_1, so as we should expect, there is a history of beriberi in such a prison, but any vegetables used in the preparation of the soup would be unlikely to retain any vitamin C during the cooking and it recorded that 448 prisoners did in fact show signs of scurvy in one year.

During the war of 1914-18, both scurvy and beriberi occurred. Soldiers of the Indian Army who ate few vegetables and little if any meat, suffered from scurvy. The British on the other hand, who ate vegetables, but while in India had rice or refined cereals, went down with beriberi. In 1916, special hospitals for the admission of cases of scurvy were operated in Baghdad and other parts of the Middle East.

A valuable study was made between 1920 and 1930 on the effect of increasing the quantities of protective foods in the meals of schoolboys. Records were kept of the heights and weights of boys between the ages of 10 and 18 at a large public

school. Conditions of housing, the amount of physical exercise taken and other possibly influential factors remained relatively constant during period and the boys were, by accepted standards of nutrition, well fed. However as a result of an increasing awareness of the importance of vitamin-containing food the authorities altered the diet, so that the boys ate less bread and more milk, vegetables and butter. The results of these changes as seen in records kept over a long period were published by the medical officer (Friend, 1935).

At all ages there was an increase in both height and weight, this being greatest, as one would expect, during the years of most rapid growth.

It may be argued that increase in size does not necessarily mean better health and we are all familiar with overweight children who are abnormally fat but, in general, and where large numbers are concerned, the rate of growth of children can be taken as one indication of physical progress.

This study as important a being the first of its kind undertaken on an adequate scale and over a long enough period to give reliable results.

With the increasing consumption of vegetables and fruits in Great Britain, scurvy, at one time described as the 'English disease,' became rare and the introduction of cod liver oil as a supplement for infants reduced the incidence of rickets, but the new knowledge of the importance of vitamins as protective factors roused great interest in nutrition and people began to wonder how present-day diets would stand up tests of adequacy.

The first report of a large-scale investigation into modern diets was published by J. Boyd Orr in 1937. He divided the community into six sections according to income and estimated the amounts of money spent on food and also the consumption of protective foods in the different income groups. He found that with rise of income, more protective foods were eaten and that in three lowest groups, the amounts of Vitamins A and C eaten were very low.

Investigations on similar lines were made in the United States and from all the data collected, it became possible to make a fairly good estimate of the amounts of the more common vitamins contained in the daily food of normal healthy people and also the amounts below which deficiency diseases were liable to occur.

Tables setting out probable human requirements were published by a Technical Commission of the League of Nations in 1937 and by the Government Bureau of Home Economics, U.S.A. (Stiebeling, 1933). This was before the adoption in the United States of international units, so the results are not comparable but the attempt to express human needs in terms of quantities of vitamins was in itself a great advance.

Later, tables using international units were prepared by the National Research Council of the U.S.A. and these gave (except for vitamin C) figures surprisingly close to the League of Nations recommendations. The most recent publication is that the British Medical Association, 1950.

	League of Nations 1937	*National Research Council 1945*	*British Medical Assoc, 1950*
Vitamin A	2,000-4,000 i.u.	5,000 (optimum)	i.u. 3,000 i.u.
Vitamin B_1	1 mg. approx.	1.5 mg.	1.4 mg.
Vitamin C	30 mg	75 mg.	20 mg.
Vitamin D	—	—	400 i.u.

During the war of 1939-45 there were many cases of dietary deficiency including beriberi in the Japanese prison camps. Some of the doctors who were among the prisoners kept record of the occurrence of beriberi and also of the vitamin B_1 of the diet and the history of the Porto Rico garrison was repeated. When the intake fell blow 0.5 mg. vitamin B_1 (or about 0.2 mg. per 1,000 nonfat Calories) men succumbed to beriberi; when it rose above these levels, the men's health improved (Smith and others, 1951).

The Register-General's Statistical Review which is published

every year gives details of the number of deaths from different diseases. Although the causes of scurvy and rickets were known in the 1920s, and the foods effective for prevention were available, the diseases did not disappear immediately. There was, however, a steady drop in incidence. This demonstrates that not only must protective foods be available, but the public must be educated to use them properly. Free supplements for the provision of vitamins A, C and D were made available for infants during the war years but all mothers did not avail themselves of the opportunity to have them. The number taking them has increased considerably in the last few years with a corresponding drop in rickets almost to the point of disappearance (Registrar-General, 1949, 1950 and 1952) and the health of children in this country has steadily improved.

Functional Disorders of Diet

Most of the vitamins, being growth-promoting substances in the sense that they are needed for full structural and functional development are needed greater amounts during childhood when growth is taking place and during pregnancy and lactation for the same reason.

Vitamin A : Vitamin A is stored in the liver and during adult life the daily needs can be supplied to a considerable extent from these reserves. Volunteers who deliberately consumed diets very low in vitamin A were able to continue doing so for several months without developing any gross signs of deficiency. But in spite of this there is general agreement that it is better to consume enough to guard against any chances of deficiency and, in the table, the recommended levels of vitamin A vary very little.

Whether taken as vitamin A itself or in the plant form carotene, this factor is always converted into vitamin A used as such in the animal body. But the two forms are not of equal physiological value, unit for unit. It is usually reckoned that the number of international units needed in the form of carotene is about twice the number needed as vitamin A itself. Even this

ratio is not the same for all foods. The carotene of green vegetables supplies to the body about half as many units of vitamin A as the number obtained by chemical estimation, but yellow vegetables like carrots supply only about one-fifth or one-quarter (Graves, 1942).

Increased Needs for the Vitamin A : Vitamin A being used by the body to maintain the health of the skin, teeth and mucous linings of the respiratory organs is needed especially in illness in which these are involved. To guard against them, supplies in the daily diet should be kept up to the standards given. In treatment higher doses are available in pharmaceutical preparations can be prescribed. Dases up to 100,000 or 200,000 i.u. are sometimes given daily.

The B Vitamins : The need for vitamin B_1 varies according to the amount of food eaten and the energy expended in movements in all kinds.

As muscular activity, which depends on the oxidation of carbohydrates, continues throughout life, vitamin B_1 must be continuously supplied. Moreover, this vitamin differs from vitamin A is not being stored for long periods. In general, the fat-soluble vitamins tend to be stored in the body, but the water-soluble vitamins, i.e. those of the B group and vitamin C, are not stored. If more is eaten than needed it is rapidly excreted again in the urine.

Because vitamin B_1 is used day by day, as it is consumed, it has been much easier to find out exactly what the daily needs are in fact, as early as 1934, Cowgill was able to suggest a mathematical formula to express the daily needs in terms of body weight and calorific intake.

Increased Needs for Vitamin B : As vitamin B_1 is concerned with carbohydrate metabolism, any conditions which involve a higher calorific intake, e.g heavy manual work, climbing, Article exploration, naturally call for higher intake.

In those diseases, too, in which the rate of metabolism is increased, more vitamin B_1 is needed.

Riboflavin and nicotinic acid also take part with vitamin B_1 in energy relations and the amounts needed are related to the amount of vitamin B_1 In foods, these vitamins of the B complex occur together and a shortage of one usually implies a shortage of the others.

It is only when the staple cereal happens to be grossly short of a particular factor as, for example, polished rice and vitamin B_1 or maize and nicotinic acid, that well-defined deficiency diseases such as beriberi and pellagra occur. In countries where mixed diets are customary, deficiencies are usually less severe but of a mixed type. For example, a mild shortage of riboflavin will produce, in some individuals, roughness and redness of the skin over the nose and some cracking of the lips, easily cured by the consumption of extra quantities of one or two foods rich in riboflavin such as eggs or liver.

Similarly a shortage of nicotinic acid, not enough to cause pellagra, may produce digestive disturbances, a reddening of the skin of the face and hands and feelings of depression.

Nicotinic acid is concerned in healthy people with the control of movements of precision and if its supply is limited subjects may find themselves unusually clumsy.

The varying responses of individuals to diets deficient in the B vitamins was puzzling until, as a result of a long series of observations, it was learnt that these factors are also synthesised within the body of man and the higher animals though not by it.

The intestinal tract contains warm semifluid material rich in nutrients and is, as we know, a medium in which there is normally an expensive bacterial flora. These organisms are not disease-producing. They merely live in and with man. Their power of synthesising vitamins is a relatively recent discovery. They seem to specialise in the factors of the B group but some can also synthesise vitamin K (Najjar and Barrett, 1945).

Ruminants, whose large rumen in the first part of the

digestive tract is the site of vitamin formation, obtain so much riboflavin and nicotinic acid from bacterial activity that they fail to develop signs of deficiency even on diets lacking these vitamins. But in man the bacteria carry on their synthetic activities much further down the digestive tract and, although undoubtedly some of the B vitamins produced in this way are available for human use, enough of each is not formed to satisfy all human needs.

It is, however easy to see that in conditions such as diarrhoea in which the contents of the alimentary tract move along quickly, less than the normal amounts may be absorbed and symptoms of deficiency may occur if the condition is not checked. The recent use of some of the new drugs, sulphonamides and antibiotics has also revealed that these, too, affect the activities of the vitamin-B-producing bacteria. The drugs inhibit not only the growth of disease-producing organisms but also the synthesis of the B vitamins in the intestine.

This interference with production may be great enough to cause true secondary deficiencies in patients who may not have been very well supplied with vitamins before. Cases have been observed with signs of peripheral neuritis, enlarged heart and swelling of the limbs, indistinguishable from true beriberi. Other patients develop a sore mouth and red scaling of the skin recalling deficiencies of nicotinic acid and riboflavin.

In the treatment of these secondary deficiencies very large doses have to be given — a hundred or more times the amount needed for normal health.

Vitamin C : Vitamin C, also water-soluble, is another vitamin which should be taken daily, although the body stores are not exhausted as rapidly as those of vitamin B_1

There is difference of opinion about the amount of vitamin C needed to maintain health. The American in which estimate is higher than the British. An experiment in which volunteers did without any vitamin C for months showed that they could

do so without developing any gross signs of scurvy, and the quantity needed for health has consequently been placed at a lower level than formerly. But, as the signs of deficiency develop in such an insidious manner, it is probable that minor degrees of vitamin C deficiency are fairly common though not always recognised as such.

It is known now that vitamin C is concerned not only in protection against scurvy but in the healing of wounds, in the prevention of shock from injury, in the tolerance of high atmospheric temperature and in the control of body fluids. These processes are common in everyday life and probably account for the daily use by the body of vitamin C.

Increased Needs for Vitamin C : When any of these, conditions are actually present or exaggerated the need is correspondingly increased. Very large doses (up to 1 g.) are given before operation to prevent shock and in the treatment of severe injuries and larger than normal doses are given to man doing heavy work in an excessively hot humid atmosphere

Large amounts of vitamin C are available in fruits and vegetables and some reason or other an individual does not eat or can not eat these foods, then tablets of the synthetic vitamin or some other pharmaceutical preparation must be taken. The massive doses are, of course, always given by means of a pharmaceutical preparation.

Vitamin D : It will be noticed that the table gives the same figure for the human requirement for vitamin D except in the case of infants under one year and for expectant and nursing mothers. This uniform level is an fact a confession that very little is known about varying needs.

At one time it was assumed that vitamin D was needed only during the period of active growth. This view is no longer held and it is thought that the fragile nature of the bones people of older may be related to their low consumption of vitamin D foods.

The figures for human needs given in the B.M.A. table

have been obtained by studies of dietary surveys, of individual diets, of diets associated with deficiency diseases and so on. The first estimates were made by the Health Committee of the League of Nations; these were followed by American estimates and finally the table quoted here was produced. Considering the magnitude and difficulty of the task, it is astonishing how closely the various estimates agree and it is probable that these recently published figures are as near the truth as we are likely to get.

The question has to be asked : to what extent do individuals eat diets which satisfy these needs? How easy or difficult is it to obtain an adequate diet?

There are several ways of approaching this problem. We can see whether there is any evidence that people are better fed now than formerly or we can study foods which are normally eaten in ordinary homes and see if they supply the needs.

Reference has been made to the effect of adding protective foods to the diet of schoolboys and to the increases in height and weight produced as a result. Data on the height and weights of London school children during the first half of this century have recently been published and these show the same kind of increase. It may not be totally due to nutrition, but it is not unreasonable to conclude that improved feeding due to increased knowledge of nutrition has also played a part in producing these effects.

If we approach the problem from the dietary angle, we can analyse representative diets and estimate their vitamin content. This is a useful exercise and a comparison between the foods eaten on several different days or by different individuals shows how wide the variation in vitamin content can be.

Following examples of meals will illustrate this point. These show that the value of a diet depends on the *choice* of foods, and that likes and dislikes are of themselves not reliable guides; there must be knowledge also.

	Amount	*Vit. A or equiv. in carotene*	*Vit. B_1*	*Ribo-flavin*	*Niacin*	*Vit C*	*Vit D*
	Oz.	*i.u.*	*µg.*	*µg.*	*µg.*	*mg.*	*i.u.*
Meal 1							
Roast beef	3	42	33	195	3,900	—	
Yorkshire pudding							
3 oz. = flour	1	_	34	23	450	-	-
milk 1½	45	18	54	-	-	0.45	
egg ½	125	19	85	-	-	8.5	
cooking fat	1/8	-	-	-	-	-	
Roast potatoes	6	-	180	26	1,200	15	-
Cabbage	4	288*	4	28	320	28	-
Apple pie= flour	½	-	171	11	225	-	
cooking fat	¼	-	-	-	-	-	
apple	2	4	6	-	50	1.6	-
sugar	¼	-	-	-		-	-
Custard= milk	2	60	24	72	-	-	0.6
powder	mg.	-	-	-	-	-	-
		564 i.u.	335 µg.= .335 mg.	564 µg.= .564 mg.	6,145 µg. 6.145 mg.	44.6 = mg	9.55 i.u.
Proportion of day's need approx.	1/8	¼	1/3	½	All 1/40		
Meal 2							
Cheese	1	369	9	140	100	-	4
Lettuce	2	727*	34	40	200	6	-
Tomato	3	540*	60	30	300	21	-
Wholemeal bread	3	-	270	150	1,800	-	-
Butter	½	568	-		-	-	8.5
Orange	3	48	60	30	300	48	-
Coffee= milk and sugar	5	150	60	180	-	1	1.5
		2,402 i.u.	493 µg.= .493 mg.	570 µg.= .57 mg.	2,700 µg.= 2.7 mg.	76 mg.	14 i.u.

contd.....

Vit. A *Amount* *Oz. i.u.*	*or equiv. in carotene µg.*	*Vit. B_1 µg.*	*Ribo-flavin µg.*	*Niacin mg.*	*Vit C i.u.*	*Vit D*	
Proportion of day's need approx		¾	1/3	1/3	1/5	All	1/90
Meal 3							
Fried white fish	4	-	40	160	2,400	-	-
Fried potatoes	6	-	180	60	1,200	12	-
Rice pudding							
4 oz.							
= rice	1/3	-	5	6	100	-	-
milk	2½	75	30	90	-	-	.75
sugar	1/6	-	-	-	-	-	-
Stewed apple, 4							
=raw	3	7	9	-	75	2.4	-
		82	264	316	3,775	14.4	.75
		i.u.	µg.=	µg.=	µg.=	mg.=	i.u.
			.264	.316	3.775		
			mg.	mg.	mg.		
Proportion of day's need approx		1/40	1/5	1/6	1/4	1/2	negli-gible
Meal 4							
Sausages, pork	3	-	210	150	1,500	-	-
Baked beans	3	35	60	30	600	-	-
Mashed potatoes	6	-	144	96	1,350	12	-
Prunes	4	103	32	68	-	-	-
Custard= milk							
powder	3	90	36	108	-	-	.9
		228	482	452	3,450	12	.9
		i.u.	µg.=	µg.=	µg.	= mg.	i.u.
			.482	4.52	3.45		
			mg.	mg.	mg.		
Proportion of day's need approx		1/13	1/3	1/4	1/4	3/5	negli-gible

For green vegetables : Carotene = 40 per cent as vitamin A

Yellow veg. = 25 per cent

Half the value of uncooked flour.

Where no value is available for vitamin B_1 in cooked food, the figure for the uncooked food is halved.

Global Standards

Vitamins A and D are usually expressed in international units (i.u.)- Human or animal requirements for other vitamins are usually given in milligrams, but the value in foods is often expressed in microgrammes (1/1000s of miligram) so as to avoid the use of decimals.

In the tables, the value of *individual* foods are given in micrograms for the same reason but the *total* values and the requirements are given in milligrammes for comparison with other published figures.

Conversions —Weight and Household Measures

5 oz. milk	= 1 teacupful
1 oz. cheese is the same size as	1 oz. butter (1/4 of 1/4- 1b. pack)
1 oz. bread	= ¼ in. slice from square loaf
1 oz. cake	= portion 3 x 1 x 1 /2 in.
1 oz. Castle pudding is the same size as 1 oz. cake	
1 oz. biscuits	= 2 biscuits (2 1 /4 in. in diameter)
1 oz. flour, rice, oatmeal, custard, powder, sugar, etc.	= 1 well heaped tablespoonful
½ oz.	= 1 level tablespoonful or 1 well heaped teaspoonful
1 oz. jam	= 1 desert spoonful
2 oz. potato (mashed), green vege. tables, root vegetables or peas	= 1 heaped tablespoonful
2 oz. tomato	= 1 egg-size tomato
4 oz. stewed fruit	= 5 small plumps (with juice) or 4 tablespoonfuls of soft fruit

3 oz. meat	= 4 thin slices (approx. 3 x 4 in.)
1 oz. bacon	= 1 rasher (approx. 1/8 x 7 x 2 in.)
4 oz. fish	= portion approx. 3 x 3 x 1/2 in.
2 oz. herring	= 1 small herring (approx.5 in. long)

Food Sufficiency

A plan for selecting an adequate diet developed around 4 *food groups* will now be discussed. There are other plans, developed on the same principle as the 4-group plan, differing only in number of groups. All of the plans are developed in accordance with food habits of the people and the foods available to them. In case the habits of eating and the food supply are different, the food patterns are of necessity also different.

The 4-group plan is flexible, permitting for a wide choice of foods within most of the groups. Meals planned in one home are likely to differ from those in another, even though the meals in each home follow the same food pattern. Foods selected by persons eating out may vary widely, and still at the same time each is in complete accordance with the recommendations of the same food pattern.

Four-Food Group Plan : The 4-group plan (milk, meat, vegetable-fruit, cereal) was developed in 1958 and is a simplified version of the 7-group plan that originated during wartime 1943. In this 4-group plan the citrus fruits, green and yellow vegetables, and other vegetables and fruits are combined into one group, and the butter fortified margarine group omitted. The daily quantities of each group of recommended to provide the *foundation* of an adequate diet and the main nutritive contribution of each group are given. Calories are the least well supplied by this pattern. However, as a usual practice, facts and sugars, which are not listed, are eaten as such or combined in baked goods and desserts, with a result that the energy intake is increased. Also to satisfy the appetite, additional servings from the food groups may be exported, which will contribute both to the energy and nutrient intake.

The 7 food groups	*Daily amounts*	*Min nutritive contributions*
Green and yellow vegetables	1 or more servings	Vitamin A value Ascorbic acid Iron
Oranges, grapefruit, tomatoes or raw cabbage or salad greens	1 serving	Ascorbic acid
Potatoes, other vegetables and fruits	2 or more servings	Vitamins and minerals in general Cellulose
Milk and milk products	Children ¾ to 1	Calcium Riboflavin quart Adults 1 pint
Protein Phosphorus		
Meat, poultry, fish and eggs	1 serving meat, fish or poultry 1 egg (at	Protein Phosphorus Iron B- Vitamins least 4 a week)
Bread, flour and cereal (whole grain enriched or restored)	3 or more servings	Thiamine Niacin Riboflavin Iron Carbohydrate Cellulose
Butter or fortified margarine	2-3 tbsp.	Vitamin A Fat

The basic 7 food groups.

The 4 food groups	*Daily amounts*	*Main nutritive contributions*
Milk group:	Children 3-4 cups	
Milk, cheese, rice	Teenagers 4 or more cups	Calcium
cream (cheese and	Adults 2 or more cups	Riboflavin
ice cream can replace part of the milk)	Pregnant 4 or more cups Women	Protein
	Nursing 6 or more cups Mothers	Phosphorus
Meal group	Protein	
Beef veal pork, lemon,		Phosphorus
Poultry, fish, eggs	2 or more serving	Iron
dry beans and peas as alternates		B-vitamins
	4 or more serving, including a dark green or deep yellow vegetable at least every other day.	
Vegetable-fruit group	A citrus fruit or other fruits or vegetable rich in ascorbic acid- daily Other fruits and vegetables including potatoes	Vitamins Minerals Cellulose
		Thiamine Niacin
Bread-cereals group (Whole grain)	4 or more servings	Riboflavin Iron Carbohydrates
ebrichesd restored)		Cellulose

The 4-food group plan.

The percentage of the Daily Recommended Allowances provided by the foundation pattern will vary with individual needs. The person with a greater requirement will have those needs less well satisfied, of course, than the person with a lesser daily need. A day's dietary planned with the use of the 4-food group is illustrated in table . This menu contains more than the minimum number of servings in three of the four groups of plan (see Fig.). The nutrients contributed by foods in the four foundations groups in this dietary are shown in Table. Butter, jelly, mayonnaise, lemon pie, and sugar are not included in the calculations. A comparison of the percentage of the Daily Recommended Allowances provided by foods in the 4-group plan is made. Young men obtained a lower percentage of their allowance from the foundation groups simply because the recommendations for men are higher for some nutritents than for women. However, the young man, in satisfying his energy need and promoted by a natural urge to eat, selects more of all food groups. The result is that in actual practice the diets of young men are, as a rule, more nearly adequate than those of young women.

The 4-group plan is useful as an education tool to show the foundation for an adequate diet. Because it is such a simple plan, some may not realise its importance. For the most efficient use of the 4-group plan, one should know the main nutrients contributed by the various foods in each group.

The Nutritive Value of Food Commodities : Each food group, and each food within the group, makes a particular contribution to the nutrient content of the diet.

Comparison of the Nutrients Sulphide by the 4-Food Groups from the Above Menu to the NRC Recommended Daily Allowances for Young Women and Men

Young women (18-35 years)

Young men (18-35 Years)

Some foods supply many more of the nutrients and in much greater concentration then do others. The nutritional

contribution of the major food commodities, milk and milk products; meat, fish, and poultry vegetables and fruits; and cereal and cereal products, will be discussed in the following paragraphs.

Menu with the Four-Group Plan as a Guide

Breakfast		*Lunch*		*Dinner*	
Orange juice	4 oz	Swedish meat balls	3 oz	Veal cutlet	3 oz
Egg. soft-cooked	1	Green beans	½ c	Baked potato	1 med
w.w. toast	2 sl	Waldorf salad	2/3 c	Brocciol	2/3 c
Butter	1 pat	Pan roll	1	Hard roll	1
Jelly	1 tbsp	Cup custard	6 oz	Butter	1 pat
Milk, whole	8 oz	Iced tea	8 oz	Lemon pie	1 sl
		Snack -1 milk shake	Iced	tea	8 oz

Milk and its Products : Milk is considered as nature's "most nearly perfect food." Most of the known essential nutrients are found in milk, but it is a much better source of some than others. Milk and milk products are excellent sources of calcium, protein and riboflavin. Vitamin A, phosphorus, and thiamine are also supplied in good amounts. Milk and milk products are poor sources of iron and of ascorbic acid.

The importance of including milk or milk products in the diet so as to obtain sufficient calcium is well known. For students, or anyone for that matter, drinking milk is one of best ways of assuming an adequate calcium intake. Each gram of milk contains about 1 mg of calcium, with a glass of milk supplying about one-third of the recommended calcium intake for the college student.

Milk contains protein of high biological value; and so complements the incomplete proteins of cereal products, fruits, and vegetables. Since milk can very well be served as a beverage at each meal of the day, a source of complete protein can therefore be assured. The protein value of milk is that each ounce contains about 1 gm of protein.

Evaporated milk contains the same nutrients as whole

milk. The process of evaporation, and pasteurisation also, reduces the thiamine and ascorbic acid content by about one-fifth to one-fourth. Evaporated milk is convenient to use and economical. The fortification of milk with vitamin D is a god public health measure. By adding vitamin D to milk the essential nutrients for bone formation, calcium, phosphorus, protein, and vitamin D, are brought together in one food.

The popularity of non-fat milk solids is increasing because of the lower cost, ease of storage, and lower Calorie intake. With the removal of cream from the milk, which is done prior to drying, the caloric value is reduced by about one-half. In addition, vitamin A and other fat-soluble vitamins are removed. If nonfat milk solids are used to replace the whole milk, care must be taken to provide for additional vitamin A value from other sources such as green and yellow vegetables and liver because milk is depended on to provide some of the vitamin A need.

An ounce of Cheddar-type cheese provides about the same nutritive value as a glass of milk. In the manufacture of cheese some of the whey is removed, taking with it some of the water-soluble nutrients including lactose, the water-soluble vitamins (B-group), and minerals. Cottage cheese made with rennet has a higher calcium content than that made entirely by the souring of the milk (acid coagulation). In this process calcium, which is soluble in an acid medium, is carried into the whey.

Food	*Measure*	*Energy, Cal.*	*Protein, gm.*	*Calcium, mg.*	*Iron mg.*	*Vit. A i.u.*	*Thiamin, mg.*	*Riboflavin, mg.*	*Ascorbic Acid mg.*
Milk group									
Milk (beverage, custard, milk shake)	2½ C	400	22.5	720	0.3	875	0.2	1.05	5
Ice cream (shake)	1/8qt	145	3	87	0.1	370	0.03	0.13	1
	Total	*545*	*25.5*	*807*	*0.4*	*1245*	*0.23*	*1.18*	*6*
Meat group									
Egg, soft, ck,	1	80	6	27	1.1	590	0.05	0.15	—
Egg (custard)	1	80	6	27	1.1	590	0.05	0.15	—
Meat balls	3 oz	245	21	9	2.7	30	0.07	0.18	—
Veal cutlet	3 oz	185	23	9	2.7	—	0.06	0.21	—
	Total	*590*	*56*	*72*	*7.6*	*1210*	*0.23*	*0.69*	—

contd........

Food	*Energy, Measure*	*Protein, Cal.*	*Calcium, gm.*	*Iron mg.*	*Vit. A mg.*	*Thiamin, i.u.*	*mg.*	*Riboflavin, mg.*	*Ascorbic Acid mg.*
Vegetable-fruit group									
Orange juice	4 oz	50	.5	13	0.3	245	0.11	0.03	64
Green beans	1/2c	15	1.0	31	0.4	340	0.04	0.06	8
Apples	1/2c	35	tr	4	0.2	25	0.02	0.01	1
Celery	1/4c	4	tr	10	0.1	60	0.01	0.01	2
Potato, baked	1 med	90	3	9	0.7	tr	0.10	0.04	20
Broccoli	1/2c	20	2.5	66	0.6	1875	0.07	0.15	68
	Total	*219*	*7.0*	*124*	*2.1*	*2435*	*0.32*	*0.28*	*150*
Bread-cereal group									
Bread,w.w.	*2sl*	*110*	*4*	*44*	*1.0*	*tr*	*0.10*	*0.06*	*tr*
Roll,pan	*1*	*115*	*3*	*28*	*0.7*	*tr*	*0.11*	*0.07*	*tr*
Roll, hard	*1*	*160*	*5*	*24*	*0.4*	*tr*	*0.03*	*0.05*	*tr*
	Total	*385*	*12*	*96*	*2.1*	*tr*	*0.24*	*0.18*	*tr*
Grad total		*1734*	*100.5*	*1108*	*12.4*	*5000*	*1.05*	*2.35*	*169*

Table Nutritive Value of Milk and Some Milk Products

Milk and Milk Products	*Energy Cal.*	*Protein, gm.*	*Calcium, mg.*	*Irons mg.*	*Vit. A. IU*	*Thiamine, mg.*	*mg.*	*Riboflavin Acid, mg.*	*Ascorbic*
Milk, whole (I c)	160	9.0	288	0.1	350	0.08	0.42	2	
Buttermilk, cultured, skim (I c)	90	9.0	298	0.1	10	0.09	0.44	2	
Evaporated, water added (I c)	173	9.0	318	0.2	410	0.05	0.42	1	
Cheese, process (I oz)	105	7.0	219	0.3	350	trace	0.12	0	
Cheese, cottage, creamed, skim milk (1/2 c)	120	15.5	106	0.3	190	0.03	0.28	0	
Ice cream (1/8 qt)	145	3.0	87	0.1	370	0.03	0.13	1	

Table Nutritive Value of Some Meat, Fish, Poultry, and Egg

Food	*Energy Cal.*	*Protein, gm.*	*Calcium, mg.*	*Irons mg.*	*Vit. A. IU*	*Thiamine, mg.*	*mg.*	*Riboflavin Acid, mg.*	*Ascorbic*
Beef, round (3 oz)	165	25.0	11	3.2	10	0.06	0.19	0	
Beef, liver, fried (3 oz)	195	22.5	9	7.5	45,200	0.22	3.55	23	
Pork, lean and fat, loin roast (2 oz)		310	21.0	9	2.7	—	0.78	0.22	0
Chicken (3 oz)	170	18.0	18	1.3	200	0.03	0.11	3	
Salmon (3 oz)	120	17.0	—	0.7	60	0.03	0.16	0	
Egg (1)	80	6.0	27	1.1	590	0.05	0.15	0	

Value depends on amount of bone consumed.

value. They provide protein and iron, with some phosphorus, thiamine, riboflavin, and niacin.

Meat, Fish, Poultry, and Eggs : Meat, fish and poultry are very similar in their nutritive.

The protein in this group of foods, as in milk, is of high biological value. A 3-oz serving of meat supplies approximately

20 gm of protein, which is about one-third the Recommended Allowances for young adult women and over one-fourth for young adult men. Iron is present in relatively high concentration in the organ meats, particularly liver and kidney, and in to a lesser amounts in muscle tissues. Meat is poor source of calcium. Some fish products, mainly shell fish and canned salmon, in which a portion of the bone is edible, are somewhat better than meat as a source of calcium.

Pork muscle is high in thiamine content whether other muscle meats contain only moderate amounts. The glandular organs, liver, kidney, heart, and tongue are higher than muscle meats in riboflavin content. Liver, the storage place for vitamin A, is only edible animal tissue containing any appreciable amount of this vitamin. Eggs provide protein, iron, phosphorus, vitamin A, and riboflavin to the diet. The protein in egg, which is located both in the yolk and the white, is of good biological value. One egg contains about 6 gm of protein, approximately two-thirds the amount in one glass of milk. Phosphorus is distributed in both the white and the yolk. Iron and vitamin A are contained only in the yolk. Table gives the nutritive value of some kinds of meat, fish, poultry, and eggs.

Vegetables and Fruits : As a group, vegetables and fruits are the source of minerals, vitamins, and cellulose to the diet. Not to be disregarded is the interest they add because of their variety in colour, flavor, and texture. The great number of fruits and vegetables available for use differ markedly in nutrient content. For the sake of clarity they are grouped according to their contribution of nutrients to the diet.

Mostly the green and yellow vegetables and fruits are outstanding in their vitamin A values. One serving of sweet potatoes, carrots, kale or cantaloup furnishes sufficient vitamin A to meet the Recommended Allowance for vitamin A for persons of all ages. In addition leafy-green vegetables provide more calcium, iron , ascorbic acid, and the B-vitamins than other vegetables. Some of the green vegetables, spinach, chard, sorrel, and beet greens contain oxalic acid, which combines

with calcium during digestion forming an insoluble salt that cannot be absorbed, and makes the calcium in those vegetables of no use to the body. However, these green vegetables, even though not supplying calcium, are invaluable in the diet for the generous amounts or iron, ascorbic acid, and vitamin A value they provide. The ascorbic acid content of broccoli, Brussels sprouts, kale, and similar vegetables is quite high.

Head lettuce, celery, cabbage, and other light green vegetables are important source for cellulose. Raw cabbage is a good source of ascorbic acid. The light green vegetables are low in vitamin A value. Citrus fruits and tomatoes are good sources of ascorbic acid. Tomato juice contains about one-third as much as the citrus fruits. Since they are available the year round and are well liked, it is customary to include citrus fruits in most dietary plans.

Other foods like strawberries, cantaloups, kale, and green peppers are also rich in ascorbic acid as citrus fruits, but due to cost, unavailability, or lack of acceptance, do not fit into the diet as a routine item as well as citrus fruits. Ascorbic acid is easily destroyed during storage, freezing, and cooking. Citrus retain their ascorbic acid content better than other foods due to the preserving action of the acids they contain.

Vegetables of the seed, root and tuber classes, like Lima beans, corn green peas, sweet potatoes, and white potatoes are higher in carbohydrate content than other varieties. Starch is stored in these areas of the plant's structure.

Vegetables and fruits as a rule provide lesser protein, fat, and total Calories. Table shows the contribution of some fruits and vegetables to the diet.

Pulses and Grains : Cereals are seeds of the grass family—wheat, corn, rice, oats, rye, and barley. From these cereal grains flours (whole wheat, white rye, barley etc.) are manufactured for the preparation of the many varieties of breads, cakes, pastries, breakfasts cereals, macaroni products, hominy corn sirup, corn meal, and other products. Whole

wheat has the following composition in terms of grams of the constituent per 100 gm of whole wheat: 72 carbohydrate, 2 fiber, 12 protein, 2 fat, 2 ash, and 10 moisture.

In many countries cereals are the main food of the diet, furnishing 50 to 90 per cent of the total calories; in the United States about one-third of calories is from cereal or cereal products. More wheat and corn is grown in U.S.A. than other cereals; most of the wheat and corn is used for human consumption, a large part of the corn fed to livestock.

The cereal grains are easy to grow, easy to store, and have good keeping qualities. Whole grain cereals contribute calories (mainly in the form of carbohydrate), thiamine, iron, riboflavin, and partially in complete protein. Cereals are easy to digest, have a blend flavor, and supply roughage or cellulose.

In the preparation of cereal products for the market, parts of the cereal grain are removed. In the case of brown rice only the outer husk is removed whereas in the preparation of white rice all of the outer coats are wheat flour is made from the entire grain except for the husk; white flour is just the endosperm portion. Most of the cellulose, minerals and B vitamins are in the outer layers or coats of the grain. The endosperm portion contains the carbohydrate and incomplete protein. The germ contains most of the fat of the grain and some of the thiamine. Because the fat tends to become rancid during storage and also seems to hold a great attraction for insects, the germ is usually removed in milling. Furthermore, the public prefers the refined product for houses.

Nutritive Value of Some Fruits and Vegetables

Food	*Energy Cal.*	*Protein, gm.*	*Calcium, mg.*	*Irons mg.*	*Vit. A. IU*	*Thiamine, mg.*	*Riboflavin mg.*	*Ascorbic Acid, mg.*
Apple (1 med)	70	tr	8	0.4	50	0.04	0.02	3
Banana (l med)	85	1.0	8	0.7	190	0.05	0.06	10
Broccoli, ck (1/2 c)	20	2.5	66	0.6	1875	0.07	0.15	68
Cabbage, ck (1/2 c)	18	1.0	38	0.3	110	0.04	0.04	28
Carrots, ck (1/2 c)	23	0.5	24	0.5	7610	0.04	0.04	5
Green beans, ck (1/2 c)	15	1.0	31	0.4	340	0.04	0.06	8

contd

Food	*Energy Cal.*	*Protein, gm.*	*Calcium, mg.*	*Irons mg.*	*Vit. A. IU*	*Thiamine, mg.*	*Riboflavin mg.*	*Ascorbic Acid, mg.*	
Kale, ck (1/2 c)	15	2.0	74	0.7	4070	—	—	34	
Lettuce, Iceberg, (50 gm)	7	0.4	10	0.2	167	0.03	0.03	3	
Lima beans, ck (½ c)	90	6.0	38	2.0	225	0.15	0.08	14	
Orange juice, fresh (1/2 c)	50	0.5	13	0.3	245	0.11	0.03	64	
Peaches, cn (1/2 c)	100	0.5	5	0.4	550	0.01	0.03	4	
Potato, white, baked (1 med)	90	3.0	9	0.7	tr	0.10	0.04	20	
Tomato juice (1/2 c)	23	1.0	9	1.1	970	0.07	0.04	20	
Oatmeal, ck (1 c)	130	5.0	21	1.4	0	0.19	.0.05	0.3	0
Rice, white, enriched ck (1 c)	185	3.0	17	1.5	0	0.19	0.12	1.6	0
Macaroni, unenriched ck (1 c)	155	5.0	11	0.6	0	0.02	0.02	0.4	0
Macaroni, enriched ck (1 c)	155	5.0	11	1.3	0	0.19	0.11	1.5	0
Cake, plain (3 x 2 x ½)	200	2.0	35	0.2	90	0.01	0.05	0.1	0
Bread, white, unenriched (1 sl)	60	2.0	16	0.2	0	0.2	0.02	0.3	0
Bread, whole wheat (1 sl)	55	2.0	23 .	0.5	0	0.06	0.03	0.7	0
Muffin, enriched (1)	140	4.0	46	0.8	50	0.08	0.11	0.7	0
Biscuit, enriched (2 1/2" diam)	140	3.0	46	0.6	0	0.08	0.08	0.7	0
Pan roll, enriched (12 per Ib)	115	3.0	28	0.7	0	0.11	0.07	0.8	0

The protein in cereal products varies in biological efficiency but as a whole is classified as partially incomplete. However, with supplementation or protein from milk or meat or eggs, the cereal proteins make a significant contribution to the diet. About 20 per cent of the protein in diet is from cereal sources; on low-cost dietaries the percentage is relatively higher. Milk and cereal make a good nutritional team, cereal supplying iron, thiamine, and partially complete protein, and milk providing calcium, vitamin A, and complete protein. The use of the milk solids (dried milk) in bread improves the nutritive contributions of this product.

It is important that whole grain, restored or enriched cereals, and cereal products be selected for the diet in order to benefit from their greater nutritive contributions. Table gives the nutritive contributions of some cereal products to the diet.

Energetic Oxidisation : Energy for physiological activity is ultimately obtained from the sun. The sun gives out energy in the form of visible light waves and invisible ultravoilet rays. When these impinge on green plants, the latter are able to

absorb the energy and build-up compounds and such as sugar and starch from the carbon dioxide of the air and the water of soil. The compounds become storehouses of energy within the green plants.

By the action of oxygen on these carbon compounds, they can again be broken down through a series of stages into carbon dioxide and water and, in the process, the stored energy is once more liberated. This breakdown happens in the plants themselves when they respire and is the source of energy used by them for growth, development and reproduction.

But far more energy is stored in starch and sugar than is used by the plants and, when animals eat the plant products, this energy becomes available to them.

Animals, too obtain oxygen from the air in breathing. This is transferred in the lungs to the blood and is carried to all parts of the body. Food which is eaten, digested and absorbed is also taken in the circulation to organs where it can be temporarily stored. Most of the starch and sugar eaten is stored in the form of another carbohydrate, glycogen, in either the liver or the muscles, and as we have seen, is broken down to provide energy.

The recognition of these facts led to a great deal of work on the quantities of food needed to produce the energy for certain tasks and it was found that the energy value of a food could be expressed in terms of a heat-unit or calorie. A calorie is the amount of heat-energy required to raise the temperature of 1 g. of water through 1 °C. This was found too small a unit to be convenient for expressing the energy values of foods, so a unit was chosen of 1,000 calories or 1 large Calorie (spelt with a capital C).

The next step was to find out how many calories were required by individuals of different weights, both resting and carrying out tasks with different energy requirements.

The following are average values:

Calories	
Metabolism of sleep (8 hours)	518
Basal metabolism while awake (16 hours)	1,152
Allowance for light work	1,200
Total	2,870

(Best and Taylor, 1950).

Daily Calorin Allowances for Adults

Requirment	*Description*	*Calorie requirement*
0	No work, almost basal (e.g. lying in bed).	1,750
1	Sedentary work (30 Cals/hr.) and little travelling (65 Cals.)	2,250
2	Light work (70 Cals. /hr.) and travelling (130 Cals.)	2,750
3	Medium work (100 Cals./hr.) and travelling (130 Cals.	3,000
4	Heavy work (200Cals. / hr.) and travelling	(130 Cals.)
	Extremely heavy work (450 Cals./hr.) and travelling (130 Cals.)	5,000
0	No work, almost basal (e.g lying in bed)	1,500
1	Sendentay work (30 Cals./hr.) and little travelling (50 Cals.)	2,000
2	Light work (70 Cals./hr.) and travelling (100 cals.)	2,250
3	Medium work (100 Cals./hr.) and travelling (100 Cals.).	2,500
4	Heavy work (200 Cals./hr.))	
5	Very heavy work (200 Cals./hr.) and travelling (100 Cals.)	3,750

The recognition that the quantity of food required could be exactly measured so captured the imagination that for many years no thought was given to the quality of the food supplied. Calories were all important and the neat way in which a man's energy needs could be expressed numerically no doubt delayed the recognition of the more subtle needs for particular factors in foods.

This is doubtless one reason why vitamins were not discovered until the present century, and why are still ignorant about certain aspects of nutritional requirements.

It is, however, established that the basal needs for calories, are best met by a balance between proteins, fats and carbohydrates: other needs are adequate amounts of fluid, certain mineral salts and vitamins.

Although this book is mainly concerned with vitamins, it is not possible to study them without reference to other nutrients.

The Table shows present-day views of human requirements of vitamins and calories.

The figures given in the table are, of course, averages. As with all physiological functions, there are great individual variations but these can be discovered only by experience. However, an example will serve to illustrate the point. Mr. A, a lean, bony man, has a hearty appetite and seems able to eat as much as he likes without showing any tendency to grow fat. By contrast Mr. B, who is also physically active, puts on weight without apparently eating

to excess. These two seem to have different basal mechanisms for the use of food, one making his calories go further, as it were, than the other.

The same seems to be true, to some extent for vitamin needs. Among people in a group living on a diet deficient in one or other of the vitamins, some will show signs of disease, others, whose needs are evidently less, will escape.

Both nicotinic acid and nicitinatmide are physiologically active as what was originally the P-P factor. To avoid confusion with nicotine (which has no vitamin activity) the term Niacin was introduced in the U.S.A. and has been used in the table compiled by the B.M.A. Committee on Nutrition.

In planning diets, however, the table given can safely be used as a guide. Any individual idiosyncrasies in the nature of special needs must be allowed for separately.

Methods of Assessing the Adequacy of Diets : How can you assess that a diet which has been planned or eaten is

adequate in nutrient content? There are various procedures for assessing the nutritive value of the diet. The most simple of these is the use of a "score card." The use of a food pattern as a guide in planning an adequate diet has already been discussed, and score cards have been developed for determining how well diet meets the suggested pattern. A more precise method of evaluating the nutritive adequacy of diets is computing the nutrient content of each food using a food composition table. The most precise method of all is actual laboratory determination of the energy, protein, mineral, and vitamin content. The choice of the procedure for evaluating the energy and nutrient content of a diet depends on the use to be made of the findings and the accuracy with which the food consumption data were collected.

Vitamins in Nutrition and Health

Report of the Committee on Nutrition- B.M.A. 1950. p. 22 and p. 23 (extract)

Age and sex	*Require- ment class*	*Calo- ries*	*Vit. A &- carotene i.u. daily*	*Vit. D i.u. daily*	*Thia min mg. daily*	*Niacint mg. daily*	*Ribo flavin mg. daily*	*Vit. C mg. daily*
Both								
0-1		1,000	3,000	800	0.4	4	0.6	10
2-6		1,500	3,000	400	0.6	6	0.9	15
7-10		2,000	3,000	400	0.8	8	1.2	20
Males								
11-14		2,750	3,000	400	1.1	11	1.6	30
15-19		3,500	5,000	400	1.4	14	2.1	30
	sedentary work	2,250	5,000		0.9	9	1.4	20
	medium work	3,000	5,000		1.2	12	1.8	20
	heavy work	4,250	5,000		1.7	17	2.6	20
Females								
11-14		2,750	3,000	400	1.1	11	1.6	30
15-19		2,500	5,000	400	1.0	10	1.5	30
	sedentary work	2,000	5,000		0.8	8	1.2	20
	medium work	2,500	5,000		1.0	10	1.5	20
	very heavy work	3,750	5,000		1.5	15	2.2	20
Pregnancy								
	1st half	2,500	6,000	400	1.0	10	1.5	40
	2nd half	2,750	6,000	600	1.1	11	1.6	40
	Lactation	3,000	8,000	800	1.4	14	2.1	50

Vitamin B_1, aneurin and thiamine are synonymous but the most recent recommendation is that Thiamine should be adopted as the chemical name of vitamin B_1.

Use of Dietary Score Cards : Dietary score cards set up for evaluating the diet on the basis of food groups are used mainly as a basis for nutrition education programs (Tables). A limitation of each score card is its failure to give the complete nutritive contribution of each food. On the other hand, it is a commendable educational tool in that it encourages a varied diet and provides a good "rule of thumb" for the selection of an adequate diet.

Use of Food Composition Tables : By far the most common method of determining the nutrient content of the diet is the use of food composition tables. In various chapters of the book of the food tables have been used in the study of the energy value and content of the different nutrients of various foods. The values are obtained by laboratory analysis with each representing many determinations.

The composition of the same food varies widely, specially in fruits and vegetables. Growing conditions, variety, fertilizer treatment, care in handling, all make a difference. Preparation of the food for the table preservation, and storage also have an influence on the composition of most foods.

A Dietary Score Card

Foods	*Each Day You Need*	*Score for Your One Day Score*
Green and yellow vegetables (raw or cooked in small amount of water)	1 serving	10
Oranges, grapefruit, tomatoes, or other vitamin-C-rich food	1 serving	10
Potatoes and other vegetables and fruits	3 servings	5
Milk and milk products	Children, 3-4 cups Adolescents, 3-4 cups Adults, 2 cups	20
Meat, poultry, or fish	1 serving	15
Meat, poultry, fish, or meat alternates (dried beans, peas, peanut butter)	1 serving	10
Eggs	1 daily (at least 4 a week)	5
Cereal, whole grain or enriched	1 serving or 2 slices bread	

contd....

Foods	*Each Day You Need*	*Score for Your One Day Score*
Bread, whole grain or enriched	1 or 2 slices at every meal	5
Butter or other fats	2 to 3 level tbsp butter or enriched margarine	5
A good breakfast, including some form of protein, as milk or egg		10
	Total	*100*

Those who have compiled food tables have had to decide on the values that they considered most representative for inclusion in the table. The necessity of exercising judgment in preparing a food table accounts for some of the differences that exist among the compilations of food composition values. As a rule the most food table includes new data add to the accuracy.

Short Methods Using a Food Composition Table : Short method for dietary calculation have been developed in which foods similar in composition are grouped, together, a representative value for each of the nutrients is calculated, and the values listed in a table. Totalling the amount of foods of like composition in a diet and computing the nutritive contribution of the group is shorter, of course, than computing each food separately. That is the manner in which the short methods save time in calculation.

The short methods are found to agree with the long or usual method if applied to a varied diet, and if the food habits do not differ widely from the usual pattern of the dietary habits of the people of a country as a whole. In regions where there is a relatively higher intake of certain foods, such as corn meal in the southern region, a revision in the cereal and bread portion of the table becomes necessary to obtain an accuracy approaching that of calculating each food individually. For survey studies the shorter method of dietary calculation has

been found satisfactory. Time required for computation can be reduced without great sacrifice to accuracy.

Chemical Analysis : In research studies designed to determine quantitatively the nutrient intake as part of a metabolic balance study, the composition of a food is determined by laboratory analysis rather than by use of a food composition table Laboratory Analysis is the most accurate of any procedure for determining the nutritive value of food, since the nutrient content of some foods varies widely. To illustrate, the ascorbic acid content the thirteen different varieties of potatoes ranged from 8.2 to 17.4 mg. per 100 gm. of fresh weight, and for thirty-five varieties of cabbage from 32.4 to 100.7 mg. per 100 gm. of fresh weight. It should be recognised, however, that the fruits and vegetables vary more in composition than most other foods. The differences between computed and analysed values for given diets have been studied. It is generally concluded that agreement is sufficiently close to warrant the use of food tables in dietary surveys, reserving laboratory analyses for controlled studies.

Diet and the Family Budget

Ultimately, daily menus, a market list, and a plan for purchasing food must be made by families who maintain households and by agencies who make allowances for families in need of assistance. The Consumer and Food Economics Research Division of the U.S. Department of Agriculture prepares materials that are of genuine assistance as guides in planning food budgets. The quantities and kinds of foods needed to adequately nourish different individuals are estimated at low-cost, moderate-cost, and more expensive diet. The Research Division has prepared plans for twenty age-sex groups and for eleven food commodity groups. These groups include milk and milk products; meat, poultry, and fish, eggs, dry beans, peas, and nuts; grain products; potatoes; citrus fruits and tomatoes; dark green and dark yellow vegetables; other vegetables and fruits; fats and oils; sugars and sweets.

To plan a food budget for a family of a certain size it is necessary to add together the estimated quantities needed for each family member. For example, a week's food plan for a family of a man and a woman both 33 years of age, a girl 8, and a boy 11 would be as summarised in Table.

The student of nutrition is aware that merely planning for kinds and amounts of food is not enough; the planning is made best with knowledge of the food preferences of the family and the availability and price of these foods. The budget plan allows for choice of foods within the commodity groups. Further adjustment can be made even between the groups with a knowledge of food values. For instance, potatoes could not substitute for eggs nutritionally, but another serving of meat would substitute very well.

The estimated cost of one week's food supply for a four-member family (a girl 8, a boy 11, man and woman each 33 years of age) as of January 1963 was $32.80 for the moderate-cost plan (Table).

It is gratifying that an adequate diet can be provided at different cost levels. The low-cost plan contains substantially more potatoes and grain products. The lower-cost meat cuts can be used in the limited-expense plan with the assurance that the nutritive value is equal to that of the higher-priced cuts. The moderate-cost and liberal plans include more milk, eggs, meat, fruits and vegetables. More expensive items such as foods out of season and the more highly processed foods are allowed in the moderate-cost and liberal plans.

Quantity of Food for a Family of Four for One Week

Food Groups	*Unit*	*Low-Cost Plan*	*Moderate Cost-Plan*	*Liberal Plan*
Milk, milk products	qt	16.5	17.5	19.0
Meal, poultry, fish	lb	11.5	17.2	20.8
Eggs	no.	25.0	29.0	29.0
Dry beans, peas, nuts	lb	1.4	.9	0.9
Grain products	lb	12.5	11.5	10.9

contd.....

Food Groups	*Unit*	*Low-Cost Plan*	*Moderate Cost-Plan*	*Liberal Plan*
Potatoes	lb	10.0	8.5	7.8
Citrus fruit, tomatoes	lb	7.5	9.0	11.5
Dark green, deep yellow vegetables	lb	3.5	3.5	3.5
Other vegetables, fruits	lb	19.8	22.5	25.2
Fats, oils	lb	3.1	2.8	2.8
Sugars, sweets	lb	3.0	3.9	4.6

Cost of one Week's Food at Home[1] Estimated for Foods Plans at Three Cost Levels, June 1964, U.S.A. Average

Sex-Age Groups[2]	*Low-Cost Plan, Dollars*	*Moderate Cost Plan, Dollars*	*Liberal Plan, Dollars*
Families			
Family of two 20-35 years[3]	14.60	19.60	22.70
Family of two, 55-75 years[3]	12.20	16.50	18.70
Family of four, preschool children[4]	21.40	28.40	32.80
Family of four, school children[5]	24.60	33.00	38.30
Individuals[6]			
Children under 1 year	2.90	3.80	4.10
1-3 years	3.70	4.80	5.50
3-6 years	4.40	5.80	6.70
6-9 years	5.20	7.00	8.30
Girls, 9-12 years	6.00	8.00	8.90
12-15 years	6.60	8.80	10.20
15-20 years	6.90	9.00	10.20
Boys 9-12 years	6.10	8.20	9.40
12-15 years	7.10	9.70	11.00
15-20 years	8.30	10.90	12.60
Women, 20-35 years	6.20	8.30	9.40
35-55 years	6.00	7.90	9.10
55-75 years	4.70	6.90	7.80

contd......

Sex-Age Groups[2]	*Low-Cost Plan, Dollars*	*Moderate Cost Plan, Dollars*	*Liberal Plan, Dollars*
75 years and over	4.70	6.20	7.20
Pregnant	7.50	9.60	10.80
Nursing	8.60	11.10	12.30
Men 20-35 years	7.00	9.50	11.20
35-55 years	6.60	8.80	10.20
55-75 years	6.00	8.10	9.20
75 years and over	5.60	7.80	8.90

Food Values and Food Costs : The buying of food is the largest single expense in the budget for most families. An average of 19 per cent capita of disposable personal income is spent for food in the United States. With increasing income, the per cent going food diminishes.

Economical purchase of food makes it possible to provide adequate diets at lower cost. Some nutrients are provided at less expense by one food or group of foods than by others. The approximate percentage of the food dollar spent for milk and milk products; meat, fish, poultry, eggs, dry beans, and nuts; vegetables; fruits; and grain products is shown in Fig. Other foods not included in Fig. are fats, sugar, and sweets, and miscellaneous items, which account for about 20 per cent of the total food expenditure. The protein group of foods is the most expensial, amounting to 37 cents of the food dollar or 37 per cent of the food expenditure. Milk and milk products account for 15 cents of the food dollar, vegetables 12 cents, grain products 11 cents, and fruits 8 cents.

Foods differ in the economy with which they provide the nutrients of the diet. Protein, calcium, vitamin A value, and thiamine were selected to illustrate this point. Using Fig. as a guide, in each food group if the nutrient bar exceeds the cost bar in length, that food group is a good buy for providing that nutrient, because the per cent of the expenditure for that group of foods is less than per cent of the nutrient obtained. The

relative lengths of the bars will vary, of course, with price. It can be clearly seen that the milk group of foods is an excellent buy for calcium but only fair for supplying thiamine; the meat, fish, poultry, eggs, dry beans, and nuts group is a good buy for protein but a poor buy for calcium; vegetables are excellent for vitamin A, poor for protein; fruits are not a good buy for the nutrients listed, but had ascorbic acid been selected as one of the nutrients in the illustration, fruits would appear as good buy for that nutrient.

Grain products are a good buy for protein and thiamine, and it may be surprising to note, for calcium, accounted for in part by the use of bread containing non-fat milk solids. Of course, little if any vitamin A value is furnished by grain products. The protein of cereals is economical to purchase and becomes of the good nutritional quality when supplemented with milk or other sources of complete protein. Grain products are a good buy for protein because, as shown in the cost bar is shorter than the protein bar.

7

Dietary Schemes

Energy is defined as the capacity for doing work. From sunlight plants store energy as carbohydrate with the help of chlorophyll. Animals and men avail their energy through foods. When food is metabolised in the body, energy is released. Body requires this energy for all its activities. For all muscular activities and for the function of vital organs like heart, lungs, alimentary canal, nervous tissues and glandular tissues energy is required.

Energy-sources and their Applications

The energy requirement of a person depends on various factors. The most important factor in assessing the energy requirement is the basal metabolic rate of the person.

Basal Metabolic Rate (BMR) : This is defined as the amount of heat or energy required by the body to do the involuntary work of the body. Functions of brain, heart, liver, kidneys, lungs, secretary activities of the glands and intestinal movements are the basal activities in our body.

Factors Affecting the BMR : Basal metabolic rate of a

person depends on the size, shape, composition, body weight, age and physiological condition of the person. Nutritional status and endocrine system of the body also exert their influence on basal metabolic rate of a person. Climatic condition of the surroundings also affects the basal metabolic rate.

1. *Size:* Since the heat loss in body is proportional to the skin surface, a tall and thin person has greater surface and thus higher basal metabolic rate compared to the short person. The body composition shows variations in energy use. With little fat deposition in the body the basal metabolic rate increases and thus a tall, thin person has higher rates of basal metabolism compared to a short, fat man. Constant muscular activities of an athlete demand about 5 per cent more basal metabolic rate.
2. *Sex:* Sex also makes a variation in energy requirement. The metabolic rate of women is 6-10 per cent lower than that of men.
3. *Growth period:* During the growth period the basal metabolic rate and thus the energy requirement are increased. The highest BMR is during the first two years. After the growth period, especially after 25 years of age, there is a decline in energy requirement.
4. *Endocrine glands:* The thyroid gland exerts influence over energy requirements. Thyroid hyperactivity will speed up basal metabolism. The pituitary gland also increases the metabolic fate if it has disturbance. Adrenaline increases the BMR.
5. *Nutritional status:* During undernourishment the basal metabolic rate also declines to an extent of thirty per cent.
6. *Pregnancy condition:* During pregnancy and lactation the basal metabolic rate is increased by about 5 per cent during the first and second trimester and 12 per cent during the third trimester.
7. *Sleep:* During sleep the BMR is less than in the waking state.

8. *Climate:* Climatic conditions of the environment also affect the energy requirement. If the temperature falls below 14°C the energy requirement increases.
9. *Body temperature:* Fever increases the BMR.
10. *Disease conditions* : Diseases like typhoid fever, medullary diseases and lymphatic leukaemia show an increase in the BMR.
11. *Physical activities:* Physical activities half an hour before BMR measurement show high rates of BMR.

Total energy requirement of 4 person varies according to the basal metabolic rate, effect of food, activity involved, age, sex, physiological conditions and climatic conditions of the surroundings. Energy requirements of an adult in India was reviewed by 'The Expert Group' of the FAO/WHO in terms of a reference man and reference woman. The reference man is in the age group of 20-39 years, with a weight of 55 kgs, without any disease and with a capacity to perform 8 hours of moderate activity. When not engaged in work a reference man spends 8 hours in bed and 4-6 hours in moving around or in a sitting position and 2 hours either walking or doing household activities.

In the case of a reference woman, the difference is only in her body weight that is, as against 55 kg of a man she weighs 45 kg. Instead of the physical activity of the occupation, the woman does household duties. Other conditions are the same in the case of a reference woman. The energy expenditure for man and woman is calculated considering their internal and external activities. The FAO/WHO expert group (1983) suggested some recommendations as given in Table.

Energy Allowance for Various Groups

Category	*Reference body weight*	*Activity*	*Energy allowance kcals.*
Man	55 kg	light	2400
		moderate	2800
		heavy	3900

contd.....

Category	*Reference body weight*	*Activity*	*Energy allowance kcals.*
Woman	45 kg	light	1900
		moderate	2200
		heavy	3000
	Pregnancy	2nd and 3rd trimester	+300
	Lactation	first six months	+550
		6-12 months	+400

Additional requirement of energy is needed for the growth of the foetus, placenta and maternal tissues during pregnancy. The BMR is also increased due to increased internal activities. Daily 150 kcals during the first trimester and 300 kcals during the rest of pregnancy is recommended. The energy cost during the term of pregnancy is 62,500 kcals. Additional energy requirement during lactation is for the secretion of milk. For a normal output of 850 ml of milk during the first 6 months 550 kcals/day is recommended.

During infancy the energy requirement is high. For 0-3 months 120 kcalsAg, 3-5 months 115 kcals/kg, 6-8 months 110 kcal/kg, 9-11 months 105 kcal/kg and at one year 112 kcals are recommended.

For children, the energy requirement varies according to their body weight The ideal weight of the children according to their age must be considered for assessing energy requirements. Thus in the latest recommendations, some suggestions were made as shown in Table.

Energy Allowance for Children

	Age group	*Body weight*	*Energy allowance kg. kcal.*
Children	1-3 years	12.03 kg	1220 kcal
	4-6 years	18.87 kg	1720 kcal
	7-9 years	26.37 kg	2050 kcal

contd.....

	Age group	*Body weight*	*Energy allowance kg. kcal.*
Boys	10-12 years	34.30 kg	2420 kcal
	13-15 years	47.03 kg	2660 kcal
	16-18 years	56.50 kg	2820 kcal
Girls	10-12 years	36.47 kg	2260 kcal
	13-15 years	45.53 kg	2300 kcal
	16-18 years	50.00 kg	2200 kcal

Specific Dynamic Action : Energy requirement is affected by the type of food ingested. The extra heat which is produced after taking food is known as the specific dynamic action (SDA). The stimulating effect of carbohydrates, fats and protein on energy metabolism is different. Protein foods produce the highest per cent increase in energy metabolism. The SDA of protein is about 30 per cent while carbohydrates and fats exert 6 per cent and 4 per cent, respectively. The average SDA of a mixed diet is about 8-10 percent.

The activities which demand maximum energy are in the following order:

> Walking very fast, severe exercises, running, swimming, sawing wood, labourer's work, carpentry, metal and industrial work, walking slowly, laundry work, typing and ironing.

Different Types of Scales

Energy is measured in terms of Kcalorie. A Kcalorie is defined as the amount of heat required to raise the temperature of 1 kilogramme (1 litre) of water by 1°C. But joule is the new term used to measure energy. A joule is defined as the amount of work done or heal generated by an electric current of 1 ampere acting for 1 second against a resistance of 1 ohm. This

unit is very small for measuring energy in nutrition and so 1 kilojoule (KJ) or megajoule (MJ) is used for practical use.

1 Kilojoule	=	1000 joules
1 Megajoule	=	10,00,000 joules
1 Calorie	=	4.184 joules (J)
1 Kcalorie	=	4.184 Kjoules
1,000 Calories	=	4.184 Mj (Megajoules)

Energy Value of Foods : Energy value of foods is otherwise known as the caloric value of foods. To measure the caloric value of food, an apparatus known as Bomb Calorimeter is used. In this apparatus a known amount of food is burnt and the rise in temperature of water in a container is recorded. The heat liberated by 1 gramme of carbohydrate is recorded as 4.1 kcals, protein 5.6 kcals and fat 9.4 kcals.

The energy value of food is otherwise known as the fuel value of foods. This is determined by means of an instrument known as Bomb calorimeter (Fig.).

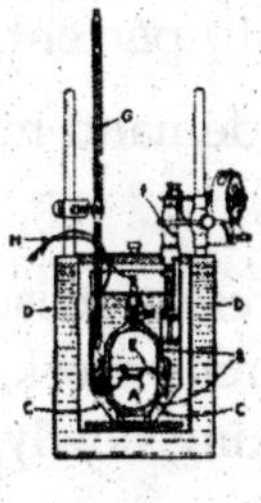

Fig. : Bomb Calorimeter

A weighed sample of dried food in pellet form is placed in a heavy steel container called a bomb. The bomb is held in place in a well insulated vessel. This is surrounded by a known volume of water.

The bomb is charged with oxygen to a pressure of about 300 pounds to the square inch. After the temperature of the

water has been determined the sample is ignited by means of an electric fuse. The heat from the sample is dissipated into the water. By noting the change in the temperature of the water the energy value of the food can be calculated.

Physiological Fuel Value of Foods : Small losses occur in digestion when carbohydrates, fats and proteins are utilised by the body. Digestibility coefficiency of western diets are available and based on these data:

the physiological fuel value of carbohydrate	4kcals
the physiological proteins	4kcals
the physiological fats	9kcals

Among these nutrients the body loses maximum energy from protein. This is due to the loss during digestion. The digestibility of various foods varies which affects the physiological fuel value.

Among the various sources of energy, foods rich in fats and dairy produce are the best Dried fruits, cereals and other starchy foods and legumes are fair sources of energy.

Determination of Energy Requirements of the Body : There are two methods generally applied for measuring the energy requirements of human body. They are direct and indirect calorimetry methods.

Direct Calorimetry : In this method the amount of heat produced by the body is measured directly. At Water-Rosa, Benedict perfected the human respiration calorimeter. To measure the heat production in an individual a specially constructed chamber, called a respiration calorimeter, is used. It is determined through the relation between energy output and oxygen consumed in a person in this calorimeter.

Respiration calorimeter consists of an airtight copper chamber insulated by wooden wall with air spaces in between. Enough provision for sitting or to lie down is made through a folding bed, chair and table. Food is provided through the opening at one end and the excreta is removed through another

opening at the other end. The chamber is very well arranged for the individual to relax, read or write. It is insulated to prevent the entry and exit of heat through the walls. The chamber is ventilated by a current of air, the CO_2 and water given off are removed by soda lime and sulphuric acid respectively. For oxygen supply, a known amount of oxygen is introduced to the chamber through a gas meter with an air current. The oxygen consumption and CO_2 production are calculated using a formula. The heat produced is also measured by circulating a current of water through copper pipes and measuring the quantity of water that has been circulated through the chamber and also the difference between the temperature of the water entering and leaving the chamber.

Respiratory Quotient: Respiratory Quotient is the ratio between the volume of CO_2 given out and the volume of O_2 consumed by the human subjects.

$$\text{Respiratory Quotient (RQ)} — \frac{\text{Volume of } CO_2 \text{product}}{\text{Volume of } O_2 \text{ consumed}}$$

The RQ varies with the type of food being oxidised. When glucose is oxidised RQ is 1, for a fatty acid it is 0.7 and for protein it is 0.8. Under resting condition with no food for 12-14 hours the RQ is 0.82. In this condition carbohydrate, fats and protein present in the body are utilised.

Indirect Calorimetry : In this method oxygen consumed and carbon dioxide excreted in a given time under basal condition is measured. Many experiments on people of different ages have shown that 1 litre of oxygen is equal to 4.825 kcals when the person is in Basal Metabolic condition. The basal metabolic rate is measured by indirect calorimetry. The person has to observe certain conditions before measuring the basal metabolic rate. They are:

1. In a post-absorptive state, i.e., 12-16 hours after the last meal- preferably in the morning.
2. In a reclining state but awake.

3. Before the test the person must take one hour rest if he is engaged in any work.
4. Relaxed and free from emotional upset, excitement or fear.
5. Body temperature must be normal.
6. Room temperature must be comfortable.

In these favourable conditions the BMR is measured. For an adult man the BMR rate is 34.2-36.7 kcals/sqm/hr. and for an adult woman 30.9-35.1/kcals/sqm/hr.; for children it is more. For 7-8 year boys it is 49.1/kcals/sqm/hr. and for girls it is 45.2/kcals/sqm/hr.

Instructions of Diet

Planning a balanced diet within the income level needs precise knowledge about the sources of nutrients, their requirements for various groups, seasonal availability of various foodstuffs and dietary habits of the group.

Blending these theoretical knowledge into a day's menu needs skill which has to be acquired through practice. There are certain basic principles which has to be observed while planning a balanced diet. As the name indicates a planned diet should be balanced. A diet is called balanced when all the nutrients required by the body are present in correct proportion. A judicious selection from the food groups will supply adequate nutrients for maintenance, repair and growth. Other factors to be considered are composition of the family and the physiological phases of family members, dietary habits of the family, food budget of the family, variety in meal patterns, appearance, texture and taste of meals, selection of food for seasons (cold and hot climate condition) and nutritional balance.

While planning a diet, the composition of the family has to be considered. Infants, children, teenagers, pregnant and lactating mothers and geriaric people need food in different amounts. There are variation in their nutritional requirements and their digestive capacity. Quality and quantity and often

the preparation need certain changes. A careful planning gives provisions for all these without making great changes in the preparation pattern of foods.

Dietary habits of the family also have to be considered while meal planning. Food habits are influenced by earlier experiences in life: sensory, aesthetic, economic, geographic, social and cultural factors. Often food habits change according to the new environment, new values and availability of items. When a child is exposed to group feeding practices slight changes in likes and dislikes of foods are common. Age and sex also influence food choices. The best example is an adolescent's dietary habits. There are vegetarian, non-vegetarian, lacto-aro-vegetarian and pesco-vegetarian dietary habits (plant foods and fish but no milk, egg, meat or poultry).

Food budget is a very important factor in meal planning. Money available to buy food items, time available for food preparation, job pattern, life-style of the family members, location of residence, and family values decide how much has to be spent on food. First of all to reduce food expenditure, one must have an idea about low cost nourishing foods for selection. Intelligent selection, good marketing, just enough bulk purchase, seasonal buying, food processing and preservation, better food preparation methods, avoiding more indulgence in ready-to-eat foods, fast foods, readymade food and hotel catering reduce food budget.

Variety in meal makes eating a pleasure. The same meal, if repeated, creates monotonous taste and reduces appetite. Variety in choice of food items, methods of preparation in colour, texture and flavour induce appetite and provide satiety value to the meal. When food is prepared in a proper form with characteristic taste and flavour, it derives satiety value.

Above all these factors, a nutritional balance in each meal is essential. Successful meal planning principles suggest that each meal must be nutritionally balanced. One-third of the total nutrients must be in the lunch, one-third in the dinner and the remaining one-third in breakfast and teatime. Skipping

a meal is not good. Many people often skip breakfast which is very bad as far as the body is concerned. Studies have showed that the performance of a person after mid-morning is very poor if they skip their breakfast

Based on these principles, if a diet is planned it will be a balanced one. The Indian Council of Medical Research has suggested the amounts of various food items to make a balanced diet for various groups.

Some Clues for Different Balanced Diet —Low Cost BD: Millets, cereals, roots and tubers, pulses and leafy vegetables can be included liberally. Cheap seasonal fruits and salad vegetables lend attraction and freshness to the diets. Balanced diets of middle income groups can include moderate amounts of protective and proteins-rich food. Milk, egg, fish and meat can be included in moderate amounts while pulses, nuts, fruits and other foods can be used in liberal amounts.

Balanced Diets for High Cost : In a high cost diet protect the foods and protein-rich foods can be included liberally. Wet surveys in India have shown that malnutrition is prevalent among high class people also. B vitamins and certain minerals are not included m enough quantities in their dietaries. Over-nutrition due to high consumption of milk, meat, fats, ghee and sweets are common.

The ultimate aim of consuming food is to maintain health. There is no single foodstuff which can contribute all the nutrients needed by the body. Only a judicious selection can provide all the nutrients in required amount. In order to select the sources of all nutrients and that also in correct proportion one must know about the basic principles of food selection. The nutritionists and food scientists all over the world have put in a lot of efforts to formulate certain guidelines in these matters. Basic food groups suitable to different countries were evolved by the nutritionists.

The US Department of Agriculture suggested a number of food group plans like the Basic-4 food group.

The Basic-7 Food Group and the Basic-11 Group Plan: The Nutrition Expert Group of I.C.M.R. (India) suggested a five-food group plan.

Based on their nutritive values foods are grouped into 11. They are:

1. Cereals and millets
2. Pulses (legumes)
3. Nuts and oil seeds
4. Vegetables
5. Fruits
6. Milk and milk products
7. Egg
8. Meat, fish and other fleshy foods
9. Fats and oils
10. Sugar and jaggery
11. Spices and condiments.

In India this classification has great significance.

Cereals and Millets : They supply the major portion of calories (70%), more than 50 per cent of B Vitamins. In a poor vegetarian diet cereals supply more than 70 per cent of protein.

Rice and wheat are abundantly used as staple cereals in the world. In India rice is used mainly in the diet of South India, whereas wheat is the main staple in all other parts.

The major millets used are jowar, ragi and bajra. Though they are cheaper they are superior in their mineral supply and protein content Ragi is a very good source of calcium.

Pulses and legumes are fair sources of proteins and they are good sources of B vitamins, especially thiamine and riboflavin. Quantitatively their protein content is high but qualitatively they are not as superior as animal proteins. But in an Indian diet cereals and pulses supplement mutually in their amino acid content and improve the quality of protein.

Pulses and legumes are comparatively cheap sources of protein. Cowgram, horsegram, soyabean, dry field beans and dried peas are some of the cheap pulses.

Nuts and oilseeds are fair sources of protein. Coconut is a good source of energy. It has also fair sources of B vitamin. Vegetables are unique in their vitamins and mineral content. Vegetables are again grouped into yellow vegetables, leafy vegetables and as other vegetables. Yellow vegetables and leafy vegetables are rich sources of vitamin 'A' (carotene), calcium, iron, folic acid, riboflavin and vitamin 'C. But the iron content in vegetables is comparatively poorly utilised by the body. Other vegetables are good sources of minerals like calcium, trace elements and vitamin 'C.

Apart from these, vegetables have a high percentage of water in them. They contribute the alkaline ash and the bulk or cellulose to the body. Beetroot, carrot and ladyfinger are high in sodium content. Fruits, especially citrus fruits, are good sources of ascorbic acid. Some starchy fruits are rich in fructose and dextrose; yellow fruits contribute carotene, the provitamin A. Fruits as a group supply water and electrolytes like potassium. Most of the fruits are rich sources of potassium and poor sources of sodium. Dates, dry raisins, grapes, peaches, lime and ripe papaya are very high in their potassium content Bananas are rich in starch content in the raw form and in sugar content in the ripe form.

Milk and milk products are the best sources of calcium and ribo-flavin and second to the meat groups in protein content. Except a iron and vitamin 'C milk is good in all other nutrients. Milk products are high in biological value and milk cream is excellent in vitamin A.

Egg has biologically very good protein. Egg white is rich in protein and egg yolk is rich in fat, iron, sodium, potassium, calcium and magnesium. Vitamins like thiamine, riboflavin, vitamins A and D are also present in egg. It plays very many roles in cookery.

Meat, fish and other flesh foods rank first in protein content.

They are also good sources of phosphorous, magnesium, iron, thiamine, niacin, B 6, and B 12. The biological value of meat protein is high. Fat is present in a good quantity in it.

Fats and oils contribute mainly calories and they are good sources of essential fatty acids and vitamins A and E. Sugar and jaggery are the common sweetening agents in our diet. They are rich sources of carbohydrate.

Spices and condiments are necessary foodstuffs used for flavouring foods. Apart from these functions some of them contribute certain nutrients. Red chillies and coriander supply carotene; green chillies vitamin 'C turmeric and tamarind iron; and garlic stimulates synthesis of certain vitamins.

Then food is classified into 11 groups for convenience in selection. But in different countries different food group plans are followed by the nutritionists to formulate a balanced diet.

In India the nutrition expert group of ICMR suggested the five-food group plans.

They are given below:

Group	*Nutrients*
1. *Milk group*	
This consists of milk and milk products and protein-rich foods like pulses, nuts, meat, fish and egg.	This group supplies proteins, minerals and vitamins.
2. *Fruits and green leafy vegetables*	
Fruits like papaya, orange, mango, Indian gooseberry, guava and green leafy vegetables like spinach, amaranthus, drumstick leaves.	Rice sources of vitamin 'C' vitamin A (carotene), minerals and roughage.
3. *Other vegetables*	
Peas, beans, drumsticks, ladyfingers, bringals, cucumber, bittergourd, ashgourd, plantain etc.	They are fair to good sources of minerals, vitamins and roughage and water content.

contd....

Group	Nutrients
4. *Cereals, roots and tubers*	
Rice, wheat, maize, ragi corn, tapioca, potato, sweet potato, yam, colocacia etc.	Rich sources of starch, fair to good sources of proteins and B vitamins. In an Indian diet this group supplies about 75-80 per cent of calories and 50-60 per cent of proteins.
5. *Fats and oils and pure carbohydrate foods*	
Butter, ghee, vegetable oils, sugar, jaggery, honey, starch and egg powders, custard powders etc.	They are rich sources of energy, sential fatty acids, vitamins A and E and cholesterol.

When these food groups were suggested they were known as Basic groups. Basic-7 food group plan, Basic-4 food group plan, and Basic-11 food group plan of US Department of Agriculture was the first in these series. Basic dietary pattern and daily food guides were suggested by different nutrition boards. In India the National Institute of Nutrition of ICMR has done remarkable work in this field. Guidelines for a balanced diet, composition of balanced diet and balanced diet for different regions of India, viz., balanced diets for the eastern region, northern region, southern region and western region were suggested by the NIN and it was periodically revised based on their advanced research in this field. Guidelines for low-cost balanced diets were prepared by them for a student of nutrition. In our Indian background this has great significance.

Selection of Food

Most of us have to plan either our own or other people's diets if only to the extent of selecting from a hotel or cafe menu what they well eat.

The housewife has to plan meals for her family, the members of which may be of different ages and have very different likes

and dislikes. Meals have to be planned for the school dining-room, the institutional canteen etc.

Everyone should therefore have some knowledge of the values of foods and of the factors that influence choice, recognising that the more costly foods are not necessarily those of the highest nutritional value. Grapes, for example, contain only traces of vitamins compared with the amount present in an orange or cabbage.

During and since the war, the science of nutrition has been applied to home economics as never before and in 1952 the Ministry of Food (U.K.) published a Report on Domestic Food Consumption and Expenditure in which data are given on nutrients consumed by different sections of the community and the extent to which these satisfy nutritional needs.

In 1950, the *mean* intake of the chief vitamins per head of the population was as follows:

Vitamin A	3,536 i.u. (a)
Vitamin B_1	1.51 mg.
Riboflavin	1.69 mg.
Nicotinic acid	13.0 mg.
Vitamin C	84 mg. (a)
Vitamin D	172 i.u. (a)

Including Welfare Foods : Except for vitamin D which is definitely low and riboflavin which seldom reaches the estimated needs, all these are at the level of adequacy. However, the mean figure is obtained by averaging some above and some below and that in the case of the vitamins of the B complex, all of which are near the requirement level, there must be as many individuals receiving less that this as there are consuming more.

There was not, in 1950, the marked difference between the intake of people in different income groups as was found by the pre-war survey. It has therefore been possible to make

estimates of the food consumption of the whole population with regard to the supply of the principal nutrients.

Vitamin A	20.7	per cent obtained from vegetables
	17.4	per cent obtained from fats, milk and cheese
	14.5	per cent obtained from meat (incl. bacon)
	7.1	per cent obtained from eggs
Vitamin B_1	43.8	per cent obtained from bread and cereals
	22.7	per cent obtained from vegetables
	14.8	per cent obtained from meats
	12.5	per cent obtained from milk
Riboflavin	40.0	per cent obtained from milk and cheese
	17.6	per cent obtained from bread and cereals
	16.5	per cent obtained from meat
Nicotinic acid	34.8	per cent obtained from meat
	31.9	per cent obtained from bread and cereals
	20.1	per cent obtained from vegetables
Vitamin C	37.2	per cent obtained from potatoes
	33.0	per cent obtained from fruit
	10.0	per cent obtained from green vegetables
Vitamin D	36.0	per cent obtained from fats
	22.7	per cent obtained from fish

The report points out that after August 1950 when the extraction rate of national flour was lowered from 85 per cent to 81 per cent, there was a corresponding drop in vitamin B intake in all groups of the population. Since bread and cereals supply over 40 per cent of the vitamin B_1, this change was to be expected. Average figures are no guide to individual differences. Everyone can do something, by the application of a knowledge of food values, to improve his or her own food intake and can do so without in any way becoming a food-faddist. It should be a natural habit to choose the better food when several choices are presented and such good habits can be developed easily in children.

If a mother or a school matron has enough knowledge to provide children with a nutritious diet then they are likely to grow up liking the nutritious foods to which they have become accustomed. Habits about food tastes are very pronounced and it is sometimes difficult training children (or adults) to like something new.

We would like to study various ways in which meals can be improved by various additions.

One of the simplest breakfasts taken in Great Britain is toast with butter and marmalade, accompanied by tea, but many people add to this either porridge or a cereal with milk or a cooked dish such as bacon or fish, or fruit, sometimes raw, sometimes cooked; and some take coffee in stead of tea.

The accompanying table shows the effect of a number of different additions to what is taken as a 'basal' breakfast.

The Effects of Various Additions to a 'Basal' Breakfast

Basal breakfast		*Vit. A or equiv. in carotene* i.u.	*Vit.* B_1 µg.	*Ribo-flavin* µg.	*Niacin* µg.	*Vit. C* mg.	*Vit. D* i.u.
Toast (81 percent edtraction)	2. oz	—	98	34	660	—	—
Butter	¼ oz.	284	—	—	—	—	—
Milk	1 oz.	30	13	40	—	—	4
Tea							
Sugar	—	—	—	—	—	—	—
		314	111 =.111 mg.	74 =.074 mg.	660 =.66 mg.	—	4.3
Additions							
(1) 5 oz. milk for cereal or coffee	150	65	200	—	—	1.5	
(2) Cornflakes	2 oz.	—	—	—	—	—	—
(3) Porridge	2 oz.	—	250	80	600	—	—
(4) Bemax	1 oz.	—	450	200	1,700	—	—

contd....

Basal breakfast		*Vit. A or equiv. in carotene i.u.*	*Vit. B_1 µg.*	*Ribo-flavin µg.*	*Niacin µg.*	*Vit. C mg.*	*Vit. D i.u.*
(5) I egg	2 oz.	500	74	340	—	—	3.4
(6) 2 x 2 oz. tomatoes	4oz.	500 Approx	80	40	400	20	—
(7) Rasher bacon	1 oz..	—	160	125	1,200	—	
(8) Herring	6 oz.	250	—	—	6,000	—	1,500
(9) Haddock	4 oz.	70	48	300	1,500	—	—
(10) Grape-fruit pulp	2 oz.	3	20	14	100	12	—
(11) Stewed prunes	4 oz.	300-400	32	68	—	1	—

The same kind of analysis can be made of the snacks obtainable at nearly every snack bar.

1. *Baked beans on toast*							
Baked beans	3 oz.	70	60	30	600	—	—
Nat. bread	1 oz.			49	17	330	
Margarine	1/8 oz.	71	—	—	—	—	7
		141	109	47	930	—	7
2. *Spaghetti on toast*							
Spaghetti (uncooked)	1 oz.	—	20	—	600	—	—
Nat. bread	1 oz.	—	49	17	330	—	—
Margarine	1/8 oz.	71	—	—	—	—	7
		71	69	17	930	—	7
3. *Sardines on toast*							
Sardines	2 oz.	150	20	160	1,300	—	560
Nat. bread	1 oz.	—	49	17	330	—	—
Margarine	1/8 oz	71	—	—	—	—	7
	221	69	177	1,630	—	567	

contd......

Basal breakfast		*Vit. A or equiv. in carotene i.u.*	*Vit. B_1* µg.	*Ribo-flavin* µg.	*Niacin* µg.	*Vit. C* mg.	*Vit. D* i.u.
4. *Grilled tomato on toast*							
Tomatoes	3 oz.	420	60	30	300	6	—
Nat. bread	1 oz.	—	49	17	330	—	—
Margarine	1/8 oz	71	—	—	—	—	7
		491	109	47	630	6	7
5. *Welsh rarebit*							
Cheese	1 oz.	369	9	140	100	—	4
Nat. bread	1 oz.	—	49	17	330	—	—
Margarine	1/8	71	—	—	—	—	7
		440	58	157	430	—	11
6. *Poached egg on toast*							
1 egg	2 oz.	500	74	340	—	—	34
Nat. bread	1 oz.	—	49	17	330	—	—
Margarine	1/8	71	—	—	—	—	7
		571	123	357	330	—	41

The value as carotene is halved to give an approximate equivalent in terms of i.u. vitamin A.

It is worth nothing that all of these (except tomatoes) are completely lacking in vitamin C, that sardines make the only appreciable vitamin D contribution and provide the only large quantity of niacin and that, except for numbers 3 and 6, not one contains as much as one-tenth of the day's requirement of vitamins of the B complex. While these meals are useful, people consuming them regularly must remember to make up for these defects at other meals.

It is sometimes assumed that any fruit will supply vitamin C. This is erroneous as can be seen by reference to the table of fruit contents and it is worth repeating the values of a few.

One orange with about 3 oz. of edible pulp will yield 45-50 mg.

One apple with about 3 oz. of edible portion yields only 10-15 mg.

the same weight of raw cherries	
grapes about	3 mg.
pear	

Fruit drinks taken with a snack lunch supply vitamin C only if they do, in fact, contain the juice of the fr·uit. Some brands do and it is worth examining the label on the bottle. Some drinks, e.g. Ribena and Rosena have a declared vitamin-C potency.

An attempt was made during the war years to popularise the salad lunch, based on the Oslo breakfast of Norway and this met with a measure of success although the general preference in Great Britain is still for 'meat and two veg.' However, the following example illustrates the value of the salad meal especially when it is combined with such highly protective foods as egg, cheese and one or other of the fat fish.

The value of these cold salad foods can be turned to practical use in the preparation of picnic meals and sandwiches of which a few examples cab be given.

From these examples it will be clear that certain foods are mentioned over and over again. These are the ones with a high vitamin content which should find a place in the diet of everyone.

Typical salad lunch	*Amount o.z.*	*Vit. A or equtv. in carotene i.u.*	*Vit. B_1 µg.*	*Ribo-flavin µg.*	*Niacin µg.*	*Vit. C mg.*	*Vit D i.u.*
Canned salmon	4	280	40	240	7,200	—	680
Hard-boiled egg	1	225	18	85	—	—	8.5
Cheese	1/3	123	3	47	33	—	1
Lettuce	1	363*	17	20	100	3	—
Tomato	2	140*	40	20	200	14	—

contd....

Typical salad lunch	*Amount o.z.*	*Vit. A or equiv. in carotene i.u.*	*Vit. B_1 µg.*	*Ribo-flavin µg.*	*Niacin µg.*	*Vit. C mg.*	*Vit D i.u.*
Beetroot	1	—	5	20	30	1.7	—
Wholemeal bread	2	-180	100	1,200	—	—	
Butter	¼	281	—	—	—	—	4.25
Orange	3	21*	60	30	300	48	—
Coffee (= Milk)	5	150	60	180	1		1.5
		1,583 i.u.	423 µg. = .423 mg.	742 µg. = .742 mg.	9,063 µg. = 9.063 mg.	67.7 mg.	695.25 i.u.
Proportion of day's approx. needs		½	1/3	½-½	3/4	all	all

Sandwich Fillings of Good Vitamin Content

	A	*B_1*	*Ribo-flavin*	*Niacin*	*C*	*D*	*E*
Grated cheese with lettuce & tomato	x		x		x		
Grated cheese with black currant jam	x		x		x		
Grated cheese with Marmite & watercress	x		x	x	x		
Grated cheese with Bemax & tomato	x	x	x	x	x	x	x
Buttered egg with parsley	x		x	x			x
Buttered egg with tomato	x	x	x		x		x
Buttered egg with lettuce & Bemax	x	x	x	x	x	x	x
Sardine with lettuce	x		x	x		x	
Salmon with tomato	x		x	x	x	x	
Ham with black currant jam.		x		x	x		
Ham with grated carrot & lettuce		x	x		x	x	

No hard and fast rules need be made about what must be included in the day's meals and what is better omitted but the examples given show how a knowledge of vitamin values can be applied to the planning of diets for a single individual or for groups of any size.

If vitamin values are forgotten it is easy to omit essentials, but the necessary information is available and no one today need eat a poor diet through ignorance. It is one function of teachers and others who have an opportunity to influence the tastes and habits of the population in dietary matters, to make sure that proper guidance is given on this subject.

The science of nutrition is a development of the present century. This has been a period during which there have been many advances in science and in medicine advances which are so closely interconnected that it is difficult to distinguish the effects of one alone.

There are, however, certain pointers indicating where nutrition has played a part.

The effect of improved diet on the heights and weights of boys at school can be compared with the increases in height and weight of London schoolboys during the half-century.

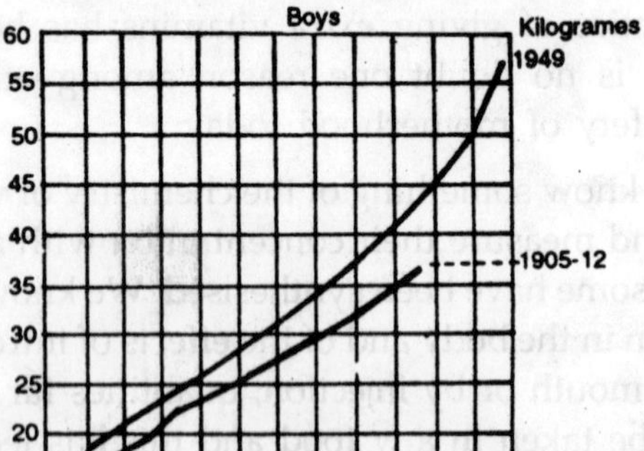

Fig. County of London. Average weights of London school children for the years 1905-12 and 1949.

Other effects known to be because of nutritional factors are

the fall in the incidence of rickets and in the death rate of children from this cause.

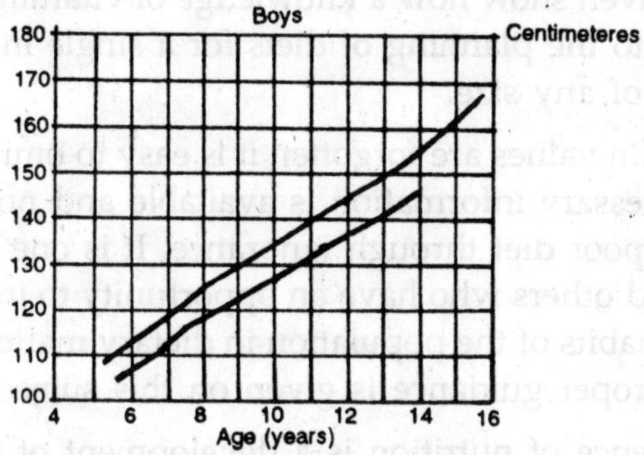

Fig. County of London. Average heights of London school children for the years 1905-12 and 1949.

Another improvement for which a better knowledge of nutrition is responsible is the reduced occurrence of complications of pregnancy, particularly the form of illness called toxaemia. Now this is partly due to improved nutrition because during 1940-42 the addition of various vitamins and minerals to the intake of over, 2, 000 pregnant women was tried out and the effects noted. It was found that the number of cases of toxaemia was reduced one-third.

The practice of giving extra vitamins has become more general and is no doubt one reason among others for the increased safety of motherhood today.

We now know something of the chemistry of vitamins, we can detect and measure their concentration with a fair degree of accuracy; some have been synthesised. We know something of their action in the body and of the effects of introducing into it, either by mouth or by injection, quantities far in excess of what could be taken in any food and this has led to the use of vitamins in massive doses in the treatment of disease. But there are still gaps in our knowledge, gaps which are gradually being filled as one discovery after another is made.

Chemists and biologists continue to work on obscure problems connected with the behaviour of vitamins. Isolated

results of their works may seem sometimes to be of little importance. It is only when a result obtained in one laboratory is linked with another discovery made somewhere else that ultimately a fact of practical value to ordinary people emerges.

Such knowledge has wide applications. It influences agricultural policy, international economics, the world supply of food. These are problems too vast and complicated, but every individual can learn enough to be able to select the most nutritious food from what is available and to play a part in influencing public opinion in favour of producing and eating the foods which are best for nutrition and health.

Proteins and their Functions

In 1838, a Dutch scientist, Mulder, suggested the name 'Protein' is complex nitrogen-bearing substance in tissues. The word 'protein' is derived from the Greek work 'proieot' meaning to take the first place[1]. Proteins play a significant role in all activities of living organisms. Due to the multiplicity of their functions the name given to them is very apt. In our skin, muscles, skeleton and in body fluids protein is present. They are essential during growth because it is the main constituent in out tissues. In a person's life the first eighteen months are very important as eighty pet cent of the brain development is over by this time. If protein is not sufficient during this time menial retardation takes place. Acute shortage of protein in early months of life brings intellectual dwarfism and it remains so for the rest of the life. That is why this nutrient is called 'protein'.

Specific Characteristics

Proteins are complex organic compounds which contain carbon, hydrogen, oxygen, nitrogen, sulphur and phosphorus. Certain proteins contain iron, iodine, copper and other inorganic elements. The presence of nitrogen distinguishes protein from carbohydrate and fat. Protein contains an average of 16 per cent nitrogen.

On hydrolysis large molecules of protein give out smaller units called amino acids. Amino acids contain a drboxyl group (CuoH) and an amino group (NH.) They are otherwise known as die acidic and basic group, respectively. There are twenty-three amino acids in proteins. Amino acids are grouped into five groups based on the number of amino groups and carbosyl group in them. They are amino-monocarboxylic acids, monoamino-dicatboylic acids, dianino-moftocarboxylic acids, sulphur containing amino acids and aromatic and hetezocyclic amino acids.

1. Monoamimo-monocarboxylic acid.

   eg., Glycine, Alanine, Viliae, Leuciw,

2. Monoamino-dicarboxylic acids

   eg., As paartic acid, Glutamitic acid.

3. Diamino-monocarboxylic acid

   eg., Arginine, Lysine

4. Sulphur containing amino acids

   eg., Cystine, Cysteine, Methionine

5. Aromatic and heterocyclic amino acids

   eg., Phenylalanine, Histidine, Tyrosine, Tryptophan, Proline, Hydroxyproline

A protein contains a number of amino acids. Two amino acids are linked together through peptide linkage. In this, the basic group of one amino acids combines with the carboiyl or acidic group of another group of amino acids. Two amino acids are linked together to form a dipeptide and it combines with another amino acid. When a number of adds are linked together it is known as polypeptide.

Classification : Proteins are biochemically classified into three groups based on their chemical composition. They are simple proteins, conjugated proteins and derived proteins.

Simple Proteins : On hydrolysis simple proteins give nothing but only amino acids. Such proteins are called simple

proteins. Examples of simple proteins ate albumins, globulins and glutelins. Prolamins, fibrous protein, histamins and protamins.

Albumins and globulins are found within all body cells and in the blood serum. Potatoes and peanut also have globulin in them. Globulin is in haemoglobin and myoglobin.

Glutelin is present in wheat glutens, zein in corn and gliadin in wheat has prolamine.

Collagen, keratin and elastin have fibrous proteins and histones are present in thymus and in globin. Protamins is present in salmine from salmon sperm.

Conjugated Proteins : On hydrolysis conjugated proteins give a protein fraction and a non-protein fraction. In our body conjugated proteins are present in various parts. Examples of conjugated proteins are haemoglobin, nucleoprotein, glycoprotein and lecithoprotein. Nucleoproteins are present in nucleus, glycoproteins in mucin and lecithoprotein in lecithin.

Derived Proteins : Derived proteins are derived from a mother protein either through hydrolysis or digestion. On digestion derived proteins are formed. Examples of derived proteins are metaproteins, coagulated proteins, peptides and peptones.

Metaproteins are products resulting from the action of acids or alkalies.

Coagulated Proteins : By the action of heat on protein or by the action of alcohol on proteins it coagulates. Cooked egg is an example.

Peptides, Proteoses and Peptones : Peptide linkage forms peptides from amino acids. Two amino acids linkage forms dipeptide and three amino acid tripeptides. Digested proteins form peptides.

Proteoses are derived from proteins by the action of digestive enzyme pepsin and precipitated by saturating their solution with ammonium sulphate.

Proteins are again classified based on their nutritive value. From the nutritional standpoint amino acids are of two types, essential and non-essential. Essential amino acids are termed essential because the body requires them through our daily food. Non-essential amino acids are also essential to the body but the body can manufacture them from the available chemicals. Thus, the nutritional quality of a protein is based on the presence of essential amino acids in them. There are eight essential amino acids in the case of adults and ten essential amino acids in the case of children and others are non-essential amino acids.

The essential amino acids are isoleucine, leucine, lysine, methionine, phenylalanine, threonine, tryptophan and valine. For infants and children arginine and histidine are also essential.

Distance : The non-essential acids are alanine, aspartic acid, cystine, cysteine, glutamic acid, glycere, hydroxyproline, proline, serine, tyrosine, citrulline and norleucine.

Proteins *at* classified into complete proteins, partially incomplete proteins and incomplete proteins based on the presence of essential amino acids in them. The presence of amino acids helps a protein to perform all the functions of proteins in our body. A complete protein food contains all essential amino acids in proper proportion. It helps the protein to promote growth, maintenance and repair: Examples of complete protein foods are milk, egg and fish.

Partially incomplete proteins lack some essential amino acids and so they will help to maintain our body but growth is not promoted. Vegetable proteins are examples of this group.

Incomplete proteins neither help maintenance nor growth. Gelatin is an example of this type of protein.

There are certain methods to evaluate the quality of a protein. Biological value of protein and growth gain or protein efficiency ratio (PER) are the important methods used for this.

Biological Value : Biological value of protein foods depends to a great extent on their amino acid composition and they are

grouped as protein of low or high biological value. The biological value of protein also depends on the digestibility of a protein food. Cooking enables the digestibility whereas overheating and frying reduces the biological value of a protein because it destroys the essential amino acid, lysine. Egg, milk, fish, poultry and meat are of high biological value whereas proteins from cereals like wheat, rice, pulses and nuts are of low biological value. Vegetable proteins are low in essential amino acid pattern, especially of lysine, tryptophan, methionine and threonine. When a mixture of proteins from different sources is consumed mutual supplementation takes place. For example, rice or wheat protein is low in lysine whereas pulse is rich in lysine and low in methionine. Rice is rich in methionine and a mutual supplementation results when rice and pulse are consumed together.

Protein Efficiency Ration (Growth Method): The ability of a protein to bring increase in weight or growth gain is considered on the efficiency of a protein or its nutritive quality. Nutritive value is expressed as growth per gram of protein consumed.

$$\text{PER} = \frac{\text{Gain in body weight (gm)}}{\text{Protein in take (gm)}}$$

Main Functions of Diet

Building Block : The most important function of protein is to supply amino acids to cells for the continuous replacement of cells throughout life. It is the most abundant organic compound in our body. It is the building block of our body. From conception to growth at various levels, foetus, infant, child, and during pregnancy and lactation period protein is very essential.

In our body, tissues are not in a static position. Degradation and resynthesis of protein and other nitrogenous compounds always take place in our cells. Tissue proteins are broken down and new substances are continually synthesised in our body.

Muscles and other tissues, bones, cartilage and trabecula contain fairly high percentage of proteins. Our hair, nails and skin also contain protein. All body fluids except bile and urine have protein in them.

Regulatory Functions : Different proteins perform highly specialised regulatory function in the body. Haemoglobin, which is the chief constituent of the red blood cells, carries oxygen to the tissues. Oxyhemoglobin and their alkaline salts permit oxygen to enter into the tissues and receive carbonic acid from the cells. This process helps the removal of 92-97 per cent of all carbon dioxide from the tissues.

Plasma proteins, especially albumin and globulin, play an important role in regulating osmotic pressure and water balance within the body. When plasma proteins are decreased the water balance is upset and accumulation of fluids in the body takes place. Proteins enable the blood to maintain their slight alkalinity.

Proteins function as buffers, thus helping various sites of chemical reaction to maintain its pH. Nucleo proteins contain the blueprint for the synthesis, of all body proteins.

Formation of Enzymes, Hormones and other Secretions: Proteins supply raw materials for the body to synthesise enzymes like trypsin and pepsin. Hormones like insulin and thyroxine are protein in nature. Digestive juices contain protein in them. Antibodies which give resistance power to the body are protein in nature. They are known as immune proteins.

Sources of Energy : Protein is generally considered as the building material of our body. But when the diet contains insufficient carbohydrate and fat for fuel, proteins are used as a fuel by the body.

Each gramme of protein yields four calories. But it is not a wise contribution because the nitrogen excretion increases along with the cost of the food.

Proteins form Part of Vital Compounds in Body: Nitrogenous compounds are present in certain substances of

immunological and antigenic reactions. Examples of these are globulin of blood serum and chromatin in nucleus. Methionine, an amino acid, provides methyl groups for the formation of creatinine and choline.

Tryptophan, an amino acid, is converted to niacin, a B-complex vitamin, in emergencies, Glutathione, a tripeptide of crystine, is present in small amounts in all our active tissues and this helps oxidation- reduction reactions. Histidine, an amino acid, is decarboxylised to histamine and it is a stimulator of gastric glands to secrete gastric juice. It is also used as a dialator of capillaries.

Threonine, another amino acid, acts as a lipotropic agent which prevents deposition of excess fat in the liver. Proteins also help the transport of drugs by binding them into protein molecules. Contractile proteins (myocin, actin) regulates muscle contraction.

Acts as a Binding Factor : Rational binding protein, lipo proteins, transferring and serum proteins are very essential to our body for transporting many chemical substances.

Various Steps of Digestion

There is no protein-splitting enzyme in saliva and so in salivary digestion protein is not affected. Hydrolysis of protein begins in the stomach. Pepsin, the enzyme secreted by gastric glands in the stomach, breaks down proteins into proteoses and peptones. Milk proteins are first converted to casein by a special enzyme called rennin.

Casein combines with calcium to form calcium casemate. Pepsin converts this into peptones. All the peptide linkages are not broken by gastric digestion. Stronger enzymes of pancreatic and intestinal juices complete the digestion of protein into peptones. Pancreatic juice contains trypsin and chymotrypsins.

They hydrolyse peptones and proteoses into polypeptides. The final breakdown of all protein fractions to amino acids is brought about by erepsin secreted by intestinal mucosa.

Amino acids are absorbed by the small intestine and they are carried to the tissues or to the liver. A small amount, that is, about eleven per cent of the total protein is absorbed by the stomach, sixty per cent by the intestine and twenty-eight per cent by the colon. The absorbed amino acids reach the tissues where then metabolism takes place.

Energy-producing Reactions in Living Beings

In the tissues amino acids undergo breakdown and synthesis. They are also transaminised, doaminised or decarboxylated. The anabolic activities consist of a formation of new cells or the repair and maintenance of existing ones and secretion of various substances. If there is excess of amino acids after these anabolic activities, they will be either used for energy purpose or are converted to fat

Deamination : One of the catabolic action is deamination. In deamination, the amino group is removed from an amino acid. This reduced amino group is used for the synthesis of new substances like simple amino acids or eliminated as urea. The carbon, hydrogen and oxygen may be used for carbohydrate or fat metabolism.

Transamination : The deaminated amino group may be utilised for transamination process. Transamination is a process whereby a new amino acid is formed using the deaminated amino group and another precursor. If all the amino group is not used for transamination, nitrogen freed from the amino group forms urea.

Formation of Urea : Amino acid is oxidised to ketoacids and ammonia. The ammonia, if it accumulates in the tissues, is toxic. So in the liver ammonia is converted into urea and excreted as urine. Therefore, the end product of protein metabolism is urea. Ammonia is converted into carbamyl phosphate.

This reacts with ornithine to form citrulline. Citrulline combines with aspartic acid and forms arginosuccinic acid

which is converted to arginine. Arginine is hydrolysed into ornithine and urea. This ornithine again combines with carbamyl phosphate and this cycle is repeated.

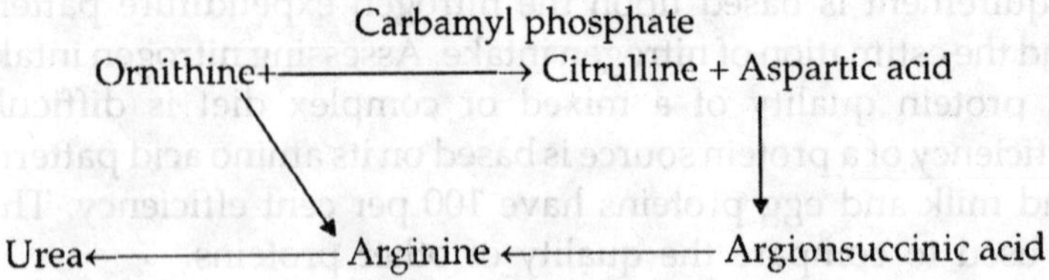

Fig. : Urea Cycle

Another catabolic change which takes place in an amino acid is decarboxylation. In this carbon dioxide splits off from carboxyl group and amines.

Nitrogen Balance : Tissue proteins are broken down at various times. When tissue proteins are utilised by the body for different functions nitrogen is excreted in urine and faeces. When protein food is taken new tissue proteins are formed. Nitrogen balance is the difference between the amount of nitrogen taken in through the diet and the sum of nitrogen and nitrogenous substances excreted from the body. When the nitrogen intake is more and the nitrogen loss is less the body is said to be in a positive nitrogen balance. When the nitrogen intake is less and nitrogen loss is more the body is said to be in a negative nitrogen balance. When the nitrogen intake equals the nitrogen lost from the body, the body is said to be in nitrogen equilibrium. During deficiency state the body shows a negative nitrogen balance.

Factors Affecting Protein Utilisation : If sufficient calories are present in the diet, protein will be effectively used for its function. Certain minerals and vitamins are essential for the proper utilisation of protein. Potassium, phosphorus, vitamin A, riboflavin, niacin and vitamin B6 are the factors which enable the utilisation of proteins. Heat treatment or processing improves the utilisation. Heat destroys inhibitors of certain

enzymes and that is how it helps digestion. But if direct heat is applied to meat it retards protein utilisation. Examples of this action is roasting meat in direct heat or frying.

Requirement of Proteins : Determination of protein requirement is based upon the nitrogen expenditure pattern and the estimation of nitrogen intake. Assessing nitrogen intake or protein quality of a mixed or complex diet is difficult. Efficiency of a protein source is based on its amino acid patterns and milk and egg proteins have 100 per cent efficiency. This is used to compare the quality of other proteins.

Covering a 30 per cent addition to the variability of different proteins 1 gm of protein/kg body weight is recommended by FAO/ WHO expert group. Protein allowance during infancy and childhood is more as there is rapid growth. Studies have shown that well nourished pre-school children can attain heights and weights close to Western children. During pregnancy and lactation there is an increase in protein content. Protein requirement during various phases of life is given in table .

Protein Allowance

Man	1 gm/kg/day...55 gm (55 kg)			
Woman	1 gm/kg/day...45 gm (45 kg)			
Pregnancy	2nd, 3rd trimester 14 gin/day			
Lactation	0-6 months + 25 gm/day			
	0-3 months - 2.3 gm/kg (for milk protein)			
	3-6 months -1.8 gm/kg			
	6-9 months -1.8 gm/kg			
	0-12 months -1.5 gm/kg			
Children	1-3 years	-	1.83 gm/kg day	- 22 gm (12.03 kg)
	4-6 years	-	1.56 gm/kg	29 gm (18.87 kg)
	7-9 years	-	1.35 gm/kg	36 gm (26.37 kg)
Boys	10-12 years	-	1.24 gm/kg	43 gm (34.30 kg)
	13-15 years	-	1.10 gm/kg	52 gm (47.03 kg)
	16-18 years	-	0.94 gm/kg	53 gm (56.50 kg)
Girls	10-12 years	-	1.17 gm/kg	43 gm (36.47 kg)
	13-15 years	-	0.95 gm/kg	43 gm (45.53 kg)
	16-18 years	-	0.88 gm/kg	44 gm (50.00 kg)

Protein Deficiency : Deficiency of protein-predominant foods during growth period is known as kwashiorkor and protein-calorie malnutrition is known as 'marasmus'. Protein calorie malnutrition is one of the largest nutritional problems in India.

Kwashiorkor : Kwashiorkor is a term used by Ga tribe in Ghana and the meaning of this word is 'the sickness the old child gets when the next baby is born'. Dr. Cicely Williams in 1933 reported the occurrence of kwashiorkor (red baby) in poorer sections of population in Africa. Internal and external changes occur in the structure and functional efficiency of various organs in the body. Starting from hair and skin to tissues, blood, water content in the cells, blood cells, organs like liver, alimentary canal and nervous tissues are affected by protein deficiency. The child will maintain an apathetic, anaemic, anorexic, ocdematous and diarrhoeic condition.

The early symptoms of kwashiorkor include loss of weight, fatigue, reduced resistance to infection, delay in recovery from illness and retarded growth. As the protein deficiency prolongs, changes in the body take place. Discolouration of the skin, oedema with ascites, peeling of skin, cracks in the skin, angular stomatitis, cheilosis and atrophy of the tongue and- anaemia occur. Degeneration of the epithelial tissues and reduced secretions of digestive juices result in diarrhoea. Absorption of nutrients is delayed and result in general weakness and mental retardation. Liver may be palpable and soft. Fatty liver, especially in the periphery, is seen. Functions of liver are affected adversely, especially the synthesis of lipoprotein is reduced, hence fat accumulates in the liver. Deficiency of all vitamins occurs. Deficiency symptoms of B-complex vitamins lead to stomatitis and ulcers in the mouth and stomach, vitamin A deficiency leads tonight blindness and xerophthalmia, severe protein deficiency results in permanent retardation of mental development, psychomotor development and behaviour.

Marasmus : Marasmus is the result of starvation in children. Marasmus occurs when the diet is low in both calories and in

protein. The term Marasmus is derived from the Greek word meaning 'to waste. Tissues are wasted and unlike kwashiorkor the body looks shrunken. The child has an emaciated look. Irritation, apathy, loss of appetite, dehydration, wasting of muscles and weight loss and incessant diarrhoea are the other symptoms of marasmus. All these result in retardation of growth and severe wasting of subcutaneous fat Deficiency of vitamins A and B complex is common in these patients.

Marasmic Kwashiorkor : This occurs when a combination of marasmus and kwashiorkor of varying degrees is present in a child. The child shows signs of both kwashiorkor and marasmus. Gastrointestinal and respiratory infection may accompany the other symptoms.

Major Sources and Needs

The best sources of protein are animal sources like meat, fish, poultry, eggs, milk and milk products. They are otherwise known as first class proteins because they are rich in essential amino acids.

Cereal grains and legumes are the important sources of plant proteins. They are also known as second class proteins. Pulses are commonly knows as "poorman's meat" because they are comparatively cheap and fair in the nutritive content. The biological value of vegetable proteins can be improved when cereals and pulses are taken together.

Mutual Supplementation of Amino Acids : Amino acid content of the diet can be improved by supplementing one protein source with the other. In Indian diet this method has a significant role to play. For example, cereals and pulses. Lysine, the limiting amino acid in cereals, is supplemented by pulses, which is rich in it Methionine, the limiting amino acid in pulse, is supplemented by cereals. Soyabeans is also rich in lysine but deficient in methionine. A small quantity of animal foods can also do the same to vegetable proteins.

8

Planned Diets

Vitamins are essential in animal diet though they are needed only in minute quantities. Most of the vitamins are essential for the metabolism of all living organisms. Some fat soluble vitamins may have metabolic roles unique to animals and they all synthesised from plants and micro-organisms. Some vitamins can be synthesised from pro-vitamins. A large proportion of vitamin requirement of humans and higher animals is supplied by the intestinal flora.

Whether a substance must be supplied intact to the cell or can be synthesised by the cell depends upon the assortment of enzymes peculiar to the cell species. A substance may therefore be a vitamin for one species but not for other. The difference lies in whether the substance is required in the diet or whether it can be synthesised by the animal. The number of substances supplied by cellular synthesis tends to vary inversely with the complexity of the organism and appears to reflect an evolutionary adaptation as more complex organisms evolved from a simple one cell origin. Thus man is dependent upon his environment for some nutrients that microorganisms and other lower form of life can synthesise for themselves.

Primitive Techniques

Very soon after vitamins began to be studied seriously it became apparent that their distribution in foods was very uneven. Cod liver oil and butter, for example, both contained vitamin D and could be used to cure rickets but the amount of butter needed was much greater than the amount of cod liver oil. Similarly wheat germ and yeast both contained the anti-beriberi factor B but some investigators found wheat germ the better and others yeast and no one had reliable information. Various attempts were therefore made to discover which were the best sources and later to find some means of measuring the extent to which one food was superior to another.

The fact that animals failed to grow when deprived of either fat-soluble vitamin A or water-soluble vitamin B provided a ready method of testing foods for the presence or absence of these two factors. Rats, when fed on purified foods containing all the known dietary essentials except vitamin A, grew for a time, but their rate of growth gradually decreased until finally they began actually to lose weight. At this stage foods to be tested for vitamin A would be given. If the effects were a restoration of growth, then it could be concluded that the food added to the deficient ration contained the missing factor A, but if no growth took place, then the food was deficient in vitamin A.

Foods found by this test to contain vitamin A were cod liver oil, halibut liver oil, butter, milk, liver, egg yolk.

The same type of experiment was used for testing the presence of vitamin B in foods, because by this time it was known that vitamin B not only protected from beriberi but was also a growth factor.

Experiments of this sort gave information not only on the presence or absence of these vitamins in a wide variety of foodstuffs but also an indication of their relative concentration. If a very small quantity of food were needed to restore or maintain normal growth, it was considered that this food was

'rich' in the vitamin; if a large amount were needed, then the concentration was taken to be 'poor.'

Other tests besides the growth test were used for some of the vitamins, for example, the cure of head retraction in pigeons for vitamin B, the prevention of scurvy in guinea pigs for vitamin C, the cure of rickets in rats for vitamin D, and the cure of sterility in rats for vitamin E.

Cure of Head Retraction : It had long been known that pigeons fed on polished rice developed avian polyneuritis, the condition in birds corresponding with beriberi in man. In this disease, the muscles of the neck and limbs become paralysed, the bird can no longer stand upright but sinks down on its hocks and the head cannot be held up. It usually flops backwards in the position known as head retraction.

In the test, birds were fed on polished rice until they developed this condition and were then given doses of the food to be tested until, by trial and error, the weight was discovered which was just enough to effect a cure. A good source of vitamin B would be indicated by a small curative dose and vice versa. Vitamin B was found in the following :

Wheat germ

yeast

oats

whole wheat

egg yolk

liver

peas and beans

Prevention of Scurvy

Animals were found to vary in their sensitivity to lack cf vitamin C. Some, like rats and birds, seemed able to live and thrive without foods containing it, but guinea pigs were very sensitive to a shortage and on diets consisting of oats, skimmed milk powder and butter fat without any vegetables, their growth

would begin to fait after about three weeks and afterwards they would develop scurvy. Their joints would become tender and sometimes bleeding under the skin could be detected. The test consisted of adding to a scurvy-producing diet small quantities of test foods and finding out (again by trial and error) the least quantity needed to prevent the onset of scurvy.

By this test it was discovered that most fresh fruits and vegetables contained vitamin C, the best sources being citrus fruits and fresh green leaves. Other foods, like meat and milk, contained only very small quantities.

Cure and Prevention of Rickets : Rickets does not readily develop in rats if they have access to sunlight, but if they are reared in darkness (a condition quite normal for a rat) and are fed a diet containing no cod liver oil or other source of vitamin D and if the calcium and phosphorus of the diet are suitably unbalanced, rickets will develop in three or four weeks. This is shown by swollen joints and a disinclination on the part of the young rats to run about. The joints, if X-rayed at this stage, show swollen ends of the long bones and poor calcification indicative of rickets. The test consisted either of feeding minimum curative doses to rats with rickets or of finding the smallest daily dose needed to prevent the onset of rickets and permit normal growth. Fish liver oils were found to be the best sources, others were animal fats like butter.

Cure of Sterility in Rats : The same principle was applied in this test. Female rats were kept on diets known to lack sources of vitamin E. On such diets they failed to produce any young, but normal fertility could be restored by feeding daily doses of wheat germ oil or another source of vitamin E. This vitamin was found in some green foods as well as in the oils of wheat germ and cotton seed. Lettuce and grass both contain it.

Different Classes

A great deal was learnt by these methods about the distribution of vitamins in foods and the two main groups

were distinguished: the fat-soluble vitamins, namely, A, D and E, and the water-soluble, B and C. But still very little was known of the actual quantities present and a great need was felt for a method of accurate measurement.

An attempt was made to express vitamin activity numerically by giving an arbitrary value of 100 to some well known and reliable source and then assigning other appropriate values to other foods. For example, if food X were chosen as the standard of comparison and given the value 100, and the dose of food Y needed to produce the same result as a given weight of X were twice as great, then Y would have half the value, i.e. 50.

Comparisons such as these were made and lists of values published. They were useful as preliminary tests but owing to variation in potency from sample to sample and variation in response from animal to animal, the resulting figures tended to be confusing and even sometimes contradictory.

The next step was the adoption of 'units' of activity. Several such units were suggested. For example, a unit of vitamin B was taken as the amount of vitamin B which, when fed as a daily allowance, resulted in maintenance of weight of a standard rat over an 8-week period.

There were obvious disadvantages in such units as these. There is, for example, no such thing as a 'standard' rat any more than there is a 'standard' baby. Moreover, 'units' failed to take into account the different conditions under which the tests were performed and thus the standards themselves varied.

It was therefore a significant advance in technique when, in 1931, the Health Organisation of the League of Nations suggested the adoption of 'International Standards' for the four principal vitamins A, B, C and D.

A suitable highly active preparation was chosen for each vitamin and it was agreed that a given weight contained so many units. Except in the case of vitamin C which is unstable, a bulk supply sufficient for several years' use throughout the

world, was put into storage and from this, allocations were made to scientific institutions where estimations were being carried out.

This measure helped greatly to dispel, at one stroke, all the difficulties caused by the use of as many standards as there were countries and workers.

Vitamin A **:** On account of the unstable nature of vitamin A, the material chosen as a provisional standard was its precursor P-carotene, which occurs as a yellow pigment in vegetables such as carrots. One unit was the activity of 0.6 mg. of pure P-carotene.

Vitamin B **:** This standard was specially prepared from a single bulk supply of rice polishings. The material was extracted with water and then absorbed on kaolin. The resulting powder was highly active. One unit was the activity of 10 mg.

Vitamin C **:** On account of the impossibility of keeping fruit-juices without loss of vitamin C activity from oxidation, fresh lemon juke was chosen as the standard substance and a unit was 0.1 ml. of this. It was assumed that this was relatively constant which of course it was not, since lemons, like all products, vary among themselves.

Vitamin D **:** For this, a good quality cod liver oil was selected. The unit was the activity of 1 mg. of this oil.

Present-day Units **:** Later on with advances in the chemistry of the vitamins all these were modified but in the meantime between the years 1931 and 1938 much very valuable work was done by their use.

People were familiar with units of length, weight, temperature, heat. The vitamin units were not given names corresponding to inches, grammes, degrees or calories, but were called merely units of activity. The ideas, however, were the same and just as there is a standard yard against which all other yardsticks are measured, so these international units of vitamins became standard and all other estimates were measured against them.

Before vitamins were chemically synthesised in a state of purity estimations of vitamins in foods were carried out by biological methods. Chemical procedures were greatly simplified when the chemical nature of the vitamins was discovered.

International Standard : The general method was to select a group of experimental animals (usually rats) all from the same stock and arrange them in several groups with eight or ten in each so that their weights were as nearly as possible distributed evenly and fairly between all the groups. They were fed with a basal diet containing everything needed for health except the one vitamin to be measured and were weighted regularly until they either ceased to grow or began to develop typical signs of deficiency. Each rat was then fed an accurately weighed dose either of the international standard or of the food being tested.

Both the standard and the test would be fed at several dose levels and from the weight measurements obtained a graph was drawn showing what was known as a 'Dose-response curve'.

From this curve it was possible to read off the weight of test substance which produced a response equal to that produced by a given number of units of standard and hence to calculate the potency of the test food in terms of international units per gramme or ounce.

The mathematical procedures were at first very simple and to attempt was made to measure how accurate they were. It v as, however, accepted that methods depending on the use if living animals were subject to considerable error but for the time being, even these simply performed tests using the international standards were a great advance and many foods were subjected to estimation or 'assayed' for their vitamin content.

From values published it became possible to calculate the approximate total vitamin content of any diet and hence the next logical step was the carrying out of dietary surveys.

The international standards now in use (Timmerman, 1951) as established by the World Health Organisation (1951) are as follows:

International Standard Preparations	*One International Unit*	
Vitamin A acetate	0.000344	mg.
Provitamin A (beta-carotene)	0.0006	mg.
Vitamin B_1 (Pure synthetic vitamin B_1)	0.003	mg.
Vitamin C (1-ascorbic acid)	0.05	mg.
Vitamin D_3 (crystalline)	0.000025	mg.
Vitamin E (alpha-tocopherol acetate)	1.0	mg.

Functional Difference

The function of each vitamin in living processes is studied in the first place by excluding it from the diet and observing the effects either on experimental animals or on man. But the study of deficiency states does not reveal the whole story because no vitamin acts alone. In the living body there are interactions between all the essential nutrients and lack of one may interfere with the actions of others.

However, to arrive at an understanding of how the vitamins work, it is logical to consider first what we know of each one individually, and then to study their relationships one with another.

Usefulness for the Life Process

Vitamins are concerned in the development of the body from its beginning until it has reached full maturity and with the maintenance of its structure throughout life. They are not structural constituents in the way that calcium phosphate is of bone or protein is of muscle but they are concerned in the formation and maintenance of the whole body of the various organs and tissues.

It is important to recognise that the body, while retaining its identity and its individual characteristics throughout life, is in a continuous state of change. The outer layers of the skin are shed and cells of the deeper layers constantly renew the covering of the body. This is a process which we can all observe. The same kind of renewal is all the time taking place even with structures as hard and apparently permanent as bone.

The initial growth and continued renewal of the bodily structures depends in the first instance on food taken. That is the source of supply. But although food consists of proteins, fats, carbohydrates, mineral salts and water, and the body consists of the same there is a vast difference between bread and butter, meat and vegetables and the human or animal body into which they are transformed. This transformation to be effective, needs the essential vitamins and each one plays some part in the camplicated process. Some have functions still only very imperfectly understood. Others we know more about.

The deficiency diseases through which vitamins were first discovered showed how lack of a single vitamin caused a disturbance of function and also structural change in organs or tissues peculiarly sensitive to that particular factor, e.g. the connection between vitamin A and the skin, vitamins B and the nervous system, vitamin C and capilaries, vitamin D and bone, but as soon as observations and experiences were followed by experiments, it was found that in all these deficiencies growth was affected. Vitamin A and Vitamin B were, in fact, known as growth factors for experimental animals.

The first investigations into the ways where by these factors actually functioned in the body and thus brought about the effects on growth, were made on vitamin B_1 which was found to be essential at certain stages in the breakdown of carbohydrates and the production of energy.

Vitamins and Carbohydrates : Carbohydrates (starches and sugars) are basically fuel foods. They supply the day-to-day needs of the body for the maintenance of body temperature and for the performance of all physiological processes, whether

involving muscular activity, glandular secretions or even mental and nervous functions. Carbohydrates do not become part of the body structure except in so far as they are stored as reserves. Glycogen is the form in which storage occurs in muscle and liver and carbohydrate eaten in excess over the daily needs may be stored as body fat, but in general these foods are eaten to provide energy and this is liberated when they are chemically broken down (or metabolised) to produce at the final stage of the process, carbon dioxide and water.

Various chemical stages are controlled by enzymes (compounds produced within the body, which accelerate or regulate chemical reactions). Lactic acid and pyruvic acid are intermediate products produced during the breakdown of carbohydrates and in deficiency of vitamin B_1 is associated with the enzyme which is concerned in the oxidation of pyruvic acid. It is, in fact, a co-enzyme and is responsible for the control of chemical oxidative processes which take place within the living cell in connection with its energy relations. Since vitamin B_1 performs such a vital function as this, it is not surprising that shortage affects many physiological processes in addition to inhibiting growth.

Vitamin B_1 Functions : Being readily soluble in water, vitamin B_1 is quickly absorbed from the digestive tract and enters the blood stream to be carried to the tissues of the body. It is used continuously in the vital physiological processes and the tissues are not able to store it except for short periods. Consequently regular supplies are required for health.

The effect on growth is seen in children, who make only subnormal gains in weight if their food is short of vitamin B_1. At the same time their appetite may be poor, the slackness of their intestinal muscles may cause constipation and they may lack the mental alertness of normally healthy children.

The same apathy and lethargy affect adults, abnormal fatigue may be because of excess of such breakdown products as lactic acid and pyruvic acids in the tissues. All stages occur

between such general signs of debility and the symptoms and signs characteristic of beriberi.

In this disease, the loss of appetite and consequent reduced food intake may produce extreme emaciation, but this may be masked by the presence of fluid in the tissues causing swelling of the limbs. The heart gets always affected in beriberi, being considerably enlarged and its rate is increased, so that when death occurs, this is frequently from failure of the circulatory mechanism.

The nervous system changes that causes abnormal sensations in the limbs, extreme weakness, wrist-drop and loss of knee-jerks.

Relations with other Vitamins : Riboflavin and nicotinic acid also participate in the same enzyme systems concerned in the intermediate breakdown of carbohydrates. It is possible in experiments with animals to make diets deficient in one vitamin only and to study the effects of deficiency of one factor at a time but deficiency diseases in man rarely occur singly. The, effects of one deficiency could be more marked than those of another but even in extreme cases like beriberi and pellagra, several factors are involved.

Riboflavin : Although riboflavin is more concerned in carbohydrate oxidation, deficiency is shown by effects quite different from those caused by vitamin B_1 deficiency. The chief characteristics are cracking of the lips with sore, red patches, scaliness with raw cracks at the corners of the mouth and a sore, red tongue. There may also be scaly red patches across the nose and lips and over the eyelids and sometimes redness of the eyes. These signs may appear in individuals even in Great Britain if their diet has been poorly supplied with such riboflavin-containing foods as milk, eggs and liver.

Nicotinic Acid : Nicotinic acid is also fundamentally a tissue co-enzyme and is essential, along with vitamins B_1 and riboflavin for the breakdown of carbohydrate, but extreme deficiency of nicotinic acid causes not beriberi but pellagra, the

disease that is still prevalent in countries where maize is the staple cereal. The word is derived from pelle=skin, agra=rough, and was used first to describe this disease in 1771 in Italy. It was Funk, who had coined the word 'vitamine', that suggested in 1914 that pellagra was a deficiency disease. Its exact cause has given rise to much controversy but can probably be explained as follows:

> Maize, like any other cereal grain, consists of both carbohydrate and protein and the protein is made up of a collection of simpler nitrogenous compounds, the amino-acids. Some of these are necessary for life and health; others can be dispensed to be very poorly represented in maize. It has been known that the occurrence of pellagra was associated with the consumption of foods poor in protein. The connection with nicotinic acid was recognised when it was found that tryptophan was a precursor of nicotinic acid. The disease disappears when food protein containing tryptophan or when nicotinic acid (or amide) is given.

Pellagra is characterised by soreness of the tongue and mouth, presistent diarrhoea and the development of pigmentation of the skin leading to dermatitis. At the same time patients become mentally depressed. The pigmentation of the skin occurs in exposed parts :

> Hands, face and neck. The skin is dark and brown, later becoming red and scaling with raw patches. The depression may develop into a severe dementia.

Pellagra is Fairly Common in Parts of America and in the Middle East : Mild degrees of nicotinic acid deficiency is found in other countries if individuals fail to get their quota of nicotinic acid and protein foods, or if some other condition prevents either absorption or use by the body of these food substances. The chief sources of nicotinic acid in human foods are meats and fish. Hence when people eat only a little of these and large quantities of maize, they lack both the protective factors, tryptophan and nicotinic acid, and pellagra is the result. In

animals other vitamins, and in particular B_1 and B_6 are needed for the full utilisation of nicotinic acid and also for its formation from tryptophan - another example of vitamin interactions.

B Vitamins and Proteins : Riboflavin and nicotinic acid are not only concerned with carbohydrates and energy relations but also with protein and the muscular structures of the body.

All proteins are made up of amino acids and there are some twenty of these latter, which, combined in different ways and in different proportions account for the characteristics of the various proteins as we know them. For example lean meat is protein but we know from our own experience that there are differences between beef and mutton and between mutton and pork; and even greater differences between meat and milk, cheese or eggs. But the body can use all these proteins as sources of the amino acids which are to be built up into human muscle. Protein eaten in any form is first digested and absorbed and then its amino acids have to be re-arranged and built up again either for the growth of the human body or for its maintenance when it is fully grown.

In this process both riboflavin and nicotinic acid play a part. Moreover, riboflavin, when eaten in food, is transported to the liver where it is temporarily stored, to be released again when required during the digestion and assimilation of food. It has been shown that for this release from the liver vitamin B_1 is needed.

Thus these three vitamins are very intimately connected not only in foods where they are found in nature, but also in their functions in the animal body, and we can see why the absence of any one of them interferes with growth and with general well-being.

B Vitamin and Fats : Just as certain vitamins have a special function in the metabolism of carbohydrates and others in that of protein, so a few seem to have close association with the body's use of fats in connection with either absorption or subsequent use.

Fats are digested in the small intestine and afterwards pass through the cells lining the tract into the lacteals and thence to the Lymphatics.

After absorption, fat is carried first to the liver and thence passes to other parts of the body, being either broken down for the production of heat and energy like carbohydrates or laid down in various fat depots for storage.

Choline : Choline is concerned in the use of fats by the body. It is required first of all for their absorption through the cells lining the intestinal wall and it is required subsequently by the liver for the conversion of fats into substances suitable for passing in the circulating blood to different parts of the body. Fats do not circulate as such but as compounds with phosphorus, phospho-lipids, and part of the choline molecule is required for this conversion. For this reason choline is known as a lipotropic agent and in this respect is similar to inositol and vitamin B_{12}.

Lack of choline affects the functions of the liver, but in the absence of choline, an amino acid, methionine, can take on some of its functions. Because of this reason some doubt has been cast on the justification for considering choline a vitamin since it is not *absolutely* essential for life. However, at present it is usual to include choline among the vitamins and its role in metabolism is certainly similar to that of other vitamins.

In the absence of choline, fat accumulates in the liver, interfering with the proper functioning of the hepatic cells.

This causes dyspepsia, and if the process continues, the liver may become permanently damaged.

Since normal metabolism of fat depends on supplies of choline, animals on a choline-deficient diet grow badly and various signs of deficiency develop, for example, in adequate lactation in the rat and slipped tendon in chicks.

Inositol : The function of inositol in the animal body is only partly understood. Like; choline, it helps in the metabolism of fats and it may have some effect on the skin. It has also been suggested that with vitamin E it affects the health of the muscles.

Inositol itself is soluble in water, but in plant products it is often found as phytin, a compound with phosphorus, the calcium salt of which is insoluble. This fact gave rise to the opinion that cereals containing inositol caused rickets by preventing the absorption of calcium. It has been found however, that this only happens in severe deficiency of calcium and that with normal supplies of calcium and vitamin D, no apprehension need be felt on this score.

B_6 or Pyridoxin : The role of this vitamin is not yet clearly understood Animals on a deficient diet develop eczema of the skin especially of the feet, but nothing corresponding closely with this seems to occur in man. However, the vitamin has been found useful in medicine for the treatment of several conditions, e.g. nausea and vomiting in pregnancy and radiation sickness and is one example illustrating the fact that a vitamin may be found of value before an explanation is forthcoming of exactly how it works. One fact that is known at present is that it is associated in the body with protein metabolism.

Biotin : Experiments with animals on restricted diets have demonstrated that this is essential for the growth of almost every animal and also of yeasts, bacteria and fungi, but because of its almost universal presence in human foods, little is known of deficiency in man or of human requirements.

We know, that raw egg white combines with it in the intestinal tract rendering it insoluble. The factor in egg white is responsible for this action is referred to as avidin due to its avidity of biotin.

There is thus a very slight danger that individuals consuming excessive amounts of raw egg white may become biotin-deficient. Actually this is a rather remote possibility, but sick children in hospital, given abundant quantities of albumin water, have been reported as developing a form of scaly dermatitis that has cleared up with small doses of biotin. These are, however, only isolated observations and have not been confirmed in general.

Pantothenic Acid : Rats and chicks fed on diets deliberately made to contain no pantothenic acid develop various signs of deficiency: e.g. dermatitis in chicks and growth failure, hemorrhages and changes in the adrenal glands in rats.

Deficiency in man is not likely to occur because the vitamin is so widely distributed in all kinds of foods.

It has been suggested from time to time that pantothenic acid affected graying of the hair but this has not been confirmed. It has been used in medicine in doses higher than those which would normally be taken in foods. These relatively high doses have been found effective in the treatment of several conditions including the sore tongue which some patients with pernicious anaemia develop. Pantothenic acid is not of any value in the treatment of the anaemia itself.

Para-aminobenzoic Acid : As a dietary factor this presents no problem because there is little risk of shortage in foods, but like several other vitamins, it has found a place in medicine given in high doses and has been used in the treatment of various forms of eczema and also of rheumatic fever.

Vitamin D and Bone : Among animals with a bony skeleton, the form, shape and size depend on the growth and perfection of this skeletal framework. The bones are chemically of calcium phosphate, but each bone has its own characteristic architecture and internal structure. The bones begin as relatively soft cartilage and gradually the hard calcium phosphate is deposited in them, being carried there in the blood stream.

The formation of bone from the calcium and phosphorus brought by the blood is controlled by vitamin D. The body cannot make good bone (or teeth) without vitamin D and conversely the vitamin cannot produce good bone without the materials from which bone is made. Calcium and phosphorus, together with traces of other elements such as fluorine are the building materials; vitamin D regulates the process.

The supply of these minerals and vitamins reaches its greatest importance during the period of rapid growth and it

is when the bones are young, cartilaginous and relatively soft that deficiencies may cause rickets with the characteristic bending of the bones of the limbs and widening at the ends. It may be noted, however, that since bones are hard, they are not of unchanging structure like a piece of wooden or metal furniture. Every bone is supplied with blood vessels and nerves and is as much a living part of the body as muscle or skin. Moreover the blood is continuously bringing new nutrients to the bones and removing waste products. There is thus a state of balance or equilibrium between the calcium and phosphorus of bones and of circulating blood and this continues throughout life.

The amount of vitamin D required during active growth is definitely greater than that needed when the full size of the body is reached, but some vitamin D continues to be required to maintain the bones in a healthy condition. Old people, whose consumption of calcium and vitamin D has been neglected, sometimes develop very brittle and fragile bones which fracture easily — and in some cases the bones of the spinal column lose their firm hard texture and collapse, causing great pain and weakness.

Vitamin D is the one vitamin that can be produced by sunlight, vitamin D is formed in the oil on the surface of the skin and in its deeper layers. This is absorbed and in sunny climates makes it less necessary for sources of vitamin D to be taken by mouth.

Vitamin D is also formed of the fur and feathers of mammals and birds and is consumed when they lick themselves or preen their feathers.

The fact that vitamin D is formed when the skin is exposed to sunlight has led to its use in high doses for certain diseases formerly treated by ultra-violet irradiation. One such disease is lupus vulgaris (tuberculosis of the skin) in which results have been encouraging.

Bones and other Vitamins : Concentration on the importance of vitamin D in the development of bones and

teeth led to neglect of other factors possibly concerned. However, vitamins A, C and E all seem to play some part in promoting normal bone formation.

In vitamin A deficiency the structure becomes disorganised—the bone becomes cancellous (less hard and compact) and overgrown, losing its fine moulding and outline.

In infantile scurvy, the ends of the long bones are abnormally brittle and fractures frequently occur. Vitamin E deficiency is thought to lead to imperfect calcification of the skull, possibly through its effect on vitamin A or on the phosphorus 'turn-over' in bone.

Vitamins A and the Skin : All parts of the body whether external or internal are covered or lined with skin. The type of skin differs according to situation, being tougher and thicker on the outside of the body and more delicate when lining internal cavities, but carrying out important functions throughout the life of any animal.

Vitamin A has special functions in relation to the skin, both that covering the outside of the body and also the mucous membrane lining the various parts which communicate with the exterior, e.g. the mouth, nose, trachea and bronchial tubes, the conjunctiva of the eyes and the linings of the eyelids.

The eyes are peculiarly sensitive to severe deficiency and the condition of xerophthalmia (dry eyes) was one of the first signs of vitamin A deficiency to be discovered.

The mucous membranes of the respiratory tract become less healthy if there is a shortage of vitamin A and due to this fact vitamin A was considered 'anti-ineffective.' We know now that it has no direct effect on infective organisms but it may help to prevent respiratory infections by maintaining the mucous membranes in a healthy condition.

The outer skin of the body shows little change in mild degrees of deficiency but subjects suffering extreme deprivation for long periods may develop a horny, dry condition of the skin—keratomalacia.

Other Functions of Vitamin A : Much has been discovered about the connection between vitamin A and vision in dim light. Deficiency causes a condition known as night-blindness in which individuals experience difficulty in adjusting their sight to distinguish objects in a poor light especially after being in a bright light or looking at bright objects. Normal persons can adapt themselves readily to changes in light intensity but vitamin A-deficient individuals have poor dark-adaptation. It can be explained as follows: seeing in semi-darkness depends on the presence in the retina of the eye of a pigment, visual purple. When bright light falls on the eye, the visual purple is bleached. If there are ample supplies of vitamin A in the body, the pigment is reformed, but if supplies are poor, this re-synthesis is delayed with the result that the clarity of vision in a dim light is reduced. Visual purple is actually a chemical derivative of vitamin A itself.

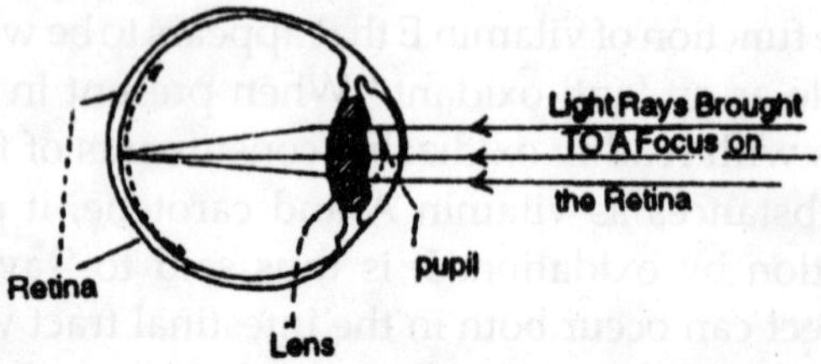

Above facts from the basis of tests used tc find out the state nutrition of groups of people in relation to vitamin A. By means of it, subjects not otherwise unhealthy have been detected as in a state of mild deficiency and it has been possible to correct the faults in their diet before any serious consequences have developed.

Vitamin E and Body Tissues : The function of vitamin E are less well understood than those of vitamins A and D there is still much conflict of opinion about its role in human nutrition.

It is absorbed from the digestive tract along with fats, and anything which interferes with fat absorption may reduce the quantity reaching the tissues of the body.

In experimental animals, e.g. rats, it has been proved that

vitamin E is necessary for normal fertility and this fact is still used in biological methods of estimation, but because of this, it cannot be assumed that the same is true of other species.

Although the role of vitamin E in connection with fertility is still not understood and seems to vary from species to species, there is no doubt at all about its importance in the physiology of some of the tissues. Just as vitamin A is concerned with the skin and mucous membranes and vitamin D with bones, so vitamin E is concerned with muscle and connective tissues. Deficiency in man does not produce such well-marked and easily recognised diseases as xerophthalmia and rickets but in animals such as rats, the young develop a paralysis of the limbs due to a muscle defect or dystrophy. This is not unlike a rather uncommon human disease but attempts to cure the latter with vitamin E have so far met with very inconclusive results.

One function of vitamin E that appears to be well established is its role as an 'anti-oxidant.' When present in the tissues or in foods with readily oxidizable constituents of fats, including such substances as vitamin A and carotene, it prevents their destruction by oxidation. It is thus said to 'save' vitamin A. This effect can occur both in the intestinal tract where vitamin A in food may be oxidized, or in the liver where vitamin A is stored, and this fact means that if vitamin A is given *with* small quantities of vitamin E, less A is needed to achieve a given result. It has been shown, for example, that vitamin E increases the growth-promoting power of vitamin A.

It may be partly because of this property and partly for other reasons not fully understood that vitmain E is as essential for the maintenance of tissue integrity and is thought to have an effect in delaying the process of ageing.

Vitamin E is one of the best examples to illustrate the complications with which a study of vitamins may be involved. There is no simple deficiency here. The vitamin itself is made up of a complex of at least four components. Their biological activity differs; their distribution varies from source to source;

they are known to interact within the body with proteins and protein constituents, the amino acids, with vitamin A and probably vitamins of the B complex and to be concerned in the metabolism of some of the most complex tissues, the muscular and fibrous tissues, of the body. It is not surprising in the circumstances that there are still many gaps in our knowledge of this particular vitamin.

Vitamins, Blood and Blood-Vessels : Vitamins C, K, P, folic acid and B_{12} are all concerned either with blood or with the smallest blood-vessels, the capillaries, which bring blood into contact with all the tissues.

Blood is made up of a fluid, plasma, of highly complex composition where in are suspended cells or corpuscles: red cells containing the pigment hemoglobin and white cells which are unpigmented. The blood, by circulating through the vessels comes into contact with all the tissues, even the hardest bone. Holes can be found in the shafts of the long bones and in the skull, through which these vessels pass to the interior.

As the fluid (plasma) circulates, it carries the cells with it and together they form the distributing and collecting system which functions for the whole body. Oxygen is carried from the lungs to all parts; carbon dioxide, formed in tissue respiration, is removed from its sites of formation and taken back to the lungs. For this, the red cells are largely responsible.

The plasma transports nutrient materials from the alimentary tract to the organs or tissues where they undergo further breakdown or assimilation. The waste products of physiological processes are similarly removed by the plasma, which may contain at any one time proteins, inorganic and organic substances in solution, nutritive and excretory materials, antibodies, hormones, vitamins and possibly compounds still unidentified.

By studying the effects of deficiencies of vitamins C, P and K, we can see how they affect the blood. All three are associated with bleeding into the tissues.

Vitamins C and P : In scurvy, there are haemorrhages into the joints and into the tissues. These appear as purplish-red patches in dependent parts, i.e. around the shoulder blades and lower part of the back when the patient is lying down. The gums may also bleed, and there is swelling due to an increase in the amount of fluid in the tissues. It is evident that the capillaries have allowed blood to escape.

The walls of these minute vessels are made up of a single layer of very delicate cells, held together by a cement substance and surrounded by a sheath of very fine fibrils.

Bleeding into the tissues occur when blood-vessels of any size are damaged. This can happen in mechanical injuries and the severity varies according to the size of the vessels which are damaged. Bruising causes injury to the walls of the capillaries and the discolouration that follows is because of the presence of blood in the tissues.

When similar escape of blood takes place without mechanical injury it means that the walls of the capillaries have become unduly permeable. This is what happens in scurvy and in another condition : purpura — a term applied to a number of conditions in which spontaneous bleeding occurs beneath the skin. The capillary sheath and the cement substance holding the cells together give way and allow the blood to escape. The haemorrages seems as purplish spots of various sizes and the colour changes which occur are similar to those of an ordinary bruise.

Purpura occurs in diseases of different kinds — many quite unrelated to scurvy, and it should not be looked upon as a disease in itself but rather as a disease sign. In the most serious varieties of measles, scarlet fever and smallpox, purpuric spots appear and the diseases are then described as 'haemorrhagic.'

Purpura may also take place in conditions of irritation of the digestive tract and associated with weals in some forms of urticaria (nettle-rash).

Scurvy can be arrived and treated by means of vitamin C; purpura by vitamin P.'

Vitamin C (ascorbic acid) is concerned not only with the prevention of tissue bleeding as in scurvy, but in the maintenance of a healthy state of equilibrium between the fluid plasma in the capillaries and the fluid in the body tissues (Hoskin, 1948). Where excess body fluid tends to collect as in diseases of the heart and circulation, vitamin C helps to reduce this.

Vitamin K : This is concerned not with the capillary walls but with the process of clotting, which is slowed down in deficiency of the vitamin.

Clotting is a complicated process. It does not normally occur within the intact, healthy blood-vessels but when the vessel or the tissues are damaged the blood which escapes, clots within a few minutes. The clot is made up of fibrin, a jelly-like protein that forms in the plasma and that entangles the blood cells in its meshes. This fibrin sets as a jelly only in the presence of another constituent of plasma, thrombin, and this in turn is formed from a precursor prothrombin, that is liberated into the circulating blood from the liver where it is formed. And this is where vitamin K comes into the picture. Without it, there is a failure in the maintenance of prothrombin in the liver and if the amount liberated into the plasma is much reduced there develops a tendency for injuries, which would otherwise be negligible, to continue to bleed because no clot forms to seal them.

In the blood of infants, the prothrombin is low for the first few days of life and consequently hemorrhage may occur from undetectable small injuries to blood-vessels. These haemorrhages may be in the digestive tract when their existence is detected by the passing of discoloured stools, or in the nose or palate or under the skin, hence the descriptive term 'haemorrhagic disease of the newborn.'

The low levels of prothrombin in the blood of infants may be due to the fact that vitamin K is obtained only to a small

extent from food but chiefly as a result of bacterial synthesis in the digestive tract. Only relatively few bacterias are associated with disease; some inhabit the digestive tracts of man and of other animals and their presence is an advantage rather than the reverse.

Although vitamin K is eaten with certain foods, there is no doubt but that much of a man's needs for this vitamin are supplied by these bacteria in the intestines. Vitamin K cannot, however, be synthesised by the intestinal bacteria until these are established and this does not happen until several days after birth, so that during this short period there is a real danger of vitamin K deficiency and subsequent haemorrhage. To prevent this, it is now a usual practice to give vitamin K either to the mother before the child is born or to the infant after birth and this has done much to reduce the incidence of this type of haemorrhagic disease.

Vitamin K, whether formed by bacteria in the intestine or eaten in foods, is fat-soluble and is so absorbed along with fats. Fat-absorption, in turn, depends on the presence of bile in the intestine since bile salts act on fats converting them into a form in which they will pass through the cells lining the intestine. Anything which interferes with the presence of bile salts in the intestine will therefore prevent the absorption of fats and with them of vitamin K. The commonest such condition is an obstruction in the bile duct, through which bile passes from the gall-bladder to the intestine. A gall-stone in this duct is a common form of obstruction and a construction from a growth can produce the same sort of result.

When the bile can no longer pass into the intestine it becomes re-absorbed into the blood stream and the pigments in it circulate in the blood and turn the skin yellow producing the condition known as obstructive jaundice.

Sometimes surgical operations are required to remove the obstruction and these operations formerly attended with some danger from haemorrhage because, with poor vitamin K absorption from the intestine, the liver had failed to produce

adequate quantities of prothrombin. Now, vitamin K is given to such patients before operation and this danger is greatly reduced.

A secondary deficiency of vitamin K may occur in any disease where in there is defective absorption of fats. Actually individuals may have low prothrombin levels in the blood with no signs to show this until some minor damage to the tissues occur such as a tooth extraction, after which bleeding may be excessive. The practice of giving vitamin K before such minor operations is becoming more widespread.

The other two vitamins in this group effect the cells of the blood rather than the plasma. Red blood cells are formed in the red marrow of bones and go through stages in development before they are fully mature and ready to be released into the circulation. They remain in circulation carrying out their various functions for about four months after which they tend to become fragmented and are disposed of by the spleen and new ones take their place.

Primitive cells in the marrow contain no haemoglobin and they stain with basic dyes, so that in stained films a cell seems blue or purple with a large nucleus. Gradually haemoglobin is formed within the cell, and as this stains with acid dyes (usually pink) the cells appear in films with both pink and bluish colouring. These cells are larger than circulating red cells and are called megaloblasts. They gain more haemoglobin, lose their nuclei and become smaller before being liberated into the circulation as normal red cells or normocytes.

In those cases where this process for some reason or other goes wrong, various things can happen; cells may be liberated with less than the normal quantity of haemoglobin, or before they are fully mature, or are of uneven size.

These abnormalities of blood cells or the amount of haemoglobin is reduced. All these are anaemias of that many varieties are known. Those types that are characterised by abnormally large red cells (macrocytes) and by a reduction of

haemoglobin are the macrocytic anaemias, the most common of which are the so-called pernicious anaemia and the anaemias that occur in the disease, sprue. There are other forms of macrocytic anaemia.

Folic Acid and Vitamin B : It has never been proved that these anaemias are directly due to dietary deficiencies, but it is now known that a normal blood picture can be restored by both folic acid and vitamin B_{12} However, the mechanism by which this is brought about is still only vaguely understood.

The story of vitamin B_{12} has been, in recent years, as dramatic as the history of the earlier vitamin discoveries. For the beginning we have to go back to 1926 when Minot and Murphy found that a form of anaemia hitherto resistant to treatment and hence described as 'pernicious' responded to treatment with liver. From that time onwards chemists sought to isolate the active factor from liver and, after nearly twenty years, the goal was thought to be reached when folic acid was synthesised. This substance restored the blood of pernicious anaemia patients to normal but this disease is characterised not only by abnormal blood formation but also changes in the nervous system causing disturbances of sensation and these were not affected by folic acid.

In 1948, workers in the laboratories of pharmaceutical firms in the U.S.A. and in Great Britain simultaneously announced the isolation from liver of a crystalline compound several thousand times more active than folic acid. The crystals were dark red needles and contained, in addition to carbon, hydrogen, oxygen, nitrogen and phosphorus, 4 per cent of cobalt — the only vitamin known to contain this element.

The new substance, known as vitamin B_{12} was found to be active in pernicious anaemia in doses as small as 20 , micro-grammes and also effective in the treatment of the nervous system complications.

It was marketed in Great Britain in 1949 and is now widely used in doses of 50 to 100 micro-grammes given by injection. It is no longer extracted from liver, twenty tons of which are

required to produce one gramme, but is produced from cultures of streptomyces griseus (an organism allied to primitive moulds).

It is stable in the solid state and also in neutral or slightly acid solution. It exists in several forms — all physiologically active. The solution, which is pale pink, is supplied in ampoules for medical use.

Standard Diet

A large majority of our people are satisfied by filling their bellies with some foodstuffs. Most of the people ignore the nutritious aspect of the food they eat. Prevalence of malnutrition is due to insufficient intake of foodstuffs. In order to assess the nutritional status dietary intake must be calculated. Diet surveys are used to collect information about the dietary intake, dietary habits, preservation methods, cooking methods and special diets for special conditions and in diseases. Diet surveys also provide information regarding the economic and social factors influencing food production and consumption patterns of families and communities.

To conduct diet surveys certain factors are to be taken into account. They are trained personnel, population sampling, various methods available for conducing diet surveys and analysis of the data and calculations of nutritive values.

The persons to be engaged in conducting diet survey must be trained in conducting such surveys. They must be trained to create rapport with the housewives and in adopting tactful methods to collect information from them. They must be aware of other principles of nutrition and dietary habits of the local community.

Population sampling must be carried out with the help of a statistician. The choice of diet survey methods depends on the purpose of the survey.

The sample size also depends on the purpose of the survey, availability of trained personnel, time and resources like money.

The duration of the survey depends on the purpose and methods adopted. Normally 7 to 10 days are preferred for diet surveys.

Different Processes : Commonly used methods in our country are:

(1) oral questionnaire method,

(2) food inventory or log book method, and

(3) weighing of raw and cooked foods.

Oral Questionnaire Method : Oral questionnaire or interview is the most commonly used method of diet survey. The investigator has to formulate a diet survey questionnaire depending upon the purpose. After conducting a pilot study with the questionnaire changes can be made on the pattern of the questions. Then the investigator goes from door to door and collects information regarding the types and qualities of foods consumed. The data is not very perfect, only approximate amounts can be collected through this method. Therefore, this method is suitable only to collect information on the general dietary patterns or an approximate picture of dietary habits of a large section of population. General ideas like choice of foods, foods avoided in certain conditions, foods included in special conditions or in festivals, food fads and fallacies, are to be collected. The advantages of the questionnaire method are that it is not a time-consuming method, therefore a large number of families can be covered within a specific time.

Food Inventory or Log Book Method : This method can be used only with a literate group because a book with question is entrusted to the head of the family or to the housewife. The person entrusted must enter all purchases in the book. Full cooperation from the householder is essential because reliability of the data depends upon the entry. Therefore at times the data may not be authentic.

Weighing of Raw and Cooked Foods : This method is the most reliable one. But it is time-consuming and therefore it is not used for a large sample. As the investigator cannot leave

the questionnaire with the householder she has to stay near and record the amount of various foods before and after cooking. The amount of food used can be recorded by weighing the leftovers after consumption. Often the housewife shows more raw foods to give a wrong impression about their food intake. Therefore the investigator must have some practical knowledge about the quantities of cooked foods of various sources.

Procedure of Consumption : In order to find out the nutritional adequacy of a family, a group adult consumption unit is calculated. In this method, the heterogenic groups of people i.e., school going, adult, geriatric, pregnant, a uniform calculation can be applied. This makes the total requirement calculation simple. For adult consumption unit a sedentary man is taken as one adult consumption unit (ACU) and the calorie requirements of other age groups and special groups are expressed in terms of proportionate adult consumption units.

The adult consumption unit coefficient laid down by the Indian Council of Medical Research for the various groups and the type of physical activity is given below:

Adult male (Sedentary worker)				1.0
Adult male (Moderate worker)				1.2
Adult male (Heavy worker)				1.6
Adult female (Sedentary worker)				0.8
Adult female (Moderate worker)				0.9
Adult female (Heavy worker)				1.2
Adolescents	12	-	21 yrs	1.0
Children	9	-	12 yrs	0.8
Children	7	-	9 yrs	0.7
Children	5	-	7 yrs	0.6
Children	3	-	5 yrs	0.5
Children	1	-	3 yrs	0.4

Calculation of Nutritive Value of the Family Diet in Terms of Adult Consumption Unit : To find out the adult consumption

unit of the family, the number of persons, their physical activities and their ages have to be recorded. Total consumption unit has to be calculated using the above table. For example, for a family of 4, the adult consumption unit is given below:

Member	*Age*	*Occupation*	*A.C.U.*
Father	57 yrs	Retired Teacher	1.2
Mother	55 yrs	Housewife	0.9
Daughter	23 yrs	Student	0.9
Son	17 yrs	Student	1.0
		Total	4.0

The adult consumption unit of the family is 4. The total calorie requirement of the family is calculated below.

The energy requirement for a sedentary = 2,800 kcals

man for 4 A.C.U. = 2,800 x 4 = 11, 200 kcals

Total cereals for the family per day = 460 gms x 4 = 1,040 gms

Preserving Methods

Food spoilage is brought about by the action of enzymes present in foods or due to the action of micro-organisms such as mould, yeast and bacteria or due to infestation with insects and worms. The environment unfavourable to the action of enzymes or to the growth of micro-organism is the main objective of food preservation. Natural and artificial methods are adopted for this purpose. Preservatives are also added to foods to preserve them. Various agents are used to bring either physical or chemical changes in food materials which are to be preserved.

As the principal spoilage agents are normally present in foods, destroying them or preventing their development

becomes the chief problem of food preservation. Any condition opposed to the development of these organisms, whether by retarding their growth or by entirely destroying them, aids in the preservation of food. Methods commonly used to this end include common or cellar storage, refrigeration, canning, freezing, use of preservatives, drying and the exclusion of air. Although not a commonly used method at present, "cold sterilisation" or irradiation by beta and gamma rays to prolong the keeping quality of foods is being widely studied experimentally. The possibilities of this method of food preservation are great. When some of the major problems now limiting its use are solved, it is probable there will be radical changes in the present accepted methods of handling foods.

Refrigeration : Micro-organisms, although not readily destroyed by severe cold, are much less active at low temperature. Refrigeration is widely used both in homes and in commercial plants, as a means of maintaining the low temperature found satisfactory in the storage of perishable foods. Fresh milk, meat and similar foods are kept just above the freezing point.

Certain fruits and vegetables also keep better when cold, although as a rule they do not require low temperature for the limited time they are stored in the home as do the protein foods cited. Under-ripe fruits usually ripen rapidly at room temperatures. The keeping of cooked foods and leftovers is greatly facilitated by refrigeration, which is a more expensive but also a more efficient method of preservation than common storage. Wrapping certain fruits and vegetables in paper or cellophane, or coating them with wax, improves the keeping quality. Pears and apples of high quality to be kept for winter use and oranges, lemons and grapefruit are often wrapped.

The fruit or the paper in which it is wrapped may be treated to retard mould formation. Green peppers, tomatoes, oranges and other citrus fruits, cucumbers and cantaloupes are among the foods whose keeping quality is improved when they are coated with wax.

Heating : Food is commonly preserved through the application of high temperatures. A temperature considerably above that of the body may result in either a pasteurised or a sterilised product. Foods are commonly pasteurised by being held at a temperature of from 60° to 66° C, for thirty to forty minutes, during which time most of the organisms, although not all, are killed. Spores and some vegetative forms of bacteria remain, but as a rule those causing diseases are destroyed. This method is used for temporary preservation of milk, and for preservation of fruit juices and other fruit products of delicate flavour. Acid foods are rendered sterile by boiling for a comparatively short time.

Canning : If the effectiveness of pasteurisation and sterilisation is to extend over a period of time of practical significance, the material thus treated must be protected from fresh contamination by micro-organisms. The fact that in canning the sterilised or pasteurised food is afforded such. Protection accounts, at least in part, for the rapid growth in popularity of the process as a means of food preservation. The first foods preserved %9 canning were sealed in glass.

All prejudice which once existed against canned foods has been removed. Experiments have shown canned foods to be as wholesome as any form of preserved food. Often the canned food is found to superior in food value to the freshly cooked. Canned foods are ready to use at a moment's notice. When food is to be kept over a long period of time, canned products are also most satisfactory and they do not require special temperatures or equipment for storage.

Freezing : Freezing, like cold storage, does not destroy the micro-organisms enzymes present in the foods. It does render them more or less inactive however, so that frozen foods, when held at the proper temperature undergo change slowly. This applies to the nutritive value as well as to other chemical and physical characteristics.

Practically all common fruits are now preserved by freezing. However, the success with which the different fruits can be

freezing varies. Most of the vegetables that are served cooked may be preserved by freezing. Scalding to destroy the enzymes is essential for successful preservation of almost all vegetables by this method. Hence vegetable to be used fresh and crisp in salads cannot be preserved by freezing. Fish, poultry and game, as well as beef, veal, pork and lamb to preserved by freezing. Frozen foods have little or no waste at the time using and little time is required for preparation. However, time must allowed for the thawing of fruits to be served raw, and the cooking period of frozen meats and vegetables not already defrosted must be sufficient to allow for both thawing and cooking.

Using Chemical Preservatives : Certain chemicals are helpful in preserving foods either by retarding preventing the growth of micro-organisms. These may be either added to the product or produced in it by fermentation. Sugar, a common he preservative, is often used in such quantities as to increase concentration of the food and make it an unfavourable medium for organisms. Salt and acid, in the form of vinegar or lemon juice, are substances added for their preservative action. Frequently, fermentation process is used to produce, within the food itself, acetic lactic that exerts the same preservative action as is obtained by addition of vinegar. Salt is used as an aid in the preservation of vegetables by fermentation. It aids both in drawing out the juices are delaying the action of certain spoilage agents. Formerly, spices often used in such quantities as to have a definite preservative effect. Today the concentration in which they are used is usually less than employed earlier and it is believed that only a few of them a0re of value as preservatives.

Certain preservatives such as borax, boric acid, sulfites, and formaldehyde, once sold as canning powders, are now considered injurious to human beings, and their use is usually prohibited.

Preserving and Pickling : Preserving and pickling are methods of preservation frequently employed. Long before canning was known as a means of preserving foods, the use

of preservatives was practised. Salting of foods was an early method of preservation. Salt is a valuable preservative, both as an antiseptic and as an agent for removing water. Placing foods in brine of certain concentration promotes fermentation and the foods develop an agreeable flavour. Sugar in large amount is likewise a favourite preservative. Today, although practically every known variety of preserve and pickle is found in the market, the distinctive flavour of the homemade product is often preserved.

Drying : Removal of moisture is of benefit in preserving food. Although it is difficult to dry foods to the point of destroying micro-organisms, it is comparatively easy to dry foods so that no spoilage takes place. All organisms must have food on which to develop and grow and before the food is available to them it must be in a solution which is rather dilute. These micro-organisms obtain food by osmotic action. Drying of foods to prevent spoilage does not necessarily mean complete removal of the water, but it does mean concentration to such a point that the liquid is denser than the body fluid of the organisms. When the liquid outside the cell wall is more dense than that inside, the liquid within tends to be removed from the cell and the body processes are delayed or prevented. Hence, although the food is present it may not be available to the organism which might cause spoilage. Drying may retard enzyme action also, but in vegetables especially, the effect of enzymes is sufficient to cause comparatively rapid deterioration. Scalding to destroy the enzymes is highly beneficial in prolonging the storage life of dehydrated vegetables. Drying alters greatly the character of the food and requires some time for preparation both before and after the process.

Bibliography

Achaya, K.T. : *Indian Dairy Products,* Asia Publishing House, Bangalore, 1974.

Adam, H. : *Food Science,* John Wiley and Sons, New York, 1982.

Austin, A. and Ram, A. : *Studies on Chapati-Making Quality of Wheat,* Indian Council of Agricultural Research, New Delhi, 1971.

Bachman, W. : *Food Chemistry,* Maclaren and Sons Ltd., London, 1955.

Balsdon, J.P. : *Life and Leisure in Ancient Rome,* Bodley Head, London, 1966.

Beersteeker, E. : *The Biochemistry of Vitamins,* Reinhold, New York, 1950.

Bharathi, V.V. : *Anxiety Behaviour in Children,* Discovery Publishing House, New Delhi, 1992.

Board, R.G. : *A Modern Introduction to Food Microbiology,* Blackwell Scientific Publications, London, 1983.

Burton, T.L. : *Recreation Research and Planning,* Allen and Unwin, London, 1970.

Cambell, A.M., Penfield, M.P. and R.M. Griswold : *Experimental Study of Foods,* Houghton, London, 1990.

Cameron, Spencer : *Food Science,* Pargamon Press, New York, 1984.

Challoway, D.H. : *Nutrition and Physical Fitness,* W.B. Saunders Co., Philadelphia, 1979.

Cherry, G.E. : *Town Planning and its Social Context,* MacGraw-Hill, London, 1970.

Cherry, J.P. : *Protein Functionality in Foods,* American Chemical Society, Washington D.C., 1981.

Chichester, C.O. : *The Chemistry of Plant Pigments,* Academic Press, New York, 1972.

Dale, E. : *Management Theory and Practice,* McGraw-Hill, New York, 1973.

Dharamraj, E. : *Food and Hotel Legislations and Policies,* New Age Publishers, New Delhi, 1998.

Dumazedier, J. : *Towards a Society of Leisure,* Free Press, New York, 1967.

Dyke, S .F. : *The Chemistry of Vitamins,* Interscience Publishers, New York, 1965.

Edmunds, F. : *Environmental Administration,* McGraw-Hill, New York, 1973.

Eley, G. : *Wild Fruits and Nuts,* E.P. Publishing Ltd., Yorkshire, 1976.

Feiffer, M. : *Giving Places,* MacMillan, London, 1985.

Furia, T.E. : *Hand Book of Food Additives,* C.R.C. Press, Ohio, 1977.

Geiss, I. : *Proteins : Structure, Function and Evolution,* Menlo Park, California, 1983.

Gell, D. L. : *International Hotel Industry Policy,* Van Nostrand Reinhold, New York, 1990.

Gyorgy, P. and Pearson, W.N. : *The Vitamin,* Academic Press, New York, 1967.

Haschemeyer, A.H. : *Proteins : A Guide to Study of Physical and Chemical Methods,* John Wiley & Sons, New York, 1973.

Hudson, F. : *Development in Food Proteins,* Applied Science Publisher, London, 1982.

Inglett, G. : *The Quality of Foods and Beverages : Chemistry and Technology,* Academic Press, New York, 1981.

Jackson, R. : *The Geography of Recreation and Leisure,* Hutchinson, London, 1972.

Jasscler, D.K. and M.A. Joslyn : *Fruit and Vegetables Juice,* The AVI Publishing Co., Connecticut, 1971.

Johnson, A.H. : *Encyclopedia of Food Science,* The AVI Publishing, Connecticut, 1978.

Kacharoo, P. : *Pulse Crops of India,* ICAR, New Delhi, 1970.

Kadans, J.M. : *Encyclopedia of Fruits, Vegetables, Nuts and Seeds,* Parker Publishing Co., New York, 1973.

Kalyani, T. : *Home and School Environment,* Discovery Publishing House, New Delhi, 2003.

Kinton, R. : *Practical Cookery,* Edward Arnold, London, 1981.

Krause, M.V. and L.K. Mahann : *Food Nutrition and Diet Therapy,* W.B. Saunders Co., Philadelphia, 1979.

Kumari, P. : *Role of Anganwadi Workers in Integrated Child Development Services Programme,* Discovery Publishing House, New Delhi, 1994.

Lal, H. : *Total Quality Management : A Practical Approach,* New Age Publishers, New Delhi, 2000.

Law, C. : *Urban Tourism : Attracting Visitors to Large Cities,* Mansell, London, 1993.

Less, R. and E.B. Jackson : *Sugar Confectionery and Chocolate Manufacture,* Leonard Hill Books, Burks, 1973.

Lickorish, L. J. : *Hotel Industry and International Balance of Payments,* International Institute of Scientific Travel Research, Geneva, 1953.

Lloydryall, A. and W.T. Pendzer : *Handling Transportation and Storage of Fruits and Vegetables : Fruits and Peanuts,* The AVI Publishing Co., Connecticut, 1974.

Mackinney, G. and A.C. Little : *Colour of Foods,* The AVI Publishing Company, Connecticut, 1972.

Malhotra, R. K. : *Food Service and Catering Management,* Anmol Publications, New Delhi, 1998.

Marton, I.D. and A.J. Macbod : *Food Flavours,* Elesvier Scientific Publishing Co., New York, 1982.

Masefield, G.B. : *The Oxford Book of Food Plants,* Oxford University Press, London, 1975.

Mate, S.A. : *Bakery Technology and Engineering*, The AVI Publishing Co., Connecticut, 1972.

——————: *Cereal as Foods and Feeds*, The AVI Publishing Co., Connecticut, 1959.

Mayo, J. : *The Psychology of Leisure Travel*, CBI Publishing Company, Boston, 1981.

Mcwilliams, M. : *Food Fundamentals*, John Wiley & Sons, New York, 1979.

Medlik, S. : *Hotel Management: Past, Present and Future*, Heinemann, London, 1976.

Mellor, J. D. : *Fundamentals of Freete-Drying*, Academic Press, London, 1978.

Meyer, L.H. : *Food Chemistry*, Reinhold Publishing Corporation, New York, 1961.

Michael Eskin, N.A. : *Plant Pigments, Flavours and Textures*, Academic Press, New York, 1979.

Murthy, R.B. : *Breeding Procedures in Pearl Millet*, ICAR, New Delhi, 1977.

Nath, P. : *Vegetables for the Tropical Region*, Indian Council of Agricultural Research, New Delhi, 1976.

Oberleas, D. : *Trace Elements in Health and Disease*, Academic Press, New York, 1976.

Palmer, T. : *Understanding Enzymes*, John Wiley & Sons, New York, 1981.

Pamela, V.P. : *Directory of Wines and Spirits*, North Wood Publication, London, 1974.

Paul, P.C. and Palmer, H.H. : *Food Theory and Application*, John Wiley & Sons, New York, 1972.

Peckham, G.G. : *Foundations of Food Preparation*, MacMillan Publishing Co., New York, 1979.

Pigman, W.W. and Horton, D. : *The Carbohydrates : Chemistry and Biochemistry Plants*, C.S.I.R, New Delhi, 1969.

Pigram, J. : *Outdoor Recreation and Resource Management*, Croom Helm, London, 1993.

Potter, N.M. : *Food Science*, The AVI Publishing Co., Connecticut, 1978.

Pudney, John : *The Thomas Cook Story*, Michael Joseph, London, 1953.

Puruthi, J.S. : *Spices and Condiments*, National Book Trust of India, New Delhi, 1976.

Ranganna, S. : *Manual of Analysis of Fruit and Vegetable Products*, Tata McGraw-Hill Publishing Co. Ltd., New Delhi, 1984.

Rani, Usha D. : *Nutrition and Religion*, Discovery Publishing House, New Delhi, 2003.

Rao, D. V. : *Community and School Nutrition Education*, Discovery Publishing House, New Delhi, 1998.

Roberts, D.V. : *Enzyme Kinetics*, Cambridge University Press, 1977.

Rogers, C.A. : *Profitable Poultry Keeping in India and the East*, Treasure House of Books, Bombay, 1966.

Ross, G. F. : *The Psychology of Hotel Management*, Hospitality Press, Melbourne, 1994.

Schapira, J. : *The Book of Coffee and Tea*, St. Martin's Press, New York, 1975.

Schultz, G.E. and R.H. Schirmer : *Principles of Protein Structure*, Springer-Verlage, New York, 1979.

Shreedevi, V. : *Nutrition Education*, Discovery Publishing House, New Delhi, 1997.

Smartt, J. : *Tropical Pulses*, Longman Group Ltd., London, 1976.

Stadlmare, W.J. and. Cotterill, O.Y. : *Egg Science and Technology*, The AVI Publishing Co., Connecticut, 1973.

Stumbo, C.R. : *Thermobacteriology in Food Processing*, Academic Press,, New York, 2000.

Swern, D. : *Bailey's Industrial Oil and Fat Products*, John Wiley & Sons, New York, 1964.

Tandon, G.L. : *Preservation of Fruits and Vegetables*, Indian Council of Agricultural Research, New Delhi, 1960.

Tandon, H.P. : *Egg and its Care*, Ministry of Agriculture, Govt. of India, New Delhi, 1973.

Thangam, P.E. : *Modern Cookery for Teaching and Trade*, Orient Longmans Ltd., Bombay, 1983.

Thankamma, J. : *Food Adulteration*, S.G. Wasani, New Delhi, 1976.

Tressler, D.K. : *Food Technology the World Over*, The AVI Publishing Co., Connecticut, 1963.

Uma, K. : *Poverty, House Hold Fold Security and Nutrition in Rural Areas*, Discovery Publishing House, New Delhi, 2000.

Underwood, E.J : *Trace Elements in Human and Animal Nutrition*, Academic Press, New York, 1977.

Veldhnis, M.K. : *Food Science and Technology*, The AVI Publishing Co., Connecticut, 1977.

Vidya, D.B. : *A Textbook of Nutrition*, Discovery Publishing House, New Delhi, 1996.

Weiss, T. J. : *Food Oils and their Uses*, The AVI Publishing, Connecticut, 1970.

Whitaker, J.R. : *Principles of Enzymology for Food Science*, Marcel Dekker, New York, 1977.

Williams, P.N. : *The Chemical Constitution of Natural Fats*, Chapman and Hall, London, 1964.

Winitz, M. : *Chemistry of Amino Acids*, John Wiley & Sons, New York, 1961.

Yasoda, P. : *Determinants of Nutritional Status of Pre-School Children*, Discovery Publishing House, New Delhi, 1998.

Yukic, T.S. : *Fundamentals of Recreation*, Harper Row, London, 1963.

Index

K

L

M

N

O

P

❑❑❑